现代分析检测技术丛书

Green Extraction Techniques:
Principles, Advances and Applications

绿色萃取技术
原理、进展与应用

［西］Elena Ibáñez Alejandro Cifuentes 主 编
杨 飞 邓惠敏 主 译

中国轻工业出版社

图书在版编目（CIP）数据

绿色萃取技术原理、进展与应用／（西）艾琳娜·依班娜（Elena Ibáñez），（西）亚历桑德罗·西弗恩特斯（Alejandro Cifuentes）主编；杨飞，邓惠敏主译．—北京：中国轻工业出版社，2021．11

ISBN 978-7-5184-3516-6

Ⅰ．①绿… Ⅱ．①艾… ②亚… ③杨… ④邓… Ⅲ．①萃取-应用-化学工业-无污染技术 Ⅳ．①X78

中国版本图书馆 CIP 数据核字（2021）第 096610 号

Green Extraction Techniques：Principles，Advances and Applications
Elena Ibáñez，Alejandro Cifuentes
ISBN：9780128110829

《绿色萃取技术原理、进展与应用》（杨飞，邓惠敏译）
ISBN：9787518435166

注意

本书涉及领域的知识和实践标准在不断变化。新的研究和经验拓展我们的理解，因此须对研究方法、专业实践或医疗方法作出调整。从业者和研究人员必须始终依靠自身经验和知识来评估和使用本书中提到的所有信息、方法、化合物或本书中描述的实验。在使用这些信息或方法时，他们应注意自身和他人的安全，包括注意他们负有专业责任的当事人的安全。在法律允许的最大范围内，爱思唯尔、译文的原文作者、原文编辑及原文内容提供者均不对因产品责任、疏忽或其他人身或财产伤害及/或损失承担责任，亦不对由于使用或操作文中提到的方法、产品、说明或思想而导致的人身或财产伤害及/或损失承担责任。

责任编辑：张　靓　王庆霖
策划编辑：张　靓　　责任终审：唐是雯　　封面设计：锋尚设计
版式设计：砚祥志远　　责任校对：宋绿叶　　责任监印：张　可

出版发行：中国轻工业出版社（北京东长安街 6 号，邮编：100740）
印　　刷：三河市万龙印装有限公司
经　　销：各地新华书店
版　　次：2021 年 11 月第 1 版第 1 次印刷
开　　本：787×1092　1/16　印张：31．25
字　　数：710 千字
书　　号：ISBN 978-7-5184-3516-6　定价：158．00 元
邮购电话：010-65241695
发行电话：010-85119835　传真：85113293
网　　址：http：//www. chlip. com. cn
Email：club@ chlip. com. cn
如发现图书残缺请与我社邮购联系调换
200499K1X101ZYW

本书翻译人员

主　译	杨　飞	国家烟草质量监督检验中心
	邓惠敏	国家烟草质量监督检验中心
副主译	叶长文	中国烟草总公司郑州烟草研究院
	鞠华波	西北烟草质量监督检测站
	熊　巍	四川省烟草质量监督检测站
	刘　欣	云南中烟工业有限责任公司
	韶济民	四川省烟草质量监督检测站
	严　俊	广西中烟工业有限责任公司
	陈玉松	山东省烟草专卖局（公司）
	蔡宪杰	上海烟草集团有限责任公司
译　者	刘珊珊	国家烟草质量监督检验中心
	闫　鼎	上海烟草集团有限责任公司
	曹亚凡	上海烟草集团有限责任公司
	陈晓水	浙江中烟工业有限责任公司
	张小涛	贵州中烟工业有限责任公司
	许成悦	上海烟草集团有限责任公司
	虞　沁	上海烟草集团有限责任公司
	朱文静	贵州省烟草质量监督监测站
	王建波	贵州省烟草质量监督监测站
	孟冬玲	广西中烟工业有限责任公司
	朱　静	广西中烟工业有限责任公司
	王　颖	国家烟草质量监督检验中心
	纪　元	国家烟草质量监督检验中心
	杨　进	国家烟草质量监督检验中心
	柳　均	湖北省烟草质量监督监测站
	边照阳	国家烟草质量监督检验中心
	范子彦	国家烟草质量监督检验中心
	唐纲岭	国家烟草质量监督检验中心

本书编写人员

Maryline Abert Vian

Avignon University, INRA, UMR 408, Avignon, France

Tamara Allaf

ABCAR-DIC, La Rochelle, France

Sergio Armenta

University of Valencia, Burjassot, Valencia, Spain

Francesco Cacciola

University of Messina, Messina, Italy

Pilar Campíns-Falcó

University of Valencia, Valencia, Spain

Soledad Cárdenas

University of Córdoba, Córdoba, Spain

Maria Celeiro

Universidade de Santiago de Compostela, Santiago de Compostela, Spain

Farid Chemat

Avignon University, INRA, UMR 408, Avignon, France

Alejandro Cifuentes

Institute of Food Science Research (CIAL-CSIC), Madrid, Spain

Meire R. da Silva

University of São Paulo, Institute of Chemistry at São Carlos, São Carlos, Brazil

Thierry Dagnac

Agronomic and Agrarian Research Centre (INGACAL-CIAM), A Coruña, Spain

Miguel de la Guardia

University of Valencia, Burjassot, Valencia, Spain

Paola Dugo

University of Messina, Messina, Italy; University Campus Bio - Medico of Rome, Rome, Italy

Francesc A. Esteve-Turrillas

University of Valencia, Burjassot, Valencia, Spain

Anne Sylvie Fabiano-Tixier

Avignon University, INRA, UMR 408, Avignon, France

José Manuel Florêncio Nogueira

University of Lisbon, Lisbon, Portugal

Bruno H. Fumes

University of São Paulo, Institute of Chemistry at São Carlos, São Carlos, Brazil

Carmen Garcia-Jares

Universidade de Santiago de Compostela, Santiago de Compostela, Spain

María J. García-Sarrió

Instituto de Química Orgánica General (CSIC), Madrid, Spain

Salvador Garrigues

University of Valencia, Burjassot, Valencia, Spain

Javier González-Sálamo

Universidad de La Laguna (ULL), San Cristóbal de La Laguna, Spain

Javier Hernández-Borges

Universidad de La Laguna (ULL), San Cristóbal de La Laguna, Spain

Rosa Herráez-Hernández

University of Valencia, Valencia, Spain

Antonio V. Herrera-Herrera

Universidad de La Laguna (ULL), San Cristóbal de La Laguna, Spain

Miguel Herrero

Institute of Food Science Research (CIAL-CSIC), Madrid, Spain

Elena Ibáñez

Institute of Food Science Research (CIAL-CSIC), Madrid, Spain

John M. Kokosa

Kettering University, Flint, MI, United States; Mott Community College, Flint, MI, United States

Fernando M. Lancas

University of São Paulo, Institute of Chemistry at São Carlos, São Carlos, Brazil

Maria Llompart

Universidade de Santiago de Compostela, Santiago de Compostela, Spain

Rafael Lucena

University of Córdoba, Córdoba, Spain

Mariarosa Maimone

University of Messina, Messina, Italy

Ciara McDonnell

Teagasc Food Research Centre, Dublin, Ireland

Yolanda Moliner-Martínez

University of Valencia, Valencia, Spain

Carmen Molins-Legua

University of Valencia, Valencia, Spain

Luigi Mondello

University of Messina, Messina, Italy; University Campus Bio-Medico of Rome, Rome, Italy

Antonio Moreda-Piñeiro

University of Santiago de Compostela, Santiago de Compostela, Spain

Jorge Moreda-Piñeiro

Grupo Química Analítica Aplicada (QANAP), University Institute of Research in Environmental Studies (IUMA), Centro de Investigaciones Científicas Avanzadas (CICA), University of A Coruña, A Coruña, Spain

Jacek Namieśnik

Gdańsk University of Technology, Gdańsk, Poland

Carlos E. D. Nazario

Federal University of Mato Grosso do Sul, Institute of Chemistry, Campo Grande, Brazil

Katarzyna Owczarek

Gdańsk University of Technology, Gdańsk, Poland

Janusz Pawliszyn

University of Waterloo, Waterloo, ON, Canada

Yolanda Picó

University of Valencia, Valencia, Spain

Merichel Plaza

Lund University, Lund, Sweden; University of Alcalá, Alcalá de Henares, Spain

Justyna Płotka-Wasylka

Gdańsk University of Technology, Gdańsk, Poland

Lourdes Ramos

Institute of Organic Chemistry, CSIC, Madrid, Spain

Miguel Á. Rodríguez-Delgado

Universidad de La Laguna (ULL), San Cristóbal de La Laguna, Spain

Ana I. Ruiz-Matute

Instituto de Química Orgánica General (CSIC), Madrid, Spain

Andrea del Pilar Sánchez-Camargo

Institute of Food Science Research (CIAL-CSIC), Madrid, Spain

María L. Sanz

Instituto de Química Orgánica General (CSIC), Madrid, Spain

Pascual Serra-Mora

University of Valencia, Valencia, Spain

Bárbara Socas-Rodríguez

Universidad de La Laguna (ULL), San Cristóbal de La Laguna, Spain

Ana C. Soria

Instituto de Química Orgánica General (CSIC), Madrid, Spain

Érica A. Souza-Silva

Instituto de Química, Universidade Federal do Rio Grande do Sul (UFRGS), Porto Alegre, Brazil

Natalia Szczepańska

Gdańsk University of Technology, Gdańsk, Poland

Brijesh K. Tiwari

Teagasc Food Research Centre, Dublin, Ireland

Charlotta Turner

Lund University, Lund, Sweden

Jorge Verdú-Andrés

University of Valencia, Valencia, Spain

Eugene Vorobiev

Université de Technologie de Compiègne (UTC), Compiègne, France

顾问委员会

译者序

工业革命以后，尤其是20世纪以来，科学技术有了突飞猛进的发展，分析化学也从传统的化学分析逐步地进入以仪器分析为主的时代。但是，日益频繁的工业生产和人类生活导致了全球性的三大危机：资源短缺、环境污染和生态破坏。想必以下这些场景大家肯定深有体会：雾霾、气候变暖、恶臭的水体、横飞的白色污染……日益严峻的环境问题呼吁我们要善待自然。

从事分析化学研究工作十余年来，每天都在思考自己工作的意义，我们每天在消耗大量有毒有害的有机溶剂，到底是在造福人类还是在污染环境？直到阅读了 *Green Extraction Techniques: Principles, Advances and Applications* 一书后，才豁然开朗，意识到自身工作的意义。善待自然就是善待人类自身，作为从事分析化学的一名研究人员，我们有责任和义务去开发更为环保、绿色的分析技术。

绿色化学又称环境无害化学、环境友好化学，即减少或消除危险物质的使用和产生的化学品和过程的设计。而绿色分析化学是把绿色化学的原理使用在新的分析方法和技术中，旨在减轻分析化学对环境的影响。

Green Extraction Techniques: Principles, Advances and Applications 一书对当前绿色萃取技术的研究热点进行了翔实地归纳，并洞悉了未来的发展趋势。本书是汇编当前所有可用的多种绿色萃取技术及其应用的第一部著作，适合从事分析化学的学生、研究人员等查阅。希望本译著的出版能为国内从事分析化学的人员提供借鉴，通过发展绿色环保的萃取技术来减少分析过程中能源的消耗、有毒有害溶剂的使用，从而为天蓝、地绿、水清的美好环境贡献一份力量。

本书的翻译工作由科研院所和企业中从事分析化学工作的研究人员共同完成，译者均在各自的领域从事分析化学工作多年，具有丰富的经验和扎实的专业知识。本书的第 1 章及第 2 章概述了绿色分析化学的现状，并介绍了加压液体萃取和超临界流体萃取在生物活性物质萃取中的应用，由刘珊珊翻译完成。第 3 章和第 4 章介绍了加压液体萃取的应用，由陈晓水翻译完成。第 5 章和第 17 章介绍了超声提取及其偶联技术的应用，由刘欣翻译完成。第 6 章介绍了微波辅助萃取，由邓惠敏翻译完成。第 7 章和第 8 章介绍了离子液体和免溶剂萃取的原理和应用，由张小涛翻译完成。第 9 章介绍了顶空技术，由王颖翻译完成。第 10 章和第 15 章介绍了微型化固相萃取和搅拌棒吸附萃取，由叶长文翻译完成。第 11 章和第 12 章介绍了 QuEChERS 方法和基质固相分散萃取，由鞠华波翻译完成。第 13 章介绍了绿色分析方法中溶剂的选择，由杨飞翻译完成。第 14 章介绍了在线固相微萃取技术，由陈玉松翻译完成。第 16 章介绍了食品基质中采用固相微萃取进行污染物分析，由韶济民翻译完成。第 18 章和第 19 章介绍了绿色样品制备的新材料和新技术，由严俊翻译完成。熊巍参与了全文校稿。孟冬玲、唐纲岭、边照阳、范子彦、纪元等为翻译工作提供了技术指导。

译者虽力求准确传达原著者的思想，但由于时间关系及水平有限，译文中难免存在疏漏或不当之处，恳请读者批评指正。

最后送读者，乃至地球上生活的我们每个人一句话：减少环境污染，你能做到！

译者

原著者主编序

作为丛书主编，非常高兴在 COAC 系列中介绍这本有关绿色分析化学的新书。这是该主题的第二部书。第一部书是第 57 卷，书名为《绿色化学，理论和实践》，由 Miguel 及其同事 Sergio Armenta 于 2011 年出版。非常感谢 Elena 和 Alejandro 接受我的邀请，编写了这本关于新兴且不断发展的绿色化学主题的新书。

本书共有 19 章，从绿色萃取技术的一般概念开始介绍，其中包括绿色萃取技术的最新进展，绿色萃取技术旨在通过在实验室中使用少量的溶剂来避免化学污染。本书涵盖了所有主要的萃取技术，例如加压液体萃取、超声萃取、微波辅助萃取、免溶剂萃取、小型化技术、搅拌棒吸附萃取、QuEChERS 和固相微萃取等。

本书各章均由在该领域工作多年的著名科学家撰写，涵盖了仪器、环境和食品领域的应用。作为丛书主编，非常感谢所有人花费大量的时间和精力完成了这部全面而独特的绿色萃取方法的书。

本书的编写目的很明确，即希望实验室研究人员和学生在日常工作中使用更环保的分析化学方法。

最后，本书主要面向分析化学和相关学科（例如生物学、环境学和食品学等）的教师、学生以及技术人员。对于行业和不同机构在专业会议上组织的高级课程而言，本书也是一本非常有用的教科书。

D. Barceló
IDAEA-CSIC, Barcelona and ICRA, Girona
《综合分析化学》丛书主编
巴塞罗那，2017 年 4 月

前　言

人们越来越关注我们的环境以及人类对环境的影响方式。如今，污染无处不在，气候变化已经成为一个至关重要的问题，我们对能源和食物的需求以及世界上的人口总数均在不断增长，这是没有任何非科学性的争论。在这种情况下，需要进一步努力寻找替代方法，为所有人提供能源和粮食，并以更绿色环保的方式与我们的星球互动。多年来，科学家一直担心化学对环境的影响。因此，多年前出现了一种被定义为“绿色化学”的新方法，试图尽可能减少化学对环境的影响。在这种情况下，绿色萃取技术的发展可能对减少环境影响具有重要作用，主要是考虑到这些萃取技术的性质、涉及的领域、行业和实验室数量众多。

《绿色萃取技术原理、进展与应用》的 19 章内容充分记录了科学界对开发新的绿色萃取技术以及热点领域中不同趋势和应用的兴趣。本书回顾了绿色萃取技术（包括加压流体萃取、超临界流体萃取、超声和微波辅助萃取、单滴微萃取、搅拌棒吸附萃取、离子液体、固相萃取、固相微萃取等）的最新分析进展，并介绍了主要工作原理及其主要应用。本书还对新兴的绿色萃取方法以及它们的小型化和偶联技术进行了深入的描述和讨论，显示了最近几年中开发的最新技术。这些技术在临床分析、环境分析、食品科学等方面的相关应用也包括在内，从而提供了最具创新性的应用场景，并洞悉了未来的趋势。本书是第一部汇编当前所有可用的多种绿色萃取技术及其应用的著作。

本书的第 1 章概述并讨论了绿色分析化学的现状，以及绿色萃取技术可以发挥的重要作用，包括使用绿色指标来衡量萃取方法的绿色程度。第 2 章和第 3 章分别关注加压流体萃取（包括加压液体和超临界流体萃取）和加压热水萃取在获得生物活性化合物方面的应用。第 4 章重点介绍加压液体萃取环境和食品中的有机污染物。第 5 章介绍了超声萃取生物活性化合物和污染物的方法。第 6 章介绍了微波辅助萃取环境中农药和新兴污染物的方法。接下来的六章分别深入介绍了样品制备中离子液体的基本原理、应用和发展趋势（第 7 章），免溶剂萃取（第 8 章），挥发性物质顶空技术（第 9 章），微型化固相萃取（第 10 章），QuEChERS 方法（第 11 章）和基质固相分散萃取（第 12 章）。第 13 章概述了如何为绿色分析方法选择合适的溶剂微萃取模式。在本书的最后几章中，读者可以找到有关在线固相微萃取（第 14 章），搅拌棒吸附萃取（第 15 章），食品基质中采用固相微萃取进行污染物分析（第 16 章），辅助萃取技术的偶联（第 17 章），绿色样品制备的新材料（第 18 章）和全二维色谱法中的绿色样品制备技术（第 19 章）的最新研究进展和发展趋势。

本书的读者面非常广，主要面向从事、应用或研究与绿色萃取技术有关的不同方向的专业人员，包括来自与控制化学、医学、生物科学、食品、营养、临床、天然产物、植物化学、环境科学、毒理学、法医学、制药科学等相关的控制实验室、大学、行业和监管机构的研究人员和学生。

作为《绿色萃取技术原理、进展与应用》一书的主编，要感谢为本书做出杰出贡献的所有作者，感谢 Damiá Barceló 邀请我们参与这份工作，感谢 Shellie Bryant 和 Vignesh Tamil 的帮助和支持，同时也感谢 Elsevier 团队的工作，他们为本书的编写做出了自己的贡献。

最后我们也要感谢那些每天挑战自我，积极为更美好的世界而努力的人们。作为研究人员，我们很幸运能够拥有这一动力，帮助我们在打破愚昧和障碍的同时建立科学。

Elena Ibáñez&Alejandro Cifuentes

Laboratory of Foodomics，CIAL，Madrid National Research Council of Spain（CSIC）

目录 CONTENTS

1　绿色分析化学：绿色萃取技术的作用

2　加压液体萃取和超临界流体萃取植物、海藻、微藻和食品副产品中的生物活性物质

3　加压热水萃取生物活性物质

4　加压液体萃取环境和食品样品中的有机污染物

5　超声波：一种用于生物活性物质和污染物的清洁、绿色萃取技术

6　微波辅助萃取环境中的农药及新兴污染物

7 离子液体在样品制备中的应用

8 免溶剂萃取技术

9 挥发性样品的顶空分析技术

10　微型固相萃取

11　QuEChERS 技术的最新研究进展

12　基质固相分散萃取

13　绿色分析方法中选择合适的溶剂进行微萃取

14 在线管内固相微萃取技术的发展趋势

15 搅拌棒吸附萃取

16 固相微萃取技术在食品污染物分析中的进展

17 辅助萃取偶联技术的进展

18 绿色样品制备技术中的新材料：研究进展与发展趋势

19 全二维色谱中的绿色样品制备技术

1　绿色分析化学：绿色萃取技术的作用

Sergio Armenta, Francesc A. Esteve - Turrillas, Salvador Garrigues and Miguel de la Guardia*
University of Valencia, Burjassot, Valencia, Spain
*通讯作者：Email：miguel. delaguardia@ uv. es

1.1 绿色分析化学：直面问题

保罗·阿纳斯塔斯（Paul Anastas）定义了绿色化学的起源和支撑这一理论的12条原则[1]。制造更环保的化学品、工艺或产品的原则包括：①预防；②原子经济；③危险性较小的化学品合成；④设计更安全的化学品；⑤更安全的溶剂和助剂；⑥设计能效；⑦使用可再生原料；⑧减少衍生物；⑨催化；⑩降解设计；⑪防止污染的在线分析；⑫更安全的化学以预防事故发生。由此可见，上述许多原则都可以直接应用于分析方法的开发和使用。

就其本身而言，绿色分析化学的历史与绿色化学并驾齐驱，始于1995年皇家化学学会（Royal Society of Chemistry）一篇题为“通过小型化、控制和试剂替换走向环境友好的分析化学”的社论[2]。从那时起，一系列致力于绿色分析化学的书籍[3-6]和文章[7-11]相继出版。

Galuzska[12]等将绿色化学的12条原则加以修改来定义绿色分析化学，包括：①采用直接分析技术避免样品处理；②以最小样本量和最少样本数为目标；③进行现场测量；④分析过程与操作相结合可节约能源并减少试剂的使用；⑤选择自动化和小型化方法；⑥避免衍生化；⑦避免产生大量分析废物，并为产生的分析废物提供适当的管理措施；⑧多分析物或多参数方法比一次仅分析一种化合物的方法更可取；⑨最大限度地减少能源消耗；⑩首选从可再生资源获得的试剂；⑪淘汰或更换有毒试剂；⑫提高操作人员的安全性。这12条原则同时也采用了有助于记忆的单词“significance”来表示（图1.1）。

分析仪器的发展，尤其是高性能实验仪器的发展，遵循了奥林匹克座右铭（更快、更高、更强），不断追求对方法的主要分析性能的改进，从而寻求最大限度的灵敏度、选择性和精密度。然而，高性能的分析仪器通常相当昂贵，因此通常无法实现直接、原位和/或实时分析。此外，它能耗很高，分析化学实验中通常需要材料和试剂的消耗，从绿色分析化学的角度来看这也是最浪费的，尤其是若将仪器生产的生命周期评估（LCA）[13]考虑在内的话。在这个中心思想下，就很容易理解“适用性”概念的重要性了；根据ISO/IEC指南2[14]，适用性是指产品、方法或服务在特定条件下服务于特定目的的能力。在分析化学中，它可以理解为选择一种适当的分析方法，以获得足够高质量的数据，以便能尽快、高效地对特定目的做出决策[15]。因此，用Koel教授的话说，“即便没有超高性能的实验室仪器，根据传感器的输出做出决策并非不可行”[13]。这意味着，当较为先进的基于传感器的方法，其分析响应足以解决问题的情况下，则不需要使用复杂的保守方法。但是，在必须要使用这些保守方法时，从绿色分析的角度出发，方法筛选的应用也可以在使用参考的保守方法分析时帮助排除阴性样本[16]。事实上，使用关注点和一般情况下的需求点筛选法作为筛选工具，通常将具有解决诊断、污染预防或在工业生产中检测故障问题的能力。相反，要获得关于各种样品或情况尽可能多的信息，就需要更多高度先进的分析方法，当然在大部分传感器的校准过程中，必须使用高性能的方法来获得参考数据。

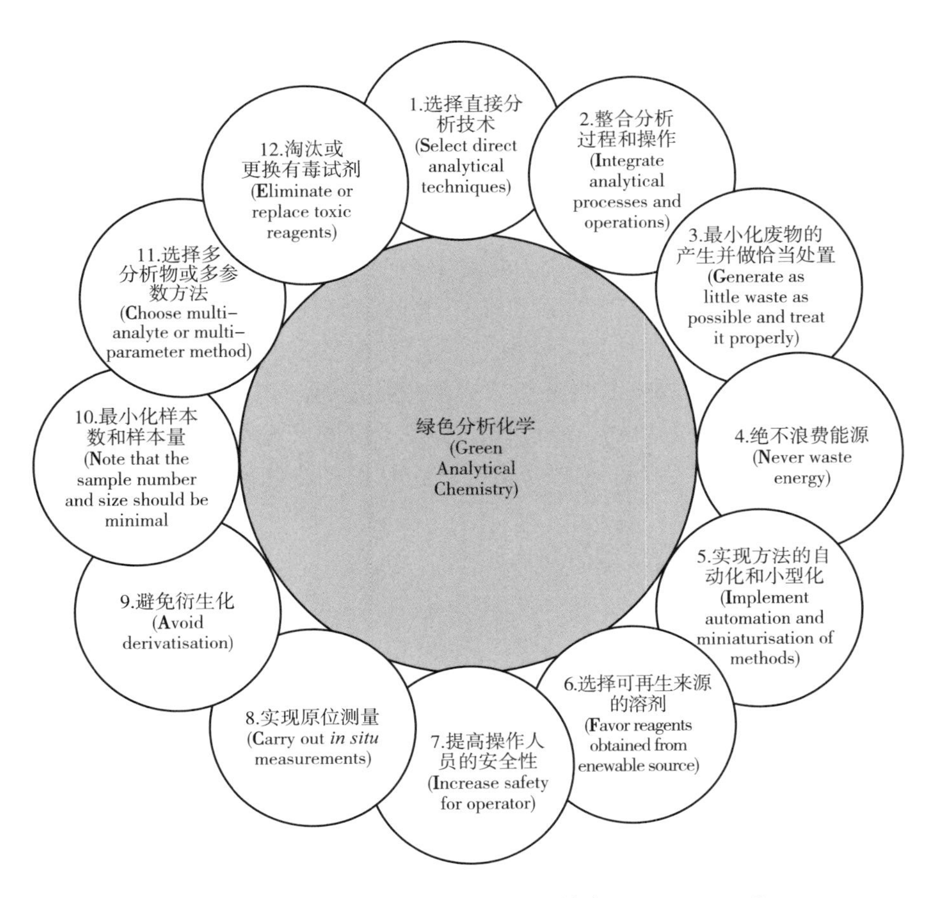

图 1.1　绿色分析化学的 12 项原则总结为 “significance”

［资料来源：A. Gałuszka, Z. Migaszewski, J. Namieśnik. The 12 principles of green analytical chemistry and the SIGNIFICANCE mnemonic of green analytical practices, TrAC Trend. Anal. Chem. 50（2013）78-84.］

因此，最绿色的分析方法必须是最适合解决问题的、具有最高准确性和对操作员及环境危害性最小的方法[17]。

1.2　从“梦想”到“绿色”

我们可以认为分析过程是一个在某特定序列中，以定义分析的问题为开端，继而发生的一系列步骤；接下来是样品的选择和采样、运输和保存、样品处理、目标分析物的分离和/或检测以及结果收集。当完成对结果的解释、方法模型的开发和对先前定义问题的解答，此过程也随之结束。显然，分析程序的主要目标应该是为起初的问题提供充分的答案，因此，分析的焦点应该是问题本身而不是样品，我们必须从问题中决定要采集的样品、要测定的分析物和要使用的方法[18]。

为了达到解决问题的目的，在分析方法中必须满足几个条件。例如：①所取样品的代表性；②所有标准、设备、仪器和操作在一定程度上的可追溯性；③所有分析方法步骤的有效性；④对样品可追溯性的持续控制。综合以上考虑，绿色分析化学的“梦想”是开发能够直接进行现场测量或遥感测量的程序，在解决起始问题的同时，亦能实现上述所有条件。换言之，绿色分析化学家的“梦想”意味着取消样品处理步骤，因为前处理是最具有分析物损失或污染风险的操作之一，同时该步骤也对环境有害，不仅因为溶解、提取或消解样品需要大量的有机溶剂和/或无机酸，而且在许多情况下还需要对样品进行热处理[19]。因此不难想象，从绿色分析化学的角度来看，遥感测量、无创检测和直接分析等策略应该被认为是最绿色的策略。然而，正如前面所述，很明显这些方法并不总是解决问题最恰当的方法，在许多情况下，仍推荐使用包含了采样、运输和预处理步骤的分析程序。总而言之，多数情况下我们的目标是“绿色”而不是“梦想”；因此，制定绿色战略以改善和减少样品处理的副作用，预期在未来几年实现有价值的绿色方法。正因为如此，绿色样品制备方法的开发是绿色分析化学最活跃的研究领域之一。在这个架构下，本书完全有理由从旧的方法转移到一种新的加强版的分析物提取方式。

1.3 绿色指标

鉴于此，我们的“绿色分析方法”需要一个工具以鉴定其绿色的程度。近年来在该方面发表了几项研究，强调了所谓的绿色指标在评估分析程序的绿色性以及选择方法时将这些指标作为附加标准的重要性。

评价方法的绿色程度的第一个准则是基于 2002 年建立的国家环境方法指南（NEMI）[20]。NEMI 是一个环境友好分析方法的数据库，是根据四个关键术语进行评估：①所用试剂的 PBT 性质（持久性、生物累积性和毒性）；②过程和产物的危险性；③所用试剂的腐蚀性；④产生的废物（图 1.2）。尽管这是衡量绿色的第一种方法，但该方法显然仅是出于定性的考虑；其更关注所涉及试剂和方法的副作用以及产生的废物，且明确禁止了能源消耗等重要方面。

Raynie 和 Driver[21] 对 NEMI 的概念进行修改，提出了绿色评估概要，增加了鉴别范围，将能源消耗作为一项重要的标准。同时也将 NEMI 危害贡献象形图分为三个不同的分区，分别表达对健康、安全和环境的危害（图 1.2）。

Garrigues 和 de la Guardia[6] 还对 NEMI 象形图进行了修改，其中，使用从红色到黄色再到绿色三个色标度评估了所用试剂和工艺的风险，即 PBT 危险和消耗，包括操作员风险和能源（图 1.2）。这个彩色的方案不仅延续了 SEMI 法的直观，并在其中结合了半定量。

2012 年，Galuszka 等[22] 将 Van Aken 等[23] 为评估有机合成的绿色性而提出的扣分制和生态坐标的概念应用于分析方法的评估。该评估是基于将扣分制应用于表达分析方法与理想的绿色分析方法之间的差异，其中理想的绿色分析方法为 100 分。扣分涉及试剂性质和能源消耗、与使用试剂和溶剂有关的危害以及产生的废物。但是，此过程并不能区分方法应用的微观、中观和宏观尺度。于是在这个层面上对生态坐标提出了修改[24]，基于加权后的扣分制，将每个颜色代码用字母 A～G 来表示，A 是最绿色的

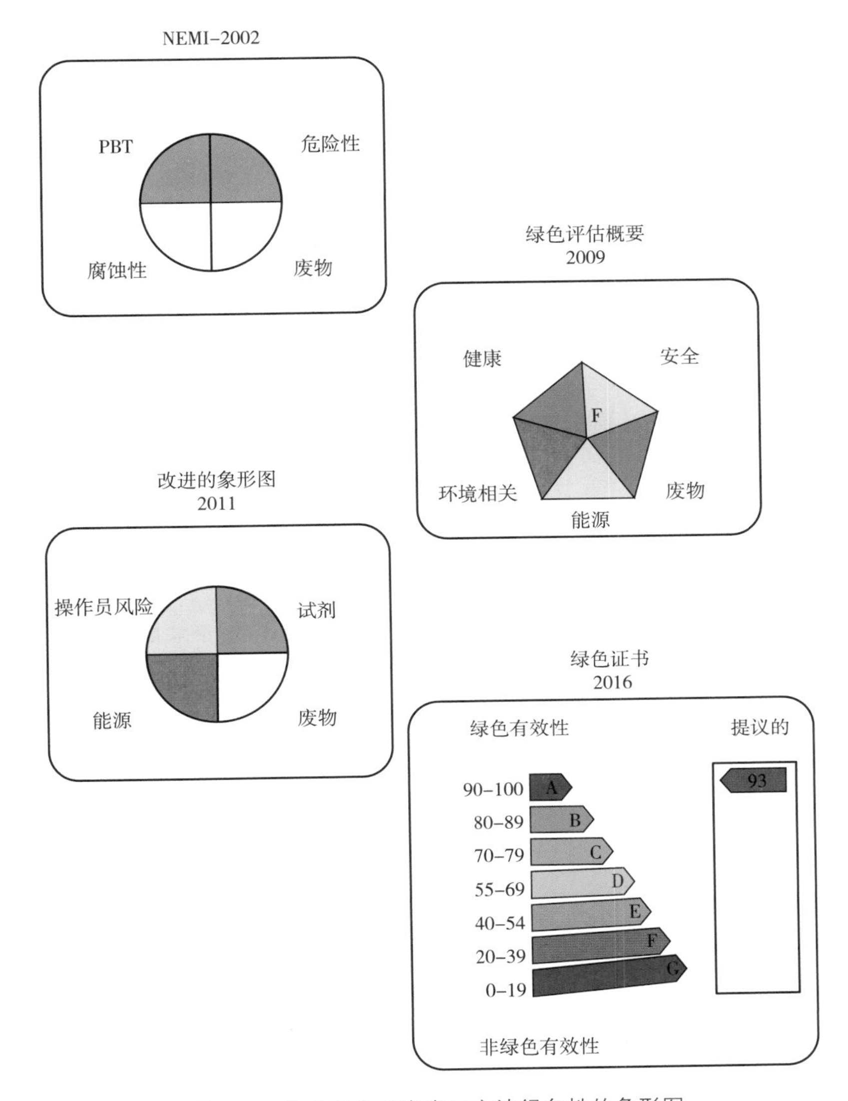

图 1.2 在文献中用来表示方法绿色性的象形图

字母，这便是绿色证书。分析方法的总分是使用数学方程式而非阶梯标准来评估扣分（图 1.2）。

此外，还提出了专门的绿色指标来评估色谱领域的危害，特别是与液相色谱中流动相的使用相关的危害。高效液相色谱-环境评估工具评估流动相中使用的每种溶剂的安全、健康和环境因素，还包括溶剂消耗[25]。它的主要优点是简单；但获得的结果是一个单一的数字，不提供其潜在威胁的相关信息。分析方法的体积强度是根据每个感兴趣的色谱峰的溶剂消耗量来评价液相色谱方法的绿色特性的过程[26]。此过程考虑了方法的多分析能力，但同样未能指明所用溶剂的性质和特征。

前面提到的所有绿色衡量程序都试图以简单直观的方式表明被评估方法的绿色特征，但有时无法明确地表明环境和健康方面的特征。多变量方法的提出，正是为了避免这个问题，即使用化学计量学工具将分析过程归纳为其绿色特征的函数，如聚类分析、主成分分析和自组织映射[27]。这个方法的主要优点是容易识别最绿色和最不绿色的对象，允许变量之间的正相关和负相关[28]。

此外，分析方法的绿色性还利用多准则决策分析进行了比较，该分析中对客观绿度进行了定义，并通过定量标准来对客观进行描述；每个定量标准通过其权重来衡量其相对重要性，并通过多种算法来确定方法涉及的参数与理想情况之间的差异或相似之处。因此测定废水中药物的几种不同的分析方法，就是按与理想溶液相似的优先顺序等技术进行评估的[29]。绿色分析指标的另一种算法是富集评价的偏好排序组织法，该算法已应用于水样中农药测定的不同方法的评价，所得结果与生态坐标[30]的分析结果相吻合。简而言之，公开推荐方法的绿色参数，形成一个简单的指南去指导如何针对分析问题选择可持续的工具，这一点对分析领域非常重要。

1.4 萃取步骤：优势与难点

如上所述，大多数样品不能直接测量或直接进入分析仪器，因此在测定之前应该对它们进行物理和/或化学前处理。样品处理步骤中可以包括不同的过程，如样品均质、分析物提取和预浓缩、样品矿化、样品和目标分析物的过滤或蒸馏、样品萃取液的净化。在任何情况下，样品前处理的选择不仅取决于待测样品和目标分析物的性质，还取决于要采用的分析技术及其能力。许多分析方法的样品前处理中都需要萃取这一步骤，以将目标分析物从固体/液体样品中分离出来，减少基体干扰，有时也会对分析物预浓缩以达到所需的灵敏度水平（图 1.3）。

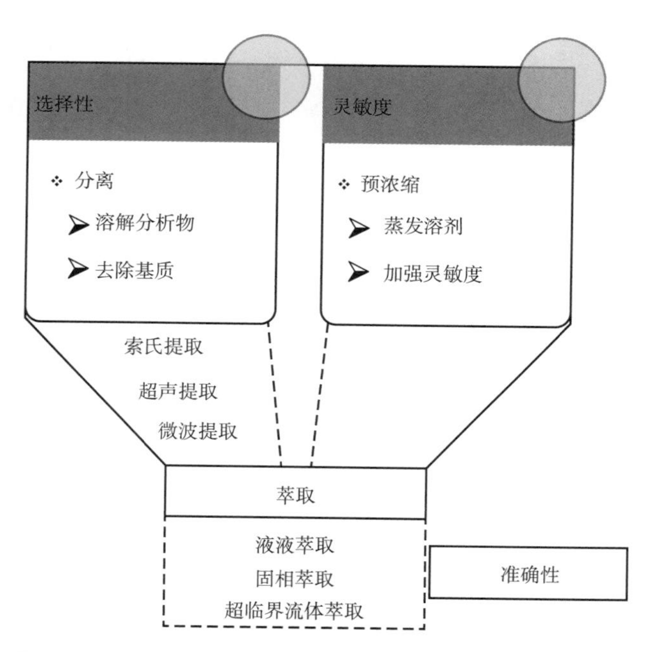

图 1.3　从样品基质中分离分析物到分析物浓缩的萃取过程

在现代分析化学中，类似于经典的湿法和干法灰化消解等强破坏性方法，已被一些适用于从样品基质中定量提取目标分析物的温和技术所取代。温和的前处理不需要将样品完全销毁，也不需要所有样品成分完全溶解。这种方法从经济、环境和基本分析的角度来看，具有很大优势。较困难的样品溶解步骤通常需要高能耗和高预算，可能会造成分析物损失或污染，而且在实际检测过程中，共溶解的杂质也会带来潜在的干扰。正因如此，温和的前处理技术是将分析物从样品中分离出来的极佳选择。

从固体样品中将分析物溶解，是索氏提取或溶剂分配等传统方法及超声辅助萃取（UAE）或微波辅助萃取（MAE）等现代方法的目标，是指在不进行样品的完全消解的前提下，能够将固体中的分析物定量转移到液相的部分溶解过程。

不要忘记，萃取过程的主要目的之一是对分析物进行预浓缩，以便能够进行痕量或超痕量分析。考虑到仪器信号和分析物浓度之间的关系为：

$$S = KC \tag{1.1}$$

式中，S 为信号，C 为分析物浓度，K 为校准曲线的斜率。IUPAC 称之为方法的灵敏度。浓度可以表达为分析物质量除以测量体积，因此方程（1.1）可以写为

$$S = K(\text{分析物质量} / \text{测量体积}) \tag{1.2}$$

根据这个方程，可以得到三种主要的提高灵敏度的方案：①放大；②增敏；③预浓缩。用于扩增 DNA 片段的聚合酶链式反应是一个通过放大来提高灵敏度的例子，它是一种增加选定样本体积中的分析物质量的方法。增敏可以定义为提高 K 值的不同策略，如长光程测量单元、测量介质的改性、表面活性剂等信号增强剂的加入以及在拉曼光谱中引入反射元素等。然而，放大和增敏都非常依赖于其存在体系，并非是每个分析物及每种测量技术都能应用的通用性策略。相反，预浓缩只减少样品体积而不损失分析物质量，因此被认为是提高灵敏度的常用策略。不仅如此，如图 1.3 的方案所示，萃取不仅仅是一种大幅减少样品体积的预浓缩策略；在某些情况下，萃取能够选择性地将目标分析物与样品中存在的其他化合物分离以消除潜在的干扰。

每种提取技术都有其优缺点，如何选择最合适的方法，主要取决于几个方面：是否容易获取；获得及操作的成本；提取溶剂是否简单；溶剂性质及使用量；是否绿色环保[31]。图 1.4 总结了从绿色分析化学的角度选择最合适的萃取方法时要考虑的主要因素。接下来会深入探讨：①所需溶剂和试剂；②能耗；③仪器设备成本和萃取能力；④产生的废物这四项因素。

必须考虑的是，传统的人工进行液-液萃取（LLE）的方法现已被商业化的设备所取代，尽可能地将萃取步骤与分析物的分离和测量结合起来。用连续前处理取代传统的批处理模式，为降低操作人员的风险和试剂的消耗带来极大的可能性。

从环保角度来看，必须考虑萃取设备制造过程中涉及的额外二氧化碳的排放。但从绿色分析化学的观点来看，关键在于操作条件。

若使分析过程更加绿色环保，那么萃取过程中使用的溶剂的性质和体积是要考虑的主要方面。为了寻求所用溶剂和能源需求之间优劣的最佳平衡点，就必须要考虑使用少量无毒或至少是低毒的溶剂。样品与萃取相之间的温度、压力和接触时间均对环境有不良副作用，因此，不能一味地采用简单方法只将试剂消耗和废物产生最小化。

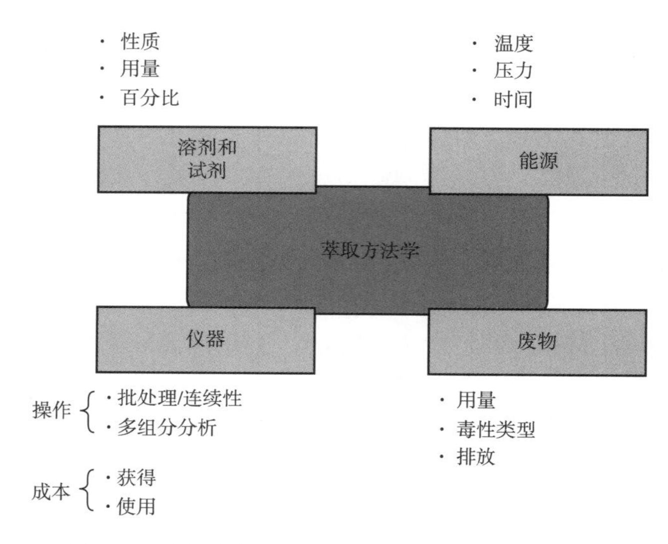

图 1.4 选择最合适的提取方法时要考虑的主要因素

上述问题将在接下来的小节里详细讨论。

1.4.1 萃取溶剂

绿色分析化学在样品提取领域要达到的主要目标之一是引入无溶剂步骤、降低溶剂消耗，并用绿色溶剂取代标准提取溶剂[25]。那么随之而来的问题是：什么是绿色溶剂？图 1.5 的金字塔示意图，给出了按照环境友好性及对操作员的环境、健康和安全（EHS）的危害进行分类的常用萃取溶剂。使用无溶剂是最环保的方法，但应用无溶剂萃取的方法仅限于使用固相微萃取法（SPME）、搅拌棒吸附萃取法（SBSE）或顶空（HS）直接进样。其他萃取方法都必须使用溶剂，这种情况下，水是最佳的绿色溶剂。

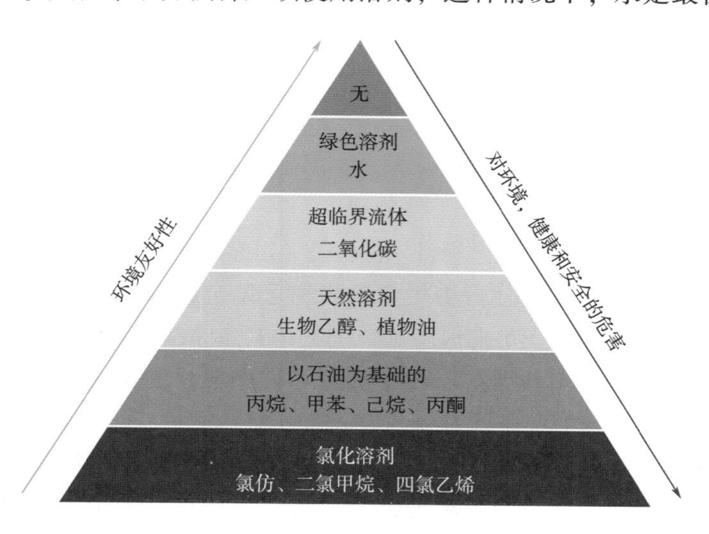

图 1.5 从绿色分析的角度对萃取过程中使用的溶剂类型采用分层方法进行分析

像二氧化碳这种超临界流体的使用，由于其低毒性而具有较大的环境优势，但这种技术通常需要使用少量的有机改进剂，例如使用甲醇或甲苯作为极性改进剂，并且超临界条件意味着需要很高的能源消耗以维持运行所需温度和压力。

溶剂的选择最主要的问题还是对绿色溶剂的识别或定义；经常会有些观点或结论采用单一的论点来识别或定义绿色溶剂，例如其是否来源于生物衍生物，亦经常忽略溶剂在其合成过程中需要的化学物质，尤其当该溶剂本身的绿色特性已被评估时。基于此，Capello 等[32]根据 EHS 特性和 LCA 生产方法的应用，以及溶剂的潜在回收和处置，提出了一种根据绿色特性对常见有机溶剂进行分类的程序。从生物乙醇或植物油等可再生资源中获得的有机溶剂因其毒性低、来源可再生而被认为是潜在的绿色溶剂。绿色方法不推荐使用从石油、油脂和天然气等化石资源中获得的溶剂，除非在浓度很低的情况下。然而，从生命周期分析的角度来看，甲醇、乙醇和乙酸甲酯等溶剂显然是首选的溶剂；而从健康和安全的角度来看，庚烷、己烷和二乙醚是首选的溶剂。无论如何，那些属于高度关注物质的溶剂必须避免使用，如二氧六环、乙腈、甲酸和乙酸、甲醛、四氢呋喃、氯化溶剂、*N*-甲基吡咯烷酮、*N*,*N*-二甲基甲酰胺和某些乙二醇醚[33]。

值得深入讨论的是过去被认为是绿色溶剂的离子液体的使用。正如 Jessop 教授在其综述《寻找绿色溶剂》中所述[34]，离子液体并不比他们取代的传统溶剂更环保。Zhang 等[35]使用 11 个描述符比较了水、丙酮、苯和 1-丁基-3-甲基咪唑四氟硼酸盐离子液体等几种溶剂对环境的影响，在几乎所有使用的描述符中，离子液体被归类为最糟糕的溶剂。Capello 教授等[32]也得出了类似的结论，他们指出，生产烷基咪唑等离子液体是非常耗能的；它的生物降解性很差，需要更多的毒理学数据来评估其绿色特性。简而言之，可以得出结论，水是萃取极性化合物最环保的溶剂；超临界 CO_2 和所谓的农业用溶剂，是相比于化石来源溶剂和含氯溶剂而言的更好选择。然而，为了正确评估所选溶剂是否为最合适的提取溶剂，需要额外考虑所需溶剂的量和使用混合溶剂的可能性。

1.4.2 能源消耗

能源的产生涉及原材料和电力的消耗，同时根据原材料的不同，也涉及废弃物的产生和对空气、水和土壤的排放。基于核能和化石燃料产生的电力，分别由于其放射性残留物的危害和大量温室气体排放，从而并不适用于绿色方法[36]。而太阳能、生物质能、风能、海洋和水力发电等可再生能源被广泛认为是绿色能源。

萃取方法通常依靠电力才能完成工作。搅拌器、振荡器、真空泵和其他小型实验室设备（分析天平、电动移液器）消耗的能量很少。然而，使用高温和（或）高压的高效萃取设备，如 MAE、加压溶剂萃取（PSE）[或加压液体（PLE）]、超临界流体萃取（SFE）和顶空进样都需要相当大的能耗。然而，考虑到提取时间的减少和所需溶剂的量的减少，从绿色分析化学的观点来看，可以预见上述方法的能量消耗高。

通常使用旋转蒸发器或氮气蒸发器通过去除溶剂来进行分析物的预浓缩，这联合了热源、真空泵和旋转电机的使用，消耗能量在中等水平。但是，诸如 SpeedVac 之类的高效蒸发系统，由于使用了冷阱冷却装置而需要很高的能耗。

基于以上的原因，显然要考虑到所用设备的消耗功率以及对电能的来源、提取步骤中涉及的能耗进行深入评估。

1.4.3 所需仪器

仪器的适用性，是选择从样品中提取目标分析物的最绿色程序的主要限制。因此，还必须考虑与设备购置、运行成本、设施条件和安全条件有关的问题。因此，从绿色分析化学的角度出发，建议使用既不需要大量的一次性物品、也不需要消耗品的、成本可承受的、容易获得的提取方法。

使用快速提取和机械化程序，能够大大提高样品通量，提高萃取步骤的绿色性；从这个意义上讲，允许多个样品同时萃取的程序比单个样品依次萃取的方法更环保。

同样，通常建议在建立针对一种或少量化合物的特定且高选择性的方法之前，先对较大范围的目标化合物进行多组分提取。但应该记住的是，我们的目标是遵循“适用性”原则去开发方法，而不是在解决现有问题之外寻找更高的选择性或敏感度。

所考虑的每个方面都对分析物或样品的总能量和溶剂消耗有很大贡献，从而证明了不同变量之间的相互联系，这些变量必须一起评估，为提取过程的可持续性提供补充观点。

1.4.4 废物和排放

溶剂和试剂消耗是确定提取方法是否绿色的一个重要变量，但也必须考虑产生的废物和潜在的排放。萃取方法通常需要使用溶剂从样品中溶解目标分析物，然后直接或在蒸发后对其进行分析，以对目标分析物进行预浓缩并提高测定的灵敏度。在这两种情况下，都必须按照良好的处置规范将一定量的溶剂视为废物。溶剂的处置通常通过蒸馏或焚烧技术进行[37]。LCA 研究可以评估这两种处置方法的利弊，它们适用于常被用作萃取剂的各种溶剂，更有利于环境保护，如通过蒸馏回收乙腈、四氢呋喃、丙酮、环己烷或甲苯；焚烧甲醇、乙醇、戊烷、己烷和庚烷以产生电能[32]。

考虑到 EHS 危害影响，分析化学实验室必须控制溶剂气体的排放；这类排放通常集中在萃取系统附近。因此，必须采用足够的通风和使用高效的排风罩，以降低操作人员的职业风险。但依然会有大量的溶剂排放到大气中；因此，在萃取过程中，溶剂蒸汽向空气中的排放通常会通过某些绿色度量指标显示出来[37]。

1.5 现代分析提取方法的绿色评价

为了清楚的描述此概念，以固体和液体样品中提取多环芳烃（PAH）为例，比较了传统提取技术到现代提取技术针对固体和液体样品的绿色差异。评价的固体萃取方法有索氏提取法、机械振荡法、UAE 法、PSE 法、MAE 法、咖啡机浓缩萃取法、SFE 法和 SPME 法。

另一方面，选择的液体萃取方法有 LLE 法、固相萃取法（SPE）、SBSE 法、SPME

法、悬浮有机液滴微萃取法、分散液液微萃取法与分散 μ-SPE 联用法（DLME-μ-SPE）。从土壤和水样中提取多环芳烃的实验条件取自文献，具体分别见表 1.1 和表 1.2。

表 1.1　从土壤中提取多环芳烃的提取方法及相应的绿色证书值

萃取方法	样品量/g	使用溶剂	溶剂体积/mL	时间/min	扣分			绿色证书值	参考文献
					溶剂	废物	能量		
索氏提取	20	正己烷/丙酮/甲苯	100/50/10	720	18.0	12.0	4	66	[38]
	30	甲苯	100	480	7.6	10.5	4	78	[41]
	1	正己烷/丙酮	125/125	360	16.4	13.7	4	66	[42]
	2	正己烷/丙酮	30/30	120	10.5	7.8	4	78	[43]
	1	甲苯	80	480	7.1	8.7	4	80	[44]
振荡提取	10	丙酮/石油醚	100/50	15	13.3	11.4	2	73	[38]
	20	水/丙酮/二氯甲烷	50/100/75	720	9.7	12.4	2	76	[39]
	20	丙酮/石油醚	100/100	60	15.3	13.0	2	70	[39]
UAE	20	正己烷/丙酮/甲苯	62/31/6	30	15.6	10.2	2	72	[38]
	5	二氯甲烷	25	120	3.3	5.8	2	89	[39]
	1	正己烷/丙酮	10/10	15	7.5	5.1	2	85	[40]
	2	甲苯	90	60	7.4	9.2	2	81	[42]
	5	乙腈	100	30	5.1	9.7	2	83	[43]
PSE	2	正己烷/丙酮/甲苯	19/9/2	15	10.7	6.0	2	81	[38]
	1	正己烷/丙酮	5/5	11	6.0	3.9	2	88	[40]
MAE	1	正己烷/丙酮	15/15	10	8.5	5.9	2	84	[40]
	2	甲苯/水	10/1	20	3.7	4.1	2	90	[44]
	10	甲苯	30	30	5.3	6.6	2	86	[42]
咖啡机浓缩萃取	5	乙腈/水	20/30	20/30	3.6	7.5	1	88	[43]
SFE	5	甲醇/二氯甲烷/甲苯	0.4/1.6/1.3	0.4/1.6/1.3	4.8	3.5	3	89	[38]
	1	甲苯	11.5	11.5	3.9	4.1	3	89	[42]
SPME	0.1	水	10	10	0.0	3.8	2	94	[45]

注：MAE—微波协同萃取；PSE—加压溶剂萃取；SFE—超临界流体萃取；SPME—固相微萃取；UAE—超声协同萃取。

表 1.2　水样中多环芳烃测定的提取方法及相应的绿色证书

萃取方法	样品量/g	使用溶剂	溶剂体积/mL	仪器	扣分				绿色证书值	参考文献
					溶剂	废物	职业危害	能量		
LLE	800	正己烷/二氯甲烷	50/50	—	12.3	9.5	2	0	76	[46]
	1000	正己烷	250	—	13.5	13.7	2	0	71	[47]
SPE	100	甲醇/丙酮/四氢呋喃	17/1.5/1.5	真空	7.9	5.0	1	1	85	[48]
	1000	二氯甲烷/异丙醇/乙醇/甲醇/乙腈	20/20/20/20/10	真空	16.4	9.1	1	1	73	[49]
SBSE	10	丙酮/乙腈	36/0.2	搅拌器	4.4	6.3	1	4	84	[50]
	20	甲醇/二氯甲烷/乙腈	5.1/5/0.1	搅拌器、超声、烘箱	5.6	3.8	1	4	86	[51]
SPME	100	—	—	烘箱	0.0	0.0	0	2	98	[52]
	10	乙腈	0.25	搅拌器	0.8	0.9	0	2	96	[50]
	12	—	—	超声、烘箱	0.0	0.0	0	4	96	[53]
FODME	20	1-十一醇	0.01	搅拌器、冷水机组	0.3	0.2	0	3	97	[54]
DLLME-u-SPE	20	正辛醇，乙腈	0.02/0.1	搅拌器	0.8	0.6	0	1	98	[55]

注：DLLME-u-SPE—分散液液微萃取与分散微固相萃取偶联；FODME—漂浮有机液滴微萃取；LLE—液液微萃取；SBSE—搅拌棒吸附萃取，SPE—固相萃取；SPME—固相微萃取。

每种所选的方法都进行了绿色分析评价，评价不是针对整个分析方法而是只考虑了萃取步骤，这样就避免了对分离和检测步骤的评价。绿色证书是所选择的绿色度量工具，因为它可以快速直观地评估要比较的方法并进行量化比较。

如前所述，绿色证书是基于与字母相关联的颜色代码的应用；从 A 到 G，A 是最环保的。A 类的分析方法是扣分少于 10 的分析方法，B 类方法的扣分总和在 11~20 分之间，C 类方法的扣分在 21~30 之间，D 类方法的扣分总和在 31~45 分之间，E 类方法的扣分总和为 46~60 分，F 类方法的扣分总和为 61~80 分，G 类方法则扣分达 81 分以上。

本评估将考虑试剂毒性、能源消耗、使用的试剂和溶剂的数量、职业暴露危害、对环境的有害排放，特别是提取步骤中产生的废物数量等参数（图 1.6）。根据使用的试剂或产生的废物量计算的扣分公式见公式（1.3）：

$$y = (ax^b) \tag{1.3}$$

式中，y 是计算出的扣分，x 是试剂/废物的量（g），试剂消耗量为 $a = 0.61 \pm 0.05$、$b = 0.31 \pm 0.02$，废物产生为 $a = 1.50 \pm 0.08$、$b = 0.40 \pm 0.02$。获得此方程的更多信息

详见参考文献［24］。此外，计算出的试剂消耗罚分应乘以《全球危险系统分类和标签》中相应试剂的象形图数量。

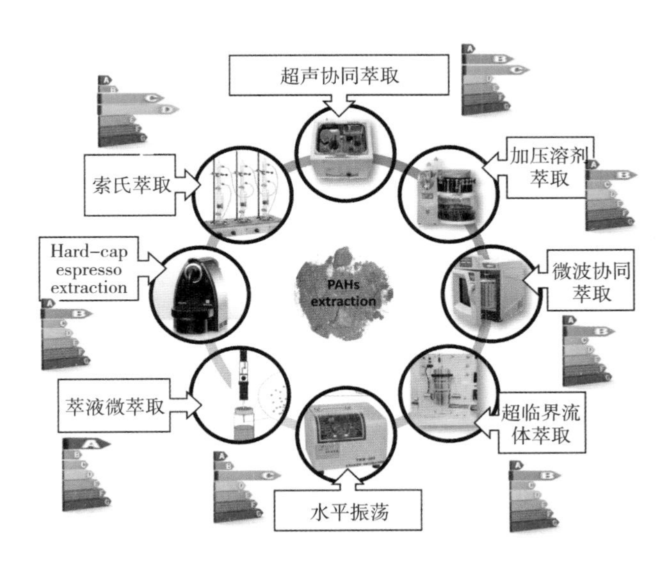

图 1.6 文献报道的关于多环芳烃萃取的绿色认证的经典和现代方法

除 LLE、SPE 和 SBSE 三种方法与职业暴露相关的扣分建议分别为 2、1 和 1，其余所有萃取方法的扣分都默认为零。

能源消耗扣分的指定依据是 Raynie 和 Driver[21] 在有关样品分析所涉及电力方面给出的值。每个样品耗电值小于或等于 0.1 kWh 时（0.04455 kg CO_2）将被扣 0 分；每个样品耗电值在 0.1~1.5kWh 之间（0.04455~0.668 kg CO_2）将被扣 1 分；每个样品超过 1.5 kWh 将被扣 2 分。这些扣分已根据所用设备的用电量和所需的提取时间进行了微调。就固体样品的提取方法而言，指定索氏提取法扣 4 分，SFE 扣 3 分，其余提取方法扣 2 分，但咖啡机浓缩萃取因其所需萃取时间非常短（0.2min）而仅扣 1 分。液体样品萃取方法中与能耗相关的扣分设置为 0 到 4 分，取决于分析用到的设备数量（如真空泵、搅拌器、超声浴、烘箱或冷冻机）及其各自的耗电量。

从表 1.1 和图 1.6 可以看出，从固体样品中提取 PAH 的绿色证书值在 66~94 之间。索氏提取，振荡和 UAE 萃取的绿色证书值最低，分别在 66~80、70~76 和 72~89 之间，取决于所使用的萃取条件；PSE，MAE，SFE 和咖啡机浓缩萃取的绿色证书值高于 81；而 SPME 的值最高，为 94。

关于液体样品中多环芳烃提取方法的绿色指标（表 1.2），LLE 和 SPE 显示的绿色证书值分别在 71~76 和 73~85 之间，最不理想；SBSE 绿色证书值处于均值水平，在 84~86 之间；而 SPME、FODME 和 DLLME-μ-SPE 方法的绿色证书值高于 96。

所评价的多环芳烃萃取技术绿色证书值覆盖了一个较宽泛的范围。因此，应选择最合适的萃取方法来解决分析问题，不能只考虑是否操作简单和分析速度快不快，还要考虑方法的绿色参数。采用传统方法的绿色证书值通常最低，但对于相同萃取技术，所获值具有很大的可变性。因此，这意味着当因为缺乏适用性、成本高或设备故障而不能采用现代方法的情况下，适当优化萃取条件也可能会增加所用传统方法的绿色效果。

1.6 绿色萃取的未来发展趋势

分析方法的灵敏度和选择性的提高在大部分情况下要通过增加萃取的步骤来实现和优化，特别是需要将样品由固体转移至液体以进行原子光谱或色谱分析的情况。由于萃取过程可能损害实验人员的健康并产生环境污染等副作用，其过程中各个方面都受到越来越多具有环境意识的分析方法的高度重视。因此，将萃取操作从毫升规模缩小到微升级别是如今的趋势，在此基础上，微萃取技术已经在涉及不同类别的分析物和所有适用的萃取策略的领域得到了广泛发展。

同样值得关注的是，在最新的有关萃取方法的文献中，均在寻找能从可再生资源获得的、污染物和有毒物质少的溶剂，来替代那些从化石燃料获得的溶剂。

然而，在不久的将来，萃取步骤的自动化一定会引起分析化学界的关注和重视，因此相关方面仍需改进研究，并结合到整个分析过程中，从而可以将液相微萃取和SPME与进样系统相结合，以避免操作员与萃取剂接触，并大幅减少所需的体积。

此外，萃取步骤中必须包含在线净化程序，以将废物转化为所谓的清洁废物，以减少环境处置和排放的毒性，并避免实验室残留的增加而带来的风险。

正如本章所建议的那样，绿色指标的进展必须与分析方法及其不同步骤的评估相协调，以定量地评估化学分析的副作用，而非只是寻求主要方法特征的优化。

总而言之，样品前处理中萃取过程可能会对环境产生较大风险。但是，对萃取条件的深入评估可以在提高分析操作的准确性和环境安全性之间取得良好的平衡，这是本书的主要关注点之一。

致谢

由衷感谢 Ministerio de Economía y Competitividad（项目 CTQ2014 - 52841）和 Generalitat Valenciana（项目 PROMETEO-II2014-077）的资金支持。

参考文献

[1] P. T. Anastas, J. C. Warner, Green Chemistry: Theory and Practice, Oxford University Press, New York, 1998.

[2] M. de la Guardia, J. Ruzicka, Towards environmentally conscientious analytical chemistry through miniaturization, containment and reagent replacement, Analyst 120 (1995) 17N.

[3] M. Koel, M. Kaljurand, Green Analytical Chemistry, RSC Publishing, 2010.

[4] M. de la Guardia, S. Armenta, Green analytical chemistry: theory & practice, in: Comprehensive Analytical Chemistry, vol. 57, Elsevier, 2011.

[5] M. de la Guardia, S. Garrigues, Handbook of Green Analytical Chemistry, John Wiley &

Sons, 2012.

[6] M. de la Guardia, S. Garrigues, Challenges in Green Analytical Chemistry, RSC, 2011.

[7] S. Armenta, S. Garrigues, M. de la Guardia, Green analytical chemistry, TrAC Trend. Anal. Chem. 27 (2008) 497-511.

[8] J. Namieśnik, Green analytical chemistry-some remarks, J. Sep. Sci. 24 (2001) 151-153.

[9] J. Wang, Real-time electrochemical monitoring: toward green analytical chemistry, Acc. Chem. Res. 35 (2002) 811-816.

[10] M. Tobiszewski, A. Mechlińska, J. Namieśnik, Green analytical chemistry—theory and practice, Chem. Soc. Rev. 39 (2010) 2869-2878.

[11] P. T. Anastas, Green chemistry and the role of analytical methodology development, Crit. Rev. Anal. Chem. 29 (1999) 167-175.

[12] A. Gałuszka, Z. Migaszewski, J. Namieśnik, The 12 principles of green analytical chemistry and the SIGNIFICANCE mnemonic of green analytical practices, TrAC Trend. Anal. Chem. 50 (2013) 78-84.

[13] M. Koel, Do we need green analytical chemistry? Green Chem. 18 (2016) 923-931.

[14] ISO/IEC Guide 2, Standardization and Related Activities—General Vocabulary, International Standards Organization, Geneva, 2004.

[15] M. Thompson, T. Fearn, What exactly is fitness for purpose in analytical measurement? Analyst 121 (1996) 275-278.

[16] M. Valcárcel, S. Cárdenas, Vanguard-rearguard analytical strategies, TrAC Trend. Anal. Chem. 24 (2005) 67-74.

[17] M. de la Guardia, S. Garrigues, Partial least squares attenuated total reflectance IR spectroscopy versus chromatography: the greener method, Bioanalysis 4 (2012) 1267-1269.

[18] H. L. Pardue, J. Woo, Unifying approach to analytical chemistry and chemical analysis, problem-oriented role of chemical analysis, J. Chem. Educ. 61 (1984) 409-412.

[19] M. de la Guardia, S. Armenta, Green Analytical Chemistry, Theory and Practice in Series Comprehensive Analytical Chemistry, Elsevier, 2010.

[20] National Environmental Methods Index (NEMI), 2002. https://www.nemi.gov/home/.

[21] D. Raynie, J. L. Driver, Green assessment of chemical methods, in: 13th Green Chemistry and Engineering Conference, Washington, DC, 2009.

[22] A. Gałuszka, P. Konieczka, Z. M. Migaszewski, J. Namieśnik, Analytical eco-scale for assessing the greenness of analytical procedures, TrAC Trend. Anal. Chem. 37 (2012) 61-72.

[23] K. Van Aken, L. Strekowski, L. Patiny, EcoScale, a semi-quantitative tool to select an organic preparation based on economical and ecological parameters, Beilstein J. Org. Chem. 2 (2006) 3.

[24] S. Armenta, M. de la Guardia, J. Namieśnik, Green microextraction, in: M. Valcárcel, S. Cárdenas, R. Lucena (Eds.), Analytical Microextraction Techniques, Bentham Science Publishers, 2016, pp. 3-27.

[25] Y. Gaber, U. Tornvall, M. A. Kumar, M. A. Amin, R. Hatti-Kaul, HPLC-EAT (environmental assessment tool): a tool for profiling safety, health and environmental impacts of liquid chromatography methods, Green Chem. 13 (2011) 2021-2025.

[26] R. Hartman, R. Helmy, M. Al-Sayah, C. J. Welch, Analytical method volume intensity (AMVI): a green chemistry metric for HPLC methodology in the pharmaceutical industry, Green Chem. 13 (2011) 934-939.

[27] M. Tobiszewski, S. Tsakovski, V. Simeonov, J. Namieśnik, Application of multivariate statistics in assessment of green analytical chemistry parameters of analytical methodologies, Green Chem. 15 (2013) 1615-1623.

[28] M. Tobiszewski, S. Tsakovski, V. Simeonov, J. Namieśnik, Multivariate statistical comparison of analytical procedures for benzene and phenol determination with respect to their environmental impact, Talanta 130 (2014) 449-455.

[29] H. Al-Hazmi, J. Namieśnik, M. Tobiszewski, Application of TOPSIS for selection and assessment of analytical procedures for Ibuprofen determination in wastewater, Curr. Anal. Chem. 12 (2016) 261-267.

[30] M. Tobiszewski, A. Orłowski, Multicriteria decision analysis in ranking of analytical procedures for aldrin determination in water, J. Chromatogr. A 1387 (2015) 116-122.

[31] M. Tobiszewski, A. Mechlińska, B. Zygmunt, J. Namiśnik, Green analytical chemistry in sample preparation for determination of trace organic pollutants, TrAC Trend. Anal. Chem. 28 (2009) 943-951.

[32] C. Capello, U. Fischer, K. Hungerbühler, What is a green solvent? A comprehensive framework for the environmental assessment of solvents, Green Chem. 9 (2007) 927-934.

[33] ECHA Candidate List of Substances of Very High Concern for Authorisation. http://echa.europa.eu/candidate-list-table.

[34] P. G. Jessop, Searching for green solvents, Green Chem. 13 (2011) 1391-1398.

[35] Y. Zhang, B. R. Bakshi, E. S. Demessie, Life cycle assessment of an ionic liquid versus molecular solvents and their applications, Environ. Sci. Technol. 42 (2008) 1724-1730.

[36] Eurelectric, Union of the Electricity Industry, Life Cycle Assessment of Electricity Generation, Eurelectric, Brussels, Belgium, 2011.

[37] M. Tobiszewski, Metrics for green analytical chemistry, Anal. Methods 8 (2016) 2993-2999.

[38] J. D. Berset, M. Ejem, R. Holzer, P. Lischer, Comparison of different drying, extraction and detection techniques for the determination of priority polycyclic aromatic hydrocarbons in background contaminated soil samples, Anal. Chim. Acta 383 (1999) 263-275.

[39] Y. F. Song, X. Jing, S. Fleischmann, B. M. Wilke, Comparative study of extraction methods for the determination of PAHs from contaminated soils and sediments, Chemosphere 48 (2002) 993-1001.

[40] V. Flotron, J. Houessou, A. Bosio, C. Delteil, A. Bermond, V. Camel, Rapid determination of polycyclic aromatic hydrocarbons in sewage sludges using microwave-assisted solvent extraction: comparison with other extraction methods, J. Chromatogr. A 999 (2003) 175-184.

[41] A. Pastor, E. Vázquez, R. Ciscar, M. de la Guardia, Efficiency of the microwave-assisted extraction of hydrocarbons and pesticides from sediments, Anal. Chim. Acta 344 (1997) 241-249.

[42] C. Miege, J. Dugay, M.-C. Hennion, Optimization and validation of solvent and supercritical-fluid extractions for the trace-determination of polycyclic aromatic hydrocarbons in sewage sludges by liquid chromatography coupled to diode-array and fluorescence detection, J. Chromatogr. A 823 (1998) 219-230.

[43] S. Armenta, M. de la Guardia, F. A. Esteve-Turrillas, Hard cap espresso machines in analytical chemistry: what else? Anal. Chem. 88 (2016) 6570-6576.

[44] A. Zouir, F. A. Esteve-Turrillas, T. Chafik, Á. Morales-Rubio, M. de la Guardia, Evaluation of the soil contamination of Tangier (Morocco) by the determination of BTEX, PCBs, and PAHs, Soil Sediment Contam. 18 (2009) 535-545.

[45] R. Doong, S. Chang, Y. Sun, Solid-phase microextraction and headspace solid-phase

microextraction for the determination of high molecular-weight polycyclic aromatic hydrocarbons in water and soil samples, J. Chromatogr. Sci. 38 (2000) 528-534.

[46] K. Farshid, H. S. Amir, M. Rokhsareh, Determination of polycyclic aromatic hydrocarbons (PAHs) in water and sediments of the Kor River, Iran, Middle East J. Sci. Res. 10 (2011) 1-7.

[47] E. Manoli, C. Samara, I. Konstantinou, T. Albanis, Polycyclic aromatic hydrocarbons in the bulk precipitation and surface waters of Northern Greece, Chemosphere 41 (2000) 1845-1855.

[48] P. Sibiya, M. Potgieter, E. Cukrowska, J. Å. Jönsson, L. Chimuka, Development and application of solid phase extraction method for polycyclic aromatic hydrocarbons in water samples in Johannesburg area, South Africa, S. Afr. J. Chem. 65 (2012) 206-213.

[49] A. K. M. Kabzinski, J. Cyran, R. Juszczak, Determination of polycyclic aromatic hydrocarbons in water (including drinking water) of Lodz, Pol. J. Environ. Stud. 11 (2002) 695-706.

[50] O. Krüger, S. Olberg, R. Senz, F. G. Simon, Comparison of stir bar sorptive extraction (SBSE) and solid phase microextraction (SPME) for the analysis of polycyclic aromatic hydrocarbons (PAH) in complex aqueous soil leachates, Water Air Soil Pollut. 226 (2015) 397.

[51] C. Margoum, C. Guillemain, X. Yang, M. Coquery, Stir bar sorptive extraction coupled to liquid chromatography-tandem mass spectrometry for the determination of pesticides in water samples: method validation and measurement uncertainty, Talanta 116 (2013) 1-7.

[52] Y. Zhang, D. Wu, X. Yan, Y. Guan, Rapid solid-phase microextraction of polycyclic aromatic hydrocarbons in water samples by a coated through-pore sintered titanium disk, Talanta 154 (2016) 400-408.

[53] M. Shamsipur, M. B. Gholivand, M. Shamizadeh, P. Hashemi, Preparation and evaluation of a novel solid-phase microextraction fiber based on functionalized nanoporous silica coating for extraction of polycyclic aromatic hydrocarbons from water samples followed by GC-MS detection, Chromatographia 78 (2015) 795-803.

[54] M. R. K. Zanjania, Y. Yaminia, S. Shariatia, J. Å. Jönsson, A new liquid-phase microextraction method based on solidification of floating organic drop, Anal. Chim. Acta 585 (2007) 286-293.

[55] Z. G. Shi, H. K. Lee, Dispersive liquid-liquid microextraction coupled with dispersive M-solid-phase extraction for the fast determination of polycyclic aromatic hydrocarbons in environmental water samples, Anal. Chem. 82 (2010) 1540-1545.

2

2　加压液体萃取和超临界流体萃取植物、海藻、微藻和食品副产品中的生物活性物质

Andrea del Pilar Sánchez-Camargo, Elena Ibáñez, Alejandro Cifuentes, Miguel Herrero *
Institute of Food Science Research (CIAL-CSIC), Madrid, Spain
* 通讯作者：E-mail：m. herrero@ csic. es

2.1 引言

随着功能性食品行业的发展，从天然产物中提取生物活性物质是食品及相关行业的研究热点。功能性食品通常被定义为除了食物必须提供的能量和营养需求之外，还能提供额外生理益处的产品[1]。通常是通过添加功能性成分获得这种额外的健康益处，这种额外的健康益处应该是其中的活性成分的作用。

商业产品中已经有很多使用功能性成分的例子，例如不饱和脂肪酸[2]、植物固醇和甾烷醇[3]、肽[4]或多酚[5]等。以消费者的角度，显然更希望添加到食品中的这些功能性成分的来源更天然。因此近年来对生物活性化合物新的天然来源进行了广泛研究探索[6,7]。本章节讨论了这些成分的一些最重要的天然来源：植物、微藻和海藻以及农业食品副产品。从自然界获得某种或某类特定组分的技术变得越来越重要。传统方法意味着通常会使用大量有机溶剂，而且往往是有毒的。由于这些方法对环境的重要影响，如今关注的焦点也大不相同[8]。随着生态意识的增强和可持续发展的重视，传统的具有侵略性的技术正在被更环保的工艺所取代。欧洲联盟和其他国际组织正在推动由刺激经济向循环经济的转变，寻求在生产链的每个环节都进行彻底的回收和再利用[9]。这也与绿色化学原则密切相关，绿色化学的主要目的是减少废物和更有效地利用资源[10]。天然产品的绿色提取有六项具体原则[11]，包括通过选择品种和使用可再生植物资源进行创新；使用以水或农用溶剂为主的替代溶剂；通过能源回收和利用创新技术减少能源消耗；在生物和农业精炼工业中以生产农副产品取代废物的产生；减少单元操作，支持安全、可靠和受控的过程；旨在获得一种不含污染物的非变性和可生物降解的提取物。这些原则的直接应用促进了生物精炼方法的发展。生物炼制意味着以可持续的方式开发将生物物质转化为能源和多种产品（主要是生物燃料和增值副产品）的集成工艺[12]。可以推断，上述某些原则与自然选择的来源有关，其余的则与所采用的提取技术密切相关。为了满足这些要求，需要高效、廉价和安全的提取技术。在当今使用的先进提取方法中，基于压缩流体的提取方法应用最广泛。

2.2 基于压缩流体的萃取过程

与常规大气压条件下操作的方法相比，大多基于压缩流体的技术除了具有更高的效率，同样也需要考虑到使用溶剂的物理化学性质以达到方法的操作条件。这些技术在一定条件下也符合绿色化学的原理。加压液体萃取法（PLE）和超临界流体萃取法（SFE）是利用压缩流体从天然来源中提取生物活性物质的两种最常用的技术。

2.2.1 加压液体萃取的原理与操作

PLE 法为了使萃取溶剂保持液态，需要在温度和压力都足够高的条件下进行萃取，但仍需要低于溶剂对应的临界温度和压力。也有文献称此技术为加速溶剂萃取、加压流体萃取或加压热溶剂萃取。而应用这些特定的提取条件，不仅能够减少萃取过程所使用提取溶剂的量，提取速度也比传统方法更快。在高温高压下溶剂物理化学的变化意味着其传质速率的增加，与此同时，分析物的溶解度会增加、溶剂表面张力和黏度则会降低，这使得溶剂能够更容易且更深地渗透到所提取的固体基质中，以上这些方面均能够提高萃取率。溶剂的选择是最重要的参数，不仅是因为涉及目标分析物，更考虑到其对环境的友好性。因此，应该首选公认的安全（GRAS）溶剂，如乙醇、乳酸乙酯，甚至 D-柠檬烯。但是，最环保的方法当属选用水作为提取溶剂。读者可参考本书的“加压热水萃取生物活性”一章[12a]，深入了解在加压条件下如何使用水作为萃取溶剂。

进行 PLE 萃取所应考虑的参数中，温度对萃取的影响目前来看是最大的。如前所述，温度的升高更利于破坏样品基质中的分子间力（范德华力、氢键和偶极引力）。显然，萃取压力也很重要，因为需要足够高的压力使溶剂保持液态。然而，当压力超过某一临界值（此值取决于所用溶剂及萃取温度）时，可能会对样品产生破坏效应；但多项工作表明，当压力高至足以使溶剂保持液态的情况下，其继续升高并不会显著影响所得结果[6,7,13,14]；因此通常采用的压力值在 5~10MPa 之间。一旦确定了萃取温度和压力，就需要考虑萃取时间了。有效萃取时间是指在要求或所选萃取条件（即选定的萃取温度和压力）下溶剂与样品基质充分接触的时间。为此，一些商用仪器会通过设置预热时间使系统在预期条件下达到平衡。一般来说，特定样品所需的萃取时间主要取决于几个参数，而其中最关键的是萃取类型。静态萃取是最为常见的萃取类型，是指一定体积的溶剂在特定压力和温度下，与样品在一定时间内接触的过程。在此过程中，一旦残留在基质中的目标成分与已被溶剂萃取的目标成分达到平衡，该萃取过程的回收率将不会再增加；因此，需要对静态萃取时间进行合理优化，以避免发生降解或其他类似情况，最大限度地提高分析物的萃取效率。采用静态萃取从天然来源中提取生物活性物质，通常需要的提取时间为 5~20min。另外一种萃取类型则是动态萃取，在萃取过程中，经过加热和加压的溶剂不断地流入萃取池。这种萃取方式可以避免平衡状态的形成，从而更有利于样品的完全萃取，但与此同时使用溶剂的量也会增加。不仅如此，由于流速直接决定了萃取时间，因此还需要对流速进行研究和优化。因此相比动态萃取，文献中报道更多的是采用耗时短的静态重复萃取循环。连续几个萃取循环可以有效地帮助样品萃取完全。除此之外，样品粒径也是 PLE 过程中需要控制的参数之一，因此粒径会影响样品接触的表面积，从而影响传质速率。有时会将分散剂和样品一起引入萃取池，以促进溶剂的均匀分布并最大程度提高萃取效率；其中硅藻土和海砂是最常用的分散剂。

原则上，进行 PLE 需要的仪器很简单［图 2.1（1）］：

①泵，能够将溶剂引入系统中，并在萃取过程完成后帮助将萃取液推出。泵应能够保持压力恒定；

②提取池，提供萃取场所，因此应能够适应高压环境，并配备至少两个开/关阀，以保持萃取条件的稳定；

③烤箱，用于放置萃取池，以便将其加热到预设值。大多数仪器的最高工作温度

在200℃左右；

④采集瓶。然而，实际使用的仪器或多或少会更复杂，这主要取决于萃取过程的要求。PLE仪器可以配备加热线圈，用于动态萃取中的溶剂加热；还可以增加一个氮气回路，能够在萃取结束后帮助排除管路中所有溶剂。市面上有很多商用仪器，尽管与实验室自制的仪器有所区别，但鉴于通常采用的工作压力和温度，两种情况均需使用耐腐蚀材料。

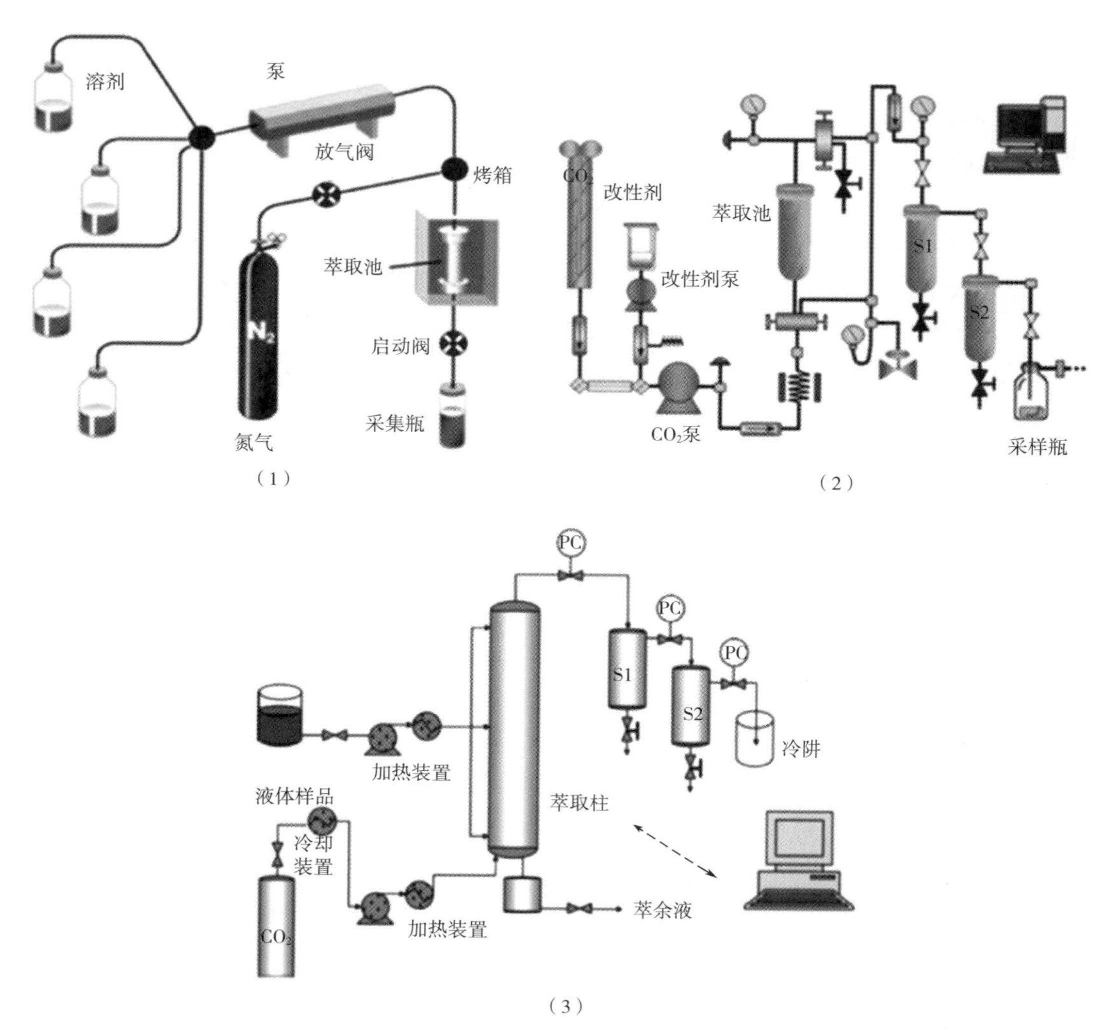

图2.1　加压液体萃取（1），用于固体样品的超临界流体萃取（SFE）（2）和用于液体样品的SFE系统（3）的仪器示意图

S1：分离器1；S2：分离器2

［资料来源：（1）改编自 Dionex Thermo Scientific；（3）在获得许可后改编自 L. Vázquez，T. Fornari，F. J. Señoráns，G. Reglero，C. F. Torres，Supercritical carbon dioxide fractionation of nonesterified alkoxyglycerols obtained from shark liver oil. J. Agric. Food Chem. 56（3）（2008）1078-1083.］

2.2.2 超临界流体萃取的原理与操作

超临界流体进行萃取主要是溶剂在高于其临界点的压力和温度下进行。一旦超过临界点，作为溶剂的流体就会因产生的物理变化而获得显著不同的性质。当溶剂的温度升高且达到其压力和临界点，就会得到两相间没有区别的均相的超临界流体（图2.2）。因此，超临界流体具有介于气体和液体之间的混合特性。它们的黏度与气体相似，但其密度接近液体的值；此外，它的扩散性介于液体和气体之间。在超临界流体中，例如表面张力和溶剂强度等其他重要特性也有所改变，这些均对于其独特的溶剂化性能更有利。此外，在超临界条件下，密度的改变有效地改变了溶剂的性能，更能够实现萃取的高度选择性[15,16]。尽管可以将不同的溶剂用作超临界流体进行萃取，但CO_2是迄今为止在生物活性物质的提取中应用最广泛的天然来源的溶剂，原因是：①具有温和的临界温度和压力（临界条件为31.2℃和7.38MPa）；②被认为是食品工业的GRAS；③廉价且容易获得；④允许重复利用其他工业过程中产生的CO_2，符合绿色化学的部分原则；⑤在室温下为气体。最后这一点非常重要，因为减压后能直接获得无溶剂的萃取物。

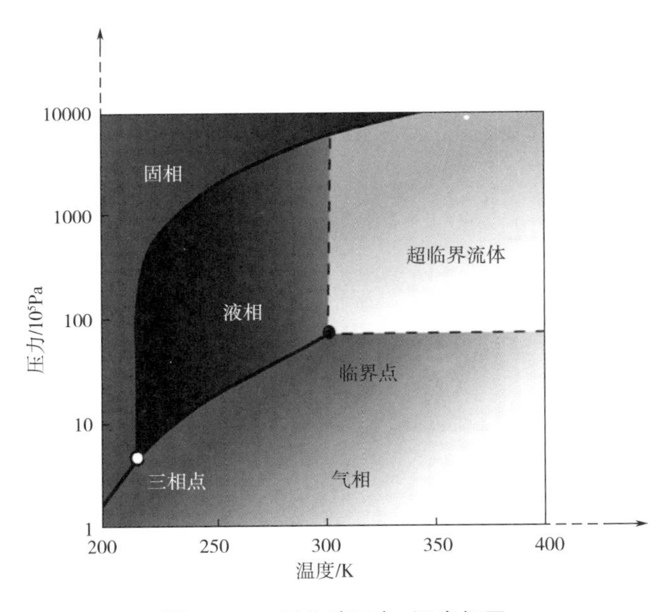

图2.2 二氧化碳压力-温度相图

在能够影响超临界流体萃取的参数中，操作压力和温度是最关键的，两者结合起来将决定物质在超临界流体中的溶解度。操作压力和温度的选择应当充分考虑实验目的及目标分析物。增加压力可以提高超临界流体的密度，从而使样品组分的溶解度增加；而萃取温度的等压升高会导致溶剂密度的降低，不过同时也会增加基质表面分析物的蒸汽压，从而促进质量转移。尽管超临界CO_2具有上述优点，但其极性低的特征仍然极大限制了它的应用，它只能用于提取极性极低的组分。从天然基质中获得的SFE

萃取物通常是同类物质的混合物，例如甘油三酯、脂肪酸、萜类、植物甾醇、生育酚、类胡萝卜素、三烯生育醇和酚类等。在图 2.3 中，给出了每种同类物质的例子。

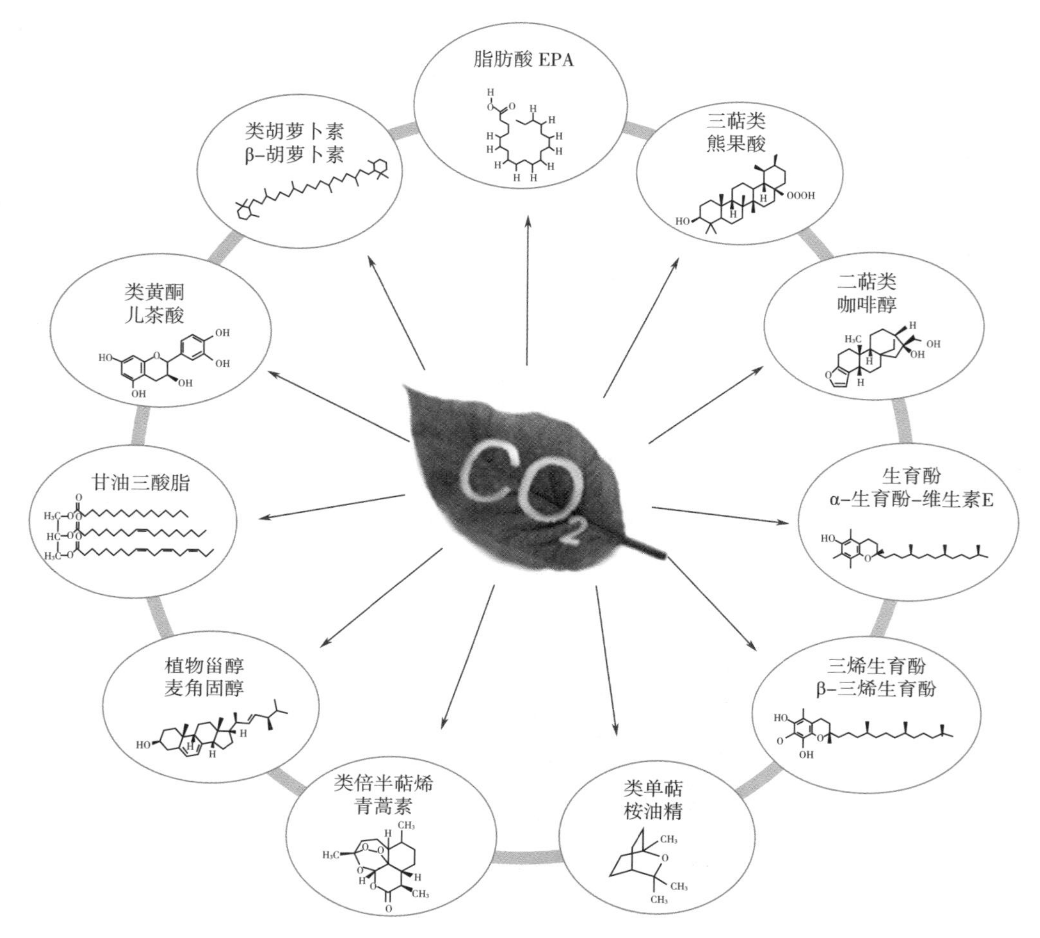

图 2.3　在天然来源的超临界流体萃取物中主要发现的同类物质

改性剂的使用可以扩展这项技术的应用范围。改性剂是一种有机溶剂，通常按给定比例（5%~10%）与超临界 CO_2 结合使用，用于增加整体极性。在天然生物活性化合物领域中常采用这种方法，通过优化改性剂的选择和使用比例，提高萃取效率，应用到更多极性组分的提取中。

与 PLE 的操作模式不同，通常 SFE 至少有一部分会涉及动态提取；一些应用做了相关改进，以便在给定时间内执行第一个静态周期，并随后执行一段持续时间较长的动态提取；在这些应用以及在那些纯动态萃取的应用中，超临界流体的流速都是一个重要的参数。一方面须控制流速、尽可能减少溶剂的使用，另一方面要使传质效率最大化。这一参数在逆流条件下萃取液体样品时更为重要，此时流速（或溶剂进料比）将直接决定样品与超临界 CO_2 之间的接触时间。操作中其他需要考虑的重要因素包括样品的物理状态（主要是粒径、孔隙率和水分含量），以及是否需要在样品中使用物理分散剂以避免流窜。

所需的仪器，与 PLE 中的同样须能承受高压（高达 50MPa 或更高）；根据要处理的是固体样品还是液体样品，所需设备会有所不同［分别参见图 2.1（2）和（3）］；不同之处主要与萃取池有关，处理液体样品时将用萃取柱取代萃取池。在任何情况下，首先都会要求有一个装提取溶剂（通常是 CO_2）的储罐，一个将气体加压到所需压力的泵，以及一个保持系统内部高压的限流器或阀门。当使用有机改性剂时，需要一个额外的泵。对于固体样品，需要高压萃取池，而萃取液体则需要逆流柱。逆流柱通常具有不同的进料水平（在不同的高度），而溶剂从柱子的底部引入，此时，目标组分分布在溶剂和逆流通过萃取柱的液体样品之间。最后，需要用于回收萃取物的捕集容器（或分离池，也称为分馏池），并可以采用固体捕集、液体捕集或冷凝捕集等不同的捕集方法。在中试或工业系统中，通过快速降低压力和/或提高温度进行萃取溶质的收集。尤其是，考虑到每个分离容器可具有特定的温度和压力，以选择性地沉淀和分离一些被提取的化合物，萃取后也可以采用几个分馏或分离池串联进行降压。

2.3 天然来源中生物活性物质的提取

2.3.1 植物中活性物质的提取

植物可能是提取生物活性化合物领域里研究最广泛的天然来源，在食品、保健品和制药工业中具有潜在的用途。PLE 和 SFE 的应用已有广泛的报道[6,13,17]。提取条件通常是根据基质以及目标成分的不同而变化的。因而统计评估不同提取参数的影响并根据某些响应变量选择最佳提取条件的实验设计非常有用[18,19]。

PLE 已被广泛用于从各种植物来源中提取生物活性物质，如黄酮类化合物[20]、花青素[21]、酚酸和其他多酚[22]、皂苷[23]或精油[24]。溶剂或溶剂混合物的选择取决于实际的应用情况，其中乙醇、甲醇、水及其混合物是最常用的溶剂。实例证明，添加酸可以提高某些成分的回收率，例如黑木莓中多酚的提取[25]。萃取温度也很大程度取决于实际应用，但实际常常采用高温条件（150℃左右），因为高温能够改善传质动力。此外，文献报道当使用水作为提取溶剂时，高温（甚至超过 200℃）可以提高提取物的生物活性，主要是抗氧化活性。这样不仅可以改善提取过程，同时有可能发生其他反应，生成可能对萃取物的整体抗氧化性有积极影响的非天然化合物。在本书中，Plza 和 Turner 编写的“加压热水萃取生物活性”部分[12a]深入讨论了高水温下加速反应。大多数采用的是基于商业仪器的静态提取过程，静态提取时间通常为 5~20min。当执行多个静态提取循环时，会加快提取速度。

还有一些连续或同时进行 PLE 萃取生物活性化合物的方法。例如，已有报道从菜籽粕中提取芥子碱并将其转化为芥子酸和甘露醇的研究，其目的是为了有效获得具有生物活性的芥子酸和甘露醇[26]。结果表明，用 60%甲醇在 200℃下提取 20min 得到总的酚类的回收率要比索氏提取 10h 高 300%以上。要将芥子碱转化为甘露醇，在提取溶剂中加入 NaOH 十分关键，最佳的萃取转化条件为：在 200℃下使用含 1% NaOH 的甲醇萃取 5min[26]。

另一方面，超临界萃取主要用于从植物中提取弱极性组分，目前已经报道了脂肪酸、植物甾醇[27]、胡萝卜素[28]、生物碱[29]或相对低极性的酚类化合物[30]、精油[31]等化合物[32]的提取。如前所述，工作压力和温度，是影响最大的参数，决定了控制 CO_2 浓度和溶解度。由于超临界二氧化碳的特殊性质，工作温度通常在 40~65℃，压力变化更大，在 10~40MPa。

在其他影响参数中，尤其是从植物中提取极性较弱的生物活性化合物时，水分含量是一个重要的因素。例如，对于含水量分别是 8%和 12%的南瓜样品中，类胡萝卜素的提取具有显著差异[33]。烘干样品（残余水分含量较低）中类胡萝卜素的回收率是冷冻干燥样品（含 12%）的 8 倍。从表 2.1 所示的参考文献中可以检索到 SFE 应用于植物中生物活性成分提取的细节和详尽列表。

过程强化是通过开发高效的技术经济系统，系统地优化能源使用、资金或其他效益的一种方式。从绿色化学的角度来看，这种方法有吸引力的地方在于它能有助于减少废物和能源消耗。在过程强化方法下，SFE 和 PLE 的结合可能有几个优点：首先，这两种技术都是在高压下操作，对仪器的要求都相同；对于给定的天然基质中，超临界 CO_2、有机溶剂、甚至是水的使用，可以选择性靶向不同的组分。例如，已有多功能仪器用于对不同的迷迭香化合物采用 SFE 和亚临界水萃取进行连续萃取[55]。第一步，在 40℃和 30MPa 下，采用超临界 CO_2，溶剂质量比为每克迷迭香 2.5g CO_2，得到 1,8-桉叶脑、樟脑和迷迭香精油的典型成分。随后，以水为溶剂，在 10MPa 的压力下以 1.1℃/min 的速率升温至 172℃，富集非挥发部分，每步萃取率分别为 2.5% 和 18.6%[55]。类似的方法针对其他植物的使用和优化，可以提高效率，降低操作成本，改善化学性质不同的生物活性化合物的提取。

表 2.1　最近发表的有关使用 PLE 和 SFE 从不同天然基质中提取生物活性物质的相关论文（2006—2016）

年份	题目	相关方法	文献
2006	Supercritical CO_2 extraction and purification of compounds with antioxidant activity	SFE	[34]
	Pressurized hot water extraction of bioactive or marker compounds in botanicals and medicinal plant materials	SWE	[35]
	Sub- and supercritical fluid extractionof functional ingredients from different natural sources: plants, food by-products, algae and microalgae: a review	SWE, SFE	[36]
2007	Supercritical fluid extraction in plant essential and volatile oil analysis	SFE	[37]
	Use of compressed fluids for sample preparation: food applications	SFE, PLE	[38]
2008	Extraction of functional substances from agricultural products or by - productsby subcritical water treatment	SWE	[39]
2009	Design and scale-up of pressurized fluid extractors for food and bio products	SFE, PLE, SWE	[40]
	Application of supercritical CO_2 in lipid extraction-a review	SFE	[41]

续表

年份	题目	相关方法	文献
2010	Supercritical fluid extraction: recent advances and applications	SFE	[42]
	Pressurized hot water extraction (PHWE)	SWE	[43]
2011	Pressurized liquid extraction as a green approach in food and herbal plants extraction: a review	PLE	[44]
	Supercritical carbon dioxide extraction of molecules of interest from microalgae and seaweeds	SFE	[45]
2012	Steps of supercritical fluid extraction of natural products and their characteristic times	SFE	[46]
	Techniques to extract bioactive compounds from food by-products of plant origin	PLE, SFE	[47]
	Application of accelerated solvent extraction in the analysis of organic contaminants, bioactive and nutritional compounds in food and feed	PLE	[14]
2013	Compressed fluids for the extraction of bioactive compounds	PLE, SFE	[7]
	Supercritical fluid extraction of plant flavors and fragrances	SFE	[48]
2014	Extraction behavior of lipids obtained from spent coffee grounds using supercritical carbon dioxide	SFE	[49]
	Recovery of biomolecules from food wastes—a review	SFE, PLE	[50]
2015	Plants, seaweeds, microalgae and food by-products as natural sources of functional ingredients obtained using pressurized liquid extraction and supercritical fluid extraction	SFE, PLE	[6]
	Pressurized hot water extraction of bioactive	PLE	[13]
	Pressurized fluid systems: phytochemical production from biomass	SWE, PLE	[51]
2016	Green alternative methods for the extraction of antioxidant bioactive compounds from winery wastes and by-products: a review	SFE, PLE	[8]
	Subcritical water extraction of bioactive compounds from plants and algae: applications in pharmaceutical and food ingredients	SWE	[17]
	Extraction of oil and carotenoids from pelletized microalgae using supercritical carbon dioxide	SFE	[52]
	Supercritical fluid extraction of bioactive compounds	SFE	[53]
	Application of non-conventional extraction methods: toward a sustainable and green production of valuable compounds from mushrooms	SWE, SFE	[54]

注：PLE—加压液体萃取；SFE—超临界流体萃取；SWE—亚临界水萃取。

虽然过程强化和过程整合十分相关，但前者使用的是相同设备，而后者可能是通过不同单元操作的组合来实现的。与本章节相关的一个典型例子是超临界流体辅助的PLE萃取和分馏的联用。这种方法已经应用到了不同植物样品的萃取中。首先，根据提取的目标化合物对 PLE 进行适当的优化。然后，用超临界抗溶剂分馏（SAF）法对在最佳条件下得到的 PLE 萃取物进行分馏。这种技术在应用中，通过加压条件下建立的 PLE 萃取物与超临界 CO_2 的连续接触，可以选择性地从 PLE 萃取物中沉淀出相对极性的化合物。这样，超临界流体溶解混合物中极性较弱的化合物和溶剂，而不溶的极性较强的化合物沉淀出来，可以单独回收。这种方法可以用来同时富集极性和非极性生物活性化合物；前一种情况，目标成分在沉淀部分[56]，后一种情况，目标成分将集中在 CO_2+溶剂部分[22]。

2.3.2 海藻和微藻中活性物质的提取

海藻和微藻是各种未被开发的生物活性化合物的来源。虽然目前这些生物体中有些成分已经应用于食品和制药行业，但仍有数千种不同的物种尚未被研究。微藻是生物燃料和二氧化碳捕集领域里大量研究的焦点；而且，食用微藻也已经商业化。事实上，这些微生物的化学组成可能会随着其生长条件的变化而改变，所以可以将它们视为能够生产特定生物活性化合物的生物工厂，这更增加了它们的受关注度。另一方面，大型藻类或海藻也在研究中。有些物种被认为侵略性很强，因此，维持它们的稳定意义很大。

随着这两类生物的应用日益增多，最近开发了基于 PLE 和 SFE 的不同提取方法，可以从微藻和海藻中获得生物活性化合物。表 2.1 列举了的已发表的综述论文对这些应用做了总结。这些应用包括使用 SFE 从微型和大型藻类中提取类胡萝卜素[52,57] 和聚不饱和脂肪酸（PUFAs）[58]，或者使用 PLE 从日本产海带中提取生物活性碳水化合物(褐藻搪胶)[59]或从钝马尾藻中提取褐藻多酚[19]等。

考虑到它们的特性，微藻和海藻的高含水量对提取过程的影响至关重要。但是去除所获藻类的水分成本较高，较实用的方法就是确定不妨碍目标生物活性化合物提取的最大含水量。此研究是以杜氏盐藻（类胡萝卜素的重要来源）为研究对象；考察了不同的含水量的样品，评估了含水量对 CO_2 超临界流体萃取类胡萝卜素的影响[60]。最终作者得出的结论是，23%的含水量有助于提高类胡萝卜素（主要是 β-胡萝卜素）的萃取率，并不会对提取过程产生负面影响。还有一种替代方案，可以直接使用生长后收获的微藻糊（含水量 70%~80%），及在萃取池内使用低成本的吸附剂作为载体[61]。对于吸附剂对富油新绿藻中类胡萝卜素的吸附效果，研究考察了不同的硅胶、壳聚糖和活性炭，结果表明使用壳聚糖基吸附剂的回收率较高，效果最好。也有考察了水分含量对不同细胞破坏机制的研究[62]。与植物不同的是，微藻通常具有坚固而厚实的细胞壁，在提取之前必须通过一些化学、酶促或机械方法来破坏细胞壁，以提高生物活性成分的回收率。

撇开传统的一步提取法不谈，PLE 和 SFE 正在作为潜在的单元操作被广泛研究，并应用于微藻及海藻的生物精炼工艺。例如，有研究建立了一个过程强化的下游平台，使用几种加压溶剂从球等鞭金藻中提取生物活性化合物[63]。相同的初始原料，按顺序先使用纯净的超临界 CO_2 进行超临界萃取，再使用 CO_2 膨胀的乙醇进行过渡萃取，最后使用加压的乙醇和水进行萃取。前两步主要萃取的是低极性到中等极性的组分，主要是类胡萝卜素（岩藻黄质），PLE 萃取的更多是极性组分。图 2.4 所示为整体优化流程的方案。

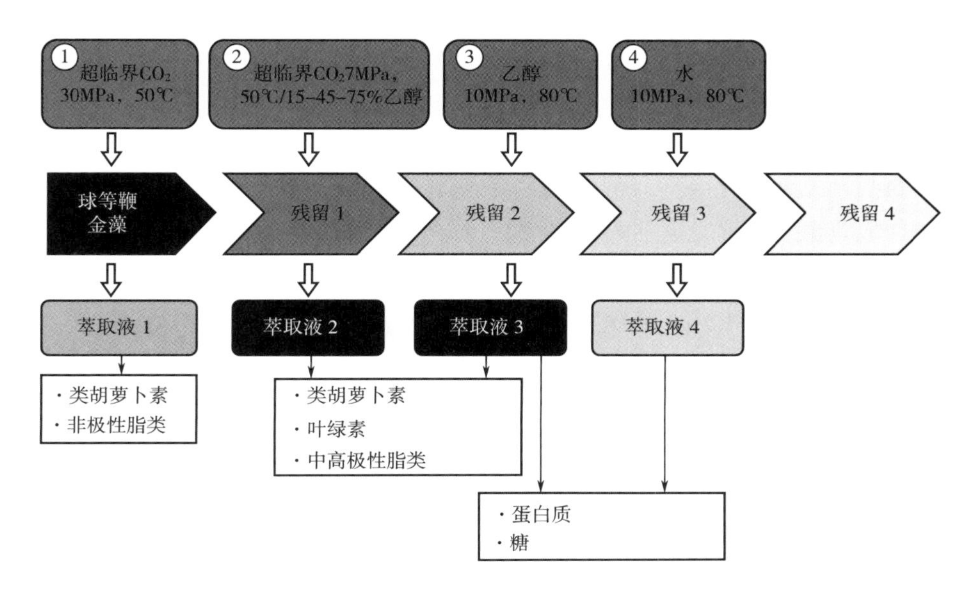

图 2.4　按照生物精炼方法，从球等鞭金藻中提取生物活性化合物的综合顺序提取工艺流程示意图

［资料来源：B. Gilbert-López, J. A. Mendiola, J. Fontecha, L. A. M. van den Broek, L. Sijtsma, A. Cifuentes, M. Herrero, E. Ibáñez, Downstream processing of Isochrysis galbana: a step towards microalgal biorefinery, Green Chem. 17 (2015) 4599-4609, http://dx. doi. org/10. 1039/C5GC01256B.］

2.3.3　食品副产品中活性物质的提取

减少浪费是对可持续性至关重要的话题。在各种食品和农产品相关的生产活动中会产生许多副产品。这些副产品传统上会重新用于饲料的制造或能源生产。如今看来，显然其中一些副产品和废物仍然是生物活性化合物的丰富来源。因此，这些副产品的再利用和增值，可能是生产其他高附加值产品的好方法，同时也可显著减少浪费。根据前面所述每种技术的要求和特殊性，已经有不同的 SFE 和 PLE 方法用于从农副产品中回收不同的物质；其中包括咖啡副产品[49]、酿酒产生的废弃物[8]、橄榄油[64]或水果工业副产品[65]等[23,66]的再利用。读者请参阅表 2.1 中列出的典型文献，以更深入地了解特定的应用方法和提取条件。

由于食品副产品的内在性质，已在生物精炼方法中使用基于压缩流体的提取技术（例如 PLE 和 SFE）对这些副产品进行了研究。例如，橄榄油行业产生的废物可能是石油的四倍。出于这个原因，已经对方法进行了优化，以进一步限定产生的副产品（如磨橄榄的废料）。通过对几种基于纯超临界 CO_2 和添加乙醇为改性剂的超临界 CO_2 连续 SFE 方法进行研究，可以得到富含多酚、多不饱和脂肪酸、单不饱和脂肪酸和角鲨烯的馏分[67]。此外，此方法还有一个额外的优点，能够有效地干燥余下物质，有利于进一步的处理加工。

为了回收生物活性成分，也有研究采用 PLE 和 SFE 联用进行副产物萃取。首先选择超临界流体萃取进行对芒果皮废弃物的连续萃取，然后再采用加压乙醇进行萃取。在整个连续过程中保持压力和温度不变。芒果皮中存在的非极性类黄酮和类胡萝卜素优先在超临界 CO_2 阶段萃取，而极性多酚的萃取在基于 PLE 的第二步[68]。然而在报道的其他方法中，萃取顺序发生了改变[69]；为了控制 Euterpe edulis 工业残留，优化了包括不同压力和温度以及不同溶剂混合物的 PLE 程序。一旦确定了最佳萃取条件，就采用优化的溶剂（乙醇和水的酸化混合物）作为第二步超临界萃取的改进剂。最终利用该方法成功地获得了富含花色苷的提取物。

至于其他天然来源，从食品副产品中提取生物活性物质的整合和/或强化过程的开发是一个热门的话题。最有前景的强化过程就是将萃取和干燥相结合，因为干燥过程需要的能耗较高。就这个层面来说，一个最显著的改进就是将 PLE 和超临界抗溶剂（SAS）沉淀进行偶联，使用相同的多功能仪器从洋葱废弃物中提取黄酮类化合物[70]。SAS 利用了超临界 CO_2 流动通过阀门与极性溶液（PLE 萃取物）混合产生的效应。混合过程中当压力突然降低，由于抗溶剂作用，超临界 CO_2 迅速溶解在有机溶液中，使得极性溶液中存在的化合物沉淀下来。该技术在概念上与之前讨论的 SAF 类似。优化后的最佳萃取条件为使用乙醇在 40℃、12MPa 下静态萃取 20min，而动态萃取/沉淀步骤的时间为 200min，压力降为 10MPa。在此条件下，可从槲皮苷含量高的干洋葱皮中获得 4%的干燥微粒。从能量角度来看，这种方法不仅新颖，而且易于产生不易氧化的干燥颗粒，因此有助于保持生物活性成分的完整性。为该应用开发的装置方案如图 2.5 所示。

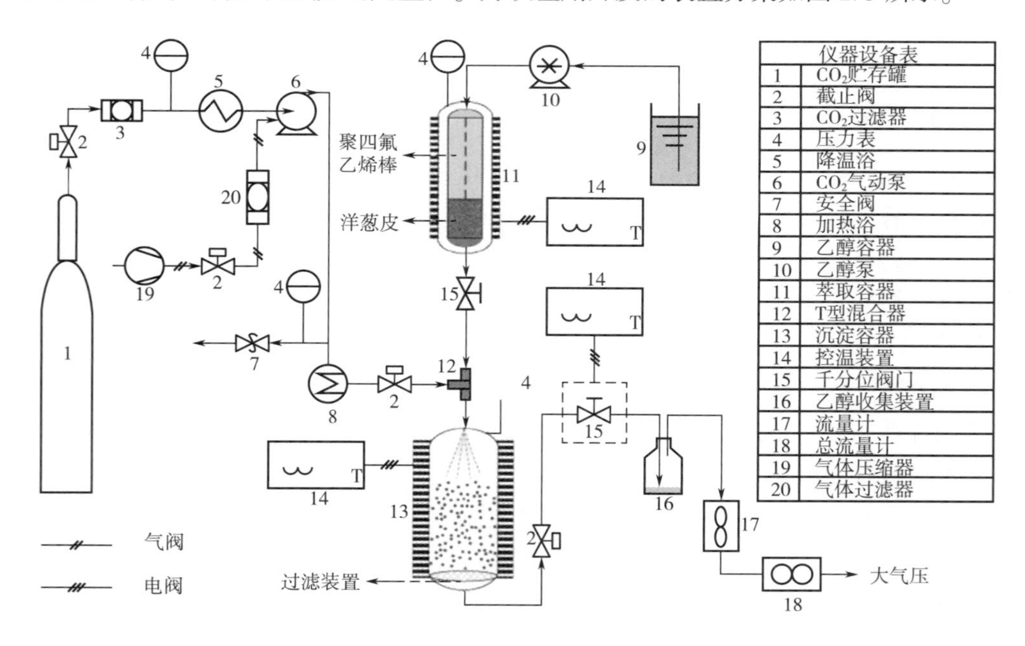

图 2.5 洋葱皮提取物在线萃取和颗粒形成（EPFO）的强化过程设备流程图

[资料来源：G. L. Zabot, M. A. A. Meireles, On-line process for pressurized ethanol extraction of onion peels extract and particle formation using supercritical antisolvent, J. Supercrit. Fluids 110 (2016) 230-239. http://dx.doi.org/10.1016/j.supflu.2015.11.024.]

2.4 结论与展望

综上所述，基于压缩流体的萃取法广泛应用于不同天然基质中生物活性化合物的萃取。为了从大量样品中得到化学性质不同的生物活性物质，所需要的萃取条件包含大量丰富的信息。尽管我们知道使用 PLE 和 SFE 的基本要求，以及不同因素对方法的影响，但该技术若想要迎合这些领域最前沿的需求，依然有很长的路要走。

单从技术应用来看，新萃取方法的开发很可能与新溶剂的研究和使用有关。即使不能预见别的溶剂能够取代超临界 CO_2 成为 SFE 首选溶剂，但这种情况在 PLE 的应用中未必不可能。事实上，PLE 法中新的食品级环保溶剂的使用在增加。一些研究已经提出了使用新型溶剂如乳酸乙酯[71]或 D-柠檬烯[72]从不同的自然来源获得生物活性物质的可能性。虽然离子液体（ILS）和低共熔溶剂（DESS）已经有了一些应用[73]，但是在加压条件下使用这些新型溶剂可能会带来更多的可能性。DESS 是由氢键受体和氢键供体这两个天然存在的组分混合而成的，它们之间可以通过氢键相互作用而结合。考虑到从天然化合物出发很容易制备 DESS，不但经济可行而且环境友好，这些化合物有望在不同的应用领域替代传统挥发性有机溶剂。因此，预计在接下来的几年内，基于 ILS 和 DES 的 PLE 的应用将会有更多。

基于 PLE 和 SFE 的方法，其未来发展必然与过程集成和过程强化的应用有关，并将进一步集成到生物炼制工业中。可以将其他辅助技术与 PLE 和/或 SFE 整合在一起，从而提高效率。超声波辅助提取或酶促提取就是同样的情况。虽然在大多数情况下偶联仍然需要优化，但在这些领域已经取得了一些进展。无论如何，这些技术的应用都可能大大有助于破坏正在提取的天然基质，从而提高萃取效率并促进目标生物活性化合物的回收。另一个创新就是提取和干燥过程的集成，并且如本章所述已有不少相关的研究应用。基于气体抗溶剂、SAS、超临界流体强化分散和超临界溶液的快速膨胀等的超临界流体的干燥和选择性沉淀技术，主要应用于药物微粉化领域。然而，将其直接应用于与萃取过程偶联的天然提取物的干燥和在线造粒则更为复杂，需要更多的开发和研究。但是鉴于这些过程偶联的固有优势，该技术在商业化的应用中很有前景。

总而言之，虽然使用 PLE 和 SFE 提取生物活性物质已经被认为是一个成熟的领域，但仍有很大的发展空间来显著提高这些技术的应用。此外，基于压缩流体的萃取机制，使得这两种技术的整合更具有可持续性和生态友好性，从而在更复杂的生物精炼方法中具有关键的核心价值。

致谢

感谢哥伦比亚大学科学技术和创新管理部门（568-2012）（哥伦比亚）的博士奖学金。同时要感谢项目 AGL2014-53609-P（MINECO，西班牙）和 I-LINK+1096（CSIC，西班牙）的资金支持。

参考文献

[1] I. Goldberg, Functional Foods, first ed., Springer US, Boston, MA, 1994. http://dx.doi.org/10.1007/978-1-4615-2073-3.

[2] E. Lopez-Huertas, Health effects of oleic acid and long chain omega-3 fatty acids (EPA and DHA) enriched milks. A review of intervention studies, Pharmacol. Res. 61 (2010) 200-207. http://dx.doi.org/10.1016/j.phrs.2009.10.007.

[3] E. De Smet, R. P. Mensink, J. Plat, Effects of plant sterols and stanols on intestinal cholesterol metabolism: suggested mechanisms from past to present, Mol. Nutr. Food Res. 56 (2012) 1058-1072. http://dx.doi.org/10.1002/mnfr.201100722.

[4] P. Patil, S. Mandal, S. K. Tomar, S. Anand, Food protein-derived bioactive peptides in management of type 2 diabetes, Eur. J. Nutr. 54 (2015) 863-880. http://dx.doi.org/10.1007/s00394-015-0974-2.

[5] M. Dueñas, I. Muñoz-González, C. Cueva, A. Jiménez-Girón, F. Sánchez-Patán, C. Santos-Buelga, M. V. Moreno-Arribas, B. Bartolomé, A survey of modulation of gut microbiota by dietary polyphenols, Biomed. Res. Int. 2015 (2015) 1-15. http://dx.doi.org/10.1155/2015/850902.

[6] M. Herrero, A. del P. Sánchez-Camargo, A. Cifuentes, E. Ibáñez, Plants, seaweeds, microalgae and food by-products as natural sources of functional ingredients obtained using pressurized liquid extraction and supercritical fluid extraction, TrAC Trends Anal. Chem. 71 (2015) 26-38. http://dx.doi.org/10.1016/j.trac.2015.01.018.

[7] M. Herrero, M. Castro-Puyana, J. A. Mendiola, E. Ibáñez, Compressed fluids for the extraction of bioactive compounds, TrAC Trends Anal. Chem. 43 (2013) 67-83. http://dx.doi.org/10.1016/j.trac.2012.12.008.

[8] F. J. Barba, Z. Zhu, M. Koubaa, A. S. Sant'Ana, V. Orlien, Green alternative methods for the extraction of antioxidant bioactive compounds from winery wastes and by-products: a review, Trends Food Sci. Technol. 49 (2016) 96-109. http://dx.doi.org/10.1016/j.tifs.2016.01.006.

[9] Commission of the European Communities, Communication from the commission to the European parliament, the council, the European economic and social committee and the committee of the region, COM (2007) (2015) 1-16.

[10] P. T. Anastas, J. C. Warner, Green Chemistry: Theory and Practice, Oxford University Press, New York, 1998.

[11] F. Chemat, M. A. Vian, G. Cravotto, Green extraction of natural products: concept and principles, Int. J. Mol. Sci. 13 (2012) 8615-8627. http://dx.doi.org/10.3390/ijms13078615.

[12] F. Cherubini, The biorefinery concept: using biomass instead of oil for producing energy and chemicals, Energy Convers. Manag. 51 (2010) 1412-1421. http://dx.doi.org/10.1016/j.enconman.2010.01.015;

[12a] M. Plaza, C. Turner, Pressurized hot water extraction of bioactives, in: E. Ibanez, A. Cifuentes (Eds.), Green Extraction Techniques: Principles, Advances and Applications, vol. 76, 2017, pp. 53-82.

[13] M. Plaza, C. Turner, Pressurized hot water extraction of bioactives, TrAC Trends Anal. Chem. 71 (2015) 39-54. http://dx.doi.org/10.1016/j.trac.2015.02.022.

[14] H. Sun, X. Ge, Y. Lv, A. Wang, Application of accelerated solvent extraction in the analysis of organic contaminants, bioactive and nutritional compounds in food and feed, J. Chromatogr. A 1237 (2012) 1-23. http://dx.doi.org/10.1016/j.chroma.2012.03.003.

[15] G. Brunner, Supercritical fluids: technology and application to food processing, J. Food Eng. 67 (2005) 21–33. http://dx.doi.org/10.1016/j.jfoodeng.2004.05.060.

[16] J. A. Mendiola, M. Herrero, M. Castro-Puyana, E. Ibáñez, Supercritical fluid extraction, in: M. A. Rostagno, J. M. Prado (Eds.), Nat. Prod. Extr. Princ. Appl., The Royal Society of Chemistry, 2013, pp. 196–230. http://dx.doi.org/10.1039/9781849737579-00196.

[17] S. M. Zakaria, S. M. M. Kamal, Subcritical water extraction of bioactive compounds from plants and algae: applications in pharmaceutical and food ingredients, Food Eng. Rev. 8 (2016) 23–34. http://dx.doi.org/10.1007/s12393-015-9119-x.

[18] K. M. Sharif, M. M. Rahman, J. Azmir, A. Mohamed, M. H. A. Jahurul, F. Sahena, I. S. M. Zaidul, Experimental design of supercritical fluid extraction-a review, J. Food Eng. 124 (2014) 105–116. http://dx.doi.org/10.1016/j.jfoodeng.2013.10.003.

[19] A. D. P. Sánchez-Camargo, L. Montero, V. Stiger-Pouvreau, A. Tanniou, A. Cifuentes, M. Herrero, E. Ibáñez, Considerations on the use of enzyme-assisted extraction in combination with pressurized liquids to recover bioactive compounds from algae, Food Chem. 192 (2016) 67–74. http://dx.doi.org/10.1016/j.foodchem.2015.06.098.

[20] D. Wianowska, A. L. Dawidowicz, Effect of water content in extraction mixture on the pressurized liquid extraction efficiency—stability of quercetin 4′-glucoside during extraction from onions, J. AOAC Int. 99 (2016) 744–749. http://dx.doi.org/10.5740/jaoacint.16-0019.

[21] A. Liazid, G. Barbero, L. Azaroual, M. Palma, C. Barroso, Stability of anthocyanins from red grape skins under pressurized liquid extraction and ultrasound-assisted extraction conditions, Molecules 19 (2014) 21034–21043. http://dx.doi.org/10.3390/molecules191221034.

[22] A. P. Sánchez-Camargo, J. A. Mendiola, A. Valdés, M. Castro-Puyana, V. García-Cañas, A. Cifuentes, M. Herrero, E. Ibáñez, Supercritical antisolvent fractionation of rosemary extracts obtained by pressurized liquid extraction to enhance their antiproliferative activity, J. Supercrit. Fluids 107 (2016) 581–589. http://dx.doi.org/10.1016/j.supflu.2015.07.019.

[23] P. S. Saravana, A. T. Getachew, R. Ahmed, Y. J. Cho, Y. B. Lee, B. S. Chun, Optimization of phytochemicals production from the ginseng by-products using pressurized hot water: experimental and dynamic modelling, Biochem. Eng. J. 113 (2016) 141–151. http://dx.doi.org/10.1016/j.bej.2016.06.006.

[24] D. Villanueva Bermejo, I. Angelov, G. Vicente, R. P. Stateva, M. Rodriguez GarcíaRisco, G. Reglero, E. Ibañez, T. Fornari, Extraction of thymol from different varieties of thyme plants using green solvents, J. Sci. Food Agric. 95 (2015) 2901–2907. http://dx.doi.org/10.1002/jsfa.7031.

[25] T. Brazdauskas, L. Montero, P. R. Venskutonis, E. Ibáñez, M. Herrero, Downstream valorization and comprehensive two-dimensional liquid chromatography-based chemical characterization of bioactives from black chokeberries (*Aronia melanocarpa*) pomace, J. Chromatogr. A 1468 (2016) 126–135. http://dx.doi.org/10.1016/j.chroma.2016.09.033.

[26] J. Li, Z. Guo, Concurrent extraction and transformation of bioactive phenolic compounds from rapeseed meal using pressurized solvent extraction system, Ind. Crops Prod. 94 (2016) 152–159. http://dx.doi.org/10.1016/j.indcrop.2016.08.045.

[27] M. Sajfrtová, I. Ličková, M. Wimmerová, H. Sovová, Z. Wimmer, β-Sitosterol: supercritical carbon dioxide extraction from sea Buckthorn (*Hippophae rhamnoides* L.) seeds, Int. J. Mol. Sci. 11 (2010) 1842–1850. http://dx.doi.org/10.3390/ijms11041842.

[28] M. Durante, M. Lenucci, G. Mita, Supercritical carbon dioxide extraction of carotenoids from

pumpkin (*Cucurbita* spp.): a review, Int. J. Mol. Sci. 15 (2014) 6725-6740. http://dx.doi.org/10.3390/ijms15046725.

[29] J. Xiao, B. Tian, B. Xie, E. Yang, J. Shi, Z. Sun, Supercritical fluid extraction and identification of isoquinoline alkaloids from leaves of *Nelumbo nucifera* Gaertn, Eur. Food Res. Technol. 231 (2010) 407-414. http://dx.doi.org/10.1007/s00217-010-1290-y.

[30] L. L. M. Marques, G. P. Panizzon, B. A. A. Aguiar, A. S. Simionato, L. Cardozo-Filho, G. Andrade, A. G. de Oliveira, T. A. Guedes, J. C. P. de Mello, Guaraná (*Paullinia cupana*) seeds: selective supercritical extraction of phenolic compounds, Food Chem. 212 (2016) 703-711. http://dx.doi.org/10.1016/j.foodchem.2016.06.028.

[31] V. Micic, Z. Lepojevic, M. Jotanoviae, G. Tadic, B. Pejovic, Supercritical extraction of *Salvia officinalis* L. J. Appl. Sci. 11 (2011) 3630-3634. http://dx.doi.org/10.3923/jas.2011.3630.3634.

[32] P. Kraujalis, P. R. Venskutonis, Supercritical carbon dioxide extraction of squalene and tocopherols from amaranth and assessment of extracts antioxidant activity, J. Supercrit. Fluids 80 (2013) 78-85. http://dx.doi.org/10.1016/j.supflu.2013.04.005.

[33] M. Durante, M. S. Lenucci, L. D' Amico, G. Piro, G. Mita, Effect of drying and co-matrix addition on the yield and quality of supercritical CO_2 extracted pumpkin (*Cucurbita moschata* Duch.) oil, Food Chem. 148 (2014) 314-320. http://dx.doi.org/10.1016/j.foodchem.2013.10.051.

[34] B. Díaz-Reinoso, A. Moure, H. Domínguez, J. C. Parajó, Supercritical CO_2 extraction and Purification of compounds with antioxidant activity, J. Agric. Food Chem. 54 (2006) 2441-2469. http://dx.doi.org/10.1021/jf052858j.

[35] E. S. Ong, J. S. H. Cheong, D. Goh, Pressurized hot water extraction of bioactive or marker compounds in botanicals and medicinal plant materials, J. Chromatogr. A 1112 (2006) 92-102. http://dx.doi.org/10.1016/j.chroma.2005.12.052.

[36] M. Herrero, A. Cifuentes, E. Ibañez, Sub- and supercritical fluid extraction of functional ingredients from different natural sources: plants, food-by-products, algae andmicroalgae: a review, Food Chem. 98 (2006) 136-148. http://dx.doi.org/10.1016/j.foodchem.2005.05.058.

[37] S. M. Pourmortazavi, S. S. Hajimirsadeghi, Supercritical fluid extraction in plant essential and volatile oil analysis, J. Chromatogr. A 1163 (2007) 2-24. http://dx.doi.org/10.1016/j.chroma.2007.06.021.

[38] J. A. Mendiola, M. Herrero, A. Cifuentes, E. Ibañez, Use of compressed fluids for sample preparation: food applications, J. Chromatogr. A 1152 (2007) 234-246. http://dx.doi.org/10.1016/j.chroma.2007.02.046.

[39] J. Wiboonsirikul, S. Adachi, Extraction of functional substances from agricultural products or by-products by subcritical water treatment, Food Sci. Technol. Res. 14 (2008) 319-328. http://dx.doi.org/10.3136/fstr.14.319.

[40] C. Pronyk, G. Mazza, Design and scale-up of pressurized fluid extractors for food and bioproducts, J. Food Eng. 95 (2009) 215-226. http://dx.doi.org/10.1016/j.jfoodeng.2009.06.002.

[41] F. Sahena, I. S. M. Zaidul, S. Jinap, A. A. Karim, K. A. Abbas, N. A. N. Norulaini, A. K. M. Omar, Application of supercritical CO_2 in lipid extraction—a review, J. Food Eng. 95 (2009) 240—253. http://dx.doi.org/10.1016/j.jfoodeng.2009.06.026.

[42] M. Herrero, J. A. Mendiola, A. Cifuentes, E. Ibañez, Supercritical fluid extraction: recent advances and applications, J. Chromatogr. A 1217 (2010) 2495-2511. http://dx.doi.org/10.1016/

j. chroma. 2009. 12. 019.

[43] C. C. Teo, S. N. Tan, J. W. H. Yong, C. S. Hew, E. S. Ong, Pressurized hot water extraction (PHWE), J. Chromatogr. A 1217 (2010) 2484 - 2494. http://dx. doi. org/10. 1016/j. chroma. 2009. 12. 050.

[44] A. Mustafa, C. Turner, Pressurized liquid extraction as a green approach in food and herbal plants extraction: a review, Anal. Chim. Acta 703 (2011) 8 - 18. http://dx. doi. org/10. 1016/j. aca. 2011. 07. 018.

[45] C. Crampon, O. Boutin, E. Badens, Supercritical carbon dioxide extraction of molecules of interest from microalgae and seaweeds, Ind. Eng. Chem. Res. 50 (2011) 8941 - 8953. http://dx. doi. org/10. 1021/ie102297d.

[46] H. Sovová, Steps of supercritical fluid extraction of natural products and their characteristic times, J. Supercrit. Fluids 66 (2012) 73-79. http://dx. doi. org/10. 1016/j. supflu. 2011. 11. 004.

[47] H. Wijngaard, M. B. Hossain, D. K. Rai, N. Brunton, Techniques to extract bioactive compounds from food by-products of plant origin, Food Res. Int. 46 (2012) 505-513. http://dx. doi. org/10. 1016/j. foodres. 2011. 09. 027.

[48] A. Capuzzo, M. Maffei, A. Occhipinti, Supercritical fluid extraction of plant flavors and fragrances, Molecules 18 (2013) 7194-7238. http://dx. doi. org/10. 3390/molecules18067194.

[49] N. A. Akgün, H. Bulut, I. Kikic, D. Solinas, Extraction behavior of lipids obtained from spent coffee grounds using supercritical carbon dioxide, Chem. Eng. Technol. 37 (2014) 1975-1981. http://dx. doi. org/10. 1002/ceat. 201400237.

[50] A. Baiano, Recovery of biomolecules from food wastes—a review, Molecules 19 (2014) 14821-14842. http://dx. doi. org/10. 3390/molecules190914821.

[51] M. D. A. Saldaña, C. S. Valdivieso-Ramírez, Pressurized fluid systems: phytochemical production from biomass, J. Supercrit. Fluids 96 (2015) 228 - 244. http://dx. doi. org/10. 1016/j. supflu. 2014. 09. 037.

[52] S. Millao, E. Uquiche, Extraction of oil and carotenoids from pelletized microalgae using supercritical carbon dioxide, J. Supercrit. Fluids 116 (2016) 223 - 231. http://dx. doi. org/10. 1016/j. supflu. 2016. 05. 049.

[53] R. P. F. F. da Silva, T. A. P. Rocha - Santos, A. C. Duarte, Supercritical fluid extraction of bioactive compounds, TrAC Trends Anal. Chem. 76 (2016) 40 - 51. http://dx. doi. org/10. 1016/j. trac. 2015. 11. 013.

[54] E. Roselló-Soto, O. Parniakov, Q. Deng, A. Patras, M. Koubaa, N. Grimi, N. Boussetta, B. K. Tiwari, E. Vorobiev, N. Lebovka, F. J. Barba, Application of nonconventional extraction methods: toward a sustainable and green production of valuablecompounds from mushrooms, Food Eng. Rev. 8 (2016) 214-234. http://dx. doi. org/10. 1007/s12393-015-9131-1.

[55] G. L. Zabot, M. N. Moraes, P. I. N. Carvalho, M. A. A. Meireles, New proposalfor extracting rosemary compounds: process intensification and economic evaluation, Ind. Crops Prod. 77 (2015) 758-771. http://dx. doi. org/10. 1016/j. indcrop. 2015. 09. 053.

[56] P. Kraujalis, P. R. Venskutonis, Ibáñez, M. Herrero, Optimization of rutin isolationfrom *Amaranthus paniculatus* leaves by high pressure extraction and fractionation techniques, J. Supercrit. Fluids 104 (2015) 234-242. http://dx. doi. org/10. 1016/j. supflu. 2015. 06. 022.

[57] J. Fabrowska, E. Ibañez, B. Łęska, M. Herrero, Supercritical fluid extraction as a tool to valorize

underexploited freshwater green algae, Algal Res. 19 (2016) 237 - 245. http: //dx. doi. org/10. 1016/j. algal. 2016. 09. 008.

[58] A. Zinnai, C. Sanmartin, I. Taglieri, G. Andrich, F. Venturi, Supercritical fluid extraction from microalgae with high content of LC-PUFAs. A case of study: Sc-CO_2 oil extraction from *Schizochytrium* sp. J. Supercrit. Fluids 116 (2016) 126-131. http: //dx. doi. org/10. 1016/j. supflu. 2016. 05. 011.

[59] P. S. Saravana, Y. J. Cho, Y. B. Park, H. C. Woo, B. S. Chun, Structural, antioxidant, and emulsifying activities of fucoidan from *Saccharina japonica* using pressurized liquid extraction, Carbohydr. Polym. 153 (2016) 518-525. http: //dx. doi. org/10. 1016/j. carbpol. 2016. 08. 014.

[60] A. Mouahid, C. Crampon, S. -A. A. Toudji, E. Badens, Effects of high water content and drying pre-treatment on supercritical CO_2 extraction from *Dunaliella salina* microalgae: experiments and modelling, J. Supercrit. Fluids 116 (2016) 271-280. http: //dx. doi. org/10. 1016/j. supflu. 2016. 06. 007.

[61] F. A. Reyes, J. A. Mendiola, S. Suárez-Alvarez, E. Ibañez, J. M. del Valle, Adsorbentassisted supercritical CO_2 extraction of carotenoids from *Neochloris oleoabundans* paste, J. Supercrit. Fluids 112 (2016) 7-13. http: //dx. doi. org/10. 1016/j. supflu. 2016. 02. 005.

[62] M. Viguera, A. Marti, F. Masca, C. Prieto, L. Calvo, The process parameters and solid conditions that affect the supercritical CO_2 extraction of the lipids produced by microalgae, J. Supercrit. Fluids 113 (2016) 16-22. http: //dx. doi. org/10. 1016/j. supflu. 2016. 03. 001.

[63] B. Gilbert - López, J. A. Mendiola, J. Fontecha, L. A. M. van den Broek, L. Sijtsma, A. Cifuentes, M. Herrero, E. Ibáñez, Downstream processing of *Isochrysis galbana*: a step towards microalgal biorefinery, Green Chem. 17 (2015) 4599-4609. http: //dx. doi. org/10. 1039/C5GC01256B.

[64] E. Roselló - Soto, M. Koubaa, A. Moubarik, R. P. Lopes, J. A. Saraiva, N. Boussetta, N. Grimi, F. J. Barba, Emerging opportunities for the effective valorization of wastes and by-products generated during olive oil production process: non - conventional methods for the recovery of high - added value compounds, Trends Food Sci. Technol. 45 (2015) 296 - 310. http: //dx. doi. org/10. 1016/j. tifs. 2015. 07. 003.

[65] J. Viganó, I. Z. Brumer, P. A. de C. Braga, J. K. da Silva, M. R. Marostica Júnior, F. G. Reyes Reyes, J. Martínez, Pressurized liquids extraction as an alternative process to readily obtain bioactive compounds from passion fruit rinds, Food Bioprod. Process. 100 (2016) 382 - 390. http: //dx. doi. org/10. 1016/j. fbp. 2016. 08. 011.

[66] L. Manna, C. A. Bugnone, M. Banchero, Valorization of hazelnut, coffee and grape wastes through supercritical fluid extraction of triglycerides and polyphenols, J. Supercrit. Fluids 104 (2015) 204-211. http: //dx. doi. org/10. 1016/j. supflu. 2015. 06. 012.

[67] A. Schievano, F. Adani, L. Buessing, A. Botto, E. N. Casoliba, M. Rossoni, J. L. Goldfarb, An integrated biorefinery concept for olive mill waste management: supercritical CO_2 extraction and energy recovery, Green Chem. 17 (2015) 2874-2887. http: //dx. doi. org/10. 1039/C5GC00076A.

[68] M. P. Garcia - Mendoza, J. T. Paula, L. C. Paviani, F. A. Cabral, H. A. Martinez - Correa, Extracts from mango peel by-product obtained by supercritical CO_2 and pressurized solvent processes, LWT—Food Sci. Technol. 62 (2015) 131-137. http: //dx. doi. org/10. 1016/j. lwt. 2015. 01. 026.

[69] M. del P. Garcia - Mendoza, F. A. Espinosa - Pardo, A. M. Baseggio, G. F. Barbero, M. R. Maróstica Junior, M. A. Rostagno, J. Martínez, Extraction of phenolic compounds and anthocyanins from juçara (*Euterpe edulis* Mart.) residues using pressurized liquids and supercritical fluids, J. Supercrit. Fluids 119 (2017) 9-16. http: //dx. doi. org/10. 1016/j. supflu. 2016. 08. 014.

[70] G. L. Zabot, M. A. A. Meireles, On-line process for pressurized ethanol extraction of onion peels extract and particle formation using supercritical antisolvent, J. Supercrit. Fluids 110 (2016) 230-239. http://dx.doi.org/10.1016/j.supflu.2015.11.024.

[71] M. Lores, M. Pájaro, M. Álvarez-Casas, J. Domínguez, C. García-Jares, Use of ethyl lactate to extract bioactive compounds from *Cytisus scoparius*: comparison of pressurized liquid extraction and medium scale ambient temperature systems, Talanta 140 (2015) 134-142. http://dx.doi.org/10.1016/j.talanta.2015.03.034.

[72] M. T. Golmakani, J. A. Mendiola, K. Rezaei, E. Ibáñez, Pressurized limonene as an alternative bio-solvent for the extraction of lipids from marine microorganisms, J. Supercrit. Fluids 92 (2014) 1-7. http://dx.doi.org/10.1016/j.supflu.2014.05.001.

[73] F. Pena-Pereira, J. Namieśnik, Ionic liquids and deep eutectic mixtures: sustainable solvents for extraction processes, ChemSusChem. 7 (2014) 1784-1800. http://dx.doi.org/10.1002/cssc.201301192.

3

3 加压热水萃取生物活性物质

Merichel Plaza[1,2], Charlotta Turner[1,*]
1. Lund University, Lund, Sweden
2. University of Alcalá, Alcalá de Henares, Spain
*通讯作者: Email: Charlotta. Turner@ chem. lu. se

缩略语

CED	内聚能密度
$C_{p,m}$	比热容（等压摩尔）
D	扩散系数
DPPH	2,2-二苯基-1-三硝基苯肼
FC	福林-酚
HSP	汉森溶解度参数
H_v	蒸发热
K_w	解离常数
MRP	美拉德反应产物
P	压力
PHWE	加压热水萃取
PLE	加压液体萃取
T	温度
USWE	超声强化亚临界水萃取
ε_r	相对静态介电常数
h	动力黏度
π^*	极化率

3.1 引言

本章是根据同一作者 Plaza 和 Turner 的综述文章[1]编写的。

加压热水萃取（PHWE）是在高于水的大气沸点（100℃/273K，0.1MPa）但低于水的临界点（374℃/647K，22.1MPa）的温度下，使用液态水作为提取剂（提取溶剂）的萃取技术，见图 3.1。加压热水萃取（PHWE），也可以称为亚临界水萃取，过热水萃取、加压液体萃取或以水为溶剂的加速溶剂萃取，其在分析化学中的应用始于 Hawthorne 及其同事在 20 世纪 90 年代中期开展的环境方面的分析[2,3]。近期有一些 PHWE 相关的综述文章[4-7]，建议读者阅读。

本章提供了详尽的关于水的基本特性的背景知识，在迄今为止有关 PHWE 分析技术的大多数综述文章和书本中几乎都忽略了这一点。因此，本章的第一部分介绍了水的化学/物理特性的基本原理，这些特性如何随温度的升高而变化，以及在不同的分析应用中这些特性对萃取性能产生的影响。

第二部分介绍了 PHWE 的技术解决方案以及如何在实际操作中开展实验。主要包括有关市场化设备以及自组装设备使用的讨论。第三部分包括 PHWE 中各个参数的方法优化。第四部分介绍了该方法的一些主要应用方向，如从植物、食品、生物和制药样品中提取生物活性化合物等。最后一部分是关于 PHWE 技术在分析化学领域设备和

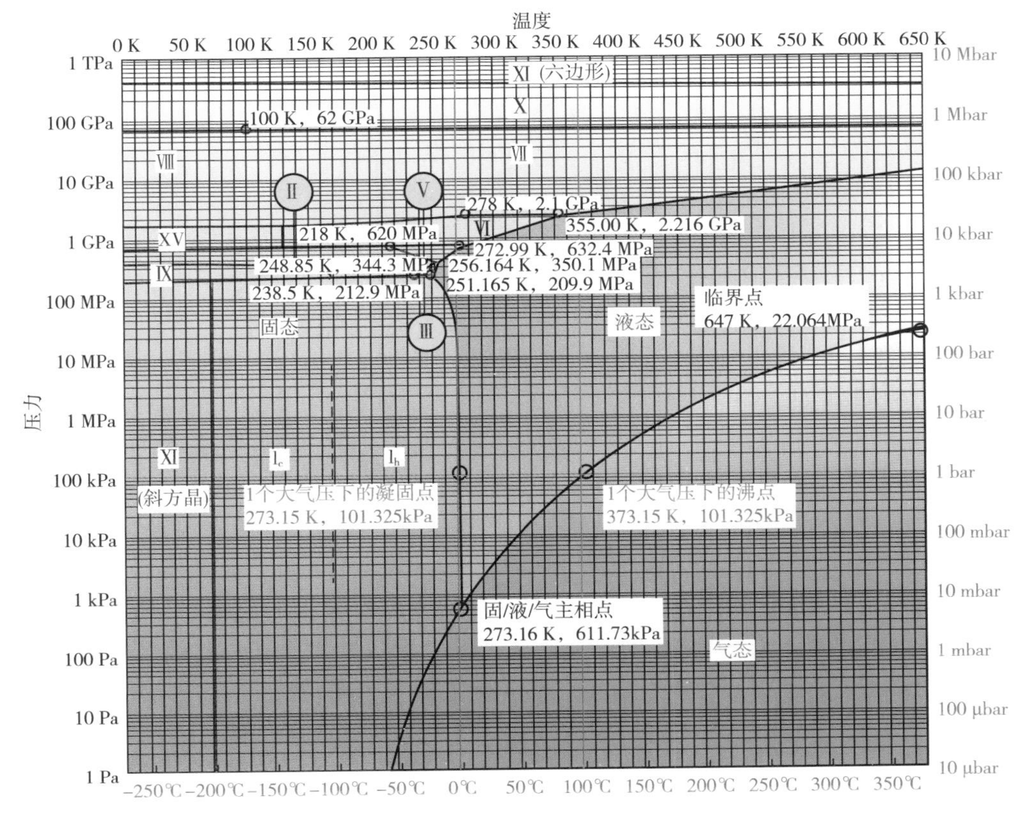

图 3.1　水的相图[8]

方法学方面的发展的总结和展望。

3.2　从环境到近临界状态水的基本特性

水也许是地球上最有趣的自然存在的液体，它对生命来说是必不可少的，且随处可以取用。考虑到生产和运输，实际上使用水作为溶剂对环境的影响几乎可以忽略不计。但是，当样品提取过程中使用水作为溶剂时，某些情况下常常需要消耗大量的能量通过蒸发去除水分。值得注意的是，加热液态水（25~250℃，5MPa）所需的能量几乎为使水汽化以产生蒸汽（25~250℃，0.1MPa）所需能量的1/3[9]。

3.2.1　环境条件下的水

水具有其他溶剂所不具备的化学和物理特性[10]。在环境温度下，水是一种极性液体，如其解离常数（K_w，25℃时为 1.0×10^{-14}）所描述的那样，它分解为水合氢离子（H_3O^+）和氢氧根离子（OH^-），这种解离通常被称为“水的自电离”[11]，见公式（3.1）：

$$K_w = \frac{[H_3O^+][OH^-]}{[H_2O]} \tag{3.1}$$

另外，水在20℃下具有高达80的相对静态介电常数（ε_r，也称为介电常数）。这意味着水会与其他分子产生静电结合，从而减少或消除周围离子之间的分子间相互作用。换句话说，水可以很好地溶解盐。介电常数（ε_r）可以通过电容器测量，如公式（3.2）所示，将真空中的电容（C_0）与液体中的电容（C_x）相比。

$$\varepsilon_r = \frac{C_0}{C_x} \tag{3.2}$$

ε_r是与极化率（π^*）有关的宏观特性，而极化率本身是一种微观特性，即在分子中建立具有分离电荷分布的感应偶极子的难易程度。π^*和ε_r一样，都是由实验确定的。对于在环境条件下的液态水，π^*和ε_r都相对较高，并且它们会随着温度的升高而降低，如下所述。

介电常数较高的水也具有较高的偶极矩（1.85D），而在20℃时ε_r为1.9的正己烷的偶极矩为0D[12]。但是，ε_r不应与分子的偶极矩相混淆。例如，二氯甲烷的ε_r相对较低，为9.1（20℃），而其偶极矩相当高（1.60D），对于乙酸乙酯也是如此（ε_r=6.0，偶极矩1.78D）。

水的高偶极矩可以用氧的负电性来解释，与典型的109°四面角相比，O—H键之间的夹角较小（104.5°）。这种电荷分布导致水分子之间以及水分子与其他偶极分子之间形成分子间吸引力，即所谓的氢键。水具有很强的氢键作用，因此具有很高的比热容［$C_{p,m}$= 75.3J/（mol·K），等压摩尔，25℃］，以及很高的蒸发热（H_v= 40.7kJ/mol，100℃）。

表3.1所示为水的一些重要化学和物理特性，在使用水作为溶剂时可以提供参考。许多参数会随温度变化而发生急剧变化。

表3.1　　不同温度和饱和压力状态下液态水的化学和物理特性

特性	T=298K（25℃，0.1MPa）	T=373K（100℃，0.1MPa）	T=473K（200℃，1.5MPa）	T=623K（350℃，17MPa）
解离常数（K_w）[10]	1.0×10^{-14}	5.6×10^{-13}	4.9×10^{-12}	1.2×10^{-12}
pK_w	13.99	12.25	11.31	11.92
介电常数（ε_r）[13]	78.5	55.4	34.8	14.1
偶极矩	1.85	1.85	1.85	1.85
比热容（C_p）/［J·（g·K）］[14]	4.18	4.22	4.51	10.1
蒸发热（H_v）/（kJ/mol）	44.0	40.7	35.0	15.6
密度/（g/cm^3）[15]	0.997	0.958	0.865	0.579
动态黏度（η）/（mPa·s）[16]	0.891	0.282	0.134	0.067
表面张力/（dyn/cm）[16]	72.0	58.9	37.6	3.7
自扩散系数（D）/（m^2/s）[17]	2.3×10^{-9}	8.6×10^{-9}	23.8×10^{-9}	N/A

3.2.2 高温高压条件下的水

水之所以能成为萃取过程中人们感兴趣且可行的溶剂的原因是，当温度变化时，水的一些化学和物理性质会发生很大变化[17]。如表 3.1 所示，饱和压力下液态水的介电常数，从 25℃时的 78 变为 350℃时的 14[13]。接近临界条件下的水（压力和温度在临界点以下）可以溶解疏水性化合物，如多环芳烃和多氯联苯等[3,18]。盐不再溶于水，即发生沉淀。发生这种现象的原因是温度的升高降低了水分子之间以及水分子与周围离子和分子之间的静电相互作用，即 ε_r 和 π^* 均随温度的升高而降低。水分子的运动/旋转也会随温度的升高而增加。在高温高压下使用液态水可以溶解极性较小的化合物，因为涉及氢键的分子间相互作用变得不那么明显，从而有利于伦敦色散力（诱导偶极之间的相互作用力）起作用。换言之，在高温（和高压）下的液态水已成为一种较低极性的溶剂。作为对比可知，温度在 200 到 275℃饱和压力下的液态水的介电常数与环境条件下的甲醇和乙醇的介电常数相似[13,19]。

溶剂性质可以使用三个溶剂致变色 Kamlet-Taft 参数[20]，通过实验进行表征和“定量”：

- 氢键贡献能力（酸度，α）
- 氢键接受能力（碱度，β）
- 极化率/极性（π^*）

每个参数都是介于 0~1 之间的表征刻度，这种表征是基于两个参比溶剂确定的，一种溶剂设置为 0，另一种设置为 1。例如，设置环己烷的极化率为 0，DMSO 的极化率为 1。大多数溶剂在每个刻度上的值都在 0~1 之间。这些参数是使用具有可见光吸收的化合物通过实验得到的，例如 4-硝基苯甲醚和 *N*，*N*-二乙基-4-硝基苯胺染料，当它们溶解在溶剂中时，相应吸光带[20]的最大吸光度会发生偏移。使用该方法，可对溶剂性质进行比对[20,21]。在表 3.2 中，列出了水与一些常见有机溶剂的溶剂致变色参数。结果表明，275℃时的高温液态水具有与甲醇相似的极化率，但其氢键接受能力与甲醇和乙醇有很大的不同。因此，在讨论加压热水的溶剂特性时，如果只考虑极化率或介电常数都是有风险的。

表 3.2　水和一些常见有机溶剂的溶剂变色参数，计算得到的平均值[20]

溶剂	α	β	π^*
水（环境）	1.20	0.37	1.12
水（275℃）	0.84	0.20	0.69
甲醇	1.01	0.71	0.59
乙醇	0.89	0.79	0.53
乙酸乙酯	0	0.36	0.61
环己烷	0	0	0

Carr 等综述了有机化合物在加压热水中的溶解度，包括一些生物活性化合物，如

D-柠檬烯和辛酸[19]。本文描述了有机化合物在水中的溶解度是有一个最小值的，这是由于溶质化合物周围的“氢键笼”导致溶解度随着温度的升高而降低。然而，在一定的温度下，溶解度反而随着温度的升高而增加，因为空穴形成的正热量大于溶解时的负热量。

然而，实验测定生物活性化合物在水中的溶解度时是有缺点的。可用的化学标准品较少，而且在溶解度测定实验的平衡过程中存在化合物降解的相关风险。因此，溶解度模型和理论估算是对实验的较好补充。汉森溶解度参数（HSP）就是评估溶剂性质的理论工具之一。该方法最初由 Hansen[22]开发，可用于描述溶剂性质或预测分析物及其他化合物在溶剂中的溶解度。该方法基于三个内聚能密度（CED）参数：

- 分子间分散力产生的能量（δ_d）
- 分子间偶极力产生的能量（δ_p）
- 分子间氢键能量（δ_h）

这三个参数分别构成分析物和溶剂分子在三维空间中的坐标。球体重叠越多，分析物在溶剂中的溶解性就越高。总溶解度参数 δ_T（平方）是三个 CED 参数的平方之和[22]，如式（3.3）所示：

$$\delta_T^2 = \delta_d^2 + \delta_p^2 + \delta_h^2 \tag{3.3}$$

CED 参数的计算基于基团贡献法[22]，例如 Srinivas 等人[23]用汉森溶解度参数表征了白桦醇在水和乙醇中的溶解度。白桦醇是一种抗真菌和抗流感化合物，大量存在于桦树皮中[24]。在图 3.2 和文献[23]中表明，水需要达到 250~325℃才能溶解该化合物。图 3.2 还表明对总溶解度参数贡献最大的是氢键 CED，而上文讨论过，氢键 CED 也是随温度升高而降低最多的。

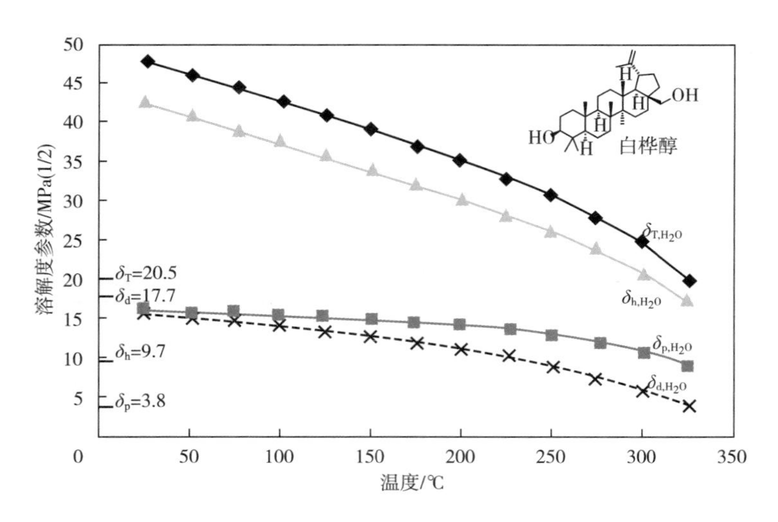

图 3.2 饱和压力下水的汉森溶解度参数与温度的关系（曲线）和白桦醇溶解度参数（标记在左侧）

［资料来源：经 K. Srinivas，J. W. King，J. K. Monrad，L. R. Howard，C. M. Hansen 许可修改和重印，Optimization of subcritical fluid extraction of bioactive compounds using Hansen solubility parameters，J. Food Sci. 74（2009）E342-E354，http：//dx. doi. org/10. 1111/j. 1750e3841. 2009. 01251. x. John Wiley & Sons.］

从上面的讨论可以看出，温度对水的介电常数、溶剂特性等有很大的影响。另一方面，压力的影响就比较小。在文献中，通常表述为压力对加压热水的溶剂强度[4]或对 PHWE 的萃取效率的影响较小[5]。压力主要用于将水保持在液态，尽管在一些参考文献中提到压力升高有助于润湿样品基质，从而提高萃取效率[25,26]。值得注意的是，饱和压力下液态水的温度升高导致极化率和浓度降低的曲线完全重叠（图 3.3[27]）。如表 3.1 和图 3.4（3）所示，饱和压力下的液态水密度从 25℃时的 0.997 g/mL 降至 350℃时的 0.579 g/mL[15]。当温度高于 200℃时，分子间相互作用力减弱导致水密度降低，而补偿所需的压力一般在 100MPa 或更大（图 3.4）。因此，需要注意的是液态水在高温高压下的密度不是恒定的，而是随着实验温度上升（100~374℃）而显著下降，这对于理解不溶性工作和萃取效率实验是非常重要的。

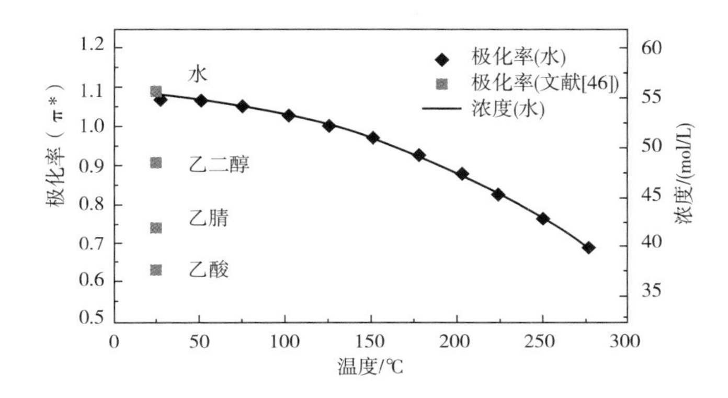

图 3.3 液态水的极化率和浓度随温度的变化

［资料来源：经 J. Lu，J. S. Brown，E. C. Boughner，C. L. Liotta，C. A. Eckert 许可转载，Solvatochromic characterization of near－critical water as a benign reaction medium，Ind. Eng. Chem. Res. 41（2002）2835－2841，American Chemical Society.］

除了介电性质和密度外，水的自电离性质也随温度而变化。表 3.1 所示为水的解离常数（K_w）会随温度变化发生两个数量级的变化，从 25℃时的 1.0×10^{-14} 增加到 350℃时的 1.2×10^{-12}，在 250℃时出现最大值为 4.9×10^{-12}。这意味着 pH 从约 7.0 变化到 5.5（图 3.5），并且氢离子和氢氧根离子的离子强度在 250℃时明显高于环境温度。这些变化可能在几个方面影响 PHWE。首先，高温时离子强度较高，pH 较低，发生水解等不必要反应的风险增加。氢离子在反应中起催化剂的作用[28]。此外，一些分析物可能发生向其他带电形式的平衡转移，例如，花青素随 pH 变化可以有五种不同形式出现[29]。

上述化学和物理性质主要与加压热水的热力学溶剂性质有关。此外，还有一些特性会影响液态水的传质特性，如黏度、扩散率和表面张力等，所有这些特性都会受到水温的影响（表 3.1 和图 3.6）。

水的黏度随温度的变化可以用阿伦尼乌斯方程式[30]来描述，方程式（3.4）如下：

$$\mu(T) = 2.414\times10^{-5}\times10^{247.8/(T-140)} \tag{3.4}$$

式中，μ 是动力黏度（$N\cdot s/m^2$），T 是绝对温度（K）。如式（3.4）所示，黏度随温度

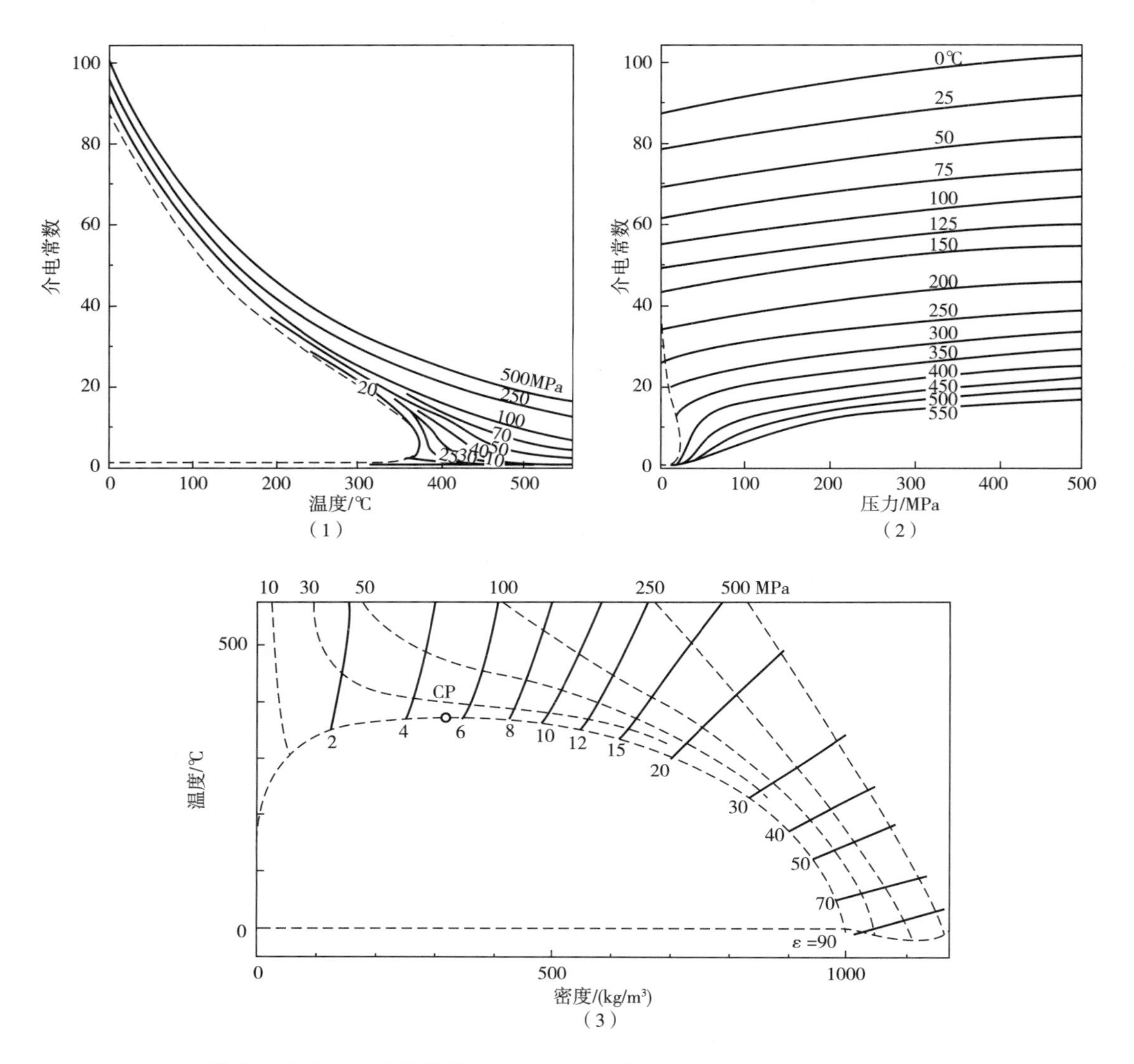

图 3.4 液态水的静态介电常数随温度（1）、压力（2）和温度/压力/密度（3）的变化

[资料来源：经 M. Uematsu, E. U. Franck 许可转载，Static dielectric-constant of water and steam, J. Phys. Chem. Ref. Data 9 (1980) 1291-1306. AIP Publishing LLC.]

的升高而降低。

斯托克斯-爱因斯坦公式（3.5）将化合物的自扩散系数描述为其黏度的函数：

$$D = \frac{k_B T}{6\pi\eta r_H} \tag{3.5}$$

式中，D 是自扩散系数（m^2/s），η 是动力黏度（mPa s），r_H 是化合物的流体力学半径（m），T 是绝对温度（K），k_B 是玻尔兹曼常数（1.38×10^{-23} J/K）。显然，自扩散系数随温度的升高和黏度的降低而增大。如表 3.1 所示，水的自扩散系数从 25～200℃增大 10 倍，在临界点（374℃），水的自扩散系数大约是环境条件下的 100 倍。

化合物在液体中的有效（二元）扩散也取决于浓度（或化学势）梯度，如菲克第

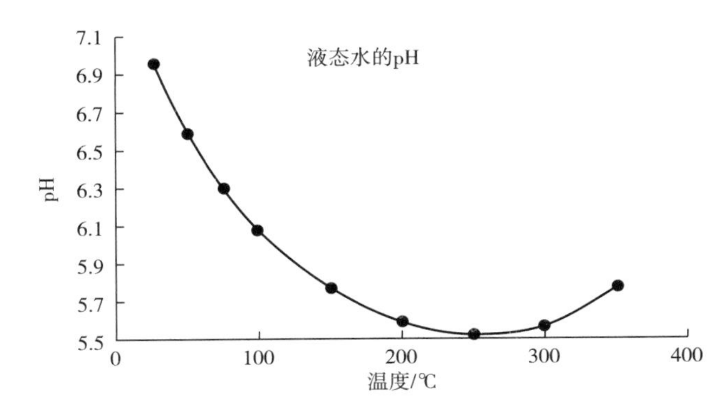

图 3.5　水电离图

在 25MPa 下，水的解离常数（K_w）负对数的一半，即 pH 与温度的函数变化

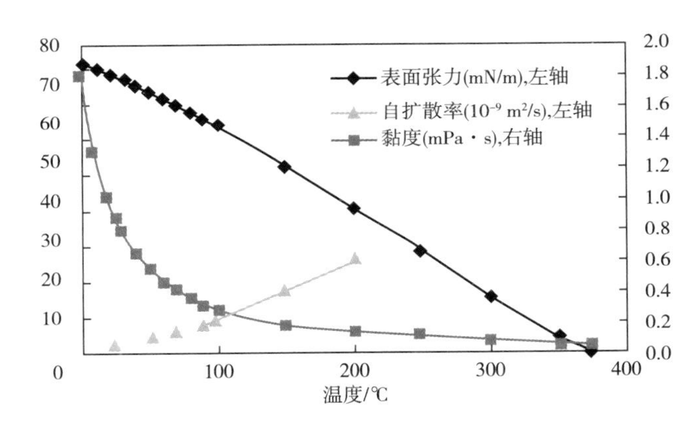

图 3.6　饱和压力下液态水的黏度、自扩散率和表面张力随温度的变化

[资料来源：数据取自蒸汽表，http：//dx. doi. org/10. 1615/AtoZ. s. steam_ tables。可从以下网址获得：http：//www. thermopedia. com/content/1150/；G. Brunner，Hydrothermal and supercritical water processes in：E. Kiran（Ed. ），Supercritical Fluid Science and Technologies Series，vol. 5，Elsevier，Amsterdam，The Netherlands，2014，p. 604；K. Krynicki，C. D. Green，D. W. Sawyer，Pressure and temperature dependence of self-diffusion in water，Faraday Discuss. Chem. Soc. 66（1978）199-208.]

二定律所述。对于分析物在球形样品颗粒内的扩散，菲克第二定律可写成公式（3.6）[31]：

$$\frac{\partial C}{\partial t}=D_{\text{eff}}\left(\frac{\partial^2 C}{\partial r^2}+\frac{2}{r}\frac{\partial C}{\partial r}\right) \tag{3.6}$$

式中，C 为浓度（mol/m^3），t 为时间（s），r 为样品颗粒半径（m），D_{eff} 为有效扩散系数（m^2/s）。

Karacabey 等人[31]报告了当温度从 105℃上升到 160℃时，反式白藜芦醇的扩散系数从 $3.3\times10^{-11}m^2/s$ 增加到 $10.4\times10^{-11}m^2/s$。但有人指出，通过实验确定化合物在 PHWE 中的扩散系数是一个挑战，因为总有化学反应发生的风险，进而可能会产生错误的结果。

与空气接触的水的表面张力随水温的升高而降低，见表3.1和图3.6。对比25℃和200℃，后者的表面张力约为前者的50%。在临界点时，表面张力为零。PHWE中萃取剂（水）的表面张力偏低将有利于样品的润湿性得到改善，即在萃取剂和样品基质之间形成较大的接触面积。这反过来可能会提高萃取效率。

总而言之，高温高压的液态水是一种具有较低的介电常数、极化率和密度的溶剂，由于其较高的扩散率、较低的黏度和表面张力，因而能够加快质量传递并改善样品的润湿性（图3.3、图3.4和图3.6）。在3.4中，我们将深入讨论如何在考虑不同变量的情况下优化PHWE，特别是影响溶解度和萃取动力学的变量。首先，对PHWE设备的实用方面进行描述。

3.3 加压热水萃取设备

如何实际应用液态水作为萃取剂进行萃取？首先，用去离子水或自来水作为萃取剂。为防止分析物氧化，水应该是无氧的。通常用于实现水脱气的方法是超声波或氮气净化，如果实验室有超声波水浴，那么超声脱气是一种低成本的方法，但比氮气净化要慢得多。超声波脱气至少需要60min，才可以确保排出水中的大部分氧气。

实施PHWE的基本设备并不复杂。有两种类型的设备：动态（连续流动）系统和静态（间歇）系统，以及两者的组合，见图3.7。

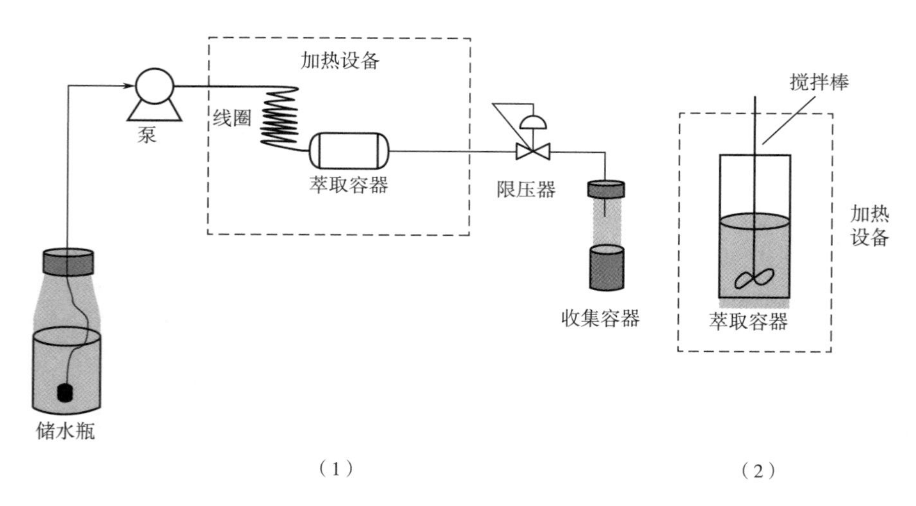

图3.7 动态（连续流动）加压热水萃取（PHWE）系统（1）和静态（间歇）PHWE系统（2）的主要部件示意图

3.3.1 动态加压热水萃取

在动态PHWE中，需要的基本部件有泵、萃取容器、加热装置、限压器和收集瓶［图3.7（1）］。系统中使用的泵可以是旧的高效液相色谱泵，因为萃取过程中的流速精度通常没有色谱法中那么严格。泵将水输送至萃取容器，并通过限压器输送至收集瓶。泵应能达到在萃取过程中使水保持液态所需的压力（通常为3.5~20MPa）。水的加热可以用烘箱，例如旧的气相色谱仪柱温箱，也可以用热交换器、加热带或加热套等。通常将不锈钢制成的管子盘绕在加热装置内，以确保水的设定温度也是进入萃取容器前的实际水温。萃取容器通常由不锈钢制成，两端应有滤芯，以避免样品损失和管道堵塞。如果使用中等的温度和压力，即典型的生物活性物质提取条件100~200℃和5~10MPa，则可用空的高效液相色谱柱作为提取容器。然而如果提取温度接近临界点，即250~374℃，那么为了防止容器腐蚀，萃取容器应选择耐腐蚀金属合金，例如哈司特镍合金。需要通过限压器来控制萃取容器内的压力，并防止萃取容器出口处的水发生沸腾。限压器可以是针阀、背压调节器、薄毛细管或简单的短管，其挤压端提供足够小的出口，以保持上游足够的压力。

3.3.2 静态加压热水萃取

在静态PHWE［图3.7（2）］中，不需要泵。但是，如果实际工作中用泵将水输送至萃取容器，则需要入口和出口处安装两个阀门，用来维持萃取过程中容器内的压力。当有泵存在时，压力可以通过在出口阀关闭的情况下向容器中加入更多的水来控制。如果不使用泵，手动向萃取容器加水，当容器关闭并加热时，压力增大，萃取在系统的饱和压力下进行。在静态PHWE中，对流是通过搅拌器加速传质来实现的。静态PHWE中的萃取容器通常为高压釜式，或至少比动态PHWE中的直径更宽，以满足搅拌器的使用。加热时，宜使用烘箱、加热套或加热带。一般不需要限压器，除非要控制从容器中取出提取物的速度。

不论用动态或静态PHWE进行实验，都要先将样品放入萃取容器（在某些情况下样品与玻璃珠或其他分散剂混合），然后将系统平衡到设定温度。之后，可以手动添加水，也可以用泵以一定的流速将水输送到萃取容器中，然后通过节流器将水排出到收集瓶中。水的流速可以在泵处进行监测，或者通过对收集到的提取物进行连续重量分析。

使用这两种系统各有利弊。由于不需要泵和压力限制器，静态PHWE更简单、更易于使用。然而，分析物在静态PHWE中的停留时间比在动态PHWE中要长得多，这可能导致热不稳定分析物降解[32]。因此，萃取时间将影响萃取效率和分析物的降解程度。此外，在静态PHWE中，由于萃取剂的体积是恒定的，分析物从样品基质到萃取剂的分配平衡将在一段时间后趋于稳定。因此在这种情况下，萃取效率取决于分析物在水中的分配比。另一方面，在动态PHWE中，由于新鲜的萃取剂不断地被泵入萃取容器并被泵出到收集瓶中，分析物在高温水中的停留时间较短。在这种情况下，提取液的流速可以控制停留时间。

因此，使用动态（连续流动）萃取装置更容易控制PHWE中的萃取和降解动力学。动态PHWE的缺点是成本更高，而且在提取过程中管道内部始终存在堵塞的风险。

一些商用PHWE仪器解决了上述问题。例如，即使萃取是在静态模式下进行的，且无流体通过容器，萃取剂也可以定期更换为新鲜溶剂（水），以防止萃取完成前发生平衡。在这种情况下，萃取包括几个静态萃取循环，其中每个循环有一定的持续时间，并且每个体积的萃取剂被替换到一定程度。另一个挑战是管道堵塞，主要涉及动态PHWE。在萃取容器的下游，水提取液冷却后，某些萃取分析物不再可溶而析出，并因此沉淀堵塞管道。有两种方法可以避免这个问题，一种方法是在从烘箱出口到收集瓶的管子周围使用加热带。另一种方法是使用一个额外的泵清洗萃取容器之后和限压器之前的管路[33]。

对于任何PHWE设备，必须考虑最大工作温度和压力、萃取容器的材料以及一般安全预防措施，如爆破片、通风和废水管等。

3.4 加压热水萃取方法的优化

从固体和半固体样品中提取分析物的过程可描述为以下五个步骤：①用萃取剂润湿样品基质；②目标物从样品基质中初步解吸；③在样品基质孔隙中扩散；④在样品基质和萃取剂之间分配；⑤通过滞流萃取剂层进行扩散，直到达到对流区为止[34]。所有这些步骤或多或少都是并行的。在PHWE中，温度是优化过程的关键参数，它会影响所有步骤的效率。此外，提取时间和/或流速也是需要优化的重要变量。

3.4.1 温度

水温越高，样品基质的润湿性越好（步骤1）。此外，温度升高也有利于被分析物和基质相互作用的破坏，特别是氢键和其他偶极-偶极力的传质动力学，从而促进被分析物从样品基质的初始解吸（步骤2）。较高的温度也会导致更快的扩散率（步骤3和5）以及改变（通常表现为提高）溶解度，后者导致样品基质和萃取剂之间分析物分配的变化（步骤4）。总之，在PHWE中提高温度可以改善萃取动力学。

在PHWE中使用高温主要有三个缺点，即降低萃取选择性、导致分析物的降解和样品基质中的其他化学反应。温度越高，将提取出越多不需要的化合物，即所谓的污染物，这将导致较低的选择性，并增加在PHWE后进一步净化的需要。例如，Vergara-Salinas等人[35]对分别在100℃和200℃下从葡萄渣中提取的两种加压热水提取物进行了表征。在200℃下获得的提取物在280和320 nm处扫描得到的HPLC色谱图显示，其含有在100℃下获得的提取物中不存在色谱峰，这些峰可能是热降解或其他反应中形成的化合物，或从样品中提取的额外化合物，而这些均会导致提取方法选择性的降低。一般来说，很难确定哪些化合物是天然的，哪些是反应生成的[32,36]。

分析物的降解和新化合物的生成是使用高温时不可避免的问题。如表3.1所示，水的离解常数（K_w）从环境温度（25℃）下的1.0×10^{-14}增加到200℃下的4.9×10^{-12}。这意味着在较高温度下，水是水合氢离子（H_3O^+）和氢氧根离子（OH^-）的强大的来源，它能催化多糖和蛋白质水解成小分子（如低聚糖、单糖、肽和氨基酸）的反应[37]。这些小分子更容易相互反应。例如，已经证明在PHWE处理时，无论是糖基化模型系统[38]还是真实的自然样品[39]中都可能发生美拉德、焦糖化和热氧化反应。这些反应的发生会形成具有不同结构和化学性质的新化合物，这可能会导致错误的分析结果。特别是，如果使用分光光度法测定植物中抗氧化酚类化合物，结果可能会错误地表述为——在 PHWE 中最高温度时，抗氧化物的回收率也最高[40]。实际上，较高的抗氧化物回收率是由于在美拉德、焦糖化和热氧化反应中形成了新的抗氧化物[38,39]。例如，全酚测试（福林酚法，FC）背后的化学作用，它依赖于电子从酚类化合物和其他还原物转移到钼，形成蓝色络合物，进而可以用分光光度法检测[41]。显然，其他非酚类化合物的还原物也能将电子转移到钼[42]。举个例子，图 3.8 给出了提取苹果副产品中黄酮醇时 PHWE 的温度和提取时间优化的响应曲面，结果表明以自来水为溶剂时温度 120℃，时间 3min 为最优条件。而其他的抗氧化剂分析均给出了错误的结果，这与高温条件下的褐变是一致的。

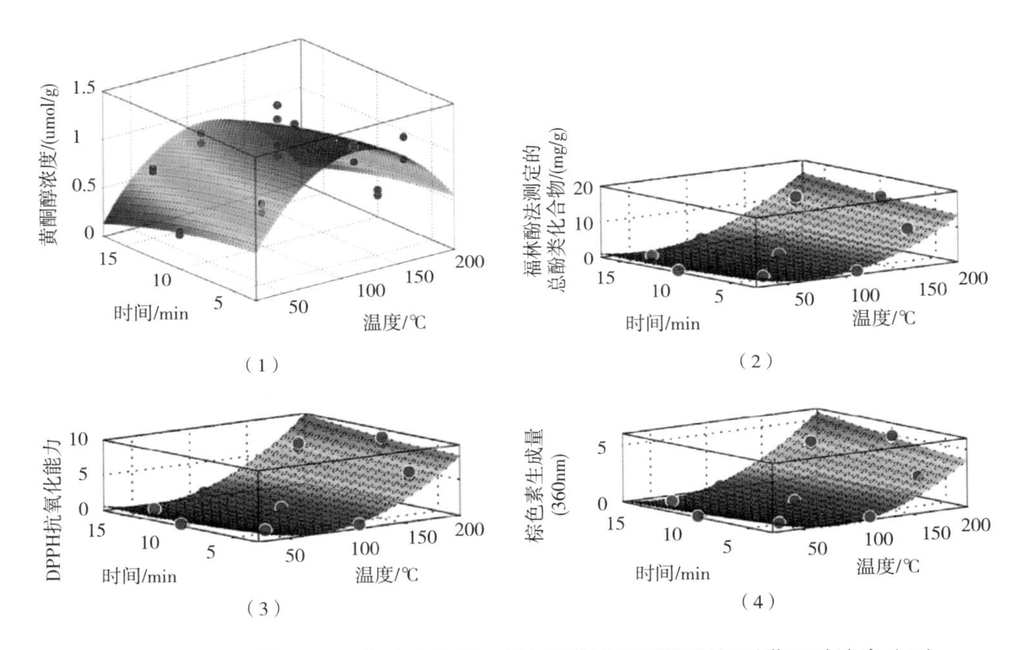

图 3.8 苹果压榨饼 PHWE 的响应面图，显示了温度和萃取时间对黄酮醇浓度（1）、通过 FC 分析测定的总酚类化合物（2）、通过 DPPH 测定的抗氧化能力（3）以及在 360nm 处测量的棕色（黑色素）（4）形成的影响

［资料来源：M. Plaza，V. Abrahamsson，C. Turner，Extraction and neoformation of antioxidant compounds by pressurized hot water extraction from apple byproducts，J. Agric. Food Chem. 61（2013）5500－5510. http：//pubs. acs. org/doi/abs/10. 1021/jf400584f. ］

在许多情况下，在提取过程中热敏感化合物的降解是不可避免的。例如，花青素对氧化反应非常敏感，这取决于溶剂的温度和 pH。有关加压热水中酚类化合物降解的文献[43-45]，表明在方法开发中应考虑温度对分析物稳定性的影响。例如，以水/乙醇/甲酸（94/5/1，体积比）为溶剂，采用静态 PHWE，在 110℃下测定了红洋葱中花青素的降解速率常数，在假定没有降解的前提下，用所得数据计算理论提取曲线[32,36]。结果表明，红洋葱中不同种类花青素在提取过程中的损失约为 21%～36%（图 3.9）。也就是说，如果能将 PHWE 过程中的降解降到最低，红洋葱中花青素的回收率可以提高

21%~36%，从而得到更准确的分析结果。

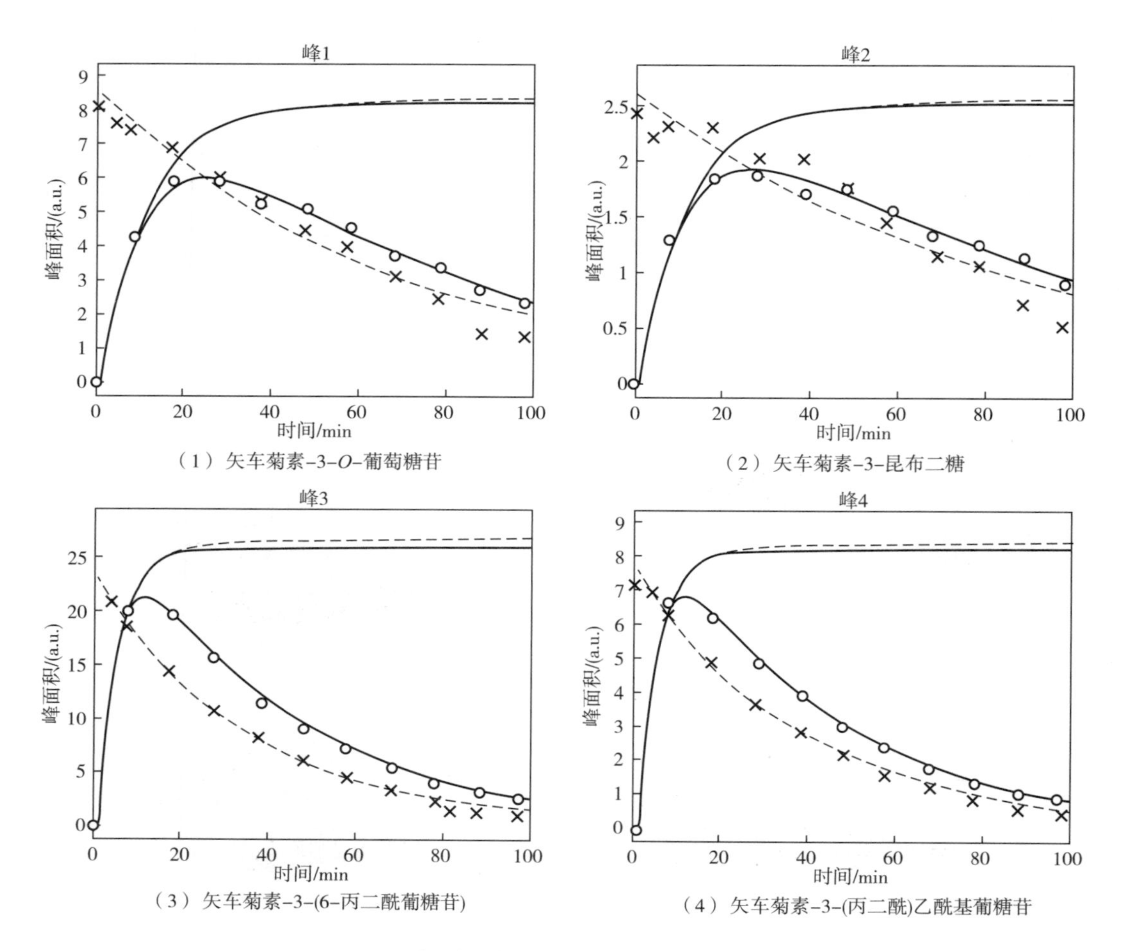

（1）矢车菊素-3-*O*-葡萄糖苷　（2）矢车菊素-3-昆布二糖

（3）矢车菊素-3-(6-丙二酰葡糖苷)　（4）矢车菊素-3-(丙二酰)乙酰基葡糖苷

图 3.9　红洋葱中四种主要花青素的提取曲线

同时降解萃取的实验值显示为圆圈（三次平均值），模型值（每分钟计算）为相邻的实心曲线。降解数据点显示为交叉点（三次平均值），相应的模型显示为相邻的虚线。假定没有降解存在时，采用两种不同的方法（方法A：右上虚线；方法 B：右上实线）计算得到的理论提取曲线

［资料来源：经 E. V. Petersson，J. Liu，P. J. R. Sjöberg，R. Danielsson，C. Turner 许可转载，Pressurized hot water extraction of anthocyanins from red onion：a study on extraction and degradation rates，Anal. Chim. Acta 663（2010）27-32 Elsevier.］

总而言之，在 PHWE 中仔细优化萃取温度是非常重要的，需要充分考虑提高溶解度、改善传质和减少降解效应。事实上，这比用除水以外的其他溶剂进行加压液体萃取（PLE）更为重要[46]。通常，适当优化后的 PHWE 比传统溶剂萃取法能提供更高、更准确的萃取率[47-49]。

3.4.2　流速和提取时间

在 PHWE 过程中减少化学反应的一种方法是使用具有足够大萃取剂流速的连续回

流系统。较高的流速不仅会减少分析物在高温水中的停留时间，而且也会提高分析物的萃取率（如果萃取动力学是受到溶剂中化合物溶解度的限制）。另一方面，萃取剂的流速过高会导致萃取液不必要的稀释，这可能需要在 PHWE 之后增加浓缩步骤。如果萃取动力学主要受样品基质孔隙内解吸和扩散的限制，那么提高流速也不会提高萃取率。在 Liu 等人对不同流速进行考察时，当流速为最高 4mL/min 时，红洋葱中花青素的萃取效率也最高，这主要是受提取过程中热降解的影响[32]。同样，在静态 PHWE 中，萃取时间需要根据溶剂与样品的比例而进行优化。

3.4.3　压力

如上所述，只要水保持液态，压力对水的特性几乎没有影响。因此，除非采用饱和压力状态下的水，否则通常采用 5~10MPa 的压力即可。

3.4.4　其他参数

除上述重要变量外，其他参数也可能对 PHWE 的提取效率有较高的影响。例如，添加一些有机和无机改性剂、表面活性剂和添加剂可以增加分析物在萃取剂中的溶解度，同时可影响样品基质的物理性质以及分析物从样品中的解吸。例如，水中添加 5%乙醇和 1%甲酸有利于从红甘蓝中提取花青素[50]。

此外，样品的粒径影响萃取动力学，因为较小的粒径导致样品和萃取剂之间的接触面增加。颗粒尺寸必须适当，以使接触面最大化，同时避免沟道效应，即颗粒团聚。在一些应用中，将分散剂（如玻璃珠）引入到萃取容器的样品中，有利于样品和萃取剂的均匀分布，以最大限度地提高萃取率。搅拌也可以用来避免结块的形成。

溶剂样品比是 PHWE 的一个重要参数。萃取剂与样品比率的增加可以在不使用新的溶剂的前提下萃取更多的分析物[51]。然而，较高的溶剂样品比需要加热更多的水。此外，分析物在提取液中的浓度较低，因此可能需要分析物的浓缩步骤，这将导致总分析时间变长，分析物降解的风险增大。因此，重要的是溶剂与样品的比例应尽可能小，但同时应能提供尽可能高的提取率[52]。

样品的含水量是另一个可能影响提取率的参数。一些研究表明，含水量高的粗样品比干燥样品具有更好的多酚提取率[33,53,54]。

3.5　加压热水萃取在复杂样品内生物活性物质化学分析中的应用

从复杂的天然来源提取得到的生物活性化合物成分，在食品和医药产品中的应用越来越广泛。同样，寻找含有感兴趣特性如抗氧化、抗炎作用的生物活性化合物的天然植物来源也备受关注。无论是用于分析目的还是工业生产，PHWE 都是从植物和其他复杂样品中分离生物活性物质的最有效的技术之一。表 3.3 所示为天然来源生物活性化合物的分析实验室规模 PHWE 的应用选择。更完整的应用综述参见文献[1]。

PHWE 主要用于提取极性的生物活性化合物（表 3.3），即酚类化合物、二萜、多糖等。一般来说，萃取酚类化合物所采用的 PHWE 条件是萃取温度为 80~150℃，萃取时间为1~60min。在许多研究中，都只分析了总抗氧化能力和总酚类化合物含量。在超过175℃的温度和更长的提取时间下获得的提取物具有更高的抗氧化能力[55,56]，这是第3.4.1节讨论的方法不足造成的。需要更先进的分析技术来量化酚类化合物以及其他生物活性物质。例如，Plaza等人[40]考虑到抗氧化能力最大化和棕色反应最小化，使用期望函数响应面优化了提取苹果副产品中黄酮醇的PHWE方法，其最佳温度为125℃、萃取时间为3min。这些最佳条件与提取物中酚类化合物的最大浓度一致。最佳提取条件随考察的酚类化合物种类而变化。例如对于极不稳定的多酚化合物如花青素，需要较温和的PHWE条件即中等温度（80~100℃）、较短的提取时间（分批提取）或较高的流速（连续提取）以及乙醇和甲酸或乙酸作为提取溶剂中的添加剂[32,50,59,60]。二萜类化合物（如甜菊糖苷）的最佳PHWE条件与酚类化合物的PHWE条件非常相似，萃取温度为100~125℃，萃取时间短（分批萃取）或流速较高（连续回流萃取）[49,61]。最后，提取多糖的最佳PHWE条件是中高提取温度（100~200℃）和较长的提取时间（20~60min）[62-64]。

表3.3　从天然来源提取生物活性化合物的分析实验室规模的加压热水萃取的应用选择

生物活性化合物	来源	温度/℃	压力/MPa	静态/动态	时间/流速	修饰/其他	参考文献
抗氧化剂							
	牛角籽	175	—	动态	180min/2.0mL/min	—	[55]
	橄榄叶	200	10.3	静态	20min	—	[56]
黄酮醇							
	苹果副产物	125	10.3	静态	3 min	—	[40]
	辣木籽	100	—	动态	20min/1.0mL/min	—	[57]
酚酮类							
6-姜根素	姜根	130	3.5	静态	60min	—	[58]
6-姜酚	姜根	170	3.5	静态	60min	—	[58]
二苯乙烯类化合物							
反式白藜芦醇；反式葡萄素	葡萄藤	85~105	5.2	动态	150min/1.0mL/min	乙醇/水（15/85，体积比）	[31]
花青素							
	红卷心菜；红洋葱	99	5.0	静态	7 min	水/乙醇/甲酸（94/5/1，体积比）	[50，59]
	红洋葱	110	1.5	动态	60min/4.0mL/min	水/乙醇/甲酸（94/5/1，体积比）	[32]

续表

生物活性化合物	来源	温度/℃	压力/MPa	静态/动态	时间/流速	修饰/其他	参考文献
原花青素							
	葡萄渣	100	10.2	静态	5 min	—	[60]
单宁和单宁花青素							
	葡萄渣	150	10.2	静态	5 min	—	[60]
二萜类化合物							
甜菊糖苷	甜叶菊叶	100	1.1~1.3	动态	15 min/1.5mL/min	—	[61]
		100	10.3	静态	4 min	—	[49]
多糖							
	宁夏枸杞	100	5.0	静态	53 min	超声波辅助（160W）	[62]
	金蚝菇	200	0.002~5.0	静态	60min	—	[63]
	伸长海条藻	100	10.3	静态	20min	—	[64]
蛋白和糖							
	豆渣	240	—	静态	5 min	—	[65]
油溶性和水溶性							
	葵花籽	130	3.0	静态	30min	—	[52]
酶促反应							
槲皮素糖苷	洋葱渣	90	5.0	静态	10min	pH5，β-葡萄糖糖苷酶 TnBgl3B	[66]
槲皮素	洋葱渣	84	5.0	动态	90min/3.0mL/min	5%乙醇，pH5.5，β-葡萄糖糖苷酶 TnBgl1A_N221S/P342 L	[67]
在线水萃取和颗粒形成						干燥条件	
抗氧化剂	迷迭香	200	8.0	动态	20min/0.2mL/min	CO_2 压力 8MPa，流速 2.5mL/min，氮气流速 0.6mL/min	[68-70]
槲皮素衍生物	洋葱	120	8.0	动态	45 min/0.3mL/min	CO_2 压力 8MPa，流速 10mL/min，氮气压力 1.22MPa	[71]

3.6 加压热水萃取联用技术

3.6.1 加压热水萃取-酶促反应

在化学分析中，对被分析物进行衍生化反应，使其更易于分析可能是一个有趣的问题。其中一个例子是酚类化合物，在自然界中会以多种糖基形式存在，由于缺乏化学标准品且同一多酚苷元的几种低丰度形式的检出限不够，因此难以准确定量。在这种情况下，通常在最终分析之前对糖苷进行水解，这一般是在用盐酸、甲醇和水的混合物进行萃取时同时实现的。

新的基于酶的方法来实现水解反应与 PHWE 的联用是最近的一个发展趋势，与传统的酸催化水解相比，这是一种更快、更准确和更环保的方法[48]。在一些研究[48,66,67,72,73]中，槲皮素苷通过 PHWE 从洋葱副产品中提取，然后 β-葡萄糖苷酶催化槲皮素苷转化为槲皮素和糖。结果表明，与使用盐酸作为催化剂的传统溶剂萃取相比，PHWE 和酶水解的结合显著地提高和精准定量了复杂样品中多酚的浓度[67]。

3.6.2 加压热水萃取-吸附剂净化

如上所述，PHWE 不是最具选择性的提取技术。事实上，在进行最终分析之前，PHWE 可能需要进行净化处理。例如，可以通过在提取容器中或样品通路下游添加单独的容器中使用部分选择性吸附剂来执行净化步骤。通过使用不同的材料，如分子印迹聚合物（MIPs）或特殊设计的聚合物，可以更加有选择性地从 PHWE 提取物中分离出感兴趣的分析物，从而提高样品通量和分析测定的质量[74]。

3.6.3 加压热水萃取与干燥步骤联用

在萃取过程中使用水作为溶剂的主要缺点是难以浓缩萃取液，因为与许多有机溶剂相比，水的蒸发热相对较高。另外，由于水提液中生物活性化合物的浓度可能较低，因此经常需要浓缩样品。此外，在某些情况下，水的存在会降低提取物的稳定性。冷冻干燥是提取液干燥的一种方法，其成本高、耗时长，还可能由于热、光和氧的作用而导致生物活性物质的降解。

为了解决这一问题，2010 年开发了一种在线干燥 PHWE 提取液的方法[68]，该方法基于 Sievers 和同事先前的一项创新，称为 CO_2 辅助雾化气泡干燥机（CAN-BD）[75]。在这种新方法中，PHWE 萃取物在小体积的三通管中与超临界 CO_2 连续混合，形成小尺寸水滴的喷雾，这些水滴被热氮气流快速干燥。这个过程被称为在线水提取和颗粒形成（WEPO）。基本上，这是一个提取和干燥结合的过程，产生微米级的含有分析物和所有其他共萃取化合物的提取物颗粒。Herrero 等人[68]和 Rodríguez-Meizoso 等人[69,70]采用 WEPO 从迷迭香叶中提取抗氧化剂，Andersson 等人[71]利用 WEPO 从洋葱中提取得到与未干燥提取物相同成分的槲皮素衍生物。

此外，已经证明，与超临界流体萃取真空干燥法和 PHWE 冷冻干燥法相比，基于门到门的生命周期评估结果，WEPO 工艺的环境影响最小[69]。

3.7 加压热水萃取技术的展望

水在高温高压下无疑是一种有吸引力的溶剂，它具有高效的传质和许多生物活性化合物的高溶解度。然而，市场上仍然没有专门针对 PHWE 的商业分析系统，即市场上可用的 PLE 系统是专为有机溶剂设计，最高温度限制在 200℃，管道和容器通常由 316 型不锈钢制成。专为 PHWE 设计的系统应由更优质的钢合金制成，可以改进溶剂的预热，可以调节萃取剂流速，以及能承受达到水临界点的温度和压力，即 374℃和 22.1MPa。但是，有一个预分离比例系统可用，设计用于在高达 300℃的温度和 100mL/min 的流速下亚临界水进行萃取操作。此外，适当的脱气系统也将是有益的。另一个方面是由于焦糖化反应而导致管道有堵塞的风险，这都是一个商业 PHWE 系统需要解决的问题。

从化学分析中涉及 PHWE 的方法学的未来发展来看，萃取与净化相结合是一种趋势。新型聚合物吸附材料包括 MIP 材料等是一个很有潜力的发展方向。此外，在 PHWE 与色谱联用方面还需要更多的努力，包括在过去 5~10 年中色谱技术的发展（主要由液体或超临界 CO_2 组成的流动相，在更高的温度和压力下的可操作性）。也许随着更耐用色谱固定相和微型化的发展，PHWE 与加压热水色谱联用分析生物活性化合物将得以实现。到目前为止，“过热水色谱法” 主要是针对热稳定化合物的研究，并且没有取得广泛的成功。

由于在萃取过程中存在水解反应和其他降解反应的风险，基于使用水作为萃取剂的 PHWE 对其他更成熟的使用有机溶剂的萃取技术进行替代的可能性，仍然存在疑问。因此，未来研究的方向应该是基于水的特性、不同萃取时间和流速时分析物、其他不需要的共萃取物和剩余样品基质在萃取过程中实际发生的事情。这些信息不仅对方法开发有价值，而且对大规模的过程开发也有价值。

致谢

这项工作得到了瑞典 Formas 研究委员会（229-2009-1527）（SuReTech）、瑞典 VR 研究委员会（2010-5439，2010-333）和瑞典隆德大学 VINNOVA VINN 卓越中心抗糖尿病食品中心的资助。

参考文献

[1] M. Plaza, C. Turner, Pressurized hot water extraction of bioactives, TrAC-Trends Anal. Chem. 71 (2015) 39-54.

[2] S. B. Hawthorne, Y. Yang, D. J. Miller, Extraction of organic pollutants from environmental solids with subcritical and supercritical water, Anal. Chem. 66 (1994) 2912-2920.

[3] Y. Yang, S. Bowadt, S. B. Hawthorne, D. J. Miller, Subcritical water extraction of

polychlorinated-biphenyls from soil and sediment, Anal. Chem. 67 (1995) 4571-4576.

[4] J. Kronholm, K. Hartonen, M. L. Riekkola, Analytical extractions with water at elevated temperatures and pressures, TrAC-Trends Anal. Chem. 26 (2007) 396-412.

[5] C. C. Teo, S. N. Tan, J. W. H. Yong, C. S. Hew, E. S. Ong, Pressurized hot water extraction (PHWE), J. Chromatogr. A 1217 (2010) 2484-2494.

[6] H. Wijngaard, M. B. Hossain, D. K. Rai, N. Brunton, Techniques to extract bioactive compounds from food by-products of plant origin, Food Res. Int. 46 (2012) 505-513.

[7] M. Herrero, M. Castro-Puyana, J. A. Mendiola, E. Ibanez, Compressed fluids for the extraction of bioactive compounds, TrAC-Trends Anal. Chem. 43 (2013) 67-83.

[8] Phase Diagram of Water. Available from: http: //en. wikipedia. org/wiki/Phase_ diagram.

[9] G. Brunner, Near critical and supercritical water. Part I. Hydrolytic and hydrothermal processes, J. Supercrit. Fluids 47 (2009) 373-381.

[10] Organisation, 2014, The International Association for the Properties of Water and Steam. Available from: http: //www. iapws. org/index. html.

[11] Wikipedia, Self-ionization of Water, 2014. Available from: http: //en. wikipedia. org/wiki/Self-ionization_ of_ water.

[12] Wikipedia, Solvent, 2014. Available from: http: //en. wikipedia. org/wiki/Solvent.

[13] M. Uematsu, E. U. Franck, Static dielectric-constant of water and steam, J. Phys. Chem. Ref. Data 9 (1980) 1291-1306.

[14] The Engineering ToolBox. Available from: http: //www. engineeringtoolbox. com.

[15] Table 3. Compressed Water and Superheated Steam. Available from: http: //www. nist. gov/srd/upload/NISTIR5078-Tab3. pdf.

[16] Steam Tables, http: //dx. doi. org/10. 1615/AtoZ. s. steam_ tables. Available from: http: //www. thermopedia. com/content/1150/.

[17] G. Brunner, Hydrothermal and supercritical water processes, in: E. Kiran (Ed.), Supercritical Fluid Science and Technologies Series, vol. 5, Elsevier, Amsterdam, The Netherlands, 2014, p. 604.

[18] M. N. Islam, Y. T. Jo, J. H. Park, Remediation of PAHs contaminated soil by extraction using subcritical water, J. Ind. Eng. Chem. 18 (2012) 1689-1693.

[19] A. G. Carr, R. Mammucari, N. R. Foster, A review of subcritical water as a solvent and its utilisation for the processing of hydrophobic organic compounds, Chem. Eng. J. 172 (2011) 1-17.

[20] P. G. Jessop, D. A. Jessop, D. Fu, P. Lam, Solvatochromic parameters for solvents of interest in green chemistry, Green Chem. 14 (2012) 1245-1259.

[21] P. G. Jessop, Searching for green solvents, Green Chem. 13 (2011) 1391-1398.

[22] C. Hansen, Hansen Solubility Parameters: A User's Handbook, CRC Press, BocaRaton, 2007.

[23] K. Srinivas, F. Montanes, J. W. King, Solubilities of flavonoids in subcritical water, Abstr. Pap. Am. Chem. Soc. 237 (2009) 70.

[24] M. Co, P. Koskela, P. Eklund-Åkergren, K. Srinivas, J. W. King, P. J. R. Sjoberg, C. Turner, Pressurized liquid extraction of betulin and antioxidants from birch bark, Green Chem. 11 (2009) 668-674.

[25] B. E. Richter, B. A. Jones, J. L. Ezzell, N. L. Porter, N. Avdalovic, C. Pohl, Accelerated solvent extraction: a technique for sample preparation, Anal. Chem. 68 (1996) 1033-1039.

[26] G. Joana Gil-Chávez, J. A. Villa, J. Fernando Ayala-Zavala, J. Basilio Heredia, D.

Sepulveda, E. M. Yahia, G. A. González-Aguilar, Technologies for extraction and production of bioactive compounds to be used as nutraceuticals and food ingredients: an overview, Compr. Rev. Food Sci. F 12 (2013) 5-23.

[27] J. Lu, J. S. Brown, E. C. Boughner, C. L. Liotta, C. A. Eckert, Solvatochromic characterization of near-critical water as a benign reaction medium, Ind. Eng. Chem. Res. 41 (2002) 2835-2841.

[28] K. Chandler, F. Deng, A. K. Dillow, C. L. Liotta, C. A. Eckert, Alkylation reactions in near-critical water in the absence of acid catalysts, Ind. Eng. Chem. Res. 36 (1997) 5175-5179.

[29] O. M. Anderson, K. R. Markham, Flavonoids: Chemistry, Biochemistry and Applications, CRC Press, 2006.

[30] T. Al-Shemmeri, Engineering Fluid Mechanics, Ventus Publishing ApS, 2012.

[31] E. Karacabey, G. Mazza, L. Bayindirli, N. Artik, Extraction of bioactive compounds from milled grape canes (*Vitis vinifera*) using a pressurized low-polarity water extractor, Food Biopr. Technol. 5 (2012) 359-371.

[32] J. Liu, M. Sandahl, P. J. R. Sjöberg, C. Turner, Pressurised hot water extraction in continuous flow mode for thermolabile compounds: extraction of polyphenols in red onions, Anal. Bioanal. Chem. 406 (2014) 441-445.

[33] J. K. Monrad, K. Srinivas, L. R. Howard, J. W. King, Design and optimization of a semicontinuous hot-cold extraction of polyphenols from grape pomace, J. Agric. Food Chem. 60 (2012) 5571-5582.

[34] J. Pawliszyn, Sample preparation: quo vadis? Anal. Chem. 75 (2003) 2543-2558.

[35] J. R. Vergara-Salinas, M. Vergara, C. Altamirano, Á. Gonzalez, J. R. Pérez-Correa, Characterization of pressurized hot water extracts of grape pomace: chemical and biological antioxidant activity, Food Chem. 171 (2015) 62-69.

[36] E. V. Petersson, J. Liu, P. J. R. Sjöberg, R. Danielsson, C. Turner, Pressurized hot water extraction of anthocyanins from red onion: a study on extraction and degradation rates, Anal. Chim. Acta 663 (2010) 27-32.

[37] I. Sereewatthanawut, S. Prapintip, K. Watchiraruji, M. Goto, M. Sasaki, A. Shotipruk, Extraction of protein and amino acids from deoiled rice bran by subcritical water hydrolysis, Bioresour. Technol. 99 (2008) 555-561.

[38] M. Plaza, M. Amigo-Benavent, M. D. del Castillo, E. Ibáñez, M. Herrero, Neoformation of antioxidants in glycation model systems treated under subcritical water extractionconditions, Food Res. Int. 43 (2010) 1123-1129.

[39] M. Plaza, M. Amigo-Benavent, M. D. del Castillo, E. Ibáñez, M. Herrero, Facts about the formation of new antioxidants in natural samples after subcritical water extraction, Food Res. Int. 43 (2010) 2341-2348.

[40] M. Plaza, V. Abrahamsson, C. Turner, Extraction and neoformation of antioxidant compounds by pressurized hot water extraction from apple byproducts, J. Agric. Food Chem. 61 (2013) 5500-5510.

[41] A Karadag, B. Ozcelik, S. Saner, Review of methods to determine antioxidant capacities, Food Anal. Meth. 2 (2009) 41-60.

[42] L. M. Magalhães, M. A. Segundo, S. Reis, J. L. F. C. Lima, Methodological aspectsabout in vitro evaluation of antioxidant properties, Anal. Chim. Acta 613 (2008) 1-19.

[43] R. Wang, T. L. Neoh, T. Kobayashi, S. Adachi, Antioxidative capacity of the degradation

products of glucuronic and galacturonic acid from subcritical water treatment, Chem. Eng. Technol. 34 (2011) 1514–1520.

[44] P. Khuwijitjaru, B. Suaylam, S. Adachi, Degradation of caffeic acid in subcritical water and online HPLC–DPPH assay of degradation products, J. Agric. Food Chem. 62 (2014) 1945–1949.

[45] P. Khuwijitjaru, J. Plernjit, B. Suaylam, S. Samuhaseneetoo, R. Pongsawatmanit, S. Adachi, Degradation kinetics of some phenolic compounds in subcritical water and radical scavenging activity of their degradation products, Can. J. Chem. Eng. 92 (2014) 810–815.

[46] J. Wiboonsirikul, S. Adachi, Extraction of functional substances from agricultural products or by-products by subcritical water treatment, Food Sci. Technol. Res. 14 (2008) 319–328.

[47] Gil–Ramírez, J. A. Mendiola, E. Arranz, A. Ruíz–Rodríguez, G. Reglero, E. Ibáñez, F. R. Marín, Highly isoxanthohumol enriched hop extract obtained by pressurized hot water extraction (PHWE). Chemical and functional characterization, Innov. Food Sci. Emerg. Technol. 16 (2012) 54–60.

[48] S. Lindahl, A. Ekman, S. Khan, C. Wennerberg, P. Börjesson, P. J. R. Sjöberg, E. N. Karlsson, C. Turner, Exploring the possibility of using a thermostable mutant of β–glucosidase for rapid hydrolysis of quercetin glucosides in hot water, Green Chem. 12 (2010) 159–168.

[49] J. B. Jentzer, M. Alignan, C. Vaca–Garcia, L. Rigal, G. Vilarem, Response surface methodology to optimise accelerated solvent extraction of steviol glycosides from *Stevia rebaudiana* Bertoni leaves, Food Chem. 166 (2015) 561–567.

[50] P. Arapitsas, C. Turner, Pressurized solvent extraction and monolithic column–HPLC/DAD analysis of anthocyanins in red cabbage, Talanta 74 (2008) 1218–1223.

[51] S. Rezaei, K. Rezaei, M. Haghighi, M. Labbafi, Solvent and solvent to sample ratio as main parameters in the microwave–assisted extraction of polyphenolic compounds from apple pomace, Food Sci. Biotechnol. 22 (2013) 1–6.

[52] M. Ravber, Ž. Knez, M. Škerget, Simultaneous extraction of oil–and water–soluble phase from sunflower seeds with subcritical water, Food Chem. 166 (2015) 316–323.

[53] J. K. Monrad, M. Suárez, M. J. Motilva, J. W. King, K. Srinivas, L. R. Howard, Extraction of anthocyanins and flavan–3–ols from red grape pomace continuously by coupling hotwater extraction with a modified expeller, Food Res. Int. 65 (2014) 77–87.

[54] H. N. Rajha, W. Ziegler, N. Louka, Z. Hobaika, E. Vorobiev, H. G. Boechzelt, R. G. Maroun, Effect of the drying process on the intensification of phenolic compounds recovery from grape pomace using accelerated solvent extraction, Int. J. Mol. Sci. 15 (2014) 18640–18658.

[55] Ö. Güçlü–Üstündağ, G. Mazza, Effects of pressurized low polarity water extraction parameters on antioxidant properties and composition of cow cockle seed extracts, Plants Foods Hum. Nutr. 64 (2009) 32–38.

[56] M. Herrero, T. N. Temirzoda, A. Segura–Carretero, R. Quirantes, M. Plaza, E. Ibañez, New possibilities for the valorization of olive oil by–products, J. Chromatogr. A 1218 (2011) 7511–7520.

[57] P. G. Matshediso, E. Cukrowska, L. Chimuka, Development of pressurised hot water extraction (PHWE) for essential compounds from *Moringa oleifera* leaf extracts, Food Chem. 172 (2015) 423–427.

[58] N. I. Anisa, N. Azian, M. Sharizan, Y. Iwai, Temperature effects on diffusion coefficient for 6–gingerol and 6–shogaol in subcritical water extraction, J. Phys. Conf. Ser. 495 (2014).

[59] E. V. Petersson, A. Puerta, J. Bergquist, C. Turner, Analysis of anthocyanins in red onion using capillary electrophoresis–time of flight–mass spectrometry, Electrophoresis 29 (2008) 2723–2730.

[60] J. R. Vergara-Salinas, P. Bulnes, M. C. Zúñiga, J. Pérez-Jiménez, J. L. Torres, M. L. Mateos-Martín, E. Agosin, J. R. Pérez-Correa, Effect of pressurized hot water extraction on antioxidants from grape pomace before and after enological fermentation, J. Agric. Food Chem. 61 (2013) 6929-6936.

[61] C. C. Teo, S. N. Tan, J. W. H. Yong, C. S. Hew, E. S. Ong, Validation of green-solvent extraction combined with chromatographic chemical fingerprint to evaluate quality of *Stevia rebaudiana* Bertoni, J. Sep. Sci. 32 (2009) 613-622.

[62] Z. Chao, Y. Ri-Fu, Q. Tai-Qiu, Ultrasound-enhanced subcritical water extraction of polysaccharides from *Lycium barbarum* L. Sep. Purif. Technol. 120 (2013) 141-147.

[63] E. K. Jo, D. J. Heo, J. H. Kim, Y. H. Lee, Y. C. Ju, S. C. Lee, The effects of subcritical water treatment on antioxidant activity of golden oyster mushroom, Food Biopr. Technol. 6 (2013) 2555-2561.

[64] S. Santoyo, M. Plaza, L. Jaime, E. Ibañez, G. Reglero, J. Señorans, Pressurized liquids as an alternative green process to extract antiviral agents from the edible seaweed *Himanthalia elongata*, J. Appl. Phycol. 23 (2011) 909-917.

[65] J. Wiboonsirikul, M. Mori, P. Khuwijitjaru, S. Adachi, Properties of extract from Okara by its subcritical water treatment, Int. J. Food Prop. 16 (2013) 974-982.

[66] C. Turner, P. Turner, G. Jacobson, K. Almgren, M. Waldeback, P. Sjoberg, E. N. Karlsson, K. E. Markides, Subcritical water extraction and [small beta] -glucosi-dase-catalyzed hydrolysis of quercetin glycosides in onion waste, Green Chem. 8 (2006) 949-959.

[67] S. Lindahl, J. Liu, S. Khan, E. Nordberg Karlsson, C. Turner, An on-line method for pressurized hot water extraction and enzymatic hydrolysis of quercetin glucosides from onions, Anal. Chim. Acta 785 (2013) 50-59.

[68] M. Herrero, M. Plaza, A. Cifuentes, E. Ibáñez, Green processes for the extraction of bioactives from rosemary: chemical and functional characterization via ultra-perfor-mance liquid chromatography-tandem mass spectrometry and in-vitro assays, J. Chromatogr. A 1217 (2010) 2512-2520.

[69] I. Rodríguez-Meizoso, M. Castro-Puyana, P. Börjesson, J. A. Mendiola, C. Turner, E. Ibáñez, Life cycle assessment of green pilot-scale extraction processes to obtain potent antioxidants from rosemary leaves, J. Supercrit. Fluids 72 (2012) 205-212.

[70] M. E. E. Ibanez, A. Cifuentes, I. Rodriguez-Meizoso, J. A. Mendiola, G. Reglero, J. Señorans, C. Turner, Spanish Patent No: P200900164, (2009).

[71] J. M. Andersson, S. Lindahl, C. Turner, I. Rodriguez-Meizoso, Pressurised hot water extraction with on-line particle formation by supercritical fluid technology, Food Chem. 134 (2012) 1724-1731.

[72] S. Khan, S. Lindahl, C. Turner, E. N. Karlsson, Immobilization of thermostable β-glucosidase variants on acrylic supports for biocatalytic processes in hot water, J. Mol. Catal. B: Enzym. 80 (2012) 28-38.

[73] S. Khan, T. Pozzo, M. Megyeri, S. Lindahl, A. Sundin, C. Turner, E. N. Karlsson, Aglycone specificity of *Thermotoga neapolitana* β-glucosidase 1a modified by mutagenesis, leading to increased catalytic efficiency in quercetin-3-glucoside hydrolysis, BMC Biochem. 12 (2011).

[74] V. Pakade, S. Lindahl, L. Chimuka, C. Turner, Molecularly imprinted polymers targeting quercetin in high-temperature aqueous solutions, J. Chromatogr. A 1230 (2012) 15-23.

[75] R. E. Sievers, E. T. S. Huang, J. A. Villa, J. K. Kawamoto, M. M. Evans, P. R. Brauer, Low-temperature manufacturing of fine pharmaceutical powders with supercritical fluid aerosolization in a Bubble Dryer®, Pure Appl. Chem. 73 (2001) 1299-1303.

4

4　加压液体萃取环境和食品样品中的有机污染物

Yolanda Picó
University of Valencia, Valencia, Spain
E-mail: Yolanda. Pico@ uv. es

4.1 引言

绿色提取技术是建立在提取工艺的发现和设计基础上的，它可以降低能源消耗，允许使用替代溶剂和可再生的天然产品，并确保安全和高质量的提取产物/产品[1]。现代样品富集技术的发展方向为样品处理快速化、自动化、高重现性以及尽可能少的使用有机有毒溶剂，其与绿色分析化学的发展理念相同。传统的萃取技术耗时长、效率低，需要使用大量的非环保有机溶剂、吸附剂和样品。传统的固-液提取方法包括索氏提取、浸渍、渗滤、涡旋振荡提取（高速混合）和超声波提取。相反，加压液体萃取（PLE）［也称为加速溶剂萃取（ASE）］，微波辅助萃取（MAE），超临界流体萃取（SFE）和不同类型的液相微萃取技术是符合绿色分析化学要求的新型高效非常规萃取方法[1]。

PLE 自从被美国环境保护局（USEPA）认可为环境固体样品中持久性有机污染物的官方方法以来，其使用变得更广泛了[2]。该方法操作简便、提取时间短、样品量要求低、性能好，已成为有机污染物痕量分析的常规方法，在许多实验室得到应用。在 PLE 方法中获得的小体积有机提取物（含有分析物和可溶性基质），特别是当使用水作为溶剂时，有助于进一步浓缩和净化，通常通过固相萃取（SPE）和凝胶渗透色谱（GPC）进行。

最近发表的关于测定复杂基质中有机污染物的样品处理方法的综述文章都会把一块重要的篇幅放在 PLE 方法[3]上。最近的几个例子包括海产品中聚溴二苯醚（PBDEs）及其代谢物[4]、海洋环境药物学[5]、不同环境基质中出现的污染物[6]，这些都很好地说明了 PLE 在测定有机污染物方面的广泛应用。

自 2002 年以来，有几篇专门关于 PLE 用于环境样品中有机污染物分析的综述已发表[7,8]。

在食品样品中，PLE 主要应用于药用植物和其他生物活性物质的提取。只有少数综述涉及有机污染物的分析，如抗菌药物[9]和其他化合物[10,11]。一般来讲，环境样品中使用的所有程序和关键步骤都可以应用于食品分析，考虑到食品基质相对复杂，可能含有一些干扰物质如脂类、蛋白质和色素等，这些物质必须在 PLE 提取过程中或提取后进行全面去除。除此之外，富含油脂的食物中含有微量的水可能导致对污染物的低估。

本章介绍了 PLE 在食品和环境基质中有机污染物提取中的应用。本章基于作者 Vazquez Roig 和 Pico[12]最近发表的一篇综述，并包括最新出版的关于仪器、方法学和最新应用等方面的相关论文。由于该领域的应用数量巨大，本文还对过去 4~5 年中发表的一些代表性著作进行了讨论。首先，着重介绍了萃取池内净化方法的发展，该方法将萃取和净化/分离结合在一起。其次，比较了 PLE 与传统提取技术和新提取技术（SFE 和 MAE）的优缺点。最后，回顾和讨论了许多环境和食品方面的应用，展示了这两个领域最重要的发展前景。

4.2 加压液体萃取

4.2.1 基本原理

PLE 是一种样品提取技术，利用高温高压下的溶剂从固体样品中提取分析物，用于进一步净化和分析分离与测定，测定时通常采用气相色谱法（GC）或液相色谱法（LC）。这项技术由 Dionex 公司于 1995 年在 Pittcon 会议上首次公开，也被称为加压流体萃取（PFE）、加速溶剂萃取、加压溶剂萃取或增强溶剂萃取。与传统的溶剂萃取法相比，该法具有溶剂用量少、萃取时间短、不暴露于氧气和光照等特点。

分析物从基质中的迁移很大程度上受萃取溶剂的极性和物理性质的影响，而溶剂极性和物理性质又与不同的工艺萃取参数相关。温度和压力是控制解吸、扩散、溶剂化和其他传输机制的最重要的参数。有关控制结合与传输的机理的详细信息不在本章的讨论范围内，感兴趣的读者可以参考其他文献[13-15]。

温度升高可以提高分析物的溶解度，打破基质与分析物之间的相互作用，当压力同时升高使溶剂保持在液态时，可以获得了较高的扩散速率。在高温高压条件下，溶剂的黏度和表面张力降低，因此它能更有效地渗透到固体样品中，有助于提取被困在基质孔隙中的分析物，并减少溶剂的使用量。

4.2.2 提取过程

PLE 可以单独在静态或动态模式下实施，也可以两者结合使用。在静态模式下，萃取由一个或多个萃取循环组成，在循环与循环之间进行溶剂更换。尽管设备通常以静态模式运行，但最新的 PLE 设备（ASE350）允许使用动态模式，以特定流速连续泵送小份溶剂通过萃取池。流速通常控制在 0.5~2.5mL/min。使用动态模式时萃取平衡一直被打破，因为新鲜溶剂被连续泵入样品并加速了样品传质速率。因此，与静态提取相比，动态模式需要更大体积的溶剂。动态加压萃取的不足之处是萃取液中分析物的稀释倍数较大，这可能需要在对目标化合物进行色谱分析之前增加浓缩步骤。动态模式的提取效率等于甚至高于静态模式，提取时间通常与静态模式相当。然而，通过在动态提取之前建立静态提取步骤，可以缩短完全提取所需的时间。由于许多实验室现有的 PLE 设备较老，不适用于动态提取，因此动态提取模式仍然限制于使用自制设备。

PLE 系统的示意图如图 4.1 所示。

为了进行 PLE 操作，通常需要通过冻干、烘箱干燥、风干、除水剂等方法干燥样品以消除水分，并增加溶剂对样品基质的渗透性，从而提高萃取效率[16]。将 1~100g 过筛后的样品与吸附剂（例如，惰性硅藻土、海砂等）混合，装入不锈钢萃取槽中（用吸附剂完全填充槽，以减少溶剂消耗），并用两个适用于高压密封的过滤端固定。

PLE 方法有许多可供选择的配置选项，可以用来建立一个完整、稳健的提取方法。系统对萃取池进行加压加温（温度最高不超过 200℃，压力最大不超过 20.67MPa），使

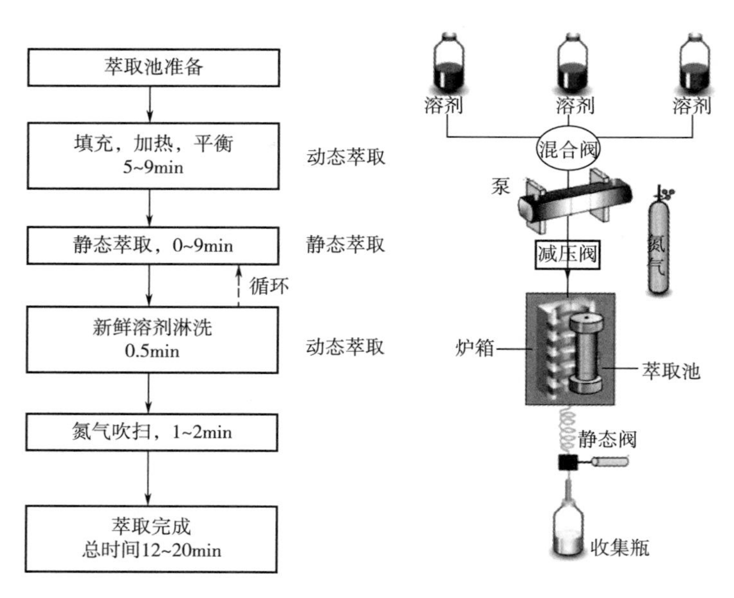

图 4.1　加压液体萃取流程图

（图片由 Thermo Fisher 科技公司提供。）

之保持在预设的优化值。方法可包括预热步骤，即在添加溶剂之前对样品进行加热（不适用于挥发性分析物）。然后，通过泵将溶剂从溶剂瓶中输送到萃取池中。

在静态模式下，通常经过 5~10min 萃取后（萃取时间需要分析人员进行优化），含有目标物和共萃取基质的提取溶剂会通过氮气泵入收集容器。大多数设备允许装载多达 24 个萃取池，所有程序都可以自动完成，因此分析人员可以在工作时间制备样品并在夜间运行样品。PLE 有许多安全传感器，以避免超压并对蒸汽和液体泄漏等进行检测。

4.2.3　方法开发

在 PLE 方法的开发过程中，需要对包括温度、萃取时间、萃取循环次数、吸附剂类型和溶剂等几个参数进行优化。寻找这些参数之间的平衡，以使大多数分析物获得可接受的回收率。当多种目标化合物具有不同的官能团和物理化学性质时，这一点尤其需要注意。

（1）吸附剂类型　为了避免萃取池中的样品颗粒聚集和溶剂通道过长，一般将样品与分散/干燥剂充分混合，以提高萃取效率。为此，Hydromatrix 通常在方法开发中作为首选吸附剂[16]，它是一种惰性硅藻土，每克可吸收 2mL 水。最近，也有用其他类型的硅藻土、硫酸钠、海砂和其他吸附剂等来去除萃取池内的水分[17-19]。部分吸附剂的使用也有萃取池内净化的效果，称为选择性加压液体萃取（SPLE）。最近，Subedi 等人[20]综述了 SPLE 技术在固体环境基质（包括生物组织、土壤、沉积物、污泥和粉尘）中持久性有机污染物和选定污染物分析中的应用。基于 SPLE 的分析方法，提高了样品通量，降低了与样品制备过程相关的固有成本（即时间、溶剂、劳动力、实验室空间、培训和分析物的潜在损失等)。文献［21］开发了一种有趣的 SPLE 方法，并对鱼类样品中的 14 种有机氯农药进行了验证。不同吸附剂（氧化铝、弗罗里硅土、酸化硅胶和硅胶等）添加到萃取池中对鱼类样品的除脂效率，如图 4.2（1）所示。图 4.2（2）和图 4.2（3）还显示，与使用弗罗里硅土吸附剂和两个萃取循环获得的提取物相比，使用氧化铝吸附剂和三个萃取循环获得的提取物中发现的未知峰更少。

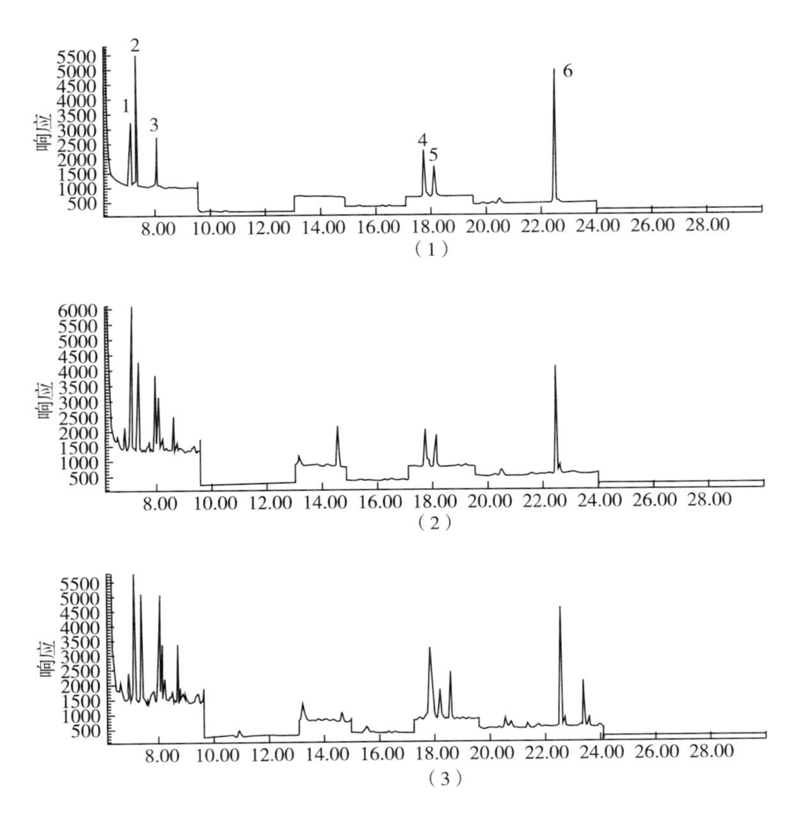

图 4.2 （1）使用氧化铝、弗罗里硅土、酸化硅胶和硅胶在线净化处理后的认证标准物质的气相色谱图；
（2）使用氧化铝吸附剂、三个萃取循环处理后的加标样品的气相色谱图；
（3）使用弗罗里硅土吸附剂、两个萃取循环处理后的加标样品的气相色谱图

标记峰分别为 1. α-HCH，2. HCB，3. γ-HCH，4. *p*，*p*’-DDE，5. 4-三联苯，6. *p*，*p*’-DDT

［资料来源：M. Choi，I-S. Lee，R-H. Jung，Rapid determination of organochlorine pesticides in fish using selective pressurized liquid extraction and gas chromatography-mass spec-trometry，Food Chem. 205（2016）1-8，经 Elsevier 许可。］

将提取和净化集成在一个步骤中，可避免对样品提取物进行额外处理。Pinto 等人考察了不同吸附剂在提取持久性农药时的萃取池内净化效果[17]。选择性依次为弗罗里硅土>酸性氧化铝>中性氧化铝>硅胶>碱性氧化铝>石墨化炭黑。如果将弗罗里硅土用于多环芳烃的分析，那么使用前需用二氯甲烷进行清洗，这是因为它本身含有这类化合物[22]。在甜味剂的提取过程中，对弗罗里硅土、C_{18}、硅胶和氧化铝的在线净化效果进行了考察。不同吸附剂得到的结果基本类似，最后选择氧化铝作为吸附剂是因为它

略微提高了某些化合物（糖精、阿力甜）的回收率，而且它是生物研究中常用的廉价材料[23]。

一个新的特点是 Dionium 萃取池（仅与 ASE 150 和 350 型号兼容），它可以处理无机酸或强碱预处理过的样品或吸附剂，而不需要对典型的不锈钢 PLE 萃取池和筛板进行老化。使用硫酸浸渍硅胶可有效去除食品基质中的脂质[24]。然而，它们的使用并不常见。

（2）温度　高温有利于提高萃取效率，这是因为提高了分析物在溶剂中的扩散速度和溶解度。高温将有助于破坏由范德华力、氢键和偶极子吸引等引起的分析物与样品基质间的相互作用，从而提高萃取效率。然而，提高萃取温度也可能会提高分析物降解率，如大环内酯类、磺胺类[25]、三嗪类、对羟基苯丙酸丁酯和三氯生、HCH（林丹）、有机磷酸酯[26]等，尤其是当萃取时间较长时。除此之外，基质组分的溶解度也随着温度的升高而增加，产生含有更多干扰的深色提取物，这些干扰可能会影响进一步测定的准确度。研究了温度对六种杂环芳香胺萃取效率的影响，温度变化范围为60~120℃，步长为20℃[27]。提取效率在80℃时达到最佳，超过80℃时回收率降低，可能是由于目标化合物的降解。另一个问题是，在高温下提取物较不纯，这可能是由于共萃取基质引起的。在80℃以下时回收率较低，主要是因为杂环芳香胺的解吸和溶解效率较低。

（3）压力　高压范围一般选择3.5~20.7MPa，主要作用是保持萃取溶剂处于液态，压力通常对 PLE 的回收率影响较小。

（4）溶剂　在传统萃取法中，理想情况下，溶剂极性必须与分析物类似（相似相溶）。与卤代化合物（二氯乙烷、氯仿等）、杂环醚（二氧六环、四氢呋喃等）和乙腈相比，醇类（甲醇、乙醇和丙醇）和烷烃类（正己烷和正庚烷）是更环保的溶剂。乙腈有一些缺点，例如在气相色谱蒸发过程中膨胀体积过大，因此如果选择气相色谱测定，则需要蒸发吹干后用更合适的溶剂进行复溶[28]。关于水作为溶剂在 PLE（加压热水提取，PHWE）中使用的基本原理，读者可参考本书 M. Plaza 和 C. Turner 的“加压热水提取生物活性物质”一章；此外，对于一些与有机污染物（极性-中等极性）提取相关的应用，见第4.2.4节。丙酮（通常与正己烷结合）能使样品颗粒和溶剂更好地接触，破碎和分解土壤团聚体，提高有机氯农药的提取率[29]。然而，丙酮的使用也可能导致从样品中提取出更多的干扰基质，从而需要进一步的净化处理。

在萃取过程中使用混合溶剂可提高萃取效率。这是由于两个方面因素的影响：极性范围的扩大和分析物解吸的增强，有助于破坏基质和分析物之间的氢键相互作用，并给出更好的回收率[19,25,30]。溶剂的选择主要取决于分析物。PLE 可用于提取非极性和极性分析物，前者主要用正己烷/二氯甲烷的不同混合物进行提取，后者则用甲醇、甲醇-丙酮或甲醇-水混合物进行提取。

对于从脂肪基质中提取极性分析物，可以在提取单元组装完成后执行在线净化步骤，即萃取前先用正己烷对样品进行脱脂处理。尽管这一步骤并没有显著提高方法的回收率，但获得了外观更干净的提取物，这有助于保护色谱柱。此外，这一步骤不需要任何样品处理，也不会显著增加分析时间。

（5）添加剂/改性剂　pH的变化可以改变分析物的离子状态，进而提高其在所选溶剂中的溶解度。有些物质会增强那些容易与金属形成络合物的化合物的提取效率。例如，氟喹诺酮类和四环素类药物是可以用柠檬酸溶液进行提取，柠檬酸除其他特性外，还具有隔离离子的能力。

（6）萃取循环　可以使用一个以上的萃取循环来实现完全提取，尽管单个萃取循环就可以达到要求，但每增加一个新的循环，都可以提高萃取效率。在某些情况下，由于干扰物质的较高共萃取率，循环次数的增加可能会导致分析物回收率下降[17]。同样，静态加热步骤后流过不锈钢萃取池的溶剂量也会影响提取效率。

（7）冲洗量　一般表示为萃取池体积的百分比（从0%到150%）。如果指定了一个以上的萃取循环，则在各循环之间对冲洗体积进行分配，这时用于冲洗样品的溶剂部分可能不足以将剩余的提取溶剂全部转移到收集瓶[17]。

所有参数必须仔细优化，以获得分析物的最大回收率。设计较差的实验可能导致不正确的结论，浪费时间和资源。可以用统计程序来完成实验设计，以尽量减少实验的数量。

PLE提取液通常需要在测定分析物之前进行浓缩和净化。SPE是常用的，具有较高重复性的解决方案。在这项技术中，提取物中的有机溶剂含量必须降低到5%以下，以避免分析物从小柱上提取流出。如果提取液是有机溶剂和水的可混溶混合物，在使用SPE进行净化操作之前，可以向提取液中添加适量的水。SPE相的范围很广（亲水-亲油平衡、C_{18}、SAX等），意味着极性或离子化可以作为保留机制。对于非常不纯的提取液，避免SPE小柱过载的一个有效策略是将两个不同的小柱串联起来使用。例如，在土壤和沉积物样品中，前者（阴离子交换柱）能去除有机杂质，主要是腐殖酸和黄腐酸，后者（亲水-亲油聚合物小柱）保留感兴趣的分析物。在分析鱼肉样品中的多溴二苯醚和多氯联苯（PCBs）时，根据脂肪含量的不同，使用酸性硅胶和超声波辅助d-SPE后、进一步串联SPE净化或使用活化的酸性硅胶和活化的碱性氧化铝小柱直接串联SPE进行净化[16]。一种三柱净化程序，包括一个高容量的酸-碱-中性（ABN）硅胶柱、碳/硅藻土柱和碱性氧化铝柱，将多氯萘分离成两个分离部分[31]。提取液流过第一个ABN柱，然后流入碳/硅藻土柱。用甲苯反冲碳柱可以收集PCNs、多氯代二苯并二噁英/多氯代二苯并呋喃（PCDD/PCDFs）和共平面PCBs等。然后，甲苯提取液在氧化铝柱上再次净化[31]。

采用在线固相萃取技术对颗粒相的PLE提取液进行预浓缩，将大量提取物注入LC-MS/MS中[32]。用泵将1mL提取液装载到在线固相萃取柱上，药物类化合物被小柱吸附，之后这些化合物用流动相进行洗脱。

作为SPE的替代方案，几种净化技术可以在线与LC或GC系统结合起来。这些技术的优点是减少了有机溶剂的消耗、成本和分析所需的时间。

4.2.4 加压热水萃取（PHWE）

当溶剂为水时，通常使用其他术语，如亚临界水萃取、热水萃取、加压热水提取、高温水萃取、过热水萃取或热液态水萃取等[33-35]。水被用作提取溶剂可以减少有机溶剂的使用，它不易燃烧、无毒、廉价，可在对环境影响最小的情况下循环使用[36]。虽然水在低温下无法提取非极性分析物，但在高温下是可行的，因为它改变了分析物在基质中的溶解度和解吸动力学。如前所述，为了更深入地讨论 PHWE 的背景和基本原理，读者可以参考 M. Plaza 和 C. Turner 的“加压热水提取生物活性物质”一章[28a]。

如前所述，PHWE 主要用于植物成分的分析，但在文献中仍有一些关于提取有机极性-中等极性污染物的报道，如苯并噻唑、苯并三唑和苯磺酰胺衍生物、烷基酚、内分泌干扰物（EDC）、甜味剂、杀虫剂和抗菌剂等[37-42]，见表 4.1。与我们之前的综述[12]对照，这项技术正在被废弃，可能是因为尽管它比使用有机溶剂更环保，但大多数有机污染物都是非极性化合物。

考察 PHWE 与 PLE、振荡和超声波提取法等四种方法，在从家禽组织中分离微量氯霉素、甲砜霉素、氟芬尼考及其主要代谢物氟芬尼考胺时的提取效果[41]。在所考察的四种方法中，PHWE 和 PLE 技术的提取效率最高，平均回收率分别为 95.3% 和 93.8%。另一方面，与 PLE、超声波和振荡萃取法中使用的非选择性有机溶剂相比，亚临界水的选择性表现出更高的灵敏度。

表 4.1　使用加压热水作为提取溶剂分析环境和食品样品中的有机污染物的研究（2013—2017 年初）

分析物	基质	温度/℃	固相吸附剂	净化	回收率/%	分析仪器	灵敏度/(ng/g)	参考文献
苯并噻唑，苯并三唑，苯磺酰胺及其衍生物	污泥	80	硅藻土	SPE（Oasis HLB）	25~107	LC-Orbitrap MS	0.25~25	[37]
烷基酚	沉积物	80	硅藻土	SPE（Oasis HLB 与弗罗里硅土小柱串联）	>80	LC-HR-MS	0.25~25	[38]
内分泌干扰物	沉积物	150	硅藻土	DLLME（氯苯）	42~135	BSTFA GC-MS	0.006~0.639	[42]
6 种甜味剂	污泥	80	沙土	SPE（Oasis HLB）	72~105	LC-MS/MS	0.3~16	[39]
农药	茶叶	80	硅藻土	无	77~107	UHPLC-MS/MS	0.7~18.6	[40]
氯霉素，甲砜霉素，氟苯尼考，氟苯尼考胺	家禽组织	150	硅藻土	SPE（Oasis HLB）	87~102	UHPLC-MS/MS	0.03~0.5	[41]

注：BSTFA—（三甲基硅烷基）三氟乙酰胺；DLLME—分散液液微萃取；GC—气相色谱；HLB—亲水亲脂平衡柱；HR—高分辨；MS—质谱；SPE—固相萃取；UHPLC—超高压液相色谱。

4.2.5 与其他技术的比较

与传统索氏提取法相比，PLE 的提取时间从几个小时缩短到小于 1h，有机溶剂消耗量从 150~500mL 减少到 5~50mL。SFE 和亚临界水的提取时间与 PLE 几乎相同。

不同分析物类别的选择性也因萃取方法而异。超声波提取法和 PLE 法在分析双酚 A 及其氯化衍生物[43]时回收率基本相同，但超声波提取法对富勒烯的回收率较低[44]。

PLE 法测定白葡萄蔗渣中 11 种杀菌剂的回收率明显优于超声波法[45]。即使在一个单一步骤中，与索氏提取法相比，SPLE 法在确定绿色海藻中优先使用的农药方面也显示出更好的性能[17]。

对于环境样品中有效农药的提取，QuEChERS 似乎是一种比 PLE 更好的方法[29]，但通常 PLE 具有更好的重现性[28]。

PLE 的主要缺点是设备的初始成本较高，即使将其作为监测分析的常规技术进行价格分摊。

分别采用三种不同的方法 PLE、UAE 和索氏提取法从海产品中提取对羟基苯甲酸酯类和内分泌干扰素等，回收率分别为 90.6%~107.8%、81.4%~85.2%、83.5%~88.7%[46]。与传统的索氏法和超声波法相比，PLE 法具有萃取效率高、自动化程度高、溶剂和时间消耗少、环境友好、高通量等优点。

然而，PLE 在参考土壤中提取多环芳烃的实验证明该方法并不适用于所有类型的土壤[47]。PLE 法是能最完全提取多环芳烃的方法，使用该方法测定时部分土壤样品的多环芳烃总量最高，但并非所有土壤样品均是如此。有机质和黏粒含量是土壤的主要特征，从土壤样品中选择最合适的多环芳烃提取方法时应考虑这两个因素。因此，有机质含量高的土壤必须采用侵蚀性程序进行提取，以确保完全提取多环芳烃。然而，黏土含量高的土壤不需要如此彻底的提取程序，固-液萃取将有助于从土壤中获得完全的多环芳烃提取率。

4.3 应用

4.3.1 环境样品

PLE 主要用于提取环境样品而非生物群中的持久性有机污染物[2,19,48]。在大多数情况下，使用正己烷和二氯甲烷作为提取溶剂，可以使有机氯、多氯联苯和多溴二苯醚得到较好的回收率。对于沉积物、土壤和生物群基质，通常需要特殊和强力的净化步骤。使用半自动 SPE 系统可以减少这种烦琐的工作。例如，文献［18］分析了沉积物中 7 种 PCDDs、10 种 PCDFs、12 种类二噁英和 6 种 PCBs、六氯苯、DDT 及其代谢物 DDE。为了避免平面化合物（PCDD/Fs）和非平面化合物（PCBs）之间的干扰，PLE 处理后的提取液用半自动 Power Prep（流体管理系统，美国）进行分段，并配有三个商用一次性柱。其他在线固相萃取装置，称为 TurboFlow，将分析物浓缩到净化装载柱中，并直接将其注入色谱系统，不仅提供可接受的回收率，而且与传统的固相萃取方法相比，基质效应显著降低。TurboFlow 已用于分析沉积物和污水污泥样品中的内分泌干扰物，包括雌激素（及其共轭物）、抗菌剂、对羟基苯甲酸酯、双酚 A、烷基酚类化合物、苯并三唑和有机磷阻燃剂[49]。沉积物中回收率在 65%～114%，中值为 91%；污泥中回收率在 55%～115%，中值为 82%。

在提取的同时进行萃取池内同步净化，省去了一般 PLE 的后处理步骤，提高了样品的处理效率。但是，不建议在没有进一步净化的情况下直接在 GC-MS 系统中注入污泥提取物，因为仪器灵敏度会显著降低[19]。在海洋沉积物中，硅胶在测定 PPBDEs、PCDDs 和 PCDFs 等非极性化合物时非常有效[18]。活性铜通常用于脱除硫，硫具有还原性，会干扰分析物的测定。铜通常与吸附剂一起放置在萃取池的底部，用于萃取池内净化。如 Aguilar 等人所证明的，当铜放在收集瓶中时，经 PLE 处理后的提取液到达收集瓶中，净化效果也较好[2]。这一新方法，如图 4. 3 所示，大大提高了用美国 EPA-1613 方法分析沉积物中 PCDD/Fs 和 dl-PCBs 的回收率。开发了一种从鲸脂中同时提取多溴二苯醚、多氯联苯和有机氯农药的 SPLE 方法[46]，其中中性硅胶放置在酸性硅胶下方，从而防止酸污染提取系统。弗罗里硅土具有去除生物大分子的作用，能有效地去除脂肪。但是它的密度非常小，而且萃取池一般体积有限制。因此与氧化铝或酸性硅胶相比较，当使用弗罗里硅土时，需要减少样品使用量。

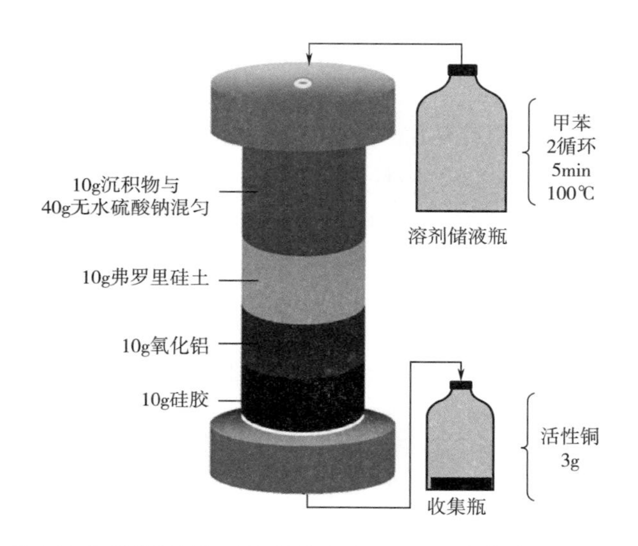

图 4. 3　带有净化吸附剂的加速溶剂萃取池，用于沉积物中 PCDDs、 PCDFs 和类二噁英物质的 SPLE 处理

［资料来源：L. Aguilar, E. S. Williams, B. W. Brooks, S. Usenko, Development and application of a novel method for high-throughput determination of PCDD/Fs and PCBs in sediments, Environ. Toxicol. Chem. 33（2014）1529-1536，经 SETAC 出版社许可。］

还有一些例子说明 PLE 在无需净化步骤的情况下分析新出现的污染物的有效性，如从污泥中提取碘化 X 射线造影剂的方法[50]。虽然观察到强烈的基质效应，但该方法仅通过一个步骤就提供了足够干净和合适浓度的甲醇提取液，易于通过 LC-MS/MS 进行分析。新出现的污染物的极性高于持久性污染物。用于极性化合物分析的其他萃取池内净化方法包括在温和条件（低温和低压）下用正己烷去除非极性物质，然后用极性溶剂提取分析物。该方法已应用于粉尘中三氯生和对羟基苯甲酸酯的分析。此外，还描述了北极熊毛发中三种关键皮质类固醇（醛固酮、皮质醇和皮质酮）的提取、净化和分析的方法；其中净化步骤如图 4.4 所示。第一步是反相 PLE（i-PLE），其中 PLE 萃取池中样品填充在硅藻土上，然后用正庚烷冲洗［图 4.4（1）］。这个步骤是为了从样品基质中除去比分析物更亲脂的化合物。第二步，一体净化（ic-PLE），是使用丙酮-甲醇实施的实际萃取步骤。一个含有净化相如活性碱性氧化铝、石墨和三种活性硅胶［图 4.4（2）］的萃取池连接在 i-PLE 萃取池下方。此装置得到的回收率较低。然后还测试了最简单的方法。

在处理非常复杂的基质如污泥时，即使经过固相萃取步骤，基质干扰的存在也不能完全消除，目标分析物的电离抑制率高达 90%。在测定 20 种滥用和非法药物及其代谢物时观察到了这一现象，文章首次报告了污泥中存在大麻素[30]。

表 4.2 所示为应用于环境样品分析时这些方法和其他几种 PLE 方法的更全面清单。

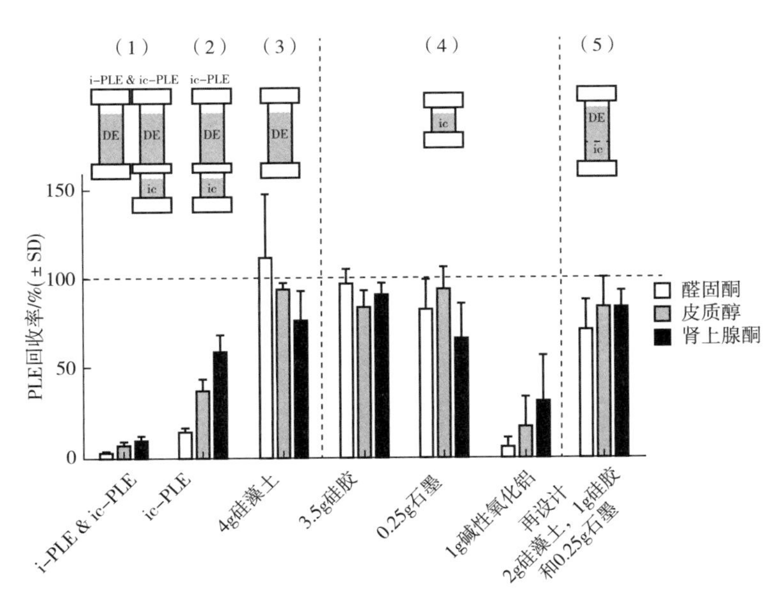

图 4.4　不同 PLE 萃取池装置的回收率（n=3）

在（1）组中，先用正庚烷冲洗硅藻土填充的 PLE 萃取池，然后在下方附着 11mL 的 ic-PLE 萃取池，用硅胶、石墨和碱性氧化铝填充，再用丙酮：甲醇进行再次萃取。在（2）组中，单独使用 ic-PLE 萃取池，用丙酮：甲醇（1：1，体积比）作为溶剂。在（3）组和（4）组中，分别对每种填充材料进行测试，其中硅藻土填充在 22mL 萃取池中，集成净化材料填充在 11mL 萃取池中，使用丙酮：甲醇（1：1，体积比）作为溶剂。

［资料来源：J. J. Weisser, M. Hansen, E. Bjorklund, C. Sonne, R. Dietz, B. Styrishave, A novel method for analysing key corticosteroids in polar bear (Ursus maritimus) hair using liquid chromatography tandem mass spectrometry, J. Chromatogr. B-Anal. Technol. Biomed. Life Sci. 1017 (2016) 45-51，经 Elsevier 许可。］

表 4.2　**使用 PLE 方法分析环境样品中的污染物的相关研究（2013—2017 年初）**

分析物	基质	固相吸附剂	温度/℃	溶剂	净化	回收率/%	检出限/(pg/g)	分析仪器	参考文献
生物基质									
17 种多环芳烃	空气	—	150	乙腈	—	43~114	0.1~5	GC-MS/MS	[51]
55 种半挥发性有机物	室内灰尘	硅藻土	100	二氯甲烷	氨玻璃柱	65~146	—	GC-MS/MS	[52]
22 种磺胺抗生素（含 5 种乙酰代谢物）	土壤、淤泥	Hydromatrix	淤泥：50 土壤：100	淤泥：乙腈-水（25∶75） 土壤：甲醇-水（90∶10）	固相萃取（HLB）	60~130	10~4190	LC-QqLIT-MS/MS	[25]
7 种溴代联苯醚	淤泥	弗罗里硅土，硫酸钠	40	正己烷-二氯甲烷（50∶50）	填充吸附剂微萃取	92~102	10~40	GC-MS/MS	[19]
5 种碘 X 射线造影剂	淤泥	Hydromatrix	100	甲醇	—	42~89	1600~3200	LC-MS/MS	[50]
5 种富勒烯（3 种衍生物）	沉积物	Hydromatrix	150	甲苯	—	70~92	0.03~160	UHPLC-APPI-MS	[44]
有机磷酸酯	沉积物	Hydromatrix	150	丙酮	SPME	82~109	9~280	GC-FPD	[26]
PCDD/Fs 和 dl-PCBs	沉积物	弗罗里硅土，氧化铝，硅胶	100	甲苯	铜	67~83	12~32	HRGC-HRMS	[2]
抗生素和其他药物残留	水的颗粒相	枫丹白露沙	80	甲醇/乙腈/0.2mol/L 柠檬酸；其中，氟喹诺酮类：pH4.5，体积比 40∶40∶20 四环素类：pH3，体积比 25∶25∶50	C_{18} 反相柱在线固相萃取	53~107	—	UHPLC-MS/MS	[32]
97 种污染物（香水、防晒剂、驱蚊剂、内分泌干扰物、杀菌剂、多环芳烃、多氯联苯、有机磷阻燃剂、农药）	沉积物	硅藻土	100	二氯甲烷	铜和氧化铝在线净化	70~100	<1	MTBSTFA 衍生后 GC-MS/MS 测定	[11]

多溴联苯醚、五溴甲苯、五溴乙苯、六溴苯、2-乙基己基-四溴苯甲酸酯、双（2，4，6-三溴苯氧基）乙烷	土壤	硫酸钠和 Hydromatrix	100	正己烷-二氯甲烷（体积比 1∶1）	活性铜、硫酸钠、硅胶、弗罗里硅土在线净化	—	0.01~4.8	GC-MS/MS	[53]
多环芳烃	土壤	硅藻土	100	二氯甲烷-丙酮（体积比1∶1）	未报道	31~108	<0.05	HPLC-PDA	[47]
双酚 A	农业土壤	未使用	70	二氯甲烷	硅胶和硫酸钠在线净化	未报道	未报道	吡啶、BSTFA 衍生后 GC-MS	[54]
三氯卡班及其转化产物	沉积物	枫丹白露沙	70	甲醇-丙酮（体积比 1∶1）	不需要	56~119	0.04~0.4	LC-HR-MS	[55]
生物样品									
多溴联苯醚、多氯联苯、有机氯农药	鲸脂	硫酸钠	100	正己烷	酸性和中性硅胶在线净化	76~74	—	GC-MS	[56]
20 种农药，15 种多氯联苯，7 种多溴联苯醚	鲸鱼耳垢	碱性氧化铝、硅胶和弗罗里硅土	100	正己烷-二氯甲烷（体积比 1∶1）	—	58~132	0.57~960	GC-MS	[57]
44 种药物，13 种内分泌干扰物	河流生物膜	硅藻土	70	甲醇	未报道	31~137	0.07~6.7	UHPLC-MS/MS	[58]
皮质类固醇（醛固酮、皮质醇、皮质酮）	北极熊毛	硅藻土	80	甲醇-丙酮（体积比 1∶1）	石墨化碳和氧化铝萃取池内净化；C_{18} 固相萃取	60~85	0.023~0.170	UHPLC-MS/MS	[59]

注：APPI—大气压光电离；BDE—溴化二苯醚；BTBPE—双（2，4，6-三溴苯氧基）乙烷；DCM—二氯甲烷；dl-PCBs—类二噁英多氯联苯；EDC—内分泌干扰物；EH-TBB—2-乙基己基-2，3，4，5-四溴苯甲酸酯；FQs—氟喹诺酮类；GC-FPD—气相色谱-火焰光度法；GC-MS—气相色谱-质谱；GPC—凝胶渗透色谱；HBB—六溴代苯；HCB—六氯苯；HLB—亲水亲脂平衡柱；HR—高分辨率；HRGC-HRMS—高分辨率气相色谱-高分辨率质谱；LC—液相色谱；MEPS—填充吸附剂微萃取；OCP—有机氯；OFRs—有机磷阻燃剂；PAHs—多环芳烃；PBDEs—多溴二苯醚；PBEB—五溴乙苯；PBT—五溴甲苯；PCBs—多氯联苯；PCDD/Fs—二苯并对二噁英/二苯并呋喃；QqLIT—四极杆线性离子阱；SAs—磺胺抗生素；SPE—固相萃取；SPME—固相微萃取；SVOCs—半挥发性有机化合物；TCs—四环素类；UHPLC—超高压液相色谱法。

4.3.2 食品

PLE 是一种全自动、可靠的提取技术，与传统的提取技术相比具有许多优点，因此它特别适用于食品和饲料中持久性和新出现的有机污染物的常规分析[22,28,57,60,61]。

PLE 被广泛用于测定各种各样的食物成分和残留物，从极性化合物（如高强度甜味剂[23]）到非极性化合物（如杀虫剂[28]、多环芳烃[22]、多溴二苯醚[16]等）。

PLE 的主要问题是伴随分析物出现的共萃取基质。食品是一种非常复杂的基质，分析时需要去除蛋白质、脂类、碳水化合物、胡萝卜素、硫、多酚等干扰物质。SPLE 单独使用或与 SPE、dSPE、GPC 联合使用是最常用的前处理方法。在富含脂质的食物中，如饲料、油和鲸脂等，干燥剂的使用非常重要，它主要是作为分散剂以降低提取微量水的风险，这可能导致样品中脂质含量过高，并在浓度与脂质含量相关时导致不准确的定量。此外，高效液相色谱柱对微量脂质高度敏感，它会影响色谱柱的分辨率，缩短色谱柱的使用寿命。在这些高脂肪样品中，正己烷通常优先于混合溶剂，因为在常温条件下正己烷能得到较干净的提取液[46,57]。

在 SPLE 中，硅胶和氧化铝被用于保留脂质，但必须仔细选择吸附剂的用量，以避免感兴趣的分析物被保留[22]。除此之外，酸性硅胶很容易堵塞，导致超压。

SPLE 中，萃取吸附剂种类以及吸附剂与样品质量比的优化需要经过多方面进行判定：目视检查（脂质沉淀）、色谱峰形状、色谱基线降低、分析物回收率和仪器维护频率[57]。

图 4.5 所示为使用 GC-MS-EI 全扫描进行监测时，不同净化剂得到的分析物色谱峰形状及基线的差异。

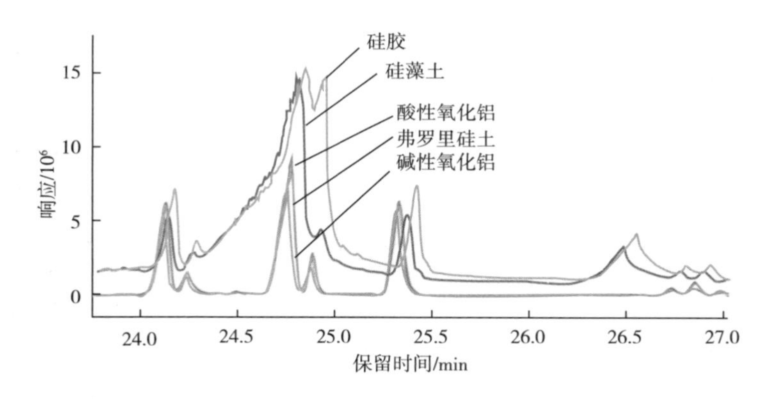

图 4.5　以 10∶1 的吸附剂与灰鲸耳垢的质量比，考察了五种吸附剂得到的提取液的 GC-MS-EI 全扫描的总离子流色谱图的叠加图，最终选择弗罗里硅土以减少基质干扰

[资料来源：E. M. Robinson, S. J. Trumble, B. Subedi, R. Sanders, S. Usenko, Selective pressurized liquid extraction of pesticides, polychlorinated biphenyls and polybrominated diphenyl ethers in a whale earplug (earwax): a novel method for analyzing organic con taminants in lipid-rich matrices, J. Chromatogr. A 1319 (2013) 14-20. 经 Elsevier 许可。]

分散在硅藻土中的 Carbopack 材料，通常用于将平面化合物（如 PCDD/Fs）从已知可以干扰它们的非平面化合物（如 PCBs）中分离出来[62]。

鱼肉和贻贝中的麝香成分已通过 PLE 进行成功测定[63]。选择二氯甲烷作为萃取剂，因为除甲醇外，正己烷和其他更环保的溶剂的回收率均较低。而在甲醇提取物中，由于麝香的亲油性，它们会保留在脂肪沉淀中。使用弗罗里硅土时，其基质效应比氧化铝和二氧化硅低。与 QuEChERS 提取法相比，PLE 法获得的方法学验证参数稍好。

表 4.3 所示为用于测定食品中有机污染物的 PLE 分析方法的相关文献。

4.4 结论与展望

随着人们对绿色技术（避免或减少使用有毒有机溶剂提取固体基质中的污染物）越来越感兴趣，PLE 已显示出其优越性。这种技术可以是替代传统提取技术的一个很好的选择，在食品污染物的提取方面有较大的潜力。它能够提供具有竞争力的重复性和回收率，可以实现自动化，提高样品检测通量，并节省分析人员的工作时间。这使其成为传统方法（如索氏提取法、超声波提取法）和其他现代固液萃取法（如 SFE、MAE 等）的良好替代品。

在可用于 PLE 的不同方法中，PHWE 不使用有毒溶剂。它已成功地用于几种污染物的分析，但由于商用 PLE 系统能达到的最高温度有限（200℃），而且非极性分析物不能用这些仪器测定，因此必须在自制仪器中进行。它们的应用通常仅限于极性分析物，如抗菌剂、胺类、三嗪类等。然而，这项技术最近被提出用于评估农药残留，这些农药残留可从茶叶中转移到其浸泡液和副产品中，随后被人类消费。

PLE 提取通常需要后续的净化步骤。SPE、GPC 和 SPME 等可以直接与 GC 或 LC 结合使用，进而节省时间、减少误差来源。SPLE 是一种将各种吸附剂组合在一起进行萃取池内净化的方案，据报道为了充分利用 PLE 提供的所有可能性，在从食物、污水、土壤和沉积物中提取有机污染物的过程中，SPLE 经常被用于保留脂肪和其他基质成分。这种萃取池内净化不仅节省了人力物力，增加了实验室样品处理通量，而且还减少了处理样品时固有的误差来源。

致谢

这项工作得到了西班牙经济和竞争力部和欧洲区域发展基金（ERDF）的 CGL2015-64454-C2-1-R 项目的支持。

参考文献

[1] S. Armenta, S. Garrigues, M. de la Guardia, The role of green extraction techniques in Green Analytical Chemistry, TrAC, Trends Anal. Chem. 71 (2015) 2-8.

[2] L. Aguilar, E. S. Williams, B. W. Brooks, S. Usenko, Development and application of a novel method for high-throughput determination of PCDD/Fs and PCBs in sediments, Environ. Toxicol. Chem. 33

表 4.3 使用 PLE 方法食品中污染物的相关研究（2013—2017 年初）

分析物	基质	固相吸附剂	温度/℃	溶剂	净化	回收率/%	检出限/（pg/g）	分析仪器	参考文献
18 种 PCB	鱼	硅藻土	100	丙酮/正己烷（1：1）	SPE（二氧化硅）	80~110	400~1100	GC-MS	[64]
12 种农药	茶叶	水基质	120	丙酮-正己烷（2：1）	GPC、SPE（Envi-Carb+LC-氧化铝-N）	96~102	20~10830	GC-HRMS	[28]
矿物油和饱和和芳香烃	谷物	沙子	100	正己烷-乙醇（1：1，v/v）	液液萃取	96	100~200	LC-GC-FID	[65]
17 种多氯二恶英和呋喃	鱼	-	100	正己烷	酸性硅胶+活化硅胶	96~108	12.5~32.2	GC-HRMS	[60]
5 种药品	海生贻贝	氧化铝+砂子	60	乙腈：水（3：1）	SPE（Stratr-X）	83~104	4000~29000	LC-MS/MS	[66]
五氯苯酚	肉、鱼	硅藻土	110	甲醇-2%三氯乙酸（3、1）	SPE（Oasis HLB）	71	160	GC-ECD，GC-MS	[61]
NUEVAS 高强度甜味剂	鱼类	硅藻土和使用氧化铝的池内净化	60	使用正己烷和甲醇：超纯水（1：1 v：v）进行在线净化	SPE	43~94	2.5~125（d. w.）	LC-Orbitrap-MS	[23]
PCN	鱼类	硅藻土	80℃加热 10min；100℃加热 20min	二氯甲烷/正己烷（10/90）	采用 3 根柱子进行 SPE：（1）ABN 硅胶，（2）碳/硅藻土，（3）碱性氧化铝	88~110	1.3~3.7pg/g	GC-HRMS	[31]
农药	茶叶	硅藻土	100	二氯甲烷	无水 Na_2SO_4 SPE GCB/PSA	77~107	0.7~18.6	UHPLC-MS/MS	[40]

酯化的3-和2-单氯丙二醇和缩水甘油脂肪酸酯	食品基质	聚丙烯酸酯和沙子	40	叔丁基甲基醚	用正己烷补偿高脂肪含量的内标法液液萃取	82~114	7~17	衍生化 GC-MS	[67]
对羟基苯甲酸酯：防腐剂；防晒剂	海产品（植物和动物）	硅藻土	70	甲醇	SPE（Oasis MCX）	81~108	10	LC-QqLIT-MS/MS	[46]
14种有机氯农药	鱼	硅藻土	100	二氯甲烷-正己烷（体积比3：7）	在线氧化铝、弗罗里硅土、酸化硅胶和硅胶净化	91~93	0.09~0.51	GC-MS	[21]
多溴联苯醚	鱼	酸洗沙土和Hydromatrix	100	二氯甲烷-正己烷（体积比1：1）	硅胶在线净化	70~124	0.0006~9	GC-MS/MS	[68]
多环芳烃	烘焙咖啡	未使用	100	正己烷-二氯甲烷（体积比85：15）	硅胶在线净化；液液萃取条件DMF-水（体积比9：1）/正己烷	87~111	0.04~0.18	GC-MS	[69]
杂环胺	烤肉	氢氧化钠预处理；硅藻土	80	二氯甲烷-乙腈（体积比1：1）	中性氧化铝与硫酸钠在线净化	69~94	1.5~3.0	LC-IT-TOF-MS	[27]

注：ABN二氧化硅—酸性-碱性-中性二氧化硅；AcN—乙腈；ECD—电子捕获检测器；FD—荧光检测器；FID—火焰离子化检测器；GC-HRMS—气相色谱-高分辨率质谱；GPC—凝胶渗透色谱；HLB—亲水亲油平衡柱；IS—内标；IT—离子阱；LC—液相色谱；LLE—液体萃取；OCPs—有机氯农药；OPPs—有机磷农药；PAHs—多环芳烃；PBDEs—多溴联苯醚；PCBs—多氯联苯；PCDD/Fs—二苯并-*p*-二噁英/二苯并呋喃；PCNs—多氯萘；PLE—加压液体萃取；PSA—*N*-丙基乙二胺；QqLIT—四极杆线性离子阱；SPE—固相萃取；TOF—飞行时间；UHPLC—超高效液相色谱法。

(2014) 1529-1536.

[3] N. Fidalgo-Used, E. Blanco-Gonzalez, A. Sanz-Medel, Sample handling strategies for the determination of persistent trace organic contaminants from biota samples, Anal. Chim. Acta 590 (2007) 1-16.

[4] R. Cruz, S. C. Cunha, A. Marques, S. Casal, Polybrominated diphenyl ethers and metabolites-an analytical review on seafood occurrence, TrAC, Trends Anal. Chem. 87 (2017) 129-144.

[5] K. Pazdro, M. Borecka, G. Siedlewicz, A. Bialk-Bielinska, P. Stepnowski, Analysis of the residues of pharmaceuticals in marine environment: state-of-the-art, analytical problems and challenges, Curr. Anal. Chem. 12 (2016) 202-226.

[6] K. M. Dimpe, P. N. Nomngongo, Current sample preparation methodologies for analysis of emerging pollutants in different environmental matrices, TrAC, Trends Anal. Chem. 82 (2016) 199-207.

[7] M. M. Schantz, Pressurized liquid extraction in environmental analysis, Anal. Bioanal. Chem. (2006) 1043-1047.

[8] L. Ramos, E. M. Kristenson, U. A. T. Brinkman, Current use of pressurised liquid extraction and subcritical water extraction in environmental analysis, J. Chromatogr. A (2002) 3-29.

[9] M. D. Marazuela, S. Bogialli, A review of novel strategies of sample preparation for the determination of antibacterial residues in foodstuffs using liquid chromatography-based analytical methods, Anal. Chim. Acta 645 (2009) 5-17.

[10] K. Ridgway, S. P. D. Lalljie, R. M. Smith, Sample preparation techniques for the determination of trace residues and contaminants in foods, J. Chromatogr. A (2007) 36-53.

[11] M. G. Pintado-Herrera, E. Gonzalez-Mazo, P. A. Lara-Martin, In-cell clean-up pressurized liquid extraction and gas chromatography-tandem mass spectrometry determination of hydrophobic persistent and emerging organic pollutants in coastal sediments, J. Chromatogr. A 1429 (2016) 107-118.

[12] P. Vazquez-Roig, Y. Pico, Pressurized liquid extraction of organic contaminants in environmental and food samples, TrAC, Trends Anal. Chem. 71 (2015) 55-64.

[13] N. Salgueiro-González, I. Turnes-Carou, S. Muniategui-Lorenzo, P. López-Mahía, D. Prada-Rodríguez, Fast and selective pressurized liquid extraction with simultaneous in cell clean up for the analysis of alkylphenols and bisphenol A in bivalve molluscs, J. Chromatogr. A 1270 (2012) 80-87.

[14] L. Do, S. Lundstedt, P. Haglund, Optimization of selective pressurized liquid extraction for extraction and in-cell clean-up of PCDD/Fs in soils and sediments, Chemosphere (2013) 2414-2419.

[15] M. Marchal, J. Beltran, Determination of synthetic musk fragrances, Int. J. Environ. Anal. Chem. 96 (2016) 1213-1246.

[16] D. Lu, Y. Lin, C. Feng, D. Wang, X. Qiu, Y. Jin, et al., Determination of polybrominated diphenyl ethers and polychlorinated biphenyls in fishery and aquaculture products using sequential solid phase extraction and large volume injection gas chromatography/tandem mass spectrometry, J. Chromatogr. B-Anal. Technol. Biomed. Life Sci. 945-946 (2014) 75-83.

[17] M. I. Pinto, C. Micaelo, C. Vale, G. Sontag, J. P. Noronha, Screening of priority pesticides in *Ulva* sp. Seaweeds by selective pressurized solvent extraction before gas chromatography with electron capture detector analysis, Arch. Environ. Contam. Toxicol. 67 (2014) 547-556.

[18] K. Conka, J. Chovancova, Z. Stachova Sejakova, M. Domotorova, A. Fabisikova, B. Drobna, et al., PCDDs, PCDFs, PCBs and OCPs in sediments from selected areas in the Slovak Republic, Chemosphere 98 (2014) 37-43.

[19] M. P. Martínez-Moral, M. T. Tena, Use of microextraction by packed sorbents following selective pressurised liquid extraction for the determination of brominated diphenyl ethers in sewage sludge by gas chromatographymass spectrometry, J. Chromatogr. A 1364 (2014) 28-35.

[20] B. Subedi, L. Aguilar, E. M. Robinson, K. J. Hageman, E. Bjorklund, R. J. Sheesley, et al., Selective pressurized liquid extraction as a sample-preparation technique for persistent organic pollutants and contaminants of emerging concern, TrAC, Trends Anal. Chem. 68 (2015) 119-132.

[21] M. Choi, I. -S. Lee, R. -H. Jung, Rapid determination of organochlorine pesticides in fish using selective pressurized liquid extraction and gas chromatography-mass spectrometry, Food Chem. 205 (2016) 1-8.

[22] M. Suranová, J. Semanová, B. Sklářšová, P. Simko, Application of accelerated solvent extraction for simultaneous isolation and pre-cleaning up procedure during determination of polycyclic aromatic hydrocarbons in smoked meat products, Food Anal. Met. 8 (2014) 1014-1020.

[23] M. Núñez, F. Borrull, E. Pocurull, N. Fontanals, Pressurised liquid extraction and liquid chromatography-high resolution mass spectrometry to determine high-intensity sweeteners in fish samples, J. Chromatogr. A 1479 (2017) 32-39.

[24] Y. P. Liu, J. G. Li, Y. F. Zhao, Y. N. Wu, L. Y. Zhu, Rapid determination of polybrominated diphenyl ethers (PBDEs) in fish using selective pressurized liquid extraction (SPLE) combined with automated online gel permeation chromatography-gas chromatography mass spectrometry (GPC-GC/MS), Food Add. Contam. A 26 (2009) 1180-1184.

[25] M. J. García-Galán, S. Díaz-Cruz, D. Barceló, Multiresidue trace analysis of sulfonamide antibiotics and their metabolites in soils and sewage sludge by pressurized liquid extraction followed by liquid chromatography-electrospray-quadrupole linear ion trap mass spectrometry, J. Chromatogr. A 1275 (2013) 32-40.

[26] J. Zheng, Z. Gao, W. Yuan, H. He, S. Yang, C. Sun, Development of pressurized liquid extraction and solid-phase microextraction combined with gas chromatography and flame photometric detection for the determination of organophosphate esters in sediments, J. Sep. Sci. 37 (2014) 2424-2430.

[27] Y. -f. Ouyang, H. -b. Li, H. -b. Tang, Y. Jin, G. -y. Li, A reliable and sensitive LCMS-IT-TOF method coupled with accelerated solvent extraction for the identification and quantitation of six typical heterocyclic aromatic amines in cooked meat products, Anal. Met. 7 (2015) 9274-9280.

[28] J. Feng, H. Tang, D. Chen, H. Dong, L. Li, Accurate determination of pesticide residues incurred in tea by gas chromatography-high resolution isotope dilution mass spectrometry, Anal. Met. 5 (2013) 4196-4204.

[28a] M. Plaza, C. Turner, Pressurized hot water extraction of bioactives, in E. Ibanez, A. Cifuentes (Eds.), Green Extraction Techniques: Principles, Advances and Applications. vol. 76, 2017, pp. 53-82.

[29] O. Svahn, E. Bjorklund, Increased electrospray ionization intensities and expanded chromatographic possibilities for emerging contaminants using mobile phases of different pH, J. Chromatogr. B-Anal. Technol. Biomed. Life Sci. 1033 (2016) 128-137.

[30] N. Mastroianni, C. Postigo, M. L. de Alda, D. Barcelo, Illicit and abused drugs in sewage sludge: method optimization and occurrence, J. Chromatogr. A 1322 (2013) 29-37.

[31] R. Lega, D. Megson, C. Hartley, P. Crozier, K. MacPherson, T. Kolic, et al., Congener specific determination of polychlorinated naphthalenes in sediment and biota by gas chromatography high resolution mass spectrometry, J. Chromatogr. A 1479 (2017) 169-176.

[32] I. Tlili, G. Caria, B. Ouddane, I. Ghorbel-Abid, R. Ternane, M. Trabelsi-Ayadi, et al., Simultaneous detection of antibiotics and other drug residues in the dissolved and particulate phases of water by an off-line SPE combined with on-line SPE-LC-MS/MS: method development and application, Sci. Total Environ. 563 (2016) 424-433.

[33] M. Plaza, C. Turner, Pressurized hot water extraction of bioactives, TrAC, Trends Anal. Chem. 71 (2015) 39-54.

[34] F. J. Barba, Z. Z. Zhu, M. Koubaa, A. S. Sant'Ana, V. Orlien, Green alternative methods for the extraction of antioxidant bioactive compounds from winery wastes and by-products: a review, Trends Food Sci. Technol. 49 (2016) 96-109.

[35] S. M. Zakaria, S. M. M. Kamal, Subcritical water extraction of bioactive compounds from plants and algae: applications in pharmaceutical and food ingredients, Food Eng. Rev. 8 (2016) 23-34.

[36] M. C. Bubalo, S. Vidovic, I. R. Redovnikovic, S. Jokic, Green solvents for green technologies, J. Chem. Technol. Biotechnol. 90 (2015) 1631-1639.

[37] P. Herrero, F. Borrull, R. M. Marcé, E. Pocurull, A pressurised hot water extraction and liquid chromatography-high resolution mass spectrometry method to determine polar benzotriazole, benzothiazole and benzenesulfonamide derivates in sewage sludge, J. Chromatogr. A 1355 (2014) 53-60.

[38] N. Salgueiro-González, I. Turnes-Carou, S. Muniategui-Lorenzo, P. López-Mahía, D. Prada-Rodríguez, Pressurized hot water extraction followed by miniaturized membrane assisted solvent extraction for the green analysis of alkylphenols in sediments, J. Chromatogr. A 1383 (2015) 8-17.

[39] E. Y. Ordoñez, J. B. Quintana, R. Rodil, R. Cela, Determination of artificial sweeteners in sewage sludge samples using pressurised liquid extraction and liquid chroma-tography-tandem mass spectrometry, J. Chromatogr. A 1320 (2013) 10-16.

[40] H. Chen, M. Pan, X. Liu, C. Lu, Evaluation of transfer rates of multiple pesticides from green tea into infusion using water as pressurized liquid extraction solvent and ultraperformance liquid chromatography tandem mass spectrometry, Food Chem. 216 (2017) 1-9.

[41] Z. Xiao, R. Song, Z. Rao, S. Wei, Z. Jia, D. Suo, et al., Development of a subcritical water extraction approach for trace analysis of chloramphenicol, thiamphenicol, florfenicol, and florfenicol amine in poultry tissues, J. Chromatogr. A 1418 (2015) 29-35.

[42] K. Yuan, H. Kang, Z. Yue, L. Yang, L. Lin, X. Wang, et al., Determination of 13 endocrine disrupting chemicals in sediments by gas chromatography-mass spectrometry using subcritical water extraction coupled with dispersed liquid-liquid microextraction and derivatization, Anal. Chim. Acta 866 (2015) 41-47.

[43] Z. R. Hopkins, L. Blaney, An aggregate analysis of personal care products in the environment: Identifying the distribution of environmentally-relevant concentrations, En-viron. Int. 92-93 (2016) 301-316.

[44] A. Astefanei, O. Nunez, M. T. Galceran, Analysis of C60-fullerene derivatives and pristine fullerenes in environmental samples by ultrahigh performance liquid chromatography-atmospheric pressure photoionization-mass spectrometry, J. Chromatogr. A1365 (2014) 61-71.

[45] M. Celeiro, M. Llompart, J. P. Lamas, M. Lores, C. Garcia-Jares, T. Dagnac, Determination of fungicides in white grape bagasse by pressurized liquid extraction and gas chromatography tandem mass spectrometry, J. Chromatogr. A 1343 (2014) 18-25.

[46] C. Han, B. Xia, X. Chen, J. Shen, Q. Miao, Y. Shen, Determination of four parabentype preservatives and three benzophenone-type ultraviolet light filters in seafoods by LC-QgLIT-MS/MS, Food

Chem. 194 (2016) 1199-1207.

[47] C. Garcia-Delgado, F. Yunta, E. Eymar, Are physico-chemical soil characteristics key factors to select the polycyclic aromatic hydrocarbons extraction procedure? Int. J. Environ. Anal. Chem. 96 (2016) 87-100.

[48] C. Lesueur, M. Gartner, A. Mentler, M. Fuerhacker, Comparison of four extraction methods for the analysis of 24 pesticides in soil samples with gas chromatography-mass spectrometry and liquid chromatography-ion trap-mass spectrometry, Talanta75 (2008) 284-293.

[49] M. Gorga, S. Insa, M. Petrovic, D. Barceló, Analysis of endocrine disrupters and related compounds in sediments and sewage sludge using on - line turbulent flow chromatography - liquid chromatography-tandem mass spectrometry, J. Chromatogr. A1352 (2014) 29-37.

[50] S. Echeverría, F. Borrull, E. Pocurull, N. Fontanals, Pressurized liquid extraction and liquid chromatography-tandem mass spectrometry applied to determine iodinated X-ray contrast media in sewage sludge, Anal. Chim. Acta 844 (2014) 75-79.

[51] C. Schummer, B. M. Appenzeller, M. Millet, Monitoring of polycyclic aromatic hydrocarbons (PAHs) in the atmosphere of southern Luxembourg using XAD-2 resin-based passive samplers, Environ. Sci. Pollut. Res. 21 (2014) 2098-2107.

[52] F. Mercier, E. Gilles, G. Saramito, P. Glorennec, B. Le Bot, A multi-residue method for the simultaneous analysis in indoor dust of several classes of semi-volatile organic compounds by pressurized liquid extraction and gas chromatography/tandem mass spectrometry, J. Chromatogr. A 1336 (2014) 101-111.

[53] T. J. McGrath, P. D. Morrison, A. S. Ball, B. O. Clarke, Selective pressurized liquid extraction of replacement and legacy brominated flame retardants from soil, J. Chromatogr. A 1458 (2016) 118-125.

[54] Z. Zhang, M. Le Velly, S. M. Rhind, C. E. Kyle, R. L. Hough, E. I. Duff, et al., A study on temporal trends and estimates of fate of Bisphenol A in agricultural soils after sewage sludge amendment, Sci. Total Environ. 515 (2015) 1-11.

[55] M. Souchier, D. Benali-Raclot, D. Benanou, V. Boireau, E. Gomez, C. Casellas, et al., Screening triclocarban and its transformation products in river sediment using liquid chromatography and high resolution mass spectrometry, Sci. Total Environ. 502 (2015) 199-205.

[56] E. M. Robinson, M. Jia, S. J. Trumble, S. Usenko, Selective pressurized liquid extraction technique for halogenated organic pollutants in marine mammal blubber: a lipid-rich matrix, J. Chromatogr. A 1385 (2015) 111-115.

[57] E. M. Robinson, S. J. Trumble, B. Subedi, R. Sanders, S. Usenko, Selective pressurized liquid extraction of pesticides, polychlorinated biphenyls and polybrominated diphenyl ethers in a whale earplug (earwax): a novel method for analyzing organic contaminants in lipid-rich matrices, J. Chromatogr. A 1319 (2013) 14-20.

[58] B. Huerta, S. Rodriguez-Mozaz, C. Nannou, L. Nakis, A. Ruhi, V. Acuna, et al., Determination of a broad spectrum of pharmaceuticals and endocrine disruptors in biofilm from a waste water treatment plant-impacted river, Sci. Total Environ. 540 (2016) 241-249.

[59] J. J. Weisser, M. Hansen, E. Bjorklund, C. Sonne, R. Dietz, B. Styrishave, A novel method for analysing key corticosteroids in polar bear (*Ursus maritimus*) hair using liquid chromatography tandem mass spectrometry, J. Chromatogr. B-Anal. Technol. Biomed. Life Sci. 1017 (2016) 45-51.

[60] D. V. Augusti, E. J. Magalhaes, C. M. Nunes, E. Vieira Dos Santos, R. G. Dardot Prates, R. ssinatti, Method validation and occurrence of dioxins and furans (PCDD/Fs) in fish from Brazil, Anal. Met.

6 (2014) 1963-1969.

[61] D. Zhao, Determination of pentachlorophenol residue in meat and fish by gas chromatography-electron capture detection and gas chromatography-mass spectrometry with accelerated solvent extraction, J. Chromatogr. Sci. 52 (2014) 429-435.

[62] B. Subedi, L. Aguilar, E. S. Williams, B. W. Brooks, S. Usenko, Selective pressurized liquid extraction technique capable of analyzing dioxins, furans, and PCBs in clams and crab tissue, Bull. Environ. Contam. Toxicol. 92 (2014) 460-465.

[63] B. Petrie, J. Youdan, R. Barden, B. Kasprzyk-Hordern, Multi-residue analysis of 90 emerging contaminants in liquid and solid environmental matrices by ultra-high-performance liquid chromatography tandem mass spectrometry, J. Chromatogr. A 1431 (2016) 64-78.

[64] G. Ottonello, A. Ferrari, E. Magi, Determination of polychlorinated biphenyls in fish: optimisation and validation of a method based on accelerated solvent extraction and gas chromatography-mass spectrometry, Food Chem. 142 (2014) 327-333.

[65] S. Moret, M. Scolaro, L. Barp, G. Purcaro, M. Sander, L. S. Conte, Optimisation of pressurised liquid extraction (PLE) for rapid and efficient extraction of superficial and total mineral oil contamination from dry foods, Food Chem. 157 (2014) 470-475.

[66] G. McEneff, L. Barron, B. Kelleher, B. Paull, B. Quinn, A year-long study of the spatial occurrence and relative distribution of pharmaceutical residues in sewage effluent, receiving marine waters and marine bivalves, Sci. Total Environ. 476e477 (2014) 317-326.

[67] V. G. Samaras, A. Giri, Z. Zelinkova, L. Karasek, G. Buttinger, T. Wenzl, Analytical method for the trace determination of esterified 3-and 2-monochloropropanediol and glycidyl fatty acid esters in various food matrices, J. Chromatogr. A 1466 (2016) 136-147.

[68] P. Wardrop, P. D. Morrison, J. G. Hughes, B. O. Clarke, Comparison of in-cell lipid removal efficiency of adsorbent mixtures for extraction of polybrominated diphenyl ethers in fish, J. Chromatogr. B-Anal. Technol. Biomed. Life Sci. 990 (2015) 1-6.

[69] R. Pissinatti, C. M. Nunes, A. G. de Souza, R. G. Junqueira, S. V. C. de Souza, Simultaneous analysis of 10 polycyclic aromatic hydrocarbons in roasted coffee by isotope dilution gas chromatography-mass spectrometry: optimization, in-house method validation and application to an exploratory study, Food Control. 51 (2015) 140-148.

5 超声波：一种用于生物活性物质和污染物的清洁、绿色萃取技术

Ciara McDonnell，Brijesh K. Tiwari*
Teagasc Food Research Centre，Dublin，Ireland
*通讯作者：E-mail：brijesh. tiwari@ teagasc. ie

5.1 引言

萃取过程的发展旨在最大化的纯化目标化合物而不影响其性质。标准的固-液萃取（SLE）技术分别依靠机械或温度增强，例如浸渍和“索氏萃取”，由于它们需要大量的溶剂和较长的处理时间，因此耗费昂贵[1]。工业上最常用的萃取方法主要是化学溶剂和高温的组合[2]。尽管该技术在促进两相之间的传质方面是有效的，但热不稳定的化合物可能在此过程中会受到损害[2]。例如，传统的萃取技术会破坏热敏性的芳香分子[3]或引入杂质化合物而污染目标分子（例如在脂质萃取过程中溶解的叶绿素）[4]。另外，萃取过程中用到的溶剂通常是有毒的并且可能致癌（例如氯化物溶剂），不符合当前安全和清洁的市场趋势[5]。

这种趋势是受消费者驱动的，因此，零售商正在向食品制造商施加压力，要求他们寻找清洁的标签成分替代化学品。零售商通常使用“食橱配料”一词，这意味着在产品开发过程中只能使用家喻户晓的成分和消费者熟悉的成分。但是，某些化学品对于特殊目的（例如提高萃取效率）非常有效。比如，尽管氯仿和甲醇可以有效地萃取脂类，但它们有毒且不符合当前的市场趋势[4]。用清洁的替代品取代化学品是一项挑战，因为这些替代品可能会导致产量下降并提高制造商的生产成本。此外，目标化合物通常会嵌入到基质中并与其他化合物发生相互作用[5]。

新型加工技术的出现与清洁溶剂提供了一种质控系统，可为制造商提供产量和标签问题的解决方案。此外，它还提供安全、低耗费且环保的萃取物[4]。许多新颖的加工技术已评估了其萃取生物活性成分与减少污染物的效果，例如超声辅助萃取（UAE），微波辅助萃取（MAE），超临界流体萃取（SFE），脉冲电场（PEF），高压技术和胶质气体泡沫（CGA）。其中一些技术已成功实现工业化。例如，PEF 已证明可通过引起细胞电穿孔提高葡萄中多酚[6]和花青素[7]的萃取效率，目前研究人员和工业界正在评估其用于商业化的可行性[2]。但是，上述技术也有一些缺点需要克服。尽管 CGA 可以充分分离大分子和小分子，但是萃取物可能在后面阶段被表面活性剂污染[3]。在上述技术中，与常规萃取技术相比，使用超声（US）萃取可提供一种环保、通用、能耗低且产生化合物或残留物毒性较小的解决方案。它可以应用于前处理或 SLE 阶段，也可以与其他萃取技术结合使用以提高效率[8]。超声辅助萃取已在多种食品基质中用于生物活性成分的提取。例如，从干燥小麦谷物蒸馏器[9]，葡萄渣[10]和黑莓[11]中的花青素中提取酚类化合物。另外，它是一种具有高度商业可行性的技术。自 20 世纪 50 年代以来，它就已经用于除萃取以外的其他应用，如清洁，去沫和均质化[12,13]。因此，随着超声波设备的日趋成熟，超声辅助萃取（UAE）技术的商业化应用并不像采用新技术那样具有挑战性。本章论述了超声辅助萃取（UAE）技术作为一种清洁、绿色的萃取技术用于生物活性成分的萃取，可有效减少污染物的最新进展，以及提供包括作用机理、设备和潜在挑战的综述。

5.2 超声辅助萃取的原理

人类认识并研究使用超声波能量来改变反应路径和提高产率的方法已有多年[14]。超声在工业中的应用最早是在 20 世纪 60 年代通过超声波清洗和塑态焊接引入的[13]。许多其他应用很快相继被认可，例如加速和提高结晶、萃取、过滤、冷冻、干燥和乳化[12]。但是，并非所有形式的声能都能有效地引起介质的物理和化学变化。随着声波的传播，它经过压缩和稀疏循环后，当声波以 20~100kHz 的频率传播时才能对介质产生破坏性作用[15]。在这个范围的频率下，介质中会发生许多机械作用，例如冲击波、剪切力、微射流、声流、自由基形成和空化作用[5]。较高频率（> 1MHz）的声能也有应用，但主要用于诊断（表 5.1）。低频高强度超声的关键机械作用是空化作用。

表 5.1　部分声音的频率范围

声音	频率范围
人耳听觉	16Hz~18kHz
功率超声	20kHz~100kHz
扩展功率超声	20kHz~1MHz
诊断超声	5MHz~10MHz

注：改编自 Mason et al.（2005）。

空化作用是高（压缩）和低（稀疏）压力交替施加到介质中微小气泡上的结果。这些气泡增长到不稳定的尺寸大小，直到最终气泡破裂，亦称为声波空化[5]。频率是气泡大小和稳定性的决定因素。对于频率约为 1MHz 的诊断性超声，在稀疏循环中气泡的生长时间仅约 0.5ms，时间太短而不会引起空化作用。另一方面，在约 20kHz 的频率下，由于稀疏循环时间较长，微气泡的生长时间约为 25ms，这会导致气泡不稳定[16]。表 5.2 所示为微气泡在不同声频下的特性。

表 5.2　微气泡在不同频率下的特性[8, 17]

频率/kHz	振幅/mm	声压/atm	波长/cm	破裂时间/μs	脉动气泡的平均直径/μm
20	2.95	5.4	7.42	10	330
500	1.1	5.4	0.29	0.4	13

空化作用有两种类型：非惯性（稳定）和惯性（瞬态）空化[5]。虽然两种气泡类型都可以在介质中引起变化，但正是惯性空化作用导致了反应和萃取过程的加速。当气泡达到不稳定的半径并破裂时，就会发生如图 5.1 所示的惯性空化作用。内爆导致局部温度和压力分别发生 5000K 和 20MPa 的变化[18]。尽管内爆的瞬间对于加速萃取最为重要，但内爆之前发生的进程在萃取过程中也起到了作用。随着气泡的振荡和脉动，

图 5.1 空化作用的示意图[20]

可能会发生微流，从而确保了介质的搅动，进而加速了反应和萃取[16]。声波的压力波动还会引起所谓的“海绵效应”，即介质的挤压和释放。高密度水平的能量还会引起分子运动，从而产生热能，这同样可以提高萃取效率[19]。但是，辅助萃取过程的关键因素是超声微喷流的发展。微喷流是气泡内爆的结果[16]，在爆炸过程中，高动能导致高速度（Suslick，1998）。因此，它会撞击诸如细胞膜之类的固体表面，从而导致机械损伤[16]。因此，超声的联合作用机制可以诱导细胞膜变薄和穿孔，从而使得溶剂流入，这有助于使用清洁溶剂提取生物活性化合物。

为了促使空化作用发生，必须要有气泡的成核位点[21]，空化作用的发生类型取决于介质的特性和多种其他因素。

5.3 超声辅助萃取的影响因素

声能是一门复杂的科学，它受多种参数影响，如波长（m）、频率（Hz）、振幅（μm）、速度（m/s）和强度（W/cm^2）。尽管在空化作用的产生中已经探讨了频率是关键参数，但是许多其他因素也会影响空化作用的效率。这些因素可能归因于声波的性质，实验装置或声波传播通过的介质。超声强度（W/cm^2）是通过功率除以发射面的表面积计算获得[5]。如前所述，随着声波的传播，它经历了高压和低压的振荡。频率描述的是振荡的速率，赫兹（Hz）是描述每秒循环数的计量单位。最后，在谈到超声设备时，对振幅的理解一定不能与声波的振幅相混淆。声波的振幅是振荡引起的压力变化程度。另一方面，超声设备的振幅，例如探针系统的喇叭，是其移动一次距离（m）的量度[22]。各种参数不是相互独立的，比如超声强度与振幅成正比[5]。传输到发射面的功率越大，振幅（mm）就越大。视传播介质而定，可能需要高功率才能进行空化作用。但是，高功率可能会导致发射探针的腐蚀加快。例如，对于高黏度的介质，需要更高的功率。可以理解，为了实现黏性液体的振幅和空化，需要更多的能量。然而，频率和介质黏度之间也存在相互作用。稀疏循环必须足以克服分子之间的内聚力，从而允许气泡生长[8]。其他介质因素如温度，蒸气压和溶解的固体，也会影响超声的效率[23]。温度对超声辅助萃取来说也是一项挑战，以致超声能量可能导致温度升高，进而影响介质或溶剂的性能。溶剂的黏度和表面张力会随着温度的升高而降低，这可能导致气泡破裂的声化学作用减弱[8]。例如，WKhemakhem 等[24]发现超声辅助萃取技术在较低温度下从橄榄叶中提取酚类化合物的效率更高，而在高温下（70℃），超声辅助萃取技术与传统技术没有发现差异。此外，溶剂的其他特性也很重要。在一项评估四种溶剂（水、甲醇、乙醇和正丁醇）对亚麻籽中多酚类化合物在超声辅助萃取中的影响研究中发现，水由于其高蒸气压、低黏度和低表面张力的特性而表现出萃取效率最佳[25]。

而且，实验装置和介质对声波的参数也有影响。实验容器的形状和尺寸将会影响能量的耗散[23]。当一部分超声波撞击固体表面时会被反射，因此建议使用比圆形容器表面积小的平底容器[8]。另外，能量在较厚的容器中也会衰减，因此应避免使用较厚的容器[4]。理解影响超声效率各因素之间的相互作用是推动技术进步和实现商业化的关键。在这种情况下，研究人员报告所有实验参数并准确测量超声的输出量是很有必要的。

在许多情况下，研究人员已应用量化模型来了解超声辅助萃取工艺的动力学过程。通过这种方式可以优化工艺，并且可以在受不同工艺参数影响的基础上量化比较工艺。各种不同的动力学模型已应用于超声辅助传质过程，例如 Naik 模型[24]，Fick 第二定律[26]和二阶模型[27]。Tiwari[5]更详细地讨论了模型的类型。但是，一般情况下这些模型都会同时考虑内在因素（基质特征）和外在因素（超声功率、时间、温度）来量化目标化合物的传质。

5.4 超声能量的量化

如 5.3 节所述，超声能量受多种因素的影响。在超声研究最初的几十年中，超声科学及其在工业中的新应用发展缓慢[14,16]。部分原因是报告单位的变化以及缺乏影响超声能量的所有其他因素（例如试验装置和归因于介质的影响因素），难以进行重复实验。就声波而言，诸如超声强度（W/cm^2），频率（Hz）和振幅（μm）等是文献中最常见的报道参数。另外，现在很常见的是找到一份超声能量的测量报告，具体到每个实验装置。超声能量的测量可以通过多种方法进行，以量热法最为常用[4]。该方法假定所有的超声能量都以热量的形式消散；因此，它仅用作预估。通过测量已知质量和比热容的液体随时间的热变化，能量测量可以将功率输出用超声功率（P）、超声强度（UI）和声能密度（AED）表征，如式（5.1）~式（5.3）所示。该技术对分析报告很有帮助，因为它使得其他研究小组可以复制实验设计并确保所有参数是正确的。同时，它对研究小组中也很有帮助，因为它可以随着时间的推移帮助研究小组检查超声设备是否有效运行，并且没有功率输出损失。

$$P = mC_p\left(\frac{\mathrm{d}T}{\mathrm{d}t}\right)_{t=0} \tag{5.1}$$

$$\mathrm{UI} = \frac{P}{A} \tag{5.2}$$

$$\mathrm{AED} = \frac{P}{V} \tag{5.3}$$

式中，(dT/dt) 是温度随时间的初始变化率（℃/s），C_p 是介质的比热容 [kJ/(kg·℃)]，P 是超声功率（W），m 是样品质量（kg），V 是样品体积（mL）。UI 是样品中释放的能量，A 是超声探头的表面积（cm^2），而 AED 则是每单位体积样品的超声能量[5]。

量热法在用于萃取过程中能量的测定存在部分局限性，因为它假设所有的超声能量都以热量的形式消散，但是这种假设是错误的，因为能量会因吸收、反射和空化而损失[5]。为此，其他方法诸如使用水听器[28]或剂量学[29]分别测量声压或声化学反应速率也被评估过。其他技术也不是没有缺点，比如水听器方法可能会由于空泡内爆导致压力波动增加，进而导致功率增加使得精确度下降[16,28]。剂量学包含测量化学反应速率，一些研究人员选择 Weissler 反应作为评估声功率的一种手段[29]。在该技术中，超声机理导致四氯化碳热解，产生游离氯，进而与碘化物反应释放出碘。通过测量游离碘的量，可以估算出超声能量的效率[29]。

在另一种剂量测定技术中，可以通过测定过氧化氢分子的量。过氧化氢分子是由于水分子的热解而形成的，这些自由基与碘发生反应，进而通过 UV/VIS 分光光度法进行定量测定（Ashokkumar 等，2008）。虽然水听器方法表明随着功率的增大，标准偏差也随之增大[28]，但 Weissler 反应却显示出与量热法具有线性关系[29]。根据超声设备和实验装置可以选择其他适合的方法。例如，由于大体积液体中温度随时间升高缓慢，在超声浴中使用量热法可能会更加困难。

5.5 超声辅助设备的设计与开发

目前存在各种类型的超声设备，具体选择哪种类型取决于实际应用。总体而言，超声设备可分为超声探头和水浴系统。在这两种情况下，换能器（转换器）都是超声能量的源头。发电机将常规电能（50~60Hz）转换为低频能量（通常为 20Hz）。换能器负责将电能转换为机械能的机械振动。换能器可以是磁致伸缩的或压电的，由于后者效率和耐用性更好，因此更为常见。在使用压电换能器的情况下，钛酸锆电致伸缩晶体会因能量供应而发生振动[19]。

换能器的布置位置很重要，因为它会影响能量的损耗，从而影响产率[5]。换能器在水浴和探头系统中最常见的布局位置如图 5.2 所示。

在不同情况下，换能器与样品之间都应避免直接接触；但是，这会导致能量损失，因为能量必须穿过设备材料的壁[5]。在水浴系统中，用于制成浴盆外壳的材料主要是不锈钢[4]，而探头通常由钛制成。金属钛是坚固的，但也允许能量传输到介质而没有高反射率[30]。尽管如此，由于发射面上的空化作用力，钛容易被腐蚀。这可能导致钛沉积到反应容器中，并因此污染产物。目前已经有一些创新方法来阻止沉积并改善能量传递。例如，石英和派热克斯玻璃之类的材料已经被用于测试[31]，并且还存在新的探针尖端形状，如统一的圆柱体、指数形、线性锥形、圆锥形或阶梯形[5]。根据探针发射面的表面积，能量可能会更加集中。虽然这引起了能量的集中，但由于分布不均匀，可能会导致处理不一致。一种可以解决这个问题的方法就是开发连续流动池[14]。

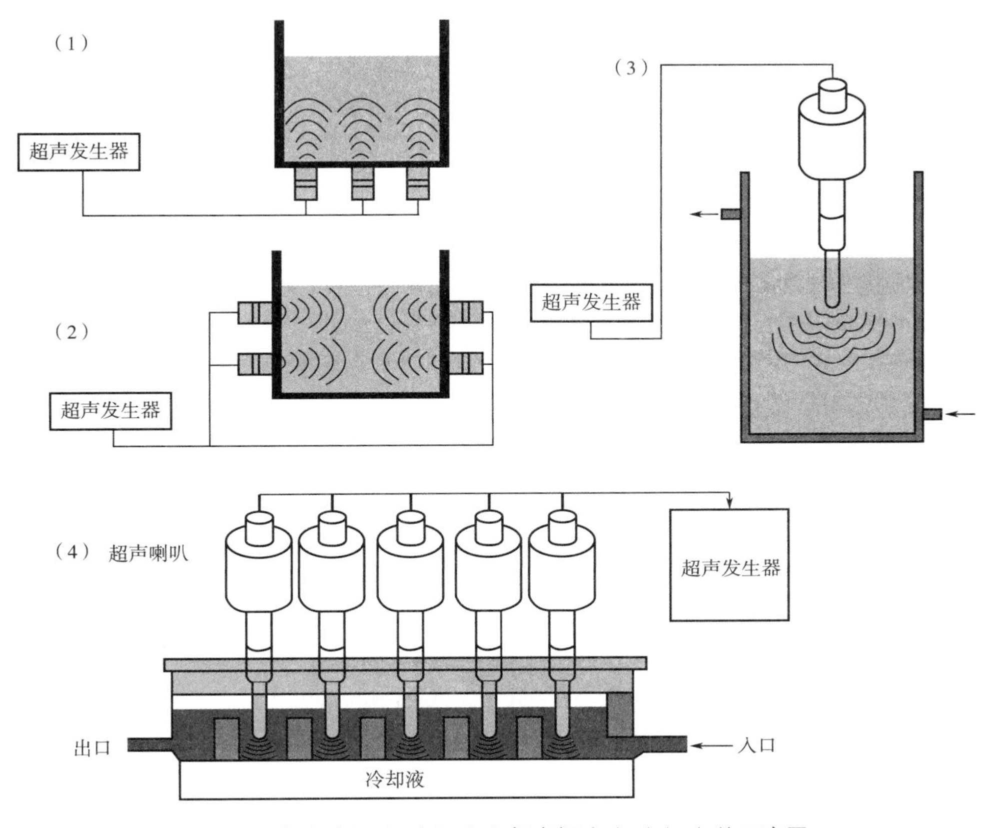

图 5.2 超声水浴（1）（2）和超声探头（3）（4）的示意图

除超声设备的物理特性外，可用设置还决定了其功能性和对各种应用的适用性。探头系统通常以一种频率运行，而部分水浴系统则具有多种设置[5]。当需要加热的时候，许多水浴系统可以进行温度设置（例如，通过加快扩散和基质分解来提高反应速率）。但是，由于高温会破坏目标分子并减弱空化作用（如前所述），许多设备还具有循环冷却系统[5]。随着振幅的增加，搅拌水平会更高，因此温度也会提升的更快。然而，由于黏性材料需要加大功率输入和振幅，因此冷却系统可能需要根据应用进行相应的调整。由于影响超声的变量和因素较多，必须根据材料和目标分子的情况，对超声辅助萃取的不同应用进行优化。

5.6 萃取机理

关于超声食品加工的动力学、系统与基质，最近有许多综述（Tiwari 等[5]；Galanakis[3]；Pingret 等[32]），而 Chemat 等[8]又对机理进行了全面的综述。

空化作用包含几个过程，这些过程可以通过诱导腐蚀、颗粒分解和表面剥落来辅助从基质中萃取分子[4]。Chemat 等[8]对许多超声辅助萃取的研究进行了综述分析，并指出“超声萃取不是通过一种机制起作用，而是通过破碎、腐蚀、毛细管作用、变形和声孔效应之间独立或不同组合的机制进行的”。在气泡破裂之前，脉动的气泡和压力的波动会引起介质的微搅拌。这种搅动确保了离子的流动，这有助于萃取[33]。然而，颗粒尺寸的减小并由此增加表面积也是提高萃取效率的关键。在一项研究中，比较了从菠菜叶中常规萃取（浸渍）和新型萃取（超声辅助萃取）叶绿素分子的过程，结果表明，经超声辅助萃取处理后的菠菜残渣的平均粒径为 200mm，而浸渍处理后的残渣平均粒径为 300mm[4]。在一个非均质系统中，例如浸在液体溶剂中的固体基质，空化的爆裂会影响固体表面并引起固体材料外表面的腐蚀或破裂。在萃取的情况下，这种机械损伤会影响细胞膜，从而使溶剂进入到细胞内并释放出目标分子。通过扫描电子显微镜在玉米面筋粉[34]、亚麻籽[25]、黑莓蔗渣[11]和肉类[26]中观察到了超声对固体天然基质造成的表面损伤和破坏（变形）的证据。这种机械损坏可能是由超声的组合机制引起的，但声孔效应很可能是由于微喷射造成的。如前所述，当气泡达到不稳定的半径时，一侧变得平整，导致液体通过气泡内部流入[35,36]。这形成了被称为微注射系统的超声微喷流[28]。微注射的动力学很复杂，但是一些研究者已经收集了相关的照片证据[16,36]。在膜和固体结构中形成孔会导致液体或溶剂进入材料的流量增加，这被称为超声毛细管效应[4]。Chemat 等[8]证明了超声辅助萃取可以改善传质，并且该过程在处理的最初几分钟内是最有效的。苹果渣在超声萃取的前 10min 其持水能力比浸渍法高 70% 也证实了这一点[4]。就超声辅助萃取在过程开始时最有效而言，这与 Khemakhem 等[24]的发现是一致的，因为 88%的多酚在处理的前 10min 中就被萃取出来了（有关超声辅助萃取的参数，请参见表 5.3）。然而，不仅是膜界面处的空化作用会引起机械损伤，压力波动和气泡内爆的组合机制也会在介质中产生大量剪切应力和能量。如前所述，其效率取决于介质的特性（Corbin 等，2015）。空化的强大作用还可能引起声致发光，从而使分子分裂，产生自由基并加速化学反应[16]。同样地，该过程可以释放目标分子并增强与溶剂的相互作用以提高萃取效率。在描述 Weissler 反应时曾探讨过超声分裂分子的能力[29]。这些机制很可能是组合在一起发生的，没有一个是独立的。但是，Tiwari[5]将其描述为两步现象，首先，溶剂进入基质后破坏了细胞壁，导致表面腐蚀和变形以及粒径减小。第二步，由于表面积增加，内部结构的萃取物被洗出[5]。

5.7 超声辅助萃取生物活性成分的萃取技术与发展

超声辅助萃取在基质中提取天然生物活性成分的应用研究已有多年。自 1949 年从鱼和鱼肝中提取鱼油的专利被授权后（Shropshire，1949 年），就在许多其他基质上进行了研究和综述。例如，表 5.3 所示为最近关于使用超声萃取生物活性成分的报道。

表 5.3　从不同基质中超声辅助萃取生物活性成分

被萃取的化合物	基质	超声参数	参考文献
抗氧化剂	补血草花	溶剂：乙醇 10%~90%（体积比）； 超声功率：400W； 频率：20kHz； 温度：30~80℃； 时间：5~18.2min[B]	[37]
抗氧化剂	石榴皮	溶剂：水； 超声功率：59.2W/cm^2； 频率：20kHz； 温度：25℃； 时间：60min[P]	[38]
抗氧化剂	枣莲果	溶剂：50%乙醇[B]； 超声功率：360W； 温度：63℃； 处理时间：min	[39]
酚类化合物	干燥小麦谷物蒸馏器	溶剂：水[P]； 超声功率：200~600W； 频率：24kHz； 萃取时间：10~40min	[40]
酚类化合物	葡萄渣	溶剂：水[P]； 超声功率：(435±5) W/L； 频率：(55±5) kHz； 温度：35℃、50℃	[10]
类胡萝卜素	马铃薯渣	溶剂：己烷/乙醇（50：50，体积比）； 超声振幅：58~145mm； 温度：25℃； 时间：>10min	[41]
酚类化合物	葡萄渣	溶剂：50%乙醇； 声能密度：6.8~47.4W/L； 频率：250kHz[P]； 温度：20~50℃	[42]
苯酚、抗氧化剂、花青素	葡萄籽	溶剂：乙醇 33%~67%（体积比）； 超声功率：250W； 频率：40kHz； 温度：33~67℃； 时间：16~34min[B]	[43]
酚类化合物	橄榄叶（油橄榄）	溶剂：水； 超声功率：(109.5±1.7) W； 温度：10~70℃； 时间：10min[P]	[24]

续表

被萃取的化合物	基质	超声参数	参考文献
花青素	野樱莓（黑果腺肋花楸）	溶剂：乙醇 0%~50%（体积比）； 超声功率：0~100W； 频率：30.8kHz[P]； 温度：20~70℃； 时间：0~240min	[44]
酚类化合物 黄酮类化合物 花青素 原花青素	野樱莓的副产物	溶剂：50%乙醇； 超声功率：72~216W； 温度：30~70℃； 时间：30~90min[P]	[45]
多酚	云杉木树皮	溶剂：50%与70%的乙醇； 超声功率：400W； 频率：35kHz； 温度：40~60℃[P]； 时间：30~60min	[46]
多酚	墨西哥原生沙漠植物	溶剂：0%~70%乙醇； 频率：40kHz； 温度：40℃、50℃[P]； 时间：20~60min	[47]

注：不同的上标表示水浴（B）或探针（P）。

表 5.4　　超声辅助萃取技术的进展[8]

类别	近期发展
萃取率	减少溶剂的用量
代谢物的类型	减少单元操作
选择性	减少萃取时间
	使用可再生植物资源
	安全
	环境影响
	投资快速回报

在 2005~2015 年间，超声辅助萃取在该领域的研究报道增长了 5 倍[5]。这导致了超声辅助萃取的快速发展，其关键进展在表 5.4 中[8]进行了总结。如前所述，超声的组合机理对它们作用于基质是有利的，并且还可以提高溶剂的效率，进而扩大溶剂的选择范围。当选择溶剂时，重要的是它对目标分子具有良好的适应性，可以溶解分子而不会引入污染。Corbin 等[25]发现，与已刊发的报道相比，用水从亚麻籽中提取酚类物质的效率要比甲醇、乙醇和正丁醇更高。同样，Pasquel Reátegui 等[11]发现，在使用

超声辅助的超临界流体萃取法中，水可以提高黑莓中酚类物质的萃取率，其原因是萃取物是极性物质，如果没有其他极性溶剂，它不会与非极性的 CO_2 发生相互作用。这表明超声辅助萃取使加工者可以使用清洁、绿色且通常被认为是安全的溶剂以提高萃取率，而且还证明了超声辅助萃取可以与其他萃取技术（如超临界流体萃取）组合使用。

许多其他的组合萃取体系已经被报道过了。对于油脂的萃取，最常用的是索氏提取技术。该技术需要使用有机溶剂，如正己烷，尽管它非常有效，但却很耗时。已经证明，在萃取室外使用超声能量可增强传质，进而减少从固体基质中萃取代谢物的时间[48,49]。尽管这表明超声可以减少脂肪酸和油脂的萃取时间，但同时也证明了超声可以提供无溶剂萃取。Adam 等人[50]使用无溶剂超声辅助萃取技术成功地从微藻中萃取出了脂类，比 Bligh-Dyer 法更快，且对脂肪酸没有影响。此外，Kimbaris 等[51]证明超声辅助萃取（35kHz，30min）相比于水蒸气蒸馏法，对从大蒜精油中萃取出热敏感分子的损伤要更小。在必须使用溶剂的情况下，有研究表明有机溶剂（己烷）可用天然替代品替换，例如用葵花籽油从胡萝卜中提取类胡萝卜素[52]。Li 等[52]将超声辅助萃取技术（20kHz，9.5~22.5W/cm^2）应用于萃取胡萝卜中的类胡萝卜素，20min 内类胡萝卜素的产量为 334.75mg/L，而使用己烷萃取在 60min 内类胡萝卜素的产量为 321.35mg/L。同样，Goula 等[27]发现在应用超声辅助萃取技术期间（20kHz，130W，10~60min），使用葵花籽油或大豆油作为溶剂时，可从石榴皮中萃取出高达 93.8%的类胡萝卜素。因此，在萃取亲脂性生物活性化合物的情况下，超声辅助萃取具有加速萃取进程、提高产量、减少溶剂使用量、除污染物以及对分子造成较小损害的应用潜力。当然，这取决于目标生物活性分子的性质。在某些情况下，不同的组合萃取体系可能更具优势。例如，当超声与酶促萃取技术结合使用时，可以提高桑树中植物素的萃取率[53]。其他组合技术包括在 Clevenger 蒸馏，MAE 和压铸工艺中组合使用超声[4]。这些都证明了该技术的多功能性以及在提高生物活性物质萃取率方面具有许多潜在应用。

5.8 结论与展望

超声是一项已经研究了几十年的技术，但是近年来它发展迅速。由于它适合当前趋势，因此受到了广泛的关注。也就是说，这是一种可持续、经济且对环境友好的工艺，能够以高效的方式得到清洁、绿色、浓缩的萃取物。同时它还扩大了天然溶剂或无溶剂萃取的选择范围，从而减少或消除了污染物。此外，它是一种通用技术，可以用于许多食物基质、目标分子及其应用。它可以作为单独的萃取工艺或与其他萃取工艺结合用于前处理，这为生产者提供了一种可以生产出含有害致癌污染物较少的萃取物从而保护人类健康的技术。该工艺安全经济，具有职业优势。超声辅助萃取最大的挑战之一是扩大其规模，这主要是由于本章中概述的影响超声能量的因素有很多。这凸显了测量声场以及准确说明实验装置和超声参数的重要性。任何扩大规模的方式还必须提供比传统技术更持续、安全、经济和绿色的优势。由于对使用小体积的超声探针装置进行了许多研究，因此扩大规模尤为困难。但是，随着超声辅助萃取的发展，连续流动池的出现为克服这一问题提供了一种手段。随着越来越多的研究和综述文章发表，对该领域的了解将继续保持稳定的增长势头，这将有助于超声设备制造商了解其工艺的基本原理，进而制造出适合不同工艺和目标分子的设备。显然，这将为克服超声辅助萃取技术商业化所带来的挑战带来许多优势。

参考文献

[1] L. Wang, C. L. Weller, Recent advances in extraction of nutraceuticals from plants, Trends in Food Sci. Technol. 17 (2006) 300-312.

[2] F. J. Barba, Z. Zhu, M. Koubaa, A. S. Sant'Ana, V. Orlien, Green alternative methods for the extraction of antioxidant bioactive compounds from winery wastes and by-products: a review, Trends in Food Sci. Technol. 49 (2016) 96-109.

[3] C. M. Galanakis, Recovery of high added-value components from food wastes: conventional, emerging technologies and commercialized applications, Trends in Food Sci. Technol. 26 (2012) 68-87.

[4] S. J. Kumar, G. V. Kumar, A. Dash, P. Scholz, R. Banerjee, Sustainable green solvents and techniques for lipid extraction from microalgae: a review, Algal Res. 21 (2017) 138-147.

[5] B. K. Tiwari, Ultrasound: a clean, green extraction technology, TrAC, Trends in Anal. Chem. 71 (2015) 100-109.

[6] R. Soliva-Fortuny, A. Balasa, D. Knorr, O. Martín-Belloso, Effects of pulsed electric fields on bioactive compounds in foods: a review, Trends in Food Sci. Technol. 20 (2009) 544-556.

[7] N. El Darra, N. Grimi, E. Vorobiev, R. G. Maroun, N. Louka, Pulsed electric field assisted cold maceration of cabernet franc and cabernet sauvignon grapes, Am. J. Enol. Vitic. (2013) ajev. 2013. 12098.

[8] F. Chemat, N. Rombaut, A. -G. Sicaire, A. Meullemiestre, A. -S. Fabiano-Tixier, M. Abert-Vian, Ultrasound assisted extraction of food and natural products. Mechanisms, techniques, combinations, protocols and applications. A review, Ultrason. Sonochem. 34 (2017) 540-560.

[9] Z. Izadifar, Ultrasound pretreatment of wheat dried distiller's grain (DDG) for extraction of phenolic compounds, Ultrason. Sonochem. 20 (2013) 1359-1369.

[10] S. Samaram, H. Mirhosseini, C. P. Tan, H. M. Ghazali, Ultrasound-assisted extraction and solvent extraction of papaya seed oil: crystallization and thermal behavior, saturation degree, color and oxidative stability, Ind. Crop. Prod. 52 (2014) 702-708.

[11] J. L. P. Reátegui, A. P. da Fonseca Machado, G. F. Barbero, C. A. Rezende, J. Martínez, Extraction of antioxidant compounds from blackberry (*Rubus* sp.) bagasse using supercritical CO_2 assisted by ultrasound, J. Supercrit. Fluids 94 (2014) 223-233.

[12] T. Mason, L. Paniwnyk, J. Lorimer, The uses of ultrasound in food technology, Ultrason. Sonochem. 3 (1996) S253-S260.

[13] T. J. Mason, Sonochemistry and sonoprocessing: the link, the trends and (probably) the future, Ultrason. Sonochem. 10 (2003) 175-179.

[14] T. Mason, Industrial sonochemistry: potential and practicality, Ultrasonics 30 (1992) 192-196.

[15] T. Leong, M. Ashokkumar, S. Kentish, The fundamentals of power ultrasound—a review, Acoust. Aust. 39 (2011) 54-63.

[16] L. A. Crum, Comments on the evolving field of sonochemistry by a cavitation physicist, Ultrason.

Sonochem. 2 (1995) S147-S152.

[17] C. Pétrier, N. Gondrexon, P. Boldo, Ultrasons et sonochimie, Techniques de l' ingénieur, Sci. Fondam. (2008).

[18] M. Kuijpers, M. Kemmere, J. Keurentjes, Calorimetric study of the energy efficiency for ultrasound-induced radical formation, Ultrasonics 40 (2002) 675-678.

[19] S. Berliner, Application of ultrasonic processors, Int. Biotechnol. Lab. 2 (1984) 42-49.

[20] C. K. McDonnell, The Use of Pulsed Electric Fields and Power Ultrasound for Accelerating the Processing and Improving the Quality of Meat, University College Dublin, 2014.

[21] W. D. O' Brien, Ultrasound-biophysics mechanisms, Prog. Biophys. Mol. Biol. 93 (2007) 212-255.

[22] M. Inc, Power vs. Intensity, Misonix Inc., 2010. http://mm.ece.ubc.ca/mediawiki/images/d/db/PowerIntensity.pdf.

[23] J. Berlan, T. J. Mason, Sonochemistry: from research laboratories to industrial plants, Ultrasonics 30 (1992) 203-212.

[24] I. Khemakhem, M. H. Ahmad-Qasem, E. B. Catalán, V. Micol, J. V. García-Pérez, M. A. Ayadi, M. Bouaziz, Kinetic improvement of olive leaves' bioactive compounds extraction by using power ultrasound in a wide temperature range, Ultrason. Sonochem. 34 (2017) 466-473.

[25] C. Corbin, T. Fidel, E. A. Leclerc, E. Barakzoy, N. Sagot, A. Falguiéres, S. Renouard, J.-P. Blondeau, C. Ferroud, J. Doussot, Development and validation of an efficient ultrasound assisted extraction of phenolic compounds from flax (*Linum usitatissimum* L.) seeds, Ultrason. Sonochem. 26 (2015) 176-185.

[26] I. Siró, C. Vén, C. Balla, G. Jónás, I. Zeke, L. Friedrich, Application of an ultrasonic assisted curing technique for improving the diffusion of sodium chloride in porcine meat, J. Food Eng. 91 (2009) 353-362.

[27] A. M. Goula, M. Ververi, A. Adamopoulou, K. Kaderides, Green ultrasound-assisted extraction of carotenoids from pomegranate wastes using vegetable oils, Ultrason. Sonochem. 34 (2017) 821-830.

[28] J. Cáarcel, J. Benedito, J. Bon, A. Mulet, High intensity ultrasound effects on meat brining, Meat Sci. 76 (2007) 611-619.

[29] T. Kimura, T. Sakamoto, J.-M. Leveque, H. Sohmiya, M. Fujita, S. Ikeda, T. Ando, Standardization of ultrasonic power for sonochemical reaction, Ultrason. Sonochem. 3 (1996) S157-S161.

[30] J. P. Clark, Update on ultrasonics, Food Technol. (2008).

[31] G. Cravotto, L. Boffa, S. Mantegna, P. Perego, M. Avogadro, P. Cintas, Improved extraction of vegetable oils under high-intensity ultrasound and/or microwaves, Ultrason. Sonochem. 15 (2008) 898-902.

[32] D. Pingret, A.-S. Fabiano-Tixier, F. Chemat, Degradation during application of ultrasound in food processing: a review, Food Control. 31 (2013) 593-606.

[33] B. Farouk, N. Hasan, Acoustic wave generation in near-critical supercritical fluids: effects on mass transfer and extraction, J. Supercrit. Fluids 96 (2015) 200-210.

[34] J. Jin, H. Ma, K. Wang, A. E.-G. A. Yagoub, J. Owusu, W. Qu, R. He, C. Zhou, X. Ye, Effects of multi-frequency power ultrasound on the enzymolysis and structural characteristics of corn gluten meal, Ultrason. Sonochem. 24 (2015) 55-64.

[35] E. Brujan, The role of cavitation microjets in the therapeutic applications of ultrasound, Ultrasound Med. Biol. 30 (2004) 381-387.

[36] K. Suslick, Kirk-othmer Encyclopedia of Chemical Technology, vol. 26, J. Wiley&Sons, New York, 1998, pp. 517-541.

[37] D.-P. Xu, J. Zheng, Y. Zhou, Y. Li, S. Li, H.-B. Li, Ultrasound-assisted extraction of natural antioxidants from the flower of *Limonium sinuatum*: optimization and comparison with conventional methods, Food Chem. 217 (2017) 552-559.

[38] Z. Pan, W. Qu, H. Ma, G. G. Atungulu, T. H. McHugh, Continuous and pulsed ultra-sound-assisted extractions of antioxidants from pomegranate peel, Ultrason. Sonochem. 18 (2011) 1249-1257.

[39] K. M. Hammi, A. Jdey, C. Abdelly, H. Majdoub, R. Ksouri, Optimization of ultra-sound-assisted extraction of antioxidant compounds from Tunisian Ziziphus lotus fruits using response surface methodology, Food Chem. 184 (2015) 80-89.

[40] C. Carrera, A. Ruiz-Rodríguez, M. Palma, C. G. Barroso, Ultrasound-assisted extraction of amino acids from grapes, Ultrason. Sonochem. 22 (2015) 499-505.

[41] E. Luengo, S. Condón-Abanto, S. Condón, I. Álvarez, J. Raso, Improving the extraction of carotenoids from tomato waste by application of ultrasound under pressure, Sep. Purif. Technol. 136 (2014) 130-136.

[42] P. Bajerová, M. Adam, T. Bajer, K. Ventura, Comparison of various techniques for the extraction and determination of antioxidants in plants, J. Sep. Sci. 37 (2014) 835-844.

[43] K. Ghafoor, Y. H. Choi, J. Y. Jeon, I. H. Jo, Optimization of ultrasound-assisted extraction of phenolic compounds, antioxidants, and anthocyanins from grape (*Vitis vinifera*) seeds, J. Agric. Food Chem. 57 (2009) 4988-4994.

[44] S. Albu, E. Joyce, L. Paniwnyk, J. Lorimer, T. Mason, Potential for the use of ultrasound in the extraction of antioxidants from *Rosmarinus officinalis* for the food and pharmaceutical industry, Ultrason. Sonochem. 11 (2004) 261-265.

[45] Y. Tian, Z. Xu, B. Zheng, Y. M. Lo, Optimization of ultrasonic-assisted extraction of pomegranate (*Punica granatum* L.) seed oil, Ultrason. Sonochem. 20 (2013) 202-208.

[46] C. Jiang, X. Li, Y. Jiao, D. Jiang, L. Zhang, B. Fan, Q. Zhang, Optimization for ultra-sound-assisted extraction of polysaccharides with antioxidant activity in vitro from the aerial root of *Ficus microcarpa*, Carbohydr. Polym. 110 (2014) 10-17.

[47] J. E. W. Paz, D. B. M. Máarquez, G. C. M. Avila, R. E. B. Cerda, C. N. Aguilar, Ultrasound-assisted extraction of polyphenols from native plants in the Mexican desert, Ultrason. Sonochem. 22 (2015) 474-481.

[48] J. Luque-Garcıa, M. L. De Castro, Ultrasound-assisted Soxhlet extraction: an expeditive approach for solid sample treatment: application to the extraction of total fat from oleaginous seeds, J. Chromatogr. A 1034 (2004) 237-242.

[49] Z. Djenni, D. Pingret, T. J. Mason, F. Chemat, Sono-Soxhlet: in situ ultrasound-assisted extraction of food products, Food Anal. Methods 6 (2013) 1229-1233.

[50] F. Adam, M. Abert-Vian, G. Peltier, F. Chemat, "Solvent-free" ultrasound-assisted extraction of lipids from fresh microalgae cells: a green, clean and scalable process, Bioresour. Technol. 114 (2012) 457-465.

[51] A. C. Kimbaris, N. G. Siatis, D. J. Daferera, P. A. Tarantilis, C. S. Pappas, M. G. Polissiou, Comparison of distillation and ultrasound-assisted extraction methods for the isolation of sensitive aroma compounds from garlic (Allium sativum), Ultrason. Sonochem. 13 (2006) 54-60.

[52] Y. Li, A. S. Fabiano-Tixier, V. Tomao, G. Cravotto, F. Chemat, Green ultrasound-assisted extraction of carotenoids based on the bio-refinery concept using sunflower oil as an alternative solvent, Ultrason. Sonochem. 20 (2013) 12-18.

[53] W. Tchabo, Y. Ma, F. N. Engmann, H. Zhang, Ultrasound-assisted enzymatic extraction (UAEE) of phytochemical compounds from mulberry (*Morus nigra*) must and optimization study using response surface methodology, Ind. Crop. Prod. 63 (2015) 214-225.

6 微波辅助萃取环境中的农药及新兴污染物

Maria Llompart[1,*], Maria Celeiro[1], Carmen Garcia-Jares[1], Thierry Dagnac[2]

1 Universidade de Santiago de Compostela, Santiago de Compostela, Spain

2 Agronomic and Agrarian Research Centre (INGACAL-CIAM), A Coruña, Spain

*通讯作者：E-mail: maria.llompart@usc.es

缩略语

2D	二维
ACN	乙腈
AMP	大气颗粒物
APEOs	烷基酚聚氧乙烯醚
BFRs	溴化阻燃剂
[BMIM][PF6]	1-丁基-1,3-甲基咪唑鎓六氟磷酸盐
BSTFA	*N*，*O*-双（三甲基甲硅烷基）三氟乙酰胺
[C_4MIM][BF4]	1-丁基-3-甲基咪唑鎓四氟硼酸盐
CPE	浊点萃取
DAD	二极管阵列检测器
DCM	二氯甲烷
DDD	二氯二苯基氯乙烷
DDE	二氯二苯基二氯乙烯
DDT	二氯二苯基三氯乙烷
DEHP	邻苯二甲酸二（2-乙基己基）酯
DLLME	分散液液微萃取
E1	雌酮
E2	17-β-雌二醇
ECD	电子捕获检测器
EDC	内分泌干扰物
EE2	17-α-炔雌醇
EP	新兴污染物
EPA	美国环境保护署
EQS	环境质量标准
ESI-ITMS	电喷雾离子阱质谱
EU	欧盟
FL	荧光检测仪
FPD	火焰光度检测仪
FRs	阻燃剂
FUSLE	聚焦超声固液萃取
GC	气相色谱
GPC	凝胶渗透色谱
HBCD	六溴环十二烷
HPLC	高效液相色谱
HPV	高产量

HTAB	十六烷基三甲基溴化铵
IL	离子液体
LAS	直链烷基苯磺酸盐
LC	液相色谱
LDTD-APCI	激光二极管热脱附-大气压化学电离
LiNTf2	双（三氟甲磺酰基）亚胺锂
LLME	液液微萃取
LOD	检出限
LOQ	定量限
LPME	液相微萃取
MA	微波萃取
MAE	微波辅助萃取
MAME	微波辅助胶束萃取
MASE	微波辅助溶剂萃取
MeOH	甲醇
MRM	多反应监测
MS	质谱
MS/MS	串联质谱
MSPD	基质固相分散
NP	壬基酚
NP1EC	壬基酚单羧酸盐
NPEO2	壬基酚二乙氧基化物
NPEOs	壬基酚聚氧乙烯醚
NSAIDs	非甾体类抗炎药
OCPs	有机氯农药
OP	辛基酚
OPEOs	辛基酚聚氧乙烯醚
OPhs	有机磷酸酯
OPPs	有机磷农药
PAEs	邻苯二甲酸酯
PAHs	多环芳烃
PBBs	多溴联苯
PBDEs	多溴二苯醚
PCP	个人护理产品
PDMS	聚二甲基硅氧烷
PDMS/DVB	聚硅氧烷/二乙烯基苯
PFOA	全氟辛酸
PFOs	全氟辛烷磺酸盐

PLE	加压液体萃取
POLE	聚氧乙烯月桂醚
POP	优先有机污染物
PUF	聚氨酯泡沫塑料
RSD	相对标准偏差
SAs	磺酰胺
SDSE	同时蒸馏溶剂
SE	溶剂萃取
SFE	超临界流体萃取
SFO	漂浮有机液滴凝固
SPE	固相萃取
SPMD	半透膜装置
SPME	固相微萃取
STP	污水处理厂
TEHP	三（2-乙基己基）磷酸酯
THF	四氢呋喃
TnBP	磷酸三正丁酯
UAE	超声辅助萃取
UAEME	超声辅助微波萃取
UHPLC	超高效液相色谱
USAEME	超声辅助乳化微萃取
UV	紫外检测器
VOCs	挥发性有机化合物
VWD	可变波长检测器
W	瓦
WWTP	污水处理厂

6.1 引言

样品制备在化学分析中起着重要的作用。大多数分析仪器不能对样品进行直接测定，需要一些预处理来萃取和分离分析物[1]。鉴于通常需要使用有机溶剂，样品制备过程被认为是分析过程中最易产生污染的步骤。采用传统溶剂萃取法来提取和分离分析物，会将溶剂释放于环境中，从而引发严重的污染。故此，目前对分析物的萃取和分离主要有以下趋势：减少溶剂消耗、提高提取通量（例如可自动化）、更高的回收率、更好的再现性。

基于上述背景，研究者开发了一些绿色环保的溶剂萃取方法，其中包括微波辅助萃取（microwave-assisted extraction，MAE），也称微波辅助溶剂萃取[2,3]。在 MAE 过程中，微波可促进溶质从固体样品基质转移至溶剂中，提取速度快、提取通量高，效率可与经典方法相媲美，甚至更高。从绿色化学的角度来看，MAE 的主要优势包括：显著减少所需溶剂和样品量，以及提取时间，从而减少废物的产生和能源消耗，减少溶剂释放到环境中，降低人类暴露的风险[4-6]。

微波在样品制备中的使用最早始于20世纪70年代，被用于高压条件下消解待进行元素分析的样品[7]。1986年，采用家用的微波设备，Ganzler等人首次将微波用于有机化合物的萃取[8]。1991年，首个采用微波萃取天然产物的专利（US 5002784 微波辅助萃取天然产物）由Pare等人提出。20世纪90年代，随着商用实验室微波萃取仪的问世，MAE技术，尤其是在污染物和天然产物萃取领域逐渐流行起来，相关文章的发表数量在2008年达到高峰。在环境分析中，采用MAE法提取环境样品中的污染物也引起了人们的广泛兴趣。起初，MAE主要被用于土壤及沉积物中的农药、多环芳烃和对氯联苯的萃取及测定[8-11]，之后逐渐用于萃取其他多种多样的环境污染物[12-14]。美国环境保护署方法3456“微波萃取”，可有效提取半挥发性有机化合物，如有机磷农药、有机氯农药、含氯除草剂、苯氧羧酸除草剂、取代苯酚、多氯联苯、多氯代二苯并二噁英、多氯代二苯并呋喃等，并将萃取液用于后续各种各样的色谱分析过程[15]。经分析人员证明，该方法也可适用于其他目标物的萃取[16]。

近来，利用微波从环境基质中提取新兴的污染物引起了人们极大的兴趣[6,17]。由于相对易于对影响提取的参数进行优化、易于自动化、效率高和样品通量大的特点，MAE已成功应用于各类有机污染物（如阻燃剂、表面活性剂、医药和个人护理产品）。

在仪器配置方面，实验室用于提取的微波系统主要有两种形式：密闭提取容器/多功能微波炉和开放式聚焦微波炉。这两种技术通常分别称为加压MAE和聚焦MAE（FMAE）。两种系统如图6.1所示。

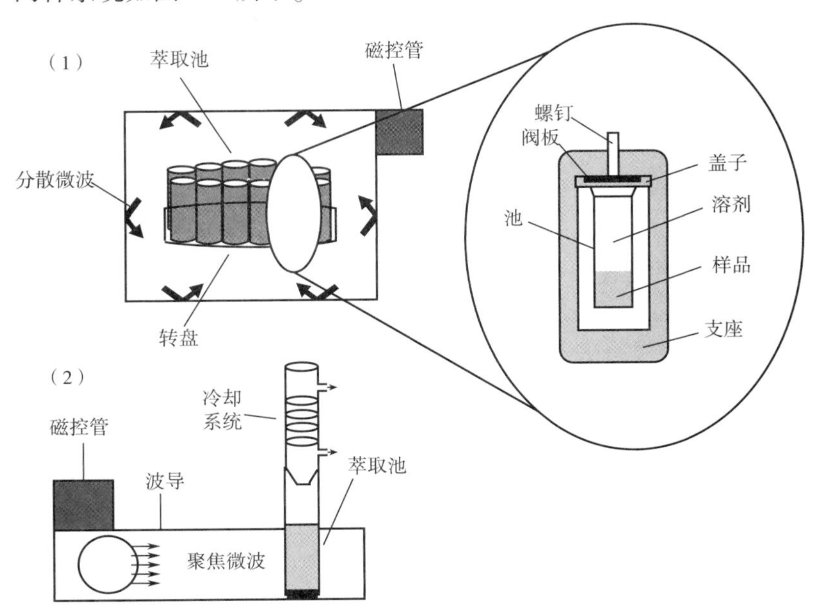

图6.1 （1）加压微波辅助萃取和（2）聚焦微波辅助萃取设备示意图

［资料来源：L. Sanchez-Prado, C. Garcia-Jares, T. Dagnac, M. Llompart, Microwave-assisted extraction of emerging pollutants in environmental and biological samples before chromatographic determination, TrAC-Trend. Anal. Chem. 71 (2015) 119—143.］

加压 MAE 过程是由施加的压力和温度控制的。因此，需要对萃取温度进行优化，一般设置在 60~120℃范围内。而压力是由萃取温度、溶剂类型和体积、样品特性和称样量大小共同决定的。在聚焦 MAE 中，只在萃取容器的含样品部分进行聚焦微波辐射，萃取温度受限于萃取溶剂的沸点，且萃取是在常压下进行的，因此减少了爆炸风险。聚焦 MAE 是在开放的系统中进行，故可使用较大的样品量，且在分析之前不需要再冷却样品。但该方法会导致挥发性物质的损失，可能受到来自空气中的潜在污染，也不能同时处理多个样品，且获得与封闭萃取体系相当的萃取效果需要的萃取时间更长[18]。

加压 MAE 在密闭体系中进行，使用的高温高压可能会导致分析物分解。商品化 MAE 设备的设计通常包括一些安全机制，以避免超压并可检测溶剂泄漏。另一方面，加压 MAE 的萃取效率总体上优于聚焦 MAE，一次可以处理 12 个样品。相比之下，聚焦 MAE 在大气压下操作，安全性更高，而且在萃取期间的任何时间都可以加入任何试剂。

MAE 使用的萃取溶剂通常局限于能够吸收微波的溶剂，即具有永久偶极的溶剂。然而，通过使用偶极和非偶极的混合溶剂，可将 MAE 的应用扩展到更广泛的分析物（极性和非极性）上。

与传统溶剂萃取相比，MAE 的使用可显著缩短萃取时间，因为微波可直接加热萃取溶液，而传统溶剂萃取在热量传递到溶液之前，需要一定的时间先对萃取容器加热。可见，微波的使用可使温度梯度最小化，并加快了加热速度。此外，如前所述，MAE 可同时处理多个样品。因此，MAE 在很大程度上都能满足现代样品处理技术所要求的最低标准，是一种可替代传统萃取技术的、颇具吸引力的萃取技术。

6.2 微波辅助萃取原理

在 MAE 中，微波被用于加热液体样品或与固体样品接触的溶剂，以促进分析物从样品基质中分配到萃取溶剂中。微波是由互相垂直的电场和磁场组成的电磁波，频率在 300~300000MHz，可通过离子迁移和偶极子旋转来引起分子的运动。磁场是波对物质的直接作用，它能吸收一部分电磁能并将其转化为热量[4,19,20]。

微波通过离子传导和偶极旋转的双重机制对分子进行加热。离子传导和偶极子旋转通常在溶剂和样品中同时发生，可有效地将微波能量转化为热能。由于介质对离子流的阻力，离子传导会产生热量。溶解离子的迁移可引起分子间的碰撞，因为离子方向可一直随着磁场方向的变化而变化。偶极旋转与极性分子的交替运动有关，这些极性分子试图与电场对齐。分子激荡产生的多次碰撞而释放能量，因此使得温度升高。样品在微波场中产热需要介电化合物的存在。介电常数越大，释放的热能就越多，给定频率下的加热速度也就越快。只有具有永久偶极子的材料或溶剂在微波下会被加热。事实上，样品中产生的热量部分取决于耗散因子（$\tan\delta$），它反映了不同溶剂在微波下加热的效率。耗散因子是样品的介电损耗（即损耗因子 ε''）与介电常数（ε'）的比值。

当微波吸收与永久偶极子被电场重定向一起发生时，吸收的能量与表示样品吸收微波能力的溶剂介电常数（ε'）成正比。损耗因子表示样品耗散吸收能量的能力。因此，微波能量的影响很大程度上取决于溶剂和样品基质的性质。大多数情况下，所选择的萃取溶剂具有较高的介电常数，因此它能强烈地吸收微波能量。具有高介电损耗的化合物主要是极性化合物。然而，在某些情况下，只有样品基质可以被加热，从而使溶质在冷溶剂中释放，这对热不稳定物质的萃取十分有利，可有效防止其降解[21]。例如，非极性溶剂如己烷和氯仿，因其不吸收微波，故不产生或产生极少的热量[22]。表 6.1 给出了几种 MAE 常用溶剂的特性参数[6,17,18]。

表 6.1　微波辅助萃取常用萃取溶剂的介电性质（25℃，2450MHz）、沸点、最高密闭温度

溶剂	ε'	ε''	$\tan\delta$（$\times 10^4$）	沸点/℃	最高密闭温度/℃
丙酮	21.1	11.5	5555	56.2	164
丙酮/正己烷（1∶1，体积比）	—	—	—	52	156
丙酮/石油醚（1∶1，体积比）	—	—	—	39	147
乙腈	37.5	2.3	620	81.6	194
四氯化碳	2.2	0.00088	4	—	—
二氯甲烷	8.9	0.042	472	39.8	140
乙醇	24.3	6.1	2500	78.3	164
乙酸乙酯	6.02	3.2	5312	77	—
正己烷	1.89	0.00019	0.10	68.7	—
甲醇	23.9	15.3	6400	64.7	151
NaCl 水溶液（0.1mol/L）	75.5	18.1	2400	—	—
NaCl 水溶液（0.5mol/L）	67.0	41.9	6250	—	—
四氢呋喃	7.58	—	—	66	—
水	76.7	12.0	1570	100	—

6.3 微波辅助萃取优化

微波辅助 MAE 的主要影响因素有：溶剂和样品基质的性质、溶剂体积、微波功率、萃取时间、样品称样量及含水率、温度[21,23,24]。通过合理的实验设计，可减少评估这些因素对 MAE 过程的影响并确定最佳萃取条件所需的实验次数。因 MAE 受到许多因素的影响，且这些因素间可能相互作用，故对 MAE 过程的优化需要更先进的优化手段。为了评估这些因素间的相互作用，也可使用多变量法，如效应面优化法（response surface methodology，RSM）。RSM 包括应用统计和数学知识来开发、改进和优化实验设计过程[13,25]。

MAE 中的温度取决于溶剂吸收微波的能力以及所使用的微波的功率。当温度升高时，溶剂对目标物的增溶能力增大，而溶剂的表面张力和黏度则下降，使溶剂黏度降低，从而改善样品润湿性和基质渗透性。密闭容器中温度的影响表明：当溶剂温度从60℃升高到120℃时，可显著提高萃取效率。萃取时间的延长可提高分析物的萃取量，但同时也面临着热不稳定组分降解的风险。

影响MAE萃取效率的另一基本因素是溶剂的选择。合适的萃取溶剂应对目标化合物具有较高的选择性，排除其他基质组分的干扰，并与进一步的色谱分析具有良好的兼容性。溶剂的选择不仅取决于目标化合物的溶解度和溶剂与基质的相互作用，还取决于决定其微波吸收性能的溶剂介电常数。

正己烷是一种不吸收微波的溶剂，而甲醇是一种能很好地吸收微波的溶剂。为了获得最优的微波辅助萃取效率，目标物的萃取机理可分为三种：

①机理Ⅰ　以吸收微波强的单一溶剂或混合溶剂作萃取剂；

②机理Ⅱ　以高、低介电损耗溶剂按不同混合比组成的混合溶剂作萃取剂；

③机理Ⅲ　以不吸收微波的溶剂萃取具有高介电损耗的样品（即含水量高的样品）。

近来，一些其他溶剂如离子液体，可有效吸收和传递微波能量，也被用于MAE中，以萃取农药、环境污染物以及天然化合物。离子液体由大量有机阳离子和无机或有机阴离子组成，室温下为液态。由于其具有蒸气压可忽略、热稳定性好、液态范围宽、黏度可调、与水和有机溶剂的互溶性好、可萃取大多数有机物等优点，离子液体在各应用领域引起了广泛的研究兴趣。因此，离子液体已被用作传统挥发性有机溶剂的绿色替代品[38]，尽管目前离子液体的绿色程度尚不清楚，但Armenta等人在《绿色分析化学：绿色萃取技术的作用》这一章中对此问题进行了进一步的讨论[38a]。

在传统的萃取技术中，较大体积的溶剂通常会增加分析物的回收率。然而，在MAE中，同样的方法可能会导致回收率降低，这可能是由于溶剂与基体的混合不充分导致的。溶剂体积的选择取决于样品的类型和大小，但一般而言，溶剂的用量可能比经典萃取中使用的低10倍左右[21]。

一般而言，溶剂与样品的比值（S/F）越高，回收率越高。通常，S/F值的范围在10∶1到50∶1之间，但必须对S/F进行适当调整和优化，以适应不同的样品。目标物所在的基质的性质对萃取效率具有重要影响。样品中含有或添加的水是另一重要影响因素，因为水分子的偶极矩较大，能有效加热样品。但相应地，也必须对含水量进行控制以获得可重复的结果。其他基质成分，如含铁成分，会因其对微波能量的吸收而引起拱效应。基质中含有的有机碳是已知的对萃取不利的因素，其可与分析物间形成强烈的相互作用，而微波能有效地破坏此类相互作用[13]。

微波功率和相应的微波辐射时间这两者间的相互影响很大，它们的选择取决于所使用的样品和溶剂的类型。原则上，使用高强度的微波时，可适当减少辐射时间。然而，有些情况下，微波功率太高时，可能会导致样品降解或溶剂（开放体系）快速沸腾，从而降低萃取效率。通常，超过最佳范围后萃取时间的延长并不会提高萃取效率，甚至可能降低热不稳定分析物的回收率。在大多数情况下，温度升高可提高萃取效率，这是由于增加了扩散到样品基质内部的溶剂，同时促进了分析物从样品基质活性位点的解吸。但是，在分析热不稳定化合物的时候，应当特别注意温度的选择。

6.4 环境中农药及新兴污染物的分析

农药包括杀虫剂、除草剂和杀菌剂，用于控制水果、蔬菜或谷类中的害虫。农药的使用可提高农作物产量，是现代农业生产质量、产量和安全的需求。然而，这些化合物在环境中的残留造成了严重的环境问题。欧盟（EU）已制定了若干指令，以确保良好的农业规范，并控制农药在环境中的使用和残留[26,27]。

新兴污染物是指现有环境质量法规未涵盖的化合物，它们尚未被充分研究，但其可能对生态系统和人类健康带来威胁[28]。新兴污染物包括许多类化合物，如药物、兽药、个人护理产品、类固醇和激素、表面活性剂、全氟化合物（全氟辛烷磺酸钠、全氟辛酸等)、阻燃剂（多溴联苯醚)、工业添加剂和制剂、汽油添加剂、纳米颗粒。

作为人类、农业和工业活动的结果，农药和新兴污染物可以通过多种途径进入环境，并残留在环境中，其过程示意图参见图 6.2。

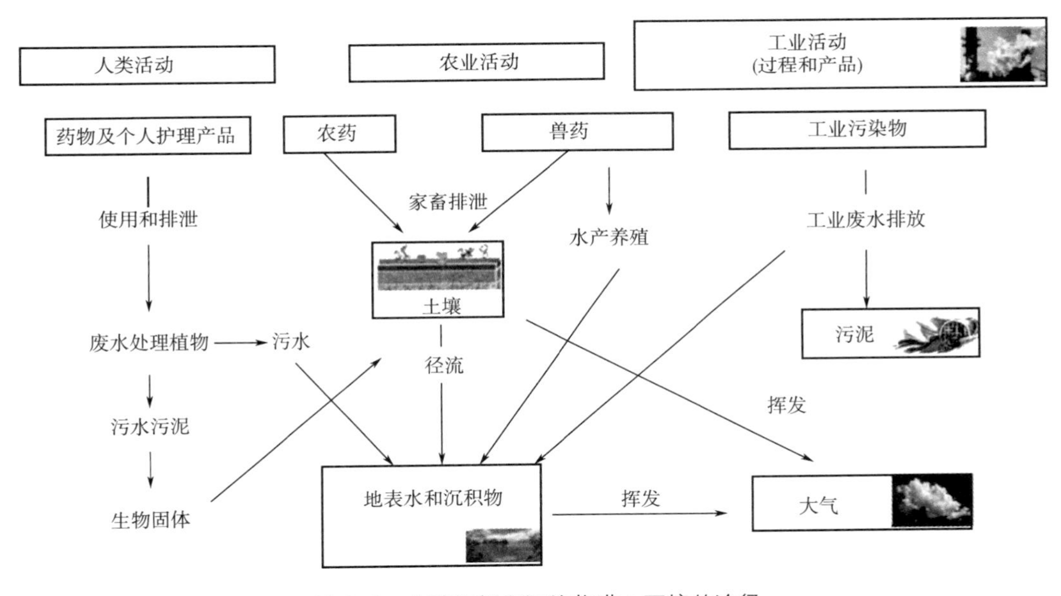

图 6.2 农药和新兴污染物进入环境的途径

由于其效率高、时间短、溶剂消耗少的优点，MAE 在萃取新兴环境污染物尤其是农药中的应用引起了人们的广泛关注。由于环境样品基质复杂，但同时又需要对痕量的农药及新兴污染物进行监控，因此需要通过前处理来对目标物进行萃取和分离。样品前处理阶段在对环境样品基质中农药和新兴污染物的检测中起着重要作用。大多数情况下，必须对从基质中分离出目标化合物进行富集，以达到所需的检测限（LOD）。

经典的分析过程通常包括各种各样的样品制备步骤，如萃取、过滤、纯化及氮吹等，如果最终进行生物测定或气相色谱法测定，可能还需要进行衍生化[28-31]。

常规萃取技术，如液液萃取、索氏萃取、Soxtec、超声辅助提取、机械振荡等，常被用于从固体或液体环境样品中萃取农药和新兴污染物。然而，冗长烦琐的分析过程、样品和试剂的大量消耗以及产生大量的废物（包括溶剂）是这些常规技术的共同缺点。

在过去的20年里，新型绿色萃取技术，如固相萃取、固相微萃取、加压液相萃取、超临界流体萃取、基质分散固相萃取及微波辅助萃取，逐渐被发展起来以克服常规萃取技术的上述缺点。这些绿色萃取技术已被成功用于从固体、半固体、液体样品中萃取各种各样的农药及新兴污染物。这些萃取技术大多数缩短了萃取时间，减少了溶剂消耗，甚至达到零溶剂，提高了萃取通量（甚至有些可自动化）。至于分析方面，这些萃取技术通常可与GC和配备紫外和二极管阵列检测器的LC联用。近年来一些选择性更好、灵敏度更高的检测器如质谱、串联质谱、离子阱质谱的使用，可实现许多环境基质中痕量农药和新兴污染物的定量分析。

6.5 微波萃取在农药分析中的应用

自20世纪90年代首次商用设备开始使用以来，微波辅助萃取（MAE）已经应用于所有种类的农药，包括有机氯农药（OCPs）、有机磷农药（OPPs）、三嗪类、拟除虫菊酯类、尿素衍生物等。最初的应用是萃取和分析单个或少数农药；后来，用于多种杀虫剂的萃取，而在过去十几年中，提出了用几种方法来同时提取多类杀虫剂。在大多数研究中，最终采用气相色谱和液相色谱结合质谱检测器或串联质谱进行分析。基于非有机溶剂的离子液体或与其他萃取技术结合的新方法也已开发。MAE在农药萃取中的应用如表6.2所示。

6.5.1 有机氯农药

有机氯农药（OCPs）是对人类和环境有害的半挥发性持久性有机污染物（persistent organic pollutants，POPs）。OCPs难于生物降解和（光）化学降解。此外，OCPs还可长距离迁移，已在未使用OCPs的地区检测到其中的一些OCPs。虽然大多数OCPs已被禁用，但他们在农业土壤中的残留水平依然较高。本小节将专门介绍基于MAE的OCPs分析方法。

DDT（Dichlorodiphenyltrichloroethane）是被研究最多的一种OCPs，它是20世纪40年代开发出的第一种合成杀虫剂，对斑疹伤寒、疟疾等疾病有奇效。后来，DDT作为一种杀虫剂被广泛使用，直到20世纪70年代，由于其对环境的影响以及对人体健康的潜在风险[32]，美国国家环境保护局（EPA）禁止了它的使用。MAE被用于从固体或水样中单独萃取DDT或其主要代谢物DDD、DDE，或从固体样品中与其他OCPs同时萃取。

在MAE之后，常需要进一步采用固相萃取（SPE）等对萃取物进行净化并浓缩目标分析物。通过此方式，MAE之后进行SPE，并以GC-MS检测，可同时分析沉积物和土壤中的OCPs[33,34]。在这两个案例中，分别以四氢呋喃/正己烷（9∶1，体积比）[32]和丙酮/正己烷（1∶1，体积比）[33]为萃取溶剂，样品分别用微波处理90min和30min。两个情况下的回收率均较为满意，且相对标准偏差较低，所有目标物的检出限均在

表 6.2 MAE 在有机氯农药（OCP）及其他农药萃取中的应用

分析物	样品	样品预处理	MAE 条件	萃取液处理	检测方法	回收率/%	LOD	RSD/%	发表年份	参考文献
OCPs										
18 OCPs	沉积物（5g）	风干、研磨、过筛	15mL 四氢呋喃/正己烷（9∶1，体积比）；MAE（800W，90℃，9min）	过滤、氮吹、正己烷复溶、SPE 净化（丙酮/正己烷作洗脱液）	GC-MS	73.4~119	1.0~2.2ng/g	< 14	2015	[33]
15 OCPs	土壤（2g）	均质、冻干、过筛	丙酮/正己烷（1∶1，体积比）；MAE（600W，30min）	SPE 净化（正己烷作洗脱溶剂）	GC-MS	70~139	0.01~0.06ng/g	< 10	2012	[34]
18 OCPs	沉积物（5g）	干燥、过筛	10mL 甲醇；MA-HS-SPME（80℃，PDMS 纤维）	—	GC-ECD 和 GC-MS	—	0.01~0.26ng/g（GC-ECD） 0.005~0.11ng/g（GC-MS）	—	2008	[35]
6 OCPs	淤泥（2g）	风干	8mLPOLE；MAE（25℃，10min）；SPME（PDMS 纤维）	甲醇解吸（8min）	LC-UV	79.8~117	28~136ng/g	< 10	2008	[36]
DDT 及其代谢物	水样（田间、海、河）（10mL）	—	LPME（4μL 正辛醇）；MAE（249W，6.5min，34℃）	—	GC-ECD	93.5~101	20~30ng/L	< 11	2011	[37]
DDT 及其代谢物	土壤（1g）	风干、过筛	5mL [BMIM] [PF6]；MAE（400W，10min，70℃）	离心、过滤	LC-UV	72.6~99.2	< 0.07μg^{-1}	< 10	2013	[38]
DDT 及其代谢物	土壤（2g）	2mL 0.2mol/L NH_4Cl（振摇）	30mL 丙酮/正己烷（1∶1，体积比）；MAE（4min）	过滤，加 100mL 水、20mL NaCl 溶液，有机相吹干并复溶	GC-ECD 和 GC-MS	73.0~116	0.005mg/kg	< 15	2015	[39]

续表

分析物	样品	样品预处理	MAE 条件	萃取液处理	检测方法	回收率/%	LOD	RSD/%	发表年份	参考文献
10 OCPs	土壤（5g）	冻干、过筛	30mL 石油醚/丙酮（1∶1，体积比）；MAE（800W，5min 升至 100℃ 并保留 10min）	过滤、Na_2SO_4 除水、SPE 净化（丙酮/正己烷作洗脱液）	GC-MS/MS	95.5~108	0.33~1.08μg/kg	< 4	2015	[40]
10 OCPs	土壤（5g）	风干、过筛	30mL 丙酮/正己烷（1∶1，体积比）；MAE（1200W，110℃，10min）	过滤、净化（氮吹至 1mL 以正己烷复溶）	GC-ECD	—	0.19~0.54ng/g	—	2007	[41]
OPPs 6 OPPs	土壤（1g）	均质	1mL 水/甲醇（1∶1，体积比）+ KH_2PO_4（0.02M）；MAE（500W，10min）	OPPs 氮吹并以正己烷复溶	GC-FPD	> 73	0.0004~0.012μg/g	< 11	2007	[42]
9 OPPs	土壤（1g）	干燥、均质	5mL 水；MAE（226W，2.5min）	USAEME（20μL 甲苯，30s US，3200r/min 离心 3min）	GC-ECD	91.4~101	0.04~0.13ng/g	< 8.2	2013	[43]
三嗪类衍生物 阿特拉津和马拉松	土壤（2g）	空气干燥	丙酮/正己烷（3∶2，体积比）	离心	GC-MS	—	< 5mg/kg	—	2011/2010	[44，45]
阿特拉津和 3 OCPs	土壤（0.5g）	均质	17.5mL 正己烷和 7.5mL 丙酮；MAE（1200W，24℃/min 升至 120℃ 并保持 20min）	过滤，氮吹、甲醇复溶	GC-ECD	70.0~85.1	12~18μg/mL（OCPs）；2μg/mL（阿特拉津）	< 5.3	2013	[46]
4 种三嗪类农药	土壤（3g）	风干、均质	50mL 甲醇/乙腈/乙酸乙酯（5∶3∶1，体积比）；MAE（850W，4min）	过滤、甲醇洗、离心	HPLC-UV	83.3~96.3	0.16~0.30μg/mL	< 8	2011	[47]
4 种三嗪类农药	水（5mL）	—	40μL [C4MIM] [BF4]；MAE（30W，90s，50℃）	DLLME	HPLC	88.4~114	0.52~1.3μg/L	< 6.2	2012	[48]

尿素类衍生物										
8 种尿素类和碳酸酯类农药	土壤（5g）	风干、过筛	20mL 乙腈；MAE（70℃，10min）	过滤、氮吹、乙腈复溶	HPLC-DAD	107~120	0.65~3.97μg/kg	< 2.9	2008	[49]
苯脲类和三嗪类农药	土壤（1g）	干燥	4mL［C4MIM］［BF4］；MAE（225W，50℃，10min）	过滤	HPLC	84.0~101	0.027~0.066mg/L	< 6	2012	[50]
拟除虫菊酯类										
8 种拟除虫菊酯类和 5 种 OCPs	沉积物（5g）	干燥、过筛	100mL 正己烷/丙酮（1∶1，体积比）；UAEME（微波功率 100W、超声功率 50W，10min）	SPE（二氯甲烷作洗脱溶剂）	GC-MS	65.2~128（OCPs）；71.0~104（拟除虫菊酯）	0.27~0.70 ng/g（干重）	<18（OCPs）；< 23（拟除虫菊酯）	2010	[51]
种拟除虫菊酯类和 OCPs	土壤（5g）	风干、破碎、过筛	25mL 乙酸乙酯+1mL 水；MAE（110℃，5min）	离心、氮吹、乙酸乙酯复溶	GC-ECD	90~110	0.4~2.0μg/g	< 10	2001	[52]
其他农药										
二甲戊灵	土壤（10g）	风干、过筛、均质	50mL 丙酮；MASE（60℃，3min）	过滤、氮吹、乙腈复溶	HPLC-UV	97	0.059μg/mL	< 3	2011	[53]
甲草胺、异丙甲草胺及其酸性代谢物	土壤（10g）	风干、过筛	50mL 甲醇/水（1∶1，体积比）；MAE（100℃，20min）	SPE（C_{18}），分两部分分析	GC-MS（甲草胺、异丙甲草胺）；HPLC-UV（代谢物）	>71	10~50μg/kg	< 10	2007	[54]
7 种杀菌剂	土壤（0.5g）	NaCl 水溶液（10%，质量分数）	5mL 乙酸乙酯；MAE（1600W，90℃，15min）	过滤、氮吹、200μL 甲醇复溶	HPLC-VWD	72.4~99.4	0.0006~0.0015μg/g	< 12	2014	[55]
噻虫嗪	土壤（10g）	干燥	40mL 水；MAE（500W，1min）	离心、SPE（乙腈作洗脱液）	HPLC-DAD	82.1~87.0	0.029μg/g	< 3	2012	[56]

注：DAD—二极管阵列检测器；DLLME—分散液液微萃取；ECD—电子捕获检测器；FPD—光火焰检测器；GC—气相色谱；HPLC—高效液相色谱；HS-SPME—顶空固相微萃取；LC—液相色谱；LOD—检出限；LPME—液相微萃取；MA—微波辅助；MAE—微波辅助萃取；MS—质谱；MS/MS—串联质谱；OCPs—有机氯农药；OPPs—有机磷农药；RSD—相对标准偏差；SPE—固相萃取；SPME—固相微萃取；UAME—超声辅助微萃取；US—超声能量；UV—紫外检测器；VWD—可变波长检测。

2ng/g 以下。

利用不需要有机溶剂的绿色萃取技术，如固相微萃取（SPME），结合微波辐射法，也被用于分析土壤和沉积物中不同的 OCPs。通过这种组合方式，采用微波辅助顶空固相微萃取（MA-HS-SPME）对河口沉积物中的 18 种 OCPs 进行提取，并分别以 GC-ECD 和 GC-MS 进行检测[35,36]。对 MAE 和 SPME 进行了优化，确定最优的条件是以甲醇做萃取溶剂、采用聚二甲基硅氧烷（PDMS）纤维、温度为 80℃。检出限分别为 0.005~0.11ng/g（GC-MS）和 0.01~0.26ng/g（GC-ECD）。MAE 结合 SPME 也被用于从泥浆样品中萃取包括 DDT 及其代谢物在内的 6 种 OCPs，并以 LC-UV 进行检测[36]。在此方法中，以一种非离子型表面活性剂，聚氧乙烯月桂醚（POLE），作萃取剂，微波辅助萃取 20min，回收率在 80%~117%范围内，RSD 小于 10%。在 LPME 分析水样中 DDT 及其代谢物时，采用微波辐射促进分析物挥发至气相区，进行顶空进样[37]。该方法以正辛醇作 LPME 溶剂，MAE 条件为 249W、34℃、6.5min，并用农田水、海水、河水样品对该方法进行了验证，回收率在 93%~101%。此外，以离子液体作为 MAE 中的萃取剂，分析土壤中的 DDT 及其代谢物的研究也有报道[38]，并以 LC-UV 进行检测。该方法对比了三种离子液体，最终选择了［BMIN］［PF_6］。

在一些研究中，将 MAE 的萃取效率与其他萃取技术进行了比较。其中，在检测土壤中的 DDT 及其代谢物的研究中，采用了以正己烷/丙酮做萃取溶剂、微波辐照 4min 的 MAE 萃取方式，并与超临界流体萃取（SFE）和溶剂萃取（SE）两种前处理方式进行了比较，检测采用 GC-MS 和 GC-ECD 分析[39]。结果显示 MAE 的萃取效率最高，耗时最短、溶剂消耗最少。在土壤中 10 种 OCPs 的分析中，采用了以石油醚/丙酮做萃取溶剂、800W、100℃下 MAE 萃取 15min，萃取液经 SPE 净化后，以 GC-MS/MS 进行分析，并与 PLE、QuEChERS、UAE 前处理技术进行了对比[40]。MAE 的回收率最高（95%~108%），相对标准偏差最小（<5%）。此外，还采用 MAE、索氏提取和 PLE 分析同类目标物[41]，尽管三种萃取方式的回收率相当，但 MAE 溶剂消耗较少、萃取耗时短，因此最终选用 MAE 技术处理实际土壤样品，表现出了很好的适用性和较高的萃取效率。

6.5.2 其他农药

6.5.2.1 有机磷农药

上世纪五十年代以来，有机磷农药（organophosphorous pesticides，OPPs）在农业中广泛应用，表现出了良好性能和较好的成本效益，有效防止、抵制或减轻虫害对各种农作物的影响。因此，这些 OPPs 有机化合物经常残留于土壤和其他环境基质中，对动物和人类健康构成危害。

文献报道采用 MAE 萃取农用土壤中的六种 OPPs，使用加了 KH_2PO_4 的水/甲醇溶液以解吸土壤样品中的 OPPs，并促进其向正己烷中的分配，最终以 GC-FPD 对萃取的 OPPs 进行检测[42]。该方法中所有目标物的回收率均高于 73%，检出限在 0.012μg/g 以下。

一种快速的无溶剂 MAE 方法与超声辅助乳化微萃取技术（ultrasound-assisted

emulsification microextraction，USAEME）相结合，被用于萃取土壤中的9种OPPs，并以GC-ECD进行检测[43]。其中，MAE采用水相萃取，微波能量226W，萃取时间2.5min，之后的USAEME条件为：甲苯20μL、超声30s、离心。采用实际土壤样品对该方法的适用性进行了评价，获得了回收率结果。

多数情况下，OPPs通常与其他农药一起分析，具体可参见6.5.3的详细介绍。

6.5.2.2　三嗪类农药

三嗪类除草剂是长期以来使用最广泛的除草剂之一，主要用于控制玉米和大豆生产中的杂草。它们是水、土壤和生物中普遍存在的环境污染物，最重要的一种是莠去津（atrazine），因其被认为是一种内分泌干扰物，现在已禁止使用。

研究报道了采用MAE（丙酮/正己烷作溶剂，3∶2，体积比）同时分析土壤中莠去津和一种许可使用的OPPs，马拉硫磷，以GC-MS检测，通过模拟迁移过程和降解过程，研究了其迁移和降解行为[44,45]。此外，采用MAE和GC-ECD，同时分析了土壤中莠去津和其他3种禁用OCPs（艾氏剂、狄氏剂、异狄氏）[46]。最优的萃取条件是采用正己烷/丙酮（2∶1，体积比）做萃取溶剂、微波能量为1200W，温度120℃、萃取时间20min。该方法的回收率在70%~85%，相对标准偏差小于5%。

MAE法也被用于分析不同土壤类型中的莠去津、赛克嗪、阿灭净和特丁净，以HPLC-UV进行检测[47]。研究优化了影响MAE效率的因素，确定最优的条件是以甲醇/乙腈/乙酸乙酯（5∶3∶1，体积比）为萃取溶剂，功率800W，萃取时间4min。三个不同加标水平的回收率较为满意，在83%~96%，相对标准偏差小于8%。

研究报道了一种基于原位离子液体-微波辅助萃取-分散液液微萃取（MAE-DLLME）分析水样中四种三嗪类除草剂的HPLC方法[48]。作者认为，这是第一次使用微波辅助原位离子液体-分散液液微萃取（IL-DLLME）方法。在水样中加入[C_4MIM]［BF_4］，当离子液体（IL）完全溶解后，再加入$LiNTf_2$分散剂。形成的悬浮液用微波辐射（30W，50℃，90s）。然后将悬浮液离心，将IL相进行HPLC分析。在三种不同类型的水（自来水、泉水和瓶装饮用水）中均获得了满意的回收率和较低的RSD，LOD小于1.3μg/L。将该方法与其他分析水中三嗪的方法进行比较，虽然回收率和LOD值相似，但由于不使用挥发性有机溶剂，且提取时间最短（1.5min），是最环保的方法。

6.5.2.3　尿素衍生物农药

尿素衍生物农药主要用作除草剂，它们与三嗪类有共同的特性，尽管它们在土壤中的持久性较短（3~6个月）。尽管如此，即便是亲脂性最强的尿素农药也可以被浸出，最终残留于环境水中。

MAE和HPLC-DAD已被用于分析土壤中的八种尿素农药和氨基甲酸盐[49]。对MAE的影响因素进行了研究，结果表明最优的条件是以乙腈为溶剂，在90℃微波辐射10min。在新鲜和老化的土壤样品对方法进行了验证。

有研究在MAE中使用IL，然后用HPLC同时分析土壤中的苯基脲除草剂和三嗪衍生物[50]。优化了IL-MAE参数，最优的条件是使用IL选择［C_4MIM］［BF_4］、微波功率225W、50℃、10min。将该方法的结果与甲醇提取法和IL-加热提取法进行了比较，

IL-MAE 最快（10min，而 IL-加热为 4h），高效且环保。

6.5.2.4 拟除虫菊酯类农药

如今，拟除虫菊酯在商用家用杀虫剂中占大多数。第一代拟除虫菊酯在光照下不稳定，从 1974 年起，第二代拟除虫菊酯进入市场，其更能抵抗光和空气的降解，从而更适合农业上使用。

超声辅助微波萃取（UAME）被提出用于同时从沉积物中提取 8 种拟除虫菊酯和 5 种有机氯农药[51]。在自制的反应器中同时向样品提供超声波和微波能量，以启动溶剂的萃取。最佳实验条件为 100mL 正己烷/丙酮（1：1，体积比）为萃取剂、微波功率为 100W、超声功率为 50W、萃取时间为 6min。提取物经固相萃取净化后，用气相色谱-质谱联用仪（GC-MS）进行分析，回收率在 65%~128%，检出限小于 0.70ng/g，与索氏提取法和超声提取法进行了比较，结果表明该方法具有较好的提取效果。

还通过 MAE 进行了拟除虫菊酯和 OCPs 的提取，然后进行 GC-ECD 分析[52]。影响 MAE 的主要实验条件表明最优的条件为：在 100℃下 5min 使用乙酸乙酯和水作为萃取溶剂。所有化合物的回收率均在 90%~110%，RSD<10%，LODs 在较低的 μg/g 水平。

6.5.2.5 其他农药

二硝基苯胺是一类化学式为 $C_6H_5N_3O_4$ 的化合物，由苯胺和二硝基苯衍生而来，是很多农药制备的中间体。二甲戊灵是此类农药中最常用的农药之一，用于一年生草本和某些阔叶杂草的出苗前和出苗后防治。研究采用 MASE 从土壤样品中萃取二甲戊灵，并以 HPLC-UV 进行检测[53]，对 MASE 参数进行了优化，最优条件是以丙酮作溶剂，60℃下萃取 3min，回收率为 97%。

甲草胺和甲氯胺是最重要的氯乙酰苯胺衍生物。其中，甲草胺在美国是第二广泛使用的除草剂，而欧盟自 2006 年起禁止其作为除草剂使用。甲氯胺的危害性存在争议，目前仍允许使用。文献报道采用 MAE 萃取甲草胺、甲氯胺和其酸性代谢物，SPE 净化后，以 GC-MS 和 HPLC-UV 分析[54]。MAE 条件是以 50mL 甲醇/水混合溶液作提取溶剂，100℃下提取 20min，然后将萃取液过 C_{18} 小柱并分成两个组分，第一组分包含甲草胺和甲氯胺，以 GC-MS 检测，第二组分为代谢物，以 HPLC-UV 检测。该方法在两种有机质含量不同的土壤中进行了验证，所有化合物的平均回收率>71%。LODs 在 10~50 μg/kg。

文献报道了采用一步 MAE 法萃取土壤中的 7 种杀菌剂[55]，并以 HPLC 在不同波长下进行检测。MAE 条件是以水/乙酸乙酯作溶剂，功率 1600W，90℃下萃取 15min。该方法与索氏提取、振荡提取和超声辅助微萃取进行了比较，方法所用的 MAE 萃取效率较高。

新烟碱是一类化学结构是与烟碱类似的新型杀虫剂，由于其水溶性较好，可施用于土壤被植物吸收，因此在病虫防治中广泛使用。噻虫嗪是使用最广泛的一种新烟碱杀虫剂，研究采用 MAE 萃取土壤中的噻虫嗪，微波能量 500W，萃取 1min，SPE 净化后以 HPLC-DAD 分析[56]。在沙质壤土、粉砂黏壤土、砂质黏壤土和壤砂土中，两个不同浓度水平下，进行了回收率研究，结果在 82%~87%，RSD 小于 3%。

6.5.3 多种农药残留分析

为了实现同时分析地下水、土壤、沉积物、室内空气和粒相物中的多种农药，研究者们做出了大量研究来开发相适用的分析方法。表 6.3 总结了 MAE 在多种农药分析中的应用。

表 6.3 MAE 在多农残分析中的应用

分析物	样品	样品预处理	MAE 条件	萃取液处理	检测方法	回收率/%	LOD	RSD/%	发表年份	参考文献
10 种	水（10mL）		35mL 正己烷/丙酮（3∶2，体积比）；MAE（120℃，20min）	过滤、吹干、正己烷复溶	GC-MS	96.2~102	0.0001~0.02mg/L	<2.6	2011	[57]
70 种	土壤（2g）	均质	20mL 正己烷/丙酮（1∶1，体积比）；MAE（1200W，100℃，10min）	离心、氮吹、正己烷复溶	GC-MS	89.3~94.7	0.5~211ng/g	<12	2012	[58]
12 种	土壤（5g）	风干、过筛	35mL 正己烷/丙酮（3∶2，体积比）；MAE（120℃，20min）	离心、氮吹、正己烷复溶	GC-MS	70~120	0.0001~5mg/L	<20	2010/2011	[59，60]
253 种	湖泊沉积物（3.5g）	干燥	25mL 正己烷/丙酮（1∶1，体积比）；MAE（1600W，60℃，10min）	氮吹、甲醇复溶	LC-MS/MS	67.0~123	0.003~0.024μg/kg（干重）	<25	2014	[61]
18 种 OCP 和 5 种 OPP	沉积物（3.5g）	真空过滤、均质、干燥	25mL 乙酸乙酯；MAE（1200W，120℃，15min）	SPE 净化［LC-Si 净化柱、正己烷/乙酸乙酯（1∶1，体积比）洗脱］	GC-μECD	74~114	0.1~1.8μg/kg（干重）	<13	2013	[62]
85 种	沉积物（10g）		DCM/甲醇（1∶9，体积比）；MAE（100℃，10min）	无水 Na_2SO_4 干燥、SPE 净化（乙酸乙酯作洗脱溶剂）	GC-ECD、GC-MS	72.0~108	0.6~8.9μg/g（ECD、干重）	<18	2008	[63]

续表

分析物	样品	样品预处理	MAE 条件	萃取液处理	检测方法	回收率/%	LOD	RSD/%	发表年份	参考文献
4 种杀生物剂	港口沉积物（1g）	冷冻干燥、过筛	10mL 甲醇；MAE（200W，6min）	过滤、SPE 净化（甲醇作洗脱溶剂）	LC-MS/MS	>75	0.1~0.3ng/g	<7	2011	[64]
40 种	室内空气	半透膜吸附（SPMD）	正己烷/丙酮（1：1，体积比）；MAE（3000W，90℃，10min）	氮吹、正己烷复溶、净化	GC-MS/MS	81.0~108	0.1~3.1ng/m	<11	2009	[65]
OPP 和拟除虫菊酯	室内空气	石英纤维过滤器/C_{18} 圆盘吸附	20mL 二氯甲烷；MAE（1200W，90℃，5min）	浓缩至 0.1mL	GC-MS	78.2~110（石英纤维过滤器）；71.0~98.8（C_{18} 圆盘）	0.03~0.20ng/m^3	—	2014	[66]
20 种	室内空气	SPMD 吸附	30mL 正己烷/丙酮（1：1，体积比）；MAE（300W，90℃，10min）	氮吹、正己烷复溶、净化	GC-MS	59.0~206	1~10 ng/device	<26	2011	[67]
30 种和 40 种	室内空气	石英纤维过滤器吸附	30mL 乙酸乙酯；MAE（1200W，50℃，20min）	浓缩、复溶［LC 分析采用水/甲醇（70：30，体积比）复溶，GC 分析采用二氯甲烷］	LC-MS/MS 和 GC-MS/MS	70~120	6.5~32pg/m^3（LC-MS/MS）；1.3~39.5pg/m^3（LC-MS/MS）LOQs	—	2009/2011	[68，69]
10 种	空气中颗粒物	玻璃纤维滤头被动采样、冻干	10mL 正己烷/丙酮（1：1，体积比）；MAE（100℃）	萃取物以环己烷/乙酸乙酯（1：1，体积比）溶解、GPC 净化	GC-MS	68.4~94.5	0.005~0.046ng/g（LOQs）	<13	2015	[70]
10 种	室内空气（0.8g）	—	正己烷+0.1M 硫酸（含 0.1%抗坏血酸）；MAE（80℃，5min）	净化（硅藻土、涡旋）	GC-ECD	84.0~103	0.22~40ng/g	<8.8	2007	[71]

注：DCM—二氯甲烷；ECD—电子捕获检测器；GC—气相色谱；LC—液相色谱；LOD—检出限；LOQ—定量限；MAE—微波辅助萃取；MS—质谱；OCPs—有机氯农药；OPPs—有机磷农药；RSD—相对标准偏差；SPE—固相萃取；SPMD—半透膜装置。

研究采用 MAE 萃取法，对不同环境水中的多种农药残留，包括 OCPs、OPPs、氨基甲酸酯类和拟除虫菊酯类，进行了 GC-MS 分析[57]。以丙酮/环己烷（3∶2，体积比）为提取溶剂，在 120℃下萃取 20min，回收率在 96%～102%，LODs 小于 0.02mg/L。以方法对 15 个不同的地表水样品进行了分析，在所分析的 10 种农药中，检出了 9 种，最高浓度达 0.181mg/L。

土壤无疑是农药多残留测定中研究最多的环境基质之一[58-60]，采用 MAE 萃取土壤中的多种类农药残留，并以 GC-MS 进行检测的研究多有报道。其中，以正己烷/丙酮为溶剂，分别提取了土壤中的 70 种农药残留（27 种 OPPs、27 种 OCPs、9 种拟除虫菊酯、5 种氨基甲酸酯）[58]和 12 种农药残留（OCPs、OPPs、氨基甲酸酯、拟除虫菊酯、杀菌剂）[59,60]。在所有情况下，均获得了良好的回收率，其回收率在 70%～120%。将 MAE 与索氏提取进行了比较，表明 MAE 回收率更高，RSD 更低，时间和溶剂消耗更低。

有研究者提出将 MAE 与 LC-MS/MS 相结合来分析湖泊沉积物中的 253 种农药残留（氨基甲酸酯，三嗪/康唑/三唑，OPPs，苯甲酰胺/嘧啶和尿素）[61]。用丙酮/己烷（1∶1，体积比）提取沉积物样品，并将 MAE 条件固定在 1600W，60℃和 10min。之后，将提取物蒸发、复溶，并通过 LC-MS/MS 进行分析。

通常，必须对提取物进行净化才能获得令人满意的分析效果。研究者建立并验证了一种基于 MAE，SPE 和 GC-ECD 的方法，可同时分析沉积物中的 18 种 OCPs 和 5 种 OPP s[62]。在最佳 MAE 条件（1200W，60℃）下，乙酸乙酯作萃取溶剂，时间为 15min。回收率在 74%～114%。也有研究提出了一种基于 MAE 的方法，随后进行两个净化步骤，即在 GC-ECD 和 GC-MS 分析之前采用凝胶渗透色谱（GPC）和 SPE 进行净化，同时测定了 85 种农药，其中包括苯胺、唑/三唑、氨基甲酸酯、氯乙酰苯胺、OCPs、OPPs、拟除虫菊酯、甲氧丙烯酸酯、硫代氨基甲酸酯和苯并三嗪/三唑[63]。此外，有研究使用二氯甲烷（DCM）/MeOH（1∶9，体积比）作萃取剂，100℃下微波辐射 10min 萃取分析物。回收率在 72%～108%，LOD 为 0.6 至 9mg/kg。另有研究采用 MAE 萃取法，通过 LC-MS/MS 同时分析了四种化学性质不同的杀菌剂[64]，对影响萃取的因素进行了优化，最佳条件是以甲醇作萃取剂，在 1200W 下萃取 6min。回收率>75%，RSD<7%，LOD 范围为 0.1～0.3ng/g。

一些 MAE 方法已经被提出用于分析空气和相关的颗粒物。多残留农药分析是在空气样本中进行的，在所有情况下，都需要使用吸附剂来吸附分析物，然后在色谱分析之前进行 MAE 和净化。研究报道了采用半透膜装置（SPMD）对室内空气中的农药残留进行采集，并结合 MAE［丙酮/正己烷（1∶1，体积比）作溶剂，3000W、90℃、10min］萃取 SPMD 吸附的目标物，萃取液经蒸发后，以正己烷复溶，最后以 GC-MS/MS 对 40 种农药进行了监测[65]。回收率在 81%～108%，7d 进样的 LOD 在 0.1～3.1ng/m^3。为了测定室内空气中的 OPP 和拟除虫菊酯农药，研究者提出了以石英纤维过滤器和 C_{18} 圆盘上进行吸附的方法，然后通过 MAE 和 GC-MS 进行分析[66]。以 DCM 为萃取溶剂，在 1200W，90℃下微波辐射 5min。回收率令人满意（石英纤维过滤器的回收率为 78%～110%，C_{18} 萃取盘的回收率为 71%～99%），LOD 范围为 0.03～0.20ng/m^3。所提出的方法与索氏提取法进行了比较，显示出相似的回收率，但是 MAE 表现出更重要的优势，例如溶剂用量少、耗时短。研究使用基于膜的装置作为被动采样器，对检疫植物的室内空气质量进行了测定[67]。用 MAE（丙酮/正己烷，300W，90℃，10min）从被动采样器中提取了 20 种多类农药和挥发性有机物。萃取物蒸发后以正己烷复溶，然后经过净化后，以 GC-MS 分析。单个装置的检出限为 1～10ng。

使用 MAE，结合 LC-MS/MS[68] 和 GC-MS/MS[69]，同时分析了空气中的细颗粒物（PM 2.5 和 PM 10）中多达 40 种农药[68,69]。研究了 MAE 参数，最佳的 MAE 条件是使用乙酸乙酯作为萃取溶剂，在 1220W、50℃下萃取 20min。在 LC-MS/MS 或 GC-MS/MS 分析之前，将萃取液浓缩并以水/甲醇或二氯甲烷中复溶，回收率为 70%～120%，定量限低于 40pg/g。

有研究提出了用 MAE 结合 GC-MS 同时从大气颗粒物（AMP）中提取 10 种有机氯农药的方法[70]。在 5d 采集时间内，使 AMP 通过玻璃纤维过滤器。然后，用正己烷/丙酮在 100℃下萃取含有 AMP 的过滤器，萃取物进行 GPC 净化，然后进行 GC-MS 分析。为了评估室内空气中 10 种 OPPs、7 种拟除虫菊酯以及新兴污染物（麝香）的存在，使用 MAE 对粉尘进行分析，然后进行瓶内净化，以 GC-ECD 进行测定[71]。以酸性正己烷为溶剂、微波辐照 5min 提取粉尘样品，通过加入硅藻土并振摇进行瓶内净化。回收率在 84%～103%，并将该方法用于多种粉尘样品，检测出的农药含量高达 5134ng/g。

6.6 微波萃取新兴污染物

如前所述，新兴污染物（EP）可以定义为现有环境质量法规当前未涵盖的化合物。新兴污染物会对生态系统和人类造成危害，其发生和安全性尚未得到全面研究[28]。EP 可能来自点污染（主要是城市和工业污染）或扩散性污染（农业污染）。

尽管现有数据仍然不足以评估 EP 在环境中的进入、分布、积累和降解，以及其环境和健康风险，一直以来，药品和滥用药物、类固醇和激素、个人护理产品、阻燃剂、增塑剂、全氟化合物、表面活性剂和其他工业添加剂和制剂等都是众多环境研究的对象。

以往在水政策领域的某些环境控制指令中几乎都未包含这些化合物。自 2008 年起，六种 PBDE 同系物，邻苯二甲酸二（2-乙基己基）酯（DEHP）和两种烷基酚（壬基酚和辛基酚）被列入水政策领域的优先物质清单及环境质量标准（EQS）中[72]。2013 年，针对优先物质清单，将全氟辛烷磺酸和溴化阻燃剂六溴环十二烷（HBCD）纳入了修订指令 2000/60/EC 和 2008/105/EC 的最新指令中，并添加至相应的水和生物群系的 EQS 中[73,74]。PBDEs 的 EQS 已在此最新指令中进行了修订，并且在生物群系的 EQS 中也具有最大价值。

除了优先物质外，欧盟委员会最近还制定了一份物质观察清单，需要收集广泛的监测数据，以支持未来的优先物质确定工作。一种药物双氯芬酸、两种激素 17-β-雌二醇（E2）和 17-α-乙炔基雌二醇（EE2）以及紫外线过滤剂 2-乙基己基甲氧基肉桂酸酯已被列入第一份观察名单，以收集监测数据，目的是协助确定适当措施，评估这些物质造成的风险[75]。

尽管前面提到的某些化合物实际上不能被认为是新兴污染物，但由于它们与环境的高度相关性以及对利用微波辅助程序进行萃取的研究十分具有价值，本章并未将它们忽略。

据报道，在水生环境（地表和地下水）和其他水生介质［污水处理厂（STP）的进水和出水］中存在大量的 EP[76,77]。然而，关于新兴污染物的环境发生、行为和风险，仍然存在许多知识空白[78,79]。

在本节中，将针对日益引起关注的几类新兴污染物，介绍应用 MAE 来开发的相应分析方法。

6.6.1　药物

超过 3000 种不同的物质被用作药物成分，包括止痛药、抗生素、抗糖尿病药、β 受体阻滞剂、避孕药、血脂调节剂、抗抑郁药和阳痿药物。药物及其代谢物随尿液和粪便排出，进入城市污水处理厂（WWTP）。污水处理厂的设计目的是去除一些有机污染物，主要是溶解的有机物、固体物质和营养物，并不包含药物。

尽管化学或生物过程可以消除部分药物化合物，但大多数对药物的去除是不完全的，需要改进废水处理和后续产生的污泥处理工艺，以防止这些微污染物进入环境[80]。将未使用的药物直接弃置于生活垃圾中，并作为兽药和饲料添加剂应用于牲畜、家禽和鱼类养殖业，也可促进其进入环境。由于药物和激素在环境中的存在，以及其可能的雌激素效应及其他不良影响，它们已成为关键的新兴污染物。虽然污水处理过程后检测到的浓度水平似乎不会对人类健康和水环境造成毒性影响，但水生生物长期暴露于药物以及药物混合产生的鸡尾酒效应仍是令人担忧的问题。

利用微波辅助萃取技术从环境基质中提取药物已经引起了研究者们的极大兴趣，表 6.4 所示为微波辐射在不同基质中类固醇、抗炎药和抗生素等药物提取中的应用。雌激素是类固醇激素中最重要的一类，由其化学结构和对性周期的影响来定义雌激素。它们是内分泌干扰物（EDC），扰乱荷尔蒙的正常生理功能[81]。此外，非甾体抗炎药（NSAID）的使用量也在增加，其残留对环境的污染也随之而来[82]。抗生素是一种抗细菌感染的药物，对环境有特殊的影响。其中，氟喹诺酮类抗生素因其广谱活性和良好的口服吸收性，可能是最重要的一类合成抗生素[83]。卡马西平是治疗癫痫和神经性疼痛的最常用药物。它已在废水中检测到，并且由于其亲脂性，可在沉积物或淤泥中积累。

6.6.1.1　类固醇

MAE 被成功地应用于沉积物[84,85]和污泥[86-89]中类固醇的测定，其应用总结如表 6.4 所示。通常以甲醇作萃取溶剂，体积在 5~25mL。在 MAE 过程之后经常使用额外的步骤（主要是固相萃取），以净化提取物并浓缩目标分析物[85,87-89]。关于测定，LC-MS/MS[84,86-89]是提供低至 ng/g 水平检出限的最常用的测定方法。

研究提出了一种中心切割-二维 LC/MS/MS 测定四种雌激素的方法[90]，与一维 LC

表 6.4 MAE 在药物分析中的应用

分析物	样品	样品预处理	MAE 条件	萃取液处理	检测方法	回收率/%	LOD	RSD/%	发表年份	参考文献
类固醇										
20 种合成及天然类固醇及其代谢物	沉积物（3g）	加甲醇、涡旋	5mL 甲醇；两次 MAE（10min 升至 90℃，保持 5min）	合并两次提取液，氮吹至 0.2mL	LC-MS	66.5～128.4	0.07～1.87ng/g	<60	2013	[84]
E1、E2、16α-羟基雌酮等	沉积物（3g）	冻干、研磨、过筛	25mL 甲醇；2g 铜粒；MAE（110℃，15min）	干燥、加水、SPE 净化、BSTFA 衍生、氮吹、正己烷复溶	GC-MS/MS	86～102	0.05～0.14ng/g	<11	2009	[85]
15 种性激素和皮质类固醇	污泥（1g）	冻干	10mL 甲醇；MAE（500W，65℃，4min）	过滤、氮吹、甲醇复溶	UPLC-MS/MS	>71	1.1～7.9ng/g	<23	2016	[86]
E2、E3、EE2 及其他	污泥（1g）		5mL 甲醇；MAE（300W，10min）	过滤、稀释、SPE 净化	LC-MS/MS	71.7～103	0.6～3.5ng/g	<10	2011	[87]
E1、E2、E3、EE2、DES 及其他	污泥（1g）	均质、风干	5mL 甲醇；MAE（300W，10min）	稀释、SPE 净化	UPLC-MS/MS	75.1～98.1	0.1～0.5ng/g	<8.3	2013	[88]
T、炔诺酮、DES、甲基炔诺酮、EE2、E2、E3、E1 及其他	污泥（1g）		5mL 甲醇；MAE（200W，6min）	过滤、稀释、SPE 净化	UPLC-MS/MS	75.1～102	0.1～0.7ng/g	<9.0	2013	[89]
E1、α-雌二醇、β-雌二醇、EE2	河流沉积物		10mL 甲醇/水（95∶5，体积比）；MAE（10min）	过滤、吹干、复溶（甲醇/水，1∶1，体积比）	1D 和 2D-LC-MS	98.8～107	90～250pg/g	<9.5	2011	[90]
E1、E2、EE2 及其他 PCP（对羟基苯甲酸酯、三氯生）、烷基酚、苯酚、双酚 A	土壤、沉积物、污泥（1g）	冻干、均质	10mL 甲醇/水（3∶2，体积比）；MAE（500W，6min）	过滤、氮吹、复溶、SPE 净化、洗脱、硅烷化	GC-MS	92～98	4.7～5.1ng/kg	<5.1	2012	[91]

抗炎药									
IBU、KET、NAP及其他PCP（麝香酮、三氯生）	土壤和沉积物（3g）	干燥、均质	二氯甲烷/甲醇（2∶1）、800W	离心、浓缩、衍生化（吡啶：BSTF，2∶1），硅胶微柱净化、氮吹、正己烷复溶	GC-MS	25	—		2010 [92]
KET、NAP、IBU、DIC	河流沉积物（5g）	破碎并过筛	40mL甲醇/丙酮（1∶1，体积比）；MAE（170℃，6min）	离心、SPE净化、MSTFA衍生	GC-MS	46~87	0.03~0.08μg/g	<11	2007 [93]
IBU、NAP、KET、DIC	沉积物（5g）	干燥、研磨并过筛	50mL水（pH = 6）；700W预加热（5min）60℃，600W（5min升至100℃），萃取（700W，100℃，20min）	萃取物和沉淀物用二甲醚洗涤，SPE净化、衍生化	GC-MS/MS	95~103	2~6ng/g（LOQs）	<12	2010[94，95]
IBU、NAP、KET、DIC、SUL及其他	土壤	消解并冻干	25mL甲醇/水（1∶1，体积比）；MAE（110℃，10min）	SPE净化（Oasis MCX）	UHPLC-MS/MS	89.6~99.7	0.05~0.54ng/mL		2016 [96]
18种药物及非法药物	废水和污泥（1g）	消解并冻干（污泥）	30mL甲醇/水（1∶1，体积比）；MAE（1200W，120℃，30min）	离心、SPE净化（甲醇洗脱）	手性LC-MS	91.1~124	0.03~1.16μg/g	<10	2015 [97]
KET、NAP、IBU	沉积物（1g）	干燥	8mL水2.75%POLE（5%，体积比）；MAE（500W，6min）	过滤、加14mL水（pH 3.0）、SPE净化（OASIS HLB）、5mL水洗两次、0.75mL甲醇洗脱两次	HPLC-DAD	>70	<46ng/g	<11	2008 [98]

续表

分析物	样品	样品预处理	MAE 条件	萃取液处理	检测方法	回收率/%	LOD	RSD/%	发表年份	参考文献
抗生素										
LEVO、NOR、CIP、ENR、SAR	沿海沉积物和污泥（2g）	风干	15mL 5%HTAB 溶液；MAME（500W，15min）	—	LC-FL LC-MS/MS	73.2~95.6	0.15~0.55ng/g	≤8	2012	[99]
25 种抗生素	含水层沉积物	风干并过筛	5min 升至 60℃（保持 25min），100W	离心、SPE 净化（Oasis HLB 柱，甲醇洗脱）	UPLC-Q-Orbitrap-MS	>70	0.1~3.8mg/kg	<21	2016	[100]
TC、DC、OTC、CTC 及其他 PCP	土壤、污泥、大气颗粒物（1g）		5mL 甲醇、μSPE、MAE（400W，60℃，20min）	用甲醇超声	HPLC-UV	70.6~110	0.1~6.3ng/g	<15	2015	[101]
8 种喹诺酮类抗生素	土壤（0.5~1g）	干燥、均质、过筛	8mL 水溶液（含 20% $Mg(NO_3)_2 \cdot 6H_2O$ 和 2%NH_3），20min，80℃	离心、酸化	HPLC-FL	69~110（MAE-LC-FL）；60~85（MAE-SPE-LC-FL）	1.4~4.8μg/kg	0.9~2.4	2010	[102]
13 种喹诺酮类抗生素	污泥（0.5g）	干燥、细磨	10mL 萃取缓冲液（pH=3）；MAE（1000W，87℃，17min）	离心、上清液吹干、复溶、离心	LC-MS/MS	97.9~105	1~5ng/g	0.3~6.7	2013	[103]
CBZ	废水及废水污泥	冻干、均质	20mL 甲醇，MAE（1200W，10min 升至 110℃并保持 10min）	浓缩并用水复溶	LDTD-APCI-MS/MS	98~113（废水）；96~107（污泥）	12ng/L（废水）；3.4ng/g（污泥）	<5	2012	[104]

注：1D——维；2D—二维；BSTFA—*N*,*O*-双（三甲基甲硅烷基）三氟乙酰胺；CBZ—卡马西平；CIP—环丙沙星；CTC—金霉素；DC—强力霉素；DES—己烯雌酚；DIC—二氯芬酸；DME—二甲醚；E1—雌酮；E2—17-β-雌二醇；E3—雌二醇；EE2—17α-乙炔雌二醇；ENR—恩诺沙星；FL—荧光检测器；GC—气相色谱；HPLC—高效液相色谱；HTAB—十六烷基三甲基溴化铵；IBU—布洛芬；KET—酮洛芬；LC—液相色谱；LDTD-APCI—激光二极管热脱附大气压化学电离；LEO—左氧氟沙星；LOD—检出限；LOQ—定量限；MAE—微波辅助萃取；MAME—微波胶束萃取；MS—质谱；MS/MS—串联质谱；MSTFA—*N*-甲基-*N*-（三甲基硅烷）三氟乙酰胺；NAP—萘普生；NOR—诺氟沙星；OTC—土霉素；PCP—个人护理产品；POLE—聚氧乙烯月桂醚；RSD—相对标准偏差；SAR—沙拉沙星；SPE—固相萃取；SUL—柳氮磺胺吡啶；T—睾酮；TC—四环素；UHPLC—超高效液相色谱；UV—紫外检测器；W—瓦。

分析相比，基质组分的含量明显减少。在此条件下，采用水/甲醇混合物作为萃取剂，微波能量作用 10min。当用 GC-MS/MS 进行分析时，需要在 MAE 之后进行衍生化步骤。选择 *N*，*O*-双（三甲基硅基）三氟乙酰胺（BSTFA）作为衍生化试剂[85]，在微波辐射下（110℃，15min）用 25mL 甲醇和铜粉从沉积物中提取 4 种酯类化合物。研究还提出了同时测定土壤、沉积物和污泥中 3 种内分泌干扰物（E1、E2、EE2）、对羟基苯甲酸酯、烷基酚、苯酚、双酚 A 和三氯生的 MAE 和 GC-MS 分析方法[91]。MAE 过程选择甲醇/水（2∶3，体积比）混合溶剂，500W 下微波萃取仅 6min。在 GC-MS 分析前，进行 SPE 净化，并进行硅烷化衍生。

6.6.1.2 抗炎药

采用 MAE 分析和 GC-MS 分析相结合的方法测定土壤和沉积物中的抗炎药[92-95]。该程序包括固相萃取净化步骤和萃取后的衍生化步骤。利用微波能量提取了 7 种结构不同的药物：3 种抗炎止痛剂（布洛芬、酮洛芬和萘普生）、一种兴奋剂和一种抗组胺药物，以及两种五氯苯酚：一种香料（麝香酮）和一种防腐剂（三氯生）[95]。以二氯甲烷/甲醇（2∶1，体积比）为萃取剂，在 800W、110℃ 下萃取 10min。在 GC-MS 分析之前，提取物经过衍生化（BSTFA）和硅胶微柱净化。

也有研究提出了几种微波萃取测定沉积物中布洛芬、萘普生、酮洛芬和双氯芬酸的方法[92-94]，采用甲醇/丙酮混合液，微波辐射仅 6min[93]，并用水在 60℃ 和 100℃ 两种不同温度下进行非有机溶剂萃取，总微波辐射时间为 20min[92,94]，建立了测定沉积物中布洛芬、萘普生、酮洛芬和双氯芬酸含量的几种 MAE 方法。GC-MS 和 GC-MS/MS 分析的检出限分别小于 80ng/g 和 6ng/g。

此外，使用 MAE 和 LC 分析包括抗炎药在内的多类药物的方法也有报道[96-98]，在分析前不需要对目标物进行衍生化。基于此，对环境基质中 90 种新兴污染物进行了多残留分析。在多种化学物质中，非甾体抗炎药（磺胺吡嗪、布洛芬、酮洛芬、萘普生和双氯芬酸）和个人护理产品（对羟基苯甲酸酯和二苯甲酮紫外线过滤剂）可用水/甲醇（1∶1，体积比）从土壤中成功提取。采用超高效液相色谱 UHPLC-MS/MS 进行分析，建立了其他多残留分析方法，用于 18 种手性药物（包括消炎药和污泥中的违禁药物）的分析[97]。在 1200W，120℃，30min 的条件下，用甲醇/水进行 MAE，然后进行净化步骤，最终以手性 LC-MS/MS 进行测定。

采用手性 LC-DAD，对沉积物种 8 种常用药物（包括酮洛芬、萘普生、布洛芬等）进行了分析，探讨了胶束介质萃取 MAE 测定这些药物的可行性[98]。该方法使用含有非离子表面活性剂 POLE 的水作溶剂，样品在 500W 下提取 6min，提供了一种绿色溶剂萃取方法。加标回收率>70%，并成功地应用于不同性质沉积物样品的分析。

6.6.1.3 抗生素

研究提出了沿海沉积物和污泥中 4 种氟喹诺酮类抗生素的微波辅助胶束萃取-液相色谱-质谱（MAME-LC-MS）分析方法[99]。用 15mL 含十六烷基三甲基溴化铵的水作为表面活性剂溶液，微波能量为 500W，时间为 15min。回收率大于 73%，RSD<8%，检出限分别为 0.15ng/g 和 0.55ng/g。该方法已用于实际固体样品中抗生素的测定，其中沿海沉积物中检出 4 种抗生素。

近来，研究报道了一种基于 MAE 和 UHPLC-Q-Orbitrape-MS 分析的高效测定含水层沉积物中 25 种抗生素的方法[100]。用甲醇在 60℃下微波萃取 5min，质量准确度好，质量偏差<2mg/kg，回收率>70%，RSD 值低。将该方法应用于不同深度的实际含水层沉积物样品，结果表明，在较深的水平（18m），金霉素和氧氟沙星的浓度较高（分别为 53mg/kg 和 19mg/kg）。

比较了同时用 MAE-μSPE 法和经典 MAE 法测定土壤、污泥和 AMP 中 4 种四环素类抗生素的结果，以 HPLC-UV 进行分析[101]。在同时萃取净化步骤（MAE-μSPE）中，将甲醇和 μSPE 装置（由封闭的微孔聚丙烯膜中的异烟酸铜组成）放入萃取容器中，并在 60℃下用微波能量（400W）照射 20min。通过这种方式，分析物从样品萃取到 μSPE 装置。然后，从溶液中取出 μSPE 装置，用甲醇超声解吸分析物。在相同条件下，将该方法与常规 MAE 方法进行了比较。两种方法均显示良好的回收率和较低的相对标准偏差。与常规 MAE 相比，MAE-μSPE 获得了更低的 LOD。

研究采用单步 MAE 成功地从农业土壤中提取了 8 种氟喹诺酮类抗生素（诺氟沙星、环丙沙星、达诺福沙星、恩诺沙星、氟罗氟沙星、左氧氟沙星、马博沙星和奥氟沙星）[102]。由于采用的是 $Mg(NO_3)_2$ 水溶液，因此定量提取避免了有机溶剂的消耗。采用反相高效液相色谱法进行荧光检测。并与 UAE 的性能进行了比较，结果表明该方法具有令人满意的回收率结果。在另一项研究中，MAE 与 UAE 和 PLE 比较了同时测定污泥样品中 13 种喹诺酮类抗生素的效率，然后采用 LC-MS/MS 分析[103]。优化的 MAE 条件是使用 10mL 缓冲液（pH = 3），100W，87℃和 17min 的提取时间。微波辅助萃取法提取率高，操作简便，分析时间短，自动化程度高，是最佳的提取方法。所有被分析化合物的 LODs 均低于 5ng/g，且定量回收率为 98%～105%。

另有研究提出了一种快速可靠的方法，利用激光二极管热解吸-大气压化学电离（APCI）-MS/MS 分析污泥中的卡马西平[104]。MAE 以甲醇为提取溶剂，微波能量为 1200W，提取时间为 10min。MAE 的效率与 UAE 和 PLE 进行了比较，结果表明，MAE 的回收率最高，溶剂消耗和时间较短。

6.6.2 兽药

在世界各地的食用动物农业中，为了治疗或预防传染病和非传染性疾病，管理生殖过程和促进生长，使用了种类繁多的兽药化合物。表 6.5 所示为为测定环境样品中兽药而开发的基于 MAE 的方法。

6.6.2.1 磺胺类药物

在欧洲国家，磺胺类药物（SAs）是畜牧业中应用最广泛的抗生素类之一。例如，在农田施用有机肥后，土壤中会出现大量的磺胺残留物。MAE 被用于土壤、沉积物、污泥和水中这些药物的萃取，应用总结如表 6.5 所示。

研究以乙腈[105,106]、甲醇[107]或乙腈/水[108,109]作为提取溶剂，采用 MAE 分析了沉积物和土壤中不同的磺胺类药物。在所有的案例中，MAE 程序之后都进行了净化步骤，最终采用 LC-MS/MS[105,107-109]和 LC-荧光（LC-FL）[106]进行分析。研究还将 MAE 与其他萃取技术（UAE 和 PLE）进行了比较[106,108,109]，证明了 MAE 在效率和速度方面的优

表 6.5 MAE 在兽药分析中的应用

分析物	样品	样品预处理	MAE 条件	萃取液处理	检测方法	回收率/%	LOD	RSD/%	发表年份	参考文献
4 种 SA	土壤（2g）	干燥	12mL 乙腈（0.8mL/min）；MAE（320W）	正相 SPE	LC-MS/MS	82.6~93.7	1.4~4.8ng/g	2.7~5.3	2009	[105]
6 种 SA	过夜老化的加标土壤（2g）	风干并过筛，黑暗条件储存，过分土器	6mL 乙腈/缓冲液（pH 9）（20：80）；MAE（115℃，15min）	体积调整、SPE 净化、衍生	LC-FL	15~64	1.0~6.0ng/g	<7.0	2010	[106]
9 种 SA	沉积物和土壤		甲醇，45min	净化、过滤	LC-MS/MS	>80	—	<20	2014	[107]
SDZ 及其两种主要代谢物	土壤（10g）	干燥、均质、过筛	乙腈/水（1：4，体积比）；MAE（150℃，15min）	—	HPLC-MS/MS	66	—	—	2018、2009	[108，109]
10 种 SA	土壤、沉积物、污泥（2g）	干燥、均质	Triton X114（1.5，体积比），1.5mL/min；MAME（800W）	—	—	69.7~102	0.42~0.68ng/g	—	2016	[110]
4 种 SA	环境水（2mL）	—	离子液体；MAE（240W，90s）	通过冷冻和离心分离离子液体	HPLC-UV	75.1~116	0.33~0.85mg/L	<12	2014	[111]
6 种 SA	河水（10mL）	调节 pH	0.75mL 甲醇 + 0.3g NaCl + 200μL 荧光胺溶液 + 100μL 离子液体；MAE（240W，90s）	离心、将离子液体相溶解并过滤	HPLC-FL	96.1~111	0.011~0.018μg/L	<5.3	2011	[112]
ABA、DOR、IVER	土壤（1g）	风干、过筛、γCo60 辐照、-20℃黑暗条件下存放	5mL 乙腈/水（90：10，体积比）；MAE（120℃，15min）	离心、体积调节、过滤、SPE 净化	UPLC-MS/MS	14.5~92.5	18~25ng/g	2~10	2011	[113]

续表

分析物	样品	样品预处理	MAE 条件	萃取液处理	检测方法	回收率/%	LOD	RSD/%	发表年份	参考文献
ENR、DAN 及其光解产物	土壤（1g）	干燥、均质、过筛	8mL 水（含 20% 0.2g/mL $Mg(NO_3)_2 \cdot 6H_2O$、2% 体积比 NH_3）；MAE（1600W，80℃，20min）	离心、酸化	LC-FL	70～130	—	1～6	2012	[114]
奥苯达唑	土壤（1g）	风干并过筛	20mL Genapol X-080（0.5%）；MAE（1000W，2min）	离心	LC-FL	87.1～95.1	0.10μg/g	<7.7	2012	[115]
8 种兽药	沉积物（1g）	干燥、研磨并过筛	15mL 甲醇/水（80∶20，体积比）；MAE（65℃，15min）	倾析、蒸发、甲醇复溶	HPLC-DAD	40～100	0.005～0.460mg/L	<5	2016	[116]
MNC、OTC、TET、ENR、CEF	沉积物、污泥	沉积物：冻干；污泥：冷冻、离心、冻干	甲醇-甲酸、40℃，20min	离心、吹干、复溶	HPLC-DAD	81	—	—	2013	[117]

注：ABA—阿维菌素；CEF—头孢硫呋喃；DAD—二极管阵列检测器；DAN—达诺沙星；DOR—多拉米汀；ENR—恩诺沙星；FL—荧光检测器；HPLC—高效液相色谱；IL—离子液体；IVER—伊维菌素；LC—液相色谱；LOD—检测限；MAE—微波辅助萃取；MAME—微波胶束萃取；MeOH—甲醇；MNC—米诺环素；MS—质谱；MS/MS—串联质谱；OTC—土霉素；RSD—相对标准偏差；SA—磺胺类；SDZ—磺胺嘧啶；SPE—固相萃取；TET—四环素；UHPLC—超高压液相色谱；UV—紫外检测器。

势，在提取样品的通量方面提供了更大的灵活性。

近年来，从环境基质中提取磺胺类药物的 MAE 多采用绿色萃取溶剂。由此，提出了一种高通量同时测定土壤、沉积物和污泥中 10 种 SA 的方法[110]。15 个样品的总制备时间为 18min，采用动态 MAME 系统，其中 Triton X-114 流经包含样品的提取容器。装满容器后，在 800W 的微波辐射下，提取液离心后用高效液相色谱-紫外分光光度计（HPLC-UV）进行分析。检出限均低至 ng/g 水平。考察了添加土壤的老化对恢复研究的影响，结果表明，10 个目标 SA 中有 8 个在土壤中存留了 3 个月。

离子液体还成功地利用微波能从水中萃取磺胺类化合物[111,112]。提出了一种基于 IL 固化的微波辅助液相微萃取（MA-LLME）技术，采用的是室温固态 IL-1-3-甲基咪唑六氟磷酸盐（1-乙基-3-甲基咪唑）。微波辐射 90s 后，冷冻离心分离 IL[111]。在此情况下，使用［C_6MIM］［PF_6］作为 IL，在微波辐射下（240W，90s）从水中提取六种磺胺类化合物。用薄层毛细管电泳法进行测定，定量回收率和检出限为 0.011～0.018μg/L。

6.6.2.2 其他兽药

MAE 也被用于从环境基质中提取其他兽药，相关应用也汇总于表 6.5。

用 MAE 从土壤中提取阿维菌素、多拉米汀和伊维菌素，使用 ACN/水（90∶10，体积比）作溶剂，温度为 120℃，萃取 15min，最终以 UHPLC-MS/MS 进行分析[113]。基于 MAE 和 HPLC-FL 的另一种方法已被提出用于土壤中恩诺沙星、达诺沙星及其光产物的测定[114]。在这种情况下，使用含 Mg（NO_3）$_2$ 的水溶液，样品在 80℃ 微波辐射（1600W）20min。两种方法均具有良好的回收率和较低的 RSD。

有些研究将 MAE 与其他萃取技术（如搅拌法、UAE 法和 PLE 法）进行比较，以测定土壤、沉积物和污泥中不同类型的兽药[115-117]，然后进行 LC-FL[115] 和 HPLC-DAD 分析[116,117]。以 Genapol X-080（0.5%）为表面活性剂，微波辐射功率 1000W，搅拌 2min，与机械振荡和 UAE 相比，MAME 对土壤中奥苯达唑的萃取效率最高[115]。采用甲醇/水（80∶20，体积比）为萃取剂，在 65℃ 微波辐射下萃取 15min，与搅拌法、UAE 法和 PLE 法相比，萃取时间最短，溶剂消耗最低，可用于沉积物中不同结构基团的 8 种兽药的测定[116]。同时测定污泥和底泥中的米诺环素、土霉素、四环素、恩诺沙星、头孢噻呋钠、氟喹诺酮的萃取效率最高。MAE 条件为甲醇-甲酸在 40℃ 下萃取 20min，对大部分化合物的萃取率最高[117]。

6.6.3 个人护理用品

近年来，个人护理产品（PCP）因其巨大的消耗量和在环境中的潜在有害浓度而受到越来越多的关注。根据已公布的数据，与药物化合物相比，PCP 的环境浓度和毒性在很大程度上被忽略了[118,119]。特别值得注意的是合成麝香香料、紫外线过滤剂和防腐剂等几类 PCP。

合成香料常被添加到盥洗用品、化妆品、家用产品和各种各样的消费品中。人们关注的问题涉及其对健康的直接影响、香味化学物质在人体组织中的生物积累以及可能对环境造成的长期影响[120]。

合成麝香被用作化妆品和家用产品的香料成分，如洗衣和洗碗剂等。尽管它们是为了取代更昂贵、更稀有的天然麝香而制造的，但它们在结构上或化学性质上与天然麝香并不相似。它们的理化特性与已知可通过食物链进行生物放大的疏水性和半挥发性有机污染物有更多的共同之处[121]。在过去的几十年里，人们认为合成麝香是惰性的，对人类和环境没有毒性作用。然而，不同的研究表明，它们会产生毒性作用，具有致癌性（如硝基麝香），并会导致雌激素失衡。合成麝香有四种类型：硝基麝香、多环麝香、大环麝香和脂环麝香[122]。硝基麝香是最早生产出来的，其中一些现在已被禁用。多环麝香香料，特别是佳乐麝香和吐纳麝香的使用量最高，佳乐麝香是一个高产量、高消耗的化学品。

有机紫外线过滤剂是专门或主要用于保护皮肤免受紫外线辐射的物质。虽然紫外线过滤剂最初设计用于防晒霜配方中，但目前许多类型的化妆品中都添加了紫外线过滤剂，以防止日常生活中暴露在紫外线下的有害影响[123]。化妆品中的紫外线过滤剂通常浓度较高，在0.1%~10%。它们还在化妆品、涂料、黏合剂、塑料、光学产品和橡胶中用作光稳定剂和防晒剂[124]。由于它们的广泛使用，大量的紫外线过滤剂可以残存于不同的环境中，这促使科学界将它们视为新兴污染物[125]。由于这些化合物在水生环境中可能会对生物的健康产生不利影响，因此引起了人们的极大兴趣。紫外线过滤剂的生态毒理学暴露评估目前仍需要进一步研究[126]。一些研究表明，在评估水生生物中紫外线过滤剂的荷尔蒙活性时，需要考虑多组分的混合物。

对羟基苯甲酸酯被广泛用作许多 PCP 的防腐剂[127,128]，与佳乐麝香和吐纳麝香一样，它们在城市污水中以相对较高的浓度持续释放。一旦它们被释放到环境中，引起的主要担忧是对羟基苯甲酸已被证实的雌激素活性[129-131]。三氯生（2,4,4′-三氯-2′-羟基二苯醚）是一种广谱杀菌剂，遍及北美和欧洲，具有多种抗菌功能[132]。研究疑这种化合物可部分转化为其他毒性更强的化合物，如氯酚和二噁英[133,134]。

人们对如 PCP 这类的有机合成化学品在环境中的行为及其潜在影响的担忧正在持续增加。其较低的环境浓度、高极性、热不稳定性以及它们与复杂的环境基质的相互作用，使得对 PCP 的分析十分具有挑战性。样品制备，然后用 GC 或 HPLC 分离，并使用各种检测器（尤其是 MS）进行定性和定量分析，已成为标准分析流程。如前所述，特别值得关注的是大量使用、在环境中存留、生物积累和具有特殊生物活性的化合物，如合成麝香剂、紫外线过滤剂和抗菌剂。表 6.6 所示为基于 MAE 的方法在环境样品中测定 PCP 的应用。微波辅助萃取技术是固体样品提取的首选技术。PCP 在城市污水中以较高的浓度持续排放，许多化合物可以在污水污泥中找到，对这种基质的研究对于评估污水处理去除这些新兴污染物的效率以及它们进入环境的方式非常重要。基于 MAE 的几种方法已被提出用于分析污泥、沉积物和土壤。在这些研究中，MAE 与其他提取技术进行了比较，证明了微波提取的优势。

从活性污泥和消化污泥中提取麝香的方法有 PLE、SFE、索氏提取、固相微萃取（SPME）、液相萃取（LLE）和微波萃取（MAE）。除了 MAE，索氏萃取、PLE 和同时蒸馏/SE（SDSE）也被应用于沉积物中这类化合物的分析[135]。与经典技术（索氏提取和液液萃取）相比，微波萃取具有溶剂消耗低、速度快和回收传统技术不易释放的紧

表 6.6 **MAE 在个人护理产品分析中的应用**

分析物	样品	样品预处理	MAE 条件	萃取液处理	检测方法	回收率/%	LOD	RSD/%	发表年份	参考文献
多环麝香和硝基麝香	污水淤泥（1~2.5g）	10g Na_2SO_4	30mL 丙酮/正己烷（1∶1，体积比）；MAE（15min 升至 110℃并保持 20min）	蒸发、硅胶柱净化	GC-MS	80~105	27~41ng/g（多环麝香）；4ng/g（硝基麝香）	—	2007	[136，137]
佳乐麝香和吐纳麝香	沉积物（1g）	冻干	15mL95%丙酮+水+5mL 正己烷；MAE（100℃，15min）	离心、硅胶柱净化、氮吹浓缩	GC-MS/MS	98（佳乐麝香）；94（吐纳麝香）	0.4，0.6μg/kg	<5	2011	[138]
7 种苯并三唑紫外线过滤剂	海洋沉积物及污水淤泥（1g）	—	2mL 乙腈；MAE（300W，5min）	过滤、在线净化	UPLC-MS/MS	50~87	53~146ng/kg	7.8~16.3	2013	[139]
5 种二苯甲酮紫外线滤光剂、酮洛芬和两种光转化产品	地表水和沉积物（4g）	干燥	30mL 含 5%甲酸的甲醇/丙酮（1∶1，体积比）；MAE（800W，10min 升至 150℃，保持 10min）	离心、吹干、稀释、SPE 净化、衍生	GC-MS	80~99	0.1~1.4ng/g	—	2014	[124]
七种对羟基苯甲酸酯，两种烷基酚	土壤（2g）	干燥、过筛、均质、冷藏	柱内：玻璃柱，层尖端含有两个过滤层，填充 Na_2SO_4（2g）和硅藻土（2.5g）；吸附剂分离+10mL ACN 和 MAE（10min 升至 80℃，保持 5min）	衍生	GC-MS	78~112	0.4~1.13ng/g	<7	2012	[140]
三氯生、双酚 A、4-叔丁基-辛基酚及其他化学品	河水沉积物（5g）	干燥、均质	30mL 二氯甲烷/乙酸乙酯/丙酮（1∶1∶1）；MAE（10min 升至 110℃，保持 10min）	氮吹、加 IS、净化（深色提取液）	GC-MS	60~101	56~114μg/kg	—	2011	[91]

续表

分析物	样品	样品预处理	MAE 条件	萃取液处理	检测方法	回收率/%	LOD	RSD/%	发表年份	参考文献
对羟基苯甲酸丁酯、烷基酚、双酚 A 等	沉积物（0.1g）	—	5mL 40mmol/L［C16MIM］［Br］溶液，FMAE 搅拌，140W，90℃，6min	离心、过滤、乙腈稀释形成的 IL 微滴	HPLC-DAD	平均值 67	0.1~0.8mg/kg（LOQ）	<19	2012	［142］
对羟基苯甲酸丁酯，烷基酚，BPA 等	沉积物（0.1g）	过筛	30mL CTAB 水溶液；FMAE	衍生化，旋流，加热，离心。取出形成的微滴并稀释	GC-MS	107	0.02~0.36mg/kg	6.4~8	2012	［143］
多环麝香	水（20mL）	用 HCl 调节 pH 至 2~3，-4℃存放	20mL 水（4g NaCl）；MA-HS-SPME：80W，5min PDMS/DVB 纤维	—	GC-MS	64~102	0.05~0.1ng/L	1~9	2009	［145］
多环麝香	污水淤泥、沉积物（5g）	搅拌均质、过滤脱水	20mL 去离子水+3g NaCl；MA-HS-SPME	—	GC-MS	85~96	0.04~0.1ng/g	<16	2010	［146，147］
硝基麝香等	室内灰尘（0.8g）	过筛	8mL 正己烷+4Ml 1M 硫酸溶液（含 0.10%抗坏血酸，质量分数）；MAE（10min）	离心、Na_2SO_4 干燥、硅藻土（100mg/mL）净化、振荡、过滤	GC-μECD	88~97	1.03~3.26ng/g	<8.5	2012	［148］
多环麝香、硝基麝香	空气（1.6m^3）	取样：活性低容量 PUF，玻璃纤维过滤器	60mL 环己烷/丙酮（1∶1，体积比）；MAE（85℃，60min）	蒸发、环已烷复溶	GC-TQMS	65~120（MAE）；62~93（取样）	0.48ng/m^3（LOQ，开司米酮）	—	2014	［122］

注：ACN—乙腈；BPA—双酚 A；CTAB—十六烷基三甲基铵；［C16MIM］［Br］—1-十六基-3-甲基咪唑溴化铵；DAD—二极管阵列检测器；ECD—电子捕获检测器；FMAE—聚焦微波辅助萃取；GC—气相色谱；HPLC—高效液相色谱；IL—离子液体；IS—内标；LOD—检测限；LOQ—定量限；MAE—微波辅助萃取；MA-HS-SPME—微波辅助顶空-固相微萃取；MeOH—甲醇；MS—质谱；MS/MS—串联质谱；PDMS/DVB—聚硅氧烷/二乙烯基苯；PUF—聚氨酯泡沫塑料；RSD—相对标准偏差；SPE—固相萃取；TQMS—三重四极杆；UHPLC—超高效液相色谱；UV—紫外检测器。

密结合残留物的潜力等优点。与MAE相比，较新技术的缺点是分析物可能在萃取过程中损失或难以从基质中释放，需要优化的参数较多，设备成本较高，存在高水分基质和基质材料共萃取的问题。

使用MAE对污水污泥中多环麝香和硝基麝香的存在进行了研究，然后用硅胶柱进行净化，回收率为80%~105%，LODs在ng/g水平。SFE和MAE的性能相当[136,137]，但SFE所需的空气干燥样品制备步骤有可能使多环麝香和硝基麝香降解和/或挥发。

研究对MAE法、SDSE法、索氏提取法和超声探针法从沉积物中提取佳乐麝香和吐纳麝香的提取条件进行了优化[138]。通过正交设计实验对MAE操作条件进行了优化，结果表明MAE和超声探针前处理技术回收率高、重现性好、准确度高、效率高、溶剂消耗少、基质效应可接受，是更为有效的前处理技术。

研究使用微波能量从沉积物和污泥中提取了用作紫外线过滤剂二苯甲酮和苯并三唑衍生物[96,124,139]。另有研究报道了微波辅助固相萃取在线净化富集-UHPLC-MS/MS联用分析污泥和海洋沉积物中7种苯并三唑类紫外线稳定剂的方法[139]。对海洋沉积物和污水污泥的实际样品分析表明，其浓度在0.18~24ng/g。最近的一项研究开发了一种测定地表水和沉积物中8种二苯甲酮衍生化合物的方法，包括5种二苯甲酮类紫外线过滤剂[124]。提出的分析方法包括用固相萃取和MAE富集水样和沉积的样品，然后在GC-MS分析之前进行衍生化。对环境样本的分析结果表明，这些紫外线过滤剂的浓度高达650ng/g。

尽管环境样品中出现对羟基苯甲酸酯防腐剂的问题得到了很好的解决，但借助微波制备样品的应用并不普遍。研究提出了一种从土壤中同时提取7种对羟基苯甲酸酯和两种烷基酚的MAE方法，并与MSPD方法[140]进行了比较。提取物用BSTFA衍生化，用气相色谱-质谱（GC-MS）分析。微波池内的玻璃样品夹持器，可用于同时提取和净化样品。在所分析的大多数样品中，对羟基苯甲酸甲酯和对羟基苯甲酸丁酯的检出量为0.5~8.0ng/g。研究已提出一种多残留方法来分析90种新兴污染物，包括对羟基苯甲酸酯、二苯甲酮紫外线过滤剂、三氯生和药物，用于液体和固体环境基质的分析[96]（第6.6.1节和表6.4）。污水淤泥的分析采用MAE和SPE，消化污泥样品中的LOQs均在1ng/g以下。

近来，研究提出了MAE和连续SPE GC-MS分析相结合的方法，用于同时测定土壤、沉积物和污泥中的EDCs，包括对羟基苯甲酸酯、烷基酚、苯基酚、双酚A和三氯生[91]。MAE与超声辅助和索氏提取结果的系统比较表明，MAE在最短的提取时间（3min）里提供了最高的提取效率（接近100%）。测定结果发现污水污泥样本中含有所有目标化合物，浓度为36~164ng/kg。

一种基于MAE的多分析物方法被提出用于分析河流沉积物中人为来源的有机污染物[141]。三氯生、4-叔丁基-辛基酚和双酚A被列为代表性的目标化合物，具有广泛的物理化学性质。三种溶剂（二氯甲烷/乙酸乙酯/丙酮）的混合物提供了最佳和最一致的回收率。

研究提出了一种新的基于离子液体和FMAE的富集和萃取沉积物中有机污染物（包括3种多环芳烃、5种烷基酚和对羟基苯甲酸丁酯）的策略[142,143]。本研究的主要目的是扩大基于IL-表面活性剂的预富集程序对复杂固体样品的适用性。首次介绍了基于IL-1-十六基-3-甲基咪唑溴化物表面活性剂的萃取/富集过程。

与固相微萃取（SPME）一样，MAE已与其他提取技术相结合。研究提出了一种用MAE-HS-SPME快速无溶剂提取水、污泥、底泥、鱼和生物群等不同基质中6种常用合成多环麝香的方法[144-147]，系统考察并优化了各提取参数对提取效果的影响。将聚二甲基硅氧烷-二乙烯基苯（PDMS-DVB）纤维置于顶空，在180W的微波辐射下，不到4min即可有效地从水样中提取分析物。固体样品（5g与20mL水混合）可按同样方法提取。研究中对方法的准确度和精密度进行了评价，并论证了该方法适用于痕量污泥和沉积物中目标分析物的测定。平均回收率为85%~96%，检出限为0.04~0.1ng/g。作者通过与文献报道的其他方法的比较，指出了MA-HS-SPME的许多优点。索氏提取和PLE需要使用溶剂（分别为400mL和50mL），LOQs较高。MA-HS-SPME一步法无须额外的净化步骤，具有较高的处理量，是提取环境样品中麝香的一种良好的、环保的替代方法。

室内粉尘是一种复杂的基质，其特征是含有较高的有机碳，这些有机碳来源于皮肤组织、毛发纤维和螨类，这给提取和进一步测定带来了困难。研究提出了一种基于析因设计优化的微波辅助萃取农药和兴环境污染物的方法，用于测定室内粉尘中的硝基麝香（以及有机氯化合物和拟除虫菊酯类杀虫剂）[71]。

固体吸附剂，如聚氨酯泡沫（PUF）、Tenax TA和XAD，通常用于空气中合成麝香的取样[148]。研究开发了一种使用微波能量测定空气中合成麝香的高灵敏度和广泛适用的方法[122]。根据美国环保局关于半挥发性有机化合物的测定方法TO-10A，采用了带有过滤取样系统的低容量PUF。使用环己烷/丙酮进行微波辅助萃取，以提取分离的分析物，并提取保留在过滤器中的颗粒物。使用具有特定多反应监测模式的气相色谱-串联质谱质谱对样品进行了测定。

6.6.4 工业污染物

工业污染物可以广泛地认为包括工业生产、制造和排放的化合物，或者由于工业发展而出现在环境中的化合物。这类物质包括工业过程和生产中使用的物质，主要是化学工业中使用的物质。增塑剂、阻燃剂和表面活性剂存在于许多产品中，在这些产品中，口服、皮肤或吸入暴露被认为是无处不在的，并且几乎是连续的。

6.6.4.1 塑化剂和阻燃剂

邻苯二甲酸酯（PAE）可用于提高工业聚合物的可塑性。由于其广泛的应用，邻苯二甲酸盐在环境中无处不在。

阻燃剂（FR）是添加到潜在易燃材料中的化学物质。为了提高效果，常常要把各种不同的化学物质混合在一起。自从上世纪80年代多溴联苯被禁用以来，多溴二苯醚一直是一种流行的阻燃剂成分。2004年，欧盟禁止使用五溴二乙醚和八溴二乙醚混合物，2008年禁止使用十溴二乙醚。然而，多溴二苯醚存在于许多往年销售的消费品中，故仍可释放到环境中。新出现的溴化阻燃剂（BFR），包括溴酚、溴苯基醚、溴化苯基酯和其他溴芳族化合物，现在正在环境中被发现。其中一些已经使用多年，但直到最近才在环境中被发现[149]。HBCD是非芳香族溴化环烷烃，具有与许多持久性有机污染物相似的物理和化学性质：持久性、生物积累性、远程传输性和毒性。几种磷酸酯［有机磷酸酯（OPh）］被广泛用作阻燃添加剂和增塑剂。有机磷农药是污水、水和室内大气中普遍存在的污染物。尽管与BFR相比，OPh的毒性相对较低，但其使用量的增加和已报道的氯化OPh的一些负面影响增加了人们对长期接触这些物种可能产生的影响的担忧[150]。

为了可靠地测定不同环境基质（空气、水、污泥、沉积物和土壤）中的PAE和阻燃剂，研究已经提出了各种微波提取方法[151,152]。表6.7总结了最近基于MAE的方法。

研究对MAE与GC-MS联用进行了优化，并将其应用于湖泊沉积物和悬浮物中多环芳烃的分析。该方法表现良好，可以在该地区邻苯二甲酸盐的分布和人类活动之间建立关系[153]。还对土壤中的邻苯二甲酸盐进行了调查。提出了同时分析土壤中34个目标的多残留方法，包括17个邻苯二甲酸酯和16个多环芳烃[154]。在GC-MS/MS分析之前，土壤用MAE提取，萃取物通过固相萃取浓缩净化。该方法对大多数分析物均获得了满意的准确度和精密度，并具有较高的选择性和灵敏度，检出限低于1mg/kg。MAE后的HPLC-DAD分析被描述为测定土壤中邻苯二甲酸酯的一种良好和可靠的替代方法[155]。

采用间歇固相萃取装置，以间烟酸铜为吸附剂，与MAE联用，HPLC-UV为测定技术，研究了AMP中邻苯二甲酸酯和双酚A等几种干扰物质的存在。吸附剂从萃取液中除去分析物，从而提高萃取效率[156]。使用小体积正己烷通过超声波将分析物从吸附剂中释放出来。与传统的MAE方法相比，该方法具有有机溶剂消耗少、灵敏度高、选择性好等优点。

研究比较了液液萃取、UAE、MAE、分散液液微萃取（DLLME）、分散液液微萃取-漂浮有机滴凝固（DLLME-SFO）和浊点萃取（CPE）对水和化妆品样品中10种多环芳烃的萃取效果，以毛细管LC-UV进行测定[157]。对于MAE，方法是先用乙酸乙酯对样品进行涡旋，然后进行微波提取。各方法比较的结论是DLLME-SFO和DLLME是提取PAE的最佳方法，虽然这一结论主要是基于化妆品样品的分析。

基于表面活性剂和离子液体的同时使用的创新微萃取方法，已经被用于水样分析。其中，有研究开发基于离子液体的微波辅助DLME方法，并以HPLC-DAD分析，适用于水中5种多环芳烃的测定[158]。研究还探索了表面活性剂在CPE中的应用，采用胶束萃取相结合微波辅助反萃取和气相色谱分离从矿泉水中萃取6种邻苯二甲酸酯[159]。将邻苯二甲酸酯包埋在非离子表面活性剂Triton X-114的胶束中，通过离心萃取本体相中的邻苯二甲酸酯，该过程是在自制的离心微萃取管中进行。用不溶于水的溶剂处理富含表面活性剂的相，用短期微波反萃取目标分析物，用GC火焰离子化检测器直接测定。

FR主要在复杂的固体样品中测定。在进行色谱分析之前，通常需要在萃取后进行净化[160-163]。

污水污泥中多溴二苯醚的分析，为污水污泥作为土地应用生物固体再利用相关风险的评估提供了有价值的信息。为此，提出了一种基于GC-ICP-MS的分析方法，用于分析污泥中的6种多溴联苯醚[164]。从污泥中提取多溴联苯醚的最有效方法是用含0.1mol/L

表 6.7　MAE 在塑化剂和阻燃剂分析中的应用

分析物	样品	样品预处理	MAE 条件	萃取液处理	检测方法	回收率/%	LOD	RSD/%	发表年份	参考文献
15 种 PAE	沉积物（1g）和悬浮物	空气干燥、研磨、均质、过筛	10mL 甲醇/乙酸乙酯（1∶1，体积比）；MAE（150℃，20min）	蒸发、甲醇/乙酸乙酯（1∶1，体积比）复溶	GC-MS	83.9~109（沉积物）；91~109（悬浮物）	—	—	2014	[153]
17 种 PAE 及其他（16 种 PAH）	土壤（10g）	滴定	20mL 二氯甲烷/丙酮（1∶1，体积比）；MAE（1200W，10min 升至 120℃，保持 10min）	用无水 Na_2SO_4 烘焙研磨、SPE 净化、浓缩和溶剂交换、离心、蒸发并复溶（正己烷）	GC-MS/MS	—	0.04~0.84μg/kg	—	2010	[154]
6 种 PAE	土壤（1g）	冻干、研磨、过筛	5mL 乙腈；MAE（100℃，30min）	过滤、蒸干、复溶（乙腈）	HPLC-DAD	84~115	1.24~3.15μg/L	—	2010	[151]
3 种 PAE 和 BPA	大气粒相物	QF20 石英纤维过滤器	10mL 甲醇+μSPE 装置（含 10mg 异烟酸铜（Ⅱ）的聚丙烯微孔膜）；MAE（400W，15min）	正己烷润洗、滤纸干燥、超声 25min、过滤、吹干、正己烷复溶	HPLC-UV	81.7~119	2.0~8.5ng/L	<10	2015	[156]
10 种 PAE	水（0.3mL）	—	150μL 乙酸乙酯、涡旋 1min；MAE（700W，6min）；离心两次后收集上清液（110μL）	合并两部分、浓缩至干、甲醇复溶	HPLC-UV	74~107	0.02~0.17μg/mL	—	2012	[157]
5 种 PAE	水	—	基于 IL 的 MADLLME；MAE（60℃，2min）	—	HPLC-DAD	85.2~103.3	0.71~1.94μg/L	<6	2013	[158]
6 种 PAE	矿泉水	邻苯二甲酸酯被包裹在 Triton X-114 的胶束中，并通过离心从样品本体中萃取	CPE 与 MABE 联用：用溶剂对富表面活性剂相进行处理，MAE 反萃取出 PAEs	—	GC-FID	89.1~96.3	11.5~19.3μg/L	2.3~5.7	2014	[159]

五溴二苯醚、八溴二苯醚混合物和十溴二苯醚	室内尘土（0.8g）	加 30ng PCB30	8mL 正己烷+4mL10% NaOH 水溶液；MAE（80℃，15min）	Na_2SO_4 干燥、硅藻土净化、过滤、浓缩至 0.2mL	GC-μECD	90~108	0.044~1.44ng/g	4~13	2007	[160]
五溴二苯醚、六溴二苯醚和八溴二苯醚混合物	粒相物（2.5）	石英纤维过滤	50mL 正己烷/丙酮（1∶1，体积比）；MAE（600W，70℃，2min）	过滤、吹干、二氯甲烷复溶、GPC 净化、吹干、正己烷复溶	GC-MS/MS	88~106	0.021~0.070pg/m^3	<18	2014	[161]
5 种 OPh	河流和海洋沉积物（5g）	冻干并过筛	40mL 丙酮；MAE（800W，120℃，20min）	过滤、浓缩、SPE 净化、过氧化铝柱、洗脱、蒸干、复溶	GC-MS	62~106	0.05~0.15	5.3~8.5	2009	[162]
8 种 BDE	污水淤泥（干重 0.25~0.5g）	均质、15g 无水 Na_2SO_4+100ng BDE 77	30mL 正己烷/丙酮（3∶1，体积比）；MAE（130℃，35min）	5mL H_2SO_4/水（1∶1，体积比）；溶剂层转移至无水 Na_2SO_4 小柱，正己烷萃取、浓缩至<1mL、硅胶柱净化	GC-MS	80~110	1~7ng/g	<10	2007	[163]
6 种 BDE	淤泥（0.5g）	干燥	10mL 含 0.1mol/L HCl 的甲醇；MAE（1200W，90℃，5min）	三柠檬酸缓冲液（pH 6）+ 2mL 异辛烷，机械振荡	GC-ICP-MS	95~104	0.2~0.3ng/g	2.2~5.7	2016	[164]
9 种 BDE	土壤（0.5g）	干燥	10mL 丙酮/正己烷；MAE（1200W，90℃，10min）	过滤、加活性铜、浓缩、净化、（硅藻土）、蒸发、正己烷复溶	GC-MS	69~130	1~5ng/g	1~30	2014	[165]
7 种 BDE	土壤和沉积物（3.0g）	干燥	30mL 正己烷/丙酮（1∶1，体积比）；MAE（1500W，115℃，15min）	离心、加入内标、蒸发、溶解、浓缩、GPC 净化、硅胶柱和无水 Na_2SO_4 净化、浓缩	GC-MS	50~96（土壤）	<24.8pg/g	2.4~20（土壤）	2010	[167]

续表

分析物	样品	样品预处理	MAE 条件	萃取液处理	检测方法	回收率/%	LOD	RSD/%	发表年份	参考文献
α，β，γ-HBCD	海洋沉积物（5g）	干燥、均质	40mL 丙酮/正己烷（1∶3，体积比）；MAE（800W，90℃，12min）	过滤、浓缩；TBA-S+异丙醇（振荡）+含50% H_2SO_4 的正己烷（振荡、离心、蒸发溶剂）；净化（Na_2SO_4 和硅胶柱）；蒸发并以甲醇复溶	LC-MS	68~91	5~10pg/g	<11	2009	[168]
TnBP 和 TEHP	地表水及废水（20mL）	加 NaCl 并调节 pH	HS-SPME（PDMS/DVB 纤维）；MAE（140W，5min）	—	GC-MS	86~106	0.2~1.5ng/L	6~15	2011	[169]
OPh 和塑化剂	河流沉积物（0.5g）	干燥、过筛	5mL 丙酮+5mL 乙腈；MAE（150℃）	离心、浓缩硅胶柱净化、乙酸乙酯洗脱、加入内标并浓缩	GC-ICP-MS	78~105	2~4ng/g	<12	2009	[171]

注：ACN—乙腈；BPA—双酚 A；CPE—浊点萃取；DAD—二极管阵列检测器；ECD—电子捕获检测器；FID—火焰离子化检测器；GC—气相色谱；GPC—凝胶渗透色谱；HBCD—六溴环十二烷；HPLC—高效液相色谱；HS—顶空；ICP—电感耦合等离子体；IL—离子液体；LC—液相色谱；LOD—检出限；MABE—微波辅助反萃取；MA-DLLME—微波辅助分散液相微萃取；MAE—微波辅助萃取；MeOH—甲醇；MS—质谱；MS/MS—串联光谱法；NaCl—氯化钠；OPh—有机磷酸酯；PAE—邻苯二甲酸酯；PAH—多环芳烃；PCB—多氯联苯；PDMS-DVB—聚硅氧烷二乙烯苯；RSD—相对标准偏差；SPE—固相萃取；TBA-S—四丁基亚硫酸铵；TEHP—三（2-乙基己基）磷酸酯；TnBP—磷酸三正丁酯；UV—紫外检测器。

盐酸的甲醇中进行机械振荡或 MAE，然后加入三柠檬酸缓冲液和异辛烷。该方法非常灵敏，检出限为 0.2~0.3ng/g。

在一个通过堆肥改良不同有机污染物暴露的土壤项目中，利用 GC-MS[165] 对蔬菜和土壤中的几种多溴二苯醚进行了分析。采用聚焦超声固液萃取（FUSLE）对样品进行预处理，并与以 EPA3546 方法为基础的 MAE 法进行了比较，结果表明采用丙酮/正己烷作为萃取溶剂，萃取物得到净化，洗脱液在正己烷中蒸发和复溶。与 FUSLE 相比，MAE 提取了更多的干扰物，几种同系物的回收率均在 100%以上。

研究还对来自中国快速工业化地区的各种环境基质（包括地表水、沉积物/土壤、植物和动物组织）中的 BDE 进行了量化，以评估具有潜在 PBDE 污染源的低压电气行业如何影响当地环境[166]。确定 MAE 与 GC-MS 相结合的分析程序，是基于先前的一项研究。在该研究中，作者比较了索氏提取、PLE 和 MAE，以评估它们在从土壤和鱼类样品中提取多氯联苯和多溴联苯醚的性能[167]。对于多溴联苯醚，MAE 和 PLE（或索氏提取）的结果之间的差异表明，前者的萃取条件需要根据基质和被分析物的特性进行仔细的优化，特别是为了避免较高溴化的 BDE 同系物的降解，以提高它们的萃取率。

研究采用 MAE-HPLC-电喷雾离子阱质谱（HPLC-ESI-ITMS）联用技术，研究了海洋沉积物中的 HBCD 非对映异构体（α-、β-和 γ-HBCD）[168]。将 MAE 提取效率与索氏提取法和 PLE 法进行了比较，发现 MAE 法和 PLE 法提取底泥中的六溴环戊二烯残留量更有效，因其时间和溶剂消耗更少。在样品制备方面，MAE 法比 PLE 法具有更高的处理通量（整个 MAE 过程需要 35min，最多可以同时提取 6 个样品）。该方法灵敏度高，5g 沉淀物的最低检出限低于 40pg/g（干重）。

研究报道了地表水和污水处理厂进/出水样品中有机磷阻燃剂的测定[169]。采用一步无溶剂 MA-HS-SPME，然后以 GC-MS（SIM）测定。MA-HS-SPME 方法简单、高效、环保，是测定水样中有机物的一种较好的替代萃取方法。选择磷酸三丁酯（TnBP）和磷酸三（2-乙基己基）酯（TEHP）作为模型化合物进行方法的开发和验证。当体系受到微波辐射时，置于顶空的 PDMS-DVB 纤维可以有效地提取化合物。TnBP 和 TEHP 的 LOQ 分别为 0.5ng/L 和 4ng/L。

MAE 萃取技术也被用于分析废水和沉积物样品中的有机磷酸酯（OPh）[170,171]。采用微波辅助萃取（MAE）和固相微萃取（SPME）对废水中的几种痕量有机磷进行提取和预富集，然后用气相色谱-质谱（GC-MS）进行磷的特异性检测。并用气相色谱-飞行时间质谱联用（GC-TOF-MS）对样品中的有机磷进行确证。OPh 的检出限低于 50ng/L，在固相微萃取前的样品制备过程中使用 MAE 可以检测到 TEHP，这在以前的研究中是很难做到的。用 SPME-GC-TOF-MS 也证实了有机磷阻燃剂（OPFR）的存在。

6.6.4.2 酚类化合物和直链烷基苯磺酸盐

酚类化合物是世界上使用最广泛的化学品之一。双酚 A 用于生产阻燃剂和许多其他产品，具有雌激素活性，属于野生动物和人类健康 EDC 优先清单的第 1 类[172]。双酚 A 在世界各地的海洋和淡水环境中被检测到相当大的浓度[173]，它经常被包括在酚类表面活性剂的测定方法中。

烷基酚聚氧乙烯醚（APEO）是应用最广泛的表面活性剂之一，特别是辛基酚聚氧乙烯醚和壬基酚聚氧乙烯醚（NPEO）是市场上最常见的两种非离子表面活性剂。APEO 被排放到废水处理设施或直接排放到环境中。在生物废水处理过程中，它们被部分转化为更持久的代谢物[174]。其中，代谢物壬基酚二乙氧基酸盐和壬基酚单羧酸盐分别属于野生动物和人类健康 EDC 优先列表的第 1 类和第 2 类[175]。

直链烷基苯磺酸盐（LA）主要用于家用洗衣、洗碗剂和万能清洁剂。LA 还在纺织加工、食品和皮革工业中用作分散剂和清洁剂；在工业过程中用作乳化剂；以及用于作物保护剂配方中引发聚合[176]。

表 6.8 所示为 MAE 应用于 APEO、酚衍生物和 LA 分析中的方法。几种酚类化合物已与其他药物（E2、E3、EE）[87]和 PCP（对羟基苯甲酸酯和三氯生）[91]一起进行了测定，如双酚 A 和其他酚类化合物。这些案例已在表 6.4 和表 6.6 中进行了总结，并分别在 6.6.1 和 6.6.3 中进行了讨论。研究还提出了使用离子液体从沉积物种同时提取烷基酚、优先污染物（PAH）和 PCP（对羟基苯甲酸丁酯），这些方法在 6.6.3 中进行了讨论，并在表 6.6 中进行了总结[142,143]。

研究比较了 UAE、MAE 和 PLE 三种萃取技术，以评价它们在复杂的污水污泥基质中测定双酚 A 及其氯化衍生物的效率[177]。通过优化了各提取参数，得到乙酸乙酯/水，1000W、90℃和 10min 条件下的萃取效率最高。使用 LC-MS/MS 对所选化合物进行检测和定量，并在负离子 APCI 和 MRM 模式下检测。经统计学比较结果表明，三种萃取技术无显著差异。

固体样品中 LA 的提取主要采用索氏法和超声法，通常以甲醇作提取溶剂。在提取污水污泥、沉积物和堆肥样品中的 LA 方面，MAE 已被证明是一种可靠、高效、可重复使用的替代技术，最终以 HPLC-FL 进行分析[178,179]，在准确性和精度方面取得了较好结果。MAE 的优点之一是时间和溶剂消耗少，不需要对提取物进行额外的净化，提取效率不依赖于 LA 烷基链的长度。微波提取也被用于从堆肥样品中提取 LA，连同其他 NPEO 和一种增塑剂（DEHP），然后进行 HPLC-FL 分析[180]。若在 MAE 后以 GC-MS 进行分析，则在分析之前要进行净化和衍生化步骤[85,181]。在所有的情况下，回收率较好，RSD 和 LOD 均较低，证明了 MAE 对从环境样品中提取这些化合物的适用性。

6.7 结论与展望

微波辅助萃取技术（MAE）已成功地应用于多种环境样品中多类化合物的萃取，其中包括农药和新兴污染物（药物、个护产品和工业污染物）。MAE 是一种易于实施和快速的技术，展示了与经典技术（如索氏提取）和其他较新技术（如 SFE 或 PLE）相似或往往更高的提取效率。MAE 的优点主要在于大大减少了提取时间、能耗和溶剂消耗，以及能同时萃取多个样品，从而增加了样品处理通量。此外，MAE 设备价格也较为适中。

由于基质水分、溶剂性质、萃取时间、施加功率和密闭容器中的温度等影响 MAE 萃取效率的这些因素是完全可控的，因此 MAE 条件的优化较为简单易行。在某些情况下，还可使用化学计量学方法进行多变量优化。

表 6.8　MAE 在酚类化合物和直链烷基苯磺酸盐分析中的应用

分析物	样品	样品预处理	MAE 条件	萃取液处理	检测方法	回收率/%	LOD	RSD/%	发表年份	参考文献
BPA 及其氯化衍生物	污水淤泥（1g）	干燥	10mL 乙酸乙酯+400μL Milli-Q 水；MAE（1000W，90℃，10min）	离心 30min、蒸发、流动相复溶、离心	LC-MS/MS	97.7~103	2~9ng/g	<6	2012	[177]
LA	污水淤泥（0.5g）	40℃干燥、研磨并过筛	5mL 甲醇；MAE（250W，10min）	玻璃棉过滤	HPLC-FL HPLC-DAD	83~102	0.33~1.83 mg/kg（FL）；0.25~2.50 mg/kg（DAD）	<5.42（FL）<5.18（DAD）	2007	[178]
LA	污水淤泥（0.5g）	干燥、研磨、过筛	5mL 甲醇；MAE（250W，10min）	玻璃棉过滤	HPLC-FL CE-DAD	>85	3.03mg/kg（HPLC-FL）；21.0mg/kg（CE-DAD）	2.98（CE）	2009	[179]
LA、NPEO、DEHP	堆肥（0.2~0.3g）	干燥	20m；MAE（600W，100℃，15min）	浓缩、10mL 甲醇/水复溶（1%HCHO 和 0.05mol/L SDS）	HPLC-FL	>92.3	46~194μg/L	<22.5	2009	[180]
NP、OP、BPA 及其他药物	沉积物（3g）	冻干、研磨并过筛	25mL 甲醇，铜粒，110℃，15min	浓缩、加水、SPE 净化、衍生	GC-MS/MS	86~114	0.08~0.14ng/g	<20	2009	[85]
OP、NP、BPA、4-CP	沉积物（4g）	干燥，加入内标、加铜粒	25mL 甲醇，110℃，20min	过滤，浓缩、稀释；SPE 净化（Oasis HLB）、浓缩、衍生	GC-MS	74~105	0.15~2.9ng/g	2.0~8.4	2011	[181]

注：BPA—双酚 A；CE—毛细管电泳；CP—异丙基苯酚；DAD—二极管阵列检测器；DEHP—邻苯二甲酸二（2-乙基己基）酯；FL—荧光检测器；GC—气相色谱；HPLC—高效液相色谱；LA—直链烷基苯磺酸盐；LC—液相色谱；LOD—检测限；MAE—微波辅助萃取；MS—质谱；MS/MS—串联质谱；NP—4-壬基酚；NPEO—壬基酚聚氧乙烯醚；OP—4-叔辛基苯酚；RSD—相对标准偏差；SDS—十二烷基磺酸钠；SPE—固相萃取。

MAE 在大多数农药和新兴污染物提取中的应用，是在封闭的容器系统中使用极性溶剂（甲醇、丙酮）在高温高压下进行萃取的。在某些情况下，也可使用水作为萃取溶剂，也是一种有效的萃取方法，从而可避免或尽量减少有机溶剂的使用。

MAE 的最新应用在于同时萃取多种分析物（农药、五氯苯酚、药品、工业污染物）。此外，还有基于使用非有机溶剂的新型萃取剂的浓缩策略方面的应用。基于此，提出了使用基于离子液体的表面活性剂或水溶液中进行 MAE，以从环境样品中提取农药和新兴污染物。对于含有高浓度、中等极性化合物的样品，简单的无机无溶剂微波萃取可为其在高温高压条件下更好地将其溶解于水中提供参考。

尽管微波萃取是一种典型的固体样品萃取技术，但与其他微萃取方法（如 μSPE、DLLME 或 LPME）相结合，也可适用于液体样品的萃取。MAE 已成功地用于从不同水样和空气中提取农药（DDT、三嗪类）、抗生素和增塑剂。MAE 发展的另一个趋势是实施无溶剂萃取，如 MAE 与固相微萃取（SPME）相结合，该方法已被开发用于快速测定水和不同固体环境和生物基质中的农药、药物、个人护理产品和工业污染物。

从技术的角度来看，最近的智能批处理微波系统现在包括预定义的方法库，它们能够计数样品的数量和识别装载的容器类型。在以前的系统中，温度传感器只插入一个装有溶剂的样品容器组件中，以创建一个反馈控制，然后调节微波输入功率。该系统的一个缺点是，这种设备假设了系统中所有样品具有等效的微波吸收能力。这项技术最近已经转换为底部温度传感器，当样品在系统内旋转时，测量样品的温度，然后根据所有测量温度的平均值来调节微波输入功率，从而改善了对整个萃取过程的控制。

此外，最近还推出了更灵活的顺序处理样品的微波系统，与批量式微波系统相比，具有许多潜在的优势。这种顺序方法使操作员能够为每个样品选择精确的温度控制，或使用不同的萃取溶剂处理不同类型的样品，所有这些样品都在同一处理批次。然而，要达到间歇式的萃取方式，主要的障碍是萃取后的样品和溶剂的分离。未来的大多数应用领域可能集中在提高最新推出的时序系统的灵活性上。例如，管理每个单独样品的提取条件的能力应该使用户能够开发出适用于不同环境基质中不同化学品的稳健、可重复性的方法。

总的来说，MAE 似乎是一种很好的绿色萃取方法，它可以准确地测定新兴有机污染物，包括杀虫剂和新兴的污染物，这使得它可以扩展到环境监管领域。

致谢

感谢欧洲区域发展基金 2007—2013（FEDER）和项目 GPC2014/035 和 R2014/013（联合研究计划，Xunta de Galicia）、UNST13-1E-2152 和 CTQ2013-46545-P（西班牙明尼科经济和竞争力部）的支持。

参考文献

[1] L. Chen, D. Song, Y. Tian, L. Ding, A. Yu, H. Zhang, Application of on-line microwave

sample-preparation techniques, TrAC-Trend. Anal. Chem. 27 (2008) 151-159.

[2] E. Destandau, T. Michel, C. Elfakir, Microwave-Assisted Extraction, Royal Society of Chemistry Publishing, Cambridge, United Kingdom, 2013, pp. 113-115.

[3] P. Tatke, Y. Jaiswal, An overview of microwave assisted extraction and its applications in herbal drug research, Res. J. Med. Plant 5 (2011) 21-31.

[4] M. Letellier, H. Budzinski, Microwave assisted extraction of organic compounds, Analusis 27 (1999) 259-270.

[5] M. Tobiszewski, A. Mechlinska, B. Zygmunt, J. Namiesnik, Green analytical chemistry in sample preparation for determination of trace organic pollutants, TrAC-Trend. Anal. Chem. 28 (2009) 943-951.

[6] L. Sanchez-Prado, C. Garcia-Jares, M. Llompart, Microwave-assisted extraction: application to the determination of emerging pollutants in solid samples, J. Chromatogr. A 1217 (2010) 2390-2414.

[7] A. Abu-Samra, J. S. Morris, S. R. Koirtyohann, Wet ashing of some biological samples in a microwave oven, Anal. Chem. 47 (1975) 1475-1477.

[8] K. Ganzler, A. s. Salgó, K. r. Valkó, Microwave extraction: a novel sample preparation method for chromatography, J. Chromatogr. A 371 (1986) 299-306.

[9] V. Lopez-Avila, R. Young, W. F. Beckert, Microwave-assisted extraction of organic compounds from standard reference soils and sediments, Anal. Chem. 66 (1994) 1097-1106.

[10] F. I. Onuska, K. A. Terry, Microwave extraction in analytical chemistry of pollutants: polychlorinated biphenyls, J. High Resolut. Chromatogr. 18 (1995) 417-421.

[11] F. I. Onuska, K. A. Terry, Extraction of pesticides from sediments using a microwave technique, Chromatographia 36 (1993) 191-194.

[12] O. Donard, B. Lalere, F. Martin, R. Lobinski, Microwave-assisted leaching of organotin compounds from sediments for speciation analysis, Anal. Chem. 67 (1995) 4250-4254.

[13] M. P. Llompart, R. A. Lorenzo, R. Cela, J. R. J. Paré, J. M. R. Bélanger, K. Li, Phenol and methylphenol isomers determination in soils by in-situ microwave-assisted extraction and derivatisation, J. Chromatogr. A 757 (1997) 153-164.

[14] M. P. Llompart, R. A. Lorenzo, R. Cela, K. Li, J. M. R. Bélanger, J. R. J. Paré, Evaluation of supercritical fluid extraction, microwave-assisted extraction and sonication in the determination of some phenolic compounds from various soil matrices, J. Chromatogr. A 774 (1997) 243-251.

[15] EPA, SW-846 Test Method 3546: Microwave Extraction, 2007. Available at: https://www.epa.gov/hw-sw846/sw-846-test-method-3546-microwave-extraction.

[16] K. Li, J. M. R. Bélanger, M. P. Llompart, R. D. Turpin, R. Singhvi, J. R. J. Paré, Evaluation of rapid solid sample extraction using the microwave-assisted process (MAP[TM]) under closed-vessel conditions, Spectr. Int. J. 13 (1997) 1-14.

[17] L. Sanchez-Prado, C. Garcia-Jares, T. Dagnac, M. Llompart, Microwave-assisted extraction of emerging pollutants in environmental and biological samples before chromatographic determination, TrAC-Trend. Anal. Chem. 71 (2015) 119-143.

[18] J. R. Dean, Microwave extraction, in: J. Pawliszyn (Ed.), Comprehensive Sampling and Sample Preparation, Elsevier, Netherlands, 2012.

[19] A. Zlotorzynski, The application of microwave radiation to analytical and environmental chemistry, Crit. Rev. Anal. Chem. 25 (1995) 43-76.

[20] L. Jassie, R. Revesz, T. Kierstead, E. Hasty, S. Metz, in: H. M. Kingston, S. J. Haswell

(Eds.), Microwave-enhanced Chemistry, Fundamentals, Sample Preparation and Applications, American Chemical Society, Washington, 1997, p. 569.

[21] V. Camel, Microwave-assisted solvent extraction of environmental samples, TrAC-Trend. Anal. Chem. 19 (2000) 229-248.

[22] W. E. Lamb Jr., R. C. Retherford, Fine structure of the hydrogen atom by a microwave method, Physiol. Rev. 72 (1947) 241.

[23] K. Madej, Microwave-assisted and cloud-point extraction in determination of drugs and other bioactive compounds, TrAC-Trend. Anal. Chem. 28 (2009) 436-446.

[24] V. Camel, Recent extraction techniques for solid matrices-supercritical fluid extraction, pressurized fluid extraction and microwave-assisted extraction: their potential and pitfalls, Analyst 126 (2001) 1182-1193.

[25] M. P. Llompart, R. A. Lorenzo, R. Cela, J. R. J. Paré, Optimization of a microwaveassisted extraction method for phenol and methylphenol isomers in soil samples using a central composite design, Analyst 122 (1997) 133-137.

[26] J. T. Zacharia, Identity, Physical and Chemical Properties of Pesticides, Pesticides in the Modern World-trends in Pesticides Analysis, Intech Publisher, Rijeka, 2011, pp. 1-18.

[27] EU-Pesticides Database, Available at:, in: E. Commission (Ed.), 2009. http://ec.europa.eu/food/plant/pesticides/eu-pesticides-database/public/? event=homepage&language=EN.

[28] B. Petrie, R. Barden, B. Kasprzyk-Hordern, A review on emerging contaminants in wastewaters and the environment: current knowledge, understudied areas and recommendations for future monitoring, Water Res. 72 (2015) 3-27.

[29] K. M. Dimpe, P. N. Nomngongo, Current sample preparation methodologies for analysis of emerging pollutants in different environmental matrices, TrAC-Trend. Anal. Chem. 82 (2016) 199-207.

[30] É. A. Souza-Silva, R. Jiang, A. Rodríguez-Lafuente, E. Gionfriddo, J. Pawliszyn, A critical review of the state of the art of solid-phase microextraction of complex matrices I. Environmental analysis, TrAC-Trends Anal. Chem. 71 (2016) 224-235.

[31] J. L. Tadeo, R. A. Pérez, B. Albero, A. I. García-Valcáarcel, C. Sánchez-Brunete, Review of sample preparation techniques for the analysis of pesticide residues in soil, J. AOAC Int. 95 (2012) 1258-1271.

[32] United States Environmental Protection Agency (EPA), DDT-A Brief History and Status. Avalilable at: https://www.epa.gov/ingredients-used-pesticide-products/ddt-brief-history-and-status.

[33] Y. Merdassa, J. F. Liu, N. Megersa, M. Tessema, An efficient and fast microwave-assisted extraction method developed for the simultaneous determination of 18 organochlorine pesticides in sediment, Int. J. Environ. Anal. Chem. 95 (2015) 225-239.

[34] C. Liao, J. Lv, J. Fu, Z. Zhao, F. Liu, Q. Xue, G. Jiang, Occurrence and profiles of polycyclic aromatic hydrocarbons (PAHs), polychlorinated biphenyls (PCBs) and organochlorine pesticides (OCPs) in soils from a typical e-waste recycling area in Southeast China, Int. J. Environ. Health. Res. 22 (2012) 317-330.

[35] P. N. Carvalho, P. N. R. Rodrigues, F. Alves, R. Evangelista, M. C. P. Basto, M. T. S. D. Vasconcelos, An expeditious method for the determination of organochlorine pesticides residues in estuarine sediments using microwave assisted pre-extraction and automated headspace solid-phase microextraction coupled to gas chromatography-mass spectrometry, Talanta 76 (2008) 1124-1129.

[36] D. Vega, Z. Sosa, J. J. Santana-Rodríguez, Application of microwave-assisted micellar extraction combined with solid-phase microextraction and high-performance liquid chromatography with UV detection for the determination of organochlorine pesticides in different mud samples, Int. J. Env. Anal. Chem. 88 (2008) 185-197.

[37] P. V. Kumar, J. F. Jen, Rapid determination of dichlorodiphenyltrichloroethane and its main metabolites in aqueous samples by one-step microwave-assisted headspace controlled-temperature liquid-phase microextraction and gas chromatography with electron capture detection, Chemosphere 83 (2011) 200-207.

[38] R. Wang, P. Su, Q. Zhong, Y. Zhang, Y. Yang, Ionic liquid liquid-based microwave-assisted extraction of organochlorine pesticides from soil, J. Liq. Chromatogr. Relat. Technol. 36 (2013) 687-699;

[38a] S. Armenta, F. A. Esteve-Turrillas, S. Garrigues, M. de la Guardia, Green analytical chemistry: the role of green extraction techniques, in: E. Ibanez, A. Cifuentes (Eds.), Green Extraction Techniques: Principles, Advances and Applications, vol. 76, 2017, pp. 1-25.

[39] M. N. U. Al Mahmud, F. Khalil, M. M. Rahman, M. I. R. Mamun, M. Shoeb, A. M. A. El-Aty, J. H. Park, H. C. Shin, N. Nahar, J. H. Shim, Analysis of DDT and its metabolites in soil and water samples obtained in the vicinity of a closed-down factory in Bangladesh using various extraction methods, Environ. Monit. Assess. 187 (2015) 1-12.

[40] S. Di, S. Shi, P. Xu, J. Diao, Z. Zhou, Comparison of different extraction methods for analysis of 10 organochlorine pesticides: application of MAE-SPE method in soil from Beijing, Bull. Environ. Contam. Toxicol. 95 (2015) 67-72.

[41] W. Wang, B. Meng, X. Lu, Y. Liu, S. Tao, Extraction of polycyclic aromatic hydrocarbons and organochlorine pesticides from soils: a comparison between Soxhlet extraction, microwave-assisted extraction and accelerated solvent extraction techniques, Anal. Chim. Acta 602 (2007) 211-222.

[42] E. Fuentes, M. E. Baez, R. Labra, Parameters affecting microwave-assisted extraction of organophosphorus pesticides from agricultural soil, J. Chromatogr. A 1169 (2007) 40-46.

[43] Y. S. Su, C. T. Yan, V. K. Ponnusamy, J. Jen, Novel solvent-free microwave-assisted extraction coupled with low-density solvent-based in-tube ultrasound-assisted emulsification microextraction for the fast analysis of organophosphorus pesticides in soils, J. Sep. Sci. 36 (2013) 2339-2347.

[44] G. Abdel-Nasser, A. M. Al-Turki, M. I. Al-Wabel, M. H. El-Saeid, Behavior of atrazine and malathion pesticides in soil: simulation of transport process using numerical and analytical models, Res. J. Environ. Sci. 5 (2011) 221.

[45] M. I. Al-Wabel, G. Abdel-Nasser, A. M. Al-Turki, M. H. El-Saeid, Behavior of atrazine and malathion pesticides in soil: sorption and degradation processes, J. Appl. Sci. 10 (2010) 1740-1747.

[46] H. Barchanska, M. Czalicka, A. Giemza, Simultaneous determination of selected insecticides and atrazine in soil by MAE-GC-ECD, Arch. Environ. Prot. 39 (2013) 27-40.

[47] J. Shah, M. R. Jan, B. Ara, Quantification of triazine herbicides in soil by microwave-assisted extraction and high-performance liquid chromatography, Environ. Monit. Assess. 178 (2011) 111-119.

[48] Q. Zhong, P. Su, Y. Zhang, R. Wang, Y. Yang, In-situ ionic liquid-based microwaveassisted dispersive liquid-liquid microextraction of triazine herbicides, Microchim. Acta 178 (2012) 341-347.

[49] P. Paiga, S. Morais, M. Correia, A. Alves, C. Delerue-Matos, A multiresidue method for the analysis of carbamate and urea pesticides from soils by microwave-assisted extraction and liquid chromatography with photodiode array detection, Anal. Lett. 41 (2008) 1751-1772.

[50] Q. Zhong, P. Su, Y. Zhang, R. Wang, Y. Yang, Ionic liquid based microwave-assisted extraction of triazine and phenylurea herbicides from soil samples, Anal. Methods 4 (2012) 983-988.

[51] H. Li, Y. Wei, J. You, M.J. Lydy, Analysis of sediment-associated insecticides using ultrasound assisted microwave extraction and gas chromatography-mass spectrometry, Talanta 83 (2010) 171-177.

[52] M.C. Hernández-Soriano, A. Peña, M.D. Mingorance, Response surface methodology for the microwave-assisted extraction of insecticides from soil samples, Anal. Bioanal. Chem. 389 (2007) 619-630.

[53] J. Shah, M.R. Jan, B. Ara, Quantification of pendimethalin in soil and garlic samples by microwave-assisted solvent extraction and HPLC method, Environ. Monit. Assess. 175 (2011) 103-108.

[54] Z. Vryzas, A. Tsaboula, E. Papadopoulou-Mourkidou, Determination of alachlor, metolachlor, and their acidic metabolites in soils by microwave-assisted extraction (MAE) combined with solid phase extraction (SPE) coupled with GC MS and HPLC-UV analysis, J. Sep. Sci. 30 (2007) 2529-2538.

[55] Y. Merdassa, J.F. Liu, N. Megersa, Development of a one-step microwave-assisted extraction procedure for highly efficient extraction of multiclass fungicides in soils, Anal. Methods 6 (2014) 3025-3033.

[56] R. Karmakar, S.B. Singh, G. Kulshrestha, Water based microwave assisted extraction of thiamethoxam residues from vegetables and soil for determination by HPLC, Bull. Environ. Contam. Toxicol. 88 (2012) 119-123.

[57] M.H. El-Saeid, A.M. Al-Turki, M.I. Al-Wable, G. Abdel-Nasser, Evaluation of pesticide residues in Saudi Arabia ground water, Res. J. Environ. Sci. 5 (2011) 171.

[58] W. Zhang, J. Lv, R. Shi, C. Liao, A rapid screening method for the determination of seventy pesticide residues in soil using microwave-assisted extraction coupled to gas chromatography and mass spectrometry, Soil Sediment Contam. 21 (2012) 407-418.

[59] M.H. El-Saeid, M.I. Al-Wabel, G. Abdel-Nasser, A.M. Al-Turki, A.G. Al-Ghamdi, One-step extraction of multiresidue pesticides in soil by microwave-assisted extraction technique, J. Appl. Sci. 10 (2010) 1775-1780.

[60] M.I. Al-Wabel, M.H. El-Saeid, A.M. Al-Turki, G. Abdel-Nasser, Monitoring of pesticide residues in Saudi Arabia agricultural soils, Res. J. Environ. Sci. 5 (2011) 269.

[61] E.C. Kalogridi, C. Christophoridis, E. Bizani, G. Drimaropoulou, K. Fytianos, Part II: temporal and spatial distribution of multiclass pesticide residues in lake sediments of northern Greece: application of an optimized MAE-LC-MS/MS pretreatment and analytical method, Environ. Sci. Pollut. Res. 21 (2014) 7252-7262.

[62] V. Alcántara-Concepción, S. Cram, R. Gibson, C.P. de León, M. Mazari-Hiriart, Method development and validation for the simultaneous determination of organochlorine and organophosphorus pesticides in a complex sediment matrix, J. AOAC Int. 96 (2013) 854-863.

[63] K.L. Smalling, K.M. Kuivila, Multi-residue method for the analysis of 85 current-use and legacy pesticides in bed and suspended sediments, J. Chromatogr. A 1210 (2008) 8-18.

[64] Á. Sánchez-Rodríguez, Z. Sosa-Ferrera, J.J. Santana-Rodríguez, Applicability of microwave-assisted extraction combined with LC-MS/MS in the evaluation of booster biocide levels in harbour sediments, Chemosphere 82 (2011) 96-102.

[65] F.A. Esteve-Turrillas, A. Pastor, M. de la Guardia, Use of semipermeable membrane devices for

monitoring pesticides in indoor air, J. AOAC Int. 92 (2009) 1557-1565.

[66] T. Otake, M. Numata, Determination of pyrethroid and organophosphorus insecticides in indoor air by microwave - assisted extraction with gas chromatography/mass spectrometry, Anal. Lett. 47 (2014) 2281-2293.

[67] D. Sanjuán-Herráez, Y. Rodríguez-Carrasco, L. Juan-Peiró, A. Pastor, M. De la Guardia, Determination of indoor air quality of a phytosanitary plant, Anal. Chim. Acta 694 (2011) 67-74.

[68] C. Coscollá, V. Yusá, M.I. Beser, A. Pastor, Multi-residue analysis of 30 currently used pesticides in fine airborne particulate matter (PM 2.5) by microwave - assisted extraction and liquid chromatography-tandem mass spectrometry, J. Chromatogr. A 1216 (2009) 8817-8827.

[69] C. Coscollá, M. Castillo, A. Pastor, V. Yusá, Determination of 40 currently used pesticides in airborne particulate matter (PM 10) by microwave-assisted extraction and gas chromatography coupled to triple quadrupole mass spectrometry, Anal. Chim. Acta 693 (2011) 72-81.

[70] S. Ding, F. Dong, B. Wang, S. Chen, L. Zhang, M. Chen, M. Gao, P. He, Polychlorinated biphenyls and organochlorine pesticides in atmospheric particulate matter of Northern China: distribution, sources, and risk assessment, Environ. Sci. Pollut. Res. 22 (2015) 17171-17181.

[71] J. Regueiro, M. Llompart, C. Garcia-Jares, R. Cela, Development of a highthroughput method for the determination of organochlorinated compounds, nitromusks and pyrethroid insecticides in indoor dust, J. Chromatogr. A 1174 (2007) 112-124.

[72] Directive 2008/105/EC of the European Parliament and of the Council on environmental quality standards in the field of water policy, Off. J. Eur. Union (2008). L 348/84.

[73] Proposal for a Directive of the European Parliament and of the Council amending Directives 2000/60/EC and 2008/105/EC as regards priority substances in the field of water policy, Off. J. Eur. Union (2000). L 327/73.

[74] Directive 2008/105/EC of the European Parliament and of the Council of 16 December 2008 on environmental quality standards in the field of water policy, amending and subsequently repealing Council Directives 82/176/EEC, 83/513/EEC, 84/156/EEC, 84/491/EEC, 86/280/EEC and amending Directive 2000/60/EC of the European Parliament and of the Council, 2008.

[75] Commission Implementing Decision (EU) 2015/495 of 20 March 2015 establishing a watch list of substances for Union-wide monitoring in the field of water policy pursuant to Directive 2008/105/EC of the European Parliament and of the Council, Off. J. Eur. Union (2015). L78/40.

[76] T. Deblonde, C. Cossu-Leguille, P. Hartemann, Emerging pollutants in wastewater: a review of the literature, Int. J. Hyg. Environ. Health 214 (2011) 442-448.

[77] M. La Farre, S. Pérez, L. Kantiani, D. Barceló, Fate and toxicity of emerging pollutants, their metabolites and transformation products in the aquatic environment, TrAC-Trend. Anal. Chem. 27 (2008) 991-1007.

[78] M. Gavrilescu, K.i. Demnerová, J. Aamand, S. Agathos, F. Fava, Emerging pollutants in the environment: present and future challenges in biomonitoring, ecological risks and bioremediation, New Biotechnol. 32 (2015) 147-156.

[79] K. Noguera-Oviedo, D.S. Aga, Lessons learned from more than two decades of research on emerging contaminants in the environment, J. Hazard. Mater. 316 (2016) 242-251.

[80] M.a.D. Hernando, M. Mezcua, A.R. Fernández-Alba, D. Barceló, Environmental risk assessment of pharmaceutical residues in wastewater effluents, surface waters and sediments, Talanta 69

(2006) 334-342.

[81] V. Gabet, C. Miège, P. Bados, M. Coquery, Analysis of estrogens in environmental matrices, TrAC-Trend. Anal. Chem. 26 (2007) 1113-1131.

[82] J. L. Oaks, M. Gilbert, M. Z. Virani, R. T. Watson, C. U. Meteyer, B. A. Rideout, H. L. Shivaprasad, S. Ahmed, M. J. I. Chaudhry, M. Arshad, Diclofenac residues as the cause of vulture population decline in Pakistan, Nature 427 (2004) 630-633.

[83] V. Andreu, C. Blasco, Y. Picó, Analytical strategies to determine quinolone residues in food and the environment, TrAC-Trend. Anal. Chem. 26 (2007) 534-556.

[84] D. D. Snow, T. Damon-Powell, S. Onanong, D. A. Cassada, Sensitive and simplified analysis of natural and synthetic steroids in water and solids using on-line solid-phase extraction and microwave-assisted solvent extraction coupled to liquid chromatography tandem mass spectrometry atmospheric pressure photoionization, Anal. Bioanal. Chem. 405 (2013) 1759-1771.

[85] A. Hibberd, K. Maskaoui, Z. Zhang, J. L. Zhou, An improved method for the simultaneous analysis of phenolic and steroidal estrogens in water and sediment, Talanta 77 (2009) 1315-1321.

[86] R. Guedes-Alonso, S. Santana-Viera, S. Montesdeoca-Esponda, C. Afonso-Olivares, Z. Sosa-Ferrera, J. J. Santana-Rodríguez, Application of microwave-assisted extraction and ultra-high performance liquid chromatography-tandem mass spectrometry for the analysis of sex hormones and corticosteroids in sewage sludge samples, Anal. Bioanal. Chem. 408 (2016) 6833-6844.

[87] T. Vega-Morales, Z. Sosa-Ferrera, J. J. Santana-Rodríguez, Determination of various estradiol mimicking-compounds in sewage sludge by the combination of microwave-assisted extraction and LC-MS/MS, Talanta 85 (2011) 1825-1834.

[88] T. Vega-Morales, Z. Sosa-Ferrera, J. J. Santana-Rodríguez, Evaluation of the presence of endocrine-disrupting compounds in dissolved and solid wastewater treatment plant samples of Gran Canaria Island (Spain), BioMed. Res. Int. 2013 (2013) .

[89] T. Vega-Morales, Z. Sosa-Ferrera, J. J. Santana-Rodríguez, The use of microwave assisted extraction and on-line chromatography-mass spectrometry for determining endocrine-disrupting compounds in sewage sludges, Water Air Soil Pollut. 224 (2013) 1-15.

[90] D. Matejicek, On-line two-dimensional liquid chromatography-tandem mass spectrometric determination of estrogens in sediments, J. Chromatogr. A 1218 (2011) 2292-2300.

[91] A. Azzouz, E. Ballesteros, Determination of 13 endocrine disrupting chemicals in environmental solid samples using microwave-assisted solvent extraction and continuous solid-phase extraction followed by gas chromatography-mass spectrometry, Anal. Bioanal. Chem. 408 (2016) 231-241.

[92] M. Varga, J. Dobor, A. Helenkár, L. Jurecska, J. Yao, G. Zaray, Investigation of acidicpharmaceuticals in river water and sediment by microwave-assisted extraction and gas chromatography-mass spectrometry, Microchem. J. 95 (2010) 353-358.

[93] J. Antonic, E. Heath, Determination of NSAIDs in river sediment samples, Anal. Bioanal. Chem. 387 (2007) 1337-1342.

[94] J. Dobor, M. Varga, J. Yao, H. Chen, G. Palkó, G. Zaray, A new sample preparation method for determination of acidic drugs in sewage sludge applying microwave assisted solvent extraction followed by gas chromatography-mass spectrometry, Microchem. J. 94 (2010) 36-41.

[95] S. L. Rice, S. Mitra, Microwave-assisted solvent extraction of solid matrices and subsequent detection of pharmaceuticals and personal care products (PPCPs) using gas chromatography-mass

spectrometry, Anal. Chim. Acta 589 (2007) 125–132.

[96] B. Petrie, J. Youdan, R. Barden, B. Kasprzyk–Hordern, Multi–residue analysis of 90 emerging contaminants in liquid and solid environmental matrices by ultra – highperformance liquid chromatography tandem mass spectrometry, J. Chromatogr. A 1431 (2016) 64–78.

[97] S. E. Evans, P. Davies, A. Lubben, B. Kasprzyk – Hordern, Determination of chiral pharmaceuticals and illicit drugs in wastewater and sludge using microwave assisted extraction, solid–phase extraction and chiral liquid chromatography coupled with tandem mass spectrometry, Anal. Chim. Acta 882 (2015) 112–126.

[98] R. Cueva – Mestanza, Z. Sosa – Ferrera, M. E. Torres – Padrón, J. J. Santana – Rodríguez, Preconcentration of pharmaceuticals residues in sediment samples using microwave assisted micellar extraction coupled with solid phase extraction and their determination by HPLC – UV, J. Chromatogr. B 863 (2008) 150–157.

[99] S. Montesdeoca–Esponda, Z. Sosa–Ferrera, J. J. Santana–Rodríguez, Combination of microwave–assisted micellar extraction with liquid chromatography tandem mass spectrometry for the determination of fluoroquinolone antibiotics in coastal marine sediments and sewage sludges samples, Biomed. Chromatogr. 26 (2012) 33–40.

[100] L. Tong, H. Liu, C. Xie, M. Li, Quantitative analysis of antibiotics in aquifer sediments by liquid chromatography coupled to high resolution mass spectrometry, J. Chromatogr. A 1452 (2016) 58–66.

[101] Z. Jiao, Z. Guo, S. Zhang, H. Chen, Microwave – assisted micro – solid – phase extraction for analysis of tetracycline antibiotics in environmental samples, Int. J. Environ. Anal. Chem. 95 (2015) 82–91.

[102] M. Sturini, A. Speltini, F. Maraschi, E. Rivagli, A. Profumo, Solvent–free microwave–assisted extraction of fluoroquinolones from soil and liquid chromatography–fluorescence determination, J. Chromatogr. A 1217 (2010) 7316–7322.

[103] N. Dorival – García, A. Zafra – Gómez, F. J. Camino – Sánchez, A. Navalón, J. L. Vílchez, Analysis of quinolone antibiotic derivatives in sewage sludge samples by liquid chromatography – tandem mass spectrometry: comparison of the efficiency of three extraction techniques, Talanta 106 (2013) 104–118.

[104] D. P. Mohapatra, S. K. Brar, R. D. Tyagi, P. Picard, R. Y. Surampalli, Carbamazepine in municipal wastewater and wastewater sludge: ultrafast quantification by laser diode thermal desorption – atmospheric pressure chemical ionization coupled with tandem mass spectrometry, Talanta 99 (2012) 247–255.

[105] L. Chen, Q. Zeng, H. Wang, R. Su, Y. Xu, X. Zhang, A. Yu, H. Zhang, L. Ding, On–line coupling of dynamic microwave – assisted extraction to solid – phase extraction for the determination of sulfonamide antibiotics in soil, Anal. Chim. Acta 648 (2009) 200–206.

[106] J. Raich – Montiu, J. L. Beltrán, M. D. Prat, M. Granados, Studies on the extraction of sulfonamides from agricultural soils, Anal. Bioanal. Chem. 397 (2010) 807–814.

[107] V. K. Balakrishnan, K. N. Exall, J. M. Toito, The development of a microwave – assisted extraction method for the determination of sulfonamide antibiotics in sediments and soils, Can. J. Chem. 92 (2014) 369–377.

[108] M. Forster, V. Laabs, M. Lamshoft, T. Putz, W. Amelung, Analysis of aged sulfadiazine residues in soils using microwave extraction and liquid chromatography tandem mass spectrometry, Anal. Bioanal. Chem. 391 (2008) 1029–1038.

[109] M. Forster, V. Laabs, M. Lamshoft, J. Groeneweg, S. Zuhlke, M. Spiteller, M. Krauss, M. Kaupenjohann, W. Amelung, Sequestration of manure-applied sulfadiazine residues in soils, Environ. Sci. Technol. 43 (2009) 1824-1830.

[110] H. Wang, J. Ding, L. Ding, N. Ren, Analysis of sulfonamides in soil, sediment, and sludge based on dynamic microwave-assisted micellar extraction, Environ. Sci. Pollut. Res. (2016) 1-12.

[111] Y. Song, L. Wu, C. Lu, N. Li, M. Hu, Z. Wang, Microwave-assisted liquid-liquid microextraction based on solidification of ionic liquid for the determination of sulfonamides in environmental water samples, J. Sep. Sci. 37 (2014) 3533-3538.

[112] X. Xu, R. Su, X. Zhao, Z. Liu, Y. Zhang, D. Li, X. Li, H. Zhang, Z. Wang, Ionic liquid-based microwave-assisted dispersive liquid-liquid microextraction and derivatization of sulfonamides in river water, honey, milk, and animal plasma, Anal. Chim. Acta 707 (2011) 92-99.

[113] J. Raich-Montiu, M. D. Prat, M. Granados, Extraction and analysis of avermectines in agricultural soils by microwave assisted extraction and ultra high performance liquid chromatography coupled to tandem mass spectrometry, Anal. Chim. Acta 697 (2011) 32-37.

[114] A. Speltini, M. Sturini, F. Maraschi, A. Profumo, A. Albini, Microwave-assisted extraction and determination of enrofloxacin and danofloxacin photo-transformation products in soil, Anal. Bioanal. Chem. 404 (2012) 1565-1569.

[115] M. del Mar Gil-Díaz, A. Pérez-Sanz, M. C. Lobo, Validation of a microwave-assisted micella rextraction method for the oxibendazole determination in soil samples, Afinidad 69 (2012) .

[116] D. Drljaca, D. Asperger, M. Ferencak, M. Gavranic, S. Babic, I. Mikac, M. Ahel, Comparison of four extraction methods for the determination of veterinary pharmaceuticals in sediment, Chromatographia 79 (2016) 209-223.

[117] P. N. Carvalho, A. Pirra, M. C. P. Basto, C. M. R. Almeida, Multi-family methodologies for the analysis of veterinary pharmaceutical compounds in sediment and sludge samples: comparison among extraction techniques, Anal. Methods 5 (2013) 6503-6510.

[118] J. M. Brausch, G. M. Rand, A review of personal care products in the aquatic environment: environmental concentrations and toxicity, Chemosphere 82 (2011) 1518-1532.

[119] R. Loos, R. Carvalho, D. C. Antonio, S. Comero, G. Locoro, S. Tavazzi, B. Paracchini, M. Ghiani, T. Lettieri, L. Blaha, EU-wide monitoring survey on emerging polar organic contaminants in wastewater treatment plant effluents, Water Res. 47 (2013) 6475-6487.

[120] B. Bridges, Fragrance: emerging health and environmental concerns, Flavour Fragr. J. 17 (2002) 361-371.

[121] A. M. Peck, Analytical methods for the determination of persistent ingredients of personal care products in environmental matrices, Anal. Bioanal. Chem. 386 (2006) 907-939.

[122] I. T. I. Wang, S. F. Cheng, S. W. Tsai, Determinations of airborne synthetic musks by polyurethane foam coupled with triple quadrupole gas chromatography tandem mass spectrometer, J. Chromatogr. A 1330 (2014) 61-68.

[123] A. Chisvert, Z. León-González, I. Tarazona, A. Salvador, D. Giokas, An overview of the analytical methods for the determination of organic ultraviolet filters in biological fluids and tissues, Anal. Chim. Acta 752 (2012) 11-29.

[124] K. Kotnik, T. Kosjek, U. Krajnc, E. Heath, Trace analysis of benzophenone-derived compounds in surface waters and sediments using solid-phase extraction and microwave-assisted extraction

followed by gas chromatography-mass spectrometry, Anal. Bioanal. Chem. 406 (2014) 3179-3190.

[125] M. P. Groz, M. J. M. Bueno, D. Rosain, H. Fenet, C. Casellas, C. Pereira, V. Maria, M. J. Bebianno, E. Gomez, Detection of emerging contaminants (UV filters, UV stabilizers and musks) in marine mussels from Portuguese coast by QuEChERS extraction and GC-MS/MS, Sci. Total Environ. 493 (2014) 162-169.

[126] D. Kaiser, A. Sieratowicz, H. Zielke, M. Oetken, H. Hollert, J. Oehlmann, Ecotoxicological effect characterisation of widely used organic UV filters, Environ. Pollut. 163 (2013) 84-90.

[127] M. Celeiro, E. Guerra, J. P. Lamas, M. Lores, C. Garcia-Jares, M. Llompart, Development of a multianalyte method based on micro-matrix-solid-phase dispersion for the analysis of fragrance allergens and preservatives in personal care products, J. Chromatogr. A 1344 (2014) 1-14.

[128] L. Sanchez-Prado, J. P. Lamas, M. Lores, C. Garcia-Jares, M. Llompart, Simultaneous in-cell derivatization pressurized liquid extraction for the determination of multiclass preservatives in leave-on cosmetics, Anal. Chem. 82 (2010) 9384-9392.

[129] T. Heberer, Occurrence, fate, and assessment of polycyclic musk residues in the aquatic environment of urban areas-a review, Acta Hydrochim. Hydrobiol. 30 (2002) 227-243.

[130] Y. Lu, T. Yuan, W. Wang, K. Kannan, Concentrations and assessment of exposure to siloxanes and synthetic musks in personal care products from China, Environ. Pollut. 159 (2011) 3522-3528.

[131] M. Clara, O. Gans, G. Windhofer, U. Krenn, W. Hartl, K. Braun, S. Scharf, C. Scheffknecht, Occurrence of polycyclic musks in wastewater and receiving water bodies and fate during wastewater treatment, Chemosphere 82 (2011) 1116-1123.

[132] R. Reiss, G. Lewis, J. Griffin, An ecological risk assessment for triclosan in the terrestrial environment, Environ. Toxicol. Chem. 28 (2009) 1546-1556.

[133] L. Sanchez-Prado, M. Llompart, M. Lores, C. Garcia-Jares, J. M. Bayona, R. Cela, Monitoring the photochemical degradation of triclosan in wastewater by UV light and sunlight using solid-phase microextraction, Chemosphere 65 (2006) 1338-1347.

[134] G. Alvarez-Rivera, M. Llompart, C. Garcia-Jares, M. Lores, Pressurized liquid extraction-gas chromatography-mass spectrometry for confirming the photo-induced generation of dioxin-like derivatives and other cosmetic preservative photoproducts on artificial skin, J. Chromatogr. A 1440 (2016) 37-44.

[135] A. M. Peck, K. C. Hornbuckle, Synthetic musk fragrances in Lake Michigan, Environ. Sci. Technol. 38 (2004) 367-372.

[136] M. L. Svoboda, J. J. Yang, P. Falletta, H. -B. Lee, A microwave-assisted extraction method for the determination of synthetic musks in sewage sludge, Water Qual. Res. J. Can. 42 (2007) 11-17.

[137] S. A. Smyth, L. Lishman, M. Alaee, S. Kleywegt, L. Svoboda, J. J. Yang, H. B. Lee, P. Seto, Sample storage and extraction efficiencies in determination of polycyclic and nitro musks in sewage sludge, Chemosphere 67 (2007) 267-275.

[138] X. Hu, Q. Zhou, Comparisons of microwave-assisted extraction, simultaneous distillation-solvent extraction, Soxhlet extraction and ultrasound probe for polycyclic musks in sediments: recovery, repeatability, matrix effects and bioavailability, Chromatographia 74 (2011) 489-495.

[139] S. Montesdeoca-Esponda, Z. Sosa-Ferrera, J. J. Santana-Rodríguez, Microwave-assisted extraction combined with on-line solid phase extraction followed by ultra-highperformance liquid chromatography with tandem mass spectrometric determination of benzotriazole UV stabilizers in marine sediments and sewage sludges, J. Sep. Sci. 36 (2013) 781-788.

[140] R. A. Pérez, B. Albero, E. Miguel, C. Sánchez-Brunete, Determination of parabens and endocrine-disrupting alkylphenols in soil by gas chromatography-mass spectrometry following matrix solid-phase dispersion or in-column microwave-assisted extraction: a comparative study, Anal. Bioanal. Chem. 402 (2012) 2347-2357.

[141] T. J. Brown, C. A. Kinney, Rapid lab-scale microwave-assisted extraction and analysis of anthropogenic organic chemicals in river sediments, Int. J. Geosci. 2 (2011) 267.

[142] B. Delgado, V. Pino, J. L. Anderson, J. H. Ayala, A. M. Afonso, V. González, An in-situ extraction-preconcentration method using ionic liquid-based surfactants for the determination of organic contaminants contained in marine sediments, Talanta 99 (2012) 972-983.

[143] B. Delgado, V. Pino, J. H. Ayala, A. M. Afonso, V. González, A novel preconcentration strategy for extraction methods based on common cationic surfactants: an alternative to classical coacervative extraction, J. Chromatogr. A 1257 (2012) 9-18.

[144] M. W. Wu, P. C. Yeh, H. C. Chen, L. L. Liu, W. H. Ding, A microwave-assisted headspace solid-phase microextraction for rapid determination of synthetic polycyclic and nitro-aromatic musks in fish samples, J. Chin. Chem. Soc. 60 (2013) 1169-1174.

[145] Y. C. Wang, W. H. Ding, Determination of synthetic polycyclic musks in water by microwave-assisted headspace solid-phase microextraction and gas chromatography-mass spectrometry, J. Chromatogr. A 1216 (2009) 6858-6863.

[146] S. F. Wu, W. H. Ding, Fast determination of synthetic polycyclic musks in sewage sludge and sediments by microwave-assisted headspace solid-phase microextraction and gas chromatography-mass spectrometry, J. Chromatogr. A 1217 (2010) 2776-2781.

[147] S. F. Wu, L. L. Liu, W. H. Ding, One-step microwave-assisted headspace solid-phase microextraction for the rapid determination of synthetic polycyclic musks in oyster by gas chromatography-mass spectrometry, Food Chem. 133 (2012) 513-517.

[148] C. Garcia-Jares, R. Barro, M. Llompart, Indoor air sampling, in: J. Pawliszyn (Ed.), Comprehensive Sampling and Sample Preparation, Elsevier, Netherlands, 2012.

[149] S. D. Richardson, S. Y. Kimura, Water analysis: emerging contaminants and current issues, Anal. Chem. 88 (2016) 546-582.

[150] M. A. E. Abdallah, G. Pawar, S. Harrad, Human dermal absorption of chlorinated organophosphate flame retardants; implications for human exposure, Toxicol. Appl. Pharm. 291 (2016) 28-37.

[151] C. Garcia-Jares, M. Llompart, P. Landin, Phthalate esters, in: L. M. L. Nollet (Ed.), Chromatographic Analysis of the Environment, CRC Press, Taylor and Francis Group, Boca Raton, 2006.

[152] S. Net, A. Delmont, R. Sempéré, A. Paluselli, B. Ouddane, Reliable quantification of phthalates in environmental matrices (air, water, sludge, sediment and soil): a review, Sci. Total Environ. 515 (2015) 162-180.

[153] X. Zheng, B. T. Zhang, Y. Teng, Distribution of phthalate acid esters in lakes of Beijing and its relationship with anthropogenic activities, Sci. Total Environ. 476 (2014) 107-113.

[154] C. Liao, P. Yang, Z. Xie, Y. Zhao, X. Cheng, Y. Zhang, Z. Ren, Z. Guo, J. Liao, Application of GC-triple quadrupole MS in the quantitative confirmation of polycyclic aromatic hydrocarbons and phthalic acid esters in soil, J. Chromatogr. Sci. 48 (2010) 161-166.

[155] P. Liang, L. Zhang, L. Peng, Q. Li, E. Zhao, Determination of phthalate esters in soil samples by microwave assisted extraction and high performance liquid chromatography, Bull. Environ. Contam.

Toxicol. 85 (2010) 147-151.

[156] Z. Jiao, Z. Guo, S. Zhang, H. Chen, H. Xie, S. Zeng, Novel extraction for endocrine disruptors in atmospheric particulate matter, Anal. Lett. 48 (2015) 1355-1366.

[157] C. H. Feng, S. R. Jiang, Micro-scale quantitation of ten phthalate esters in water samples and cosmetics using capillary liquid chromatography coupled to UV detection: effective strategies to reduce the production of organic waste, Microchim. Acta 177 (2012) 167-175.

[158] R. Wang, P. Su, Y. Yang, Optimization of ionic liquid-based microwave-assisted dispersive liquid-liquid microextraction for the determination of plasticizers in water by response surface methodology, Anal. Methods 5 (2013) 1033-1039.

[159] Y. K. Lv, W. Zhang, M. M. Guo, F. F. Zhao, X. X. Du, Centrifugal microextraction tube-cloud point extraction coupled with gas chromatography for simultaneous determination of six phthalate esters in mineral water, Anal. Methods 7 (2015) 560-565.

[160] J. Regueiro, M. Llompart, C. Garcia-Jares, R. Cela, Factorial-design optimization of gas chromatographic analysis of tetrabrominated to decabrominated diphenyl ethers. Application to domestic dust, Anal. Bioanal. Chem. 388 (2007) 1095-1107.

[161] M. I. Beser, J. Beltrán, V. Yusá, Design of experiment approach for the optimization of polybrominated diphenyl ethers determination in fine airborne particulate matter by microwave-assisted extraction and gas chromatography coupled to tandem mass spectrometry, J. Chromatogr. A 1323 (2014) 1-10.

[162] H. W. Chung, W. H. Ding, Determination of organophosphate flame retardants in sediments by microwave-assisted extraction and gas chromatography-mass spectrometry with electron impact and chemical ionization, Anal. Bioanal. Chem. 395 (2009) 2325-2334.

[163] M. Shin, M. L. Svoboda, P. Falletta, Microwave-assisted extraction (MAE) for the determination of polybrominated diphenylethers (PBDEs) in sewage sludge, Anal. Bioanal. Chem. 387 (2007) 2923-2929.

[164] P. Novak, T. Zuliani, R. Milacic, J. Scancar, Development of an analytical method for the determination of polybrominated diphenyl ethers in sewage sludge by the use of gas chromatography coupled to inductively coupled plasma mass spectrometry, Anal. Chim. Acta 915 (2016) 27-35.

[165] E. Bizkarguenaga, A. Iparragirre, I. Zabaleta, A. Vallejo, L. A. Fernández, A. Prieto, O. Zuloaga, Focused ultrasound assisted extraction for the determination of PBDEs in vegetables and amended soil, Talanta 119 (2014) 53-59.

[166] J. Wang, Z. Lin, K. Lin, C. Wang, W. Zhang, C. Cui, J. Lin, Q. Dong, C. Huang, Polybrominated diphenyl ethers in water, sediment, soil, and biological samples from different industrial areas in Zhejiang, China, J. Hazard. Mater. 197 (2011) 211-219.

[167] P. Wang, Q. Zhang, Y. Wang, T. Wang, X. Li, L. Ding, G. Jiang, Evaluation of Soxhlet extraction, accelerated solvent extraction and microwave-assisted extraction for the determination of polychlorinated biphenyls and polybrominated diphenyl ethers in soil and fish samples, Anal. Chim. Acta 663 (2010) 43-48.

[168] H. H. Wu, H. C. Chen, W. H. Ding, Combining microwave-assisted extraction and liquid chromatography-ion-trap mass spectrometry for the analysis of hexabromocyclododecane diastereoisomers in marine sediments, J. Chromatogr. A 1216 (2009) 7755-7760.

[169] Y. C. Tsao, Y. C. Wang, S. F. Wu, W. H. Ding, Microwave-assisted headspace solid phase

microextraction for the rapid determination of organophosphate esters in aqueous samples by gas chromatography-mass spectrometry, Talanta 84 (2011) 406-410.

[170] J. Ellis, M. Shah, K. M. Kubachka, J. A. Caruso, Determination of organophosphorus fire retardants and plasticizers in wastewater samples using MAE - SPME with GC - ICPMS and GC - TOFMS detection, J. Environ. Monit. 9 (2007) 1329-1336.

[171] M. García - López, I. Rodríguez, R. Cela, K. K. Kroening, J. A. Caruso, Determination of organophosphate flame retardants and plasticizers in sediment samples using microwave-assisted extraction and gas chromatography with inductively coupled plasma mass spectrometry, Talanta 79 (2009) 824-829.

[172] J. R. Rochester, A. L. Bolden, Bisphenol S and F: a systematic review and comparison of the hormonal activity of bisphenol A substitutes, Environ. Health Persp. 123 (2015) 643.

[173] J. C. Anderson, B. J. Park, V. P. Palace, Microplastics in aquatic environments: implications for Canadian ecosystems, Environ. Pollut. 218 (2016) 269-280.

[174] J. Zhang, Y. Min, Y. Zhang, C. Meixue, Biotransformation of nonylphenol ethoxylates during sewage treatment under anaerobic and aerobic conditions, J. Environ. Sci. 20 (2008) 135-141.

[175] S. Esteban, M. Gorga, M. Petrovic, S. Gonzúlez-Alonso, D. Barceló, Y. Valcúrcel, Analysis and occurrence of endocrine-disrupting compounds and estrogenic activity in the surface waters of Central Spain, Sci. Total Environ. 466 (2014) 939-951.

[176] H. Hera, Environmental risk assessment on ingredients of European household cleaning products: alcohol ethoxylates (2009).

[177] N. Dorival-García, A. Zafra-Gómez, A. Navalón, J. L. Vílchez, Analysis of bisphenol A and its chlorinated derivatives in sewage sludge samples. Comparison of the efficiency of three extraction techniques, J. Chromatogr. A 1253 (2012) 1-10.

[178] M. Villar, M. Callejón, J. C. Jiménez, E. Alonso, A. Guirúum, Optimization and validation of a new method for analysis of linear alkylbenzene sulfonates in sewage sludge by liquid chromatography after microwave-assisted extraction, Anal. Chim. Acta 599 (2007) 92-97.

[179] M. Villar, M. Callejón, J. C. Jiménez, E. Alonso, A. Guirúum, New rapid methods for determination of total LAS in sewage sludge by high performance liquid chromatography (HPLC) and capillary electrophoresis (CE), Anal. Chim. Acta 634 (2009) 267-271.

[180] C. Pakou, M. Kornaros, K. Stamatelatou, G. Lyberatos, On the fate of LAS, NPEOs and DEHP in municipal sewage sludge during composting, Bioresour. Technol. 100 (2009) 1634-1642.

[181] B. Wang, X. Wan, S. Zhao, Y. Wang, F. Yu, X. Pan, Analysis of six phenolic endocrine disrupting chemicals in surface water and sediment, Chromatographia 74 (2011) 297-306.

7　离子液体在样品制备中的应用

Rafael Lucena and Soledad Cárdenas*
University of Córdoba, Córdoba, Spain
*通讯作者：E-mail：qa1caarm@ uco. es

7.1 引言

样品制备是分析过程的第一步。一般来说，在明确的采样方案获得的实验室样品通常由于以下三个主要原因而与所选的仪器分析技术进行直接分析不兼容：①分析物浓度太低；②存在潜在干扰，主要是由于基质成分；③样品的聚集状态或因仪器要求而需要更换溶剂。

理想情况下，任何样品处理所涉及的步骤都应简单、自动化、小型化、廉价和安全。因此，有几项研究工作旨在实现这些要求。在现有的方法中，使用比传统方法更有效的新萃取相是一个非常相关的话题。纳米结构的固体和新的液体介质，如超分子溶剂、可切换溶剂或离子液体（ILs），可以作为微固相萃取（μ-SPE）和液相微萃取或分散液-液微萃取（LPME 或 DLLME）的最有价值的形式之一。

ILs 是一类由 180~600K 温度范围内的液态离子组成的半有机盐，至少有一个非定域电荷组分的存在，这是其不可能形成稳定晶格的原因[1]。有些 ILs 在室温下也是液态的。所谓的室温 ILs 在微萃取环境中表现出更大的适用性，因为它们可以用作 μ-SPE 或 LPME 中的液体萃取剂，而无须任何附加要求。尽管目前正在讨论 ILs 的绿色特性(进一步的信息见 Armenta 等人的“绿色分析化学：绿色提取技术的作用”一章[1a])，离子液体作为新的溶剂在分析过程中提供了新的特点，将其列入本书是正确的。ILs 在微萃取领域发挥突出作用的特殊性质是其可忽略蒸汽压、热稳定性、可调黏度和与其他溶剂的混溶性。阴阳离子的性质决定了离子液体的最终物理化学性质，从而决定了他们在微萃取中的可能应用[2]。

本章将对液相和固相微萃取（SPME）中的 ILs 进行概述。作为萃取剂溶剂，它们以分散的形式使用，在中空纤维上或以单个液滴的结构使用。本章还包括外部能量［温度和超声波（USs）］的使用、原位溶剂形成或磁性 ILs（MILs）等内容。在固相（微）萃取中，ILs 与纳米粒（NPs）成功结合，设计出了特殊的吸附剂。此外，聚合物 ILs 在传统的 SPME 中有着广泛的应用。

7.2 离子液体作为萃取溶剂

7.2.1 离子液体在分散微萃取中的应用

由于分析物从基质转移到萃取相的效率极高，在过去的十年中固体和液体形式的分散萃取技术得到了广泛的应用。这一事实归因于分散过程中相间接触表面的膨胀，从而产生更快的动力学性能。在 DLLME 领域，Rezaee 等[3]首次提出将萃取剂和分散溶剂的混合物注入水样中，一团细小的水滴立刻形成。它有利于在萃取剂中富集分析物，最后通过离心回收并使用最合适的仪器技术进行分析。

ILs 已被用作 DLLME 的萃取剂，利用外部辅助能量可以提高萃取效率，例如温度或者超声。此外，利用 MILs 作为萃取剂，并利用外磁场对样品进行回收极大地简化了前处理过程。最后，通过选择合适的阴离子可以调节 IL 在样品中的溶解度。这些方法将在下面的小节中讨论。

7.2.1.1　离子液体作为分散液液微萃取中的溶剂

ILs 被用于提取不同基质中的有机[4-7]或无机化合物[8,9]，尽管后者需要对分析物（Co、Pb 或 Cd）进行衍生化。一般来说，1-烷基-3-甲基咪唑是首选。其他 ILs，如 1，3-二戊咪唑啉也被推荐[10]。除了这些常规用途外，ILs 还被用作分散溶剂，替代甲醇或丙酮。在这种情况下，主要条件是所用的 IL 必须是与水样基质可混溶[11]。

关于基于 IL 的 DLLME 的其他替代品，可使用塑料一次性注射器实现分离。所谓的注射器内 DLLME 可以利用 IL 和水样之间的密度不同在提取后不经过离心而分离 IL。在这种情况下，通过柱塞的简单且可控的运动来分离 IL[12]。通过使用流量配置可以很轻松地实现这种注射器 DLLME 自动化。使用此自动化工作流程可以从环境水样中提取紫外线过滤剂和 T1[13,14]。

7.2.1.2　外部能量辅助的离子液体分散液-液微萃取

众所周知，分散溶剂直接增加了分析物在样品中的溶解度，使其很难转移到萃取溶剂中，从而对 DLLME 的热力学有负面影响。为了减少甚至消除 DLLME 中对分散溶剂的需求，对几种替代方案进行了评估。其中，利用外部能量如温度或超声似乎是最有用的。

基于温度的 DLLME 可以调节 IL 在水样基质中的溶解度。升高和降低温度分别用于溶解或诱导萃取后富含分析物的 IL 的分离。Zhou 等采用温度控制型 IL-DLLME 测定水样中的拟除虫菊酯农药[15]。他们用不溶于水的 IL 来提取。分散后，混合物的温度随着 IL 完全溶解而升高。接下来，使用冰浴冷却混合物的温度，在冰浴条件下由于 IL 溶解度较低而使两相得以分离。富集分析物的 IL 通过离心进行分离并用液相色谱法和紫外检测器进行分析。环境样品中的酚类化合物[16]、碳氢化合物[17]和紫外线过滤剂[18]已用液相色谱法与不同检测器联用进行了定性和定量，结果非常好。

2008 年，Baghadadi 和 Sheminari 提出采用冷诱导聚集微萃取法测定水中汞[19]。与前面描述的工作流程的主要区别在于他们使用化学分散剂来协助 IL 在样品中的分散。其他一些作者使用类似的方法来测定环境样品中的有机[20,21]和无机污染物[22,23]。

使用超声（US）来帮助任何 LPME 过程的有益效果是众所周知的。US 的应用不仅影响萃取过程中不互溶相的均匀分散，而且还影响萃取过程中可能发生的反应动力学。关于第一个效果，US 促进萃取剂的分散，减少甚至避免了对分散溶剂的需要。由于分析物在样品中的溶解度没有改变（通常增加），这对萃取效率有积极影响。空化作用导致萃取溶剂形成表面体积比增大的细小液滴。US 辅助的 IL-DLLME 已被用于提取环境样品中的不同污染物，尽管它可以扩展到其他分析物和样品。通常，需要向样品中添加 100μL 的 IL。然后在优化的时间内对混合物进行超声波处理（在水浴中或通过探针），以确保 IL 完全分散。根据 IL 的性质、所用的时间和功率，需要随后的冷却步骤以促进提取后 IL 的回收，通常是通过离心。参照这一程序，经过稍加修改，从环境水样中提取出了芳香胺[24]、药物[25]和汞[26]，并取得了良好的结果。

7.2.1.3　原位离子液体

另一种改变 IL 在液体样品中溶解度的方法是其结构中包含的阴离子的性质。这种性质可以作为［bmim］阳离子的例证，它以氯化物的形式在水中高度溶解，而当与[PF6]阴离子结合时几乎不溶。阴离子在 IL 中的取代反应称为复分解反应，它是所谓的原位溶剂形成微萃取（ISFM）的原因。该技术于 2009 年首次提出[22]，并明确指出其相对于其他方法的优势[27]。ISFM 不需要使用分散溶剂，因为 IL 以其可溶性形式添加到样品中。接下来，仅几秒钟后（例如，30s）添加适当的试剂，通过复分解反应诱导不溶性 IL 的形成。通过离心（通常 5min）进一步回收富含目标化合物的 IL。与其他微萃取技术相比，ISFM 的竞争力体现在相对较短的时间内（不到 10min）获得的富集因子（高出 10~20 倍）上。

一般情况下，ILs 离心后很难回收。有时，可以使用注射器或通过离心后冷却溶液的方法凝固，以便于 ILs 回收。注射器的使用不能阻止低样本量与 ILs 的分离，从而降低了方法的精密度。第二种方法允许将样品从固体提取物中完全分离出来，这些固体提取物可以进一步重新溶解，用于仪器分析[28]。

一些环境样品的高盐含量也会影响复分解反应，这阻碍了 IL 的定量回收。然而，可以通过增加复分解试剂的浓度来克服这一限制[29]。

ISFM 方法已经被用于有机[30-32]和无机化合物[28,33]的富集和分析检测。

7.2.1.4　磁性离子液体分散液-液微萃取

在 DLLME 环境中，使用具有超顺磁特性的萃取剂，萃取剂相在通过简单使用位于萃取容器壁上的外部磁铁分离分析物后很容易回收。这避免了诸如离心或过滤等烦琐步骤。MILs 是 IL 的一个特例，它成功地结合了 ILs 可调的物理化学性质和对外部磁场的强敏感性。这种协同作用使 MILs 成为 DLLME 的理想溶剂。Fe（Ⅲ）、Gd（Ⅲ）或 Dy（Ⅲ）已被用于 MIL 的合成[34-36]。四氯高铁酸三己酯（十四烷基）磷被用于从水中分离和富集酚类化合物[37]。此外，MILs 似乎在提取 DNA 等生物分子方面起着至关重要的作用[38]。与磁性 NPs 相比，它的优点是提高了提取效率，因为 IL 结构可以调整以最大限度地与 DNA 的磷酸骨架相互作用，而不会失去超顺磁性。这个分析问题的挑战是合成 MIL，其疏水性足以使萃取后的相分离。作者深入讨论了使用疏水性阴离子或功能化阳离子来提高 MIL 的疏水性，同时允许加入 $FeCl_4$ 等顺磁性阴离子的影响。最后，对三种不同的 MILs 从含有金属和其他蛋白质的简单基质中提取 DNA，以及从更复杂的基质（如细菌细胞裂解物）中提取 DNA 进行了评估。根据 DNA 链的长度，观察到了 DLLME 中使用 MIL 的明显影响。同时，基质的组成通过竞争机制影响萃取回收率。

7.2.2　中空纤维离子液体液相微萃取

DLLME 的一种替代方法是将 IL 固定在固体载体上。这种方法被称为中空纤维离子液体液相微萃取（HF-IL-LPME）。在这种配置中，IL 浸入样品（DI-HF-IL-LPME）或暴露于其顶空（HS-HF-ILLPME）中的中空纤维的腔和/或孔中。中空纤维中 IL 的存在改变了膜的物理结构，影响了萃取行为。

当 IL 同时存在于中空纤维的腔和孔中时，提取过程称为两相 HF-IL-LPME（2D-

HF-ILLPME)。这种方法特别适用于非极性或中等非极性化合物的提取，因为它们在使用的离子液体中高度可溶。紫外线过滤剂[39]和苯系物[40]已采用这种方式从水中提取，且具有较高的提取回收率。如 DLLME 所述，在 2D-HF-IL-LPME 中提取无机化合物(例如钯或镍）需要使用衍生化反应来促进分析物从水相转移到 IL[41]。

IL 只能位于纤维的孔中，在所谓的三相 HF-IL-LPME（3P-HF-IL-LPME）中，腔中充满了受体（水相或有机相)。这种结构包括双重提取，从样品到 IL 和反向提取到受体相。这增加了其应用领域，因为可以通过在膜两侧建立方便的 pH 梯度，从水相中提取可离子化的有机化合物，以促进分析物以中性形式转移到 IL 相（即中空纤维的孔）并充入受体水相（中空纤维内部)[42]。在二维和三维 HF-IL-LPME 中极性化合物的提取仍然是一个挑战，因为极性化合物在 IL 中的低溶解度阻碍了传质。此外还提出了一些替代方法，例如使用载体溶解 IL 中的分析物，以从具有良好分析价值的环境样品中提取磺胺类药物[43]。

IL 可以在 3P-HF-IL-LPME 中发挥额外的作用。例如，它们可用于减小或避免与水相混溶的有机受体相的损失。它允许从暴雨水样中提取脂肪族和芳香族碳氢化合物，与所使用的仪器技术相兼容，并且不需要蒸发再溶解步骤[44]。

它们作为 3P-HF-IL-LPME 中的假相，通过增加它们在受体相中的溶解度来提高中性化合物的萃取效率。两相之间的相容性是这个过程中的关键因素[45]。这可以提高从不同的样品中萃取铬（Ⅵ）和铬（Ⅲ）的效率[46]。

7.2.3　离子液体单滴微萃取

减小萃取相的体积是提高测量灵敏度的关键因素。Cantwell[47]和 He[48]同时提出使用单液滴萃取相。尽管 SDME 在提高灵敏度方面具有明显的优势，但它的实际发展却暴露出一些缺点，如液滴的不稳定性或在样品基质中蒸发或溶解造成的损失。因此，这一领域的研究一直致力于解决这些局限性。离子液体的物理性质主要与其高密度、高黏度和低蒸气压有关，使得其在前处理领域有很高的应用前景。ILs 生成的液滴比有机溶剂生成的液滴更稳定。此外，适当选择萃取过程中使用的 IL 离子可降低与水样的混溶性。离子液体-单滴微萃取（IL-SDME）的典型应用包括在 DI 或 HS 中分离和预富集有机化合物。在 DI 的情况下，可以通过将液滴悬浮在针尖中[49]或直接将液滴添加到样品中[50]。在测定挥发性和半挥发性化合物时，首选 HS。在这种情况下，ILs 的使用尤其重要，因为液滴即使在相对高温下也具有较高的稳定性（即 90℃)，无蒸发损失。尽管分析物易挥发，但是由于 IL 与气相色谱不相容，通常选择与不同检测器耦合的液相色谱作为检测仪器。我们课题组设计了一个可移动的接口（图 7.1)，允许在 SDME 后重复注射 IL，同时防止其进入色谱系统[51]。棉球被用作物理屏障，通过在适当温度下加热界面，保留 IL 并允许分析物蒸发。在气相色谱中使用 ILs 作为萃取溶剂的主要优点是可以在短的分析时间内获得挥发性分析物的峰，但是没有溶剂峰（通常是一个加宽的峰)。赵等[52]提出了另一个有趣的方案。作者对气相色谱的衬管进行了改进，使其能够从衬管内的 IL 液滴中解吸分析物。然后，从注射口将 IL 移出注射器。解吸管[53]或全自动配置[54]的使用也是这一领域的宝贵贡献。

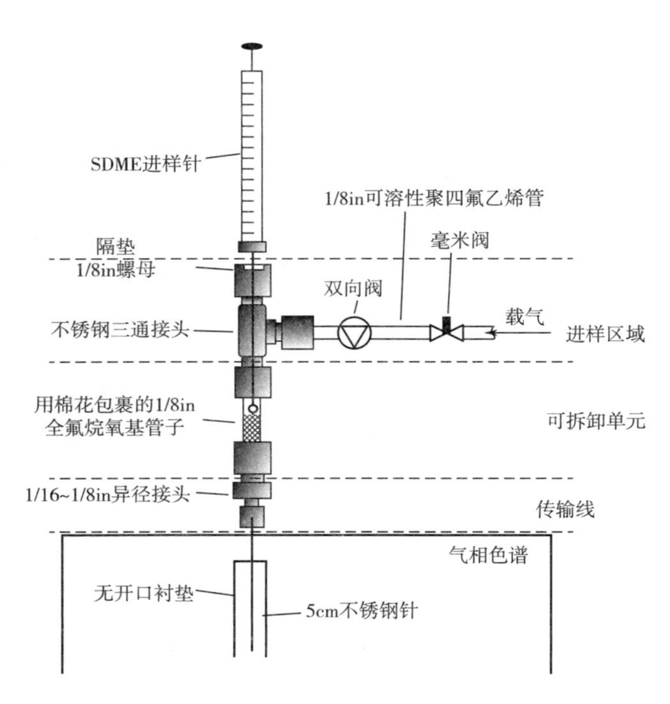

图 7.1 离子液体微萃取–气相色谱直接耦合接口示意图

PFA：可溶性聚四氟乙烯；SDME：单滴微萃取；SS：不锈钢

［资料来源：E. Aguilera-Herrador, R. Lucena, S. Cárdenas, M. Valcárcel, Direct coupling of ionic liquid based single-drop microextraction and GC/MS, Anal. Chem. 80 (2008) 793-800.］

若先前形成可溶于 IL 的中性螯合物，则 DI-IL-SDME 也可用于测定无机化合物，主要是金属。可选择分子或原子吸收技术测定分析物。通过对形成的不同螯合物的色谱分离可进行形态分析。在此背景下，Pena Pereira 等人提出了水中汞的形态的分析，通过相应的二硫腙螯合物提取，并进一步用液相色谱分离紫外进行检测[55]。如果只需要元素的总量，则原子吸收是首选，因为在分析过程中 IL 会被破坏，并且由于火焰原子吸收光谱法中可能出现的吸入流量或雾化步骤效率的变化而产生不可逆性，因此在火焰原子吸收光谱法中不建议使用 ILs。

7.3 离子液体在固相微萃取中的应用

7.3.1 经典的纤维内固相微萃取

大容量 ILs 由于其聚集状态在 LPME 环境中有很大的适用性，但它们也被提出作为 SPME 中的固定相。在早期的方法中，ILs 被涂覆到裸二氧化硅纤维上[56]，从而产生不稳定的涂层。这种涂层有吹出的趋势，特别是在气相色谱进样口中，使得纤维在萃取后必须重新涂覆，并且由于 IL 的热降解而增加色谱图中的鬼峰出现的可能性。使用特殊载体，如 nafion 膜[57]，提高了性能，但主要挑战仍然存在。聚合物离子液体（PILs）

的发展是这方面的一个里程碑，为开发 SPME 技术中的 ILs 特性打开了一扇门。PILs 是由含有典型 ILs 部分（通常是阳离子）的单体控制反应得到的聚合物，其结果是固体材料保持 ILs 的萃取能力。这些固体比大块 ILs 具有更高的机械和热稳定性，从而扩大了它们的应用范围。

Anderson 教授的研究小组首次提出在 SPME 中使用 PILs[58]。为此，PIL 先前是通过自由基聚合合成的。纯化后，将 PIL 溶解在适当的溶剂中形成溶液，将熔融石英纤维在溶液中浸泡几次。溶剂蒸发后在纤维上形成 PIL 膜，该涂层的厚度直接取决于浸渍循环的次数。图 7.2 所示为熔融石英纤维在 PIL 涂层沉积前后的扫描电子显微图。增强的机械稳定性不仅允许在顶空模式下使用 SPME，还允许在直接浸入模式下使用 SPME[59]，提供比传统 PDMS 涂层更好的灵敏度。PIL 与大体积 ILs 一样具有很大的通用性，因为可以与目标分析物发展多种相互化学作用。事实上，PILs 也可以通过引入特殊的部分来调节，包括聚阳离子[60,61]或阴离子[62]，以增强它们与目标分析物的相互作用。此外，建议将不同的 PIL 组合起来以提高萃取性能[63]。

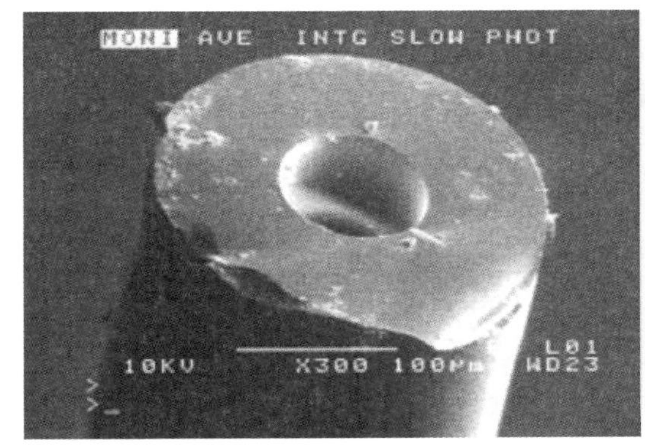

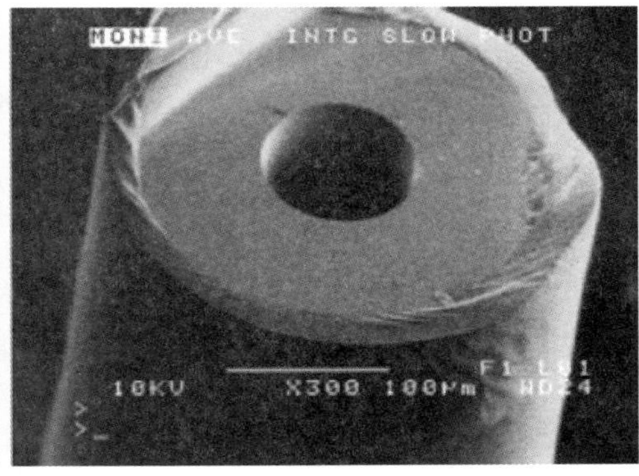

图 7.2 熔融石英纤维（左边）和聚合物离子液体涂层后的相同纤维（右边）的扫描电子照片

［资料来源：F. Zhao, Y. Meng, J. L. Anderson, Polymeric ionic liquids as selective coatings for the extraction of esters using solid-phase microextraction, J. Chromatogr. A 1208 (2008) 1-9.］

虽然在二氧化硅毛细管上沉积 PILs 的可能性不在讨论的范围，但是已经做了一些努力来提高涂层的机械强度。PIL 与纤维载体的共价键合是主要的策略，可与其他替代方案相补充，如载体的选择（不同于传统的硅基载体）或聚合物的交联。冯等[64,65]建议使用不锈钢（SS）作为 PIL 共价键合的支撑物。由于 SS 的惰性，必须对表面进行预处理以固定 PIL。从这个意义上说，SS 以前是涂银改性的，不仅是因为其表面的化学成分，而且由其面积决定。使用硫醇单分子层对银层进行改性，该硫醇单分子层最终与 PIL 单体通过表面自由基链转移聚合原位参与，得到聚合物涂层。SS 也可以通过其他方式进行改性。Pang 和 Liu 提出用黄金预涂 SS，最后涂上二氧化硅层，作为 PIL 生长的支撑[66]。此外，磁控溅射形成的硅层也被认为是不锈钢改性的一种方法[67]。

PIL 的交联也提高了涂层的机械和热稳定性[68]。交联剂的配比对涂层的最终状态起着重要的作用。它改变了从液体驱动到硬半固体的系统状态。它还增加了纤维的寿命，这是由于减少了涂层的热渗漏，但根据冯等的研究发现在较高的交联比例下，提取的重现性降低。

支撑是最终纤维获得高机械稳定性的关键，但确定涂层覆盖率也很重要。在这个意义上，钛丝尤其引人注目，因为钛丝的表面可以被修饰以增加表面积。阳极氧化为钛丝提供了一个粗糙的表面，增加了可以固定的 PIL 的数量[69]。阳极氧化也能制造出二氧化钛纳米结构[70]来增强表面积。

从这个意义上讲，一些研究人员提出了将纳米结构与 PILs 结合使用来提高纤维的萃取能力。在这种情况下，NPs 可能起到双重作用，因为它们可在纤维上引入新的相互作用化学物质，并且它们可能改变纤维的内部结构，增加孔隙率，因为纳米颗粒的加入改变了正常的聚合物生长。碳纳米管由于其纳米尺寸和与芳香族化合物相互作用的能力，在这方面得到了应用[71-73]。图 7.3 显示了纯 PIL 涂层（上边）和纳米管 PIL 涂层（下边）之间的区别。可以观察到涂层的表面结构发生了巨大的变化，形成了一个更粗糙的表面。

（1） （2） （3） （4）

图7.3 纯聚合物离子液体（PIL）涂层［（1），（2）］和纳米管 PIL 涂层［（3）、（4）］的扫描电子照片

［资料来源：Zhang, J.L. Anderson, Polymeric ionic liquid bucky gels as sorbent coatings for solid-phase microextraction, J. Chromatogr. A 1344 (2014) 15-22.］

使用整体材料[74,75]可进一步增加 PIL 涂层的孔隙率。孔隙率的增加，包括表面积的增加，对吸附容量和萃取动力学都有非常积极的影响。厚 PILs 的使用稍微改变了 SPME 单元的几何结构，如图 7.4 所示[76]。在这种方法中，浸水的 PIL 被不锈钢丝刺穿，使其直接浸入样品中。在规定的提取时间后，从样品中提取出来并进行化学洗脱，以进行随后的高效液相色谱分析。这些单体可以很容易地用氧化石墨烯修饰以提高提取性能[77]。

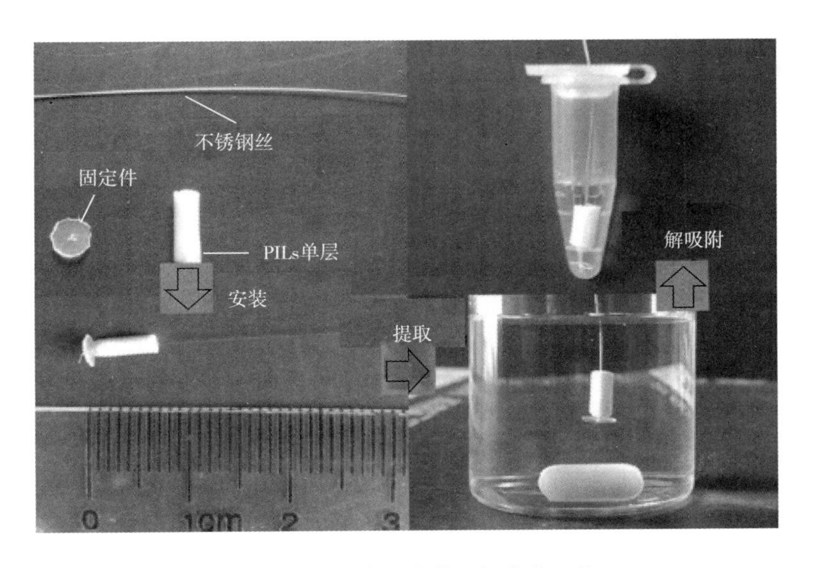

图 7.4 聚合物离子液体固相微萃取装置

［资料来源：J. Feng, M. Sun, Y. Bu, C. Luo, Development of a functionalized polymeric ionic liquid monolith for solid-phase microextraction of polar endocrine disrupting chemicals in aqueous samples coupled to high-performance liquid chromatography, Anal. Bioanal. Chem. 407 (2015) 7025-7035.］

不同于传统 PIL 涂层在纤维载体上的合成，对其他的方法也进行了探索。与 PIL 接枝的二氧化硅颗粒被粘在传统的 SPME 载体上[78]或被塞进针头中[79]用于微萃取，并取得了良好的效果。此外，中空纤维也可以用作合成基于 PIL 的 SPME 的支撑物[80]。在这种情况下，将 PVDF 中空纤维浸入预聚物中，使孔和腔湿润。聚合反应导致 PIL 胶囊的形成，其中吸附相受到中空纤维的保护，提高了萃取的选择性，并允许处理复杂的样品。图 7.5（1）为中空纤维保护的 PIL 涂层，而 PIL 的表面结构如图 7.5（2）所示。

7.3.2 固相微萃取领域中离子液体与纳米粒子的结合

ILs 与 NPs 的结合是一条不断发展的研究路线。NPs 可以定义为具有纳米范围内的一个或多个尺寸的材料（以 100nm 为基准），并且与在宏观尺度上观察到的材料比具有新的特性[81]。NPs 在样品处理中的巨大影响[82]有以下几个方面的原因。一方面，无论是无机的还是碳源的 NPs 种类繁多，它们提供了不同的相互作用，使得它们能够为研究中的问题提供了适当的选择。另一方面，可以对 NPs 进行化学修饰，以提高分析物

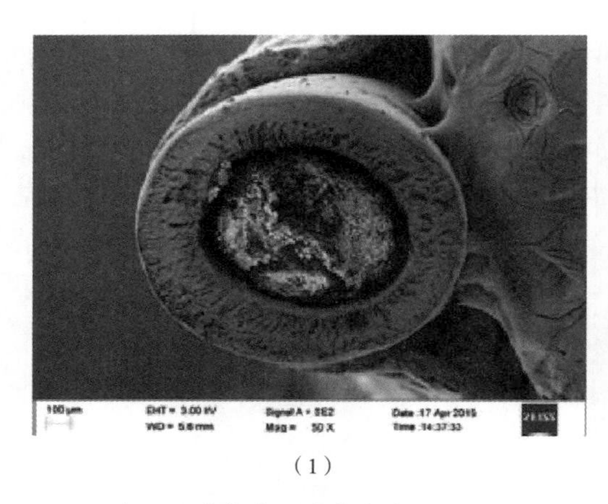

（1）

（2）

图 7.5 包覆聚合物离子液体中空纤维膜（PIL）胶囊（1）和 PIL 吸附剂（2）的 500 倍扫描电子照片

［资料来源：J. Feng, M. Sun, Y. Bu, C. Luo, Hollow fiber membrane-coated functionalized polymeric ionic liquid capsules for direct analysis of estrogens in milk samples, Anal. Bioanal. Chem. 408（2016）1679-1685.］

提取的选择性和/或效率。此外，NPs 可赋予所得到的材料特殊性能。在这些性质中，较大的表面体积比使吸附容量和萃取动力学显著增加。

顺磁性纳米颗粒（MNPs）由于受到外界磁场的吸引，已被广泛应用于与 ILs 结合，以辅助从液体样品中提取目标分析物。这些 MNPs 作为 IL 的惰性载体，IL 是真正的吸附相，可以以不同的方式固定在纳米粒子表面。MNPs 的表面电荷可以通过静电力来保持 IL 离子。Cheng 等[83]使用此方法在高于 NPs 等电点的 pH 下工作的 Fe_3O_4 NPs 表面上保留 1-十六烷基-3-甲基咪唑（未涂 Fe_3O_4 的为 6.3）。IL 的浓度起着关键作用，因为它定义了 NPs 表面上的 IL 阳离子层的数量，从而定义了与目标分析物的相互化学作用（非极性、离子性或混合性）。这个工作流程可以扩展到其他的 MNPs，只需要对 MNP 等电点进行先前的评估[84]。MNP-IL 相互作用一直持续到洗脱步骤被破坏，将 IL 释放到洗脱介质中。共价键合避免了这种释放，因为在整个提取过程中 IL 被强烈固定在 MNPs 表面。这种方法很容易与 LC[85,86]兼容，但由于最终洗脱组分中没有 IL 残留，因此它也可以与 GC[87]兼容。共价键也被提议用于在 MNPs 上锚定 PIL[88-90]。ILs 和 MNPs 的组合超出了 SPME 的范围。事实上，李等[91]在离子液体 DLLME 中提出了这种组合，以帮助萃取后溶剂的回收。

虽然 MNPs-ILs 被广泛用于微萃取，但也有人提出了其他组合。Ríos 等[92]使用涂有 IL 的金纳米粒子从水样中分离和预浓缩磺酰脲类除草剂。由于这些 NPs 不能通过外部磁场回收，作者更倾向于将它们装在传统的固相萃取盒中，以便于样品的处理。Polo Luque 等还介绍了碳纳米管离子液体的潜力[93,94]。IL 增强了纳米管的分散性，纳米管趋向于聚集，由此产生的软材料结合了两种成分的优异性能。

7.4 总结

ILs 在液相和固相微萃取新方法的开发中起着至关重要的作用。它们的理化性质以及它们的可调谐合成是最突出的特征。它们克服了 LPME 中常见有机溶剂的一些限制，提高了测量的重现性，同时保持或提高了微萃取步骤的灵敏度和选择性。包含磁性的可能性也简化了分析过程，而不需要额外的合成步骤。就 SPME 而言，它们也是竞争相，PILs 显然有助于提高涂层的坚固性。

就 LPME 而言，ILs 的主要作用将在分散的环境中发挥。原位溶剂的形成和磁性离子液体的使用是主要的替代方法。使用复分解反应来改变 IL 与样品基质的混溶性，消除了对分散溶剂的需要，分散溶剂在分离步骤中保持分析物的分布常数不变，提高了微萃取的得率。磁性 ILs 与混合 NPs-IL 核壳结构具有竞争性。当萃取剂表现出超顺磁性时，合成过程大大减少，简化了萃取后的回收过程。MILs 的可用性降低是目前应用领域的一个主要缺点。

对于微观 μ-SPE，毫无疑问，由纳米粒子或聚合物与 ILs 合成的新型杂化纳米材料具有特殊的作用。不同于 SPME，聚合 ILs 的巨大潜力将为其在这种微萃取方式中的广泛应用打开一扇门。

致谢

感谢西班牙经济和竞争力部（CTQ 2014—52939R）的资金支持。

参考文献

[1] M. Palacio, B. Bhushan, Ultrathin wear-resistant ionic liquid films for novel MEMS/NEMS applications, Adv. Mater. 20 (2008) 1194-1198;

[1a] S. Armenta, F. A. Esteve-Turrillas, S. Garrigues, M. de la Guardia, Green analytical chemistry: the role of green extraction techniques, in: E. Ibanez, A. Cifuentes (Eds.), Green Extraction Techniques: Principles, Advances and Applications, vol. 76, 2017, pp. 1-25.

[2] E. Aguilera Herrador, R. Lucena, S. Cárdenas, M. Valcárcel, The roles of ionic liquids in sorptive microextraction techniques, Trends Anal. Chem. 29 (2010) 602-616.

[3] M. Rezaee, Y. Assadi, M. R. Milani Hosseini, E. Aghaee, F. Ahmadi, S. Berijani, Determination of organic compounds in water using dispersive liquid-liquid microextraction, J. Chromatogr. A 1116 (2006) 1-9.

[4] Y. Liu, E. Zhao, W. Zhu, H. Gao, Z. Zhou, Determination of four heterocyclic insecticides by ionic liquid dispersive liquid-liquid microextraction in water samples, J. Chromatogr. A 1216 (2009) 885-891.

[5] L. He, X. Luo, H. Xie, C. Wang, X. Jiang, K. Lu, Ionic liquid-based dispersive liquid-liquid microextraction followed high-performance liquid chromatography for the determination of organophosphorus pesticides in water sample, Anal. Chim. Acta 655 (2009) 52-59.

[6] M. T. Pena, M. C. Casais, M. C. Mejuto, R. Cela, Development of an ionic liquid based dispersive liquid-liquid microextraction method for the analysis of polycyclic aromatic hydrocarbons in water samples, J. Chromatogr. A 1216 (2009) 6356-6364.

[7] J. I. Cacho, N. Campillo, P. Viñas, M. Hernández-Córdoba, Improved sensitivity gas

chromatography-mass spectrometry determination of parabens in waters using ionic liquids, Talanta 146 (2016) 568-574.

[8] P. Berton, R. G. Wuilloud, Highly selective ionic liquid-based microextraction method for sensitive trace cobalt determination in environmental and biological samples, Anal. Chim. Acta 662 (2010) 155-162.

[9] I. López-García, Y. Vicente-Martínez, M. Hernández-Córdoba, Determination of lead and cadmium using an ionic liquid and dispersive liquid-liquid microextraction followed by electrothermal atomic absorption spectrometry, Talanta 110 (2013), 46-52.

[10] B. Socas-Rodríguez, J. Hernández-Borges, M. Asensio-Ramos, A. V. Herrera-Herrera, J. A. Palenzuela, M. A. Rodríguez-Delgado, Determination of estrogens in environmental water samples using 1, 3-dipentylimidazoliumhexafluorophosphate ionic liquid as extraction solvent in dispersive liquid-liquid microextraction, Electrophoresis 35 (2014) 2479-2487.

[11] R. S. Zhao, X. Wang, F. W. Li, S. S. Wang, L. L. Zhang, C. G. Cheng, Ionic liquid/ionic liquid dispersive liquid-liquid microextraction, J. Sep. Sci. 34 (2011) 830-836.

[12] M. Cruz-Vera, R. Lucena, S. Cárdenas, M. Valcárcel, One-step in-syringe ionic liquid-based dispersive liquid-liquid microextraction, J. Chromatogr. A 1216 (2009) 6459-6465.

[13] R. Suárez, S. Clavijo, J. Avivar, V. Cerdà, On-line in-syringe magnetic stirring assisted dispersive liquid-liquid microextraction HPLC-UV method for UV filters determination using 1-hexyl-3-methylimidazolium hexafluorophosphate as extractant, Talanta, 148 (2016) 589-595.

[14] A. N. Anthemidis, K. I. Ioannou, Sequential injection ionic liquid dispersive liquid-liquid microextraction for thallium preconcentration and determination with flame atomic absorption spectrometry, Anal. Bioanal. Chem. 404 (2012) 685-691.

[15] Q. Zhou, H. Bai, G. Xie, J. Xiao, Temperature-controlled ionic liquid dispersive liquid phase micro-extraction, J. Chromatogr. A 1177 (2008) 43-49.

[16] Q. Zhou, T. Gao, J. Xiao, G. Xie, Sensitive determination of phenols from water samples by temperature-controlled ionic liquid dispersive liquid-phase microextraction, Anal. Methods 3 (2011) 653-658.

[17] A. Zhao, X. Wang, M. Ma, W. Wang, H. Sun, Z. Yan, Z. Xu, H. Wang, Temperature-assisted ionic liquid dispersive liquid-liquid microextraction combined with high performance liquid chromatography for the determination of PCBs and PBDEs in water and urine samples, Microchim. Acta 177 (2012) 229-236.

[18] Y. Zhang, H. K. Lee, Determination of ultraviolet filters in environmental water samples by temperature-controlled ionic liquid dispersive liquid-phase microextraction, J. Chromatogr. A 1271 (2013) 56-61.

[19] M. Baghdadi, F. Shemirani, Cold-induced aggregation microextraction: a novel sample preparation technique based on ionic liquids, Anal. Chim. Acta 613 (2008) 56-63.

[20] H. Zhang, M. Cheng, X. Jiang, Determination of benzoic acid in water samples by ionic liquid cold-induced aggregation dispersive LLME coupling with LC, Chromatographia, 72 (2010) 1195-1199.

[21] H. Zhang, X. Chen, X. Jiang, Determination of phthalate esters in water samples by ionic liquid cold-induced aggregation dispersive liquid-liquid microextraction coupled with high-performance liquid chromatography, Anal. Chim. Acta 689 (2011) 137-142.

[22] M. Gharehbaghi, F. Shemirani, M. D. Farahani, Cold-induced aggregation microextraction based on ionic liquids and fiber optic-linear array detection spectrophotometry of cobalt in water samples, J. Hazard.

Mater. 165 (2009) 1049-1055.

[23] M. Zeeb, M. Sadeghi, Modified ionic liquid cold-induced aggregation dispersive liquid-liquid microextraction followed by atomic absorption spectrometry for trace determination of zinc in water and food samples, Microchim. Acta 175 (2011), 159-165.

[24] Q. Zhou, X. Zhang, J. Xiao, Ultrasound-assisted ionic liquid dispersive liquid-phase micro-extraction: a novel approach for the sensitive determination of aromatic amines in water samples, J. Chromatogr. A 1216 (2009) 4361-4365.

[25] M. M. Parrilla Vázquez, P. Parrilla Vázquez, M. Martínez Galera, M. D. Gil García, A. Uclés, Ultrasound-assisted ionic liquid dispersive liquid-liquid microextraction coupled with liquid chromatography-quadrupole-linear ion trap-mass spectrometry for simultaneous analysis of pharmaceuticals in wastewaters, J. Chromatogr. A 1291 (2013) 19-26.

[26] E. Stanisz, J. Werner, H. Matusiewicz, Mercury species determination by task specific ionic liquid-based ultrasound-assisted dispersive liquideliquid microextraction combined with cold vapour generation atomic absorption spectrometry, Microchem. J. 110 (2013) 28-35.

[27] C. Yao, J. L. Anderson, Dispersive liquid-liquid microextraction using an in situ metathesis reaction to form an ionic liquid extraction phase for the preconcentration of aromatic compounds from water, Anal. Bioanal. Chem. 395 (2009) 1491-1502.

[28] L. Hu, P. Zhang, W. Shan, X. Wang, S. Li, W. Zhou, H. Gao, In situ metathesis reaction combined with liquid-phase microextraction based on the solidification of sedimentary ionic liquids for the determination of pyrethroid insecticides in water samples, Talanta 144 (2015) 98-104.

[29] S. Mahpishanian, F. Shemirani, Preconcentration procedure using in situ solvent formation microextraction in the presence of ionic liquid for cadmium determination in saline samples by flame atomic absorption spectrometry, Talanta 82 (2010), 471-476.

[30] J. Zhang, Z. Liang, S. Li, Y. Li, B. Peng, W. Zhou, H. Gao, In-situ metathesis reaction combined with ultrasound-assisted ionic liquid dispersive liquid-liquid microextraction method for the determination of phenylurea pesticides in water samples, Talanta 98, (2012) 145-151.

[31] H. Yu, K. D. Clark, J. L. Anderson, Rapid and sensitive analysis of microcystins using ionic liquid-based in situ dispersive liquid-liquid microextraction, J. Chromatogr. A 1406 (2015) 10-18.

[32] F. Galán-Cano, R. Lucena, S. Cárdenas, M. Valcárcel, Ionic liquid based in situ solvent formation microextraction coupled to thermal desorption for chlorophenols determination in waters by gas chromatography/mass spectrometry, J. Chromatogr. A 1229 (2012) 48-54.

[33] E. Stanisz, A. Zgo1a-Grześkowiak, In situ metathesis ionic liquid formation dispersive liquid-liquid microextraction for copper determination in water samples by electrothermal atomic absorption spectrometry, Talanta 115 (2013) 178-183.

[34] S. Hayashi, H. O. Hamaguchi, Discovery of a magnetic ionic liquid [bmim] $FeCl_4$, Chem. Lett. 33 (2004) 1590-1591.

[35] R. E. Del Sesto, T. M. McCleskey, A. K. Burrell, G. A. Baker, J. D. Thompson, B. L. Scott, J. S. Wilkes, P. Williams, Structure and magnetic behavior of transition metal based ionic liquids, Chem. Comm. (2008) 447-449.

[36] B. Mallick, B. Balke, C. Felser, A. V. Mudring, Dysprosium room-temperature ionic liquids with strong luminescence and response to magnetic fields, Angew. Chem. Int. Ed. 47 (2008) 7635-7638.

[37] N. Deng, M. Li, L. Zhao, C. Lu, S. L. de Rooy, I. M. Warner, Highly efficient extraction of

phenolic compounds by use of magnetic room temperature ionic liquids for environmental remediation, J. Hazard. Mater. 192 (2011) 1350-1357.

[38] K. D. Clark, O. Nacham, H. Yu, T. Li, M. M. Yamsek, D. R. Ronning, J. L. Anderson, Extraction of DNA by magnetic ionic liquids: tunable solvents for rapid and selective DNA analysis, Anal. Chem. 87 (2015) 1552-1559.

[39] D. Ge, H. K. Lee, Ionic liquid based hollow fiber supported liquid phase microextraction of ultraviolet filters, J. Chromatogr. A 1229 (2012) 1-5.

[40] X. Ma, M. Huang, Z. Li, J. Wu, Hollow fiber supported liquid-phase microextraction using ionic liquid as extractant for preconcentration of benzene, toluene, ethylbenzene and xylenes from water sample with gas chromatography-hydrogen flame ionization detection, J. Hazard. Mater. 194 (2011) 24-29.

[41] J. Abulhassani, J. L. Manzoori, M. Amjadi, Hollow fiber based-liquid phase microextraction using ionic liquid solvent for preconcentration of lead and nickel from environmental and biological samples prior to determination by electrothermal atomic absorption spectrometry, J. Hazard. Mater. 176 (2010) 481-486.

[42] J. F. Peng, J. F. Liu, X. L. Hu, G. B. Jiang, Direct determination of chlorophenols in environmental water samples by hollow fiber supported ionic liquid membrane extraction coupled with high-performance liquid chromatography, J. Chromatogr. A 1139, (2007) 165-170.

[43] Y. Tao, J. F. Liu, X. L. Hu, H. C. Li, T. Wang, G. B. Jiang, Hollow fiber supported ionic liquid membrane microextraction for determination of sulfonamides in environmental water samples by high-performance liquid chromatography, J. Chromatogr. A 1216, (2009) 6259-6266.

[44] C. Basheer, A. A. Alnedhary, B. S. Madhava Rao, R. Balasubramanian, H. K. Lee, Ionic liquid supported three-phase liquid-liquid-liquid microextraction as a sample preparation technique for aliphatic and aromatic hydrocarbons prior to gas chromatography-mass spectrometry, J. Chromatogr. A 1210 (2008) 19-24.

[45] W. Liu, Z. Wei, Q. Zhang, F. Wu, Z. Lin, Q. Lu, F. Lin, G. Chen, L. Zhang, Novel multifunctional acceptor phase additive of water-miscible ionic liquid in hollow-fiber protected liquid phase microextraction, Talanta 88 (2012) 43-49.

[46] C. Zeng, Y. Lin, N. Zhou, J. Zheng, W. Zhang, Room temperature ionic liquids enhanced the speciation of Cr (Ⅵ) and Cr (Ⅲ) by hollow fiber liquid phase microextraction combined with flame atomic absorption spectrometry, J. Hazard. Mater. 237-238 (2012) 365-370.

[47] M. A. Jeannot, F. F. Cantwell, Solvent microextraction into a single drop, Anal. Chem. 68 (1996) 2236-2240.

[48] Y. He, H. K. Lee, Liquid-phase microextraction in a single drop of organic solvent by using a conventional microsyringe, Anal. Chem. 69 (1997) 4634-4640.

[49] L. Vidal, A. Chisvert, A. Canals, A. Salvador, Ionic liquid-based single-drop microextraction followed by liquid chromatography-ultraviolet spectrophotometry detection to determine typical UV filters in surface water samples, Talanta 81 (2010), 549-555.

[50] A. Sarafraz-Yazdi, F. Mofazzeli, Ionic liquid-based submerged single drop microextraction: a new method for the determination of aromatic amines in environmental water samples, Chromatographia 72 (2010) 867-873.

[51] E. Aguilera-Herrador, R. Lucena, S. Cárdenas, M. Valcárcel, Direct coupling of ionic liquid based single-drop microextraction and GC/MS, Anal. Chem. 80 (2008), 793-800.

[52] F. Zhao, S. Lu, W. Du, W. Zeng, Ionic liquid-based headspace single-drop microextraction coupled to gas chromatography for the determination of chlorobenzene derivatives, Microchim. Acta 165 (2009) 29-33.

[53] A. Chisvert, I. P. Román, L. Vidal, A. Canals, Simple and commercial readily-available approach for the direct use of ionic liquid-based single-drop microextraction prior to gas chromatography determination of chlorobenzenes in real water samples as model analytical application, J. Chromatogr. A 1216 (2009) 1290-1295.

[54] L. Vallecillos, E. Pocurull, F. Borrull, Fully automated ionic liquid-based headspace single drop microextraction coupled to GC-MS/MS to determine musk fragrances in environmental water samples, Talanta 99 (2012) 824-832.

[55] F. Pena-Pereira, I. Lavilla, C. Bendicho, L. Vidal, A. Canals, Speciation of mercury by ionic liquid-based single-drop microextraction combined with high-performance liquid chromatography-photodiode array detection, Talanta 78 (2009) 537-541.

[56] J. -F. Liu, N. Li, G. -B. Jiang, J. -M. Liu, J. A. Jönsson, M. -J. Wen, Disposable ionic liquid coating for headspace solid-phase microextraction of benzene, toluene, ethylbenzene, and xylenes in paints followed by gas chromatography-flame ionization detection, J. Chromatogr. A 1066 (2005) 27-32.

[57] Y. -N. Hsieh, P. -C. Huang, I. -W. Sun, T. -J. Whang, C. -Y. Hsu, H. -H. Huang, C. -H. Kuei, Nafion membrane-supported ionic liquid-solid phase microextraction for analyzing ultra trace PAHs in water samples, Anal. Chim. Acta 557 (2006) 321-328.

[58] F. Zhao, Y. Meng, J. L. Anderson, Polymeric ionic liquids as selective coatings for the extraction of esters using solid-phase microextraction, J. Chromatogr. A 1208 (2008), 1-9.

[59] J. López-Darias, V. Pino, J. L. Anderson, C. M. Graham, A. M. Afonso, Determination of water pollutants by direct-immersion solid-phase microextraction using polymeric ionic liquid coatings, J. Chromatogr. A 1217 (2010) 1236-1243.

[60] Y. Meng, J. L. Anderson, Tuning the selectivity of polymeric ionic liquid sorbent coatings for the extraction of polycyclic aromatic hydrocarbons using solid-phase microextraction, J. Chromatogr. A 1217 (2010) 6143-6152.

[61] J. López-Darias, V. Pino, Y. Meng, J. L. Anderson, A. M. Afonso, Utilization of a benzyl functionalized polymeric ionic liquid for the sensitive determination of polycyclic aromatic hydrocarbons; parabens and alkylphenols in waters using solid-phase microextraction coupled to gas chromatography-flame ionization detection, J. Chromatogr. A 1217 (2010) 7189-7197.

[62] Q. Zhao, J. L. Anderson, Selective extraction of CO_2 from simulated flue gas using polymeric ionic liquid sorbent coatings in solid-phase microextraction gas chromatography, J. Chromatogr. A 1217 (2010) 4517-4522.

[63] C. M. Graham, Y. Meng, T. Ho, J. L. Anderson, Sorbent coatings for solid-phase microextraction based on mixtures of polymeric ionic liquids, J. Sep. Sci. 34 (2011), 340-346.

[64] J. Feng, M. Sun, L. Xu, J. Li, X. Liu, S. Jiang, Preparation of a polymeric ionic liquid coated solid-phase microextraction fiber by surface radical chain-transfer polymerization with stainless steel wire as support, J. Chromatogr. A 1218 (2011) 7758-7764.

[65] J. Feng, M. Sun, J. Li, X. Liu, S. Jiang, A novel aromatically functional polymeric ionic liquid as sorbent material for solid-phase microextraction, J. Chromatogr. A 1227 (2012), 54-59.

[66] L. Pang, J. -F. Liu, Development of a solid-phase microextraction fiber by chemical binding of

polymeric ionic liquid on a silica coated stainless steel wire, J. Chromatogr. A 1230 (2012) 8-14.

[67] L. Xu, J. Jia, J. Feng, J. Liu, S. Jiang, Polymeric ionic liquid modified stainless steel wire as a novel fiber for solid-phase microextraction, J. Sep. Sci. 36 (2013) 369-375.

[68] J. Feng, M. Sun, X. Wang, X. Liu, S. Jiang, Ionic liquids-based crosslinked copolymer sorbents for headspace solid-phase microextraction of polar alcohols, J. Chromatogr. A, 1245 (2012) 32-38.

[69] J. Jia, X. Liang, L. Wang, Y. Guo, X. Liu, S. Jiang, Nanoporous array anodic titanium supported co-polymeric ionic liquids as high performance solid-phase microextraction sorbents for hydrogen bonding compounds, J. Chromatogr. A 1320 (2013) 1-9.

[70] C. Chen, X. Liang, J. Wang, Y. Zou, H. Hu, Q. Cai, S. Yao, Development of a polymeric ionic liquid coating for direct-immersion solid-phase microextraction using polyhedral oligomeric silsesquioxane as cross-linker, J. Chromatogr. A 1348 (2014) 80-86.

[71] J. Feng, M. Sun, L. Li, X. Wang, H. Duan, C. Luo, Multiwalled carbon nanotubes doped polymeric ionic liquids coating for multiple headspace solid-phase microextraction, Talanta 123 (2014) 18-24.

[72] C. Zhang, J. L. Anderson, Polymeric ionic liquid bucky gels as sorbent coatings for solid phase microextraction, J. Chromatogr. A 1344 (2014) 15-22.

[73] J. Feng, M. Sun, Y. Bu, C. Luo, Facile modification of multi-walled carbon nanotubes-polymeric ionic liquids-coated solid-phase microextraction fibers by on-fiber anion exchange, J. Chromatogr. A 1393 (2015) 8-17.

[74] M. Mei, J. Yu, X. Huang, H. Li, L. Lin, D. Yuan, Monitoring of selected estrogen mimics in complicated samples using polymeric ionic liquid-based multiple monolithic fiber solid-phase microextraction combined with high-performance liquid chromatography, J. Chromatogr. A 1385 (2015) 12-19.

[75] K. Liao, M. Mei, H. Li, X. Huang, C. Wu, Multiple monolithic fiber solid-phase microextraction based on a polymeric ionic liquid with high-performance liquid chromatography for the determination of steroid sex hormones in water and urine, J. Sep. Sci. 39 (2016) 566-575.

[76] J. Feng, M. Sun, Y. Bu, C. Luo, Development of a functionalized polymeric ionic liquid monolith for solid-phase microextraction of polar endocrine disrupting chemicals in aqueous samples coupled to high-performance liquid chromatography, Anal. Bioanal. Chem. 407 (2015) 7025-7035.

[77] M. Sun, Y. Bu, J. Feng, C. Luo, Graphene oxide reinforced polymeric ionic liquid monolith solid-phase microextraction sorbent for high-performance liquid chromatography analysis of phenolic compounds in aqueous environmental samples, J. Sep. Sci. 39, (2016) 375-382.

[78] E. Wanigasekara, S. Perera, J. A. Crank, L. Sidisky, R. Shirey, A. Berthod, D. W. Armstrong, Bonded ionic liquid polymeric material for solid-phase microextraction GC analysis, Anal. Bioanal. Chem. 396 (2010) 511-524.

[79] M. M. Rahman, X. L. Osorio Barajas, J. L. Hermosillo Luján, M. A. Jochmann, C. Mayer, T. C. Schmidt, Core-shell hybrid particles by alternating copolymerization of ionic liquid monomers from silica as sorbent for solid phase microextraction, Macromol. Mater. Eng. 300 (2015) 1049-1056.

[80] J. Feng, M. Sun, Y. Bu, C. Luo, Hollow fiber membrane-coated functionalized polymeric ionic liquid capsules for direct analysis of estrogens in milk samples, Anal. Bioanal. Chem. 408 (2016) 1679-1685.

[81] M. T. García-Valverde, R. Lucena, S. Cárdenas, M. Valcárcel, Titanium-dioxide nanotubes as

sorbents in (micro) extraction techniques, Trends Anal. Chem. 62 (2014) 37-45.

[82] R. Lucena, B. M. Simonet, S. Cárdenas, M. Valcárcel, Potential of nanoparticles in sample preparation, J. Chromatogr. A 1218 (2011) 620-637.

[83] Q. Cheng, F. Qu, N. B. Li, H. Q. Luo, Mixed hemimicelles solid-phase extraction of chlorophenols in environmental water samples with 1-hexadecyl-3-methylimidazolium bromide-coated Fe_3O_4 magnetic nanoparticles with high-performance liquid chromatographic analysis, Anal. Chim. Acta 715 (2012) 113-119.

[84] H. He, D. Yuan, Z. Gao, D. Xiao, H. He, H. Dai, J. Peng, N. Li, Mixed hemimicelles solid-phase extraction based on ionic liquid-coated Fe_3O_4/SiO_2 nanoparticles for the determination of flavonoids in bio-matrix samples coupled with high performance liquid chromatography, J. Chromatogr. A 1324 (2014) 78-85.

[85] M. Bouri, M. Gurau, R. Salghi, I. Cretescu, M. Zougagh, A. Rios, Ionic liquids supported on magnetic nanoparticles as a sorbent preconcentration material for sulfonylurea herbicides prior to their determination by capillary liquid chromatography, Anal. Bioanal. Chem. 404 (2012) 1529-1538.

[86] F. A. Casado-Carmona, M. C. Alcudia-León, R. Lucena, S. Cárdenas, M. Valcárcel, Magnetic nanoparticles coated with ionic liquid for the extraction of endocrine disrupting compounds from waters, Microchem. J. 128 (2016) 347-353.

[87] F. Galán-Cano, M. C. Alcudia-León, R. Lucena, S. Cárdenas, M. Valcárcel, Ionic liquid coated magnetic nanoparticles for the gas chromatography/mass spectrometric determination of polycyclic aromatic hydrocarbons in waters, J. Chromatogr. A 1300 (2013) 134-140.

[88] X. Zheng, L. He, Y. Duan, X. Jiang, G. Xiang, W. Zhao, S. Zhang, Poly (ionic liquid) immobilized magnetic nanoparticles as new adsorbent for extraction and enrichment of organophosphorus pesticides from tea drinks, J. Chromatogr. A 1358 (2014) 39-45.

[89] R. Zhang, P. Su, L. Yang, Y. Yang, Microwave-assisted preparation of poly (ionic liquids)-modified magnetic nanoparticles for pesticide extraction, J. Sep. Sci. 37 (2014) 1503-1510.

[90] W. Bi, M. Wang, X. Yang, K. H. Row, Facile synthesis of poly (ionic liquid)-bonded magnetic nanospheres as a high-performance sorbent for the pretreatment and determination of phenolic compounds in water samples, J. Sep. Sci. 37 (2014) 1632-1639.

[91] M. Li, J. Zhang, Y. Li, B. Peng, W. Zhou, H. Gao, Ionic liquid-linked dual magnetic microextraction: a novel and facile procedure for the determination of pyrethroids in honey samples, Talanta 107 (2013) 81-87.

[92] M. J. Lerma-García, E. F. Simó-Alfonso, M. Zougagh, A. Ríos, Use of gold nanoparticle-coated sorbent materials for the selective preconcentration of sulfonylurea herbicides in water samples and determination by capillary liquid chromatography, Talanta 105 (2013) 372-378.

[93] M. L. Polo-Luque, B. M. Simonet, M. Valcárcel, Coiled carbon nanotubes combined with ionic liquid: a new soft material for SPE, Anal. Bioanal. Chem. 404 (2012), 903-937.

[94] M. L. Polo-Luque, B. M. Simonet, M. Valcárcel, Solid-phase extraction of nitrophenols in water by using a combination of carbon nanotubes with an ionic liquid coupled in-line to CE, Electrophoresis 34 (2013) 304-308.

8 免溶剂萃取技术

Farid Chemat[1*], Anne Sylvie Fabiano-Tixier[1], Maryline Abert Vian[1], Tamara Allaf[2], Eugene Vorobiev[3]

1. Avignon University, INRA, UMR 408, Avignon, France
2. ABCAR-DIC, La Rochelle, France
3. Université de Technologie de Compiègne (UTC), Compiègne, France

*通讯作者：E-mail：Farid. Chemat@ univ-avignon. fr

8.1 引言

天然产物是维生素、糖、蛋白质和脂类、纤维、芳香物、色素、抗氧化剂以及其他有机和矿物化合物的复杂混合物。在使用或分析这些物质之前，必须从食物基质中提取它们。为此，可采用不同的方法，例如索氏提取、浸渍、洗脱、煎煮、渗透等，这都需要使用石油溶剂。

需要更经济和环境友好的工艺，在提取制药、食品、化妆品、香水、生物燃料或精细化工等天然产品时，使用溶剂可能会遇到溶剂残留、某些提取化合物损失、溶剂萃取效率低、费时费能（在沸腾溶剂中长时间加热和搅拌、使用大量溶剂）等问题。这些缺点催生了新的可持续“绿色”技术在萃取中的应用，这些技术通常涉及“免溶剂”和降低能量，例如微波萃取、挤压、控制压降过程和脉冲电场。极端或非经典条件下的提取技术是当前应用研究和工业中一个动态发展的领域。取代传统提取工艺的方法通过消除溶剂的使用、减少化石能源和有害物质的产生，提高了生产效率并有助于环境保护。

本章全面介绍了免溶剂萃取技术的最新知识，提供了免溶剂萃取技术、工艺、机制、应用及环境影响必要的理论背景和一些细节。

8.2 脉冲电场

大约 100 年前，连续或交替电场首次应用于食品的加热和液体食品中微生物的灭活。20 世纪中叶，脉冲电场（PEF）加工食品在苏联和德国相继推出[1]。20 世纪 70-90 年代，生物物理学家从理论上发现了细胞膜电穿孔的概念。从 20 世纪 90 年代开始，PEF 在食品加工和生物炼制中的应用越来越广泛。

8.2.1 原理

在 PEF 的作用下，生物膜被电刺穿，暂时或永久地失去其半透性[2]。生物膜的电渗透性（称为电穿孔）可以是可逆的或不可逆的。生物物理文献中提出了许多模型来描述电穿孔机理[2-3]。跨膜电位 U_m 取决于电场强度 E 、细胞大小 d_c 、外电场 E 方向与膜表面法向矢量的夹角 θ 、膜内外电导和膜充电时间 t_c 。跨膜电位可由下式给出[3]：

$$U_m = 0.75 f d_c E \cos\theta [1 - \exp(-t/t_c)] \tag{8.1}$$

式中，f 是一个参数，取决于细胞、膜和周围介质的电生理和尺寸特性。f 和 t_c 值随细胞大小和胞外/胞内电导率 δ_e/δ_i 的变化而变化[4]。当细胞完全损坏时，$\delta_e/\delta_i \approx 1$ 和 $f \approx 1$。当外加电场强度 E 增大时，细胞膜上的电位差（跨膜电位 U_m）也随之增大。如果 U_m 超过某一阈值（通常为 0.7~2.2V），膜的半渗透性就会暂时丧失。植物组织中典型的细胞大小为 $d_c \approx$ （50~100）μm。因此，食品植物组织电穿孔所需的阈值 PEF 强度（对于 $f \approx 1$、$\theta = \pm\pi$ 和 $t_c \ll t$）可根据式（8.1）估算为 $E_t \approx 130 \sim 260$V/cm[5]。值得注意的是对于小型微生物细胞（dc 约为 1μm），E_t 的估计值为 20-50kV/cm。实际上，电渗透程度还取决于脉冲的数量及其持续时间（图 8. 1）。

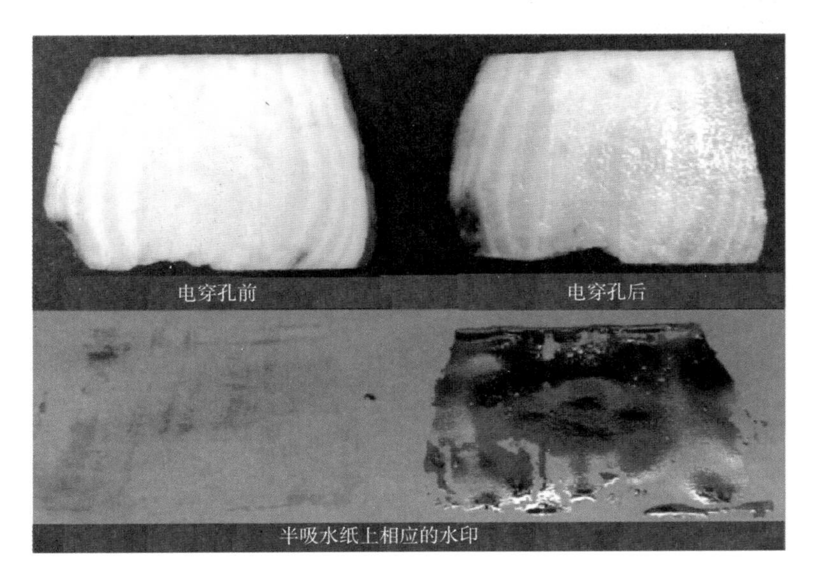

图 8. 1 电穿孔甜菜组织

［资料来源：S. Mahnič－Kalamiza，E. Vorobiev，D. Miklavčič，J. Membrane Biol. 247（12）（2014）1279－1304.］

8.2.2 仪器

高压脉冲发生器的原理是建立在储能器的基础上，储能器在相对较低的充电功率下缓慢充电，并通过激活开关迅速放电。通过这个过程，可以得到一个大的功率倍增。电容器（C）储存电能，它们的并联可以增加储存的能量。储存磁能的电感（L）延迟电流上升。许多 PEF 发生器产生方脉冲。为了产生这样的脉冲，LC 元件的各种排列是必要的。这些排列被称为脉冲网络。开关是高压脉冲发生器中继储能装置之后最重要的元件。大功率开关系统是存储装置和处理室之间的连接元件。脉冲参数（幅度、形状、上升时间）很大程度上取决于开关的特性。Vorobiev 和 N. Lebovka[4] 总结了可用于 PEF 发电机的不同类型的开关。可使用不同的处理室配置（具有平行电极、共线、同轴、非轴向等）来连续处理液体和固液混合物。文献[6,7]概述了产生电场脉冲最常用的技术。

功耗可估算为[4]：

$$w = \int_0^{t_{\mathrm{PEF}}} E^2 \sigma \mathrm{d}t / \rho \approx E^2 \sigma_d t_{\mathrm{PEF}} / \rho \tag{8.2}$$

式中，σ_{d} 是完全电穿孔产品的导电性，ρ 为产品密度，t_{PEF} 为有效电穿孔组织所需的 PEF 处理总时间。对于一些果蔬的电穿孔，典型的电耗值在 2～16kJ/kg[4] 范围内，微生物灭活为 50～200kJ/kg（图 8. 2）。

图 8.2　法国研制的 250kW 脉冲电场发生器

8.2.3　应用

PEF 为液体和固体食品的处理、生物质原料的转化、生物组织传热传质的增强、压制、扩散、渗透浸渍、干燥、冷冻、生物处理强化等提供了巨大的机遇。PEF 辅助加工糖类作物、果蔬、绿色生物质、酵母、微藻等不同原料的例子很多[4]。PEF 用于提取蔗糖、蛋白质、酚类、脂类等多种高附加值物质。中等电场（0.5~5kV/cm）的电穿孔对植物细胞网络没有破坏作用。许多植物材料经 PEF 处理后的微观和流变学研究证实了这一假设。具有受损细胞膜但具有保守细胞壁网络的植物组织具有选择渗透性，能够更好地保留不同的细胞内化合物。在萃取过程中，细胞网络作为一些不需要的化合物通过的屏障是一个很大的优势，可以提高萃取选择性。在甜菜加工、葡萄酒和苹果酒生产、淀粉和菊粉加工等领域，PEF 在免溶剂条件下或使用绿色溶剂处理果汁和溶质时获得了非常好的结果（图 8.3）。表 8.1 所示为 PEF 与免溶剂加压萃取相结合的一些实例。

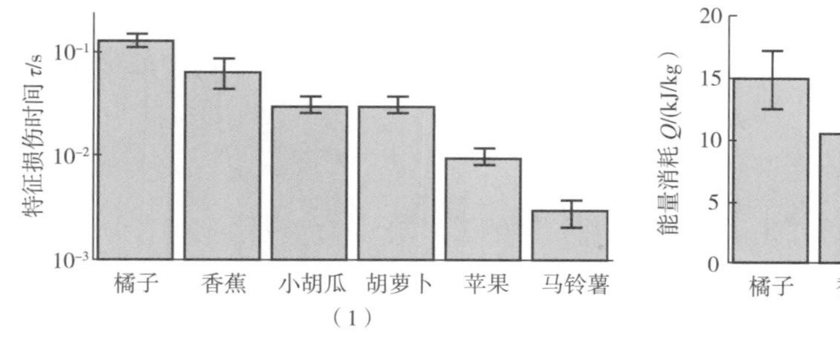

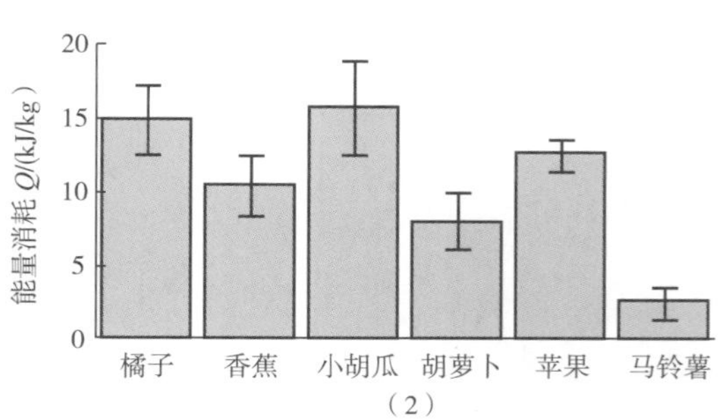

图 8.3　固体食品的特征损伤时间（1）和能量消耗（2）

［资料来源：N. Lebovka，E. Vorobiev，Food and Biomaterials Processing Assisted by Electroporation，in：Advanced Electroportion Techniques in Biology and Medicine，CRC Press，2010，pp. 463-491.］

表 8.1 脉冲电场与免溶剂加压萃取相结合的应用实例

原材料	萃取化合物	参考文献
甜菜片	甜菜汁	[8-11]
菊苣	菊粉	[12]
橘子，橘子皮	橘子汁	[13]
葡萄	葡萄汁，酒	[14-17]
葡萄渣	总多酚、花青素、黄烷醇	[18]
苹果（磨碎或碾碎）	苹果汁	[19-22]
胡萝卜	胡萝卜汁	[23]
苜蓿	苜蓿汁，蛋白质	[24]
嫩玉米	玉米汁	[25]
油菜	菜籽油	[26]
油菜茎叶	蛋白质、多酚	[26]
橄榄	橄榄油	[27]
蘑菇	蘑菇汁、蛋白质、多酚、多糖	[28]

8.3 瞬时控制压降技术辅助压榨提取

瞬时控制压降（DIC）技术是三种操作的紧密结合：自动蒸发、变形和玻璃化转变[29]。一方面，织构导致植物的孔隙率增加，打开其结构，甚至达到细胞壁的破裂[30]。

一方面，变形导致植物的孔隙率增加，结构打开，甚至达到细胞壁的破裂[30]。它意味着在这样一个基质中有更高的扩散率和更多的非挥发性液体，如植物油[31,32]。与干燥、提取、压榨和/或酶或化学反应技术相比，它显然能产生更好的材料特性。因此，传统的植物油加压提取方法变得更加有效，通常可以多 10%以上的出油量。由此产生的油饼/油料可能含有微弱的残余脂质，因此不需要通过溶剂进一步提取。与传统的植物油提取方法相比，DIC 首先需要高温短时处理。水蒸气压力的值通过向真空的突然压降达到，这意味着液体/固体混合物的瞬间冷却。这样做的好处是可以很好地保存化学成分，有效地防止提取油中的脂质和油饼/粕中的蛋白质发生任何热降解[33]。

另一方面，高强度的突然自蒸发触发了产品内部水和其他挥发物的蒸汽混合物的巨大总压力，而外部周围介质保持在约 5kPa 的真空水平。它导致了蒸汽转移，允许在很短的时间内“萃取”不同的精油[34,35]。因此，它还可以去除在加工前（甚至在收获前）添加到材料中的化学物质的挥发痕迹。与标准的精油提取工艺（即水蒸馏或水蒸气蒸馏）相比，DIC 的效率很高，而且操作时间很短（几分钟对几小时甚至几天）。如前所述，DIC 技术应突出高温短时处理和瞬间冷却两个方面。这确保①相关的化学成分保存和有效克服热降解问题；②固体残渣的价值进一步得到提升[36]。

此外，当高压瞬间下降时，营养型和孢子型微生物的细胞壁都会破裂。DIC 则是一种超高温处理（UHT）的去污手术。由于热效应和机械效应的结合，DIC 可以应用到大范围的热敏产品。提取和净化的整个操作的优化通常取决于一种多标准的方法，包括保留产品质量。多循环 DIC 的提取和净化在各个工业部门变得非常相关和方便。

冷机械压榨应用于许多油籽种子被认为是保存高天然脂质质量的最佳榨油方法[37]。对于含油量约为 40%、含水量约为 12%、纤维含量约为 18%、蛋白质含量小于 30%的油类，许多压榨系统可压榨出高达 70%的油[38,39]。压榨后的油饼通常含有超过 17%的油。这意味着它的脂肪含量太高，不能用作动物饲料。不同类型的压榨机通常是根据原材料及其机械性能来定义的。一项具体的研究和工程工作使压榨机的效率更高。其他工业上提高榨油产量的方法，已系统地与原料加热等适当预处理相结合，通常在 80~120℃，在 45~90min 内。这种系统的预处理诱导细胞壁的降解；它意味着增加了高达 73%的产量。然而，这样的加热水平也意味着脂质的重要热降解和较低的经济价值。因此，相关行业有理由使用溶剂萃取法（主要是正己烷）。

充分优化的 DIC 处理条件需要优化饱和蒸汽压力为 600kPa，处理时间为 30s。这意味着更高的产油能力。因为细胞壁破坏和一些值得注意的膨胀，油性压榨允许通过冷压榨提高油的提取率至 81%。此外，没有脂质的降解。事实上，由于 DIC 是一种高温/短时操作，显然热反应可以忽略不计[40]。

此外，热处理可以使膳食蛋白质更适合多种最终用途。尤其是对于鱼肉，在使用双螺杆挤压机之前，经过优化的 DIC 预处理，可以去除高达 93%的油。从饲料中提取的最终动物面粉含水量约为 5%（表 8.2 和表 8.3）。

表 8.2 从油料中提取植物油：常规和瞬时控制压降（DIC）技术辅助加压比较研究

湿基含量			植物油萃取效率		
水分	蛋白质	油脂		出油率	粕中残油率
5%	59%	36%	传统压榨	72%	10.6%
			DIC 辅助压榨	80%	7.70%

表 8.3 从鲱鱼中提取鱼油：双挤压机与瞬时控制压降（DIC）技术结合的影响[13]

湿基含量			油萃取效率		
水分	蛋白质	油脂		出油率	粕中残油率
68%	21.2%	10.8%	传统压榨	7.7%	3.1%
			DIC 辅助压榨	10.4%	0.4%

通过多循环“DIC”提取挥发性物质（图 8.4~图 8.6）本质上是一种特定的瞬间自蒸发。饱和蒸汽压力水平和 DIC 循环次数（在相同的总加工时间内，向真空方向的各种压降）被确定为最重要的操作参数。一旦每个循环的高温处理时间超过了允许交

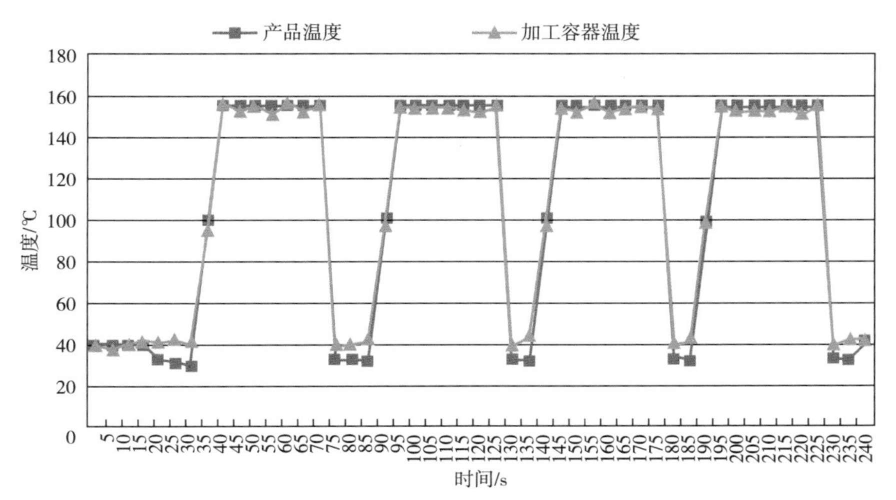

图 8.4 一个简单的单周期瞬时控制压降处理的不同阶段

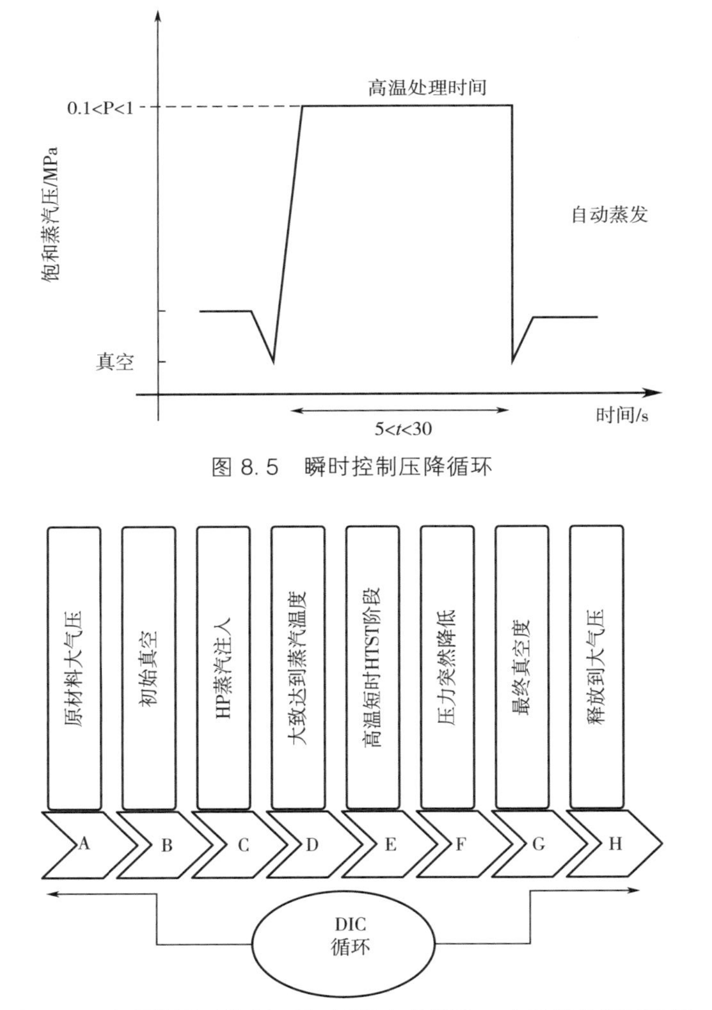

图 8.5 瞬时控制压降循环

图 8.6 多周期瞬时控制压降（DIC）处理和一个周期的不同阶段

换表面的热和冷凝蒸汽在材料内部扩散并在材料内部深度均匀的极限，总的处理时间对操作几乎没有影响。这对缩短总的运行时间（将在几分钟内完成）有直接和间接的影响，从而保持萃取化合物和残余固体的质量[41,42]。

每个周期的定义和每个阶段的影响识别都是基于基础分析。在注入高压饱和蒸汽之前进行初始真空步骤，目的是加热，从而缩短周期处理时间。事实上，这样的初始真空阶段确保蒸汽和产品之间的下一次相互作用在交换表面更接近，产品在几秒钟内达到蒸汽温度。向真空方向的 DIC 持续约 20/80ms 的时间范围。在达到平衡水平之前，根据减压速率（通常高于 5MPa/s），突然出现一个非常低的温度，这与真空水平密切相关（通常为 33℃，5kPa）。

水蒸气蒸馏中最具惩罚性的一个方面是内部传导传热、蒸汽生成和质量扩散的“自相矛盾”行为。DIC 自汽化作为挥发性萃取的最有效的特异性是其在产品核心和周围介质之间建立了非常相关的总压梯度[43]。与水蒸气相比，许多精油分子的相对压力小于 0.01。在外压为 5kPa 的情况下，必然要求有内水作为输运相。

由于这些温和的操作条件（蒸汽压力约为 600kPa；处理时间小于 6min；真空度为 5kPa），能量消耗计算为 1.10kW/kg 时，比常规程序低 20～50 倍。由于能量消耗非常低，没有带电荷的残余水，以及提高固体残渣附加值可能性很大，DIC 自蒸发是一种非常环保的操作[44,45]。法国工程师采用 ABCAR-DIC 工艺设计制造了具有 30mL 处理能力的微型 DIC 装置，可将其插入适当的光谱分析或色谱中[29]。

8.4 免溶剂微波萃取

微波能作为一种非接触式的替代热源在萃取领域得到了有效的利用。与传统的提取方法相比，几种化合物，如精油、芳香、色素、抗氧化剂等，只需消耗很少的能量就可以被连续提取出来。由于人们越来越担心石油溶剂对自然和环境的危害，一种更绿色的技术，称为免溶剂微波萃取（SFME），取得了相当大的成功，其原理与微波辅助萃取（MAE）相同。

8.4.1 微波加热原理

微波（MWs）是频率在 100MHz～3GHz 的电磁波。MWs 由电场和磁场组成，因此传播电磁能。这种能量作为一种非电离辐射，引起离子的分子运动和偶极子的旋转，但不影响分子结构。

当含有永久偶极子或感应偶极子的介电材料置于微波场中时，偶极子在交变磁场中的旋转产生热量。更准确地说，外加的微波场使分子花更多的时间将自己定向到电场的方向，而不是其他方向。当电场被移除时，热搅拌使分子在弛豫时间内回到无序状态，并释放热能。因此，微波加热是由于电磁波在被辐射介质中的耗散而产生的。介质中的耗散功率取决于材料的复介电常数和均电场强度。

在常规加热中，热量从加热介质传递到样品内部；在微波加热中，热量在被辐射的介质中以体积的方式散失。MWs 是体积分布加热，热量从样品转移到较冷的环境。

这是传统加热和微波加热的一个重要区别。

8.4.2 免溶剂微波萃取技术与仪器

SFME是实现绿色分析化学目标的关键技术。作为从天然植物资源中提取和分离高附加值化合物的一种有吸引力的技术，得到了迅速的发展。

在MWs的帮助下，免溶剂萃取可以在几分钟内完成，而不是几小时，具有多种优点，例如重复性高，能耗低，工序短，最终产品纯度高。

在SFME的情况下，水溶性植物细胞在微波辐射下被刺激旋转，瞬时的内部变化导致植物细胞内随后的压力和温度升高，从而导致细胞壁的破裂和目标分子的释放，见图8.7。

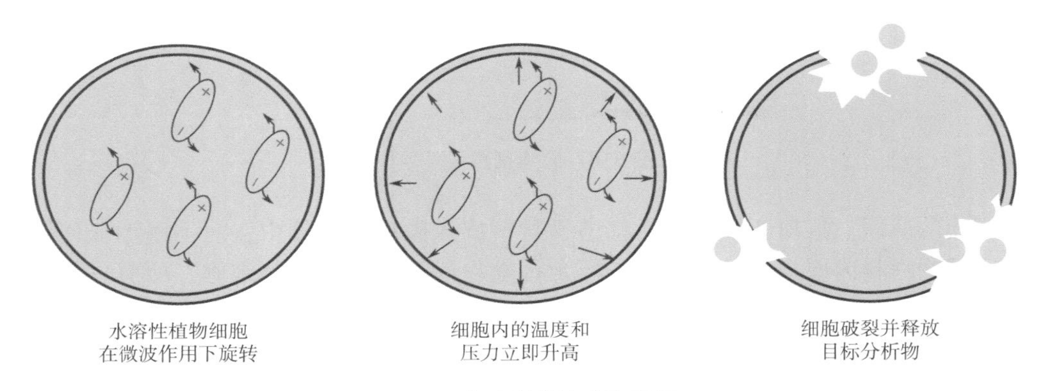

图8.7 免溶剂微波萃取机理

第一种提取天然产物的SFME技术已由Chemat等人开发并申请了专利[46,47]。该工艺是微波辅助提取精油的最新技术之一，免溶剂，且常压无水。基于一个相对简单的原理，这种方法包括将新鲜植物材料放置在微波反应器中而不添加任何溶剂或水。SFME装置如图8.8所示。水溶性植物细胞的内部加热使细胞壁膨胀，从而导致细胞壁破裂。因此，该过程释放含有生物活性化合物的精油，这些精油被植物材料的原位水蒸发。微波炉外的冷却系统可以使蒸馏物持续冷凝，蒸馏物由水和精油组成。多余的水回流到反应器中，以保持植物材料的适当湿度。

所有实验室规模的实验都可以在常压下进行，无须溶剂或加水。需要优化的操作参数易于控制。此外，它还可以帮助从不同的天然芳香植物原料中获得高质量的提取物，并且可直接用色谱法和光谱法分析，无须任何净化、溶剂交换或离心步骤。

里程碑式的“ETHOS X”微波实验室装置（最大功率1000W）被用于进行SFME。这是一个2.45 GHz的多模微波反应器。温度由外部红外传感器监测。一个典型的实验是在大气压下进行的，在没有溶剂或水的情况下，用250g基质，在500W的条件下提取30min。然后在100℃下继续提取，直到不再提取出精油。

2008年，Chemat等[48]设计了一种新的绿色提取技术，称为微波水扩散重力法（MHG）。这种绿色提取技术是一种将微波加热和大气压下地球重力相结合的“倒置”技术。MHG已被用于从不同种类的植物中提取色素、香气成分和抗氧化剂的实验室和

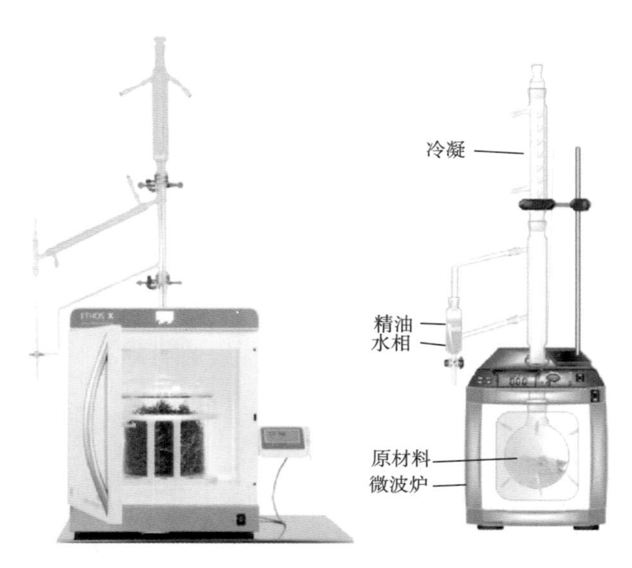

图 8.8 免溶剂微波萃取

工业规模的应用。这项新技术包括将植物材料放入微波反应器中，而不添加任何溶剂或水。植物材料内部水的加热可以破坏含有生物活性成分的植物细胞。所有可能的萃取物，包括植物的内部水都将被释放，并从植物材料的内部转移到外部。这就是物理水扩散现象，它使提取物在地球引力的作用下从微波反应器中脱落，并通过穿孔的特氟龙圆盘落入微波炉外的冷却系统中，提取物在那里不断冷凝，粗萃取物收集在接收瓶中以供进一步分析（图 8.9）。

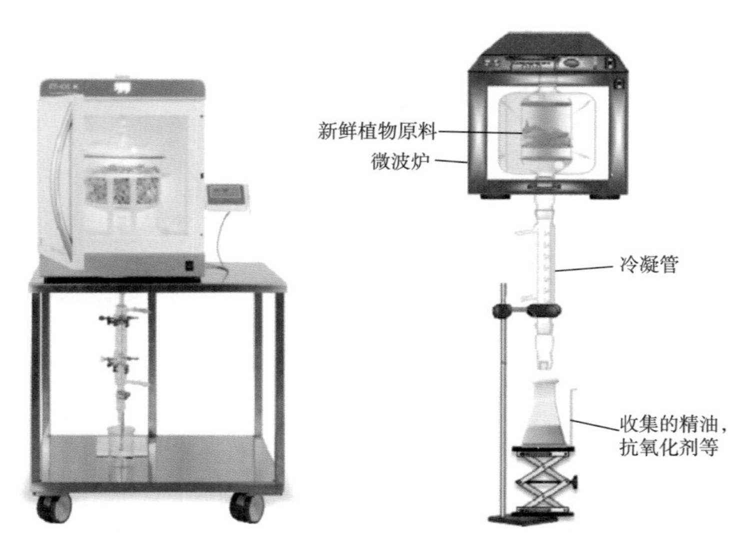

图 8.9 微波水扩散重力法

之后，MHG 系统通过引入真空泵在系统中产生真空而进行了改进。真空泵安装在冷凝器和用于收集天然植物提取物的烧瓶之间。在进一步分析之前，收集粗提取物并冷冻干燥[49]。值得注意的是，这种绿色方法允许在不蒸馏和蒸发的情况下提取生物活性成分，这是单元操作中最耗能的过程。MHG 既不是使用有机溶剂的改良微波辅助萃取，也不是使用原位水蒸发生物活性提取物的 SFME，也不是使用大量水进行能量消耗的改良水蒸气蒸馏。

对于 SFME，使用里程碑式的“ETHOS X”微波实验室装置（最大功率 1000W）来进行 MHG 提取。一个典型的实验是在大气压下进行的，500g 的基质，在没有溶剂和水的情况下，在 500W 的条件下提取 20min。

8.4.3 免溶剂微波萃取的应用

SFME 工艺被应用于几种新鲜和干植物，如香料、芳香草本植物和柑橘类水果[50-54]。干的植物在提取之前需要加湿。表 8.4 所示为采用 SFME 工艺提取精油的植物清单。第一个例子是 *Ocimum basilicum L.*[50]，在大气压下，用 500W 的功率在 30min 内提取 250g *Ocimum basilicum L.* 植物，可获得 0.029%的精油。第二个例子是 *Mentha crispa L.*[50]，在大气压下 500W 下加热 30min，250g 薄荷提取了 0.095%的精油。第三个例子是 *Thymus vulgaris L.*[50]，在大气压下 500W 下加热 30min，250g *Thymus vulgaris L.* 提取了 0.16%的精油。在大多数情况下，SFME 分离出的精油中含氧单萜和单萜烃的含量明显高于传统方法提取的精油[51]。

表 8.4　免溶剂微波萃取植物精油实例

植物物种	萃取条件	参考文献
Ocimum basilicum L.	250 g，P（atm），500W，T=30min，R=0.029%	[50]
Mentha crispa L.	250 g，P（atm），500W，T=30min，R=0.095%	[50]
Thymus vulgaris L.	250 g，P（atm），500W，T=30min，R=0.160%	[50]
Carum ajowan L.	250 g，P（atm），500W，T=60min，R=1.41%	[51]
Cuminum cyminum L.	250 g，P（atm），500W，T=60min，R=0.63%	[51]
Illicium verum	250 g，P（atm），500W，T=60min，R=1.38%	[52]
Nigella sativa L.	150g，P（atm），850W，T=10min，R=0.20%	[52]
Melissa officinialis L.	280g，P（atm），85W，T=50min，R=0.15%	[53]
Laurus nobilis L.	140g，P（atm），85W，T=50min，R=0.42%	[53]
Calamintha nepetaL. Savi	60g，P（atm），250W，T=40min，R=0.38%	[54]

MHG 工艺被应用于多种植物，如芳香植物、柑橘、洋葱和水果副产品[55-59]。表 8.5 列出了 MHG 工艺提取的植物清单。第一个例子是薄荷叶提取物[55]，通过在大气压下、500W 加热 500g 基质 20min 获得 0.95%的精油。对于 Citrus limon L.[56]，在大气压下 500g 基质在 500W 下处理 15min，分别获得了 0.7%和 1.6%的精油产率。以

Rosmarinus officinalis L.（迷迭香）为例，通过15min内以500W的功率提取500g植物，精油得率为0.33%。MHG萃取技术也在水果和蔬菜中得到了应用。

表8.5　　微波水扩散重力法萃取植物名录

植物物种	萃取条件	参考文献
Menthe pulegium L.	500g，P（atm），500W，T=20min，R=0.95%	[55]
Menthe spicata L.	500g，P（atm），500W，T=20min，R=0.6%	[55]
Citrus limon L.	500g，P（atm），500W，T=15min，R=0.7%	[56]
C. limon L.	500g，P（atm），500W，T=15min，R=1.6%	[56]
Citrus aurantifolia（Chrism.）Swing	500g，P（atm），500W，T=15min，R=0.8%	[56]
Citrus paradise L.	500g，P（atm），500W，T=15min，R=1%	[56]
Citrus sinensis L.	500g，P（atm），500W，T=15min，R=1.2%	[56]
C. sinensis L.	500g，P（atm），500W，T=15min，R=1%	[56]
C. sinensis L.	500g，P（atm），500W，T=15min，R=0.9%	[56]
Citrus paradisi Macf.	500g，P（atm），500W，T=15min，R=1.2%	[56]
Rosmarinus officinalis L.	500g，P（atm），500W，T=15min，R=0.33%	[57]
Red，yellow，white and grelot onions（Allium cepa）	500g，P（atm），300~900W，T=5~70min	[58]
Sea buckthorn（Hippophae rhamnoides）副产物	400g，P（atm），1W/g，T=0~25min	[59]

这一创新技术也被应用于抗氧化剂的提取。Zill-e-Huma等[58]报道了MHG提取洋葱黄酮的方法。通过对提取物的显微镜观察，微波辐射强烈破坏了植物组织，目标化合物可以有效地提取和通过高效液相色谱（HPLC）等分析手段进行检测。MHG还被用于沙棘副产品中特定的抗氧化剂的提取，与经典方法相比，在很短的时间内（15min）产生少量的黄酮醇，但MHG提取物中含有较高含量的还原化合物[59]。采用高效液相色谱法（HPLC）对MHG法和常规溶剂提取法提取的黄酮类化合物进行了含量测定，并用Foline-Ciocalteu方法和2，2-二苯基-1-吡啶酰肼（DPPH）自由基还原法测定相应的酚含量。

8.5　压榨

对于糖、葡萄酒和果汁的生产，或生物废料的脱水，在植物油工业中，提取是通过压榨。压榨阶段包括从多孔基质中渗出含有目标代谢物的液体的压缩步骤。果蔬组织的细胞被膜包围，并被嵌入中间层的细胞壁所封闭。刚性壁的组成部分防止了对膜的损坏，因此它们限制了通过挤压/压榨提取的效率。提取效率取决于许多参数，如机械压力、温度、材料阻力、滤饼适合度和孔隙率以及液体的黏度。使用的设备必须考虑到所有这些限制。主要使用两种压榨机：液压压榨机和螺旋压榨机。这种区别反映了压榨的发展历史。

8.5.1　压榨及过程

8.5.1.1　间歇压榨机

使用液压机，对材料施加单轴力进而从基质中分离提取物。施加在原料上的压力可能非常高（高于 50MPa，可可高达 100MPa，橄榄油高达 40MPa）[60]。在实验室中，液压压榨的研究通常是在压榨几克到几百克种子的设备上进行的。将要进行压榨的材料包裹在帆布中，并封装在圆柱形罐中，然后用从上方进入的液压锤对其进行压缩。液压压榨机仍然用于压榨可可以提取可可脂。它们的作用是施加单向力，迫使液体排出。在传统油料的压榨中，这些压榨机逐渐被螺旋压榨机所取代。第一台螺旋压榨机是 Anderson 于 1902 年发明的，他们一直在不断改进。另一种最常用于芳香和香水中精油提取的技术是无须加热和免溶剂的冷压。从某种意义上说，用冷榨油（CPO）提取法提取精油是最天然的油脂提取方法。事实上，几乎不可能获得原油成分未改性的天然精油。在任何精油提取过程中，成分的损失、化学变化等在一定程度上都是不可避免的。香气或香味成分通常是挥发性的，可转化为异构体。为了减少这些变化，有更好或更温和的提取条件，如减压和低温下的水蒸气蒸馏，利用低沸点溶剂进行提取。在从柑橘类水果中提取精油的各种方法中，冷榨法的提取步骤最少，因为该技术基本上不伴随浓缩、蒸馏和溶剂的去除。对于天然香气或风味成分的分析，低温手压技术是方便可靠的。水果的实验室 CPO 提取程序的示例如图 8.10 所示[61]。首先，将果实纵向切成 6~8 片，将果皮和反照面层剥落并丢弃。用手压榨出油，油被收集在冰上的盐水溶液中。在 4000×g、5℃下离心 15min 后，将油取出。然后，用无水硫酸钠在 5℃下干燥原油 24h，以除去蜡和其他不挥发成分。CPO 储存在 −21℃ 直到分析。该过程不包括任何升温或加热、真空或蒸发和溶剂萃取过程；CPO 在性质上接近油。对各种柑橘精油的比较研究表明该方法是可靠和重要的。我们用这种 CPO 方法进行了柑橘风味研究[61]。这些成分数据可以直接在柑橘类水果中进行比较。

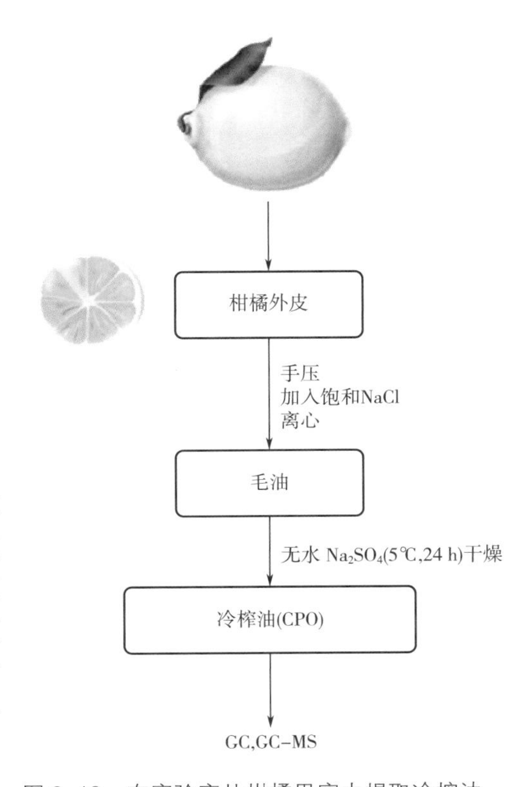

图 8.10　在实验室从柑橘果实中提取冷榨油

8.5.1.2　连续压榨机

因为间歇性生产从经济的角度来看并不

合算，所以不再使用。对连续压榨的研究兴趣是向工业规模的转变。实际上，工业上应用的压榨机都是连续压榨机。现代螺旋压榨机取代了许多液压机，因为它是一个连续的过程，具有更大的压榨能力，且需要更少的劳动力。但是液压还是有趣的，因为它可以更容易地模拟观察到的现象，它也没有在连续压力机中遇到的一个基本影响：剪切。连续压榨机可分为不同的类别：膨胀型；驱逐型；双螺杆系统。在压榨过程中，一个螺旋形的螺杆，在一个桶里朝着一个方向，把原料从压榨机的入口输送到出口。在限制区域，剪切力沿着螺杆发展，并允许压榨出种子中所含的油。Savoire 等[62]对液压和螺旋压榨机的影响因素进行了大量的回顾。在对材料施加特定压力的液压压制过程中，使用活塞是必要的。

螺旋压榨机（图 8.11）是工业螺旋压榨机中最常见的。螺旋榨油机由一个水平螺杆组成，该螺杆在一个由规则间隔的金属棒形成的穿孔筒内旋转，以允许排出的油流动。排出油和液体所需的压力增加是通过阻塞在螺钉头上的圆锥体获得的。在这种类型的压榨机中，压力上升至 110MPa，可获得 3.5%左右的最低含油量[60]。

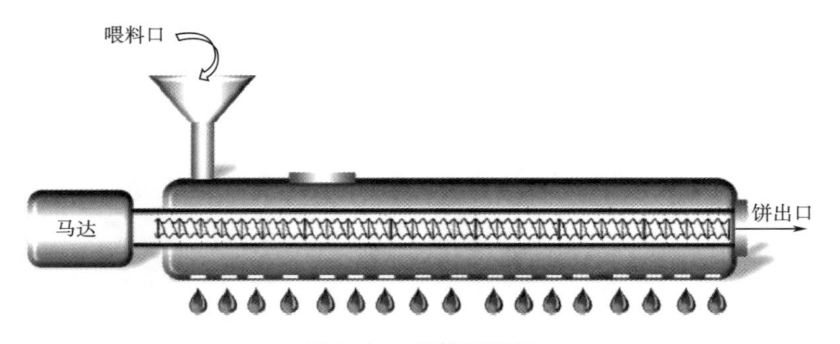

图 8.11　螺旋压榨机

膨胀压榨机和双螺杆压榨机有专门的应用。第一种主要用于低含油量材料（大豆）的预处理或压榨，第二种方法的主要优点是根据螺杆结构可以跳过预处理步骤。然而，双螺杆压榨机并没有在工业生产过程中大规模使用。

膨胀压榨机与挤压压榨机相似之处在于种子是在不榨油的情况下粉碎的。螺钉被锁定在一个封闭的系统中，在该系统中，常规喷嘴被制成，以允许水或蒸汽的注入。这种方法更多的是针对种子制备前的第二步机械提取或溶剂提取。实际上，在螺杆的末端是一个多孔板，通过它挤压材料。这种或多或少发泡的材料更适合于溶剂萃取。主要是因为这个原因，这种方法主要用于低含油量的种子（大豆或棉花）。然而，一些模型在采油区的末端有一个螺杆，这使得对富油种子的处理成为可能，由此得到的颗粒含油量为 30%~35%。

双螺杆压榨机在一步转化材料的物理和化学过程中发挥着重要作用。这种方法似乎是制备分析用样品的最常用方法，因此提出了一种更精确的操作模式。双螺杆压榨机是基于正螺距螺杆实现的输送作用和单螺杆实现的径向压缩和剪切作用[62]。这种装置的优点在于螺钉的排列可以避免预处理步骤，从而允许种子的热机械处理。然而，它们的使用主要局限于实验室研究。

位置或间距（螺距、错开角、长度）等螺杆元件特性的选择决定了螺杆的外形或配置（图 8.12）。压榨过程中的结构安排（产品转化、停留时间分布和机械能输入）是影响压榨机性能的主要因素。理论上讲，螺旋压榨机和液压压榨机的操作是可比的。液体渗出（代谢物）所需的压力是由于堵塞物的形成，堵塞物是通过逐渐减少沿压榨机可用的体积而产生的。这些压力是施加在种子质量上的两种力的结果：径向（管道内推动）；轴向（挤压样品）[63-66]。

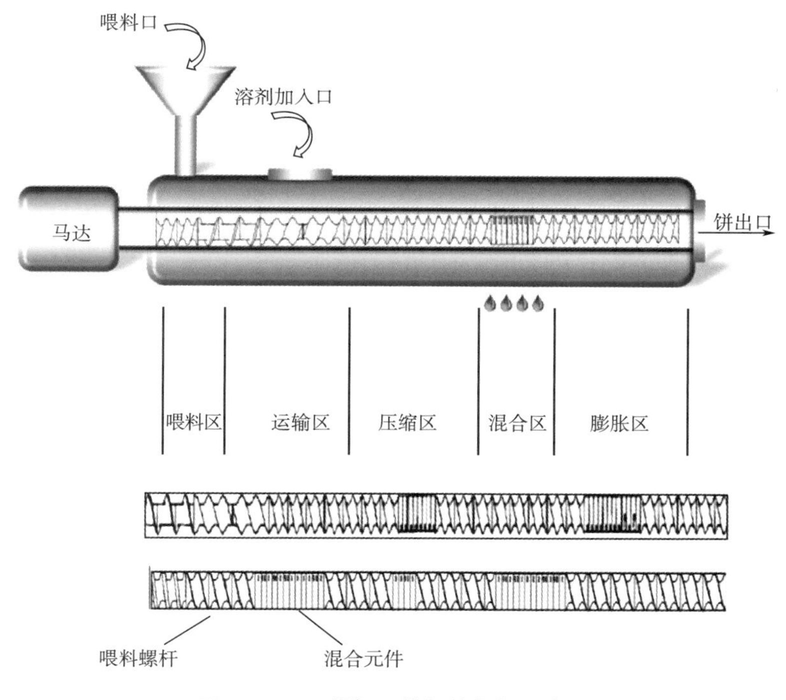

图 8.12　双螺杆压榨机的螺杆示意图

连续压榨机主要用于从纤维状食品原料，如甘蔗和一些水果中榨汁。它们通常由三个滚筒组成，当物料被迫在它们之间连续通过时，它们会挤压物料。

带式压榨机将过滤和压榨动作组合在一个连续操作中。固体被封闭在两条蛇形带之间，它们被一系列的辊逐渐挤压，迫使液体流出。产生的压力相对较低，仅限于容易去除的溶质，如果汁。

压榨是果汁加工中的一项重要操作，直接关系到产品的经济（产量）和质量。苹果、柑橘和葡萄三种不同类型水果的压榨设备有着特殊的商业重要性。

柑橘榨汁机：简单的家用铰刀在工业上用于压榨橙汁和其他果汁。水果被一把锋利的刀切开，旋转的铰刀把果汁榨出。铰刀是工业柑橘榨汁机的基本元件。柑橘类水果通常被挤压成商业果汁产品。在这项技术中，精油被认为是副产品。世界上最流行的榨汁机是食品机械和化学公司（FMC，图 8.13）的在线榨汁机，广泛应用于柑橘汁行业。

在 FMC 榨汁机中，果汁是从整个水果中提取出来的，不需要扩孔。将水果放入下方的提取杯中，上面的杯子中放入待压榨的水果，下方的圆形切割器切出一个孔，孔从底部取出。水果被挤压，果汁从水果残渣中分离出来。

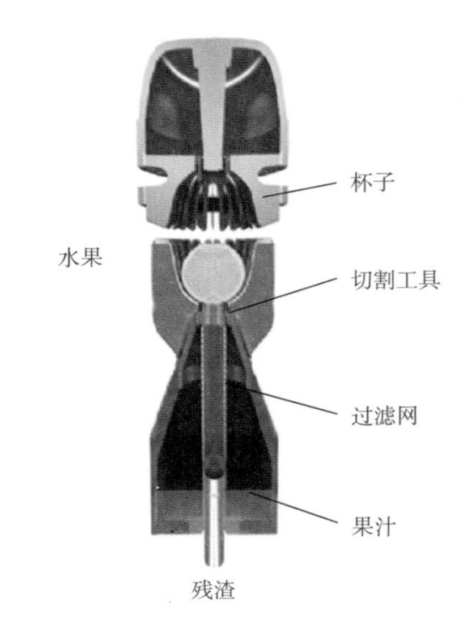

图 8.13　食品机械化工公司的在线榨汁机

图 8.13 所示为榨汁机的示意图。一旦水果被放在下杯上，上杯就下降，水果被压缩进过滤管。当下一个水果放在下一个杯子上时，孔口管上升以压榨水果。果汁收集在罐的底部，废弃的果皮、果汁囊和种子从孔管的出口丢弃。

在线榨汁机适用于各种柑橘类水果，如橙子、柚子、柠檬等。果汁中的精油含量约为 0.03%～0.04%。该系统的萃取能力约为 800 个/min。在这个压榨过程中，也可以得到精油。把废弃的果皮被切成小块。将含有果皮油和水的乳液离心得到 CPO。

苹果和葡萄汁的压榨：苹果汁的压榨可以通过快速压榨来完成，使果汁从包装材料中缓慢移动的移除，一个出口用来收集压榨出的果汁。压榨操作受到果品质量的影响，成熟果品出汁的质量最好和数量最多。水果碎片和颗粒应该大小合适（不要太小或太大）。过程中应用到带刀具的专用磨床。果胶分解酶和纤维素或淀粉分解酶被添加到果汁中，促进果汁的压榨。酶处理应在最佳的 pH、温度和时间条件下进行。压榨助剂通过增加果汁的渗透性来帮助果汁的压榨。常用的压榨助剂是木纤维、纸纤维和稻壳的混合物。每种压榨助剂在果实质量中所占的比例约为 3%～4%。通常在压榨饼中加水溶解残余固体，反复压榨，提高果汁的产量。

葡萄在粉碎机中进行预处理，这是一个孔径为 2.5cm 的旋转滚筒。葡萄从茎上摘下来，经过洞压碎，茎从滚筒中心丢弃。

压碎的葡萄被压在两个旋转的圆柱体之间以榨取果汁。例如，从红葡萄中提取葡萄色素，可以通过将压碎的葡萄加热到 60℃ 左右提取到果汁中。

威姆斯压榨机主要用于葡萄汁和葡萄酒加工。它是一个气动系统，由一个穿孔的、旋转的、水平的气缸和一个充气橡胶管组成。瓶子里装满了葡萄块，气囊压缩迫使果汁流出。然后袋子瘪了，气缸缩回。随着压力的增加，捣碎的葡萄旋转和气压压缩重复多次。

8.5.2　应用

免溶剂萃取技术-压榨在食品领域的许多工业中都有应用，例如：从葡萄和番茄中提取果汁；榨油（橄榄、可可脂、含油种子）；干物质浓缩（糖浆、污泥、食品工业废物）；乳酪生产中凝乳和血清的分离。

压榨与其他工艺的结合也进行了一些相关研究，以在工艺强化阶段获得更好的提取物（质量和数量随时间的变化）。例如，De Haan 引入了气体辅助的机械压榨以提高产油量，并使用 CO_2 辅助来降低施加在种子上的有效机械压力。新的研究工作[67]研究了用 PEF 预处理甜菜以促进甜菜油产量。在不同基质上研究了压榨工艺制备样品的可行性（表 8.6）。

表 8.6　压榨技术应用实例和实验条件

基质	分析物	实验条件	分析方法	参考文献
红葡萄	花青素	螺旋压榨机	HPLC-UV	[68]
木头	多酚	双螺旋压榨机	UV	[69]
腰果	类胡萝卜素	螺旋压榨机	HPLC-DAD-MS	[70]
向日葵	植物油	双螺旋压榨机	碘值和酸值	[71]
米糠	植物油	螺旋压榨机	GC-MS	[72]
麻枫树种子	植物油	螺旋压榨机	碘值和酸值	[73]
胡桃木地板	植物油	液压压榨机	光谱实验室	[74]

Monrad 等[68]研究了用螺旋压榨法从红葡萄渣中提取花青素和黄烷-3-醇，结果表明，这一过程可以提取出 68%的花青素（湿基）和干基质量的 41%，总黄酮-3-醇的提取率分别为 58%和 38%，并且提高了氧自由基吸的抗氧化活性，光度法的抗氧化活性比常规方法提高了 58%和 38%。

Celhay 等[69]使用双螺杆压榨机从木材副产品中提取多酚。由于其多酚含量高和高抗氧化活性，可作为食品和健康应用的高价值物质。

Pinto de Abreu 等[70]研究了从压榨后的腰果中提取类胡萝卜素。这项研究表明压榨使类胡萝卜素总含量增加了 10 倍。

压榨主要用于从各种植物中提取油脂。目前已经对葵花籽[71]、米糠[72]、麻疯树籽[73]和核桃粉[74]进行了免溶剂压榨提取的研究。油料种子的油脂提取通常是通过机械压榨来实现的，这样可以获得高质量的油脂。

8.6 展望

理想的提取溶剂就是免溶剂。首次发现的萃取技术便是不使用溶剂的。最为人所知的提取技术是用机械压榨法提取橄榄油，以及几个世纪以来使用的冷压榨法提取橙子精油。在这一章中，我们讨论了免溶剂萃取技术如何成为萃取天然产物的一个重要问题。在分子水平上，对免溶剂萃取技术相关的过程的理解还没有达到分析化学中其他课题所具有的成熟程度。这种挑战有点雄心勃勃，需要采取特殊的方法。

参考文献

[1] W. Sitzmann, E. Vorobiev, N. Lebovka, Applications of electricity and specifically pulsed electric fields in food processing: historical backgrounds, Innov. Food Sci. Emerg. 37 (2016) 302-311.

[2] A. G. Pakhomov, D. Miklavčič, M. S. Markov, Advanced Electroporation Techniques in Biology and Medicine, CRC Press, Boca Raton, 2010.

[3] G. V. Barbosa-Canovas, M. M. Gongora-Nieto, U. R. Pothakamury, B. G. Swanson, Preservation of Foods with Pulsed Electric Fields, Academic Press, London, 1998.

[4] E. Vorobiev, N. Lebovka, Pulsed electric field induced effects in plant tissues: fundamental aspects and perspectives of application, in: E. Vorobiev, N. Lebovka (Eds.), Electrotechnologies for Extraction From Food Plants and Biomaterials, Springer, 2008, pp. 39-82.

[5] E. Vorobiev, N. Lebovka, Pulse electric field assisted extraction, in: N. Lebovka, E. Vorobiev, F. Chemat (Eds.), Enhancing Extraction Processes in the Food Industry, CRC Press, 2011.

[6] S. Schilling, T. Alber, S. Toepfl, S. Neidhart, D. Knorr, A. Schieber, R. Carle, Effects of pulsed electric field treatment of apple mash on juice yield and quality attributes of apple juices, Innov. Food Sci. Emerg. 8 (2007) 127-134.

[7] S. Schilling, S. Toepfl, M. Ludwig, H. Dietrich, D. Knorr, S. Neidhart, A. Schieber, R. Carle, Comparative study of juice production by pulsed electric field treatment and enzymatic maceration of apple mash, Eur. Food Res. Technol 226 (2008) 1389-1398.

[8] H. Bouzrara, E. Vorobiev, Beet juice extraction by pressing and pulsed electric fields, Int. Sugar J. CII (1216) (2000) 194-200.

[9] M. N. Eshtiaghi, D. Knorr, High electric field pulse pretreatment: potential for sugar beet processing, J. Food Eng. 52 (2002) 265-272.

[10] A. B. Jemai, E. Vorobiev, Pulsed electric field assisted pressing of sugar beet slices: towards a novel process of cold juice extraction, Biosyst. Eng. 93 (2006) 57-68.

[11] H. Mhemdi, O. Bals, E. Vorobiev, Combined pressing-diffusion technology for sugar beets pretreated by pulsed electric field, J. Food Eng. 168 (2016) 166-172.

[12] Z. Zhu, O. Bals, N. Grimi, E. Vorobiev, Pilot scale inulin extraction from chicory roots assisted by pulsed electric fields, Int. J. Food Sci. Tech. 47 (2012) 1361-1368.

[13] E. Luengo, I. Álvarez, J. Raso, Improving the pressing extraction of polyphenols of orange peel by pulsed electric fields, Innov. Food Sci. Emerg. 17 (2013) 79-84.

[14] I. Praporscic, N. I. Lebovka, E. Vorobiev, M. Mietton-Peuchot, Pulsed electric field enhanced expression and juice quality of white grapes, Sep. Purif. Technol. 52 (2007) 520-526.

[15] N. Grimi, M. Lebovka, E. Vorobiev, J. Vaxelaire, Effect of a pulsed electric field treatment on expression behaviour and juice quality of Chardonnay grape, Food Biophys. 4 (2009) 191-198.

[16] E. Puértolas, P. Hernández-Orte, G. Sladaña, I. Álvarez, J. Raso, Improvement of winemaking process using pulsed electric fields at pilot-plant scale. Evolution of chromatic parameters and phenolic content of Cabernet Sauvignon red wines, Food Res. Int. 43 (2010) 761-766.

[17] F. Donsì, G. Ferrari, M. Fruilo, G. Pataro, Pulsed electric fields-assisted vinification, Proc. Food Sci. 1 (2011) 780-785.

[18] S. Brianceau, M. Turk, X. Vitrac, E. Vorobiev, Combined densification and pulsed electric field treatment for selective polyphenols recovery from fermented grape pomace, Innov. Food Sci. Emerg. 29 (2015) 2-8.

[19] M. Bazhal, E. Vorobiev, Electrical treatment of apple cossettes for intensifying juice pressing, J. Sci. Food Agric. 80 (2000) 1668-1674.

[20] M. Bazhal, N. Lebovka, E. Vorobiev, Pulsed electric field treatment of apple tissue during compression for juice extraction, J. Food Eng. 50 (2001) 129-139.

[21] M. Turk, C. Billaud, E. Vorobiev, A. Baron, Continuous pulsed electric field treatment of French cider apple and juice expression on the pilot scale belt press, Innov. Food Sci. Emerg. 14 (2012) 61-69.

[22] M. Turk, E. Vorobiev, A. Baron, Improving apple juice expression and quality by pulsed electric field on an industrial scale, LWT-Food Sci. Technol. 49 (2012), 245-250.

[23] N. Grimi, I. Praporscic, N. Lebovka, E. Vorobiev, Selective extraction from carrot slices by pressing and washing enhanced by pulsed electric fields, Sep. Purif. Technol. 58 (2007) 267-273.

[24] T. Gachovska, M. Ngadi, G. Raghavan, Pulsed electric field assisted extraction from alfalfa, Canadian Biosys. Eng. 48 (2006) 333-337.

[25] M. Sack, C. Eing, W. Frey, C. Schultheiss, H. Bluhm, F. Attmann, R. Stängle, A. Wolf, G. Müller, J. Sigler, L. Stukenbrock, S. Frenzel, J. Arnold, T. Michelberger, Research on electroporation devices on industrial scale, in: Proceedings of the International Conference on Bio and Food Electrotechnologies (BFE2009), 2009, pp. 265-271.

[26] M. Guderjan, P. Elez-Martínez, D. Knorr, Application of pulsed electric fields at oil yield and content of functional food ingredients at the production of rapeseed oil, Innov. Food Sci. Emerg. 8 (2007) 55-62.

[27] E. Puértolas, I. Martínez de Marañón, Olive oil pilot-production assisted by pulsed electric field: impact on extraction yield, chemical parameters and sensory properties, Food Chem. 16 (2015) 497-502.

[28] O. Parniakov, N. Lebovka, E. Van Hecke, E. Vorobiev, Pulsed electric field assisted pressure extraction and solvent extraction from mushroom (*Agaricus bisporus*), Food Bioprocess Technol. 1 (2013) 1-10.

[29] T. Allaf, K. Allaf, Instant Controlled Pressure Drop (D. I. C.) in Food Processing. Food Engineering Series, Springer, New York, 2014.

[30] J. Haddad, K. Allaf, A study of the impact of instantaneous controlled pressure drop on the trypsin inhibitors of soybean, J. Food Eng. 79 (2007) 353-357.

[31] T. Allaf, F. Fine, V. Tomao, C. Nguyen, C. Gines, F. Chemat, Impact of instant controlled pressure drop pre-treatment on solvent extraction of edible oil from rapeseed seeds, OCL 21 (2014) 1-10.

[32] K. Bouallegue, T. Allaf, C. Besombes, R. Ben Younes, K. Allaf, Phenomenological modeling and intensification of texturing/grinding-assisted solvent oil extraction: case of date seeds (*Phoenix dactylifera* L.), Arab. J. Chem. (2016) (in press).

[33] K. Bouallegue, T. Allaf, R. Ben Younes, K. Allaf, Texturing and instant cooling of rapeseed as pretreatment prior to pressing and solvent extraction of oil, Food Bioprocess Technol. 9 (2016) 1521-1534.

[34] C. Besombes, B. Berka-Zougali, K. Allaf, Instant controlled pressure drop extraction of lavandin essential oils: fundamentals and experimental studies, J. Chromatogr. A 1217 (2010) 6807-6815.

[35] M. Kristiawan, V. Sobolik, M. Al Haddad, K. Allaf, Effect of pressure-drop rate on the isolation of cananga oil using instantaneous controlled pressure - drop process, Chem. Eng. Process. Process Intensification 47 (2008) 66-75.

[36] T. Allaf, V. Tomao, K. Ruiz, K. Bachari, M. Elmaataoui, F. Chemat, Deodorization by instant controlled pressure drop autovaporization of rosemary leaves prior to solvent extraction of antioxidants, LWT-

Food Sci. Technol. 51 (2013) 111-119.

[37] S. Azadmard-Damirchi, F. Habibi-Nodeh, J. Hesari, M. Nemati, B. Fathi Achachlouei, Effect of pretreatment with microwaves on oxidative stability and nutraceuticals content of oil from rapeseed, Food Chem. 121 (2010) 1211-1215.

[38] N. Li, G. Qi, X. S. Sun, D. Wang, S. Bean, D. Blackwell, Isolation and characterization of protein fractions isolated from camelina meal, T. ASABE 57 (2014) 169-178.

[39] J. Zubr, Oil-seed crop: *Camelina sativa*, Ind. Crops Prod. 6 (1997) 113-119.

[40] C. Nguyen van, in: K. Allaf (Ed.), Maîtrise de l'aptitude technologique des oléagineux par modification structurelle: applications aux opérations d'extraction et de transestérification in-situ, Université de La Rochelle: La Rochelle, France, 2010.

[41] I. Pérez, C. Bals, I. Martinez de Maranon, K. Allaf, DIC intensification of the mechanical extraction of lipids by pressing, in: T. Allaf, K. Allaf (Eds.), Instant Controlled Pressure Drop (D. I. C.) in Food Processing, Springer, New York, 2014, pp. 163-176.

[42] T. Allaf, V. Tomao, C. Besombes, F. Chemat, Thermal and mechanical intensification of essential oil extraction from orange peel via instant autovaporization, Chem. Eng. Process: Process Intensification 72 (2013) 24-30.

[43] K. Allaf, C. Besombes, B. Berka, M. Kristiawan, V. Sobolik, T. Allaf, Instant controlled pressure drop technology in plant extraction processes, in: N. V. Lebovka, Eugene, F. Chemat (Eds.), Enhancing Extraction Processes in the Food Industry, CRC Press Taylor & Francis Group, USA, 2012.

[44] B. Berka-Zougali, C. Besombes, T. Allaf, K. Allaf, Extraction of essential oils and volatile molecules, in: T. Allaf, K. Allaf (Eds.), Instant Controlled Pressure Drop (D. I. C.) in Food Processing, Springer, New York, 2014, pp. 97-126.

[45] T. Allaf, Instant Controlled Pressure Drop for Green Extraction of Natural Products: Intensification & Combination, in UMR 408-INRA, University of Avignon, France, 2013.

[46] F. Chemat, J. Smadja, M. E. Lucchesie, Solvent Free Microwave Extraction of Volatile Natural Compound, European Patent, 2004. EP 1 439218 B1.

[47] F. Chemat, M. E. Lucchesie, J. Smadja, Solvent Free Microwave Extraction of Volatile Natural Substances, American Patent, 2004. US 0187340 A1.

[48] F. Chemat, M. Abert Vian, F. Visinoni, Microwave Hydro-Diffusion for Isolation of Natural Products, European Patent, EP 1 955 749 A1, 2008. US patent, US 2010/0062121, 2010.

[49] Zill-e-Huma, M. Abert-Vian, M. Elmaataoui, F. Chemat, A novel idea in food extraction field: study of vacuum microwave hydrodiffusion technique for by-products extraction, J. Food Eng. 105 (2011) 351.

[50] M. E. Lucchesie, F. Chemat, J. Smadja, Solvent-free microwaves extraction of essential oil from aromatic herbs: comparison with conventional hydro-distillation, J. Chromatogr. A 1043 (2004) 323-327.

[51] M. E. Lucchesie, F. Chemat, J. Smadja, Solvent-Free microwaves extraction of essential oil from spices, Flavour Frag. J. 19 (2004) 134-138.

[52] F. Benkaci-Ali, A. Baaliouamer, B. Y. Meklati, F. Chemat, Chemical composition of seed essential oils from Algerian *Nigella sativa* extracted by microwave and hydrodistillation, Flavour Frag. J. 22 (2007) 148-153.

[53] B. Uysal, F. Sozmen, B. S. Buyuktas, Solvent-free microwave extraction of essential oils from *Laurus nobilis* and *Melissa officinalis*: comparison with conventional hydrodistillation and ultrasound extraction,

Natural Product Communications 5 (2010) 111-114.

[54] S. Riela, M. Bruno, C. Formisano, D. Rigano, S. Rosselli, M. L. Saladino, F. Senatore, Effects of solvent-free microwave extraction on the chemical composition of essential oil of *Calamintha nepeta* (L.) Savi compared with the conventional production method, J. Sep. Sci. 31 (2008) 1110-1117.

[55] M. Abert Vian, X. Fernandez, F. Visioni, F. Chemat, Microwave hydrodiffusion and gravity, a new technique for extraction of essential oils, J. Chromatogr. A 1190 (2008) 14-17.

[56] N. Bousbia, M. AbertVian, M. A. Ferhat, B. Y. Meklati, F. Chemat, A new process for extraction of essential oil from citrus peels: microwave hydrodiffusion and gravity, J. Food Eng. 90 (2009) 409-413.

[57] N. Bousbia, M. AbertVian, M. A. Ferhat, E. Peticolas, B. Y. Meklati, F. Chemat, Comparison of two isolation methods for essential oil from rosemary leaves: hydrodistillation and microwave hydrodiffusion and gravity, Food Chem. 14 (2009) 355-362.

[58] M. Zill-e-Huma, A. S. Abert-Vian, M. Fabiano-Tixier, O. Elmaataoui, F. Dangles, Chemat, A remarkable influence of microwave extraction: enhancement of antioxidant activity of extracted onion varieties, Food Chem. 127 (2011) 1472.

[59] S. Périno-Issartier, Zill-e-Huma, M. Abert-Vian, F. Chemat, Solvent free microwaveassisted extraction of antioxidants from sea buckthorn (*Hippophae rhamnoides*) food by-products, Food Bioprocess Tech. 4 (2010) 1020-1028.

[60] H. G. Schwartzberg, Expression of fluid from biological solids, Sep. Purif. Methods 26 (1997) 1-213.

[61] M. Sawamura, T. Kuriyama, Quantitative determination of volatile constituents in the pummel (*Citrus grandis* Osbeck forma Tosa-buntan), J. Agric. Food Chem. 36 (1988) 567-569.

[62] R. Savoire, J.-L. Lenoisellé, E. Vorobiev, Mechanical continuous oil expression from oilseeds: a review, Food Bioprocess Technol. 6 (2012) 1-16.

[63] B. K. Gogoi, G. S. Choudhury, A. J. Oswalt, Effects of location and spacing of reverse screw and kneading element combination during twin-screw extrusion of starchy and proteinaceous blends, Food Res. Int. 29 (1996) 505-512.

[64] G. S. Chaudhury, B. K. Gogoi, A. J. Oswalt, J. Aquat, Twin-screw extrusion pink salmon muscle and rice flour blends: effects of kneading elements, Food Prod. Technol. 7 (1998) 69-91.

[65] I. Amalia Kartika, P. Y. Pontalier, L. Rigal, Oil extraction of oleic sunflower seeds by twin screw extruder: influence of screw configuration and operating conditions, Ind. Crops. Prod. 22 (2005) 207-222.

[66] I. Amalia Kartika, P. Y. Pontalier, L. Rigal, Extraction of sunflower oil by twin screw extruder: screw configuration and operating condition effects, Bioresour. Technol. 97 (2006) 2302-2310.

[67] H. Mhemdi, O. Bals, N. Grimi, E. Vorobiev, Filtration diffusivity and expression behaviour of thermally and electrically pretreated sugar beet tissue and press-cake, Sep. Purif. Technol. 95 (2012) 118-125.

[68] J. K. Monrad, M. Suarez, M. J. Motilva, J. W. King, K. Srinivas, L. R. Howard, Extraction of anthocyanins and flavan-3-ols from red grape pomace continuously by coupling hot water extraction with a modified expeller, Food. Res. Int. 65 (2014) 77-87.

[69] C. Celhay, C. E. Mathieu, L. Candy, G. Vilarem, L. Rigal, Aqueous extraction of polyphenols and antiradicals from wood by-products by a twin-screw extractor: feasibility study, CRC Chimie 17 (2014) 204-211.

[70] F. Pinto de Abreu, M. Dornier, A. P. Dionisio, M. Carail, C. Caris-Veyrat, C. Dhuique-Mayer, Cashew apple *Anacardium occidentale* L. extract from by-product of juice processing: a focus on carotenoids, Food Chem. 138 (2013) 25-31.

[71] I. Amalia Kartika, P. Y. Pontalier, L. Rigal, Twin-screw extruder for oil processing of sunflower seeds: Thermo-mechanical pressing and solvent extraction in a single step, Ind. Crops Prod. 32 (2010) 297-304.

[72] S. Sayasoonthorn, S. Kaewrueng, P. Patharasathapornkul, Rice bran oil extraction by screw press method: optimum operating settings, oil extraction level and press cake appearance, Rice Sci. 19 (2012) 75-78.

[73] R. C. Pradhan, S. Mishra, S. N. Naik, N. Bhatnagar, V. K. Vijay, Oil expression from Jatropha seeds using a screw press expeller, Biosystems Eng. 109 (2011) 158-166.

[74] D. Labuckas, D. Maestri, A. Lamarque, Effect of different oil extraction methods on proximate composition and protein characteristics of walnut *Juglans regia* L. flour, LWT-Food Sci. Technol. 1 (2014) 794-799.

9 挥发性样品的顶空分析技术

Ana C. Soria, María J. García-Sarrió, Ana I. Ruiz-Matute and María L. Sanz*
Instituto de Química Orgánica General (CSIC), Madrid, Spain
*通讯作者：E-mail：mlsanz@ iqog. csic. es

9.1 引言

气相色谱-质谱联用（GC-MS）通常是分析实际样品中挥发性化合物的首选技术。在这项技术使用之前，从基质中将这些化合物分离预浓缩出来都是必需的步骤。基于顶空（HS）技术在简单性和多功能性方面的优势，通常应用这项技术来完成上述步骤。此外，这些技术是环境友好技术，因为其减少，甚至不使用平时萃取挥发物所需的溶剂或其他化学试剂。

术语“顶空”是指处于平衡状态的气相或气相，或没有置于封闭容器（通常是密封的小瓶）中的固体或液体基质。

有两种类型的顶空取样技术可以使用：静态（S-HS）和动态（D-HS）。在S-HS中，系统在给定温度下进行一定时间的加热，挥发性成分分配在样品相和气相中［图9.1（1）］；一部分的顶空相被分离出来，采用GC直接进行分析。对操作参数需要认真设定，以获得良好的重复性。

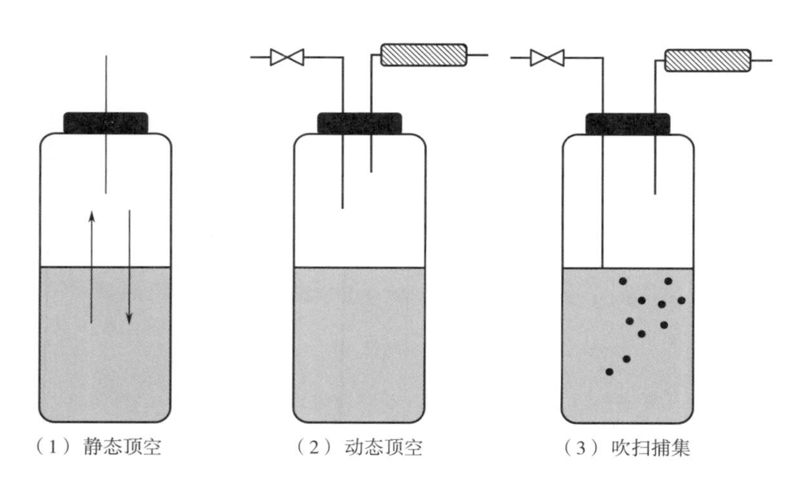

图9.1 HS的取样类型

D-HS技术使用惰性气体流对样品中的挥发性化合物连续提取。这些化合物被预浓缩进入吸附剂中或低温阱中，然后进一步解吸并转移到色谱系统中进行分析［图9.1（2）］。D-HS的最普遍应用之一是所谓的“吹扫和捕集（P&T）”技术［图9.1（3）］。在这种技术中，气体流会产生气泡，通过大部分样品，以增加挥发物的回收率，然后将其吹扫到吸收/吸附挥发物的分析阱中。通过升高温度，从该阱中释放出挥发性化合物。

对于其他取样和萃取技术，在单个萃取步骤后，大多数挥发物仅能部分回收。对于某些应用（如对样品进行表征或分类），此信息可能就足够了。但是，当需要准确测定样品中存在的分析物浓度时，可使用基于逐步HS提取的多步顶空萃取（MHE）方法[1]。使用这种方法，可以在无数次提取步骤之后从基质中将所有挥发物彻底提取出来。在实践中，MHE方法依赖于有限数量的连续提取。

如前所述，HS技术通常与GC-MS联合，用于挥发性化合物的分析；但是，在文献中也可以找到其与全二维气相色谱飞行时间质谱（GC×GC-TOF MS）[2]结合进行分析的应用。这些仪器联结具有以下优势：如通过2D（二维）GC可以增强分离度；使用TOF MS分析仪收集的数据可用于挥发物的阳性判定；检测限低（LOD）；无基质干扰；所需制备的样品量最少。

另外，一些不含分离步骤的分析方法，如基于将S-HS（静态顶空）或D-HS（动态顶空）提取与MS直接耦合的技术，这种技术对于开发快速和廉价的样品分类方法而言，变得越来越重要。样品HS中存在的挥发性化合物可以直接引入MS系统的电离源中，所得到的光谱可被认为是样品挥发性成分的代表性“指纹”图谱[3,4]。

9.2 基本原理

如前所述[1,5]，S-HS中的平衡是采用分配系数（K）进行表征的，它表示样品中气相（C_g）和凝聚相（C_s）的分析物浓度之比［式（9.1）］：

$$K = \frac{C_s}{C_g} \tag{9.1}$$

相比（β）则定义为气相体积（V_g）和样品相体积（V_s）之比：

$$\beta = \frac{V_g}{V_s} \tag{9.2}$$

公式（9.3）给出了K和β，以及HS浓度（C_g）的关系，其中C_0是样品中初始存在的分析物浓度：

$$C_g = \frac{C_0}{K + \beta} \tag{9.3}$$

在给定的系统，特定的条件下，K和β是常数，C_g与C_0成正比。此外，色谱中的峰面积（A）与分析物浓度（C_g）成正比。值得注意的是，HS和GC的很多参数都会对这一关系产生影响，因此，分析条件应保持不变，以保证分析的重复性。

另一方面，根据热力学基本定律，K与分析物i的蒸汽压力（p_i^0）成反比，蒸汽压取决于温度及摩尔浓度（x_i）和活性系数（γ_i），活性系数则取决于化合物的性质，并反映了它与样品基质的相互作用［式（9.4）］。因此，可以通过降低K，使平衡状态下的C_g增加，从而提高HS的灵敏度。

$$K \propto \frac{1}{p_i^0 x_i \gamma_i} \tag{9.4}$$

通过结合上述一些等式，并考虑响应因子（RF）对响应的仪器贡献，提出了一个新的表达式［式（9.5）］，来描述GC峰面积与样品中分析物浓度（C_0）之间的关系，其中p_{total}是小瓶中的总压力。

$$A = \frac{(RF)C_0}{\frac{p_{total}}{p_i^0 \gamma} + \beta} \tag{9.5}$$

在D-HS中，是基于式（9.6）对挥发性样品取样的，WKolb和Ettre在他们的书中写到[5,6]：

$$C_i = C_0 \times e^{-qt} \tag{9.6}$$

其中，连续吹扫过程后残留在样品中的分析物的浓度（C_i）取决于原始浓度（C_0），并随时间（t）呈指数下降，q 为与回收率相关的比例常数。色谱峰面积 A_i 和 A_0，分别代表与 C_i 和 C_0 相关的总的峰面积，与其所对应的分析物的浓度成正比。

当选择逐步模式，并考虑峰面积而不是浓度时，对于每一个连续步骤（i），式（9.6）变成式（9.7）：

$$A_i = A_1 \times e^{-k(i-1)} \tag{9.7}$$

其中 A_1 是在第一提取步骤中获得的峰面积，如果在自动仪器中进行提取，则 k 表示与某些固定的仪器参数相关的常数。

一般来说，A_0 值可以根据式（9.8）和式（9.9）中所示的几何级数之和来估计：

$$A_0 = \sum A_i = A_1 \times (1 + e^{-k} + e^{-2k} + \ldots) \tag{9.8}$$

$$A_0 = A_1/(1 - e^{-k}) \tag{9.9}$$

回收率模型应提供样品中待测成分解吸量与其总量之间的关系，这可以从 GC 数据进行估计，可以认为是为实验峰面积（A_i）与分析物总量所对应的峰面积之比（A_0）。第一步提取步骤的回收率（R）以这样的方式计算：

$$R = A_1/A_0 = 1 - e^{-k} \tag{9.10}$$

由 Kolb 和 Ettre[5,6]提出的模型采用式（9.7）的对数形式：

$$\ln A_i = -k \times (i - 1) + \ln A_1 \tag{9.11}$$

通过对式（9.11）线性回归和实验值 A_1 得到 k，然后得到 A_0 和 R。

如果数据是线性的，式（9.9）可以简化，只使用前两次提取［式（9.12）］所得到的数据可获得分析物总量。虽然对实际目的有效，但更准确的结果是通过使用更多的数据点获得的。

$$A_0 = \frac{A_1}{1 - \left(\frac{A_2}{A_1}\right)} = \frac{A_1^2}{A_1 - A_2} \tag{9.12}$$

这个 MHE 模型在理论上是正确的，在实验上是有用的，并且允许使用改进的统计方法来估计回收率和总挥发量。

9.3 相关仪器

用于 S-HS 的仪器通常由一个密封容器（通常是一个带有隔垫的封闭小瓶）和一个加热系统组成，用于温度控制。取样通常是手工进行的，采用一定体积的气密注射器，但配备有温控系统的低温注射系统，或者是阀门回路注射系统自动装置也是可用的。在最近的情况下，样品瓶是加压的（通常是 155.1～206.8kPa），蒸汽被收集在一个具有固定容量带有阀门的样品环中，样品环所处温度高于样品瓶的温度。取样后，阀门旋转，特定流量的载气通入环中，将挥发物扫向传输线，传输线与 GC 系统联结[7]。图 9.2 所示为一个商品化 HS 回路设置的示意图，包括加压、硅烷化和带温控的钢毛细管和阀[8]。

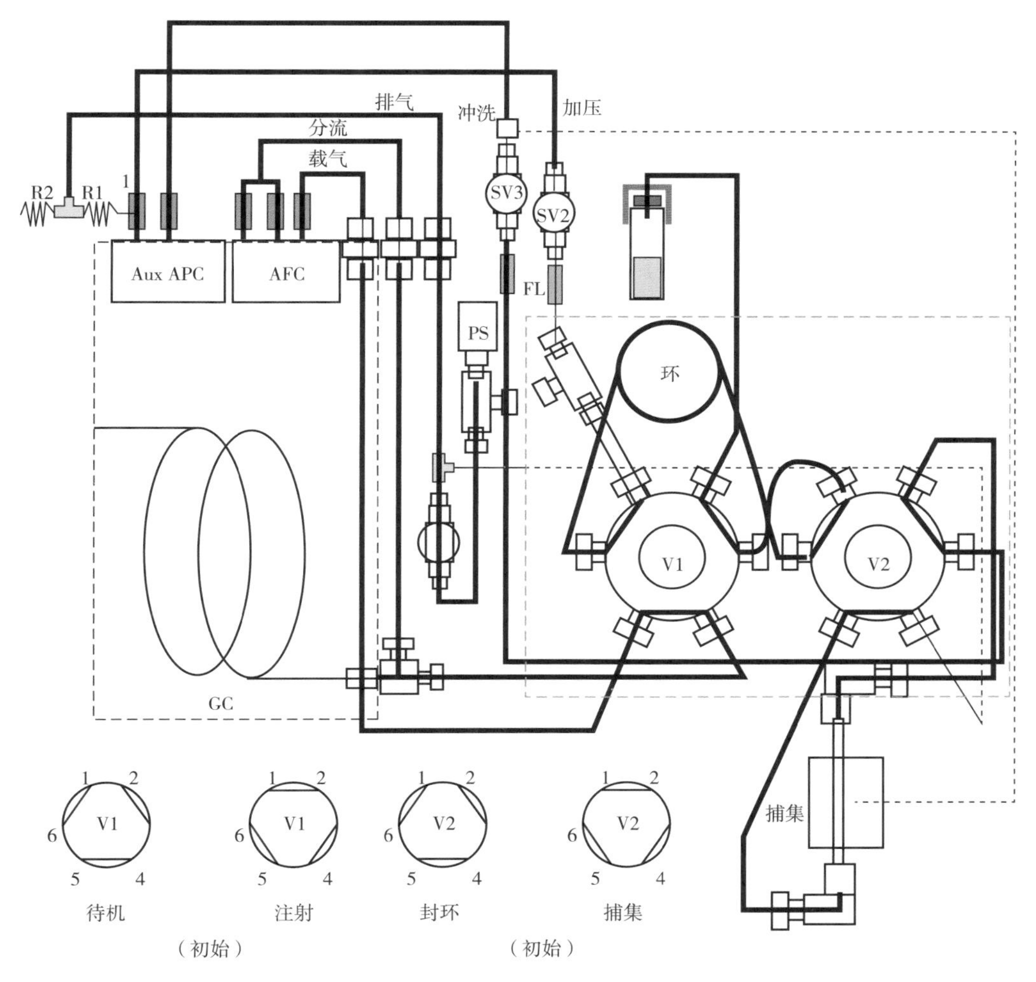

图 9.2 基于环和阀门的 S-HS 系统示意图

AFC，高级流量调节模块； APC，高级压力控制模块； FL，过滤器；

PS，压力传感器； R，节流器； SV，电磁阀阀门； V，六通阀

［资料来源：A. Kremser, M. A. Jochmann, T. C. Schmidt, Systematic comparison of static and dynamic headspace sampling techniques for gas chromatography. Anal. Bioanal. Chem. 408（2016）, 6567-6579，经 Springer 许可］

高浓度容量顶空（HCC-HS）技术，如固相微萃取（SPME）、单滴微萃取（SDME），顶空吸附萃取（HSSE）等也使用不同的设备进行 S-HS 操作。然而，有关这些技术使用的文献很多[9,10]，并且在本书的其他章节中也会对此进行介绍，出于比较的目的，在下面章节中将会提及关于此技术的少量实例。

关于 D-HS，对挥发性产物进行提取/浓缩的不同设计目前都可以商品化购买。它们包括一个样品容器，一个阱和用于调节和控制气体吹扫和温度的不同装置（流量压力调节器，阀门，探头等）。此外，某些设备使用氮气流来清除残留的水，这些水可能是在吹扫过程中被目标挥发物捕获的。水是我们希望去除的，因为它会阻止气体流动

（如果在冷阱中冷凝），破坏 GC 固定相的稳定性等[11,12]。一些仪器还包括第二个捕集阱，通常由熔融石英制成，可以在注入 GC 前将挥发物富集起来。挥发物被保留在由液氮或 CO_2 冷却的阱中，它们的快速注入减少了峰加宽，并提高了分离度和灵敏度[13]。

微型化系统目前也被用于 D-HS，首先是用一个微型吹扫提取器（μPE），与微型气相色谱（μGC）系统[14]芯片化集成。该 μPE 装置由两个入口组成，一个用于样品，另一个用于纯惰性气体对目标分析物进行吹扫。该芯片还包含两个出口，一个用于废水，一个用于将吹扫出来的待测成分引入至微热预浓缩器（μTPC）。通过电阻加热解吸，被捕集的化合物在 μGC 柱上分离，并使用与色谱柱整体集成的微热导检测器（μTCD）进行鉴定。该系统已用于水环境有机化合物（WOCs）的提取和分析。

尽管大多数 D-HS 萃取器与气相色谱系统是在线连接的，但也有一些其他萃取器设计用于离线操作。在后面这种情况下，带有吹扫挥发物的滤芯/捕集阱在附加设备［热脱附（TD）单元］上脱附，此设备与 GC-MS 是在线结合的[1]。例如，管内萃取（ITEX）系统通过填充到不锈钢毛细管中的吸附剂将 HS 气体多次循环，毛细管的一端变细，成为一个侧口注射器。另一端连接到抽吸样品的 HS 注射器。对于热脱附和吹扫步骤，将不锈钢毛细管置入一个加热单元[15]中。

种类繁多，具有不同尺寸，组成，热稳定性和解吸特性的单个和混合捕集阱和滤筒（Tenax、硅胶、硅藻土、石墨化碳、碳分子筛等），目前都是常见易得的[16]。Tenax 吸附剂制成的单个捕集阱由于适用的化合物挥发性范围较广，高温稳定性强，低水亲和力和较长的保质期，因此目前最常见。此外，具有液态 N_2 或 CO_2 的冷阱捕集技术可以改善易挥发/热不稳定化合物的分离。

9.4 实验条件的优化

对于 S-HS 取样，应优化不同的参数，如样品的温度、取样时间和 HS-样品体积比，这些取决于分析物的性质（挥发性、极性、基质亲和性等）和样品基质特征（例如，可用性）。从这些参数来看，最重要的是样品的温度和样品体积，因为 HS 灵敏度取决于 K 和 β 的综合效应（详细解释见参考文献[5]）。

最常用的温度范围为 45～150℃，这取决于目标成分和/或样品基质的稳定性。平衡时间（如果平衡没有达到，则为取样时间）也需要优化，以保证 HS 进样的重复性。

也可以采用一些技巧，如采用低的顶空-样品体积比，采用溶剂溶解样品，这些都可以提高灵敏度，特别是对于待测成分溶解性较低的情况下。当盐被添加到溶液中（所谓的“盐析效应”）时，疏水分析物在水中的溶解度随着溶液离子强度的增加而降低，同时这些化合物在 HS 中的浓度较高。

还应提及的是新型替代溶剂的使用，例如室温离子化液体（ILs）。IL 被定义为熔点低于 100℃的有机盐，由与有机/无机阴离子相关的有机阳离子组成[17]。由于它们的低挥发和低黏度，高选择性和可回收性等优点，可以认为它们是传统有机挥发性溶剂的良好和安全替代品。

挥发性化合物的进样也受基质组成的影响。众所周知，在水性体系中，芳香化合物与蛋白质，多糖或脂质之间所产生的物化作用，可以对它们的保留带来改变[18,19]。

除了前面提到的那些参数外，在D-HS采样中还应考虑其他因素，例如吹扫体积和脱附温度。根据样品的性质和所选择的提取方式（表面清扫或P&T)，可以使用不同设计的样品容器，例如针状喷射容器，带玻璃料的吹扫管或无玻璃料的吹扫容器，在某些情况下还需要进行稀释或添加消泡剂。还有一些其他的，必须考虑的参数，例如惰性气体流量（15~40mL/min）和吹扫时间（2~15min）。在这种方法中，还必须考虑突破体积（定义为每克吸附剂可采样的最大体积，从而保证捕集阱中不会有大量样品损失)。

9.5 应用

由于顶空分析技术的简单性和多功能性，在多个应用领域内，挥发性化合物进色谱分析之前，都常规性地使用HS技术进行采样。尽管可以在不同领域中找到许多应用[1]，但是食品、环境和制药是最常见领域。因此，在本章中，仅在这些领域内选定一些文献，并将其作为本技术当前水平的代表性示例进行讨论。

9.5.1 食品

挥发性样品在通过GC-MS分析以确定食品质量之前，广泛采用D-HS和S-HS技术来提取。作为D-HS的一个例子，Manzini等人[20]优化了HS采样方案，用于提取醋样品中的糠醛，特别是对Aceto Balsamico Tradizionale di Modena（摩德纳传统工艺制造的香醋）的测定。考虑到这些成分对人类安全的影响，确定其浓度具有特殊意义。这些作者使用了D-HS系统和热解吸单元与GC-MS（D-HS-TDU-GC-MS）耦合。他们使用了实验设计程序来优化与萃取和解吸步骤有关的仪器参数。最佳条件：加热温度，40℃，10min；吹扫体积，800mL；干燥体积，1500mL；热脱附时间：5min。他们还分析了两种用于糠醛捕集的吸附剂材料：Tenax TA和Tenax GR。两种吸附剂都具有相似的捕集能力，尽管Tenax TA显示出更高的重现性。通过优化方法，可以对多个醋样品中具有相对较高蒸气压的糠醛进行定量分析。

为了实现实际样品中挥发性成分的准确测定，食品基质效应的消除是另外一个需要考虑的重点。为此，沈等人[21]优化了S-HS-GC-MS方法，用于测定不同加工食品中的呋喃和烷基呋喃，包括调味料、饮料、婴儿配方食品、巧克力和肉制品等。为了避免基质效应，一种技巧是可以使用市售的同位素标记的标准化合物，例如*d*-4-呋喃。然而，这些作者观察到，呋喃和烷基呋喃具有不同的基质效应。因此，将该化合物用作唯一的内标并不适合纠正该影响。而氘代烷基呋喃相当昂贵，无法商品化购买。因此，在这项工作中，他们选择了外部基质匹配的标准校准方法来分析这些化合物，从每个食品组中选择不含待测成分的样品，对其进行加标，以获得基质匹配的校准曲线。此外，还优化了温度和平衡时间，选择60℃和15min作为最佳条件，并避免了在较高温度和更长平衡时间的顶空进样过程中目标化合物的意外形成。

出于定性和定量的目的，精密度也是验证色谱方法的一个重要参数。Soria等人[22]对一个P&T-GC-MS方法的精密度进行了评估，这个方法是来研究不同植物源蜂蜜的挥发性成分的，对通过内标法得到的相对结果（占总挥发性成分的百分比）和数据进行了比较。相对数据显示最低的分散度，这可能是由于内标在蜂蜜基质中的不完全均匀，或该标样和蜂蜜中挥发物的特性不同所致。Soria 等人[23]还提出了使用统计分析来对 P&T-GC-MS 测定中蜂蜜挥发物的百分比数据的离散度进行评估，目的是提高其精度。通过不同的统计参数［相关系数，主成分分析（PCA）特征值和载荷］对实验数据和随机模拟数据进行比较，结果表明，非随机数据对整体分散性的影响较为显著。此外，PCA 证明，当分析多个具有类似分散行为的挥发物时，对百分浓度比进行考察，而不是单个百分数据，会显著提高精度。

为了获得橄榄油中挥发性化合物的化学特征，建立纯橄榄油质量差异辨别[24]，近来，有人建议采用基于 P&T 提取系统的新方法，这个方法在 P&T 后连接 GC-MS，并且采用大气压化学电离源（APCI）和四极杆飞行时间质量分析器（QTOF）进行检测。在 P&T 中对两种不同的吸附剂材料进行了测试：Tenax TA SPE 和非多孔碳 Envi-Carb 盒；它们都是样品中主要挥发物有效的提取方法。所开发的方法灵敏度很高，油性挥发性分析物在测定中具有良好的分离度。

当前，人们对简化香气分析，开发产品质量快速检测的方法有着广泛的兴趣。实现此目的的最常用办法之一就是使用 HS-MS。这项技术已广泛应用于橄榄油[25]、奶酪[26]、葡萄酒[27]等的分析中，并且常与化学计量学工具结合使用，可以对有关产品真伪的分类进行鉴定。这些建立在“基于 MS 的传感器”或“HS-MS 电子鼻”的方法，可在几分钟的时间内（1~5min/样品）提供样品的总挥发性成分的分析响应。Vera 等[28]提出使用基于 HS-MS 的电子鼻，根据其生产场所和化学组成对一系列啤酒进行分类和表征。在该系统中，从啤酒的顶空中（65℃，1h）提取的挥发性化合物通过传输线（90℃）直接引入质谱仪的电离室。图 9.3 为分析得到的 67 个啤酒的电子鼻质谱图（m/z 范围为 50~150）。主成分分析和线性分析判别的结果表明，啤酒的生产厂地不同，香气特征也不同。此外，利用 HS-MS 电子鼻数据，可以将工厂之间的香气差异与啤酒中典型化合物的特征性 m/z 离子的存在（和丰度）相关联。

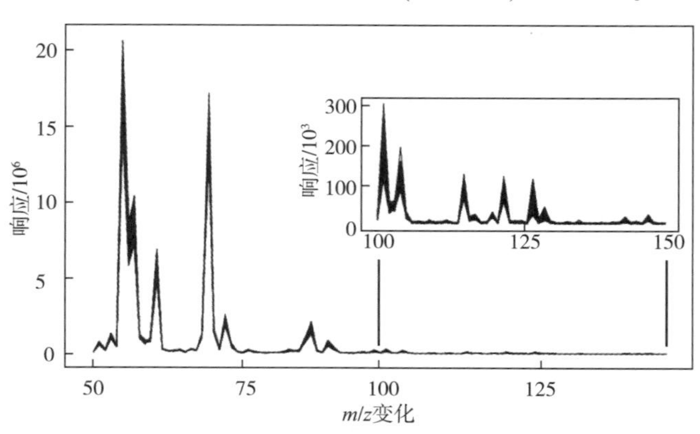

图 9.3　通过 HS-MS 电子鼻分析得到的 67 种啤酒的质谱图

［资料来源：L. Vera，L. Aceña，J. Guasch，R. Boqué，M. Mestres，O. Busto，Characterization and classification of the aroma of beer samples by means of an MS e-nose and chemometric tools，Anal. Bioanal. chem. 399（2011），2073-2081，经 Springer 许可］

另一方面，一些研究将 S-HS 或 D-HS 取样与其他 HCC-HS 技术（主要是 SPME）进行比较，以确定奶酪[29]、葡萄酒[12]、蔬菜[30]等食品样品的挥发性成分。通常，使用不同技术获得的图谱是不同的。而 D-HS 特别适合对挥发性极强的化合物进行分馏，SPME 对于萃取中等和低挥发性化合物更有效。

9.5.2　环境

S-HS 和 D-HS 技术都广泛应用于评估植物、工业过程、农药、交通中，从不同基质（空气，水，垃圾渗滤液等[31-34]）中排放的挥发性有机化合物（VOC）的存在。

MHE 也适用于 S-HS 和 D-HS 技术，用来从环境基质中对 VOC 定量取样。回收分析物所需的循环次数因其挥发性而异，应针对每个特定程序进行优化。例如，Ruiz Bevia 等[32]通过 MHE 方法对使用 P&T-GC-MS 技术分析水样品中存在的三卤甲烷的方法进行了评估。$CHCl_3$（挥发性最大的化合物）的定量净化仅需 2~4 个萃取周期，而 $CHBr_3$（挥发性最小的化合物）则需要 7~19 个周期（取决于样品浓度）。

一些文献对比了环境中挥发性化合物的 HS 采样技术。Florez-Menéndez 等[35]对两种不同的，基于 S-HS 和 HS-SPME 的操作进行了优化，并比较了其结果的灵敏度和精密度。这两个方法都是用来对未处理垃圾渗滤液和生物净化渗滤液中的氯仿、1,1,1-三氯乙烷、四氯化碳、三氯乙烯和四氯乙烯的回收率进行测定的。尽管两种方法都提供了低于 μg/kg 的检测限（LOD），但 HS-SPME 的分析速度比 HS 快（提取时间为 2vs. 15min），而后一种技术提供了更高的分析精密度（2.5%~3.5%相对于 10%~16%）。两种技术都为所研究的分析物提供了良好的回收率（±5%差异）。

Kremser 等[8]对 6 种不同的仪器 HS 方法（静态和动态）的使用进行了系统的评估，从萃取量、重现性和灵敏度方面进行了考察，这些方法是从水中萃取挥发性分析物。这些作者根据其作用方式将不同的 HS 技术分为 3 个代表性类别：S-HS 采样技术（通过注射器或“顶空环”方法，如图 9.2 所示），HS-SPME 技术（SPME 纤维和 PAL SPME 箭头）和动态富集-HS 技术［管内萃取（ITEX）和捕集阱采样］。静态采样技术的提取率约为 10%~20%，而富集技术显示的值高达 80 %。表 9.1 总结了通过不同技术分析的 41 种化合物的方法检测限（MDL）和相对标准偏差（RSD）数据。可以看出，所有技术的再现性都很好（平均 RSD 值低于 8.5%），而 S-HS 和富集 HS 技术获得了不同的 MDL 值。注射器法和定量环法（S-HS）的平均 MDL 结果分别为 64ng/L 和 56ng/L，而正如所料，富集法具有较低的检测限（平均 MDL 值介于 1.8~6.6ng/L，分别与捕集阱和 PAL SPME 箭头采样技术相对应）。在富集技术中，对于所选的分析物，RSD 和 MDL 值的差异可以忽略。但是，当分析具有不同特性（极性或特定分子作用力）的分析物时，特定吸附相材料的选择可能会很重要。

表 9.1　41 种挥发性有机化合物的 5 次重复测定得到的相对标准偏差（RSD,%）和方法的检出限（MDL，ng/L）的最小值（min），最大值（max）和平均值

	S-HS（注射器）		S-HS（定量环）		S-HS 富集（SPME）		S-HS 富集（PAL SPME 箭头）		D-HS 富集（捕集阱）		D-HS 富集（ITEX）	
	RSD	MDL	RSD	MDL	RSD	MDL	RSD	MDL	RSD	MDL	RSD	MDL
最小	4.3	40	6.1	42.1	0.2	2.9	0.9	1.4	0.9	1.7	0.5	3
最大	17.9	122	11	73.2	27	4.1	8.8	2.5	4.5	15	7	6.9
平均	7.6	64.5	8.3	56.5	4.5	3.5	4.4	1.8	2.1	6.6	3.6	4.7

资料来源：A. Kremser, M. A. Jochmann, T. C. Schmidt, Systematic comparison of static and dynamic headspace sampling techniques for gas chromatography, Anal. Bioanal. chem. 408（2016），6567-6579，经 Springer 许可。

在对环境样品中的 VOC 进行 GC-MS 分析之前，还可将 HS 技术与萃取技术相结合。从这一点出发，Min Liew 等[36]提出了一种新的样品制备方法，用于水样品中几种多环芳烃（PAHs）（萘嵌戊烯、萘己环、芴、菲、蒽、荧蒽和苯并芘）的提取，这些水样都来自工业区和自来水。这些作者将使用低分子质量萃取溶剂（正己烷）的微型液液萃取（msLLE）技术与自动全蒸发 D-HS 萃取（FEDHS）技术结合起来。对于 FEDHS，将 Tenax TA 吸附剂放在试管中，该试管自动转移到热脱附装置中（TDU），便于将 PAH 注入 GC-MS。该系统的示意图如图 9.4 所示。这种吸附剂和 PAH 之间的 π-π 强烈作用使这些化合物得到有效浓缩，同时消除了正己烷。在优化的条件下（msLLE：溶剂体积 1.75mL，混合时间：4min；FEDHS：采样温度 60℃，持续 20min，然后 150℃

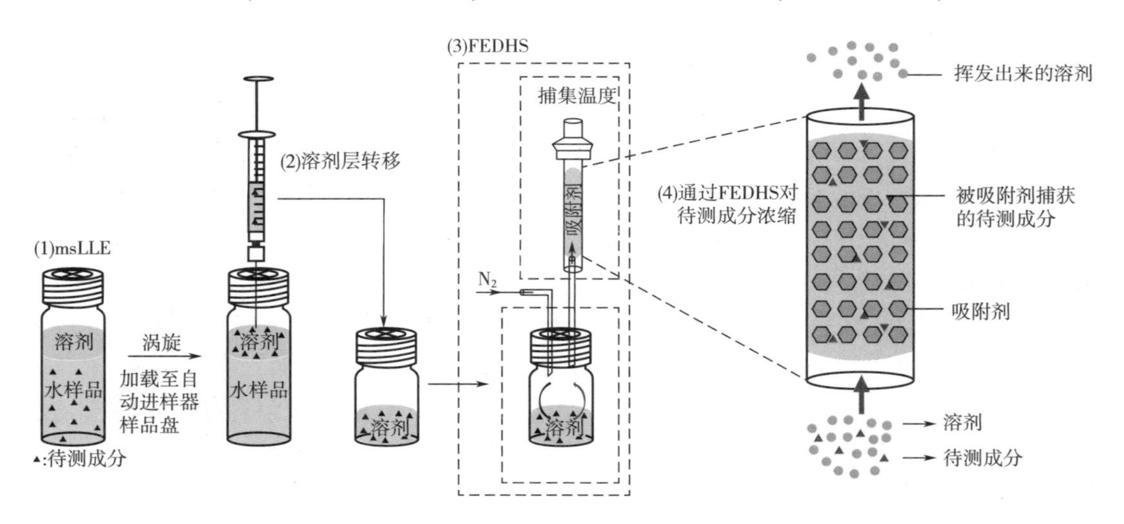

图 9.4　使用 TDU-GC-MS 系统的 msLLE-FEDHS 示意图

［资料来源：C. S. Min Liew, X. Li, H. K. Lee, Miniscale liquid-liquid extraction coupled with full evaporation dynamic headspace extraction for the gas chromatography/mass spectrometric analysis of polycyclic aromatic hydrocarbons with 4000-to-14000-fold enrichment, Anal. Chem. 88（2016），9095-9102。美国化学会版权所有（2016）］

持续5min，捕获温度40℃），得到了较低的检测限（LOD）值（1.85~3.63ng/L）和良好的线性度（R^2> 0.9989）。将该方法与搅拌棒吸附萃取法（SBSE）进行水样分析的比较表明，这种组合的萃取浓缩方法可以使低浓度的PAHs得到更快，更有效的测定。而且，msLLE FEDHS可以完全自动化。

Deng等[37]对微波辅助吹扫捕集（MAPTE）设备操作参数进行了优化，获得了良好的结果（良好的线性和可重复性，以及低的LOD值），该设备与GC-MS在线结合，可对采矿沉积物、鱼组织和藻类细胞中的二甲基三硫醚、2-甲基异冰片酚、土臭素、β-环柠檬醛和β-紫罗兰酮进行同时测定。

Cai等[38]还将磁性固相萃取（MSPE）和S-HS结合起来，用于饮用水样品中多环芳烃的GC-MS分析。图9.5所示为该系统的示意图。此技术选择胆固醇官能化的磁性纳米颗粒（MNPs）作为MSPE吸附剂。对主要的S-HS采样参数进行了优化，例如吸附剂量，萃取时间，柱箱温度和平衡时间。此过程简单，快速且对环境友好，对PAHs的分析，具有高灵敏度（LOD 0.20~7.8ng/L）和良好的精密度（RSDs≤ 9.9 %）等优点。

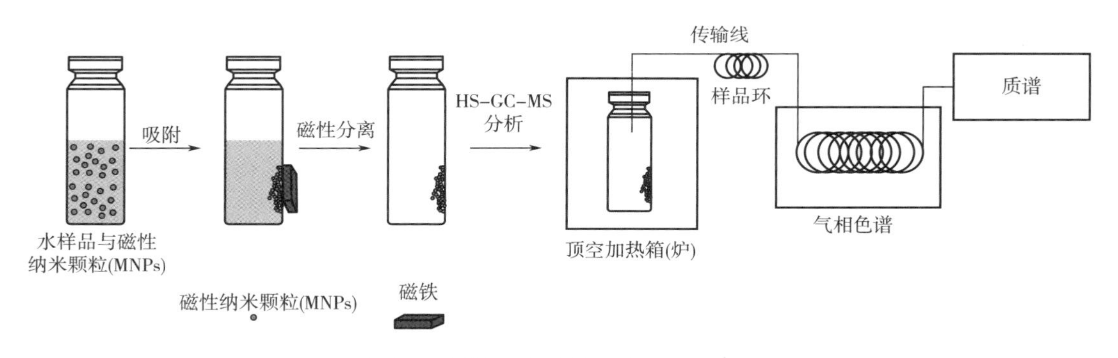

图9.5　MSPE-S-HS-GC-MS示意图

［资料来源：Y. Cai, Z. Yan, L. Wang, M. NguyenVan, Q. Cai, Magnetic solid phase extraction and static headspace gas chromatography-mass spectrometry method for the analysis of polycyclic aromatic hydrocarbons. J. Chromatogr. A, 经Elsevier许可, 2016（1429）, 97-106］

9.5.3　药物

HS与GC结合的技术常用于测定药物和药品中的残留溶剂（RS），不同的药典和《国际人类药物注册技术要求统一会议》（ICH）[39,40]也接受这种使用。残留溶剂即使存在的量非常少，也是不希望的，因为它们会影响药物产品的稳定性，功效和安全性[41]。

S-HS-GC和D-HS-GC这两种分析模式都可以用于药物分析。这些技术优先分析那些可溶于水或在水或有机溶剂中均匀分布的样品，如N,N-二甲基甲酰胺（DMF）、二甲基亚砜（DMSO）、二甲基乙酰胺（DMA）、1,3二甲基-2-咪唑啉酮（DMI）或苄醇（BA）。D-HS分析主要用于浓度非常低的溶剂的测定，因为该技术可以将分析物吸附在捕集阱上，以此得到较低的检测限，从而提高了方法的灵敏度[39,42]。但是，这种操作对残留溶剂的分析也造成了一些缺点，如采样过程中，有环境污染物的干扰，以及与静态模式相比，其精密度较低[41]。

S-HS-GC 是药品中残留溶剂测定应用最广泛的技术，其具有样品制备简单、可以全自动化、对于在 ICH 指南中提到的几乎所有溶剂，其分析的灵敏度均可以覆盖[42]。如前所述，通过添加无机盐、pH 控制、增加平衡温度或控制顶空样品体积比[1,39]，可以提高灵敏度。此外，还可以通过使用适当的溶剂作为溶解介质，最大程度降低基质效应对分析物分配系数的影响，从而增加灵敏度。分配系数影响被分析物从样品液体转移到气相能力，而那些值较低的样品在顶空的浓度更高，从而导致更高的灵敏度[43]。水是为大多数分析物提供较低分配系数值的溶剂。但是，许多药物显示出较低的水溶性，因此需要使用有机溶剂来改善样品的溶解度。因此，在 HS-GC 分析中选择合适的稀释介质至关重要，理想的介质应该能够溶解多种化合物，从而为目标分析物提供低分配系数值。Urakami 等[44]研究了在使用 S-HS-GC 测定药物成分中不同 RS 时，基质介质对测定的影响。虽然每个分析物的峰形似乎不会被基质介质改变，但所有溶剂的峰强度都受到较大的影响。例如，甲醇测定中，选择合适的溶剂，检测灵敏度可提高 4 倍。为了得到良好的回收率和增强的方法灵敏度，一些工作致力于寻找不同的溶剂作为稀释介质，例如液体石蜡[45]，二甲基亚砜和氯化钠溶液[43]或水与其他溶剂（如 DMF 或磷酸钠缓冲液[46,47]）的混合物。

考虑到它们的理化特性，尤其是在环境温度下的低蒸气压，ILs（室温离子液体）是药物产品中 RS 的顶空分析的理想基质介质，可以认为它们是许多应用的替代品。Ni 等[48]研究了 13 种含有不同阳离子或阴离子的 ILs 与 2 种有机溶剂（烷基和质子溶剂）之间的相互作用。选择 HS-GC 用于测定顶空中的分析物浓度。对不同组的 ILs 的极性与 lgK 之间的关系进行了研究。ILs 的极性对所研究的有机溶剂的分配行为有影响，ILs 中，阳离子对顶空效率的影响比阴离子的强。在这项研究提供的指导下，这些作者选择了合适的 ILs 作为基质介质，通过 HS-GC 成功的测定了酮康宁中的有机 RS。最好的顶空溶剂是 1-丁基-3-甲基咪唑六氟磷酸盐［BmIm］［PF_6］，对于不同的 RS，其回收率在 89.8 %~98.2 %，RSD 小于 4.0 %。

但是，由于关于 ILs 的理化性质的信息很少，目前，基于将这些溶剂用于 HS-GC 分析的研究主要来自于经验方法，应进行进一步研究以提供指导，为每种 RS 选择理想的 ILs[39,48]。

在 Tankiewicz 等人[39]的最新评论中，随着测定药物中 RS 的分析方法的发展，可以发现 ILs 用于 HS 进行 RS 分析的不同应用。表 9.2 所示为文献中提出的，在此类测定用作基质介质的不同 ILs。但是，ILs 的使用还有一些局限性，如 ILs 会产生热分解，造成起色谱图上干扰峰的存在，在用作 HS 分析基质介质之前需要对其进行纯化。所谓的全蒸发技术（FET）是 S-HS 的一种变体，它是基于在密封小瓶中从极少量样品中所有挥发性化合物的蒸发，这个蒸发过程是在高于基质熔点或去溶剂化温度 20℃ 的平衡温度下进行的。如果将这个小瓶加热至高于基质沸点 20℃，这时这个方法就称作全挥发法（VTV）[55]。在这些过程中，考虑到所有化合物都被转移到挥发相中，包括分析物，样品基质等，可以从粉末固体样品中直接进行取样。然而，在实践中，基质成分并不总是完全挥发性的，它还会残留在小瓶中，这可能与分析物产生作用；因此，建议进行回收率测试[56]。

表 9.2 用作药物残留溶剂顶空分析的基质介质的离子液体

名称	简写	平均质量/Da	熔点/℃	20℃下的密度/(g/mL)	水溶性	描述	参考文献
1-丁基-3-甲基咪唑四氟硼酸盐	[bmim][BF_4]	226.023	-71	1.2	混溶	溶解性很好，热稳定性高，专用于高沸点溶剂和水不溶物质	[49-51]
1-丁基-3-甲基咪唑二酰胺/双酰胺	[bmim][DCA]	205.26	-6	1.06	混溶	亲水性；热稳定性高（分解温度300℃）；不溶解药用产品的辅料	[48，52]
1-烯基-3-丁基咪唑氯化物	[abim][Cl]	200.708	65	—	混溶	亲水性；基质在室温下微溶	[48，49，52]
乙酸丁酯-3-甲基咪唑	[bmim][Ac]	198.262	<-20	1.02	混溶	亲水性能；高温下的稳定性极低（范围160~200℃）	[48，52]
1-乙基-3-甲基咪唑乙酸	[emim][Ac]	70.209	<-20	1.03	混溶	分解温度220℃	
1-丁基-3-甲基咪唑二甲基磷酸	[bmim][DMP]	264.259	<-20	1.08	混溶	溶解常见的碳水化合物药物辅料，热稳定性高，专门用于蒸汽压极低的溶剂	[48，52]
1-丁基-3-甲基咪唑六氟磷酸	[bmim][PF_6]	284.18	6.5	1.36	不混溶	疏水性，室温下基质微溶；高温下稳定性低	[48，52-54]

资料来源：M. Tankiewicz，J. Namiesnik，W. Sawicki，关于残留溶剂含量的药品质量控制分析程序：挑战和最新进展，TrAC-Trends Anal. Chem. 80（2016）328-344，经 Elsevier 许可。

FET 也被认为是对低沸点基质中高沸点挥发性有机化合物进行分析的有意义的工具。Kialengila 等[56]开发了一种用于测定不同残留溶剂的 FET 方法，对 DMF、DMSO 和 BA 在水中的混合物进行了测定，其线性、灵敏度、特异性、精密度和准确性均显示了良好的结果。该方法的优点是，与常规 S-HS 方法相比，所研究的可与水混溶的高沸点挥发性有机化合物的灵敏度大大提高。DMA、DMSO 和 BA 的检测限和定量限远低于其各自的官方限量浓度。

9.6 结论与展望

S-HS 和 D-HS 都是挥发性成分分析应用较为广泛的技术。文献中关于其在食品，环境和制药领域等不同领域的大量应用不断涌现。这些技术的主要优点是高重现性、灵敏度，以及简单、快速。此外，在 HS 技术中，避免了使用对人体健康或环境有害的溶剂或其他化学物质。这些优势使得这种环保技术可以替代其他多种应用中的，传统的蒸馏技术或基于溶剂萃取的技术。

重要的是要指出，挥发性化合物的提取在很大程度上取决于操作条件和相应的基质。因此，应针对每个特定应用认真优化 HS 参数。

近年来，已开发出各种仪器设备，主要是为了提高仪器的自动化程度和小型化程度。此外，它们能与不同分析技术（主要是 GC-MS，最近为 GC×GC-MS，以及直接与 MS 联结）的在线结合，以及与其他预浓缩或萃取技术结合也增强了其在更快，更灵敏的测定中的应用。预计在不久的将来这些方法将得到持续发展。

致谢

这项工作的资金来源是：阿雷斯基金会，马德里的自治协会和来自 FEDER 计划（项目 S2013/ABI-3028，AVANSECAL）的欧洲资金以及西班牙 A. C. S. 集团的经济、工业和竞争部（MINECO）的资助（项目 AGL2016-80475-R；AEI/FEDER，UE）。也感谢 MINECO 与 Ramony Cajal 的合同。

参考文献

[1] A. C. Soria, M. J. García-Sarrió, M. L. Sanz, Volatile fractionation by headspace techniques, TrAC-Trends Anal. Chem. 71 (2015) 85-99.

[2] S. Herrera López, M. J. Gómez, M. D. Hernando, A. R. Fernández-Alba, Automated dynamic headspace followed by a comprehensive two-dimensional gas chromatography full can time-of-flight mass spectrometry method for screening of volatile organic compounds (VOCs) in water, Anal. Methods 5 (2013) 1165-1177.

[3] S. López-Feria, S. Cárdenas, J. A. García-Mesa, M. Valcárcel, Usefulness of the direct coupling headspace-mass spectrometry for sensory quality characterization of virgin olive oil samples, Anal. Chim. Acta 583 (2007) 411-417.

[4] D. Cozzolino, H. E. Smyth, W. Cynkar, L. Janik, R. G. Dambergs, M. Gishen, Use of direct headspace-mass spectrometry coupled with chemometrics to predict aroma properties in Australian Riesling wine, Anal. Chim. Acta 621 (2008) 2-7.

[5] L. Kolb, S. Ettre, Static Headspace-Gas Chromatography, Theory and Practice, first ed., Wiley-VCH, New York, 1997.

[6] L. Kolb, S. Ettre, Theory and practice of multiple headspace extraction, Chromatographia 32 (1991) 505-513.

[7] S. Román, M. L. Alonso, I. Bartolomé, R. M. Alonso, R. Fañanás, Analytical strategies based on multiple headspace extraction for the quantitative analysis of aroma components in mushrooms, Talanta 123 (2014) 207-217.

[8] A. Kremser, M. A. Jochmann, T. C. Schmidt, Systematic comparison of static and dynamic

headspace sampling techniques for gas chromatography, Anal. Bioanal. Chem. 408 (2016) 6567-6579.

[9] A. Spietelun, L. Marcinkowski, M. de la Guardia, J. Namiesnik, Recent developments and future trends in solid phase microextraction techniques towards green analytical chemistry, J. Chromatogr. A 1321 (2013) 1-13.

[10] A. Spietelun, L. Marcinkowski, M. de la Guardia, J. Namiesnik, Green aspects, developments and perspectives of liquid phase microextraction techniques, Talanta 119 (2014) 34-45.

[11] P. R. Lozano, M. Drake, D. Benitez, K. R. Cadwallader, Instrumental and sensory characterization of heat-induced odorants in aseptically packaged soy milk, J. Agric. Food Chem. 55 (2007) 3018-3026.

[12] J. Liu, T. B. Toldam-Andersen, M. A. Petersen, S. J. Zhang, N. Arneborg, W. L. P. Bredie, Instrumental and sensory characterization of Solaris white wines in Denmark, Food Chem. 166 (2015) 133-142.

[13] M. Lakatos, Measurement of residual solvents in a drug substance by a purge-and-trap method, J. Pharm. Biomed. Anal. 47 (2008) 954-957.

[14] M. Akbar, S. Narayanan, M. Restaino, M. Agah, A purge and trap integrated microGC platform for chemical identification in aqueous samples, Analyst 139 (2014) 3384-3392.

[15] J. Laaks, M. A. Jochmann, B. Schilling, T. C. Schmidt, In-tube extraction of volatile organic compounds from aqueous samples: an economical alternative to purge and trap enrichment, Anal. Chem. 82 (2010) 7641-7648.

[16] C. F. Ross, Headspace analysis, in: J. Pawliszyn (Ed.), Comprehensive Sampling and Sample Preparation, Elsevier, Amsterdam, 2012, pp. 27-50.

[17] M. D. Joshi, J. L. Anderson, Recent advances of ionic liquids in separation science and mass spectrometry, RSC Adv. 2 (2012) 5470-5484.

[18] I. D. Fisk, M. Boyer, R. S. T. Linforth, Impact of protein, lipid and carbohydrate on the headspace delivery of volatile compounds from hydrating powders, Eur. Food Res. Technol. 235 (2012) 517-525.

[19] A. M. Seuvre, E. Philippe, S. Rochard, A. Voilley, Retention of aroma compounds in food matrices of similar rheological behavior and different compositions, Food Chem. 96 (2006) 104-114.

[20] S. Manzini, C. Durante, C. Baschieri, M. Cocchi, S. Sighinolfi, S. Totaro, A. Marchetti, Optimization of a dynamic headspace-thermal desorption-gas chromatography/mass spectrometry procedure for the determination of furfurals in vinegars, Talanta 85 (2011) 863-869.

[21] M. Shen, Q. Liu, H. Jia, Y. Jiang, S. Nie, J. Xie, C. Li, M. Xie, Simultaneous determination of furan and 2-alkyl furans in heat processed foods by automated static headspace gas chromatography mass spectrometry, Food Sci. Technol. -Leb 72 (2016) 44-54.

[22] A. C. Soria, I. Martínez-Castro, J. Sanz, Some aspects of dynamic headspace analysis of volatile components in honey, Food Res. Int. 41 (2008) 838-848.

[23] A. C. Soria, I. Martínez - Castro, J. Sanz, Study of precision in the purge - and - trap - gas chromatography-mass spectrometry analysis of volatile compounds in honey, J. Chromatogr. A 1216 (2009) 3300-3304.

[24] C. Sales, M. I. Cervera, R. Gil, T. Portolés, E. Pitarch, J. Beltran, Quality classification of Spanish olive oils by untargeted gas chromatography coupled to hybrid quadrupoletime of flight mass spectrometry with atmospheric pressure chemical ionization and metabolomics-based statistical approach, Food

Chem. 216 (2017) 365-373.

[25] S. Lopez-Feria, S. Cardenas, J. A. Garcia-Mesa, M. Valcarcel, Classification of extra virgin olive oils according to the protected designation of origin, olive variety and geographical origin, Talanta 75 (2008) 937-943.

[26] M. Bergamaschi, A. Cecchinato, F. Biasioli, F. Gasperi, B. Martin, G. Bittante, From cow to cheese: genetic parameters of the flavour fingerprint of cheese investigated by direct - injection mass spectrometry (PTR-ToF-MS), Genet. Sel. Evol. 48 (2016) 89, 1-14.

[27] M. P. Marti, O. Busto, J. Guasch, Application of a headspace mass spectrometry system to the differentiation and classification of wines according to their origin, variety and aging, J. Chromatogr. A 1057 (2004) 211-217.

[28] L. Vera, L. Aceña, J. Guasch, R. Boqué, M. Mestres, O. Busto, Characterization and classification of the aroma of beer samples by means of an MS e-nose and chemometric tools, Anal. Bioanal. Chem. 399 (2011) 2073-2081.

[29] S. Mallia, E. Fernández-García, J. Olivier Bosset, Comparison of purge and trap and solid phase microextraction techniques for studying the volatile aroma compounds of three European PDO hard cheeses, Int. Dairy J. 15 (2005) 741-758.

[30] C. Murat, K. Gourrat, H. Jerosch, N. Cayot, Analytical comparison and sensory representativity of SAFE, SPME and Purge and Trap extracts of volatile compounds from pea flour, Food Chem. 135 (2012) 913-920.

[31] D. Tholl, W. Boland, A. Hansel, F. Loreto, U. S. Röse, J. P. Schnitzler, Practical approaches to plant volatile analysis, Plant J. 45 (2006) 540-560.

[32] F. Ruiz - Bevia, M. J. Fernández - Torres, M. P. Blasco - Alemany, Purge efficiency in the determination of trihalomethanes in water by purge - and - trap gas chromatography, Anal. Chim. Acta 632 (2009) 304-314.

[33] O. Yilmazcan, E. T. Ozer, B. Izgi, S. Gucer, Optimization of static head - space gas chromatography- mass spectrometry conditions for the determination of benzene, toluene, ethyl benzene, xylene, and styrene in model solutions, Ekoloji 22 (2013) 76-83.

[34] S. X. Zhang, X. S. Chai, D. G. Barnes, Determination of the solubility of low volatility liquid organic compounds in water using volatile-tracer assisted headspace gas chromatography, J. Chromatogr. A 1435 (2016) 1-5.

[35] J. C. Flórez Menéndez, M. L. Fernández Sánchez, E. Fernandez Martínez, J. E. Sánchez Uría, A. Sanz- Méndel, Static headspace versus head space solid - phase microextraction (HS - SPME) for the determination of volatile organochlorine compounds in landfill leachates by gas chromatography, Talanta 63 (2004) 809-814.

[36] C. S. Min Liew, X. Li, H. K. Lee, Miniscale liquid-liquid extraction coupled with full evaporation dynamic headspace extraction for the gas chromatography/mass spectrometric analysis of polycyclic aromatic hydrocarbons with 4000-to-14000-fold enrichment, Anal. Chem. 88 (2016) 9095-9102.

[37] X. Deng, P. Xie, M. Qi, G. Liang, J. Chen, Z. Ma, Y. Jiang, Microwave assisted purge-and-trap extraction device coupled with gas chromatography and mass spectrometry for the determination of five predominant odors in sediment, fish tissues, and algal cells, J. Chromatogr. A 1219 (2012) 75-82.

[38] Y. Cai, Z. Yan, L. Wang, M. NguyenVan, Q. Cai, Magnetic solid phase extraction and static headspace gas chromatography-mass spectrometry method for the analysis of polycyclic aromatic hydrocarbons,

J. Chromatogr. A 1429 (2016) 97-106.

[39] M. Tankiewicz, J. Namiesnik, W. Sawicki, Analytical procedures for quality control of pharmaceuticals in terms of residual solvents content: challenges and recent developments, TrAC-Trends Anal. Chem. 80 (2016) 328-344.

[40] Q3C (R5), Impurities: guideline, for residual solvents, step 4 version, International Conference on Harmonisation (ICH) of Technical Requirements for Registration of Pharmaceutical for Human Use, February 2011, http: //www. ich. org/fileadmin/Public_ Web_ Site/ICH_ Products/Guidelines/Quality/Q3C/Step4/Q3C_ R5_ Step4. pdf.

[41] C. Camarasu, Ch Madichie, R. Williams, Recent progress in the determination of volatile impurities in pharmaceuticals, TrAC-Trends Anal. Chem. 25 (2006) 768-777.

[42] K. Grodowska, A. Parczewski, Organic solvents in the pharmaceutical industry, Acta Pol. Pharm. 67 (2010) 3-12.

[43] Y. Sitaramaraju, A. van Hul, K. Wolfs, A. Van Schepdael, J. Hoogmartens, E. Adams, Static headspace gas chromatography of (semi) volatile drugs in pharmaceuticals for topical use, J. Pharm. Biomed. Anal. 47 (2008) 834-840.

[44] K. Urakami, A. Higashi, K. Umemoto, M. Godo, Matrix media selection for the determination of residual solvents in pharmaceuticals by static headspace gas chromatography, J. Chromatogr. A 1057 (2004) 203-210.

[45] W. D' Autry, C. Zheng, J. Bugalama, K. Wolfs, J. Hoogmartens, E. Adams, B. Wang, A. Van Schepdael, Liquid paraffin as new dilution medium for the analysis of high boiling point residual solvents with static headspace-gas chromatography, J. Pharm. Biomed. Anal. 55 (2011) 1017-1023.

[46] R. Otero, G. Carrera, J. F. Dulsat, J. L. Fábregas, J. Claramunt, Static headspace gas chromatographic method for quantitative determination of residual solvents in pharmaceutical drug substances according to European Pharmacopoeia requirements, J. Chromatogr. A 1057 (2004) 193-201.

[47] Z. Li, Y. Han, G. P. Martin, Static headspace gas chromatographic analysis of the residual solvents in gel extrusion module tablet formulations, J. Pharm. Biomed. Anal. 28 (2002) 673-682.

[48] M. Ni, T. Sun, L. Zhang, Y. Liu, M. Xu, Y. Jiang, Relationship study of partition coefficients between ionic liquid and headspace for organic solvents by HS-GC, J. Chromatogr. B Analyt. Technol. Biomed. Life Sci. 945-946 (2014) 60-67.

[49] N. H. Snow, G. P. Bullock, Novel techniques for enhancing sensitivity in static headspace extraction-gas chromatography, J. Chromatogr. A 1217 (2010) 2726-2735.

[50] L. Hong, H. R. Altorfer, A comparison study of sample dissolution media in headspace analysis of organic volatile impurities in pharmaceutical, Pharm. Acta Helv. 72 (1997) 95-104.

[51] T. Gao, J. M. Andino, J. R. Alvarez-Idaboy, Computational and experimental study of the interactions between ionic liquids and volatile organic compounds, Phys. Chem. 12 (2010) 9830-9838.

[52] G. Laus, M. Andre, G. Bentivoglio, H. Schottenberger, Ionic liquids as superior solvents for headspace gas chromatography of residual solvents with very low vapor pressure, relevant for pharmaceutical final dosage forms, J. Chromatogr. A 1216 (2009) 6020-6023.

[53] M. C. Kroon, W. Buijs, C. J. Peters, G. J. Witkamp, Quantum chemical aided prediction of the thermal decomposition mechanisms and temperatures of ionic liquids, Thermochim. Acta 465 (2007) 40-47.

[54] G. VonWald, D. Albers, H. Cortes, T. McCabe, Background vapor from six ionic liquids and the partition coeff5 1icients and limits of detection for 10 different analytes in those ionic liquids measured using

headspace gas chromatography, J. Chromatogr. A 1201 (2008) 15-20.

[55] A. Brault, V. Agasse, P. Cardinael, J. C. Combret, The full evaporation technique: a promising alternative for residual solvents analysis in solid samples, J. Sep. Sci. 28 (2005) 380-386.

[56] D. M. Kialengila, K. Wolfs, J. Bugalama, A. Van Schepdael, E. Adams, Full evaporation headspace gas chromatography for sensitive determination of high boiling point volatile organic compounds in low boiling matrices, J. Chromatogr. A 1315 (2013) 167-175.

10

10　微型固相萃取

Justyna Płotka-Wasylka*, Natalia Szczepańska,
Katarzyna Owczarek and Jacek Namieśnik
Gdańsk University of Technology, Gdańsk, Poland
＊通讯作者：E-mail：plotkajustyna@gmail.com

10.1 分析化学的发展趋势：分析系统微型化

毫无疑问，分析实验室在环境保护方面发挥着重要作用，因其涉及监测空气、水和土壤中的污染物。另一方面，当使用非生态友好型试剂和溶剂时，分析试验也可能产生有毒废物。因此，为了消除/减少分析检测对操作人员和环境的不良影响[1]，2000年提出了绿色分析化学（GAC）。GAC 理念重点是减少化学分析对环境的负面影响，开发仪器和方法，并使分析实验室在成本和能源方面实现可持续性。为帮助分析人员按照 GAC 理念开展实验室操作，业内已制定了绿色化学的指导方针和原则[2]。绿色分析的组成已广为人知，有许多评论性综述可供查阅。读者可参考“绿色分析化学：绿色萃取技术的作用”一章[2a]，以便更深入地了解绿色萃取技术在 GAC 中的作用。

随着 GAC 理念和原则的出现，分析化学出现了几种趋势，其中简单化、自动化、便携性和微型化趋势得到人们的普遍认可。毫无疑问，高度的简单化、便携性和自动化开始成为微型化系统的内在要求。

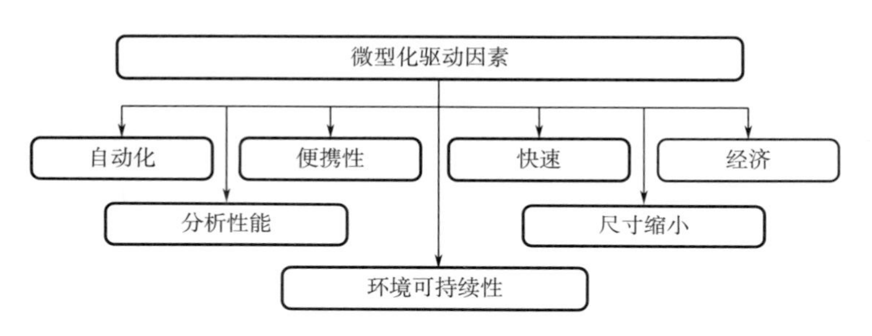

图 10.1 分析系统微型化的驱动因素

分析过程中的许多步骤已经实现了自动化、微型化和便携性。通过整合样品制备、分析分离和检测，使得开发全微型化系统成为可能。分析系统微型化也给分析化学带来了几个方面的挑战，包括分析程序的自动化、便携性和绿色化。此外，快速、经济、分析性能的提高以及分析系统尺寸的缩小都是微型化的驱动因素（图 10.1）[3]。

尽管分析过程的每一步几乎都实现了微型化，但各个步骤并未实现相同程度的微型化。例如，尽管已有一些自主和遥感分析微系统的报道，但微型化却未给样品的采集和保存带来多大益处。而数据的采集和处理已经实现极高程度的微型化。此外，人们普遍认为，经过努力实现分离系统和检测系统的微型化后，可以开发出精简的样品制备方法[3]。

已有文献提到全微型化分析系统。1990 年引入了第一个微型化系统概念，即全分析系统（TAS）[4]，其目的是提高 TAS 的分析性能，而不是简单地缩小尺寸。这项开创性工作旨在增强在线和自动化分析，提高分析性能。另一方面，它也有一些明显的缺点，例如样品导入和对一组样品的连续分析、样品输送缓慢、不同元件之间需要有装配界面等[5]。因此，通过精简和整合多个分析步骤对 TAS 概念进行修改，引入了 μ-TAS。这一概念通过精简 TAS 并将其多个分析步骤整合到单个单片器件上发展而来[5]。

如今，μ-TAS 系统已成为一个强大而令人兴奋的跨学科研究领域。

微型化系统的第二个概念称为“阀上实验室”(LOV)，这是一种将基于试剂的（生物）化学分析缩放至微升/次微升水平的通用方法，于 2000 年引入[6]。在该系统中，溶液计量、混合、稀释、孵育和监测，都可以以任何顺序在与多用途流动单元集成的通道系统中执行。

除紧凑性外，这种 LOV 系统的优点是样品处理通道位置永久刚性，确保由传统尺寸外围设备控制的微流体操作具有可重复性。由于多种原因，微顺序注射-阀上实验室［(μSI)-LOV］系统正迅速被世界各国所接受。首先，其可精简当前的流动注射分析和顺序注射（SI）技术。其次，其还支持新分析技术，如微珠注射、微亲和色谱法和 SI 色谱法。此外，(μSI)-LOV 系统有助于促进分光光度法、原子光谱法、化学发光法、荧光法和电化学技术的自动化[7]。

如前所述，分析各步骤的微型化带来了若干获益。各简化分析步骤带来的主要优点见表 10.1。

表 10.1　　不同分析步骤带来的主要优点

优点	备注
减少样品消耗	通过精简化相关的样品制备、分离和检测技术，可大大减少分析所需的样品量。这在样品量较少的情况下尤其重要
减少试剂、溶剂的消耗	为了大幅减少化学品和有机溶剂的消耗，有必要使每个分析步骤微型化，这在涉及昂贵试剂（如酶）的分析方法和涉及有毒试剂和/或有机溶剂的免疫化学和分析方法中尤其需要 目前，许多样品制备技术可去除有机溶剂和其他试剂。对于分离技术，需要减少流动相或电解质以及固定相材料。至于检测技术，可以大大减少试剂和气体消耗
减少废物生成	只有在减少样品、有机溶剂和其他试剂时才能实现。此外，有必要回收废物中的化学品和有机溶剂，同时在线生成清洁废弃物
快速	通过开发微型化样品制备、分离和检测系统，可显著缩短单次分析所需时间。毫无疑问，通过微型化，有助于改进两个最耗时的步骤，即样品预处理和分析分离。此外，缩短分析时间可带来间接获益，如减少化学品和溶剂消耗、降低能源需求和减少废物生成
提高灵敏度	通过使用微型化样品制备技术和检测系统，可以提高灵敏度。需要注意的是，如果样品体积与萃取剂相比增加，可能导致富集因子较高，从而导致检测限较低。检测系统进步可以最大限度地减少灵敏度损失
降低功耗	应用微型化分析系统可降低功耗需求。此外，微型化仪器可使用电池供电，从而提高便携性
便携性	毫无疑问，整体甚至部分分析步骤微型化均有助于提高将分析系统带到采样点的便携性。然而，便携式分析系统需要满足一定要求，如重量轻、整体尺寸小、适应多变环境和电池电源高效等

从上述获益来看，某些优点是相互关联的。因此，毫无疑问，在分析过程的常规步骤中引入微型化会带来新的挑战，而这些挑战有待解决。

10.2 传统样品制备技术的微型化必要性

在痕量/超痕量水平下监测样品中的化合物时，通常需要初步分离和/或富集分析物，这是因为分析技术的灵敏度不足以直接测定复合材料中的痕量化合物。因此，分析化学制备样品的目的是[3,8]：

（1）将目标分析物预浓缩至高于分析仪器检测限（LOD）的水平；

（2）从原始样品基质中分离目标分析物和/或基质简化；

（3）在仪器分析前净化样品（去除干扰，并消除色谱柱中强烈吸附的样品成分）；

（4）获取与所用分析技术相容的萃取物。

众所周知，80%以上的分析时间用于样品采集和制备，因此可以明确地说，样品制备步骤是分析过程的重要组成部分，应当纳入分析化学教学课程[9]。毫无疑问，样品制备技术或方法还未适应仪器的发展，无法满足定性和定量分析对目标分析物的要求。因此，越来越多的工作集中在改进样品制备上，分析化学领域的当前趋势也是提高分析样品制备阶段的绿色化。有几种方法可实现样品制备“绿色化”[10]；然而，其中最重要的一个方法是缩小分析操作规模，实现仪器微型化。微型化是如此重要，因为解决了该问题，就可实现 GAC 原则的下一个目标（图 10.2）。

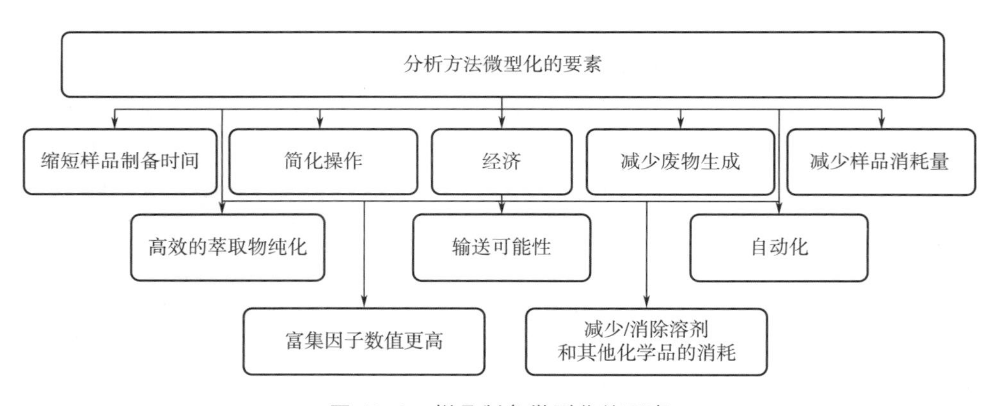

图 10.2 样品制备微型化的要素

最常用的样品制备技术是固相萃取（SPE）和溶剂萃取（包括液-液萃取，LLE）；然而，在分析物预浓缩和基质去除中，SPE 正变得比 LLE 更受欢迎。由于 LLE 存在固有缺点，SPE 获得了广泛的认可。LLE 的缺点包括[11]：①费时费力；②不能萃取极性化合物；③昂贵；④需要蒸发大量溶剂；⑤易形成乳状液；⑥需要处置有毒或易燃化学品。因此 SPE 是一种比 LLE 更有效的分离方法，SPE 在洗脱步骤中使用少量的溶剂更容易提高分析物的回收率[12]。鉴于这些优点，SPE 无疑成了改进和开发更优、更现代化和更绿色化的样品制备解决方案的首选技术，从而引入了许多具有特色的绿色技术。另一方面，SPE 并非无缺点，其缺点包括：①化学成分范围广、操作溶剂和 pH 条件选择多，使其难以掌握；②其使用方法（方法开发）难以掌握；③每份样品的成本较高；④通常需要若干步骤（和额外时间）。由于这些局限，引入了现代样品制备技术，而这些技术具有微型化和无溶剂操作的共同优点。通过采用这些微萃取技术，可整合若干步骤（包括采样、萃取、将分析物富集至高于方法 LOD 的水平、从无法直接导入测量仪器的样品基质中分离分析物等）[13]。本章简单介绍了这些技术，而关于一些最相关技术的更深入讨论，详见“在线管内固相微萃取的趋势”[13a][管内固相微萃取（SPME）]、“搅拌棒吸附萃取”[13b][搅拌棒吸附萃取（SBSE）] 和“固相微萃取用于食品基质污染物分析的最新进展”[13c]（SPME）等章节。

10.3 微型固相萃取技术

如前所述，为克服 SPE 的缺点，开发了微型化 SPE 技术，并将其引入分析领域。与 SPE 的多步操作相比，这些技术需要的时间和人力都更少。本章中描述的微型化 SPE 技术可视为绿色技术，原因见后文。本章中，对微型化 SPE 技术的讨论基于最相关、最具代表性和最新的科学参考文献。表 10.2 总结了绿色 SPE 技术的应用信息。

10.3.1 固相微萃取

固相微萃取（SPME）是目前分析化学中制备样品最常用的绿色技术之一。1990 年，Pawliszyn 和 Arthur 将 SPME 引入分析实践，试图重新调整 SPE 和 LLE 固有的局限性。从那时起，为在世界各地许多研究机构中阐述新的方法学解决方案，集中展开了一系列研究，这可能会扩大这项技术的应用范围[14,16]。SPME 是 SPE 的一种。该技术克服了其两个最主要缺点，即萃取时间长以及（更重要的是）必须使用有机溶剂[14]。

许多分析方法均认可 SPME 法制备样品，这主要归功于它具有如下的优点[15,17]：

（1）能够同时上样、浓缩和标记分析物，显著缩短分析时间；

（2）灵敏度高［能够在万亿分之一水平上标记物质］；

（3）样本量小；

（4）分析简单快速，无须使用复杂的设备、工具和装置或执行精确操作；

（5）不使用昂贵和毒性有机溶剂，SPME 纤维可多次使用，所以成本极低；

（6）小型纤维，使设备能够在原位条件下下载样品；

（7）可实现自动化；

（8）能够与其他仪器技术结合，最常见的是线上或线下模式结合气相色谱（GC）、液相色谱（LC）、高效液相色谱（HPLC）和毛细管电泳（CE）。

由于优点众多，其几乎是通用技术，因其可分析不同物理状态的多种样品，不管是液体、气体还是固体，即使基质成分非常复杂，可测定痕量和超痕量水平的分析物[18-23]。所有上述特点使 SPME 成为分析化学发展的热门，是样品制备和分析物富集最常使用的技术之一。因此，我们可以轻易查到大量关于该方法新解决方案、纤维涂层新材料以及在食品、生物和制药领域新应用的论文[18]。

表 10.2 微萃取技术的详细信息

技术	吸附剂	分析物	样品	样品体积/mL	解吸附方法	定量技术	浓度	LOD	回收率/%	参考文献
管内/纤维管内固相微萃取	结构经硝酸化学处理的铜丝/铜管	PAH	河水、雨水、煤灰	—	溶剂	HPLC	0.56~6.5μg/L	0.001~0.01μg/L	8.9~83.2	[27]
	Zylon 纤维、熔融二氧化硅毛细管、DB-5 毛细管	三环类抗抑郁药	尿液	0.2~1	溶剂	CE	—	44~153ng/mL	—	[28]
	离子液体涂层纤维和管	BPA、雌酮、17a-炔雌醇、己烯雌酚、己烷雌酚	标准储备溶液、瓶装水	35	溶剂	HPLC	10μg/L	0.02~0.05μg/L	88~108	[29]
	熔融二氧化硅、CP Sil 5CB、CP-Wax 52CB、SUpel-Q PLOT	8-羟基-2′-脱氧鸟苷、肌酐	Urine 尿液	—	—	LC-MS	0.377~22.21ng/mL	0.32~0.69ng/mL	91~95.4	[30]
EE-SPME	聚吡咯薄膜	多巴胺谷氨酸高氯酸盐 MMA	水溶液	50	电化学	FI-HPLC-MS	—	—	—	[31]
	多壁碳纳米管/全氟磺酸	甲基苯丙胺、安非他明	尿液	10	热处理	GC-FID	0.25~1μg/mL	2ng/mL	—	[32]
	活性炭	苯胺	水	—	热处理	GC-FID	5~50μg/L	0.02~0.05μg/L	87~93	[33]
	单壁碳纳米管	氟化物、氯化物、溴化物、硝酸盐、硫酸盐阴离子	水	—	—	IC-ED	4.5~45.7μg/L	0.06~0.26μg/L	77~105	[34]
薄膜-SPME	聚乙二醇，PDMS	极性酚醛树脂化合物	模型解决方案	10	热处理	GC-FID	—	7~180μg/L	—	[35]
		VOC	水	12	—	—	237~2390μg/L	0.017~0.031μg/L	—	[36]
	聚己内酯，PDMS	三嗪类	水	15	—	—	—	—	—	[37]

		雌酮								
微固相萃取	氧化石墨烯	17α-雌二醇、17α-炔雌醇、己烯雌酚	水	10	超声波促进溶剂	HPLC-UV	0. 18μg/L	0. 24~0. 52μg/L	91~111	[38]
	表面活性剂模板介孔二氧化硅	全氟羧酸	人血浆	5	超声波促进溶剂	LC-MS/MS	135. 61~196. 35ng/L	21. 23~65. 07ng/L	87. 58~102. 45	[39]
	CTAB-MCM-41、C_{18}、SDB-RPS、SDB-XC 和石墨化炭黑膜	TNT，RDX，TNT、RDX、硝甲铵基三硝基苯	土壤	1. 5	溶剂	DESI-HRMS	(4. 8 ± 0. 2) ~ (10. 6 ± 1) μg/kg	0. 008~0. 064μg/kg	96 ± 8~98 ± 9	[40]
	功能单体：EDMA Cross-linker：DVB 交联剂：DVB	可卡因、苯甲酰芽子碱、可卡乙碱	尿液	5	超声波促进溶剂	HPLC-MS/MS	(218 ± 10) ng/mL (550 ± 66) ng/mL <0. 25ng/mL	0. 16~1. 7ng/L	89~100	[41]
吸附微萃取	n-乙烯基吡咯烷酮、二乙烯基苯聚合物	双氯芬酸	尿液	25	溶剂	CE-DAD		0. 3μg/L	83. 3 ± 2. 8~97. 8 ± 6. 5	[42]
	活性炭粉	嘧菌环胺	水溶液	25	超声波促进溶剂	GC-MS	(1. 27 ± 0. 05) ng/mL	(4. 0~30. 0) ng/mL	100~107. 8	[43]
	PD-DVB	磺胺塞唑 磺胺甲恶唑 磺胺地索辛	水	25	溶剂	HPLC-DAD	—	0. 08~0. 16μg/L	63. 8~84. 2	[44]
	活性炭粉	DEET	水溶液	—	Solvent 溶剂	LVI-GC-MS	—	8~20ng/L	96. 479. 9	[45]
SPDE	PDMS/活性炭	正庚烷	血	2	热处理	GC-MS	0. 279~2. 903mg/L	0. 011mg/L	100~117	[22]
	PDMS/活性炭	BTEX	水	1	热处理	GC-MS	0. 281~1. 891μg/L	19~30ng/L	—	[46]

续表

技术	吸附剂	分析物	样品	样品体积/mL	解吸附方法	定量技术	浓度	LOD	回收率/%	参考文献
SCSE	聚合物离子液体	F^-，Br^-，NO_2^-，NO_3^-，PO_4^{3-}	河水	—	溶剂	IC-CD	19.8~105μg/L	0.061~0.73μg/L	71~111	[47]
	3-丙烯酰胺基苯硼酸与二乙烯基苯的共聚物	硝基酚类	自来水	2500	溶剂	HPLC-DAD	7.86~109μg/L	0.097~0.28μg/L	71.2~115	[48]
	聚合物离子液体	BPA EE2	湖水	2500	溶剂	HPLC-DAD	1~87.4μg/L 0.93~85.2μg/L	0.024~0.057μg/L	71.2~108	[49]
RDSE	PDMS	林丹、氯氰菊酯	河水	25	溶剂	GC-MS	(1.8 ± 0.2) μg/L (27 ± 2) μg/L	<3.1μg/L	76~101	[50]
	PDMS	六氯苯	自来水	50	溶剂	GC-EDC	—	0.08μg/L	85 ± 3	[51]
	二乙烯基苯与N-乙烯基吡咯烷酮共聚物	氟苯尼考	血浆	0.25	溶剂	HPLC-DAD	—	48.1ng/mL	91.5	[52]
SRSE	石墨烯聚合物复合材料	蒽 荧蒽 芘	湖水	—	溶剂	GC-MS	0.71ng/mL 1.14ng/mL 1.04ng/mL	0.005~0.429ng/mL	79.5 ± 10.2 55.3 ± 14.7 54.5 ± 15.5	[53]
	(AMPS-co-OCMA-co-EDMA) 聚合物	氟喹诺酮类	蜂蜜	5	溶剂	LC-MS	—	0.06~0.14ng/g	70.33~122.6	[21]
	聚 (VP-co-EDMA)	酮洛芬 芬布芬 布洛芬	湖水	—	溶剂	HPLC-UV	15.12ng/mL 20.42ng/mL 14.44ng/mL	0.09~0.25ng/mL	75.6~112.3	[15]
SBSE	分子印迹聚合物	莱克多巴胺 异克舒令 克仑特罗	动物组织（肝脏、猪肉）		溶剂	HPLC-UV	5~10μg/kg	—	73.6~90.5	[54]

	活性炭/PDMS	挥发性/极性化合物	食品和环境样品	1	热处理	GC-MS	—	—	—	[55]
	PDMS/APTES-OH-TSO	防腐剂（苯甲酸、对羟基苯甲酸甲酯、抗坏血酸）	饮品样品	—	溶剂	HPLC-UV	—	0.6~2.7μg/L	83~119.1	[56]
MIP-SPME/CS-NR-Mag-MIP	MIP	抗病毒药物（硫酸阿巴卡韦）	环境水	—	溶剂	LC-MS	—	10.1~13.6ng/L	88~99	[57]
	MIP	抗生素药物（达托霉素、阿莫西林）	人血浆、合成体液	1.5	溶剂	HPLC-UV	0.96~14.4μg/mL	0.029μg/mL	85~115	[58]
	温度敏感型MIP	氧氟沙星	牛乳	—	—	—	—	—	89.7~103.4	[59]
MSPE	二氧化硅负载的 Fe_3O_4 NP	毒虫畏、毒死蜱	废水	3	磁处理	LC-DAD	—	100ng/L 50ng/L	94 ± 5~97 ± 6	[60]
IASPME	抗体	氟西汀	血清		溶剂	LC-MS	—	5ng/mL	—	[61]
	抗体	7-氨基氟硝西泮	尿液	15	溶剂	LC-MS/MS	—	0.02ng/mL	—	[62]
DMSPE	离子液体功能化氧化石墨烯	溴布特罗	湖水	—	溶剂	HPLC-DAD	3.8μg/L	7~23ng/mL	91 ± 2	[64]
	玉米醇溶蛋白纳米微粒	氯酚	蜂蜜	5	热处理	GC-EDC	9.4770.81~61.3073.01ng/mL	0.08~0.6ng/mL	92~109	[65]
	COU-2	苯唑西林、氯洒西林、双氯西林	牛乳	1	溶剂	HPLC-UV	—	2.0~3.3μg/L	80.3~99.5	[66]

注：BPA—双酚 A；CD—电导率检测器；CE—毛细管电泳；DAD—二极管阵列检测器；DDET—*N*—*N*-二乙基间甲苯胺；DESI—解吸附电喷雾离子化；DMSPE—分散式微固相萃取；DVB—二辛苯；ED—电化学检测器；EDMA—乙二醇二甲基丙烯酸酯；EE2—炔雌醇；FI—流动注射；FID—火焰电离检测；GC—气相色谱分析；HPLC—高效液相色谱；HRMS—高分辨率质谱；IASPME—免疫组化固相微萃取；IC—离子色谱；LC—液相色谱；LVI—大容量注射；MIP—分子印迹聚合物；MS—质谱；PAH—多环芳烃；PDMS—聚二甲基硅氧烷；RDSE—转盘吸附剂萃取；TNT—三硝基甲苯；UV—紫外线；VOS—挥发性有机化合物。

如前所述，读者可参考[13a]“管内固相微萃取”和[13c]“食品基质中污染物分析固相微萃取的最新进展”章节，这两章节分别对管内固相微萃取和SPME的基本原理和最新应用进行了更深入的讨论。

10.3.1.1 管内固相微萃取

替代涂层纤维的一种方法是使用内部涂层的毛细管或毛细针，这是管内技术的基础。管内固相微萃取使用开放管式毛细管柱保留分析物。该技术主要用于为FIBRE SPME-HPLC提供自动化方案。该模式可克服一些与传统纤维SPME有关的问题，例如脆弱性、低吸附能力和厚膜纤维涂层的流失[24]。管内技术可分为使用萃取涂层的方法（将涂层用作固定在针头或毛细管壁中的内部萃取相）和使用萃取填充的方法（将吸附剂填充材料作为萃取相）。管内系统可在静态模式或动态模式下使用。在静态模式下，分析物通过扩散进行转移；在动态模式下，分析物通过泵送或在样品的重力流下通过针或管进行转移[18]。然而，传统的微萃取方法基于固定相，管内SPME和纤维SPME的基本前提也相似。然而，存在一个显著差异。分析物萃取，是在管内固相微萃取的内毛细管柱上，和纤维SPME的纤维外表面上进行。管内固相微萃取需要在萃取前通过过滤或离心去除微粒，以防止萃取毛细管堵塞[25]。而SPME纤维不需要使用此类方法去除微粒。在插入SPME-HPLC界面的解吸附室之前，简单地用水清洗纤维即可[18]。管内萃取法的缺点在于，与传统的SPME相比，其需要更复杂的仪器。然而，随着所用管的长度和吸附剂量的增加，预计该方法可以提高灵敏度。有关该技术在环境、临床、法医学和食品分析中应用，请参考文献。此外，实践证实，管内固相微萃取技术是测定环境样品中金属物质（包括砷、铬、铅、汞和硒）的实用工具[26]。萃取管的示意图见图10.3（1）。

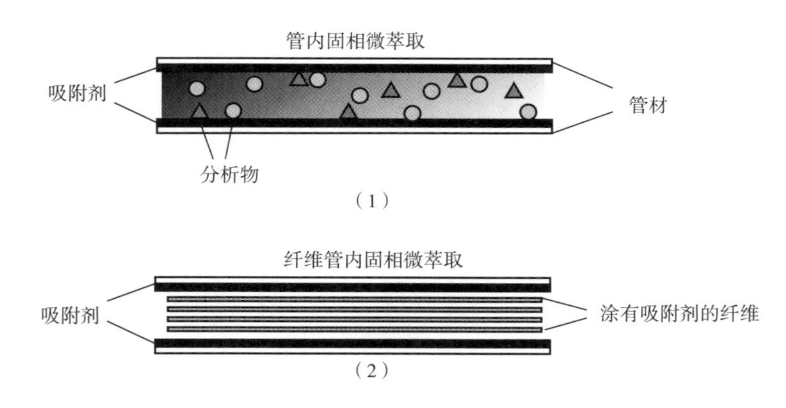

图10.3 金纤维管内和管内固相微萃取（SPME）中所用萃取管示意图

如前所述，[13a]的“管内固相微萃取”章节，对管内固相微萃取（IT-SPME）的原理进行了更全面的讨论，侧重于主要模式及其相应的配置、萃取相、相关应用及其碳足迹。

10.3.1.2 纤维管内固相微萃取

纤维管内固相微萃取技术有助于克服管内固相微萃取的主要缺点，即萃取能力低。

结合管内的纤维可以减少管内的死体积，并促进谱峰加宽[21]。传统的纤维管内固相微萃取方法基于插入管内的纤维。纤维是吸附元件，而管仅仅是支撑物。近年来，管和纤维的内侧均涂有吸附性聚合物或其他材料。该方法通过增加萃取装置的活性表面显著提高了萃取效率[21,27]。该技术已成功应用于许多研究，包括环境或生物样品。有关纤维管内固相微萃取应用的更多详细信息，参见表 10.2。萃取装置的方案见图 10.3（2）。

10.3.1.3 电吸附（电化学）增强固相微萃取（EE-SPME）

近年来还开发了一种利用电化学过程促进萃取的方法。电吸附增强 SPME 是一种电吸附和 SPME 相结合的方法，适用于水样中的电离分析物。这种技术可以消除某些问题（例如衍生化过程），并将净化和预富集合并为一个步骤。在 EE-SPME 领域的首次尝试中，萃取机理基于电荷平衡的导电聚合物（例如聚吡咯）薄膜中的反离子运动[31]。通过在萃取和解吸附过程中适当调整涂层电位，可以提高选择性和萃取效率。EE-SPME 技术适用于离子萃取和水样制备。

萃取装置通常基于三电极系统：带 SPME 纤维的工作电极（WE）、参比电极和反电极（图 10.4）[32,33]。该过程的基本原理十分简单：从水溶液中萃取正电荷物质时，将适当的负电位施加在 WE 上。通过电泳将分析物推向 SPME 纤维，并吸附在 SPME 涂层的表面。根据分析物（阳离子或阴离子）的性质，正负电位可能很容易发生改变。

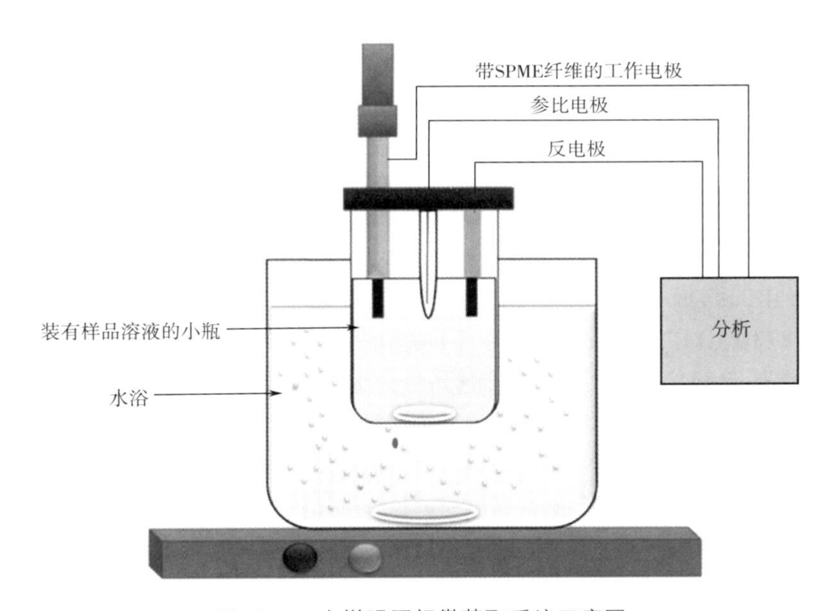

图 10.4 电增强固相微萃取系统示意图

现有多种涂层适用于电吸附增强 SPME 纤维。其中一种是由聚吡咯制成的导电薄膜[31]。如今，基于活性炭[33]单壁和多壁碳纳米管以及纳米复合物[32,34,67]的涂层在各种分析物的萃取中越来越普及。

10.3.1.4 膜固相微萃取

从环境角度看，水样中极性有机污染物的测定至关重要。从分析角度看，该方法存在一定的问题，因为此类分析物对样品基质依赖性高，需要高极性保留介质[35]。分子印迹聚合物（MIP）等特异吸附剂的使用受到极大限制，因为每一组分析物均必须有相应的聚合物，聚合物通过应用适当模板才可获得。

为避免上述问题，可使用液体或准液体保留材料。不幸的是，由于萃取介质在水中可溶，许多吸附剂材料无法用于萃取。解决方案是开发一种技术，通过聚合物膜将液体或准液体萃取剂与样品进行物理分离[34,35]，如图 10.5 所示。

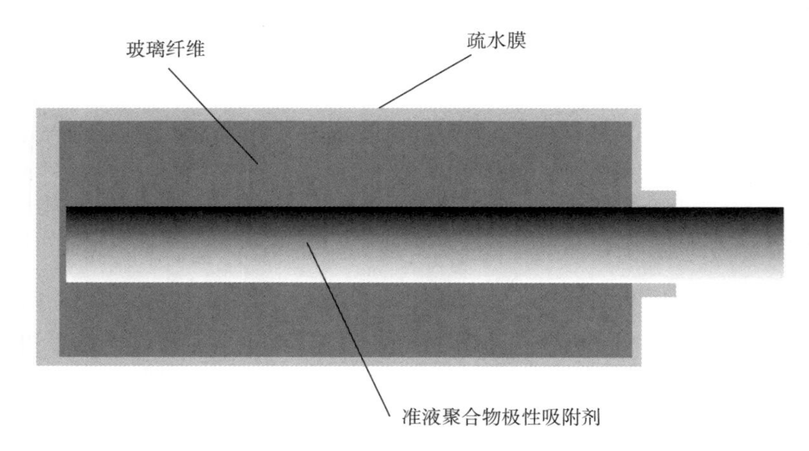

图 10.5　薄膜固相微萃取示意图（准液吸附剂通过薄膜与样品实现物理分离）

使用这种膜时，目标分析物的萃取分为如下几个阶段[35]：①可溶于膜的分析物进入膜；②膜上的分析物通过扩散作用进入萃取剂接收相；③分析物溶解于 SPME 纤维上的萃取介质。可通过质量平衡方程定量描述该萃取过程[36]。在膜 SPME 技术中，极性接收相的膜和介质所用的疏水性材料选择至关重要。最常用的市售聚合物材料包括：聚二甲基硅氧烷（PDMS）、聚乙二醇/二乙烯基苯（CW/DVB）或聚丙烯酸酯（PA）。使用 PDMS 作为膜材料，能够溶解分析物，并在纤维上保持液体或准液萃取剂的形状。以前用于从水样中萃取极性分析物的极性吸附剂是聚己内酯（PCL）[36]或聚乙二醇（PEG）[35,36]。这些物质在室温下均为准液体，具有较低的熔点和良好的吸附性能[35-37]。

10.3.1.5　分子印迹固相微萃取（MIP-SPME）

在某些情况下，由于市售纤维涂层由非选择性吸附剂制成，因此 SPE 或 SPME 的主要缺点是缺乏选择性。使用 MIP，让人们可以自行制定样本处理过程。最近，人们对 MIP 和微萃取技术的结合使用愈发关注。这类材料具有一些突出优点：①识别能力可预定；② 机械和化学稳定性；③制备过程相对容易和简化；④制备成本较低。

MIP 是具有特殊设计识别位点的合成聚合物，能够重新结合分析物分子以及密切相关的化合物[68,69]。MIP 的合成过程及其对靶分子的选择性识别示意图见图 10.6。

可用三种不同的方法制备 MIP。即通过共价键、非共价键和半价键，让单体与模板形成复合物。

1972 年报道了共价键法[68,70]，该方法基于聚合前模板和单体之间形成可逆的共价键[68]。之后，通过裂解相应的共价键去除模板，结合分析物后重新形成共价键[68,71]。

该方法的主要缺点是模板与单体的结合较缓慢，去除过程也较缓慢[72]。

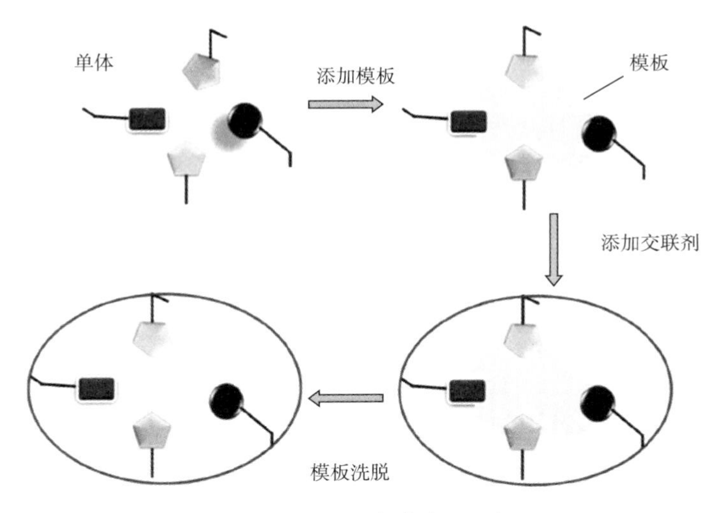

图 10.6 MIP 的合成过程示意图

半共价键法基于单体和模板之间通过共价键相互作用，但分析物通过非共价键与聚合物重新结合[68,73]。该方法的优点是模板与单体结合较快，但在合成过程中需要研究人员具有相关经验。

而使用最广的合成方法是非共价键印迹法。该方法将模板与功能性单体、致孔溶剂、交联剂和催化剂混合。模板和单体的自结合形成识别位点。只需用溶剂清洗即可除去模板[73]。该方法的主要优点是聚合物合成简单，模板易于去除，且模板与聚合物结合较快。但必须谨慎选择合成条件，以尽量减少非特异性结合位点的数量[72]。作为人工合成接收相，MIP 已广泛应用于分离、传感器、催化和药物开发与筛选等领域[69]。在这些应用领域中，使用最广的是 SPE 和 SPME，而且已有相应的市售 MIP。

2001 年，人们首次尝试将 MIP 应用于 SPME，方法还相当原始[74]。即采用管内 SPME 技术［AE，AB］，将 MIP 填充到毛细管中，以从血清样本中萃取普萘洛尔。该方法的主要缺点是，解吸附分析物的溶剂与通常使用的流动相不相容。

目前，SPME 技术通常有两种：MIP 涂层纤维和 MIP 整体纤维。2001 年，首次报道了采用 MIP 涂层二氧化硅纤维从尿液中萃取溴布特罗[75]。在这项研究中，二氧化硅纤维通过硅烷化活化，然后浸入聚合溶液中。该过程在 4℃ 和 350nm 辐射条件下持续 12h。制备的纤维成功用于选择性萃取溴布特罗。该方法主要问题是难以控制 MIP 涂层厚度。影响 MIP 涂层形态和宽度的因素有[68]：①聚合时间，其可影响涂层厚度，聚合过程据报道以 6h 为宜；②致孔剂类型。

采用化学键合策略，在不锈钢纤维涂层过程中以甲草胺为模板，可解决二氧化硅纤维基材的脆性问题[73,76]。此外，研究还证明高交联 MIP 涂层可以重复使用 200 次以上[76]。

MIP 涂层纤维也可使用以下制备技术：

(1) 可逆加成-断裂链转移聚合（RAFT）[77]，其机制涉及一系列链转移反应；

（2）单体电聚合，其主要优点是可以调整涂层厚度[68]；

（3）有机-无机杂化溶胶-凝胶印迹。

近年来，MIP-SPME 的应用范围得到了显著扩展。最近的一些应用示例，见表 10.2。

核-壳纳米环磁性分子印迹聚合物（CS-NR-Mag-MIP）是基于核-壳纳米环磁性分子印迹聚合物的全新技术，该技术开发于 2013 年。功能性聚合物合成采用超声辅助悬浮聚合法。纳米颗粒具有高吸附性、高选择性和非常快的结合性。其首先用于水样中双酚 A 的富集。CS-NR-Mag-MIP 不必填充于管柱内或固定于 SPME 纤维上，而是可以自由悬浮在样本中。萃取后，使用外部磁场将聚合物颗粒与样本快速分离[58]。应用示例见表 10.2，CS-NR-Mag-MIP 颗粒示意图见图 10.7。

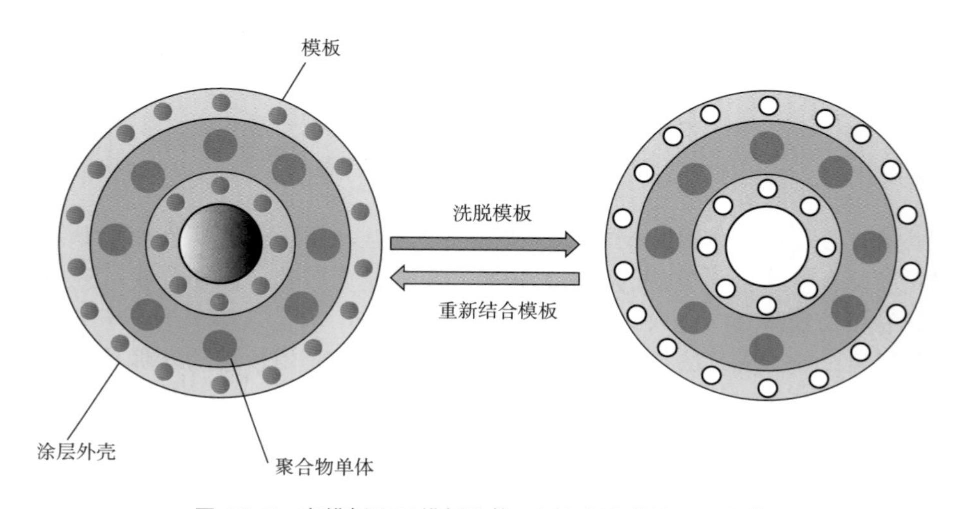

图 10.7　有模板和无模板的核-壳纳米环磁性 MIP 颗粒

10.3.1.6　磁性固相微萃取技术

利用某些材料的磁性，是当今分析化学领域的趋势之一。在化学磁性纳米颗粒的诸多领域中，杂化纳米材料和磁性复合材料得到了应用。其主要用于传感器和生物传感器的开发、分离技术和样本制备[78]。在样本制备过程中，使用磁性吸附剂可以显著减少有机溶剂的用量，并缩短分析周期。由于具备这些特点，该方法成为一种典型的环保萃取技术[79]。很明显，要想获得符合要求的富集系数和分析物回收率，主要是应选择合适的吸收材料类型和尺寸。目前，多种类型的磁性纳米颗粒（MNP）已用作各种分析物的吸附剂。然而，MNPS 核最常用的涂层材料是含有各种官能团的二氧化硅。为选择性萃取分析物，二氧化硅涂层可以用有机硅烷和/或亲和配体进行功能化。最有前景和最先进的复合材料之一是含有烷基 C_{18}（Fe_3O_4@SiO_2-C_{18}）涂层的 Fe_3O_4@二氧化硅纳米粒子（Fe_3O_4@SiO_2），由于采用纳米化技术和含有大量 C_{18} 基团，该材料具有较强的富集能力[72]。这些材料主要用于 SPE 技术中的分析样本制备阶段。但是，相关专业文献也报道过使用磁性吸附剂涂布毛细管柱内部的示例。由于该方法基于 IT-SPME 技术的典型选择性，因此称为磁性 IT-SPME。有关磁性 IT-SPME 的更多详细信息，请参见[13a]中的“管内固相微萃取”一章。

10.3.2 微固相萃取

微固相萃取（μSPE），又称多孔膜保护微固相萃取，是一种简单、有效地分离和富集复杂样本中化合物的方法。该技术开发于2006年，是多步固相萃取的替代方法[23]。该技术向多孔膜（通常为聚丙烯）制成的小型吸附剂袋（1~4cm^2）中填充几毫克粒径为3μm的吸附剂[80]，但这种吸附剂袋也可由聚酰胺和带导电聚合物涂层的尼龙孔复合材料通过静电纺丝制成[24]。然后将膜置于含有样本的小瓶中，采用磁力搅拌，使μSPE装置自由翻转，以富集分析物。

与常规SPME相比，μSPE具有几大优点[25]：

（1）吸附表面积更大；

（2）可从悬液或半固体/固体样本中有效萃取分析物，由于多孔膜可防止颗粒污染吸附剂相，因此较少存在基质效应，可避免SPE柱中通常遇到的堵塞现象；

（3）富集因子（EF）较高；

（4）成本更低，有机溶剂用量少，每个装置最多可重复使用100次；

（5）操作简便，比常规SPME耗时更少，表明更便于日常操作。

该技术的主要优点是可以一步实现分析物的净化和预浓缩[38]。μSPE还解决了SPME的一些缺点（例如，分析物残留和纤维脆性）。现有文献数据表明，不同类型的吸附剂，如反相吸附剂C_{18}、碳纳米管、纳米纤维[81]、氧化石墨烯[38]、聚合物吸附剂、氨基和尿素接枝的硅胶、合成沸石咪唑酯骨架结构8（ZIF-8）[41]已成功应用于μSPE的样本制备阶段。

μSPE已经成功地从水体中萃取有机氯农药（OCPs）、多氯联苯（PCBs）和杀菌剂等几种持久的有机污染物[26]，以及从尿样中萃取雌激素和大麻素及其代谢物[82]。该技术也应用于一些食品和酒类样本中的污染物研究。此外，μSPE还适用于检测各种极复杂基质环境样本中的痕量金属（如Cd、Pb、Se和Cr）[83]。

10.3.3 吸附微萃取

分析样本无溶剂制备技术的最新文献综述显示，分析实验室对AμE（吸附微萃取）的关注明显增加。吸附微萃取技术已经应用于实际分析操作，主要目的是消除SBSE技术萃取高$K_{o/w}$分析物所遇到的限制[84]。除了可以从具有复杂基质的各类样本中分离和富集极性化合物外，该技术还具有高效性、极高的选择性和灵敏性等特点；以及操作简便和性价比高等优点。实践中，分析物的AμE有两种方式（图10.8）。

一般而言，吸附剂的制备简单且成本不高。在棒吸附微萃取（BAμE）中，将由聚丙烯涂布并带有一层黏合剂薄膜的中空圆柱形管置于粉末状吸附剂中，并振摇。在（微球吸附微萃取）MSAμE中，将聚苯乙烯微球置于粉末吸附剂中，然后通过热处理进行固定。将吸附剂涂布的微球拧在一根线上。粉末状吸附剂的类型可很大程度上决定吸附元件的稳定性和机械强度，同时也影响测量的回收系数（RF）[85]。因此，许多研究人员目前正努力寻找合适的吸附材料。在已经应用于吸附元件制备的材料中，可确定的有以下几种：活性炭（AC）、聚苯乙烯-二乙烯基苯（PS-DVB）、硅烷、铝、金属氧化物、离子交换树脂和沸石[84]。在上述材料中，AC 和 PS-DVB 具有最佳性能和最高 RF 值。目前，还对混合吸附剂相的可行性进行了研究。使用两种或两种以上类型吸附剂似乎有希望在未来得到应用[42]。AμE 的最后一阶段可通过两种方式进行：

(1) 吸附的分析物在使用少量溶剂解吸后，通过 HPLC-二极管阵列检测器（HPLC-DAD）来测定；

(2) 通过大体积进样（LVI）和 GC-MS（LVI-GC-MS）来测定。如上所述，AμE 已应用于实验室实际操作，以消除 SBSE 技术分离和富集极性分析物所遇到的限制[86]。因此，该技术的应用领域主要包括从环境样本和法医样本中萃取和富集强极性分析物。

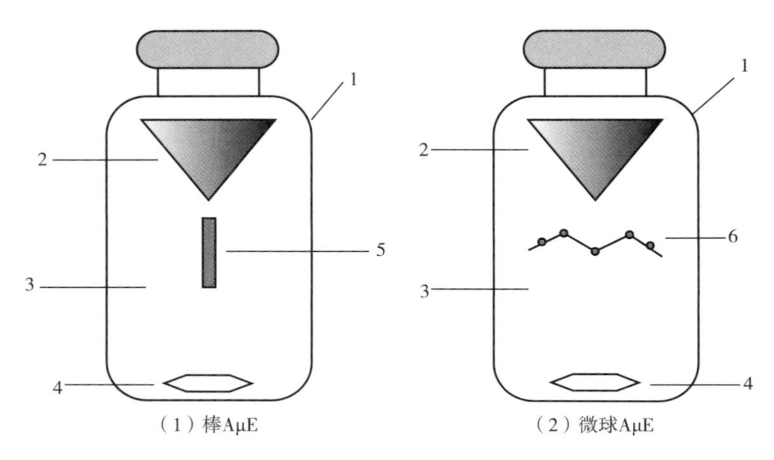

图 10.8 采用吸附微萃取技术检测分析物的装置

1—样本瓶；2—涡旋仪；3—样本；4—特氟龙磁力搅拌棒；5—棒 μ-装置；6—微球 μ-装置

10.3.4 固相动态萃取（SPDE）

在 2001 年，实验室的实际操作中首次使用了固相动态萃取（SPDE）技术。SPDE 是 SPME 的一种替代解决方案，其差别在于：在 SPME 中，吸附剂层涂布于针外侧；在 SPDE 中，吸附剂层涂布于针内侧[20]。SPME 和 SPDE 中吸附剂层典型设置方式示意图，见图 10.9。

通过注射器的反复抽吸和排出，实现在固定萃取相中富集分析物。大量研究结果表明，吸附剂层设置方式的变化可以明显地提高该技术的环保特性。样本分析所需的采样体积明显减少，而且萃取过程的总时长也显著缩短。值得注意的是，得到的富集系数远高于 SPME 技术。将吸附剂层涂布于针内侧，也可增加整个吸附元件的机械抗性，从而延长装置的有效期。与 SPDE 应用相关的主要限制在于，SPDE 纤维保留和洗脱过程复杂，同时也可能存在残留，因为分析物在热解吸后倾向于保留在针内侧[72]。该技术已成功应用于从水样[87]、各种样本、植物挥发性分成、食品和生物体液（血液和血浆）以及法医样本中分离和富集多环芳烃[22]。

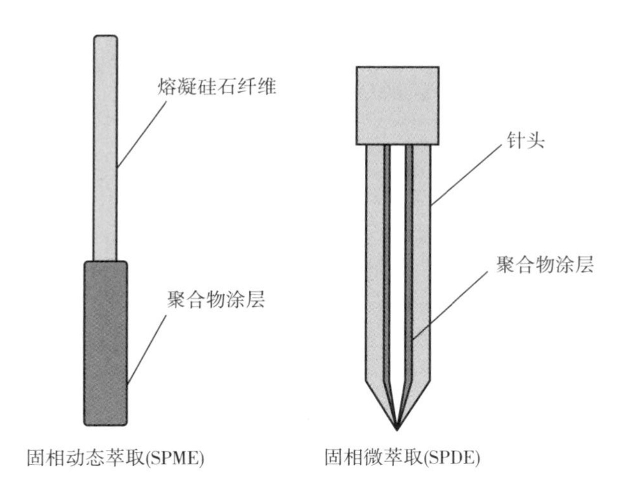

图 10.9 SPME 和 SPDE 中吸附剂层的设置

10.3.5 搅拌饼吸附萃取（SCSE）

如前所述，吸附微萃取技术中，SCSE 是对 SBSE 的一种改进。这种技术最近才被分析人员有限地使用。在 2011 年，首次报道了在分析样本制备阶段采用此类方法[88]。在上述方法中，将整体式材料（例如：萃取介质）放入专门准备的夹持器中。将具有玻璃保护涂层的铁混合物置于搅拌饼夹持器的下部。图 10.10 分别显示了空置和充满吸附剂的搅拌饼夹持器[72]。

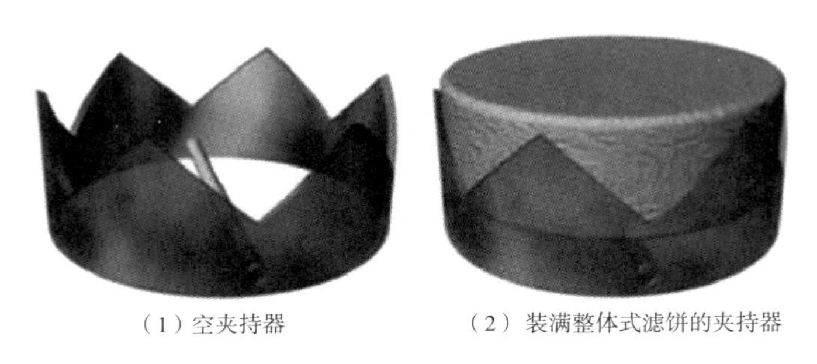

（1）空夹持器　　（2）装满整体式滤饼的夹持器

图 10.10 使用搅拌饼吸附萃取技术的分析物萃取装置

该技术的一个重要优点是将萃取、浓缩和净化过程合并成一个步骤。图 10.11 给出了使用 SCSE 萃取分析物的示意图[49]。

该技术的另一个优点是成本效益。此外，与 SBSE 相比，SCSE 具有预期的萃取能力，因为其可以利用更多的吸附剂，并且萃取介质具有明显更长的使用寿命（因为在搅拌过程中整体式滤饼不与容器芯接触，吸附剂不会摩擦损失，因此其可以重复使用很多次）。值得指出的是，SCSE 非常灵活。根据目标分析物的特性，可以轻松设计和制备吸附剂介质，以实现有效的萃取[89]。文献中最常使用的萃取相是聚（4-乙烯基苯甲酸-二炔苯），这种吸附剂的制备基于聚合离子液体和 MIP 制成的整体式圆盘。SCSE 已成功应用于从不同基质中萃取和测定分析物。SCSE 的大多数应用都集中在从环境和生物样品中萃取化合物[72]。

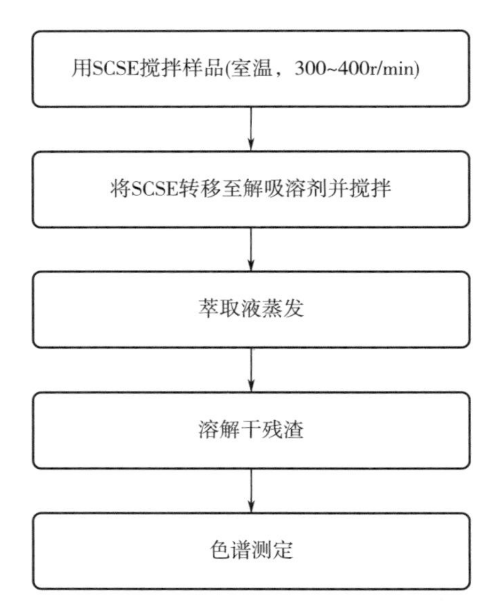

图 10. 11　SCSE 萃取工艺的步骤（SCSE：搅拌饼吸附萃取）

10. 3. 6　转盘吸附剂萃取

转盘吸附剂萃取（RDSE）是萃取技术发展的下一阶段，与 SBSE 和 SCSE 一样，其也用作磁力搅拌器。同时，如在设计阶段使用整体式滤饼分离和富集分析物的技术一样，应特别注意装置中减少使用损伤萃取介质，这是 SBSE 技术的典型特征。2009 年，提出了这种萃取和预富集半极性或非极性分析物的方法；自此以来，分析化学家对该技术的兴趣一直在不断增长。本质上，其装置由一个特氟龙圆盘组成，圆盘表面沉积一层吸附剂[90]。有两种制备萃取盘的方法。一种是使用溶胶凝胶技术将固定相（PDMS）沉积在特氟龙圆盘上；第二种是，将固定相沉积在圆盘凹槽（空腔）中。空腔用玻璃纤维过滤材料覆盖，并用特氟龙环密封[72]。转盘吸附剂萃取的示意图见图 10. 12。

作为 SCSE，转盘装置可实现更有效的分析物质量传递和更快的萃取，因为该装置提供较大的吸附相暴露面积与体积比，并且可以高速旋转[91]。其分析物的萃取及解吸类似于 SCSE 所用的方法。

萃取后，将圆盘干燥，并用少量溶剂解吸分析物。分析物的测定采用色谱法或分光光度法。这种类型的圆盘无疑具有一个优点，即可以多次使用。重复使用吸附材料肯定会降低整体分析成本。在最近发表的研究中，建议使用涂有一层尼龙、聚酰胺膜[72]、十八烷基化学改性的二氧化硅[92]和二乙烯基苯/正乙烯基吡咯烷酮共聚物的圆盘来萃取具有高富集因子和高回收价值的分析物，用于不同类型的化合物，包括废水和水中的双氯芬酸和甲芬那酸[91]，废水样品中的非甾体抗炎药[93]、OCPs[50]、六氯苯、个人护理产品以及水和尿液样品中的一些多环芳烃[72]。

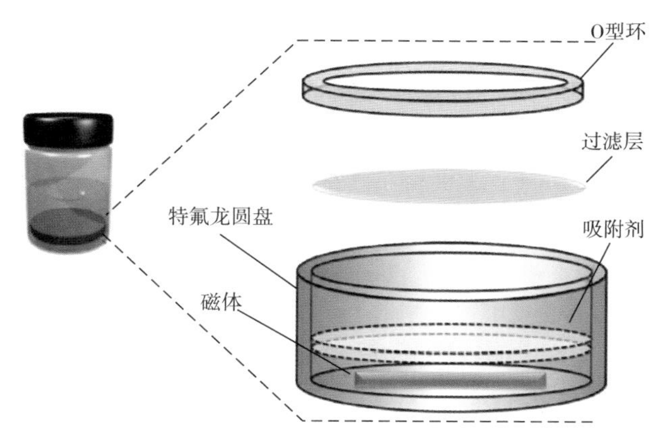

图 10.12 转盘吸附剂萃取装置

10.3.7 搅拌杆吸附萃取

在文献中，您还可以找到有关 SPE 技术的信息，该技术称为搅拌杆吸附萃取（SRSE）。该技术于 2010 年开发，用于分离和富集环境样品（具有非常复杂的基质）中具有不同理化特性的分析物[53]。SRSE 装置由一根金属线组成，该金属线的一端有一个磁铁，并在其上连接了涂有一层整体聚合物的玻璃末端。采用搅拌杆吸附元件进行萃取的装置和过程见图 10.13。

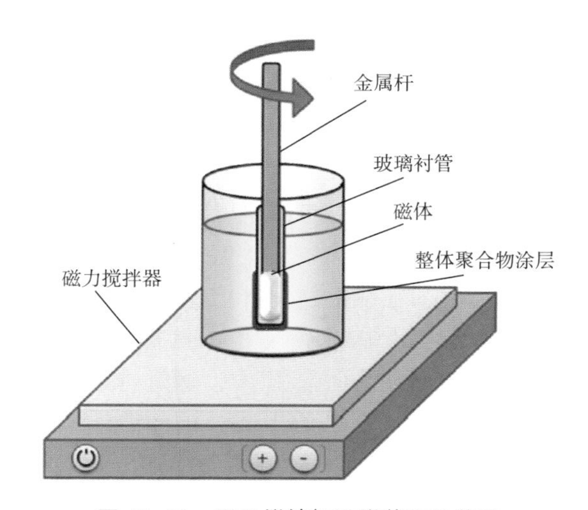

图 10.13 用于搅拌杆吸附萃取的装置

在萃取过程中，装置通过磁力搅拌器旋转；值得注意的是，其搅拌速度远低于 SBSE 的搅拌速度。

该技术已成功应用于从蜂蜜样品中分离和测定氟喹诺酮类化合物[21]，以及从环境水性样品中分离和测定非甾体类抗炎药[15]。

10.3.8 搅拌棒吸附萃取

搅拌棒吸附萃取（SBSE）这种技术，由于消除了对有机溶剂的依赖并减少了劳动密集和费时的样品制备步骤，从而满足了 GAC 的要求。SBSE 与 SPME、MLLE 和分散液液微萃取（DLLME）一样，是近年来为减少浪费而引入的最流行的技术[94]。该技术由 Baltussen 等于 1999 年引入[95]，他们提出了涉及将 PDMS 聚合物作为 SPE 吸附剂的这种新应用。SBSE 的原理与 SPME 相同，但是，不同于后者那样的涂覆聚合物纤维，其搅拌棒是将 PDMS 涂覆在玻璃棒上，这种搅拌棒是用于与目标分子进行疏水相互作用的非极性聚合物相[13b]。“搅拌棒吸附萃取”一章，对 SBSE 进行了更深入的讨论，并介绍了其最新趋势。

SBSE 虽然不是一种新技术，但通常用于所有分析领域，如环境研究、临床分析和食品分析[94]。多年来，其已用于从各种基质中萃取各种环境污染物（持久性有机污染物、农药、多溴联苯）（表 10.2）。在食品分析领域，SBSE 主要用于萃取污染物和毒素[94]。在制药和临床领域，也采用了这种技术，但是应用数量明显减少。在这些领域，样品量通常很少（至多几毫升），因此 SBSE 技术不是最佳选择。

近年来，新型涂层的开发是一个重要问题。这些涂层包括[54,94,96,97]：①聚氨酯泡沫；②有机硅材料；③聚（二甲基硅氧烷/聚吡咯）；④分子印迹聚合物；⑤聚乙烯醇和；⑥聚丙烯酸酯等。

10.3.9 免疫亲和固相微萃取（IASPME）

毫无疑问，近年来，在分析化学领域中，越来越重要的技术是使用在抗体和抗原之间发生特定反应的技术。文献提供了有关在生物分析研究中（生物传感器中的活性元素，免疫分析）以及用于分析和分离分析物的样品准备阶段（亲和色谱）使用天然抗体的广泛信息[98]。免疫吸附剂（IS）对其互补抗原（分析物）表现出很高的亲和力和非凡的特异性识别能力。因此，与传统的吸附材料相比，其具有更高的选择性，因此可以进行更可靠和更灵敏的分析[99]。IS 主要构成 SPE 柱的填料；但是，还有一个更环保的方法，其可以使用相当少量的吸附材料。免疫亲和 SPME 这种技术，将吸附材料置于诸如玻璃棒、熔融石英毛细管或不锈钢杆之类的固相载体材料上。2001 年首次报道了这种方法用于测定血清样品中的茶碱[62]。从那时起，免疫亲和固相微萃取（IASPME）已成功地主要用于临床和法医应用[78]。通过将相对匹配的抗体（单倍体或多倍体抗体）固定在固相载体物表面上来产生 IS。抗体的应用条件基于抗体和载体的理化性质，载体要加以严格控制，以避免抗体失活的风险[72]。

这些策略可以大致分为三类[63]：①非共价地固定在一种基质上；②通过截留在诸如溶胶-凝胶或丙烯酰胺的交联基质中而固定化；③利用 lincer 分子共价固定在基质上。但是，文献综述表明，上述最常用的方法是共价固定。

印迹抗体对目标分析物的特异性很高，这意味着免疫萃取对于分离和富集具有非常复杂基质的环境和生物样品中浓度非常低的分析物特别有用。迄今为止，免疫亲和 SPME 和免疫亲和 IT-SPME 已成功应用于从牛乳中分离喹诺酮类药物，和从尿液中分离苯二氮类化合物[62]。图 10.14 显示了免疫亲和 SPME 的装置和萃取过程。

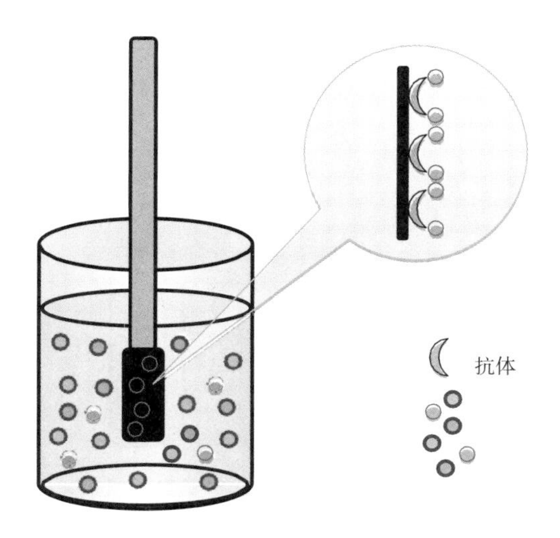

图 10.14 免疫亲和固相微萃取的装置

与抗体相反，适体对严格测定的分子（包括蛋白质、核酸、肽、氨基酸、细胞、病毒、小分子和离子）具有高度亲和力和特异性的优势[72]。适体是 DNA、RNA 或多肽的单链寡核苷酸。近年来，其引起了分析化学家越来越多的兴趣，这主要是因为其易于以低成本进行制备和修饰，并且对目标分析物显示出高选择性、高特异性和亲和力，并具有很高的再现性和出色的稳定性[100]。适体已用于覆盖纤维。文献综述表明，适体功能化 SPME 已用于从生物流体中分离和富集极高极性的分析物[101]。

10.3.10 注射器内微萃取

注射器内微萃取（MEPS）是一种最新开发的将常规 SPE 微型化的样品预处理技术[102]。已使用的大多数吸附剂是基于二氧化硅的吸附剂（C_2、C_8 和 C_{18}）。而且，基于聚酯聚合物的吸附剂也引起了许多研究小组的兴趣。根据专业文献中公布的数据，MIP 和基于聚合物静电纺丝的吸附剂已成功用于从水样中分离农药。在这些方法的解决方案中，将大约 2mg 的吸附剂填充入注射器内（100~250μL）作为塞子，或者填充在针筒和针头之间作为柱体[102]。当样品通过固相支持物时，分析物被吸附到萃取介质上。然后用有机溶剂洗脱分析物[103]。注射器内微萃取示意图如图 10.15 所示。

在文献中，有大量方法解决方案的可用信息，尤其是：MEPS-非水毛细管电泳-质谱法检测人血浆中的麻醉药，气-质联用和 MEPS 连接以检测水样中的多环芳烃。值得注意的是，在过去的十年中，MEPS 形式已从手动（在注射器或 BIN 内）改进为半自动化和全自动[104]化。与其他样品制备技术相比，注射器内微量萃取的萃取时间显著缩短，回收率较高，样品量和溶剂用量明显较少（约 10μL）[105]。在表 10.3 中列出了表明萃取技术数据特征的样品量和萃取时间信息[102]。

与用于分析物分离和富集的其他传统解决方案相比，使用 MEPS 耗时和必要的样

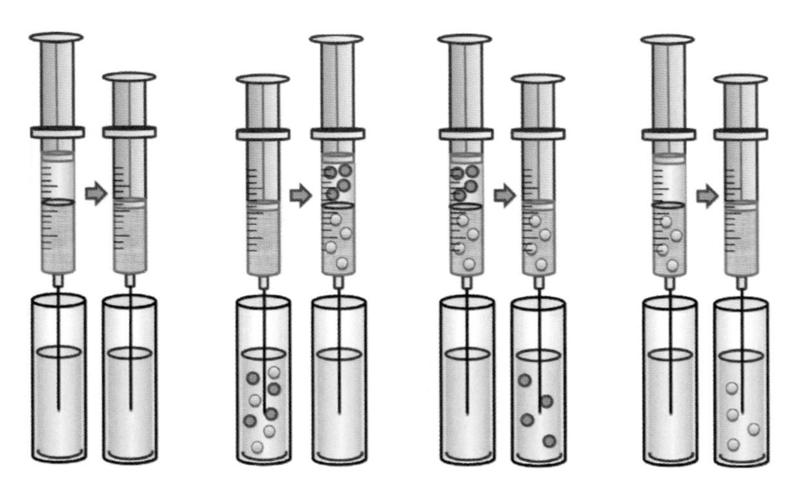

图 10.15 MEPS（注射器内微萃取）萃取步骤

品量更少。另外，其毫无疑问的优势是带有吸附剂的注射器可以多次使用，这有助于显著地降低整体分析成本[102]。

表 10.3　　使用不同样品制备技术的样品量和萃取时间比较

方法	样品体积/mL	萃取时间/min
SPE	≤ 1	10~15
SPME	0.5~20	10~20
SBSE	1~100	10~60
MEPS	0.01~1	1~4

该技术已成功应用于药物和环境分析以及生物分析[105]。MEPS 已用于分离和定量分析各种分析物，包括各种环境样品（水样、废水、土壤）和生物样品（尿液、唾液、血液、头发、血浆、血清）中的药物及其代谢产物、农药和有机污染物[102]。在最近的文献中，常见有关在食品分析中使用这种样品制备技术的信息，例如，用于测定葡萄酒中的生物活性黄酮[103]和啤酒中的异戊二烯类黄酮[106]。

10.3.11 分散式微固相萃取

为提高传统样品制备技术的环保性而开发技术的另一个方法是分散式微固相萃取（DMSPE）。这是一种基于常规 SPE 或 DSPE 微型化的样品预处理技术[65]。自 2003 年以来，该技术已成功用于分离和富集生物和环境样品中的分析物。与传统技术相比，其主要优点是显著减少所用吸附材料和有机溶剂的用量（微范围）[65]。值得注意的事实是，其样品制备过程只分为两个阶段：将固体吸附剂颗粒分散到样品溶液中，以及通过离心或摇动进行相分离[107]。这种方法可促进分析物与吸附剂之间的即时相互作用，并缩短样品制备时间。此外，由于其目标分析物与载体之间的接触比传统 SPE 更充分，因此可以提高均衡率并获得更高的萃取效率。在许多情况下，为了进一步加强质量传输，会使用辅助能量，例如超声辐射[108]。分散模式可以利用各种吸附剂来捕获或吸附目标分析物。但是，固体吸附剂需要满足几个要求。首先，需要具有高吸附能力和大表面积以保证快速，定量的吸附和洗脱，并且还必须具有在液体样品中的高分散性的特征[88]。考虑到这些要求，可以认为纳米颗粒类材料适用于 DMSPE 技术，包括碳基化合物（碳纳米管、石墨烯、富勒烯）和无机纳米颗粒类。使用磁性纳米颗粒似乎也是一种非常值得关注的解决方案。磁性颗粒可利用磁体从液体介质中分离颗粒，而不是通过离心分离。由于其简单，回收率高，速度快，可以与不同检测技术结合使用，以及有机溶剂消耗量低等诸多优势，DMSPE 激发了科学家研究用其进行分离和预浓缩的兴趣。以磁性纳米颗粒（MNP）为固体吸附剂（包括磁性碳纳米颗粒和分子印迹聚合物）的 DMSPE 已成功地用于测定水性样品中的合成代谢类固醇和 β 阻滞剂[64]，水和水果茶中的多环芳烃[108]以及猪肉样品中的雌激素[109]。

10.4 总结与展望

毫无疑问，过去几十年来，由于技术的进步，分析仪器的体积明显减小。然而，使分析系统微型化的重点不只是减小其体积，而是更着重于提高其分析性能、环境的可持续性、快速性和经济效益。尽管分析过程的每一步几乎都实现了微型化，但各个步骤并未实现相同程度的微型化。例如，尽管有一些自主和遥感分析微系统的报道，但微型化却并未给样品的采集和保存带来多大益处。此外，人们普遍认为，经过努力实现分离系统和检测系统的微型化后，已开发出微型的样品制备方法。

对传统样品制备技术（如固相萃取或溶剂萃取）微型化发展出了对环境无害的分析方法。如为克服 SPE 的缺点，引入的新型微萃取技术比 SPE 的多步过程需要更少的时间和劳动。这些微萃取技术可整合若干步骤，包括采样、萃取、将分析物富集至高于方法 LOD 的水平、从无法直接导入测量仪器的样品基质中分离分析物等。为了加速样品预处理并减少样品预处理对环境的不利影响，若干年前开发出了微型化的吸附萃取技术（例如 SPME、SBSE、MSPE、IASPME 和 MISPE），目前有多种绿色萃取方法用于处理气态、液态和/或固态样品。

这些技术具有许多优点，包括操作简单，仪器成本相对较低，多功能性，易于结合色谱系统以及萃取时间短。有关这些技术的研究仍然十分活跃，包括开发改进选择性、负载量或保留效率的新型吸附剂，尤其是集中在研究先前可用材料中保留量很低的分析物。

参考文献

[1] J. Namieśnik, Trends in environmental analytics and monitoring, Crit. Rev. Anal. Chem. 30 (2000) 221-269.

[2] A. Gałuszka, Z. Migaszewski, J. Namieśnik, The 12 principles of green analytical chemistry and the SIGNIFICANCE mnemonic of green analytical practices, Trends Anal. Chem. 50 (2013) 78-84;

[2a] S. Armenta, F. A. Esteve-Turrillas, S. Garrigues, M. de la Guardia, Green analytical chemistry: the role of green extraction techniques, in: E. Ibanez, A. Cifuentes (Eds.), Green Extraction Techniques: Principles, Advances and Applications, vol. 76, 2017, pp. 1-25.

[3] F. Pena-Pereira, From conventional to miniaturized analytical systems, in: F. Pena-Pereire (Ed.), Miniaturizing Steps in the Analytical Process, De Gruyter Open Ltd., Warszawa, 2014, pp. 1-28.

[4] A. Manz, N. Graber, H. M. Widmer, Miniaturized total chemical analysis systems: a novel concept for chemical sensing, Sens. Actuators B-Chem. 1 (1990) 244-248.

[5] R. Angel, A. Escarpa, B. Simonet, Miniaturization of Analytical Systems: Principles, Designs and Applications, John Wiley & Sons, Ltd, 2009.

[6] J. Ruzicka, Lab-on-valve: universal microflow analyzer based on sequential and bead injection, Analyst 125 (2000) 1053-1060.

[7] S. M. Decuir, H. M. Boden, A. D. Carroll, J. Ruzicka, Principles of micro sequential injection analysis in the lab-on-valve format and its introduction into a teaching laboratory, J. Flow Inject. Anal. 24 (2007) 103-108.

[8] J. Płotka-Wasylka, N. Szczepańska, J. Namieśnik, Modern trends in solid phase extraction: new sorbent media, Trends Anal. Chem. 77 (2016) 23-43.

[9] J. Pawliszyn, H. L. Lord, Handbook of Sample Preparation, John Wiley & Sons, Inc., New Jersey, 2010.

[10] J. Płotka, M. Tobiszewski, A. M. Sulej, M. Kupska, T. Góreckib, J. Namieśnik, Green chromatography, J. Chromatogr. A 1307 (2013) 1-20.

[11] Y. Picó, M. Fernández, M. J. Ruiz, G. Font, Current trends in solid-phase-based extraction techniques for the determination of pesticides in food and environment, J. Biochem. Biophys. Methods 70 (2007) 117-131.

[12] R. E. Majors, Solid phase-extraction, in: J. Pawliszyn, H. L. Lord (Eds.), Handbook of Sample Preparation, John Wiley & Sons, Inc., New Jersey, 2010, pp. 53-79.

[13] A. Spietelun, Ł. Marcinkowski, M. de la Guardia, J. Namieśnik, Recent developments and future trends in solid phase microextraction techniques towards green analytical chemistry, J. Chromatogr. A 1321 (2013) 1-13;

[13a] P. Serra-Mora, Y. Moliner-Martínez, C. Molins-Legua, R. Herráez-Hernández, J. Verdú-Andrés, P. Campíns-Falcó, Trends in online intube solid phase microextraction, in: E. Ibanez, A. Cifuentes (Eds.), Green Extraction Techniques: Principles, Advances and Applications, vol. 76, 2017, pp. 427-461;

[13b] J. M. F. Nogueira, Stir bar sorptive extraction, in: E. Ibanez, A. Cifuentes (Eds.), Green Extraction Techniques: Principles, Advances and Applications, vol. 76, 2017, pp. 463-481;

[13c] E. A. Souza-Silva, J. Pawliszyn, Recent advances in solid-phase microextraction for contaminant analysis in food matrices, in: E. Ibanez, A. Cifuentes (Eds.), Green Extraction Techniques: Principles, Advances and Applications, vol. 76, 2017, pp. 483-517.

[14] C. J. Welch, N. Wu, M. Biba, R. Hartman, T. Brkovic, X. Gong, et al., Greening analytical chromatography, Trends Anal. Chem. 29 (2010) 667-680.

[15] Y.-B. Luo, H.-B. Zheng, J.-X. Wanga, Q. Gao, Q.-W. Yu, Y.-Q. Feng, An anionic exchange stir rod sorptive extraction based on monolithic material for the extraction of non-steroidal anti-inflammatory drugs in environmental aqueous samples, Talanta 86 (2011) 103-108.

[16] J. Namieśnik, A. Spietelun, Ł. Marcinkowski, Green sample preparation techniques for chromatographic determination of small organic compounds, Int. J. Chem. Eng. Appl. 6 (2015) 215-219.

[17] B. J. Silva, F. M. Lanc, M. E. C. Queiroz, In-tube solid-phase microextraction coupled to liquid chromatography (in-tube SPME/LC) analysis of nontricyclic antidepressants in human plasma, J. Chromatogr. B 862 (2008) 181-188.

[18] H. Kataoka, Automated sample preparation using in-tube solid-phase microextraction and its application-a review, Anal. Bioanal. Chem. 373 (2002) 31-45.

[19] A. K. Malik, V. Kaur, N. Verma, A review on solid phase microextraction-high performance liquid chromatography as a novel tool for the analysis of toxic metal ions, Talanta 68 (2006) 842-849.

[20] K. Ridgway, S. P. D. Lalljie, R. M. Smith, Sample preparation techniques for the determination of trace residues and contaminants in foods, J. Chromatogr. A 1153 (2007) 36-53.

[21] Y.-B. Luo, Q. Ma, Y.-Q. Feng, Stir rod sorptive extraction with monolithic polymer as coating and its application to the analysis of froquinolones in honey sample, J. Chromatogr. A 1217 (2010) 3583-3589.

[22] B. Rossbach, P. Kegal, S. Letzel, Application of headspace solid phase dynamic extraction gas chromatography/mass spectrometry (HS-SPDE-GC/MS) for biomonitoring of n-heptane and its metabolites in blood, Toxicol. Lett. 25 (2012) 232-239.

[23] M.-Y. Wong, W.-R. Cheng, M.-H. Liu, W.-C. Tian, C.-J. Lu, A preconcentrator chip employing m-SPME array coated with in-situ-synthesized carbon adsorbent film for VOCs analysis, Talanta 101 (2012) 307-313.

[24] M. Abbasghorbani, A. Attaran, M. Payehghad, Solvent-assisted dispersive micro-SPE by using aminopropyl-functionalized magnetite nanoparticle followed by GC-PID for quantification of parabens in aqueous matrices, J. Sep. Sci. 36 (2013) 311-319.

[25] C. Basheer, A. A. Alnedhary, B. S. M. Rao, S. Valliyaveettil, H. K. Lee, Development and application of porous membrane-protected carbon nanotube micro-solid-phase extraction combined with gas chromatography/mass spectrometry, Anal. Chem. 78 (2006) 2853-2858.

[26] M. T. Garcia-Valverde, R. Lucena, S. Cardenas, M. Valcarcel, Titanium-dioxide nanotubes as sorbents in (micro) extraction techniques, TrAC 62 (2014) 37-45.

[27] M. Sun, J. Feng, Y. Bu, C. Luo, Highly sensitive copper fiber-in-tube solid phase microextraction for online selective analysis of polycyclic aromatic hydrocarbons coupled with high performance liquid chromatography, J. Chromatogr. A 1408 (2015) 41-46.

[28] K. Jinno, M. Kawazoe, Y. Saito, T. Takeichi, M. Hayashida, Sample preparation with fiber in tube solid phase microextraction for capillary electrophoretic separation of tricyclic antidepressant drugs in human urine, Electrophoresis 22 (2001) 3785-3790.

[29] M. Sun, J. Feng, Y. Bu, C. Luo, Ionic liquid coated copper wires and tubes for fiber-in-tube solid phase microextraction, J. Chromatogr. A 1458 (2016) 1-8.

[30] H. Kataoka, K. Mizuno, E. Oda, A. Saito, Determination of the oxidative stress biomarker urinary 8-hydroxy-2′-deoxyguanosine by automated on-line in-tube solid-phase microextraction coupled with liquid chromatography-tandem mass spectrometry, J. Chromatogr. B Analyt. Technol. Biomed. Life Sci. 1019 (2016) 140-146.

[31] J. Wu, W. M. Mullet, J. Pawliszyn, Electrochemically controlled solid-phase microextraction based on conductive polypyrrole films, Anal. Chem. 74 (2002) 4855-4859.

[32]J. Zeng, J. Zou, Song, J. Chen, J. Ji, B. Wang, Y. Wang, J. Ha, X. Chen, A new strategy for basic drug extraction in aqueous medium using electrochemically enhanced solid phase microextraction, J. Chromatogr. A 1218 (2011) 191-196.

[33]X. Chai, Y. He, D. Ying, J. Jia, T. Sun, Electrosorption-enchanced solid phase microextraction using activated carbon fiber for determination of aniline in water, J. Chromatogr. A 1165 (2007) 26-31.

[34]Q. Li, Y. Ding, D. Yuan, Electrosorption-enhanced solid phase microextraction of trace anions using a platinum plate coated with single-walle carbon nanotubes, Talanta 85 (2011) 1148-1153.

[35]A. Kloskowski, M. Pilarczyk, Membrane solid phase microextraction-a new concept of sorbent preparation, Anal. Chem. 81 (2009) 7363-7367.

[36]A. Spietelun, Ł. Mrcinkowski, A. Kloskowski, J. Namieśnik, Determination of volatile organic compounds in water samples using membrane-solid phase microextraction (M-SPME) (headspace version), Analyst 138 (2013) 5099-5106.

[37]Ł. Marcinkowski, A. Kloskowski, A. Spietelun, J. Namieśnik, Evaluation of polycaprolactone as a new sorbent coating for determination of polar organic compounds in water samples using membrane-SPME, Anal. Bioanal. Chem. 407 (2015) 1205-1215.

[38]N. N. Naing, S. F. Y. Lia, H. K. Lee, Evaluation of graphene-based sorbent in the determination of polar environmental contaminants in water by micro-solid phase extraction-high performance liquid chromatography, J. Chromatogr. A 1427 (2016) 29-36.

[39]M. Lashgari, H. Kee Lee, Micro-solid phase extraction of perfluorinated carboxylic acids from human plasma, J. Chromatogr. A 1432 (2016) 7-16.

[40]F. Bianchi, A. Gregori, G. Braun, C. Crescenzi, M. Careri, Micro-solid-phase extraction coupled to desorption electrospray ionization-high-resolution mass spectrometry for the analysis of explosives in soil, Anal. Bioanal. Chem. 407 (2015) 931-938.

[41]J. Sanchez-Gonzalez, M. J. Tabernero, A. M. Bermejo, P. Bermejo-Barrera, A. Moreda-Piñeiro, Porous membrane-protected molecularly imprinted polymer microsolid-phase extraction for analysis of urinary cocaine and its metabolites using liquid chromatography-tandem mass spectrometry, Anal. Chim. Acta. 898 (2015) 50-59.

[42]S. M. Ahmad, C. Almeida, N. R. Neng, J. M. F. Nogueira, Bar adsorptive microextraction (BAμE) coated with mixed sorbent phases-enhanced selectivity for the determination of non-steroidal anti-inflammatory drugs in real matrices in combination with capillary electrophoresis, J. Chromatogr. B 1008 (2016) 115-124.

[43]C. Almeida, J. M. F. Nogueira, Comparison of the selectivity of different sorbent phases for bar adsorptive microextraction-application to trace level analysis of fungicides in real matrices, J. Chromatogr. A 1265 (2012) 7-16.

[44]A. H. Ide, S. M. Ahmad, N. R. Neng, J. M. F. Nogueira, Enhancement for trace analysis of sulfonamide antibiotics in water matrices using bar adsorptive microextraction (BAμE), J. Pharm. Biomed. Anal. 129 (2016) 593-599.

[45]C. Almeida, R. Strzelczyk, J. M. F. Nogueira, Improvements on bar adsorptive microextraction (BAμE) technique-application for the determination of insecticide repel-lents in environmental water matrices, Talanta 120 (2014) 126-134.

[46]K. Sieg, E. Fries, W. Puttmann, Analysis of benzene, toluene, ethylbenzene, xylenes and n-aldehydes in melted snow water via solid-phase dynamic extraction combined with gas chromatography/mass

spectrometry, J. Chromatogr. A 1178 (2008) 178-186.

[47] L. Chen, X. Huang, Y. Zhang, D. Yuan, A new polymeric ionic liquid-based magnetic adsorbent for the extraction of inorganic anions in water samples, J. Chromatogr. A 1403 (2015) 37-44.

[48] Y. Zhang, M. Mei, X. Huang, D. Yuan, Extraction of trace nitrophenols in environmental water samples using boronate affinity sorbent, Anal. Chim. Acta 899 (2015) 75-84.

[49] L. Chen, M. Mei, X. Huang, D. Yuan, Sensitive determination of estrogens in environmental waters treated with polymeric ionic liquid-based stir cake sorptive extraction and liquid chromatographic analysis, Talanta 152 (2016) 98-104.

[50] A. Giordano, P. Richter, I. Ahumada, Determination of pesticides in river water using rotating disk sorptive extraction and gas chromatography-mass spectrometry, Talanta 85 (2011) 2425-2429.

[51] A. Cañas, P. Richter, Solid-phase microextraction using octadecyl-bonded silica immobilized on the surface of a rotating disk: determination of hexachlorobenzene in water, Anal. Chim. Acta 743 (2012) 75-79.

[52] A. Cañas, S. Valdebenito, P. Richter, A new rotating-disk sorptive extraction mode, with a copolymer of divinylbenzene and *N*-vinylpyrrolidone trapped in the cavity of the disk, used for determination of florfenicol residues in porcine plasma, Anal. Bioanal. Chem. 406 (2014) 2205-2210.

[53] Y.-B. Luo, J.-S. Cheng, Q. Ma, Y.-Q. Feng, J.-H. Li, Graphene-polymer composite: extraction of polycyclic aromatic hydrocarbons from water samples by stir rod sorptive extraction, Anal. Methods 3 (2011) 92-98.

[54] Z. Xu, Y. Hu, Y. Hu, G. Li, Investigation of ractopamine molecularly imprinted stir bar sorptive extraction and its application for trace analysis of B_2-agonists in complex samples, J. Chromatogr. A 1217 (2010) 3612-3618.

[55] C. Bicchi, C. Cordero, E. Liberto, P. Rubiolo, B. Sgorbini, F. David, P. Sandra, Dualphase twisters: a new approach to headspace sorptive extraction and stir-bar sorptive extraction, J. Chromatogr. A 1094 (2005) 9-16.

[56] J. Xu, B. Chen, M. He, B. Hu, Analysis of preservatives with different polarities in beverage samples by dual-phase dual stir bar sorptive extraction combined with high-performance liquid chromatography, J. Chromatogr. A 1278 (2013) 8-15.

[57] Z. Terzopoulou, M. Papageorgiou, G. Kyzas, D. Bikiaris, D. Lambropoulou, Preparation of MIP-SPME, fiber for the selective removal and extraction of the antivirial drug, Anal. Chim. Acta 913 (2016) 63-75.

[58] M. Szultka, J. Szeliga, M. Jackowski, B. Buszewski, Development of novel molecularly imprinted solid phase microextraction fibers and their application for the determination of antibiotic drugs in biological samples by SPME-LC/MS, Anal. Bioanal. Chem. 403 (2012) 785-796.

[59] T. Zhao, X. Guan, W. Tang, Y. Ma, H. Zhang, Preparation of temperature sensitive MIP for the SPME coatings stainless steel fiber to measure ofloxacin, Anal. Chim. Acta 853 (2015) 668-675.

[60] Y. Moliner-Martinez, Y. Vitta, H. Prima-Garcia, R. A. Gonzúlez-Fuenzalida, A. Ribera, P. Campíns-Falcó, E. Coronado, Silica supported Fe_3O_4 magnetic nanoparticles for magnetic solid-phase extraction and magnetic in-tube solid-phase microextraction: application to organophosphorous compounds, Anal. Bioanal. Chem. 406 (2014) 2211-2215.

[61] M. Euĝenia, C. Queiroz, E. B. Oliveira, F. Breton, J. Pawliszyn, Immunoaffinity intube solid phase microextraction coupled with liquid chromatography-mass spectrometry for analysis of fluoxetine in serum samples,

J. Chromatogr. A 1174 (2007) 72-77.

[62]K. Yao, W. Zhang, L. Yang, J. Gong, L. Li, T. Jin, C. Li, Determination of 11 quinolones in bovine milk using immunoaffinity stir bar sorptive microextraction and liquid chromatography with fluorescence detection, J. Chromatogr. B 1003 (2015) 67-73.

[63]H. L. Lord, M. Rajabi, S. Safari, J. Pawliszyn, Development of immunoaffinity solid phase microextraction probes for analysis of subng/mL concentrations of 7-aminoflu-nitrazepam in urine, J. Pharm. Biomed. Anal. 40 (2006) 769-780.

[64]M. Serrano, T. Chatzimitakos, M. Gallego, C. D. Stalikas, 1-Butyl-3-aminopropyl imidazolium-functionalized graphene oxide as a nanoadsorbent for the simultaneous extraction of steroids and β-blockers via dispersive solid-phase microextraction, J. Chromatogr. A 1436 (2016) 9-18.

[65]K. Farhadi, A. A. Matin, H. Amanzadeh, P. Biparva, H. Tajik, A. A. Farshid, H. Pirkharrati, A novel dispersive micro solid phase extraction using zein nanoparticles as the sorbent combined with headspace solid phase micro-extraction to determine chlorophenols in water and honey samples by GC-ECD, Talanta 128 (2014) 493-499.

[66]N. Yahaya, M. M. Sanagi, T. Mitome, N. Nishiyama, W. A. Wan Ibrahim, H. Nur, Dispersive micro-solid phase extraction combined with high-performance liquid chromatography for the determination of three Penicillins in milk samples, Food Anal. Methods 8 (2015) 1079-1087.

[67]J. Zeng, J. Chen, X. Song, Y. Wang, J. Ha, X. Chen, X. Wang, An electrochemically enhanced solid phase microextraction approach based on a multi-walled carbon nanotubes/Nafion composite coating, J. Chromatogr. A 1217 (2010) 1735-1741.

[68]A. Martin-Esteban, Molecularly imprinted polymers as a versatile, highly selective tool in sample preparation, TrAC 45 (2013) 169-181.

[69]C. He, Y. Long, J. Pan, K. Li, F. Liu, Application of molecularly imprinted polymers to solid-phase extraction of analytes from real samples, J. Biochem. Biophys. Methods 70 (2007) 133-150.

[70]G. Wulff, A. Sarchan, Enzyme models based on molecularly imprinted polymers with strong esterase activity, Angew. Chem. Int. Ed. Engl. 11 (1972) 341.

[71]E. Turiel, A. Martin-Esteban, Molecularly imprinted polymers for sample preparation: a review, Anal. Chim. Acta 668 (2010) 87-99.

[72]J. Ptotka-Wasylka, N. Szczepańska, M. de la Guardia, J. Namieśnik, Miniaturized solid-phase extraction techniques, TrAC 73 (2015) 19-38.

[73]A. Sarafraz-Yazdi, N. Razavi, Application of molecularly imprinted polymers in solid phase microextraction techniques, TrAC 73 (2015) 81-90.

[74]W. M. Mullet, P. Martin, J. Pawliszyn, In-tube molecularly imprinted polymer solid phase microextraction for the selective determination of propranolol, Anal. Chem. 73 (2001) 2383.

[75]E. Koster, C. Crescenzi, W. Den Hoedt, K. Ensing, G. J. de Jong, Fibers coated with molecularly imprinted polymers for solid-phase microextraction, Anal. Chem. 73 (2001) 3140.

[76]X. Hu, G. Dai, J. Huang, T. Ye, H. Fan, T. Youwen, Molecuarly imprinted polymer coated on stainless steel fiber for SPME of chloroacetanilide herbicides in soybean and corn, J. Chromatogr. A 1217 (2010) 7461-7470.

[77]X. Hu, Y. Fan, Y. Zhang, G. Dai, Q. Cai, Y. Cao, C. Guo, Molecularly imprinted polymer coated solid phase microextraction fiber prepared by surface reversible addition-fragmentation chain transfer polymerization for monitoring of Sudan dyes in chilli tomato sauce and chilli pepper samples, Anal. Chim. Acta

731 (2012) 40-48.

[78]A. Mehdinia, M. Ovais Aziz-Zanjani, Advances for sensitive, rapid and selective extraction in different configurations of solid-phase microextraction, TrAC 51 (2013) 13-22.

[79] G. Giakisikli, A. N. Anthemidis, Magnetic materials as sorbents for metal/metalloid preconcentration and/or separation. A review, Anal. Chim. Acta 789 (2013) 1-16.

[80]M. -M. Wona, E. -J. Chaa, O. -K. Yoona, N. -S. Kimb, K. Kima, D. -S. Lee, Use of headspace mulberry paper bag micro solid phase extraction for characterization of volatile aromas of essential oils from Bulgarian rose and Provence lavender, Anal. Chim. Acta 631 (2009) 54-61.

[81]F. Qi, L. Qian, J. Liu, X. Li, L. Lu, Q. Xu, A high-throughput nanofibers mat-based micro-solid phase extraction for the determination of cationic dyes in wastewater, J. Chromatogr. A 1460 (2016) 24-32.

[82]C. Montesano, M. Sergi, S. Odoardi, M. C. Simeoni, D. Compagnone, R. Curini, A μ-SPE procedure for the determination of cannabinoids and their metabolites in urine by LC-MS/MS, J. Pharm. Biomed. Anal. 91 (2014) 169-175.

[83]M. Miro, E. H. Hansen, On-line sample processing involving microextraction techniques as a front-end to atomic spectrometric detection for trace metal assays: a review, Anal. Chim. Acta 782 (2013) 1-11.

[84]N. R. Neng, A. R. M. Silva, J. M. F. Nogueira, Adsorptive micro-extraction techniques -novel analytical tools for trace levels of polar solutes in aqueous media, J. Chromatogr. A 1217 (2010) 7303-7310.

[85]N. R. Neng, A. S. Mestre, A. P. Carvalho, J. M. F. Nogueira, Powdered activated carbons as effective phases for bar adsorptive micro-extraction (BAμE) to monitor levels of triazinic herbicides in environmental water matrices, Talanta 83 (2011) 1643-1649.

[86]W. Liu, L. Zhang, L. Fan, Z. Lin, Y. Cai, Z. Wei, et al., An improved hollow fiber solvent-stir bar microextraction for the preconcentration of anabolic steroids in biological matrix with determination by gas chromatography-mass spectrometry, J. Chromatogr. A 1233 (2012) 1-7.

[87]H. Bagheri, E. Babanezhad, F. Khalilian, An interior needle electropolymerized pyrrole-based coating for headspace solid-phase dynamic extraction, Anal. Chim. Acta 634 (2009) 209-214.

[88]X. Huang, L. Chen, F. Lin, D. Yuan, Novel extraction approach for liquid samples: stir cake sorptive extraction using monolith, J. Sep. Sci. 34 (2011) 2145-2151.

[89]Y. Wang, J. Zhang, X. Huang, D. Yuan, Preparation of stir cake sorptive extraction based on polymeric ionic liquid for the enrichment of benzimidazole anthelmintics in water, honey and milk samples, Anal. Chim. Acta 840 (2014) 33-41.

[90]A. Canas, P. Richter, G. M. Escandar, Chemometrics-assisted excitation-emission fluorescence spectroscopy on nylon-attached rotating disks. Simultaneous determination of polycyclic aromatic hydrocarbons in the presence of interferences, Anal. Chim. Acta 852 (2014) 105-111.

[91]Y. Corrotea, N. Aguilera, L. Honda, P. Richter, Determination of hormones in water using rotating disk sorptive extraction and gas chromatography-mass spectrometry, Anal. Lett. 49 (2016) 1344-1358.

[92]V. Manzo, O. Navarro, L. Honda, K. Sánchez, M. I. Toral, P. Richter, Determination of crystal violet in water by direct solid phase spectrophotometry after rotating disk sorptive extraction, Talanta 106 (2013) 305-308.

[93] L. Jachero, B. Sepulveda, I. Ahumada, E. Fuentes, P. Richter, Rotating disk sorptive extraction of triclosan and methyl-triclosan from water samples, Anal. Bioanal. Chem. 405 (2013) 7711-7716.

[94] F. J. Camino - Sanchez, R. Rodriguez - Gomez, A. Zafira - Gomez, A. Santos - Fandila, J. I. Vilchez, Stir bar sorptive extraction: recent application, limitation and future trends, Talanta 130 (2014) 388-399.

[95] E. Baltussen, P. Sandra, F. David, C. Cramers, Stir bar sorptive extraction (SBSE), a novel extraction technique for aqueous samples : theory and principles, J. Microcolumn Sep. 10 (1999) 737-747.

[96] N. R. Neng, M. L. Pinto, J. Pires, P. M. Marcos, M. F. Nogueira, Development, optimisation and application of polyurethane foams as new polymeric phases for stir bar sorptive extraction, J. Chromatogr. A 1171 (2007) 8-14.

[97] M. Schellin, P. Poop, Application of a polysiloxane-based extraction method combined with large volume injection - gas chromatography - mass spectrometry of organic compounds in water samples, J. Chromatogr. A 1152 (2007) 175-183.

[98] A. E. Prince, T. S. Fan, B. A. Skoczenski, R. J. Bushway, Development of an immunoaffinity-based solid-phase extraction for diazinon, Anal. Chim. Acta 444 (2001) 37-49.

[99] F. Brothier, V. Pichon, Immobilized antibody on a hybrid organic-inorganic monolith: capillary immunoextraction coupled on-line to nanoLC-UV for the analysis of microcystin-LR, Anal. Chim. Acta 792 (2013) 52-58.

[100] F. Du, L. Guo, Q. Qin, X. Zheng, G. Ruan, J. Li, G. Li, Recent advances in aptamer-functionalized materials in sample preparation, TrAC 67 (2015) 134-146.

[101] L. Mua, X. Hub, J. Wena, Q. Zhoub, Robust aptamer sol-gel solid phase microextraction of very polar adenosine from human plasma, J. Chromatogr. A 1279 (2013) 7-12.

[102] M. M. Moein, A. Abdel-Rehim, M. Abdel-Rehim, Microextraction by packed sorbent (MEPS), Trends Anal. Chem. 67 (2015) 34-44.

[103] C. L. Silva, J. L. Goncalves, J. S. Camara, A sensitive microextraction by packed sorbentbased methodology combined with ultra-high pressure liquid chromatography as a powerful technique for analysis of biologically active flavonols in wines, Anal. Chim. Acta 739 (2012) 89-98.

[104] S. Fu, J. Fan, Y. Hashi, Z. Chen, Determination of polycyclic aromatic hydrocarbons in water samples using online microextraction by packed sorbent coupled with gas chromatography-mass spectrometry, Talanta 94 (2012) 152-157.

[105] A. El-Beqqali, A. Kussak, M. Abdel-Rehim, Fast and sensitive environmental analysis utilizing microextraction in packed syringe online with gas chromatography - mass spectrometry. Determination of polycyclic aromatic hydrocarbons in water, J. Chromatogr. A 1114 (2006) 234-238.

[106] J. L. Goncalves, V. L. Alves, F. P. Rodrigues, J. A. Figueira, J. S. Camara, A semiautomatic microextraction in packed sorbent, using a digitally controlled syringe, combined with ultra - high pressure liquid chromatography as a new and ultra-fast approach for the determination of prenylflavonoids in beers, J. Chromatogr. A 1304 (2013) 42-51.

[107] D. Chen, Y. Zhao, H. Miao, Y. Wu, A novel dispersive micro solid phase extraction using PCX as the sorbent for the determination of melamine and cyromazine in milk and milk powder by UHPLC-HRMS/MS, Talanta 134 (2015) 144-152.

[108] P. Rocío - Bautista, V. Pino, J. H. Ayala, J. Pasán, C. Ruiz - Pérez, A. M. Afonso, A magnetic-based dispersive micro-solid-phase extraction method using the metal-organic framework HKUST-1

and ultra - high - performance liquid chromatography with fluorescence detection for determining polycyclic aromatic hydrocarbons in waters and fruit tea infusions, J. Chromatogr. A 1436 (2016) 42-50.

[109] J. Wang, Z. Chen, Z. Li, Y. Yang, Magnetic nanoparticles based dispersive micro-solid-phase extraction as a novel technique for the determination of estrogens in pork samples, Food Chem. 204 (2016) 135-140.

11　QuEChERS 技术的最新研究进展

Bárbara Socas-Rodríguez, Javier González-Sálamo, Antonio V. Herrera-Herrera, Javier Hernández-Borges and MiguelÁ. Rodríguez-Delgado*

Universidad de La Laguna (ULL), San Cristóbal de La Laguna, Spain

*通讯作者：E-mail：mrguez@ull.edu.es

缩略词

ACN	乙腈
Al-N	中性氧化铝
AOAC	美国分析化学家协会
ASE	加速溶剂萃取
BSA	N,O-双（三甲基硅烷基）三氟乙酰胺
C18	十八烷基硅烷
CCα	决定限
CCβ	检测能力
CE	毛细管电泳
CEN	欧洲标准化委员会
CLC	毛细管液相色谱
DAD	二极管阵列检测器
DART	实时直接分析
DCM	二氯甲烷
Di-Na	柠檬酸氢二钠
DLLME	分散液液微萃取
d-DPE	分散固相萃取
ECD	电子捕获检测器
EN	欧洲规范
EPA	环保署
EtOAc	乙酸乙酯
EU	欧盟
EURL	欧盟参考实验室
FA	脂肪酸
FD	荧光检测器
FIA	流动注射分析
FID	火焰离子化检测器
GAC	绿色分析化学
GC	气相色谱
GCB	石墨化炭黑
HLB	亲水亲油平衡
HPLC	高效液相色谱
IAC	免疫亲和柱
IL	离子液体
IS	内标

LC	液相色谱
LLE	液液萃取
LOD	检出限
LPGC	低压气相色谱
MeOH	甲醇
MRL	最大残留限量
MS	质谱
MSPD	基质固相分散
MWCNT	多壁碳纳米管
Na2EDTA	乙二胺四乙酸二钠
NaOAc	醋酸钠
NH4OAc	醋酸铵
NPD	氮磷检测器
OCP	有机氯杀虫剂
OH-PAH	单羟基化多环芳烃
PAH	多环芳烃
PBDE	多溴二苯醚
PBS	磷酸盐缓冲溶液
PCB	多氯联苯
PFC	全氟化合物
PLE	加压液相萃取
PSA	*N*-丙基乙二胺
Q	单四级
QAC	季铵化合物
QqQ	三重四极杆
QTOF	四极杆-飞行时间
RSD	相对标准偏差
SAX	强阴离子交换
SFE	超临界流体萃取
SFO	漂浮的有机液滴固化
SLE	固液萃取
SPE	固相萃取
SVHC	高度关注的物质
TFA	三氟乙酸。
TMCS	三甲基氯硅烷
TMSI	*N*-三甲基甲硅烷基咪唑
TOF	飞行时间
TPP	磷酸三苯酯

tri-Na	柠檬酸三钠盐二水合物
TSL	铽敏化发光
UAE	超声协同萃取
UFLC	超快液相色谱
UHPLC	超高效液相色谱
UV	紫外线

11.1 引言

“绿色化学”这一术语源于人们对化学制品和工艺发展的日益关注，主要为了最大限度减少或者消除有害物质使用或产生。它起源于美国参议院1990年《联邦污染预防法》[1]。在所有化学学科中，绿色化学主要是从分子水平的角度来防止污染，并且应用创新科学方法来解决环境问题[1]。那么减少固有危害的化学产品和工艺设计对人类健康和环境都至关重要。起初，绿色化学显然是针对有机合成的。但是，此概念随后被应用于其他化学领域，包括分析化学，正如本书的第1章所述的，绿色分析化学：绿色提取技术的作用[1a]。

绿色分析化学（GAC）可以定义为在绿色化学的原理下指导，开发更清洁或更环保方法（污染物的消除或最小化）来分析一个复杂基质组分中的低浓度分析物，而不影响其准确性，灵敏度，和重现性[2,3]。除了环境优势外，绿色化学原理应用到分析化学中可以降低分析成本，提高了分析效率和操作员的安全性[3]。

为了减少环境污染，整个分析过程的不同步骤（例如样品的收集和储存，样品的预处理，分析和数据处理）是允许改进的[2,4]。然而，为了消除基质干扰和浓缩目标物，样品制备过程中需要消耗大量化学试剂和其他化学物质，是所有分析过程中产生污染物最多的步骤之一。这样说来，许多研究人员集中精力研究分析方法就不足为奇了，这在许多文献中都可以看出[2,3,5,6]。从这样的角度来看，QuEChERS方法（快速、简便、廉价、有效、稳定和安全）已成为传统样品制备步骤的绿色环保的选择。

QuEChERS方法于2003年由Michelangelo Anastassiades（德国斯图加特省的化学和兽医访问科学家）、Steven J. Lehotay（美国农业部，费城，宾夕法尼亚州，美国）、Darinka Stajnbaher（公共卫生研究所，马里博尔，斯洛文尼亚）和Frank J. Schenck（美国食品与药物管理局，亚特兰大，佐治亚州）[7]作为提取蔬菜和水果中农药残留的替代方法。由两个不同的步骤组成：①具有盐析效应的固液萃取/分离；②用于样品净化的分散固相萃取（d-SPE）。尽管这两个步骤都被广泛用于分析不同复杂基质中各种各样的化合物，但这一方法是一次对污染物和残留物分析的真正革命。几乎关于所有类型的化合物和样品的大量文章和评论清楚地证明了这一事实[8-12]，尽管许多文章只使用了其中一个步骤却声称使用了整个方法[11,13]。除此之外，该方法的超越性使得它的两个版本现在是用以确定水果和蔬菜中农药残留的国际标准组织（欧盟[14]和AOAC国际[15]）的官方分析方法。

该方法的重要性在于在相对较短的时间内巧妙地将步骤、溶剂、盐和吸附剂结合起来，以一种非常有效的方式和较低的成本为我们提供合适的结果。事实上，作者在原文章中已经提到过，六个切碎的样品可以由一个分析人员在不到 30min 内制备，每个样品成本低于 1 美元，并且这一事实后来也得到了证明。

作者的初衷不仅是介绍一种更环保的方法，而且是一种更有效的提取方法。到目前为止，不同的多类多残留方法得到了广泛的应用。然而这些方法并不能满足农产食品工业和健康监测方案日益增长的需求。作者对影响两个萃取步骤效率的不同因素进行了系统和深入的研究：样品大小和粉碎、样品组成（pH 和基质成分量）、溶剂类型、样品/溶剂比、搅拌方式（混合或摇动）、温度、助溶剂和/或盐的加入、萃取时间和净化吸附剂。值得注意的是，最后选择合适的条件最大限度实现简化、适用、速度、选择性和分析物的回收率。为此，需要仔细考虑共萃取物质的量、水的量和萃取物的色泽、回收率、质谱（MS）色谱图中的基质背景和基质诱导的色谱效应。

在其中一篇文章[7]中作者最初研究了样品的大小和粉碎情况。众所周知，通过减少样品到最小均匀量，这个量必须满足可以提供可靠结果，这样可以提高特定方法的分析效率（如溶剂体积、安全问题、储存、时间和成本方面）。需要注意的是，这种研磨必须提供完整的具有代表性的样品（农药残留没有损失）。以前开发的多类多残留方法使用的样品量在 50~100g 之间，但如果使用其他萃取技术［如超临界流体萃取、基质固相分散（MSPD）或加压液体萃取（PLE）］，5~15g 就足够了。根据文献中的经验和证据，建议取 10g 作为代表性样本。因为在提取过程中没有使用搅拌机，他们还建议进行彻底粉碎，最大限度地扩大表面积来确保更好的接触，同时使用干冰以避免农药损失。

传统上多类多残留方法提取农药的萃取溶剂主要是非极性溶剂和氯化物［石油醚、二氯甲烷（DCM）等］或极性溶剂［丙酮、乙腈（ACN）和乙酸乙酯（EtOAc）］。考虑到它们固有的问题（石油醚和 DCM 具有剧毒，丙酮和 ACN 很难与水分离，EtOAc 对高极性农药的亲和力降低[7]），之所以最初选择丙酮、ACN 和 EtOAc 是因为它们的回收率较高且在一个合理的范围。在实验中其余参数保持不变，作者对这三种溶剂进行了比较发现，通过使用 *N*-丙基乙二胺（PSA）的 d-SPE 后，ACN 从水果和蔬菜中提取物质的种类比丙酮和 EtOAC 要少，在未进行净化步骤的情况下，数量也比较少（图 11.1）。而且，ACN 可以通过添加盐（丙酮需要非极性助溶剂）从水相中分离出来，不能提取大量亲脂物质，可以与非极性溶剂一起使用以进行额外的净化（如果需要），与丙酮和 EtOAc 相比挥发性较低。再者，可以使用干燥剂（如 $MgSO_4$）去除 ACN 中的残余水，该溶剂适用于与气相色谱（GC）和液相色谱（LC）。鉴于这些原因，选择了这种溶剂进行下一步的实验。

溶剂与样品的比值也是保证所有农药有效提取的最基本影响因素。因为果蔬含水量很高（80%~95%），与之前提取非极性农药理想方法相比，样品/ACN 比例 1/1 具有更高的含水量[16]。这样两种溶剂之间的密切接触使 ACN 的萃取作用更强。因此选择了这一比例，同时也显然减少了有机溶剂的使用量。

关于初始萃取的搅拌方法，作者比较了振荡与匀浆。振荡被认为是可接受的，即使残留物可能不容易提取。因此考虑到这些结果和振荡胜过匀浆的其他优点（样品没

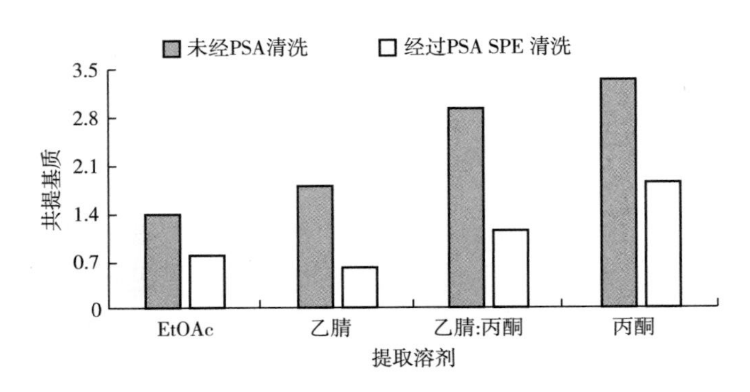

图 11.1 用 PSA 作为吸附剂，不同溶剂在有无 d-SPE 步骤处理后水果和蔬菜中混合物提取结果

y 轴表示从每克原始样品最终提取物中萃取的基质量（mg）。乙腈丙酮比为 1∶1。d-SPE，分散固相萃取；PSA，N-丙基乙二胺。

［资料来源：M. Anastassiades，S. J. Lehotay，D. Stajnbaher，F. J. Schenck，Fast and easy multiresidue method employing acetonitrile extraction/partitioning and "dispersive solid-phase extraction" for the determination of pesticide residues in produce，J. AOAC Int. 86（2003）412-431.］

有暴露在搅拌机的活性表面，样品之间不需要清洗搅拌机，更多的样品可以平行提取，不产生摩擦热），选择使用涡旋混合器。

根据先前的方法，通过盐析效应引入盐诱导相分离似乎可以提高回收率[17~19]。一方面，盐的引入通常增加了极性农药的回收率，并控制了有机相中水的百分比。另一方面，这种引入避免了使用助溶剂及其相关的缺点（稀释和提取物极性不足）。在该方法的首次应用中，对果糖、$MgCl_2$、Na_2NO_3、Na_2SO_4、LiCl、$MgSO_4$ 和 NaCl 进行了测试。其中 $MgSO_4$ 的测试结果最好，因为它结合大量的水，从而促进了分析物在 ACN 层中的分配。然而，在加入 $MgSO_4$ 后应进行剧烈振荡，避免团聚物的形成。$MgSO_4$（40~45℃）水合放热也有利于萃取，特别是对非极性农药来说。此外，使用 NaCl 对 $MgSO_4$ 进行结合或取代作用进行了评价，取得了满意的结果。在回收率方面，$MgSO_4$ 比 NaCl 效果更好。然而加入 NaCl 后，减少了极性基质化合物在有机相中的分配，这不是我们所期望的。此外 NaCl 对不同农药的峰形和面积有很大的积极影响，所以在萃取步骤中也需要加入 NaCl，但应仔细优化它的浓度来促使两种相反的作用达到最优结果。

QuEChERS 方法早期研究的另一个重要因素是样品的 pH，因为一些农药在高 pH 下迅速降解，而另一些农药在低 pH 下回收率较差。水果和蔬菜的 pH 在酸性范围内（2.5~6.5），但在分离后的 ACN 相中仍然存在大量的水，作者假设 pH 不会有显著的影响。实验证明这种酸性条件对大多数农残的影响微乎其微。但萃取溶剂如 EtOAc 的影响却很明显。然而，为了确保一些碱敏农药（碱性条件下稳定性差或易分解的农药）定量回收，需要把各种蔬菜的 pH 调至 4 以下，除此之外，pH 也会对一些产生基质干扰的共萃物有很强烈的影响。如图 11.2 所示，在较低的 pH 下，共萃物中脂肪酸和其他酸等有所增加。必须注意的是最终提取物的酸度也会影响碱敏农药的回收率。在 pH 高于 5 的样品提取过程中，以及在用 PSA（含有影响 pH 的伯胺和仲胺）d-SPE 后，这些农药都会损失。作者发现当在 ACN 提取物中加入 0.05%~0.1%（体积比）乙酸时，农药可以稳定保存一天以上。虽然他们避免使用任何酸，但这些发现为目前使用 QuEChERS 缓冲溶液法提供了合适的证据，这也是目前官方采用的方法[14,15]，此部分后面再进行讨论。

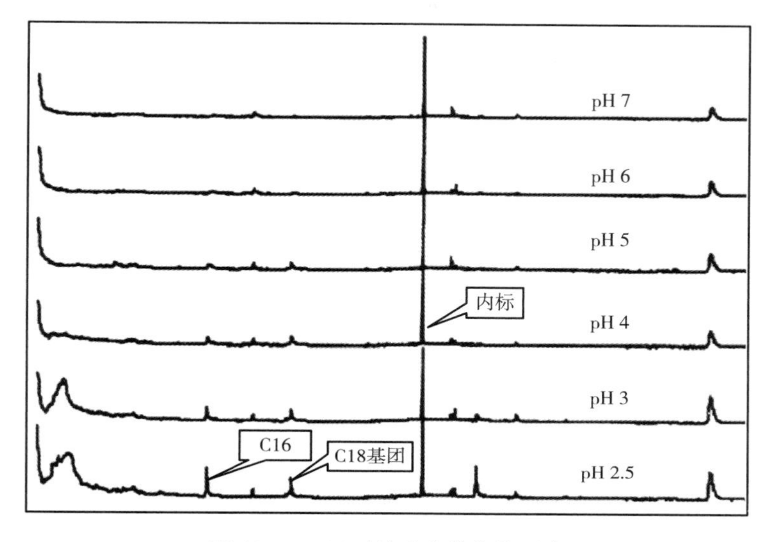

图 11.2 pH 对基质共萃物的影响

（用 1g NaCl 和 5g $MgSO_4$ 诱导分配得到 ACN 苹果汁提取物的 GC-MS 色谱图（全扫描模式））

［资料来源：M. Anastassiades, S. J. Lehotay, D. Stajnbaher, F. J. Schenck, Fast and easy multiresidue method employing acetonitrile extraction/partitionin gand "dispersive solid-phase extraction" for the determination of pesticide residues in produce, J. AOAC Int. 86（2003）412-431．经 AOAC International 许可转载］

值得注意的是，得到的 ACN 相是经过除水和净化处理的。水的存在会影响 d-SPE 和 GC 的测定，所以必须进行除水干燥。从这个意义上说，$MgSO_4$ 比 Na_2SO_4 更适合作为干燥剂，因为 $MgSO_4$ 能够更有效地去除水，提取的极性萃取物较少，这些极性萃取物会导致基质中某些极性干扰物沉淀。在 d-SPE 吸附剂方面，作者对 PSA、甲基丙烯酸酯-二乙烯基苯聚合吸附剂、石墨化炭黑（GCB）、中性氧化铝（Al-N）、强阴离子交换器（SAX）以及氰丙基、氨基丙基和十八烷基硅烷（C_{18}）进行了测试。其中，PSA 和 GCB 的混合物可以去除 ACN 相中大量的基体物质。然而，由于 GCB 对平面分子的高亲和力而保留了某些农药。其他吸附剂组合没有额外净化效果。因此，选择 PSA 作为吸附剂。

最后，作者通过基质匹配校准溶液和非基质匹配校准溶液进行了定量，基质匹配校准溶液是指在进行 d-SPE 之前和之后，将标准物添加到空白提取物中，而非基质匹配校准溶液是指将标准物加入到分析物保护剂中，这两种方式定量的结果没有任何差异。使用分析物保护剂是为不能使用基质匹配校准的实验室提供一种纠正基质效应的替代工具。因此，具有多羟基、氨基或羧基的化合物能够形成氢键，选取这样的化合物作为分析物保护剂。山梨醇和 3-*O*-乙基甘油的混合物对所有的分析物都是有效的。

此外，磷酸三苯酯（TPP）作为内标物质（IS），消除了该方法的固有误差（在第一次分配之后添加内标物质）。

概括来说，最初的 QuEChERS 方法主要是将 10mL 的 ACN 加入到 10g 切碎均匀样品中进行提取。样品涡旋摇动 1min 后，加入 4g $MgSO_4$ 和 1g NaCl，再涡旋 1min（避免团聚物的形成）。然后加入 IS 后涡旋 30s，然后 5000r/mim 下离心 5min。然后将 25mg PSA 和 150mg $MgSO_4$ 加入到 1mL 的 ACN 相进行混合。得到的混合物涡旋 30s，在 6000r/min 下离心 1min，取 0.5mL 进行 GC-MS 分析。值得一提的是，LC 分析（或更快的色谱方法）和脂质基质以及大容量进样等方面都需有待进一步研究和探讨。

11.2 原始方法的改进

尽管原始 QuEChERS 方法可以从各种各样基质中提取数百种分析物，但自创建以来，由于需要获得高回收率、避免农药降解和减少基质效应，已经进行了一些改进。这种改进（图 11.3）使得原始 QuEChERS 方法性能得到了更大的提高，特别是在分析复杂基质时。这些改进保留了液液萃取（LLE）和 d-SPE 步骤，但改进了这些步骤中使用的溶剂、盐和吸附剂。

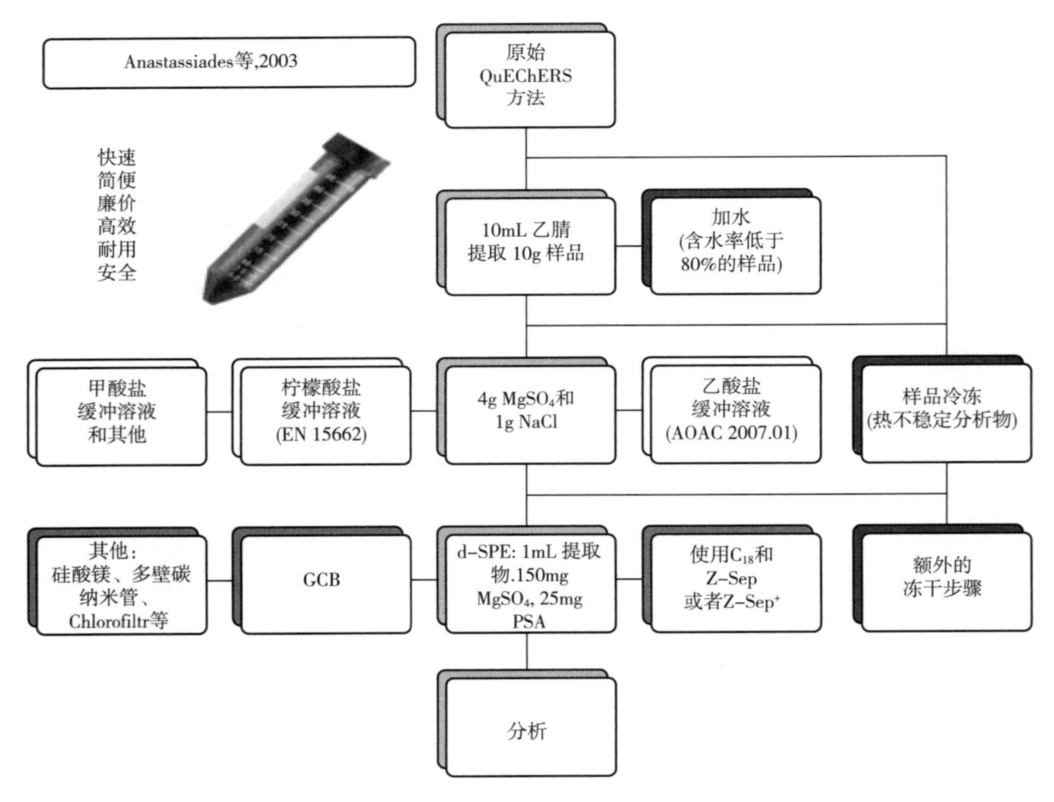

图 11.3 原始 QuEChERS 方法的原理图摘要及其相关的改进

11.2.1　pH 调节

众所周知，根据基质 pH，某些受 pH 影响较大的分析物会经历电离甚至降解。如果发生了这些情况，分析物可能无法到达有机层，也可能在最初的 LLE 步骤中就降解了，从而导致了较低回收率。因此，在提取和分层过程中要严格控制 pH。

关于前面提到的方面，从最初的无缓冲版本[7]中提出了两种方法试图将该方法扩展到这些分析物。第一种是由 Lehotay 及其同事提出的，他们利用高浓度乙酸缓冲液来获得更大的缓冲强度（AOAC 官方方法 2007.01[15]）。然而原来的方法中，部分缓冲液分层进入有机相，这样 ACN 相可以形成恒定的 pH，乙酸盐具有强缓冲能力，PSA 净化能力差[20]。第二个版本是由 Anastassiades 及其同事提出的，他们使用较低缓冲能力的柠檬酸盐缓冲液［欧洲标准化委员会（CEN）标准］，在 PSA 净化过程中没有产生负面作用。这两种缓冲方法都表明 pH 在 5 左右酸敏或者碱敏农药（例如灭菌丹、二氯氟胺、吡咯烷酮、百菌清）提取结果令人满意。由于其固有的优点，这两个版本被许多实验室都当作常规方法，但在此 pH 下，PSA 的保留能力降低，因此不建议对某些特定基质（脂质含量较高的基质）添加缓冲溶液，一些最近的研究[21]发现这种情况下产生了更多的共萃物。

这些 QuEChERS 版本最初是用于农药分析的，但它们也被应用于各种分析物如霉菌毒素的提取[22-24]。此外，还使用了其他缓冲盐如甲酸盐[25,26]（其特定方面将在 11.2.4 中描述）或磷酸盐缓冲盐水（PBS）。

最后，必须提到另一个与之相关的典型问题，一些受 pH 影响较大的农药在处理结束时会发生降解，由于最终提取物的 pH 通常在 8～9 左右，就会导致碱敏农药降解[10,20]。当样品进样前需要储存一段时间，一般在 ACN 中加入 5%（体积比）甲酸，调整 pH 约为 5，这样就可以避免碱敏农药降解。因此缩短样品处理后与进样前之间的时间也可以避免这一问题[10,27]。

11.2.2　萃取溶剂

一般情况下样品的特征决定了使用的溶剂。因此最简单的改进之一是在样品处理[28,29]开始时添加一定体积的水（除了萃取溶剂外），从而有可能将该方法扩展应用于分析含水量低于 80%的商品（谷物、面粉等）。水分可以削弱分析物和基质之间的相互作用，确保在第一步中形成适当的分层。

除此之外，选择合适的有机溶剂对于保证从水相中提取目标分析物至关重要。从这个意义上说，偶尔也可以尝试使用其他有机溶剂来替代乙腈，如丙酮[30]或正己烷[31]。也可以将溶剂混合后使用，如乙腈：甲醇（MeOH）[32,33]，乙腈：正己烷[34]或 DCM：正己烷[35]。许多这种情况下也可以加入一些盐，如 $MgSO_4$ 和 NaCl 等。

11.2.3　分散固相萃取吸附剂

除了避免分析物在 LLE 过程中电离或降解外，我们还做出了许多努力来减少共萃取化合物的数量，防止干扰分析物的准确测定。为此，其他不同于 PSA 的吸附剂已被作为 d-SPE 净化步骤的一部分，用于选择性地去除某些共萃取物。

为了在分析前获得较纯净提取物，作为目前 QuEChERS 通用方法的一部分，需要考虑两种改进。第一种改进涉及使用不同 PSA[21,36] 的数量或其与 C_{18}[23,29,37,38] 的组合。与 PSA 相反，C_{18} 不仅不会降低农药回收率，而且当对脂质含量高的基质（如谷类、鳄梨或牛乳）进行分析时特别有用。由于这些原因，一些改进后的方法不使用 PSA，只使用 C_{18} 就可以去除脂肪[39-41]。第二种改进是使用 PSA 与 GCB 的结合，这样可以从有色基质中成功地去除平面色素共萃物，获得浅色提取物[28,42]。在平面农药的特定情况下，可降低回收率高达 25%[43,44]。

虽然 C_{18} 和 GCB 分别与 PSA 结合使用或者不结合使用被认为是当前 QuEChERS 方法的重要组成部分，但在 d-SPE 净化步骤中也可以考虑使用其他吸附剂。从这个意义上说，基于氧化锆的新材料已经被用于脂质去除[45-48]，并在不同领域里提供了有趣的结果。这些类型的吸附剂比 PSA 和 C_{18} 能够更有效地去除脂肪和色素，并且得到更高的回收率。现在有可能找到两种锆基物质，商业上可用的称为 Z-Sep 和 Z-Sep$^+$。第一种是 ZrO_2 包覆的 Si，而第二种还含有 C_{18}，其与 ZrO_2 的比例含量为 2∶5。在色素去除方面，ChloroFiltr 作为一种聚合物吸附剂，已成为 GCB 的重要替代品[49]，在不损失平面分析物的情况下，它可以选择性降低最终提取物中的叶绿素含量。此外，还使用了其他新型吸附剂，包括氧化铝（用于亲脂化合物消除）[50]、氟硅酸镁（用于分离非极性或低极性分析物）[31,51-53]、纳米材料如多壁碳纳米管（MWCNTs）[54] 和磁性纳米粒子[55-57]，它们都具有较高的比表面积和较高的萃取能力。还有其他一些有趣的替代材料也已成功被用作净化吸附剂，包括硅藻土[58]、SAX（一种基于三甲基氨丙基硅颗粒的强阴离子交换吸附剂）[59] 或聚合物吸附剂，如苯乙烯-二乙烯基苯[60]，但这些材料还没有得到广泛应用。

最后还应该提到的是，一些其他方法已经出现，其中 d-SPE 程序已被传统的 SPE 步骤取代[61-63]。尽管 SPE 具有更大的净化能力，但与 d-SPE 相比，该过程本身不能达到相同简单性和速率，因此 d-SPE 仍然是目前主要使用的方法[44]。

11.2.4 加盐

如前所述，最初的 LLE 过程需要添加 NaCl 来增加盐析效果，添加 $MgSO_4$ 去除水分达到干燥有机层的目的。然而，应该考虑到这些盐也会影响后续需要使用的仪器。

从这个意义上说，GonzAlez-Curbelo 等人提出了一种有趣的替代方法，使用氯化铵、甲酸铵和乙酸铵缓冲液来提取代表性农药[25,26]。众所周知，在 QuEChERS 处理后，$MgSO_4$ 和 NaCl 容易在仪器（GC-MS or LC-MS）的某些部分沉淀为固体，从而降低仪器性能。使用铵盐可以避免这一问题，同时促进了分析物离子的形成，而在单独进行基质萃取过程中甲酸盐缓冲液可以保证 pH 的稳定。虽然这三种方法的性能优于 AOAC 官方方法中的 QuEChERS[15]，但使用甲酸盐缓冲液［在 ACN 中添加 7.5g 甲酸铵和 15mL 5%（体积比）甲酸提取 15g 水果或蔬菜样品］可以确保基质中的大部分农药在合适 pH 下得到高回收率。

在 d-SPE 清理过程中，还可以用 $CaCl_2$ 替代 $MgSO_4$，因为 $CaCl_2$ 可以更好地去除水分，这样就增强了基质组分与 PSA 之间的相互作用，从而获得纯度更好的目标物[64]。然而这种替代方案的前提是基质中没有极性农药，因为 $CaCl_2$ 降低了这类农药的回收率[27,64]。

11.2.5 样品冷冻或引入冷冻过程

在 QuEChERS 过程中，温度也会影响该方法的性能。从这个意义上说，在样品中加入无水 $MgSO_4$ 会发生水合放热反应，从而导致不稳定的分析物发生热降解[44]。为了减少热降解，在初提取前冷冻样品[65,66]或需要加水时则加入冷水（<4℃）[67]，但冷冻样品在萃取后就能达到软化温度，因而冻结样品可以更有效减少热降解。

在 d-SPE 处理前或处理后，进行所谓冷冻步骤时会发生脂质沉淀[36,68,69]。这一过程不需要额外的吸附剂，但冷冻提取物需要 1~2 小时，进而明显增加了样品处理时间[69]，但可以使用干冰浴来减少样品冷冻时间[68]。如果 d-SPE 与 PSA 和 C_{18} 联用，那么在农药分析中这一步骤是不必要的[70]。

11.3 应用领域

11.3.1 农药分析

众所周知，农药是一类令人特别关切的污染物。它们广泛应用于现代农业来保护作物免受不同疾病、杂草或昆虫的侵害[29]。然而，它们对消费者构成了重大的健康风险，因为即使收获后它们也会在农业和加工食品以及与这一领域有关的其他环境基质中持续残留[10,29]。因此，用于测定它们的分析方法要能够通过快速简单的程序分析大量的化合物同时也要能够提供良好的回收率，这是最基本的，也是最重要的。从这个意义上说，QuEChERS 方法是一个很好的替代方法，能够提供足够的灵敏度来降低国际组织建立的不同基质中污染物的最大残留限量（MRLs）。事实上，现在已有 650 多种农药及其代谢物被纳入欧盟参考实验室（EURL）数据库，从而可以使用这种方法来验证其数据的准确性[8,71]。

正如之前所论述的[9,11,72]，农药分析成了 QuEChERS 方法的主要应用领域，不仅适用于最初所研究的水果和蔬菜，而且也适用于其他各种各样的基质，包括其他不同性质的食品、环境样品，甚至生物性液体或非食用植物。表 11.1 对这一领域发表的一些文章进行了总结。可以看出，该方法的原始版本和改进版本都已应用于不同类别农药的提取，广泛进行了多残留分析[29,31,52,54,76-78,81-83]。Lozowicka 等[83]开发了一种 QuEChERS 方法来分析甜菜和甜菜糖蜜中的 400 种农药，在用 GC-MS/MS 和高效液相色谱（HPLC）-MS/MS 测定之前，提取步骤中使用柠檬酸缓冲液以及 PSA 和 GCB 作为净化吸附剂进行净化，检出限（LODs）在 5~10mg/kg 内，回收率在 60%~140%，因而可以作为一种常规多残留的分析方法[85]。

正如前面提到的，在这一领域一个共同的趋势是引入替代的净化吸附剂来提高回收效率和干净的基质，如 Florisil（弗罗里硅土）[31,52,53]、锆基吸附剂[47,48]或者

表 11.1　QuEChERS 方法分析农残的一些实例

分析物	样品（数量）	提取溶剂	提取盐类	d-SPE 中的吸附剂	分析方法	回收率/%	LODs/（μg/kg）	备注	参考文献
58 种农残	土壤（5g）	10mL 1%乙酸乙腈溶液（体积比）	4g $MgSO_4$，1g NaCl	900mg $MgSO_4$，150mg PSA，150mg C_{18}	GC-MS/MS	69~119	0.03~1.5	先加入 10mL 水，测试不同的吸附剂，环氧七氯作内标	[29]
8 种拟除虫菊酯类农残	红绿花椒（10g），干红花椒（1g）	10mL 1%乙酸乙腈溶液（体积比），2mL 己烷	2g NaCl	150mg PSA，50mg C_{18}	GC-ECD	79~104	1.2~150	先向干红花椒中加入 10mL 水，再分析试样	[73]
11 种有机氯农药	豆类，卷心菜，牛肉，鱼（10g）	10mL 乙腈	3g $MgSO_4$，1.5g NaCl	1.5g $MgSO_4$，27.5mg PSA	GC-ECD	80~92	0.25~19.29	用 GC×GC-TOF-MS 对样品中目标化合物进行了验证。对实际样品进行了分析	[74]
11 种农残	椰子肉（10g），椰子水（10mL）	10mL 1%乙酸乙腈溶液（体积比）	4g $MgSO_4$，1.5g NaOAc	600mg $MgSO_4$，100mg PSA，500mg C_{18}	UHPLC-MS/MS	70~120	3	在净化之前，在干冰中冷却上清液	[75]
88 种农残	人乳 1mL	10mL 1%乙酸乙腈溶液（体积比）	0.4g $MgSO_4$，0.1g NaOAc	157mg $MgSO_4$，9mg PSA，9mg C_{18}	GC-MS/MS	70~120	0.2~2	在净化之前，-20℃ 冻结上清液 2h，TPP 为内标物	[76]
120 种农残	苹果，黄瓜（10g）	10mL 乙腈	4g $MgSO_4$，1g NaCl，0.5g 二钠，1g 三钠	900mg $MgSO_4$，150mg PSA	HPLC-MS/MS	70~120	1.2~100	—	[77]
21 种农残	橄榄油和葡萄籽油（6g）	20mL 乙腈	8g $MgSO_4$，2g NaCl	0.5~3.5g Z-Sep	HPLC-DAD	50~130	—	最初加入 10mL 水。在 d-SPE 之前，使用 C_{18} 小柱进行过滤	[47]

4种农残	玉米（5g），玉米秆（2g），土壤（5g）	10mL 1%乙酸乙腈溶液（体积比）	4g $MgSO_4$，1g NaCl	200mg $MgSO_4$，25mg C_{18}	HPLC-MS/MS	80~110	1.8	最初加入2mL水	[41]
74种农残	橙汁（10mL）	10mL 1%乙酸乙腈溶液（体积比）	4g $MgSO_4$，1.7g NaOAc	150mg $MgSO_4$，40mg PSA	UHPLC-MS/MS	70~118	3.0~7.6	TPP作内标物	[78]
116种农残	蜂蜜（5g）	10mL 1%乙酸乙腈溶液（体积比）	4g $MgSO_4$，1g NaOAc	150mg $MgSO_4$，50mg PSA，50mg Florisil	UHPLC-MS/MS	70~120	5	最初加入10mL水	[52]
24种农残	水果，谷物，淀粉和婴幼儿食品（15g）	15mL 1%乙酸乙腈溶液（体积比）	6g $MgSO_4$，1.5g NaOAc	150mg $MgSO_4$，50mg PSA，50mg C_{18}	GC-MS	70~120	10~50	d-SPE净化与DLLME的比较。TPP作内标物	[79]
28种氨基甲酸酯农残	香草（1g）	10mL乙腈	4g $MgSO_4$，1g NaCl	200mg $MgSO_4$，200mg C_{18}	UHPLC-MS/MS	72~99	0.6	与其他d-SPE吸附剂比较	[80]
30种农残	牛乳（20mL）	16mL乙腈	8g $MgSO_4$，2g NaCl	125mg PSA，25mg Z-Sep，5mg Z-Sep+	UHPLC-DAD	70~100	0.02~0.06μg/L	—	[48]
74种农残	药用植物（2g）	20mL己烷	3g NaOAc，2g NaCl	50mg $MgSO_4$，50mg C_{18}，50mg Florisil	GC-MS/MS	70~120	3	最初加入10mL水，Chlorpyrifos-d_{10}作内标物	[31]
223种农残	烟草（2g）	20mL乙腈	4g $MgSO_4$，1g NaCl，0.5g二钠，1g三钠	150mg $MgSO_4$，25mg PSA，2.5mg GCB	GC-μECD/NPD	71~120	1~1.2	最初加入2mL水。与MSPD和SLE方法进行比较。应用GC-MS/MS进行确认。TPP作内标物	[81]

续表

分析物	样品（数量）	提取溶剂	提取盐类	d-SPE 中的吸附剂	分析方法	回收率/%	LODs/(μg/kg)	备注	参考文献
66 种农残	葡萄（10g）	10mL 1% 乙酸乙腈溶液（体积比）	4g $MgSO_4$，1.7g NaOAc	300mg $MgSO_4$，100mg PSA，	UHPLC-MS/MS	70~120	3	TPP 作内标物	[82]
171 种农残	豇豆（10g）	10mL 乙腈	4g $MgSO_4$，1g NaCl	150mg $MgSO_4$，5mg MWCNTs	GC-MS/MS	74~129	1~3	与 PSA 和 C_{18} 净化吸附剂进行比较	[54]
400 种农残	甜菜，糖蜜（10g）	10mL 1%甲酸乙腈溶液（体积比）	4g $MgSO_4$，1g NaCl，0.5g 二钠，1g 三钠	150mg $MgSO_4$，25mg PSA，2.5mg GCB	GC-MS/MS，UHPLC-MS/MS	64~140	5~10	最初甜菜糖蜜样品加入 5mL 水。在净化之前，在 -60℃ 冻结上清液 30min。TPP 作为 GC-MS/MS 分析的内标物，阿特拉津-d5 多菌灵-d3 和异丙酮-d6 作为 HPLC-MS/MS 分析的内标物。与 MSPD 和传统 QuEChERS 方法进行比较	[83]
13 种有机氯	鱼肉（5g）	10mL 乙腈	4g $MgSO_4$，1g NaCl	50mg PSA	GC-ECD	88~121	0.65~1.58	在净化中将 DLLME-SFO 过程集成	[84]
19 种农残	葡萄（10g）	10mL 1%甲酸乙腈溶液（体积比）	4g $MgSO_4$，1g NaOAc	300mg $MgSO_4$，400mg Florisil	GC-MS	95~102	6~12	与传统 QuEChERS 进行比较。五氯硝基苯和咖啡因分别作为相应方法的内标物	[53]

注：ACN—乙腈；C_{18}—十八烷基硅烷；DAD—二极管阵列检测器；di-Na—柠檬酸氢二钠水合物；DLLME—分散液微萃取；d-SPE—分散固相萃取；ECD 电子捕获检测器；GC—气相色谱；GCB—石墨化炭黑；HPLC—高效液相色谱法；IS—内部标准；LOD—检出限 MS—质谱；MSPD—基质固相分散；MWCNT—多壁碳纳米管；NaOAC—乙酸胺；NPD—氮磷检测器；OCP—有机氯农药；PSA—N-丙基乙二胺；SFO—固化悬浮的有机液滴；SLE—固液萃取；TOF—飞行时间；TPP—磷酸三苯酯；tri-Na—柠檬酸钠二水合物；UHPLC—超高效液相色谱。

MWCNTs[54,86,87]。Volpatto 等[53]通过醋酸缓冲溶液的传统方法从葡萄中提取了 19 种不同的农药，包括具有 pH 依赖性的农药，如西普罗地尼、米克洛布他尼和特布考那唑，结果表明最后一种方法比传统方法具有更好、更一致的回收率。同样通过对 GCB、C_{18}、PSA 和 Florisil 等不同的净化吸附剂进行了评价表明 Florisil 也能去除绝大部分色素，其效果类似于 PSA，但提取的农药数量较多，回收率在 70%~120%。因此在 GC-MS 分析测定 19 种农药前，将 Florisil 净化步骤与醋酸盐缓冲溶液萃取一起应用，其回收率较好，在 95%~102%，检出限也较低，在 6~12μg/kg 范围内。

在文献中可以看到的另一个重要研究是 QuEChERS 与其他方法的结合来进一步优化结果。Wang 等人[84]先用 10mL ACN、4g $MgSO_4$ 和 1g NaCl 从鲶鱼样品中粗萃取后，再使用 50mg PSA 后，将漂浮有机液滴分散液液微萃取固化法（DLLME-SFO）与 QuEChERS 的 d-SPE 净化步骤相结合，分析了鲶鱼样品中 13 种有机氯农药。DLLME-SFO 方法以 ACN 为分散剂，十一醇为萃取溶剂，加入 6mL 水促使分析物转移到有机层，从而增大了富集因子，提高了技术的灵敏度。此外，在这种方法中，有机层被固化在冰浴中，从而简化收集，避免损失。该方法回收率在在 88%~121%，相对标准偏差低于 15%，LODs 为 0.65~1.58μg/kg。与此类似的例子也可以参考相关一些文献[88,89]。

关于 QuEChERS 联用分析方法分析农药，LC[41,52,75,78,80,82,83]和 GC[29,31,53,54,76,79,83]与 MS 联用是最常用的技术，也可以与低压 GC[25]和 GC×GC 联用[69]，这些分析器都是最简单的常用的四极杆和三重四极杆。然而，LC 与其他检测器的联用，例如二极管阵列检测器（DAD）[47,48]以及 GC 与电子捕获检测器[73,74,84]或氮磷检测器的耦合[81]以及毛细管电泳（CE）等其他分离技术已被用于多项研究中[90,91]。在 CE 的情况下，该技术还没有像 LC 或 GC 那样广泛应用，尽管已经证明 CE 也可以用于这一目的，但第一批发表的研究在某种程度上被推迟了，这可能是由于最终提取物的电导率不足，阻碍了 CE 系统中进样。

TPP 是由 Anastassiades 等人在他们的第一项工作中（也在更多的情况下[76,79,82,83]）提出并成功应用的，但对目标农药采用同位素标记是一种常见的做法[25,31,70,83,92]。然而，其他如七氯环氧[29]、4-溴-3,5-二甲基苯基-*N*-甲基氨基甲酸酯[93]、4,4-二氯二苯甲酮[94]、重氮酮[95]、乙醇[96]或五氯硝基苯等偶尔被用于测定特定的某些农药。

11.3.2 药物分析

药物是人类和动物用于预防或治疗不同疾病的化合物。然而，这些分析物是一类污染物，主要与动物饲养、兽医治疗和生长促进有关。因此，它们可以出现在动物产品、食用组织和环境基质中，对人类产生不同的副作用[33]，如血液学、胃肠道和神经系统疾病[97]；肌肉震颤；心脏心悸；紧张、头痛或肌肉疼痛[98]等。由于所有这些方面，已经确定了不同的条例来进行管控。例如，欧盟已在 0.12~20000μg/kg 范围内建立了 MRLs，用于检测动物性食品中的药理活性物质[99]。考虑到这些数据，作为 QuEChERS 方法开发便捷方法的必要性就显得尤为重要，尤其是在环境和食品安全领域。

如表 11.2 所示，QuEChERS 在药物分析中的应用不仅集中在特定基团的测定上，

表 11.2　**QuEChERS 方法分析药物的一些实例**

分析物	样品（数量）	提取溶剂	提取盐类	d-SPE 中的吸附剂	分析方法	回收率/%	LODs/（μg/kg）	备注	参考文献
阿托品，东莨菪碱	荞麦，小麦，大豆，荞麦粉，荞麦面条苋属谷物，奇异籽，小米（5g）	10mL 1%甲酸乙腈溶液（体积比）	4g $MgSO_4$，1g NH_4OAc	25mg PSA，25mg GCB	UHPLC-MS/MS	50~92	0.04~0.2	最初加入 10mL 水	[28]
10 种非类固醇抗炎药物	牛乳（5g）	10mL 5%乙酸乙腈溶液（体积比），4mL 0.02mol/L 维生素 C，0.24mol/L 盐酸	1g NH_4OAc，5g Na_2SO_4	1mg $MgSO_4$，150mg C_{18}	HPLC-MS/MS	78~97	CC_α：0.4~1.5，CC_β：0.8~1.9	最初加入 6mL 水。美洛昔康-d_3、硝氟酸-$^{13}C_6$、氟芬桂酸-$^{13}C_6$ 和苯丁酮 $^{13}C_{12}$ 为内标物。采用 QqQ 和 Q-Orbitrap MS 进行比较	[40]
除虫脲	蚌类（10g）	10mL 乙腈	4g $MgSO_4$，1g NaCl，0.5g di-Na，1g tri-Na	900mg $MgSO_4$，150mg PSA，150mg C_{18}	HPLC-DAD	101	30	二氟苯脲在蚌类中半衰期的研究	[23]
3 种离子载体抗菌药物	家禽粪（10g）	10mL 乙腈	4g $MgSO_4$，1.0g NaCl	150mg $MgSO_4$，25mg Florisil	HPLC-MS/MS	70~120	~10	最初加入 10mL 水。对实际样品进行了分析。Nigericin 作内标物	[51]
14 种 β-兴奋剂，2 种 β-阻滞剂	猪肌肉（3g）	10mL 1%乙酸乙腈溶液（体积比）	3g $MgSO_4$，0.5g NaCl	150mg C_{18}	UHPLC-MS	67~121	0.17~1.67	最初加入 2mL 水。对实际样品进行了分析	[98]

22 种磺胺类药物	鸡肉，牛肉和羊肉（5g）	10mL 1%乙酸乙腈溶液（体积比）	4g $MgSO_4$，1g NaCl，0.5g di-Na，1g tri-Na	900mg $MgSO_4$，150mg PSA	LC-MS/MS	—	—	最初加入 2mL 水。磺胺甲恶唑-d_4 作内标物	[97]
90 种兽药	Royal 果酱（1g）	5mL 0.1mol/L 柠檬酸，0.2mol/L 8/5（体积比）Na_2HPO_4，20mL 5%乙酸乙腈溶液（体积比）	2.0g NaCl，2.0g Na_2SO_4	200mg NH_2 吸附剂	UHPLC-MS/MS	70~120	0.07~6	对实际样品进行了分析	[100]
乌洛托品	猪肌肉，猪肝和猪肾（2g）	10mL 乙腈，5mL 己烷	4g Na_2SO_4	50mg PSA	HPLC-MS/MS	87~110	1.5	甲酚胺-$^{13}C_6{}^{15}N_4$ 作内标物	[34]
阿维菌素，伊维菌素，多拉菌素，莫西菌素	牛肝（10g）	10mL 乙腈	4g $MgSO_4$，1g NaCl，0.5g di-Na，1g tri-Na	1g $MgSO_4$，25mg PSA，200mg C_{18}	HPLC-FD	85~90	CC_α：23~127	分析物衍生后通过 FD 分析	[101]
雷托帕明（动物生长促进药）	猪饲料（5g）	10mL 乙腈	4g $MgSO_4$，1g NaCl，0.5g di-Na，1g tri-Na	900mg $MgSO_4$，150mg PSA，150mg C_{18}	HPLC-MS/MS	96~107	1.91	样品是用蛋白酶和 β-葡萄糖醛酸酶水解的。盐酸异辛普利作内标物	[102]
26 种兽药	猪粪（2g）	20mL 甲醇/乙腈/0.1mol/L EDTA Mcilvaine 缓冲溶液 = 12.5/37.5/50（体积比）	4g $MgSO_4$，1g NaCl	400mg PSA，150mg C_{18}	HPLC-MS/MS	61~106	0.01~1.86	d-SPE 步骤与常规 HLB 进行比较	[32]

续表

分析物	样品（数量）	提取溶剂	提取盐类	d-SPE 中的吸附剂	分析方法	回收率/%	LODs/（μg/kg）	备注	参考文献
8 种磺胺类药物	鸡肉和鸡蛋（5g）	10mL 1%乙酸乙腈溶液（体积比）	4g $MgSO_4$，1g NaOAc	肌肉：300mg Z-Sep+，蛋：300m gPSA	HPLC-FD	66~81	4.2~25.5	—	[45]
地西泮，诺地西泮，替马西泮，奥沙西泮	鲤鱼（2g）	10mL 乙腈	2g $MgSO_4$，1g NaCl	100mg PSA	HPLC-MS/MS	89~110	0.5	与 d-SPE 吸附剂 MWCNTs 进行比较	[103]
3 种兽用抗生素	牛乳，蜂蜜（2g）	15mL 1%乙酸乙腈溶液（体积比）	4g $MgSO_4$，1g NaCl	900mg Na_2SO_4，500mg Z-Sep+，500mg C_{18}	CLC-MS/MS	96~100	0.02~0.045	—	[16]
11 种药物	芹菜，莴笋（500mg）	7mL Na_2EDTA 150mg/L/乙腈/甲醇=28.6/46.4/25.0（体积比）	2g Na_2SO_4，0.5g NaCl	225mg Na_2SO_4，12.5mg PSA，12.5mg C_{18}	HPLC-MS/MS	70~119	0.7~8	与 ASE 和 UAE 方法进行比较	[33]

注：ACN—乙腈；ASE—加速溶剂萃取；C_{18}—十八烷基硅烷；CC_α—决定范围；CC_β—检测能力；CLC—毛细管液相色谱；DAD—二极管阵列检测器；di-Na—柠檬酸氢二钠水合物；d-SPE—分散固相萃取；EDTA—乙二胺四乙酸盐；FD—荧光检测器；HLB—亲水亲脂平衡柱；HPLC—高效液相色谱法；IS—内标；LOD—检测限；MeOH—甲醇；MS—质谱；MWCNT—多壁碳纳米管；NaOAc—乙酸钠；NH4OAC—乙酸铵；PSA—N-丙基乙二胺；Q—单四极杆；QqQ—三重四极杆；tri-Na—柠檬酸钠二水合物；UAE—超声辅助提取；UHPLC—超高效液相色谱。

如生物碱[28]、非甾体化合物[40]、酰基尿素衍生物[23]、聚醚羧酸[51]、β-兴奋剂剂和β-阻滞剂[98,102]、磺酰胺[45]、杂环胺[34]、大环内酯[101]、苯二氮卓类[103]或苯丙胺类[46]，而且也可以在多残留分析上[32,33,100]。其中一些药物可用于治疗血糖、低血压、低胆固醇和低血糖调节因子[28]以及用作抗炎[40]、抗寄生虫[23,101]、抗微生物[51]、生长促进剂[98,102]、抗菌[45,46,97]、抗感染[34]或镇静剂药物[103]。

由于用于这一目的的化合物种类繁多，以及所研究基质的不同性质，所以还存在大量应用于药物分析的改良版本 QuEChERS。其中可以发现各种各样的环境样本（如家禽排泄物[51]和猪粪[32]）或者食品样本（如蔬菜[33]，谷物[28]，牛乳[40,46]，蜂蜜和蜂王浆[46,100]，软体动物[23]，鱼[103]，家禽，猪和可食用牛组织[34,45,97,98,101]，动物饲料[102]或鸡蛋[45]）。

在萃取步骤过程中主要调节不同酸的 pH，如乙酸[40,97,98]、甲酸[28,45,46]、抗坏血酸[40]或柠檬酸[100]、缓冲液[23,28,40,45,97,101,102,104]。

净化步骤中 C_{18} 和 PSA 是最常见的吸附剂，二者可以混合使用，也可以单独应用，但其他吸附剂，如 Florisil[51] 或 Z-Sep+[45,46] 已被引入药物研究。此外，还将 d-SPE 步骤与其他常规清洗程序进行了比较[32]。为此，Guo 等人[32]把 40mgPSA 和 20mgC_{18} 作为猪粪提取物的净化吸附剂，用 QuEChERS-HPLC-MS/MS 方法对 26 种兽药进行了分析，并将结果与亲水亲油平衡（HLB）SPE 柱方法进行了比较。此外，HLB 柱的使用需要更多时间，如图 11.4 所示，d-SPE 比传统的 HLB SPE 的回收率更高。而且 d-SPE 稳定性（4.4%~15%）比 SPE 稳定性（3.3%~21.2%）更高。

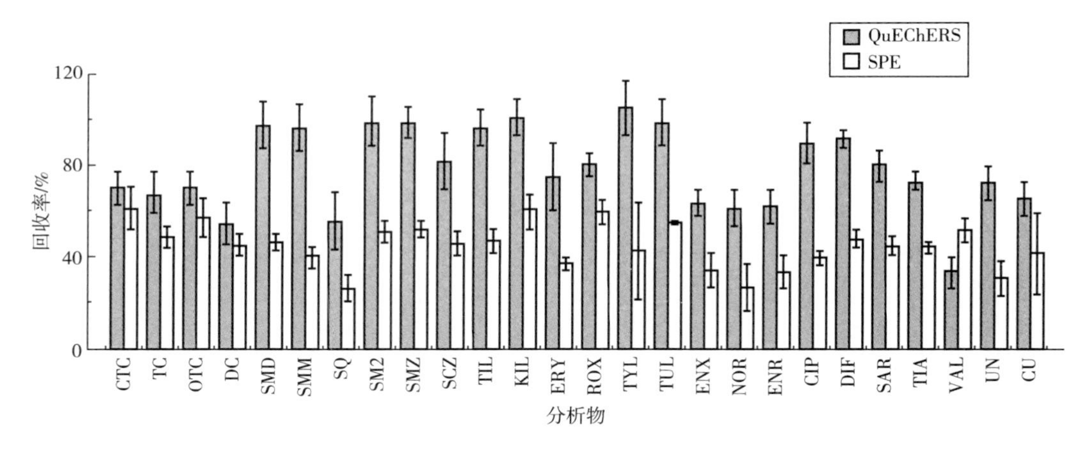

图 11.4 d-SPE 法和传统的 HLB SPE 法对猪粪中 26 种不同兽药的回收率比较（包括磺胺类、大环内酯类和氟喹诺酮类）d-SPE，分散固相萃取；HLB，亲水亲油平衡

[资料来源：C. Guo, M. Wang, H. Xiao, B. Huai, F. Wang, G. Pan, X. Liao, Y. Liu, Development of a modified QuEChERS method for the determination of veterinary antibiotics in swine manure by liquid chromatography tandem mass spectrometry, J. Chromatogr. B 1027（2016）110-118. 经 Elsevier 许可引用。]

除了与其他净化方法比较外，整个方法还与替代方法进行了比较。在这个意义上，Chuang 等人[33]使用 7mL Na_2EDTA、ACN 和 MeOH 混合物以及 2g Na_2SO_4 和 0.5g NaCl

从芹菜和生菜中提取 11 种药物，然后用 225mg Na_2SO_4、12. 5mg PSA 和 12. 5mg C_{18} 进行清洗，然后用 HPLC-MS/MS 分析，再与加速溶剂萃取（ASE）和超声辅助萃取（UAE）进行比较。UAE 中某些分析物的回收率低于 40%，ASE 和 QuEChERS 方法得到了满意结果，回收率均高于 70%。然而，QuEChERS 方法比其他方法更容易，所消耗的溶剂和时间更少，整体回收率在 70%～119%范围内，LODs 为 0. 7～8μg/kg。

在药物分析中 QuEChERS 方法与其他分析技术联用最常见的是 LC-MS/MS，HPLC[32-34,40,51,102,103]或超高效液相色谱（UHPLC）[28,100]。然而，其他技术也能提供满足要求的结果，如 GC（不太常见）[107]、毛细管 LC[46]或不需要提前色谱分离[106]的实时直接分析（DART）。

11. 3. 3 真菌毒素分析

真菌毒素是多种真菌的次生代谢物，主要产生于作物的生长、收获、转化和储存过程中[22]。真菌毒素是重要的污染物，因为很多可能是有毒的，可导致人类和动物产生各种疾病，包括激素紊乱、免疫抑制或致畸、致突变性甚至致癌[108]。

正如以前所论述的[9,10,109]，因为食物是人类接触的最直接途径，在提取真菌毒素方面的应用上 QuEChERS 方法几乎完全集中在对食品基质的分析上，如谷物[37,39,110-114]、牛乳[115]、蔬菜[42]、鸡蛋、水果[67]、饮料或其他可食用植物[117]。然而，如表 11. 3 所示，其中介绍了 QuEChERS 在提取这组分析物方面的应用，同时也对其他几种基质进行了研究，如药用植物[22]、蚯蚓[118]或膳食补充剂[119]。Veprikova 等[119]分析了用于肝病治疗、减少更年期效应和一般健康支持的植物膳食补充剂中 57 种真菌毒素。他们用 10mL 1%（体积比）乙酸、10mL ACN、4g $MgSO_4$ 和 1g NaCl 进行萃取，通过 300mg $MgSO_4$ 和 100mg C_{18} 清洗后，然后用 UHPLC-MS/MS 测定。结果表明，回收率在 40%～122%，LODs 在 1. 5～300mg/kg。

通常在该方法的第一步中添加酸性添加剂和柠檬酸盐[9,23,28,40,97,98,100-102]，但还会添加一些其他盐来提高萃取效率。Michlig 等[115]将 1mL 1mol/L Na_2EDTA 加入到 10mL 0. 1%（体积比）甲酸乙腈溶液中，从 10mL 牛乳中提取黄曲霉毒素 M_1，得到了较好结果。黄曲霉毒素 M_1 与蛋白质结合，Na_2EDTA 可以使酪蛋白胶束断裂。Xing 等[22]将 5mL PBS 和 2. 5mL ACN 混合，通过乙酸 5%（体积比）来调节建立足够的 pH 和水相有机相比，从中草药中提取 21 种具有不同疏水性的真菌毒素，获得良好的回收率（75%～104%）。

在净化步骤中，最常使用的吸附剂是 PSA 和 C_{18}[22,37,39,67,108,110,115,116,119]，但其他的一些吸附剂在一些特定的基质中也取得了满意的效果，例如用于谷物样品的 Al-N[113]用于动物治疗[118]、蔬菜[42]或食品植物样品的 GCB。此外，还有几项研究比较了 d-SPE 和其他净化程序的有效性[110,115]。在进行 GC-MS 测定前，Pereira 等[110]从谷类婴儿食品中提取 12 种真菌毒素，通过用 PSA 作为吸附剂净化和 MultiSep 柱净化和免疫亲和柱（IAC）进行对比。结果表明，IAC 方法因其特异性只能提取 12 种化合物中的 3 种，不能应用于多种真菌毒素的分析。MultiSep 柱中填充吸附剂混合物，专为真菌毒素分析而设计，虽然回收率为 21%～103%较为满意，但该方法相对更复杂、更昂贵和更耗时。

表 11.3 QuEChERS 方法分析霉菌毒素的一些实例

分析物	样品（数量）	提取溶剂	提取盐类	d-SPE 中的吸附剂	分析方法	回收率/%	LODs/（μg/kg）	备注	参考文献
21 种霉菌毒素	中草药（1g）	2.5mL 5%乙酸酸乙腈溶液（体积比），5mL PBS	2g $MgSO_4$，0.5g NaCl，0.5g tri-Na，0.25g di-Na	150mg $MgSO_4$，150mg C_{18}	UHPLC-MS/MS	75~104	0.031~5.4	对不同净化吸附剂进行研究。对实际样品进行了分析	[22]
棒曲菌素	小麦玉米粉饼（10g）	10mL 乙腈	4g $MgSO_4$，1g NaCl	1.2g $MgSO_4$，400mg PSA，400mg C_{18}	HPLC-MS/MS	—	0.2	—	[37]
22 种霉菌毒素	药用蚯蚓（1g）	15mL 15%甲酸乙腈溶液（体积比）	4g $MgSO_4$，1g NaCl	900mg $MgSO_4$，300mg PSA，900mg C_{18}，60mg GCB	UHPLC-MS/MS	73~105	0.05~10	在净化之前，将样品放置在冰浴中 10min。以阿特拉津-d_5 和^{13}C-玉米烯酮为内标物	[118]
黄曲霉毒素 M1	脱脂全牛乳（10mL）	10mL 0.1%甲酸乙腈溶液（体积比），1mL 1Mna_2EDTA	4g $MgSO_4$，1g NaOAc	200mg $MgSO_4$，67mg PSA，180mg C_{18}	UHPLC-MS/MS	70~95	0.002ug/L	与 IAC 净化相比	[115]
7 种霉菌毒素	鸡蛋（2g）	10mL 0.1% 甲酸乙腈溶液（体积比）	4g $MgSO_4$，1g NaCl	100mg PSA，100mg C_{18}	UHPLC-MS/MS	85~115	1~5	最初加入 2mL 水，对实际样品进行了分析	[108]
9 种霉菌毒素	甜菜（6g）	5mL 1%甲酸溶液（体积比），10mL 0.1%甲酸乙腈溶液	4g $MgSO_4$，1g NaCl，1g tri-Na，0.5g di-Na	150mg $MgSO_4$，25mg PSA，7.5mg GCB	HPLC-MS/MS	64~168	—	对实际样品进行了分析	[42]
10 种霉菌毒素	小麦，玉米，水稻（7.5g）	15mL 1% 乙酸乙腈溶液（体积比）	5.5g $MgSO_4$，1g NaCl，1g tri-Na，0.5g di-Na	2g $MgSO_4$，337mg C_{18}	HPLC-MS/MS	87~106	0.062~199	最初加入 10mL 水。对实际样品进行了分析	[39]

续表

分析物	样品（数量）	提取溶剂	提取盐类	d-SPE 中的吸附剂	分析方法	回收率/%	LODs/（μg/kg）	备注	参考文献
14 种霉菌毒素	啤酒（10mL）	5mL 乙腈	4g $MgSO_4$，1g NaCl	900mg $MgSO_4$，300mg C_{18}	GC-MS/MS	70~110	0.05~8μg/L	分析物衍生后，通过 GC-MS/MS 进行分析。对实际样品进行了分析	[116]
3 种霉菌毒素	石榴（2g），石榴汁（2mL）	15mL 1% 乙酸乙腈溶液（体积比），7mL 冷水	4g $MgSO_4$，1g NaCl	600mg $MgSO_4$，200mg PSA	UHPLC-DAD	82~108	15~20	—	[67]
12 种霉菌毒素	谷物婴儿食品（2.5g）	10mL 乙腈	4g $MgSO_4$，1g NaCl	1.350g $MgSO_4$，450mg PSA	GC-MS	44~135	0.37~19.19	最初加入 15mL 的水。采用 α-氯醛糖和 $^{13}C_{15}$-脱氧雪腐烯醇作内标物，与 MultiSep 柱和 IAC 进行比较	[110]
3 种霉菌毒素	草莓植株（10g），玉米植株（5g）	10mL 乙腈	6g $MgSO_4$，1g NaCl，1g tri-Na，0.5g di-Na	855mg $MgSO_4$，150mg PSA，45mg GCB	UHPLC-QTQF-MS	55~106	0.6~0.96	最初加入 10mL 水。罗红霉素 4 作内标物	[117]
57 种霉菌毒素	植源节食补充（1g）	10mL 1% 甲酸溶液（体积比），10mL 乙腈	4g $MgSO_4$，1g NaCl	300mg $MgSO_4$，100mg C_{18}	UHPLC-MS/MS	40~122	1.5~300	—	[119]

注：ACN—乙腈；C_{18}—十八烷基硅烷；DAD—二极管阵列检测器；di-Na— 柠檬酸氢二钠水合物；d-SPE—分散固相萃取；GC—气相色谱；GCB—石墨化炭黑；HPLC—高效液相色谱法；IAC—免疫亲和柱；IS—内标；LOD—检测极限；MS—质谱；NaOAc—醋酸钠；PBS—磷酸盐缓冲盐水；PSA—N-丙基乙二胺；QTOF—四极飞行时间；tri-Na— 柠檬酸钠二水合物；UPHLC—超高效液相色谱。

最常用的分离技术是 GC-MS[110,116] 和 LC-MS/MS[22,37,39,42,108,115,117-119]，对于 GC-MS 来说，在某些情况下分析物需要用双（三甲基硅基）乙酰胺/三甲基氯硅烷/N-三甲基硅咪唑进行衍生化。还有些情况不太常见，如 LC-DAD[67]、荧光检测器的流动注射分析或 terbium-sensitized luminescence[120] 和 DART-Orbitrap-MS[114] 也被应用于 QuEChERS 样品预处理后的真菌毒素分析。

11.3.4 多环芳烃分析

多环芳烃（PAHs）是一组环境有机污染物，其结构中至少含有两个缩合芳香环，并由不完全燃烧过程产生。由于其已被证明的致突变性和致癌活性，必须对其进行监测，并确定其来源以控制其向环境释放。从这个意义上说，尽管这些化合物有数百种，但美国环境保护局（EPA）已经建立了一份含有 16 种多环芳烃的优先污染物清单，其中包括苊、蒽、苯并 [a] 蒽、苯并 [a] 芘、苯并 [b] 氟蒽、苯并 [g、h、i] 芳烃、苯并 [k] 氟蒽、䓛、二苯并 [a、h] 蒽、氟蒽、氟、茚并 [1.2.3-cd] 芘、萘、菲和芘[121]。

表 11.4 概述了 QuEChERS 方法在 PAH 分析中的应用。可以看出，这一领域大多数应用都集中在获得优先清单中[58,123,126-128,130,132,134] 16 种多环芳烃的良好回收率，其他文章只包含了其中一些作为目标分析物[30,38,59,122,124,125,131,133]，有时甚至还取代了这些优先多环芳烃的变体[125,129,131,133]。

除了前面提到的 16 种管控的多环芳烃外，WKnobel 等人还研究了单羟基多环芳烃（OH-PAHs），它们被用作牛乳生产过程中的消毒剂[129]。因此它们的测定是重要的，因为它们可以作为生物标志物来确定牛接触多环芳烃的情况，并防止人类摄入受污染的牛乳。少数文献将 QuEChERS 和 CE 联合起来采用 CE-紫外（UV）来测定其中五种 OH-PAHs[129,136,137]。

在工业化地区，大量的 PAH 污染源使它们成为环境中最广泛的污染物之一，污染物浓度非常高[138,139]。它们的高亲脂性导致脂肪含量高的动物食品被污染的风险更高。因此，在文献中可以发现各种应用 QuEChERS 方法的基质。其中鱼类[38,122,133,135]、海鲜[123,125,126,130,131,134]、肉类[124,126,128,132] 和牛乳等动物性食品居多。然而也对土壤进行了分析[58]，因为这种基质是多环芳烃的主要蓄积源之一，它们往往与土壤中有机质有很强的结合。Cvetkovic 等[58] 已经对原始方法进行了修改，由相同的 LLE 步骤组成，但以硅藻土为 *d*-SPE 吸附剂从土壤样品中提取美国国家环境保护局所列出的 16 种多环芳烃。这种提取方法具有较高的回收率和良好的洗脱能力。对标准样品、空白样品和加标样品进行了分析得到 GC-MS 色谱图如图 11.5 所示。

此外在热处理过程中产生多环芳烃（例如吸烟、烘焙、烧烤、干燥）可能导致食物污染。因此对进行过干燥或焙烧的基质进行了研究，包括茶叶[59]、米粒[127] 或面包[30]，其原料可能受到这些加工方法的影响，同时对烧烤、烟熏或烘烤动物性食物也进行了分析[38,126]。

对于 QuEChERS 方法提取多环芳烃所用的溶剂、盐和吸附剂，应该强调的是对原始方法的改进很少。因此在萃取步骤[38,58,59,122,123,125,129,130,134] 中，几乎所有的文献都使用

表 11.4 QuEChERS 方法分析 PAHs 的一些实例

分析物	样品（数量）	提取溶剂	提取盐类	d-SPE 中的吸附剂	分析方法	回收率/%	LODs/（μg/kg）	备注	参考文献
16 种 PAHs	土壤（10g）	30mL 乙腈：水 2/1（体积比）	8g $MgSO_4$，2g NaCl	150mg $MgSO_4$，50mg 硅藻土	GC-MS/MS	81~110	0.39~1.53	还测试了正己烷和 H_2O 混合物。PSA、C_{18}、Florisil 和沸石也被用作净化吸附剂。丙烯-d_{10} 和二甲苯-d_{12} 作内标物	[58]
苯并［a］芘	面包（5g）	10mL 丙酮	6g $MgSO_4$，1.5g NaCl	1.8g $MgSO_4$，400mg PSA	GC-MS/MS	97~120	0.3	最初加入 5mL 去离子水。蒽-d_{10} 作内标物	[30]
13 种 PAHs	烟熏鱼（5g）	10mL 乙腈	4g $MgSO_4$，1g NaCl	900mg $MgSO_4$，300mg PSA，150mg C_{18}	GC-FID	72~90	1.1~5.5	—	[38]
5 种 PAHs	鲤鱼（2.5g）	10mL 乙腈	2g $MgSO_4$，0.5g NaCl	2g $MgSO_4$，150mg PSA	HPLC-FD	—	—	与 Soxhlet 进行了比较。尝试了一些不同的吸附剂	[122]
16 种 PAHs	野生和商业蚌类（10g）	10mL 乙腈	4g $MgSO_4$，1g NaCl	900mg $MgSO_4$，150mg PSA	GC-MS/MS	89~112	0.01~0.99μg/L	5 种氘化为内标物	[123]
12 种 PAHs	火腿（8g）	10mL 乙酸乙酯	4g $MgSO_4$，1g NaCl	900mg $MgSO_4$，150mg PSA，300mg C_{18}	GC-MS	72~111	0.1~1	蒽-d_{10} 作内标物	[124]
14 种 PAHs	马尼拉蛤，硬蛤，牡蛎，乌蛤（2g）	5mL 乙腈	2g $MgSO_4$，0.5g NaCl	150mg $MgSO_4$，50mg PSA	HPLC-FD	87~116	0.05~0.5270	提取步骤进行了两次	[125]

12 种 PAHs	茶叶（1g）	10mL 乙腈	4g $MgSO_4$，1g NaCl	900mg $MgSO_4$，150mg PSA，150mg SAX	GC-MS	50~120	—	最初加入 10mL 沸水。其中一个氘化作内标物。在 QuEChERS 后进行 LLE	[59]
16 种 PAHs	炭火烤禽肉，红肉，海鲜（5g）	10mL 乙腈	6g $MgSO_4$，1.5g NaOAc	1200mg $MgSO_4$，400mg PSA，400mg C_{18} 硅胶颗粒	GC-MS	71~104	0.1~2μg/L②	最初加入 10mL 水。商业 QuEChERS 套件	[126]
16 种 PAHs	大米（10g）	10mL 1% 乙酸乙腈溶液	6g $MgSO_4$，1.5g NaOAc	150mg $MgSO_4$，50mg PSA	GC-MS	70~122	—	最初加入 10mL 水。六种氘化后作为内标物	[127]
16 种 PAHs	肉（5g）	10mL 乙腈	6g $MgSO_4$，1.5g NaOAc	1200mg $MgSO_4$，400mg PSA，400mg C_{18}	GC-MS	71~104	0.1~2μg/L②	最初加入 10mL 水。商业 QuEChERS 套件	[128]
5 种 OH-PAHs	牛乳（1200μL）	300μL 乙腈	480mg $MgSO_4$，120mg NaCl	30mg PSA，30mg C_{18}	CE-UV	80~105	0.98~3.72	在 QuEChERS 之前先进行水解	[129]
16 种 PAHs	牡蛎（10g）	15mL 乙腈	6g $MgSO_4$，1.5g NaCl	150mg $MgSO_4$，50mg PSA	UHPLC-MS	71~110	13~129	最初加入 7mL 水。商业 QuEChERS 套件	[130]
17 种 PAHs	海胆（1g）	1mL 乙腈	600mg $MgSO_4$，150mg NaOAc	90mg $MgSO_4$，30mg PSA，15mg C_{18}	GC-MS/MS	70~120①	0.7-1.5②	—	[131]
16 种 PAHs	鸡肉和鸭肉（5g）	10mL 乙腈	6g $MgSO_4$，1.5g NaOAc	1200mg $MgSO_4$，400mg PSA，400mg C_{18}	GC-MS	71~104	0.1~2μg/L②	最初加入 10mL 水。商业 QuEChERS 套件	[132]

续表

分析物	样品（数量）	提取溶剂	提取盐类	d-SPE 中的吸附剂	分析方法	回收率/%	LODs/（μg/kg）	备注	参考文献
33 种 PAHs	三文鱼（1g）	2mL 丙酮：乙酸乙酯：异辛烷 2/2/1（体积比）	方法 1：6g $MgSO_4$，1.5g NaOAc 方法 2：4g $MgSO_4$，1g NaOAc，1g tri-Na，0.5g di-Na	150mg $MgSO_4$，50mg PSA，50mg C_{18}	GC-MS	70~120①	2~10	2 种氘化后作内标物。商业 QuEChERS 套件，与 EN 和 AOAC 方法进行比较	[133]
16 种 PAHs	虾（10g）	10mL 乙腈	6g $MgSO_4$，1.5g NaCl	150mg $MgSO_4$，50mg PSA	UFLC-MS/MS	70~120①	20~510	商业 QuEChERS 套件	[134]
16 种 PAHs	鱼（5g）	8mL 乙腈	6g $MgSO_4$，1.5g NaOAc	900mg $MgSO_4$，300mg PSA，150 C_{18}	HPLC-FD	70~120①	0.09~1.4ug/L	商业 QuEChERS 套件	[135]

注：ACN—乙腈；AOAC—分析协会；C_{18}—十八烷基硅烷；CE-UV—毛细管电泳-紫外；di-Na—柠檬酸氢二钠水合物；EN—欧洲规范；EtOAc—乙酸乙酯；FD—荧光检测器；FID—火焰离子检测器；GC—气相色谱；HPLC—高效液相色谱法；IS—内标；LLE—液液萃取；LOD—检测极限；MS—质谱；NaOAc—乙酸钠；OH-PAH—单羟基化多环芳烃；PAH—多环芳烃；PSA—*N*-丙基乙二胺；SAX—强阴离子交换剂；tri-Na—柠檬酸钠二水合物；UFLC—超快速液相色谱。

①对于大多数测定的分析物，回收率在 70%~120%。

②仪器检测限。

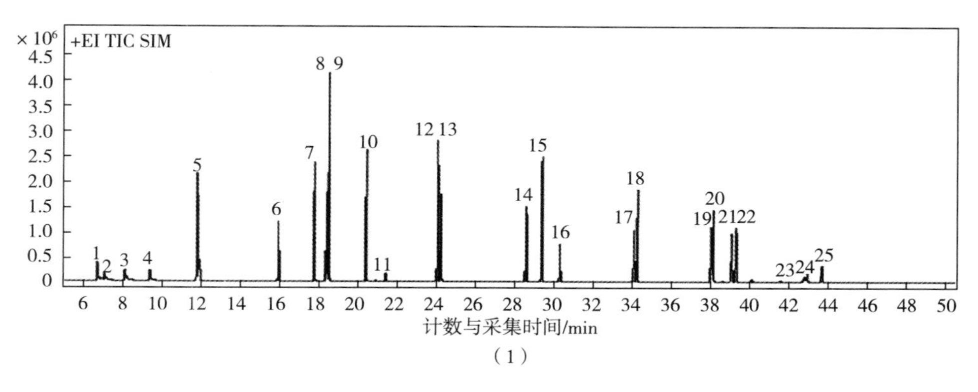

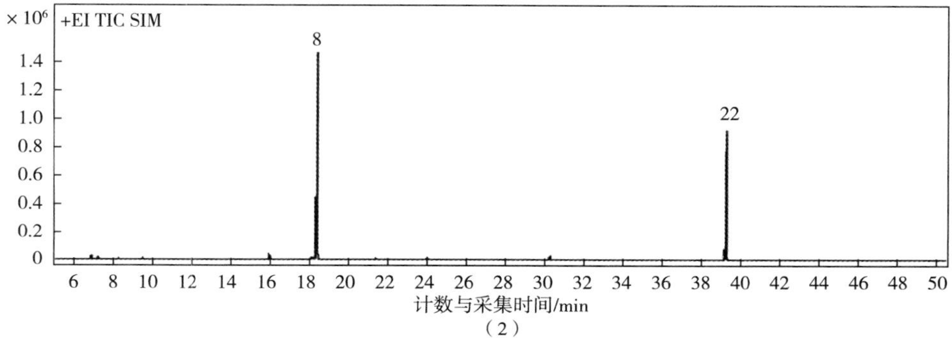

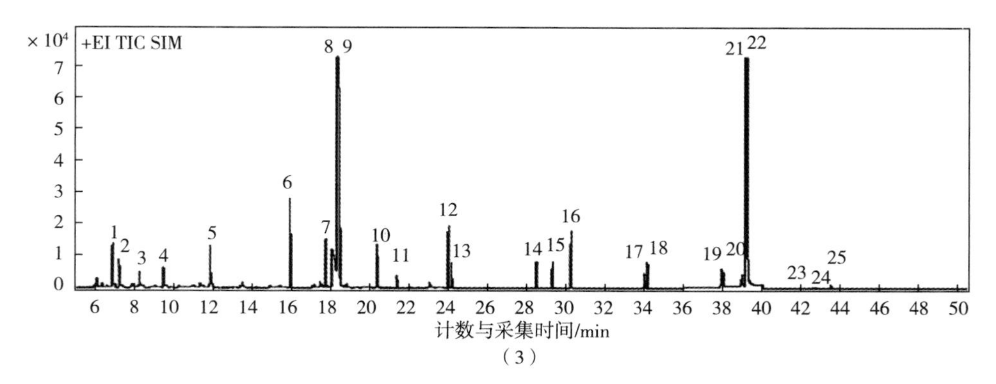

图 11.5 19.23mg/mL 标准溶液的 GC-MS 色谱图（1），空白样品的 GC-MS 色谱图（2），加标样品的 GC-MS 色谱图（3）（ACN：水：硅藻土）

1—苯酚-d_6；2—2-氯苯酚-3，4，5，6-d_4；3—1，2-二氯苯-d_4；4—硝基苯-d_5；5—萘；6—2-氟联苯；7—亚萘；8—亚萘-d_{10}；9—亚萘；10—氟烯；11—2，4，6-三溴苯酚；12—菲；13—蒽；14—氟蒽；15—芘；16—对叔苯基-d_{14}；17—黄烯；18—苯并［a］蒽；19—苯并［b］氟蒽；20—苯并［k］氟蒽；21—苯并［a］芘；22—过烯 d_{12}；23—吲哚［1，2，3-cd］芘；24—二苯并［a］蒽；25—苯并［g，h，i］乙腈，气相色谱，质谱

［资料来源：J. S. Cvetkovic, V. D. Mitic, V. P. S. Jovanovic, M. V. Dimitrijevic, G. M. Petrovic, S. D. Nikolic-Mandic, G. S. Stojanovic, Optimization of the QuEChERS extraction procedure for the determination of polycyclic aromatic hydrocarbons in soil by gas chromatographymass spectrometry, Anal. Methods 8（2016）1711-1720 with permission of Royal Society of Chemistry.］

了原溶剂（ACN）和盐（$MgSO_4$ 和 NaCl）。然而一些文章提出了一些轻微的修改，如使用柠檬酸盐[133]或改变盐的类型如乙酸钠（NaOAc）[126-128,131,132,135]。一些作者还提出了使用丙酮或 EtOAc 等替代溶剂以及 ACN 水或丙酮 EtOAc 异辛烷等溶剂的混合物来提高方法的回收率[30,124,133]。Forsberg 等[133]使用溶剂混合物（丙酮、EtOAc 和异辛烷）与 AOAC 和 CEN 官方方法相结合从鱼中提取 33 种多环芳烃，比官方方法的提取效果更好。作者认为丙酮和 EtOAc 与水的相容性较高，有利于 PAH 从水封基质孔转移到异辛烷层。

在存在 $MgSO_4$ 的情况下，净化步骤中主要使用的吸附剂是 PSA[30,38,59,122-135]，并且经常与 C_{18} 结合以减少最终提取物的脂肪含量。只有少数的文章建议吸附剂使用不同的材料。Cvetkovic 等[58]第一次使用硅藻土（主要由 87%~91%的 SiO_2 组成，其余为 Al_2O_3 和 Fe_2O_3）作为净化吸附剂，而 Sadowska-Rociek 等人[59]把 PSA 与 SAX 结合作为吸附剂，适用于羧酸等化合物的萃取。最后值得注意的是 PAHs 是无极性平面化合物，不能使用 GCB，因极易被该吸附剂高度保留。

除了 QuEChERS 外，GC 和 LC 也被用来测定 PAH，但最广泛使用的分离技术还是 GC-MS 联用[30,58,59,123,124,126-128,131-133]。Knobel 等人[129]通过毛细管区电泳电离来测定 OH-PAHs。

所有基于 QuEChERS 方法和用于多环芳烃分析的方法都得到了满意结果，回收率一般在 70%~120%之间，许多研究在这一过程中都使用了 IS。除了较高回收率外，这些方法提供的 LODs 在几微克/千克或几微克/升范围内。如表 11.4 所示的数据来看，在 PAH 分析中 QuEChERS 方法具有较突出的优越性。

11.3.5 其他

除了前面提到的分析物外，QuEChERS 方法还被用于分析个人护理产品[140,141]、阻燃剂[140,142]、工业化学品[140,143]、多溴二苯醚[144]和多氯联苯[140,145]、紫外线过滤剂[140]、激素[146,147]、海洋毒素[148,149]、农业食品工业[150]（可能的污染物[150,151]和天然物质[152]、植物提取物[60,153]）、硅氧烷[35]、脂类[154]和邻苯二甲酸盐[155]。

从表 11.5 可以看到这些应用的一些示例。值得一提的是，这种方法的稳健性能够满足同时提取不同种类的化合物[24,140,157]（包括前面提到的化合物以及农药、药品、真菌毒素和多环芳烃），并得到了满意的结果。Plassmann 等开发了一种方法来测定 14 个不同种族中的 64 种化合物[140]。在萃取过程中，没有使用缓冲溶液而使用 5mL ACN、2g $MgSO_4$ 和 0.5g NaCl，d-SPE 萃取使用 PSA 和 $MgSO_4$ 的混合物。将该方法与仅使用 LLE 步骤（回收率类似但色谱背景更高）或使用甲酸或柠檬酸盐缓冲的 QuEChERS（彩色混浊提取物）进行了比较。尽管 Quechers 方法具有通用性，但猪和人类血液中的 47 种化合物回收率高于 70%，加热蒸发和 PSA 吸附导致了其余 17 种化合物的损失。

几乎所有类型的液体样品[140,156,158]和固体样品[24,35,144,146,148,151,153,156]中上述分析物都被测定过，主要包括生物样品（血液[140,153,156]、尿液[156]、人体组织、血浆[154]和脑脊液[159]）、工业产品（卫生用品[141]、食品用纸[151]）、环境基质（沉积物和土壤[35,144,146]、植物[35,60]和水[158]）和食品（大豆产品[24]、饮料[160]、肉类[150]、鱼类和海鲜[148]、牛乳及其衍生物[161,162]、蜂蜜和蜂巢产品[143]和水果、蔬菜及其衍生物[155,157]）。

表 11.5 QuEChERS 方法分析其他化合物的一些实例

分析物	样品（数量）	提取溶剂	提取盐类	d-SPE 中的吸附剂	分析方法	回收率/%	LODs	注意事项	参考文献
64 种多残留化合物（10 种潜在引起变态反应的化妆品原料，5 种芳香胺，10 种阻燃剂，5 种工业化学添加剂，4 种杀虫剂，3 种 PAHs，3 种 PCBs，8 种 PFCs，苯酚，2 种增塑剂，4 种紫外过滤剂）	血液（5mL）	5mL 乙腈	2g $MgSO_4$，0.5g NaCl	375mg $MgSO_4$，62.5mg PSA	HPLC-MS/MS，GC-MS	47 种化合物 ≥70	1~128μg/L	加入乙腈后再加入不锈钢球来打破凝块。比较了使用其他盐组合（2g $MgSO_4$ + 0.5g NaCl+5g tri-Na+0.25g di-Na 或 2g $MgSO_4$+0.5g NaCl+0.5g 甲酸钠 + 340mg 甲酸）的 QuEChERS，LLE（类似于 QuEChERS 第一步）。用所选的 QuEChERS 方法获得了更好的结果（在效率和更清洁的提取物方面）	[140]
257 种多残留化合物（农药和霉菌毒素）	大豆异黄酮营养产品：胶囊、片剂和粉末（2g）	10mL 1% 乙酸乙腈溶液（2 倍）	1g $MgSO_4$，0.5g NaOAc	200mg $MgSO_4$，100mg C_{18}	UHPLC-MS/MS	70%的化合物为 70~120	0.5~5μg/kg	最初加入 8mL 水。对 PSA、GCB、C_{18}、Z-Sep+和它们的混合物进行了净化吸附剂的测试。Florisil 也用于清理过程中。对 LLE 进行了测试［用 7.5mL 1%甲酸乙腈溶液萃取（体积比）］。用 LLE 和 Florisil 联用萃取结果最好	[24]
3 种合成大麻素类	尿液（0.10mL）、血液（0.10mL）、脑、心肌、肺、肝、脾、肾、胰腺、脂肪组织（1g）	液体样品：1mL 乙腈 固体样品：9mL 乙腈	—	150mg $MgSO_4$，25mg PSA，25mg C_{18}	UHPLC-MS/MS	85~109	0.1μg/L	QuEChERS 提取物经过脂质捕获器小柱过滤	[156]

续表

分析物	样品（数量）	提取溶剂	提取盐类	d-SPE 中的吸附剂	分析方法	回收率/%	LODs	注意事项	参考文献
6 种 PBDEs	沉积物（1g）	5mL 己烷：二氯甲烷 1/1（体积比）（3 倍）	—	1.5g $MgSO_4$，300mg PSA，50mg C_{18}	GC-MS/MS	86~113	0.03~0.05μg/kg	超声辅助提取。C_{18}、PSA 和 GCB 为吸附剂。对比 QuEChERS 和 PLE。PLE 的 LODs 较低	[144]
6 种类固醇激素	沉积物（2.5g）	10mL 乙腈：异丙醇 90/10（体积比）	6g $MgSO_4$，1.5g NaOAc	900mg $MgSO_4$，150mgPSA	LC-MS/MS	63~123	0.03~0.2 μg/kg	最初加入 7.5mL 水。对比醋酸盐缓冲液和柠檬酸盐缓冲液（1g 柠檬酸盐，4g $MgSO_4$，1g NaCl，0.5g 二钠）。醋酸缓冲液的结果较好。以 PSA、C_{18} 和 GCB 为吸附剂	[146]
12 种合成麝香	个人洗护用品：沐浴露，洗发水，肥皂，剃须产品，牙膏，除臭剂，止汗剂，润肤露和香水（500mg）	3mL 乙腈	2.4g $MgSO_4$，750mg NaOAc	180mg $MgSO_4$，60mg PSA，30mg C_{18}	GC-MS	50~112	0.01~500μg/kg	超声辅助提取	[141]
13 种致瘫贝类毒素	蚌类（1g）	1mL 0.1%甲酸水溶液（2 倍）	—	10mg ABS Elut-NEXUS 吸附剂（具有非极性保留机制的聚合物吸附剂）	HPLC-MS/MS	83~113	3~708 μg/kg	d-SPE 前，在-20℃处加入甲醇和冷冻来消除蛋白质	[148]
5 种亚硝胺	烤培根（5g）	10mL 乙腈：水 1/1（体积比）	4g 甲酸铵盐	300mg $MgSO_4$，100mg PSA，100mg C_{18}，100mg Z-Sep	GC-MS/MS	70~120	0.1 μg/kg	在 d-SPE 步骤中加入 2mL 饱和正己烷乙腈溶液去除脂肪	[150]

2 种全氟化合物	蜂蜜（5g）	10mL 0.15% 甲酸乙腈溶液	4g $MgSO_4$，1g NaCl	900mg $MgSO_4$，150mg 苯乙烯基二乙烯苯	UHPLC-MS/MS	82~86	0.016~0.040 μg/kg	初加入 5mL 水。以 PSA、SAX、氨基丙基、C_{18} 和 Florisil 为吸附剂	[143]
6 种多酚类	人体血液细胞（2mL）	5mL 1% 乙酸乙腈溶液	4g $MgSO_4$，1g NaCl，	600mg $MgSO_4$，100mg PSA	HPLC-MS/MS	1~64	0.12~48.40 μg/L*	样品在进行缓冲保护，提取前水解共轭分析物。对 18 种不同的蛋白质沉淀、LLE 和 SPE 相结合的方法进行了试验	[153]
2 种甜菜红碱	红甜菜根（2g）	10mL 甲醇：水 90/10（体积比）	4g $MgSO_4$，1g NaCl，0.01g 二钠，1g 三钠	900mg $MgSO_4$，150mg SAX	HPLC-MS	—	1.063~1.166 mg/kg	以 PSA、C_{18}、氨基吡啶、Florisil、硅胶和苯乙烯基二乙烯苯为吸附剂	[60]
9 种硅氧烷	松针，土壤（2.5g）	10mL 二氯甲烷：己烷 1/1（体积比）	6g $MgSO_4$，1.5g NaOAc	900mg $MgSO_4$，300mg PSA 150mg C_{18}	GC-MS	56~63	1.845~19.853ng/kg	—	[35]

注：ACN—乙腈；C_{18}—十八烷基硅烷；DCM—二氯甲烷；di-Na—柠檬酸氢二钠水合物；d-SPE—分散固相萃取；GC—气相色谱；GCB—石墨化炭黑；HPLC—高效液相色谱法；LLE—液液萃取；LOD—检测限；MeOH—甲醇；MS—质谱；NaOAc—醋酸钠；PAH—多环芳烃；PBDE—多溴联苯醚；PCB—多氯联苯；PFC—全氟化合物；PLE—加压液体萃取；PSA—N-丙基乙二胺；QAC—季铵化合物；SAX—强阴离子交换剂；SVHC—高度关注的物质；tri-Na— 柠檬酸钠二水合物；UHPLC—超高效液相色谱；UV—紫外线。

*：检测限。

绝大多数文献都使用 ACN 作为萃取溶剂，添加不同比例的 NaCl 和 $MgSO_4$。然而，某些萃取过程需要特定的溶剂（如 MeOH[60]）或它们的混合物（如正己烷：DCM1/1（体积比）[35,144]，ACN：异丙醇 90/10（体积比）[146]，$CHCl_3$：MeOH 2/1（体积比）[154]）。这一步骤经常在缓冲介质中进行。因此在这种应用中可以通过添加乙酸盐[24,141,146]或柠檬酸盐缓冲液[60]和甲酸[143,148]或乙酸[24,153]来控制 pH。

在 d-SPE 过程中首选的吸附剂是 PSA，其次是 C_{18}。然而在净化步骤中还使用了其他吸附剂，包括经典吸附剂 GCB[24,144,146]、硅胶[60]、3-氨丙基三甲氧基硅烷[60,143]、SAX[60] 和 Florisil[60,143] 或替代材料 Z-SEP/Z-SEP$^+$[24]、苯乙烯-二乙烯基苯和聚合物吸附剂[148]。通过比较每种特定材料或及其混合物的结果后（效率、回收率或清洁方面），再选择合适的吸附剂或其组合。

一般来说，QuEChERS 方法能够提供高回收率的干净提取物。然而在 d-SPE 步骤前后，根据分析样品和化合物的性质，增加一个步骤来去除蛋白质或脂肪[148,150,156]。同样不仅仅需要去除蛋白质和脂肪，而且还需要去除其他基质干扰，所以一些文献增加了额外的提取步骤（通常是 DLLME 或 SPE）。Faraji 等[163]通过 QuEChERS 方法（10mL ACN、4g 无水 $MgSO_4$ 和 1g NaCl 萃取 10g 样品，50mg PSA 和 300mg $MgSO_4$ 进行洗脱，酱汁样品需要再通过 25mg C_{18} 处理）进行萃取后，选用离子液体 1-己基-3-甲基咪唑六氟磷酸酯（[HMIm][PF6]）作为萃取剂，再通过 DLLME 进行处理。使用 50μL [HMIm] [PF6]（以水为分散溶剂），可在 0.1μg/L 的条件下检测罐装食品（扁豆、番茄酱中的平托豆和罐装意大利面）中的双酚 A，回收率高于 90%。

在这些情况下最常用的分离技术和测定方法主要是 LC（HPLC[60,140,146,148,153,156] 或 UHPLC[24,143] 模式）和 GC[35,140,141,144,150] 与 MS[35,60,140,141] 或 MS/MS[24,140,143,144,146,148,150,153,156] 联用。然而较小程度上也需要使用 CE[136,137]。

值得注意的是，一些文献[137,140,154,157,164]将优化的 QuEChERS 方法与其他成熟的提取技术［如 LLE、Folch 方法（用于脂类提取）、未改进 QuEChERS（避免 d-SPE 步骤）、微波辅助溶剂萃取、经典 SPE 和 PLE（与凝胶渗透色谱或 SPE 耦合）］的萃取效率进行了比较。通常 QuEChERS 在效率、速度、溶剂和试剂消耗以及经济成本方面更显优势。然而也可以找到 QuEChERS 不是首选方法的一些例子[137,157]，因为通过替代方法获得了更好结果。在这方面，Fan 等[157]通过萃取茶叶中 201 种农药和化学污染物比较了三种不同的方法：

（1）双酸性 ACN 提取后用经典 SPE 小柱（由三种材料制成，采用不同的机理，并与着色剂、有机酸、碱、多酚以及非极性干扰物质相互作用）进行洗脱；

（2）以 GCB 和 PSA 为 d-SPE 吸附剂醋酸盐为缓冲溶液的 QuEChERS 方法；

（3）用 NaCl 和 ACN 双萃取，用 $MgSO_4$ 除水，用 SPE 进行清洗。结果表明第三种方法没有作用，只有不到 4%的分析物的回收率较高。另外两种方法在回收率方面表现出相似的效率，但经典方法在去除色素方面表现出较好的效果。

11.4 结论与展望

在 GAC 原则下，分析化学的当前趋势是向着高通量多残留方法的方向发展。从这个角度来说，方法应该易于处理，成本低，所需要的样品、溶剂体积和试剂最小。此外，他们为广泛的分析物提供了一个高选择性，不需要复杂的洗脱方法。

在这种观点下，QuEChERS 方法作为一种替代的样品制备主流方法来测定水果和蔬菜中的农药残留。然而其固有的优势，已超出其最初的适用范围。事实上，QuEChERS 在使用不同溶剂、缓冲盐（以增加 pH 敏感分析物的回收率）和吸附剂的实验过程中迅速发展起来了，从而增加了分析物的数量和类型以及所研究基质的复杂性。

目前该技术适合从各种基质中提取极性和碱性化合物，特别是水果、蔬菜和其他食品。此外，除了农产品中的农药外，QuEChERS 还是从食品和饮料、生物样品和环境基质中提取不同有机污染物的有效方法，如真菌毒素、挥发性有机化合物、多环芳烃和药物化合物。尽管取得了这些积极的研究成果，但某些应用尚未得到充分的开发和研究。

虽然 QuEChERS 方法大多数都是面向污染物分析的，但有一些例子表明它也可以用于测定天然化合物。由于这类物质的潜在利益和 QuEChERS 的处理效率，未来这种方法在这一领域得到更广泛的应用就不足为奇了。

如正在开发的大多数分析方法一样，待分析的样品数量在急剧增加的情况下，QuEChERS 方法另一个重要的未来展望可以朝着自动化方向发展，而 QuEChERS 基本上是多步骤的手动程序。目前自动化主要面向预称盐、缓冲液和吸附剂的商业可用性，进而减少这一步骤所需的时间。实际上有许多由一些公司提供的不同组合。然而其他步骤或完整步骤的自动化还不容易实现。特别是已经开发了不同的方法，如使用一次性吸管提取[165,166]（可以实现全自动化的 d-SPE 技术替代传统的离心技术）、带有自动采样器的微型 SPE 小柱[167]用于实现不同步骤的自动化、在线 SPE 小柱[168]或全自动化设备[169-172]，这些设备主要用于进行液体分配或移液、涡旋混合、样品瓶振荡、打开或关闭样品瓶、添加固体试剂、识别液体水平、倒液、离心、基质加标和 d-SPE 净化。在接下来的几年里，将开发关于这一课题相关的新方法，在不同的实验室这些设备将作为常规技术而存在（尤其是对人员不足的实验室）。

致谢

B. S. R 和 J. G. S 感谢加那利群岛政府经济，工业，贸易和知识局提供的 FPI 奖学金（由欧洲社会基金提供 85%的共同资助）。

参考文献

[1] US Environmental Protection Agency, Basics of Green Chemistry. https://www.epa.gov/greenchemistry/basics-green-chemistry;

[1a] S. Armenta, F. A. Esteve-Turrillas, S. Garrigues, M. de la Guardia, Green analytical chemistry: the role of green extraction techniques, in: E. Ibanez, A. Cifuentes (Eds.), Green Extraction Techniques: Principles, Advances and Applications, vol. 76, 2017, pp. 1-25.

[2] R. L. Pérez, G. M. Escandar, Experimental and chemometric strategies for the development of green analytical chemistry (GAC) spectroscopic methods for the determination of organic pollutants in natural waters, Sustain. Chem. Pharm. 4 (2016) 1-12.

[3] A. B. Eldin, O. A. Ismaiel, W. E. Hassan, A. A. Shalaby, Green analytical chemistry: opportunities for pharmaceutical quality control, J. Anal. Chem. 71 (2016) 861-871.

[4] M. Koel, Do we need green analytical chemistry? Green Chem. 18 (2016) 923-931.

[5] A. Gałuszka, P. Konieczka, Z. M. Migaszewski, J. Namieśnik, Analytical eco-scale for assessing the greenness of analytical procedures, TrAC-Trend. Anal. Chem. 37 (2012) 61-72.

[6] S. Armenta, S. Garrigues, M. de la Guardia, The role of green extraction techniques in green analytical chemistry, TrAC-Trend. Anal. Chem. 71 (2015) 2-8.

[7] M. Anastassiades, S. J. Lehotay, D. štajnbaher, F. J. Schenck, Fast and easy multiresidue method employing acetonitrile extraction/partitioning and "dispersive solid-phase extraction" for the determination of pesticide residues in produce, J. AOAC Int. 86 (2003) 412-431.

[8] M. L. Schmidt, N. H. Snow, Making the case for QuEChERS-gas chromatography of drugs, TrAC-Trend. Anal. Chem. 75 (2016) 49-56.

[9] M. Á. González-Curbelo, B. Socas-Rodríguez, A. V. Herrera-Herrera, J. González-Sálamo, J. Hernández-Borges, M. Á. Rodríguez-Delgado, Evolution and applications of the QuEChERS method, TrAC-Trend. Anal. Chem. 71 (2015) 169-185.

[10] T. Rejczak, T. Tuzimski, A review of recent developments and trends in the QuEChERS sample preparation approach, Open Chem. 13 (2015) 980-1010.

[11] M. C. Bruzzoniti, L. Checchini, R. M. De Carlo, S. Orlandini, L. R. M. Del Bubba, QuEChERS sample preparation for the determination of pesticides and other organic residues in environmental matrices: a critical review, Anal. Bioanal. Chem. 406 (2014) 4089-4116.

[12] S. J. Lehotay, QuEChERS sample preparation approach for mass spectrometric analysis of pesticide residues in foods, Methods Mol. Biol. 747 (2011) 65-91.

[13] G. Martínez-Domínguez, P. Plaza-Bolaños, R. Romero-González, A. Garrido-Frenich, Analytical approaches for the determination of pesticide residues in nutraceutical products and related matrices by chromatographic techniques coupled to mass spectrometry, Talanta 118 (2014) 277-291.

[14] European Committee for Standardization (CEN) Standard Method EN 15662, Food of plant origin - determination of pesticide residues using GC-MS and/or LC-MS/MSfollowing acetonitrile extraction/ partitioning and clean-up by dispersive SPE -QuEChERS method.

[15] AOAC Official Method 2007.01, Pesticide residues in foods by acetonitrile extraction and partitioning with magnesium sulfate, AOAC Int., Gaithersburg, USA, 2007.

[16] P. A. Mills, J. H. Onley, R. A. Guither, Rapid method for chlorinated pesticide residues in non fatty food, J. Assoc. Off. Anal. Chem. 46 (1963) 186-191.

[17] M. Luke, J. E. Froberg, H. T. Masumoto, Extraction and cleanup of organochlorine, organophosphate, organonitrogen, and hydrocarbon pesticides in produce for determination by gas-liquid chromatography, J. Assoc. Off. Anal. Chem. 58 (1975) 1020-1026.

[18] W. Specht, S. Pelz, W. Gilsbach, Gas-chromatographic determination of pesticide residues after clean-up by gel-permeation chromatography and mini-silica gel-column chromatography, Fresenius J. Anal. Chem. 353 (1995) 183-190.

[19] S. J. Lehotay, A. R. Lightfield, J. A. Harman-Fetcho, D. A. Donoghue, Analysis of pesticide

residues in eggs by direct sample introduction/gas chromatography/tandem mass spectrometry, J. Agric. Food Chem. 49 (2001) 4589-4596.

[20] M. Anastassiades, E. Scherbaum, B. Tasdelen, D. Stajnbaher, Recent developments in QuEChERS methodology for pesticide multiresidue analysis, in: H. Ohkawa, H. Miyagawa, P. W. Lee (Eds.), Pesticide Chemistry. Crop Protection, Public Health, Environmental Safety, Wiley-VCH Verlag GmbH & Co. KGaA, Weinheim, 2007.

[21] K. Mastovska, K. J. Dorweiler, S. J. Lehotay, J. S. Wegscheid, K. A. Szpylka, Pesticide multiresidue analysis in cereal grains using modified QuEChERS method combined with automated direct sample introduction GC-TOF-MS and UPLC-MS/MS techniques, J. Agric. Food Chem. 58 (2010) 5959-5972.

[22] Y. Xing, W. Meng, W. Sun, D. Li, Z. Yu, L. Tong, Y. Zhao, Simultaneous qualitative and quantitative analysis of 21 mycotoxinsin Radix Paeoniae Alba by ultra-high performance liquid chromatography quadrupole linear ion trap mass spectrometry and QuEChERS for sample preparation, J. Chromatogr. B 1031 (2016) 202-213.

[23] L. Norambuena-Subiabre, M. P. González, S. Contreras-Lynch, Uptake and depletion curve of diflubenzuron in marine mussels (*Mytilus chilensis*) under controlled conditions, Aquaculture 460 (2016) 69-74.

[24] G. Martínez-Domínguez, R. Romero-González, F. J. Arrebola, A. Garrido-Frenich, Multi-class determination of pesticides and mycotoxins in isoflavones supplements obtained from soy by liquid chromatography coupled to orbitrap high resolution mass spectrometry, Food Control 59 (2016) 218-224.

[25] M. Á. González-Curbelo, S. J. Lehotay, J. Hernández-Borges, M. Á. Rodríguez-Delgado, Use of ammonium formate in QuEChERS for high-throughput analysis of pesticides in food by fast, low-pressure gas chromatography and liquid chromatography tandem mass spectrometry, J. Chromatogr. A 1358 (2014) 75-84.

[26] L. Han, Y. Sapozhnikova, S. J. Lehotay, Streamlined sample clean-up using combined dispersive solid-phase extraction and in-vialfiltration for analysis of pesticides and environmental pollutants in shrimp, Anal. Chim. Acta 827 (2014) 40-46.

[27] M. Anastassiades, Crl-srm 1st Joint CRL Workshop, 2006. Stuttgart, http://www.eurl-pesticides.eu/library/docs/srm/1stws2006_ lecture_ anastassiades_ quechers.pdf.

[28] H. Chen, J. Marín-Sáez, R. Romero-González, A. Garrido Frenich, Simultaneous determination of atropine and scopolamine in buckwheat and related products using modified QuEChERS and liquid chromatography tandem mass spectrometry, Food Chem. 218 (2017) 173-180.

[29] Y. Yu, X. Liu, Z. He, L. Wang, M. Luo, Y. Peng, Q. Zhou, Development of a multiresidue method for 58 pesticides in soil using QuEChERS and gas chromatographytandem mass spectrometry, Anal. Methods 8 (2016) 2463-2470.

[30] S. Eslamizad, F. Kobarfard, K. Javidnia, R. Sadeghi, M. Bayat, S. Shahanipour, N. Khalighian, H. Yazdanpanah, Determination of benzo [a] pyrene in traditional, industrial and semiindustrial breads using a modified QuEChERS extraction, dispersive SPE and GC-MS and estimation of its dietary intake, Iran, J. Pharm. Res. 15 (2016) 165-174.

[31] X.-Q. Liua, Y.-F. Li, W.-T. Meng, D.-X. Li, H. Sun, L. Tong, G.-X. Suna, Amulti-residue method for simultaneous determination of 74 pesticides in Chinese material medica using modified QuEChERS ample preparation procedure and gas chromatography tandem mass spectrometry, J. Chromatogr. B

1015–1016 (2016) 1–12.

[32] C. Guo, M. Wang, H. Xiao, B. Huai, F. Wang, G. Pan, X. Liao, Y. Liu, Development of a modified QuEChERS method for the determinationof veterinary antibiotics in swine manure by liquid chromatography tandem mass spectrometry, J. Chromatogr. B 1027 (2016) 110–118.

[33] Y.-H. Chuang, Y. Zhang, W. Zhang, S. A. Boyd, H. Li, Comparison of accelerated solvent extraction and quick, easy, cheap, effective, rugged and safe method for extraction and determination of pharmaceuticals in vegetables, J. Chromatogr. A 1404 (2015) 1–9.

[34] X. Xu, X. Zhang, E. Duhoranimana, Y. Zhang, P. Shu, Determination of methenamine residues in edible animal tissues by HPLC–MS/MS using a modified QuEChERS method: validation and pilot survey in actual samples, Food Control 61 (2016) 99–104.

[35] S. Ramos, J. A. Silva, V. Homem, A. Cincinelli, L. Santos, A. Alves, N. Ratola, Solvent-saving approaches for the extraction of siloxanes from pine needles, soils and passive air samplers, Anal. Methods 8 (2016) 5378–5387.

[36] S. S. Herrmann, M. E. Poulsen, Clean-up of cereal extracts for gas chromatographytandem quadrupole mass spectrometry pesticide residues analysis using primary secondary amine and C_{18}, J. Chromatogr. A 1423 (2015) 47–53.

[37] F. Saladino, L. Manyes, F. B. Luciano, J. Mañes, M. Fernandez-Franzon, G. Meca, Bioactive compounds from mustard flours for the control of patulin production in wheat tortillas, LWT-Food Sci. Tech. 66 (2016) 101–107.

[38] M. T. Ahmed, F. Malhat, N. Loutfy, Residue levels, profiles, emission source and daily intake of polycyclic aromatic hydrocarbons based on smoked fish consumption, an Egyptian pilot study, Polycycl. Aromat. Comp. 36 (2016) 183–196.

[39] P. J. Fernandes, N. Barros, J. L. Santo, J. S. Câmara, High-throughput analytical strategy based on modified QuEChERS extraction and dispersive solid-phase extraction cleanup followed by liquid chromatography-triple-quadrupole tandem mass spectrometry for quantification of multiclass mycotoxins in cereals, Food Anal. Method 8 (2015) 841–856.

[40] A. Rúbies, L. Guo, F. Centrich, M. Granados, Analysis of non-steroidal anti-inflammatory drugs in milk using QuEChERS and liquid chromatography coupled to mass spectrometry: triple quadrupole versus Q-orbitrap mass analyzers, Anal. Bioanal. Chem. 408 (2016) 5769–5778.

[41] N. Pang, T. Wang, J. Hu, Method validation and dissipation kinetics of four herbicides in maize and soil using QuEChERS sample preparation and liquid chromatography tandem mass spectrometry, Food Chem. 190 (2016) 793–800.

[42] H. Boudra, B. Rouillé, B. Lyan, D. P. Morgavi, Presence of mycotoxins in sugar beet pulp silage collected in France, Anim. Feed Sci. Tech. 205 (2015) 131–135.

[43] L. Li, W. Li, D. M. Qin, S. R. Jiang, F. M. Liu, Application of graphitized carbon black to the QuEChERS method for pesticide multiresidue analysis in spinach, J. AOAC Int. 92 (2009) 538–547.

[44] S. J. Lehotay, M. Anastassiades, R. E. Majors, The QuEChERS revolution, LCGC Europe 23 (2010) 418–429.

[45] J. F. Huertas-Pérez, N. Arroyo-Manzanares, L. Havlíková, L. Gámiz-Gracia, P. Solich, A. M. García-Campaña, Method optimization and validation for the determination of eight sulfonamides in chicken muscle and eggs by modified QuEChERS and liquid chromatography with fluorescence detection, J. Pharmaceut. Biomed. 124 (2016) 261–266.

[46] H. -Y. Liu, S. -L. Lin, M. -R. Fuh, Determination of chloramphenicol, thiamphenicol and florfenicol in milk and honey using modified QuEChERS extraction coupled with polymeric monolith-based capillary liquid chromatography tandem mass spectrometry, Talanta 150 (2016) 233-239.

[47] T. Tuzimski, T. Rejczak, Application of HPLC-DAD after SPE/QuEChERS with ZrO_2-based sorbent in d-SPE clean-up step for pesticide analysis in edible oils, Food Chem. 190 (2016) 71-79.

[48] T. Rejczak, T. Tuzimski, QuEChERS-based extraction with dispersive solid phase extraction clean-up using PSA and ZrO_2-based sorbents for determination of pesticides in bovine milk samples by HPLC-DAD, Food Chem. 217 (2017) 225-233.

[49] S. Walorczyk, D. Drożdżyński, R. Kierzek, Two-step dispersive-solid phase extraction strategy for pesticide multiresidue analysis in a chlorophyll-containing matrix by gas chromatography-tandem mass spectrometry, J. Chromatogr. A 1412 (2015) 22-32.

[50] P. Kaczyński, B. Łozowicka, M. Jankowska, I. Hrynko, Rapid determination of acid herbicides in soil by liquid chromatography with tandem mass spectrometric detection based on dispersive solid phase extraction, Talanta 152 (2016) 127-136.

[51] J. S. Munaretto, L. Yonkos, D. S. Aga, Transformation of ionophore antimicrobials in poultry litter during pilot-scale composting, Environ. Pollut. 212 (2016) 392-400.

[52] P. A. Souza Tette, F. A. da Silva Oliveira, E. N. Corrêa Pereira, G. Silva, M. B. de AbreuGlória, C. Fernandes, Multiclass method for pesticides quantification in honey by means of modified QuEChERS and UHPLC-MS/MS, Food Chem. 211 (2016) 130-139.

[53] F. Volpatto, A. D. Wastowski, G. Bernardi, O. D. Prestes, R. Zanella, M. B. Adaime, Evaluation of QuEChERS sample preparation and gas chromatography coupled to mass spectrometry for the determination of pesticide residues in grapes, J. Braz. Chem. Soc. 27 (2016) 1533-1540.

[54] Y. Han, L. Song, N. Zou, R. Chen, Y. Qin, C. Pan, Multi-residue determination of 171 pesticides in cowpea using modified QuEChERS method with multi-walled carbon nanotubes as reversed-dispersive solid-phase extraction materials, J. Chromatogr. B 1031 (2016) 99-108.

[55] Y. Chen, S. Cao, L. Zhang, C. Xi, Z. Chen, Modified QuEChERS combination with magnetic solid-phase extraction for the determination of 16 preservatives by gas chromatography-mass spectrometry, Food Anal. Method. (2016), http://dx.doi.org/10.1007/s12161-016-0616-1.

[56] Y. -F. Li, L. -Q. Qiao, F. -W. Li, Y. Ding, Z. -J. Yang, M. -L. Wang, Determination of multiple pesticides in fruits and vegetables using a modified quick, easy, cheap, effective, rugged and safe method with magnetic nanoparticles and gas chromatography tandem mass spectrometry, J. Chromatogr. A 1361 (2014) 77-87.

[57] H. -B. Zheng, Q. Zhao, J. -Z. Mo, Y. -Q. Huang, Y. -B. Luo, Q. -W. Yu, Y. -Q. Feng, Quick, easy, cheap, effective, rugged and safe method with magnetic graphitized carbon black and primary secondary amine as adsorbent and its application in pesticide residue analysis, J. Chromatogr. A 1300 (2013) 127-133.

[58] J. S. Cvetkovic, V. D. Mitic, V. P. S. Jovanovic, M. V. Dimitrijevic, G. M. Petrovic, S. D. Nikolic-Mandic, G. S. Stojanovic, Optimization of the QuEChERS extraction procedure for the determination of polycyclic aromatic hydrocarbons in soil by gas chromatography-mass spectrometry, Anal. Methods 8 (2016) 1711-1720.

[59] A. Sadowska-Rociek, M. Surma, E. Cieślik, Comparison of different modifications on QuEChERS sample preparation method for PAHs determination in black, green, red and white tea, Environ. Sci. Pollut.

Res. 21 (2014) 1326-1338.

[60] T. Sawicki, M. Surma, H. Zieliński, W. Wiczckowki, Development of a new analytical method for the determination of red beetroot betalains using dispersive solid-phase extraction, J. Sep. Sci. 39 (2016) 2986-2994.

[61] B. D. Morris, R. B. Schriner, Development of an automated column solid-phase extraction cleanup of QuEChERS extracts, using a zirconia-based sorbent, for pesticide residue analyses by LC-MS/MS, J. Agric. Food Chem. 63 (2015) 5107-5119.

[62] F. J. Schenk, A. N. Brown, L. V. Podhorniak, A rapid multiresidue method for determination of pesticides in fruits and vegetables by using acetonitrile extraction/partitioning and solid-phase extraction column cleanup, J. AOAC Int. 91 (2008) 422-438.

[63] R. E. Hunter Jr., A. M. Riederer, P. B. Ryan, Method for the determination of organophosphorus and pyrethroid pesticides in food via gas chromatography with electroncapture detection, J. Agric. Food Chem. 58 (2010) 1396-1402.

[64] A. Lozano, Ł. Rajski, N. Belmonte-Valles, A. Uclés, S. Uclés, M. Mezcua, A. R. Fernández-Alba, Pesticide analysis in teas and chamomile by liquid chromatography and gas chromatography tandem mass spectrometry using a modified QuEChERS method: validation and pilot survey in real samples, J. Chromatogr. A 1268 (2012) 109-122.

[65] T. D. Nguyen, B. S. Lee, B. R. Lee, D. M. Lee, G. Lee, A multiresidue method for the determination of 109 pesticides in rice using the quick easy cheap effective rugged and safe (QuEChERS) sample preparation method and gas chromatography/mass spectrometry with temperature control and vacuum concentration, Rapid Commun. Mass Spectrom. 21 (2007) 3115-3122.

[66] L. Geis-Asteggiante, S. J. Lehotay, H. Heinze, Effects of temperature and purity of magnesium sulfate during extraction of pesticide residues using the QuEChERS method, J. AOAC Int. 95 (2015) 1311-1318.

[67] C. K. Myresiotis, S. Testempasis, Z. Vryzas, G. S. Karaoglanidis, E. Papadopoulou-Mourkidou, Determination of mycotoxins in pomegranate fruits and juices using a QuEChERS-based method, Food Chem. 182 (2015) 81-88.

[68] P. Parrilla-Vázquez, A. Lozano, S. Uclés, M. M. Gómez-Ramos, A. R. FernándezAlba, A sensitive and efficient method for routine pesticide multiresidue analysis in bee pollen samples using gas and liquid chromatography coupled to tandem mass spectrometry, J. Chromatogr. A 1426 (2015) 161-173.

[69] S. Niell, V. Cesio, J. Hepperle, D. Doerk, L. Kirsch, D. Kolberg, E. Scherbaum, M. Anastassiades, H. Heinzen, QuEChERS-based method for the multiresidue analysis of pesticides in beeswax by LC-MS/MS and GCxGC-TOF, J. Agric. Food Chem. 62 (2014) 3675-3683.

[70] U. Koesukwiwat, S. J. Lehotay, K. Mastovska, X. K. J. Dorweiler, N. Leepipatpiboon, Pesticide multiresidue analysis in cereal grains using modified QuEChERS method combined with automated direct sample introduction GC-TOF-MS and UHPLCMS/MS techniques, J. Agric. Food Chem. 58 (2010) 5950-5972.

[71] EURL DataPool. EU Reference Laboratories for Residues of Pesticides. http://www.eurl-pesticides-datapool.eu.

[72] A. Wilkowska, M. Biziuk, Determination of pesticide residues in food matrices using the QuEChERS methodology, Food Chem. 125 (2011) 803-812.

[73] Y. Zhang, D. Hu, S. Zeng, P. Lu, K. Zhang, L. Chen, B. Song, Multiresidue determination of

pyrethroid pesticide residues in pepper through a modified QuEChERS method and gas chromatography with electron capture detection, Biomed. Chromatogr. 30 (2016) 142-148.

[74] Y. Nuapia, L. Chimuka, E. Cukrowska, Assessment oforganochlorine pesticide residues in raw food samples from open markets in two African cities, Chemosphere 164 (2016) 480-487.

[75] J. Alves Ferreira, J. M. Santos Ferreira, V. Talamini, J. de Fátima Facco, T. Medianeira Rizzetti, O. D. Prestes, M. B. Adaime, R. Zanella, C. B. G. Bottoli, Determination of pesticides in coconut (Cocos nucifera Linn.) water and pulp using modified QuEChERS and LC-MS/MS, Food Chem. 213 (2016) 616-624.

[76] J. Du, Z. Gridneva, M. C. L. Gay, R. D. Trengove, P. E. Hartmann, D. T. Geddes, Pesticides in human milk of Western Australian women and their influence on infant growth outcomes: a cross-sectional study, Chemosphere 167 (2017) 247-254.

[77] G. Ramadan, M. Al Jabir, N. Alabdulmalik, A. Mohammed, Validation of a method for the determination of 120 pesticide residues in apples and cucumbers by LC-MS/MS, drug test, Analysis 8 (2016) 498-510.

[78] T. M. Rizzetti, M. Kemmerich, M. L. Martins, O. D. Prestes, M. B. Adaime, R. Zanella, Optimization of a QuEChERS based method by means of central composite design for pesticide multiresidue determination in orange juice by UHPLC-MS/MS, FoodChem. 196 (2016) 25-33.

[79] M. H. Petrarca, J. O. Fernandes, H. T. Godoy, S. C. Cunha, Multiclass pesticide analysis in fruit-based baby food: a comparative study of sample preparation techniques previous to gas chromatography-mass spectrometry, Food Chem. 212 (2016) 528-536.

[80] E. A. Nantia, D. Moreno-González, F. P. T. Manfo, L. Gámiz-Gracia, A. M. García-Campaña, QuEChERS-based method for the determination of carbamate residues in aromatic herbs by UHPLC-MS/MS, Food Chem. 216 (2017) 334-341.

[81] B. Lozowicka, E. Rutkowska, I. Hrynko, Simultaneous determination of 223 pesticides in tobacco by GC with simultaneous electron capture and nitrogen-phosphorous detection and mass spectrometric confirmation, Open Chem. 13 (2015) 1137-1149.

[82] M. L. Martinsa, T. M. Rizzetti, M. Kemmerich, N. Saibt, O. D. Prestes, M. B. Adaime, R. Zanella, Dilution standard addition calibration: a practical calibrationstrategy for multiresidue organic compounds determination, J. Chromatogr. A 1460 (2016) 84-91.

[83] B. Lozowicka, G. lyasova, P. Kaczynski, M. Jankowska, E. Rutkowska, I. Hrynko, P. Mojsak, J. Szabunko, Multi-residue methods for the determination of over four hundred pesticides in solid and liquid high sucrose content matrices by tandem mass spectrometry coupled with gas and liquid chromatograph, Talanta 151 (2016) 51-61.

[84] X.-C. Wang, B. Shu, S. Li, Z.-G. Yang, B. Qiu, QuEChERS followed by dispersive liquid-liquid microextraction based on solidification of floating organic droplet method for organochlorine pesticides analysis in fish, Talanta 162 (2017) 90-97.

[85] Document SANTE/11945/2015. https://ec.europa.eu/food/sites/food/files/plant/docs/pesticides_ mrl_ guidelines_ wrkdoc_ 11945.pdf, 2016.

[86] X. Hou, S. R. Lei, S. T. Qiu, L. A. Guo, S. G. Yi, W. Liu, A multi-residue method for the determination of pesticides in tea using multi-walled carbon nanotubes as a dispersive solid phase extraction absorbent, Food Chem. 153 (2014) 121-129.

[87] M. Á. González-Curbelo, M. Asensio-Ramos, A. V. Herrera-Herrera, J. Hernández-Borges,

Pesticide residue analysis in cereal-based baby foods using multi-walled carbon nanotubes dispersive solid-phase extraction, Anal. Bioanal. Chem. 404 (2012) 183-196.

[88] B. Bresina, M. Piol, D. Fabbro, M. A. Mancini, B. Casetta, C. Del Bianco, Analysis of organo-chlorine pesticides residue in raw coffee with a modified "quick easy cheap effective rugged and safe" extraction/clean up procedure for reducing the impact of caffeine on the gas chromatography-mass spectrometry measurement, J. Chromatogr. A 1376 (2015) 167-171.

[89] Y. Zhang, X. Zhang, B. Jiao, Determination of ten pyrethroids in various fruit juices: comparison of dispersive liquid-liquid microextraction sample preparation and QuEChERS method combined with dispersive liquid-liquid microextraction, Food Chem. 159 (2014) 367-373.

[90] D. S. Bol'shakova, V. G. Amelina, A. V. Tret'yakova, Determination of polar pesticides in soil by micellar electrokinetic chromatography using QuEChERS sample preparation, Anal. Chem. 69 (2014) 89-97.

[91] J. Wei, J. Cao, K. Tian, Y. Hu, H. Su, J. Wan, P. Li, Trace determination of five organophosphorus pesticides by using QuEChERS coupled with dispersive liquid-liquid microextraction and stacking before micellar electrokinetic chromatography, Anal. Methods 7 (2015) 5801-5807.

[92] Y. Sapozhnikova, Evaluation of low-pressure gas chromatography_ tandem mass spectrometry method for the analysis of 140 pesticides in fish, J. Agric. Food Chem. 62 (2014) 3684-3689.

[93] S. W. C. Chung, B. T. P. Chan, Validation and use of a fast sample preparation method and liquid chromatography-tandem mass spectrometry in analysis of ultra-trace levels of 98 organophosphorus pesticide and carbamate residues in a total diet study involving diversified food types, J. Chromatogr. A 1217 (2010) 4815-4824.

[94] V. C. Fernandes, V. F. Domingues, N. Mateus, C. Delerue-Matos, Multiresidue pesticides analysis in soils using modified QuEChERS with disposable pipette extraction and dispersive solid-phase extraction, J. Sep. Sci. 36 (2013) 376-382.

[95] I. -S. Jeong, B. -M. Kwak, J. -H. Ahn, S. -H. Jeong, Determination of pesticide residues in milk using a QuEChERS-based method developed by response surface methodology, Food Chem. 133 (2012) 473-481.

[96] S. J. Lehotay, K. Maštovská, A. R. Lightfield, Use of buffering and other means to improve results of problematic pesticides in a fast and easy method for residue analysis of fruits and vegetables, J. AOAC Int. 88 (205) (2005) 615-629.

[97] A. Hiba, A. Carine, A. R. Haifa, L. Ryszard, J. Farouk, Monitoring of twenty-two sulfonamides in edible tissues: investigation of new metabolites and their potential toxicity, Food Chem. 192 (2016) 212-227.

[98] Z. Zhang, H. Yan, F. Cui, H. Yun, X. Chang, J. Li, X. Liu, L. Yang, Q. Hu, Analysis of multiple β-agonist and β-blocker residues in porcine muscle using improved QuEChERS method and UHPLC-LTQ orbitrap mass spectrometry, Food Anal. Method. 9 (2016) 915-924.

[99] Official Journal of the European Union, L15 20 January 2010, Commission Regulation (EU) No 37/2010 of 22 December 2009 on Pharmacologically Active Substances and Their Classification Regarding Maximum Residue Limits in Foodstuffs of Animal Origin, 2009. Brussels, Belgium.

[100] Y. Zhang, X. Liu, X. Li, J. Zhang, Y. Cao, M. Su, Z. Shi, H. Sun, Rapid screening and quantification of multi-class multi-residue veterinary drugs in royal jelly by ultra performance liquid chromatography coupled to quadrupole time-of-flight mass spectrometry, Food Control 60 (2016) 667-676.

[101] D. Pimentel-Trapero, A. Sonseca-Yepes, S. Moreira-Romero, M. Hern-ández-Carrasquilla,

Determination of macrocyclic lactones in bovine liver using QuEChERS and HPLC with fluorescence detection, J. Chromatogr. B 1015-1016 (2016) 166-172.

[102] V. Gressler, A. R. L. Franzen, G. J. M. M. de Lima, F. C. Tavernari, O. A. Dalla Costa, V. Feddern, Development of a readily applied method to quantify ractopamine residue in meat and bone meal by QuEChERS-LC-MS/MS, J. Chromatogr. B 1015-1016 (2016) 192-200.

[103] J. Li, J. Zhang, H. Liu, L. Wu, A comparative study of primary secondary amino (PSA) and multi-walled carbon nanotubes (MWCNTs) as QuEChERS absorbents for the rapid determination of diazepam and its major metabolites in fish samples by high-performance liquid chromatography-electrospray ionisation-tandem mass spectrometry, J. Sci. Food Agric. 96 (2016) 555-560.

[104] M. Lombardo - Agüí, L. Gámiz - Gracia, C. Cruces - Blanco, A. M. García - Campaña, Comparison of different sample treatments for the analysis of quinolones in milk by capillary - liquid chromatography with laser induced fluorescence detection, J. Chromatogr. A 1218 (2011) 4966-4971.

[105] A. Martínez - Villalba, E. Moyano, M. T. Galcerán, Ultra - high performance liquid chromatography - atmospheric pressure chemical ionization - tandem mass spectrometry for the analysis of benzimidazole compounds in milk samples, J. Chromatogr. A 1313 (2013) 119-131.

[106] A. Martínez-Villalba, L. Vaclavik, E. Moyano, M. T. Galcerán, J. Haj-šslova, Direct analysis in real time high-resolution mass spectrometry for high-throughput analysis of antiparasitic veterinary drugs in feed and food, Rapid Commun. Mass Spectrom. 27 (2013) 467-475.

[107] F. Plössl, M. Giera, F. Bracher, Multiresidue analytical method using dispersive solidphase extraction and gas chromatography/ion trap mass spectrometry to determine pharmaceuticals in whole blood, J. Chromatogr. A 1135 (2006) 19-26.

[108] Y. Li, S. Wen, Z. Chen, Z. Xiao, M. Ma, Ultra-high performance liquid chromatography tandem mass spectrometry for simultaneous analysis of aflatoxins B1, G1, B2, G2, zearalenone and its metabolites in eggs using a QuEChERS based extraction procedure, Anal. Methods 7 (2015) 4145-4151.

[109] O. Núñez, H. Gallart - Ayala, C. P. B. Martins, P. Lucci, New trends in fast liquid chromatography for food and environmental analysis, J. Chromatogr. A 1228 (2012) 298-323.

[110] V. L. Pereira, J. O. Fernandes, S. C. Cunha, Comparative assessment of three cleanup procedures after QuEChERS extraction for determination of trichothecenes (type A and type B) in processed cereal-based baby foods by GC-MS, Food Chem. 182 (2015) 143-149.

[111] Y. Rodríguez - Carrasco, G. Font, J. C. Moltó, H. Berrada, Quantitative determination of trichothecenes in breadsticks by gas chromatography-triple quadrupole tandem mass spectrometry, Food Addit. Contam. Part A 31 (2014) 1422-1430.

[112] Y. Rodríguez - Carrasco, J. C. Moltó, H. Berrada, J. Mañes, A survey of trichothecenes, zearalenone and patulin in milled grain - based products using GC - MS/MS, Food Chem. 146 (2014) 212-219.

[113] U. Koesukwiwat, K. Sanguankaew, N. Leepipatpiboon, Evaluation of a modified QuEChERS method for analysis ofmycotoxins in rice, Food Chem. 153 (2014) 44-51.

[114] L. Vaclavik, M. Zachariasova, V. Hrbek, J. Hajšlova, Analysis of multiple mycotoxins in cereals under ambient conditions using direct analysis in real time (DART) ionization coupled to high resolution mass spectrometry, Talanta 82 (2010) 1950-1957.

[115] N. Michlig, M. R. Repetti, C. Chiericatti, S. R. García, M. Gaggiotti, J. C. Basílico, H. R. Beldoménico, Multiclass compatible sample preparation for UHPLC-MS/MS determination of aflatoxin M1 in

raw milk, Chromatographia 79 (2016) 1091-1100.

[116] Y. Rodríguez-Carrasco, M. Fattore, S. Albrizio, H. Berrada, J. Mañes, Occurrence of fusarium mycotoxins and their dietary intake through beer consumption by the European population, Food Chem. 178 (2015) 149-155.

[117] J. Taibon, S. Sturm, C. Seger, H. Strasser, H. Stuppner, Quantitative assessment of destruxins from strawberry and maize in the lower parts per billion range: combination of a QuEChERS based extraction protocol with a fast and selective UHPLC-QTOF-MS assay, J. Agric. Food Chem. 63 (2015) 5707-5713.

[118] S. Zhang, J. Lu, S. Wang, D. Mao, S. Miao, S. Ji, Multi-mycotoxins analysis in Pheretima using ultra - high - performance liquid chromatography tandem mass spectrometry based on a modified QuEChERS method, J. Chromatogr. B 1035 (2016) 31-41.

[119] Z. Veprikova, M. Zachariasova, Z. Dzuman, A. Zachariasova, M. Fenclova, P. Slavikova, M. Vaclavikova, K. Mastovská, D. Hengst, J. Hajšlova, Mycotoxins in plant-based dietary supplements: hidden health risk for consumers, J. Agric. Food Chem. 63 (2015) 6633-6643.

[120] E. J. Llorent-Martínez, P. Ortega-Barrales, M. L. Fernández-de Córdova, A. Ruiz Medina, Quantitation of ochratoxin A in cereals and feedstuff using sequential injection analysis with luminescence detection, Food Control 30 (2013) 379-385.

[121] United States Environmental Protection Agency (U. S. EPA), Appendix A to 40 CFR Part 423, U. S. EPA, Washington, D. C, 2010. http: //www. epa. gov/waterscience/methods/pollutants. html.

[122] A. O. Oduntan, N. T. Tavengwa, E. Cukrowska, S. D. Mhlanga, L. Chimuka, QuECh-ERS method development for bio-monitoring of low molecular weight polycyclic aromatic hydrocarbons in South African carp fish using HPLC-fluorescence: an initial assessment, S. Afr. J. Chem. 69 (2016) 98-104.

[123] T. V. Madureira, S. Velhote, C. Santos, C. Cruzeiro, E. Rocha, M. J. Rocha, A step forward using QuEChERS (quick, easy, cheap, effective, rugged, and safe) based extraction and gas chromatography-tandem massspectrometry-levels of priority polycyclic aromatic hydrocarbons in wild and commercial mussels, Environ. Sci. Pollut. Res. 21 (2014) 6089-6098.

[124] M. Surma, A. S. Rociek, E. Cieślik, The application of d-SPE in the QuEChERS method for the determination of PAHs in food of animal origin with GC-MS detection, Eur. Food Res. Technol. 238 (2014) 1029-1036.

[125] M. Yoo, S. Lee, S. Kim, S.-J. Kim, H.-Y. Seo, D. Shin, A comparative study of the analytical methods for the determination of polycyclic aromatic hydrocarbons in seafood by high-performance liquid chromatography with fluorescence detection, Int. J. Food Sci. Tech. 49 (2014) 1480-1489.

[126] T. H. Kao, S. Chen, C. W. Huang, C. J. Chen, B. H. Chen, Occurrence and exposure to polycyclic aromatic hydrocarbons in kindling-free-charcoal grilled meat products in Taiwan, Food Chem. Toxicol. 71 (2014) 149-158.

[127] A. L. V. Escarrone, S. S. Caldas, E. B. Furlong, V. L. Meneghetti, C. A. A. Fagundes, J. L. O. Arias, E. G. Primel, Polycyclic aromatic hydrocarbons in rice grain dried by different processes: evaluation of a quick, easy, cheap, effective, rugged and safe extraction method, Food Chem. 146 (2014) 597-602.

[128] S. Chen, T. H. Kao, C. J. Chen, C. W. Huang, B. H. Chen, Reduction of carcinogenic polycyclic aromatic hydrocarbons in meat by sugar-smoking and dietary exposure assessment in Taiwan, J. Agric. Food Chem. 61 (2013) 7645-7653.

[129] G. Knobel, A. D. Campiglia, Determination of polycyclic aromatic hydrocarbon metabolites in

milk by a quick, easy, cheap, effective, rugged and safe extraction and capillary electrophoresis, J. Sep. Sci. 36 (2013) 2291-2298.

[130] S. -S. Cai, J. Stevens, J. A. Syage, Ultra high performance liquid chromatography-atmospheric pressure photoionization-mass spectrometry for high-sensitivity analysis of US Environmental Protection Agency sixteen priority pollutant polynuclear aromatic hydrocarbons in oysters, J. Chromatogr. A 1227 (2012) 138-144.

[131] A. Angioni, L. Porcu, M. Secci, P. Addis, QuEChERS method for the determination of PAH compounds in sardinia sea urchin (*Paracentrotus lividus*) Roe, using gas chromatography ITMS-MS analysis, Food Anal. Method. 5 (2012) 1131-1136.

[132] T. H. Kao, S. Chen, C. J. Chen, C. W. Huang, B. H. Chen, Evaluation of analysis of polycyclic aromatic hydrocarbons by the QuEChERS method and gas chromatography - mass spectrometry and their formation in poultry meat as affected by marinating and frying, J. Agric. Food Chem. 60 (2012) 1380-1389.

[133] N. D. Forsberg, G. R. Wilson, K. A. Anderson, Determination of parent and substituted polycyclic aromatic hydrocarbons in high-fat salmon using a modified QuEChERS extraction, dispersive SPE and GC-MS, J. Agric. Food Chem. 59 (2011) 8108-8116.

[134] M. Smoker, K. Tran, R. E. Smith, Determination of polycyclic aromatic hydrocarbons (PAHs) in shrimp, J. Agric. Food Chem. 58 (2010) 12101-12104.

[135] M. J. Ramalhosa, P. Paíga, S. Morais, C. Delerue-Matos, M. B. P. P. Oliveira, Analysis of polycyclic aromatic hydrocarbons in fish: evaluation of a quick, easy, cheap, effective, rugged, and safe extraction method, J. Sep. Sci. 32 (2009) 3529-3538.

[136] M. Bustamante-Rangel, M. M. Delgado-Zamarreño, L. Pérez-Martín, R. Carabias-Martínez, QuEChERS method for the extraction of isoflavones from soy-based foods before determination by capillary electrophoresis-electrospray ionization-mass spectrometry, Microchem. J. 108 (2013) 203-209.

[137] J. Domínguez-Álvarez, E. Rodríguez-Gonzalo, J. Hernández-Méndez, R. Carabias-Martínez, Programed nebulizing-gas pressure mode for quantitative capillary electrophoresis-mass spectrometry analysis of endocrine disruptors in honey, Electrophoresis 33 (2012) 2374-2381.

[138] J. de Boer, M. Wagelmans, Polycyclic aromatic hydrocarbons in soil - practical options for remediation, Clean-Soil Air Water 44 (2016) 648-653.

[139] K. Srogi, Monitoring of environmental exposure to polycyclic aromatic hydrocarbons: a review, Environ. Chem. Lett. 5 (2007) 169-195.

[140] M. M. Plassmann, M. Schmidt, W. Brack, M. Krauss, Detecting a wide range of environmental contaminants in human blood samples - combining QuEChERS with LCMS and GC-MS methods, Anal. Bioanal. Chem. 407 (2015) 7047-7054.

[141] V. Homem, E. Silva, A. Alves, L. Santos, Scented tracese dermal exposure of synthetic musk fragrances in personal care products and environmental inputassessment, Chemosphere 139 (2015) 276-287.

[142] Y. Sapozhnikova, S. J. Lehotay, Multi-class, multiresidue analysis of pesticides, polychlorinated biphenyls, polycyclic aromatic hydrocarbons, PBDEs and novel flame retardants in using fast, low-pressure gas chromatography-tandem mass spectrometry, Anal. Chim. Acta 758 (2013) 80-92.

[143] M. Surma, W. Wiczkowski, E. Cie_ slik, H. Zieli_ nski, Method development for the determination of PFOA and PFOS in honey based on the dispersive solid phase extraction (d-SPE) with micro-UHPLC-MS/MS system, Microchem. J. 121 (2015) 150-156.

[144] S. Song, X. Dai, W. Wang, Y. He, Z. Liu, M. Shao, Optimization of selective pressurized

liquid extraction and ultrasonication-assisted QuEChERS methods for the determination of polybrominated diphenyl ethers in sediments, Anal. Methods 7 (2015) 9542-9548.

[145] Z. He, L. Wang, Y. Peng, M. Luo, W. Wang, X. Liu, Determination of selected polychlorinated biphenyls in soil and earthworm (Eisenia fetida) using a QuEChERS-based method and gas chromatography with tandem MS, J. Sep. Sci. 38 (2015) 3766-3773.

[146] J. Camilleri, E. Vulliet, Determination of steroid hormones in sediments based on quick, easy, cheap, effective, rugged, and safe (modified-QuEChERS) extraction followed by liquid chromatography-tandem mass spectrometry (LC-MS/MS), Anal. Methods 7 (2015) 9577-9586.

[147] R. Shuiying, W. Hongfei, F. Shun, W. Jide, L. Yi, Determination of 21 plant growth regulators in tomatoes using an improved ultrasound-assisted QuEChERS technique combined with a liquid chromatography tandem mass spectrometry method, Anal. Methods 8 (2016) 4808-4815.

[148] M. Mattarozzi, M. Milioli, F. Bianchi, A. Cavazza, S. Pigozzi, A. Milandri, M. Careri, Optimization of a rapid QuEChERS sample treatment method for HILIC-MS2 analysis of paralytic shellfish poisoning (PSP) toxins in mussels, Food Control 60 (2016) 138-145.

[149] A. Rúbies, E. Muñoz, D. Gibert, N. Cortés-Francisco, M. Granados, J. Caixach, F. Centrich, New method for the analysis of lipophilic marine biotoxins in fresh and canned bivalves by liquid chromatography coupled to high resolution mass spectrometry: a quick, easy, cheap, efficient, rugged, safe approach, J. Chromatogr. A 1386 (2015) 62-73.

[150] S.J. Lehotay, Y. Sapozhnikova, L. Han, J.J. Johnston, Analysis of nitrosamines in cooked bacon by QuEChERS sample preparation and gas chromatography-tandem mass spectrometry with backflushing, J. Agric. Food Chem. 63 (2015) 10341-10351.

[151] Z. Xun, J. Huang, X.Y. Li, S. Lin, S. He, X. Guo, Y. Xian, Simultaneous determination of seven acrylates in food contact paper products by GC/MS and modified QuEChERS, Anal. Methods 8 (2016) 3953-3958.

[152] Y. Ruiz-García, C.L. Silva, E. Gómez-Plaza, J.S. Câmara, A powerful analytical strategy based on QuEChERS-dispersive solid-phase extraction combined with ultrahigh pressure liquid chromatography for evaluating the effect of elicitors on biosynthesis of trans-resveratrol in grapes, Food Anal. Method. 9 (2016) 670-679.

[153] M. Mülek, P. Högger, Highly sensitive analysis of polyphenols and their metabolites in human blood cells using dispersive SPE extraction and LC-MS/MS, Anal. Bioanal. Chem. 407 (2015) 1885-1899.

[154] D.Y. Bang, S.K. Byeon, M.H. Moon, Rapid and simple extraction of lipids from blood plasma and urine for liquid chromatography-tandem mass spectrometry, J. Chromatogr. A 1331 (2014) 19-26.

[155] Y. Ma, Y. Hashi, F. Ji, J.M. Li, Determination of phthalates in fruit jellies by dispersive SPE coupled with HPLC-MS, J. Sep. Sci. 33 (2010) 251-257.

[156] K. Hasegawa, A. Wurita, K. Minakata, K. Gonmori, I. Yamagishi, K. Watanabe, O. Suzuki, Postmortem distribution of AB-CHMINACA, 5-fluoro-AMB, and diphenidine in body fluids and solid tissues in a fatal poisoning case: usefulness of adipose tissue for detection of the drugs in unchanged forms, Forensic Toxicol. 33 (2015) 45-53.

[157] C.L. Fan, Q.Y. Chang, Z.Y. Li, J. Kang, G.Q. Pan, S.Z. Zheng, W.W. Wang, C.C. Yao, X.X. Ji, High-throughput analytical techniques for determination of residues of 653 multiclass pesticides and chemical pollutants in tea, part II: comparative study of extraction efficiencies of three sample preparation techniques, J. AOACInt. 96 (2013) 432-440.

[158] E. Vulliet, A. Berloiz-Barbier, F. Lafay, R. Baudot, L. Wiest, A. Vauchez, F. Lestremau, F. Botta, C. Cren-Olivé, A national reconnaissance for selected organic micropollutants in sediments on French territory, Environ. Sci. Pollut. Res. Int. 21 (2014) 11370-11379.

[159] A. Wurita, K. Hasegawa, K. Minakata, K. Gonmori, H. Nozawa, I. Yamagishi, O. Suzuki, K. Watanabe, Postmortem distribution of α-pyrrolidinobutiophenone in body fluids and solid tissues of a human cadaver, Leg. Med. 16 (2014) 241-246.

[160] A. R. Fontana, R. Bottini, QuEChERS method for the determination of 3-alkyl-2-methoxypyrazines in wines by gas chromatography-mass spectrometry, Food Anal. Method 9 (2016) 3352-3359.

[161] R. P. Z. Furlani, F. F. G. Dias, P. M. Nogueira, F. M. L. Gomes, S. A. V. Tfouni, M. C. R. Camargo, Occurrence of macrocyclic lactones in milk and yogurt from Brazilian market, Food Control 48 (2015) 43-47.

[162] W. Jia, Y. Ling, Y. Lin, J. Chang, X. Chu, Analysis of additives in dairy products by liquid chromatography coupled to quadrupole-orbitrap mass spectrometry, J. Chromatogr. A 1336 (2014) 67-75.

[163] M. Faraji, M. Noorani, B. N. Sahneh, Quick, easy, cheap, effective, rugged, and safe method followed by ionic liquid-dispersive liquid-liquid microextraction for the determination of trace amount of bisphenol A in canned foods, Food Anal. Method (2016), http://dx.doi.org/10.1007/s12161-016-0635-y.

[164] H. Gallart-Ayala, O. Núñez, E. Moyano, M. T. Galcerán, Analysis of UV ink photoinitiators in packaged food by fast liquid chromatography at sub-ambient temperature coupled to mass spectrometry, J. Chromatogr. A 1218 (2011) 459-466.

[165] O. G. Cabrices, A. Schreiber, W. E. Brewer, Automated sample preparation and analysis workflows for pesticide residue screenings in food samples using DPXQuEChERS with LC/MS/MS, Gerstel AppNote 8/2013. http://www.gerstel.com/pdf/p-lc-an-2013-08.pdf.

[166] P. Kaewsuya, W. E. Brewer, J. Wong, S. L. Morgan, Automated QuEChERS tips for analysis of pesticide residues in fruits and vegetables by GC-MS, J. Agric. Food Chem. 61 (2013) 2299-2314.

[167] S. J. Lehotay, L. Han, Y. Sapozhnikova, Automated mini-column solid-phase extraction cleanup for high-throughput analysis of chemical contaminants in foodsby low-pressure gas chromatography-tandem mass spectrometry, Chromatoraphia 79 (2016) 1113-1130.

[168] S. Ferhi, M. Bourdat-Deschamps, J.-J. Daudin, S. Houot, S. Nélieu, Factors influencing the extraction of pharmaceuticals from sewage sludge and soil: an experimental design approach, Anal. Bioanal. Chem. 408 (2016) 6153-6168.

[169] Teledyne Tekman. http://www.teledynetekmar.com/products/semi-volatile-organic-compounds-(svoc)/automate-q40.

[170] T. Trent, Pesticide analysis using the AutoMate-Q40: an automated solution to QuEChERS extractions, LC-GC N. Am. 25 (June 1, 2013). The application notebook.

[171] T. Trent, Veterinary drug residue analysis using the AutoMate-Q40: an automated solution to QuEChERS, LC-GC N. Am. 24 (June 1, 2014). The application notebook.

[172] T. Trent, Determination of carbendazim in orange juice using an automated QuEChERS solution, LC GC N. Am. 26 (September 1, 2014).

12 基质固相分散萃取

Lourdes Ramos
Institute of Organic Chemistry, CSIC, Madrid, Spain
E-mail: l. ramos@ iqog. csic. es

12.1 引言

样品制备，此处理解为内源或外源目标化合物在样品均质化、二次采样和仪器检测之间所进行的几种操作，仍然被认为是许多现代分析程序的瓶颈[1]。对于有机化合物，常规样品处理方案的第一步包括对基质中的分析物的完全提取，在痕量组分情况下尤为重要。如果最终分析测定确实没有一种非常有选择性的分离检测技术，那么后续首先清除非化学相关的萃取物，然后再清除与目标分析物化学相关的成分，此步骤本质上是非选择性的。在大多数情况下，从共提物中提取分析物以及随后的逐步分离等几种处理和操作都是离线进行。这不仅增加了劳动强度，还需要花费大量时间来处理多个程序，通常涉及大量样品、吸附剂和大体积有机溶剂等。持续的手动提取导致分析物容易损耗和污染，同时也使分析人员持续暴露在有害化学物质的环境里。因此就时间、材料和能源消耗而言，这些方法花费太高，同时还产生大量废物，此外样品处理能力通常也是相当有限。在这些特定的应用领域中，使用复杂强大的联用系统，例如液液萃取（LLE）或索氏提取，用于提取物最终的分析净化，与基于大规模且劳动密集型的常规提取（尽管通常稳健、完善且受到全球认可）形成鲜明对比。为了消除这些最急需修改的缺点，在过去的几十年中进行了许多努力，已经形成了更快、更强大且更通用的绿色提取技术，其中一些在前面的章节中已经进行了改进。但是，尽管有了这些进步和创新，目前的发展水平仍然非常依赖于样品的物理状态。对于水性样品和（某些）生物学样品液体，使用固相萃取（SPE）、固相微萃取或基于透析的专用工作站在线处理并自动进样完全是可以实现的（参见第 10 章[1a]和第 13~17 章[1b-1f]以获取更完整的信息）。然而就半固体和固体样品而言，则受到更多限制[1,2]。

当今增强型流体/溶剂萃取技术，例如超临界流体萃取，加压液体萃取（PLE），亚临界水提取，微波辅助提取和超声辅助萃取（UAE），在一定程度上进行了改进且受到人们的追捧，在不降低处理效率的情况下，其开发和商业化有助于加快（半）固体样品的萃取。将其与更常规的程序（通常是索氏提取）进行比较并在本章的相应章节中进行了讨论。

但是，对于复杂基质，相当复杂的多步方法仍然是必需的，而不是例外。尽管在最近的文献中可以找到一些例子说明几种处理方法进行整合的可能性，但在实践中，这些方法相当匮乏，并且常常仅限于理论层面[2]。与这些增强的萃取技术（其中萃取在高压和/或高温下或借助外部能量的帮助下）相比，基质固相分散（MSPD）萃取是在环境条件下进行，不需要任何类型的专用设备。尽管提取条件比较温和，但该技术已证明可提供令人满意的结果，该结果与使用更复杂，更昂贵的增强技术以及许多应用领域中的常规提取方法所获得的结果相似。

12.1.1 基质固相分散萃取的发展

MSPD 是一项受专利保护的处理方法，由 Barker 等人在 1989 年同时破碎萃取半固体和固体样本时首先提出[3]。在开发 MSPD 时，SPE 被认为是完美的液体样品处理技术。SPE 可以用最少的溶剂一步实现目标化合物的提取和预浓缩。SPE 还避免了一些与当时用于液体样品处理的其他技术相关的主要限制，例如通常在 LLE 或逆流提取中形成的乳状液。鉴于其优势，Barker 和同事们考虑开发与 SPE 类似的方法用于处理固体样品。初始方法（最直观的）包括将均质组织直接倒入 SPE 小柱顶部。但是，这种方法总是由于玻璃或上层色谱柱堵塞导致不可修复的色谱柱塌陷[3,4]。还有另一种策略：将均质的组织和适当的吸附剂混合。在这种情况下，吸附剂充当研磨材料，并且用研钵和研杵或其他机械设备通过混合过程产生的剪切力破坏了总体样品的结构，并提供了细小的提取材料[5]。最早用作分散吸附剂材料之一的是硅藻土。样品与这种吸附剂的混合产生半干物质，可以很容易地装在色谱柱中并作为 SPE 的吸附剂进行洗脱。另外，因为整个基质分散在吸附剂表面上，暴露于萃取溶剂的样品表面积大大增加。

实际上，整个样本（即蛋白质、脂质、结缔组织等）暴露于洗脱液中不需要再沉淀细胞成分和离心杂质成分。与当时在用的更常规的技术相比，例如 LLE，简化和加快了样品处理过程。新方法的实用性通过成功地将其应用于中性和/或中等极性的化合物的分离（即芬苯达唑及其代谢物）得到了证实，例如肝脏组织[6]、血浆、尿液、粪便和组织匀浆[4]。然而，由于所研究基质的复杂性，极性分析物仍然不可逆地保留在色谱柱中，因此在仪器分析前对提取物进行了额外的净化是必需的（通常使用配备了紫外可见光检测器 LC-UV 的液相色谱）。

为了提高该技术分离更大极性分析物的整体性能，建议采用完全多残留的方法，作者用溶解脂质的吸附剂，如 C_{18} 包裹的二氧化硅球，替换了最初考虑的吸附剂（硅藻土或任何其他干燥的惰性试剂）。C_{18} 可以通过溶解细胞表面上的磷脂和胆固醇促进细胞膜破裂。更多的极性分子（例如水）与吸附剂表面上的（剩余）极性基团（如硅烷醇基团）以及基质中具有氢键结合能力的组分缔合。其他极性较小的分子可根据它们的极性分布在此范围内多相结构中[7]。为避免在基于表面活性剂对样品进行直接处理的过程中观察到的不便，通过与固体载体结合固定该分散剂，确保其在洗脱过程中有效保留在色谱柱上。

当通过扫描电子显微镜检查混合材料表面的变化时，新的 MSPD 技术对样品在吸附剂表面上的破坏和分散的效率变得显而易见[5]。（图 12.1）

图 12.1（1）表明在样品机械混合过程中起到剪切作用的原始 C_{18}-材料具有锋利边缘和粗糙表面。在图 12.1（2）中观察到的呈团块状分布在未衍生二氧化硅颗粒上的实体表明，所研究的牛肝组织的混合导致了样品结构的真正破坏，尽管细胞成分本身并未受到破坏。相反，使用 C_{18} 衍生的材料会导致样品组分的完全分解，这些组分均匀地分散在吸附剂表面［图 12.1（3）］，厚度约为 100μm［图 12.1（4）］[5]。

1989 年发表的研究表明通过选择一种合适的洗脱方法，可以选择性提取单一样品中的特定类别药物，单一药物，或者几种药物[3]。早期的 MSPD 应用研究是将 0.5g 的复杂基质牛肌肉加入到 2g 预洗过亲脂性固相填料中（40μm C_{18} 硅胶）。将混合物在玻璃研钵中用玻璃棒彻底研磨混合 30s 后，在 10mL 自制 SPE 柱管中装入 0.50g 干净的 C_{18}，将半干状和表面均匀的粉末状混合物加入到 C_{18} 层上，再在 MSPD 混合物的顶部添加 0.25g 干净的 C_{18}。在溶剂洗脱分析物前，整个柱子用注射器柱塞稍微压缩一下以

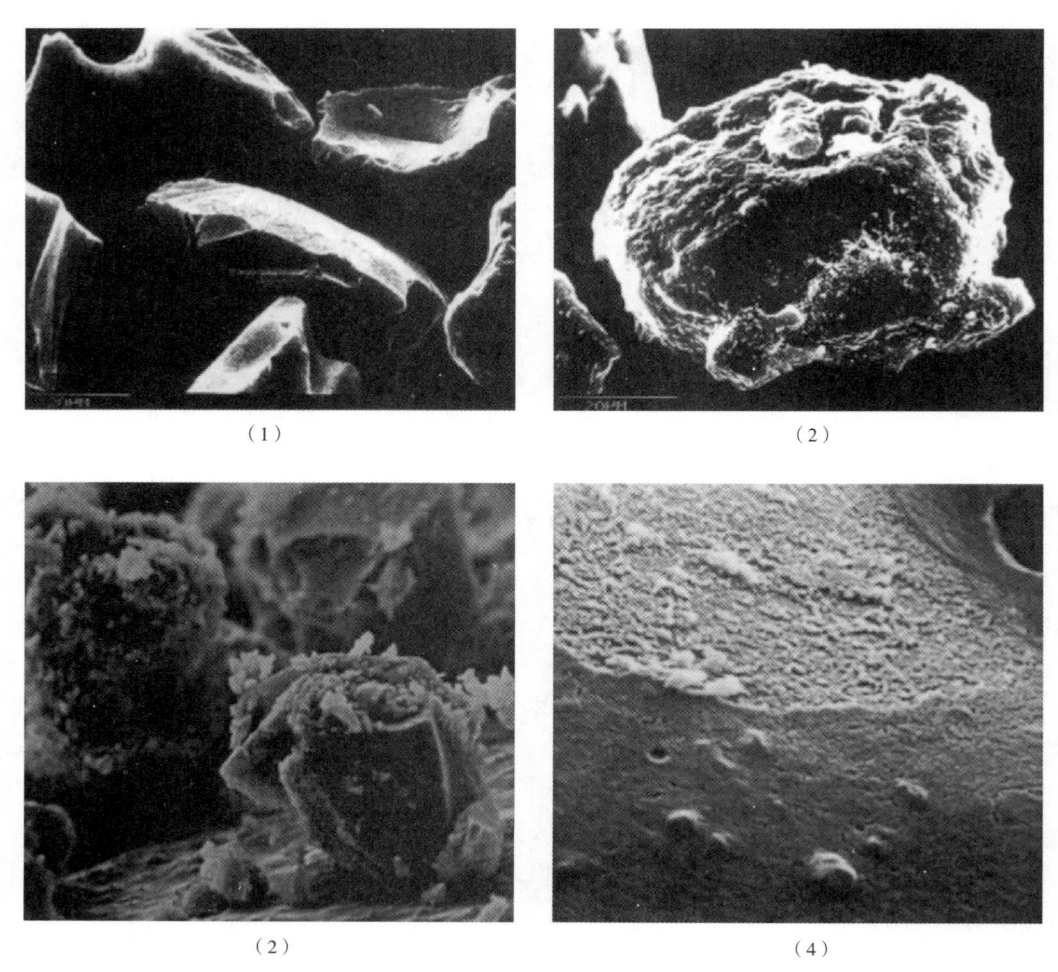

图 12.1　扫描电子镜图（分辨率 20μm）
（1）C_{18} 修饰的二氧化硅颗粒，（2）组织团块和混有牛肝的未衍生二氧化硅颗粒，
（3）使用 C_{18} 修饰的二氧化硅的相同过程，（4）C_{18} 修饰二氧化硅
表面上形成的样品薄层的详细信息（分辨率 2μm）[5]

确保去除空气。MSPD 处理过程涉及的几个步骤如图 12.2 所示[8]。

在此特定研究中，三类化合物分别收集为四个独立的部分，依次通过 MSPD 柱，分别用已烷、苯、乙酸乙酯和甲醇（每种溶剂 8mL）进行洗脱。倍硫磷和香豆磷在第一部分中进行选择性和定量洗脱（回收率高于 77%），而另外两个有机磷酸酯，即克芦磷脂和伐灭氧磷（用作该类分析物的内标），则用苯选择性洗脱（回收率高于 82%）。尽管在前两部分中发现有少量的脂质，使用气相色谱仪氮磷检测器分析时，它们似乎并没有影响目标化合物测定。因此，这两种提取液无须进行额外净化。随后苯并咪唑（7 种化合物）用乙酸乙酯洗脱，回收率达到 63%~86%。几个未确定的组分暂时指定为基质组分和背景干扰物，在此部分中也检测到了。然而根据作者的说法，利用液相色谱-二极管阵列（LC-DAD）分析时，它们没有影响目标分析物的检测，因此没有对

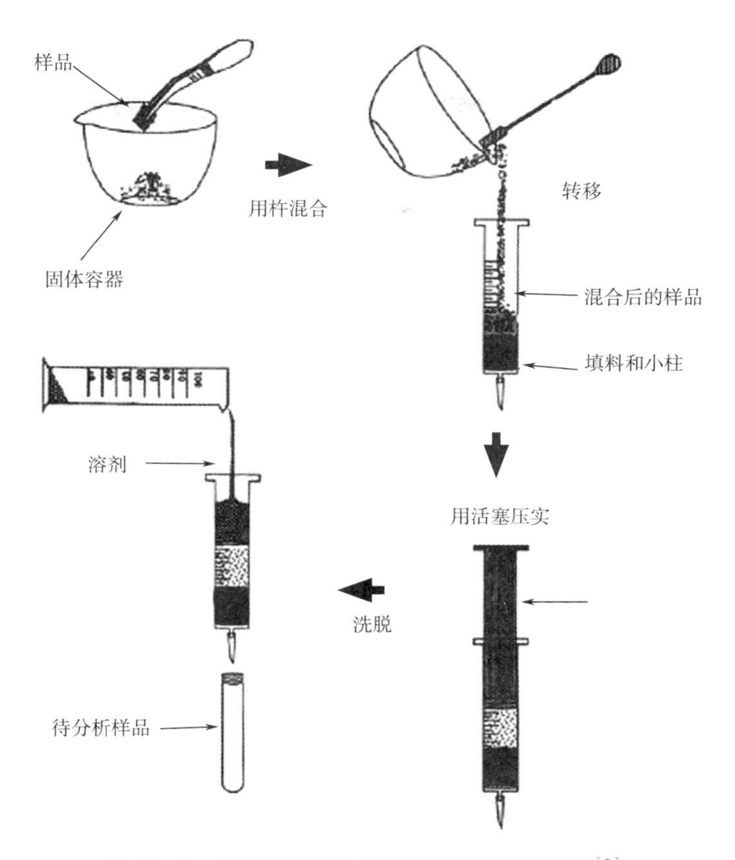

图 12.2 典型基质固相分散萃取程序的示意图[8]

收集的萃取物进行额外的净化。最后，研究中选取三种 β-内酰胺抗生素头孢哌啶、青霉素和氨苄西林用甲醇进行选择性洗脱。这一部分中蛋白质的存在阻碍了萃取物直接进行液相色谱（LC）分析。但是，用 0.017mol/L 磷酸溶液去除蛋白质后，头孢氨苄和青霉素的回收率分别为 72%和 86%，虽然有点低，但仍然可以接受（60%），对氨苄西林来说重复性相对较好［相对标准偏差（RSD），9%，$n=5$］。虽然该方法未经所有样品分析验证，但是所研究的几类化合物的成功分离和令人信服满意的优点证明了 MSPD 具有作为一类分析动物组织中药物残留及其代谢物技术的潜力。

最初建议用于 MSPD 的材料是 C_{18} 包裹的二氧化硅，但如 C_3，C_8，C_3-磺酸和其他聚合物的吸附剂应该也能够提供同样令人满意的结果。实践也证明了，样品的破坏效率和 MSPD 处理效率取决于所选取的组织-基质的组合和比率。有趣的是，在混合过程中可以通过添加溶剂、表面活性剂或其他试剂调节以上效率[9,10]。应该注意在样品混合过程中施加的机械力可能会破坏一些二氧化硅涂层。在开发 MSPD 技术时，作者推测此过程可能会影响溶剂在柱中的流动速率并有助于增加活跃点的数量，在这些活跃点上分析物不可洗脱，因此限制了 MSPD 在痕量分析中的适用性（即浓度小于 100ng/g）[3]。但是，最近的研究证明在实践中这些因素均未对该技术的性能产生影响，其应用范围比发明者最初预期的要宽得多[11-13]。

Backer 等最初研究了一些药物应用分析，然后又对相关的微污染物和天然化合物进行了研究[5,7,8]，还有一些不太常见的，包括肺结核病菌株 19698 和大肠杆菌 DH5a 菌株[14]的裂解及后续分离，随后证明了 MSPD 用于制备各种液体、半固体和固体样品的可行性。

自该技术引入以来，早期研究成果已在有关的论文[5,7,8,15-17]和书籍[18]中进行了总结，同时在更多以主题为导向的论文[1,19-21]中讨论了最新趋势和新颖方法。一般而言，所有这些研究都表明 MSPD 作为一种简单、通用、快速、廉价和环保的技术，可以轻松适应各种应用。MSPD 还可以有效简化样品制备过程。从这个意义上讲，MSPD 可以减少样品的如下复杂处理步骤，例如样品彻底均质化（实际上，所研究的子样本代表整个初始实体）、组成组分的沉淀、离心、pH 调节、样品转移以及在某些情况下萃取物额外净化和浓缩收集。此外，缩小 MSPD 流程可以促进其随后的样品处理，实际上就是最终的仪器分离和检测步骤[22,23]，因此有助于提高样品通量以及商业实验室的潜在利益。

12.2 基质固相分散萃取的一般原理

12.2.1 基质固相分散萃取的基本原理

按照 Barker 等人最初描述，MSPD 的基本程序包括三个主要步骤。如前所述，第一步由机械搅拌的方式混合所研究的带有分散吸附剂的液体、半流体、半固体或固体样品（图 12.2）。通常混合以手工方式用玻璃杵在玻璃杯或玛瑙研钵中进行，因为多孔材料（例如瓷器）会导致分析物和/或样品损失[5]。这种处理的目的是剪切破坏粗样品结构并分解成小块。当使用亲脂性键合相材料，根据分散吸附剂性质，将细胞成分完全破坏并以薄层形式分散在吸附剂颗粒表面上也是可行的［图 12.1（4）］。

涉及简易设备的机械混合过程看似简单又直接，可以在实验室或现场进行。但是，在随后的洗脱步骤中应该适当和广泛地分解样品，以尽可能扩大暴露于溶剂中的样品表面，这也是 MSPD 流程的一个关键方面。吸附剂的使用数量取决于样品的性质。但是，对于大多数涉及药物和有机微污染物分析的应用研究，样品：吸附剂的比例在（1：1）～（1：4）范围内可以形成所需的干燥、均匀、粉状混合物[5,11,18,20]。获得这种混合物所需的时间是变化的，大约需要 30s[3]～1h[24]。这种可变性可能依赖所研究基质的数量和性质，样品数量越大，均质所需的时间越长。混合时间还应确保结缔组织和其他可能存在基质中刚性生物聚合物的有效破坏[5]。因此，所研究样品中有大量这些物质时，通常需要更长的混合时间。作为固体支持物，所用吸附剂的性质取决于样品成分、所研究化合物的性质和所选的洗脱方法。然而，在迄今为止的大多数应用指出目标分析物是直接从 MSPD 柱上洗脱出来，亲脂键合相衍生材料如 C_8 或 C_{18} 等已被用来促进吸附剂表面上的细胞成分稀释并均匀分散。所得的双相键合相/分散样品脂质结构[8]既有助于洗脱时增加干扰基质组分在 MSPD 柱中的保留，又有助于增大样品和溶剂相互作用的表面积，这两个关键方面决定了 MSPD 的处理效率。

混合步骤完成后，通常将均质样品填充到空柱中或 SPE 吸附剂顶部，洗脱前不进行任何额外干燥或处理。大多数情况下，MSPD 柱是一个空的注射器筒，或者带不锈钢或者聚丙烯玻璃料的 SPE 柱，或者是纤维素过滤器，也或底部塞有硅烷化玻璃棉。应仔细包装样品并压实以避免在柱上形成通道或气泡。用注射器柱塞压实前，在样品顶部一般再放一些玻璃棉或者玻璃料。大型玻璃柱也可用于数量较大的样品（例如提取痕量多氯联苯和 PCDD/Fs[24]或农药[25]）。同时小型不锈钢柱已用于涉及微小样品和联用系统的小型化方法中[11,13,22,23]。

MSPD 的最后一步是用适当的溶剂或一系列溶剂从柱上将目标分析物洗脱下来。这里有两种可能的方法。在最简单的情况下，干扰成分有选择的保留在柱上，目标分析物直接从 MSPD 柱上洗脱下来。或者在用另一种溶剂选择性洗脱目标分析物之前，干扰基质化合物从柱中洗脱出来而目标化合物保留在柱里。

对相对简单的基质，载体吸附剂和洗脱方案适当组合可以提供直接用于分析的提取液（到目前为止最终用于仪器分析的技术可提供所需的灵敏度）[26]。对于复杂的基质，通常需要额外净化来收集需要提取液。一般通过将 MSPD 柱与 SPE 柱[12]直接连接进行额外的净化，或如前所述将适当的 SPE 吸附剂装在 MSPD 柱[3,12]的底部进行纯化（参阅后面的讨论）。

尽管 MSPD 技术具有许多积极的功能，但应该注意，在所有情况下将样品与分散剂混合然后将所得混合物填充到柱中都需要手动操作，这必定失去了无人操作的可能性。

12.2.2 影响基质固相分散萃取的可变因素

与 SPE 一样，MSPD 过程涉及的三个方面是基质、分散吸附剂和萃取溶剂。但在这几个方面 MSPD 性能与 SPE 有所不同[8]。首先，在 MSPD 过程中样品基质被破坏并分散成大表面，为后续溶剂萃取提供了有效的表面接触。同时，在 MSPE 过程中样品破碎应单独进行，并且要在上柱前进行。其次，在 MSPD 中，样品是均匀分布并保留在整个萃取柱中。相反，在 SPE 中，样品主要保留在柱的上部。第三，在基质成分、固相支撑物以及萃取溶剂之间存在物理和化学相互作用，这些相互作用在许多方面与 SPE 不尽相同。但是，SPE 和其他色谱分离的通用原理是适用于 MSPD。换句话说，固体支撑物的特征和化学性质可能会影响目标分析物（和基质成分）在 MSPD 柱上的保留和洗脱效果。因此方法开发和优化过程中应考虑到分散吸附剂和萃取溶剂这两个参数的影响。尽管 MSPD 的选择性严格取决于所使用的吸附剂/溶剂的组合，但要意识到分散剂和基质样品之间不同的相互作用，所研究基质的性质在 MSPD 过程中也发挥着重要作用。

12.2.2.1 吸附剂选择

通常 MSPD 中使用的吸附剂材料与在 SPE 中使用的材料是相似的。总体而言，为了增大表面积来改善样品的分散性和溶剂接触性、降低固体支持物的成本以及避免因较小的粒径（3~10μm）导致的流量受限或柱堵塞，粒径在 40~100μm 范围的吸附剂提供了一个很好的折中方案，较小粒径的吸附剂主要用于小型 MSPD[11,27]。

样品和吸附剂比率是影响 MSPD 性能的关键参数。吸附剂的量应足以确保样品充分破碎并均匀分散在材料表面。文献中大多数 MSPD 应用涉及的样品与吸附剂之比为（1∶1）～（1∶4）[8,15,17,20,21]。然而也有一些例外的情况[28]，表明该参数对 MSPD 效率有着重要的影响，需要仔细优化样品和吸附剂的比例。

在柱填充之前，根据所选的洗脱方案（非极性溶剂用作洗脱液时）需要将样品和吸收剂的最终混合物完全干燥。不像其他技术单独进行样品干燥（如处理前加热或冷冻干燥），在 MSPD 中可以通过加入适量的干燥剂（如钠硫酸盐）轻松实现在线干燥生成 MSPD 混合物[12]。

目前为止，反相材料是 MSPD 中使用最多的吸附剂，特别是与 C_8 和 C_{18} 硅胶结合的相[8,15,17,20]。可能是最初 MSPD 在处理生物样品和食品时表现出的优势导致了这样的趋势。但是，应该注意的是在涉及环境非脂质样品时，如室内灰尘[29]，C_{18} 还是首选的吸附剂。

反相材料可以选择性保留低极性和中等极性基质成分。但是基质成分在分散和洗脱过程中有着显著作用，对某一种特定应用而言，不同键合相的可行性最好通过实验来确定。Kristenson 等人[11]通过萃取柑橘中某些农药对两家商业公司生产的 C_8 键合硅胶与 C_{18} 键合硅胶和二氧化硅进行了比较。基质成分和三种分散剂之间的相互作用的差异性显而易见。如图 12.3 所示，当使用二氧化硅作为吸附剂载体时，许多农药被强烈地保留在 MSPD 柱中。同时，用 C_{18} 和 C_8 获得了相似（总体上令人满意）的结果，在气相色谱质谱（GC-MS）分析中后者提供了更干净干扰较少的提取物。

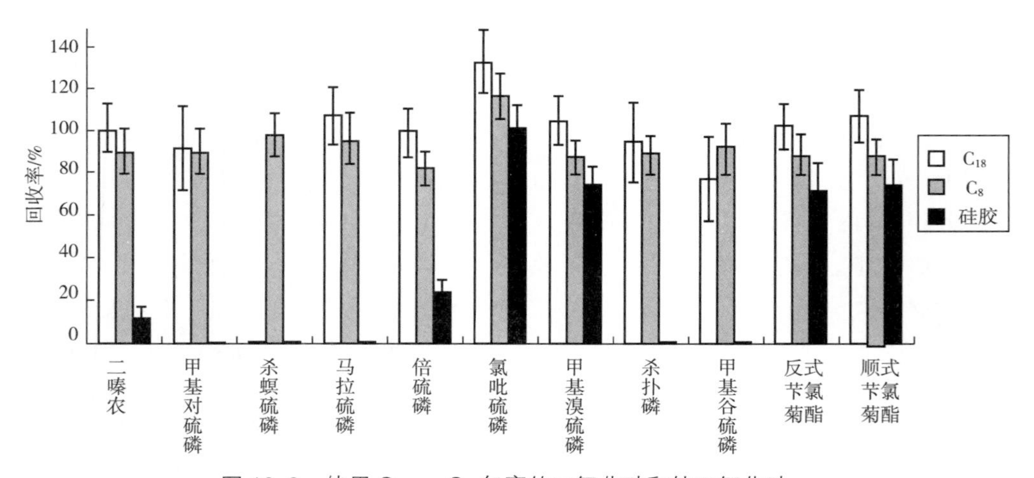

图 12.3　使用 C_{18}、C_8 包裹的二氧化硅和纯二氧化硅，
乙酸乙酯洗脱时从橙子中提取选定农药的回收率 ±SD（n=4，添加水平，0.5mg/kg）[11]

使用不同公司的 C_8 吸附剂获得的结果之间没有观察到差异，这一发现与其他作者先前的观察一致[5]，尽管额外净化的要求取决于最终用于分离检测的仪器，结合适当的洗脱方案，反相材料通常可以提供现成的分析提取物。根据干扰的性质，可以采取不同的净化方法（除了随后的离线纯化收集的提取物[12,30,31]）。例如极性分析物在用甲醇、乙腈、温水或其混合物等溶剂进行分析物提取之前，可以很容易地（而且更有选

择性地）从 MSPD 柱中洗出[32]。其他较小极性分析物可以通过柱上化学降解[33]或 MSPD 柱底部填充的吸收剂[12]来消除。

通过 MSPD 分析不同基质中多种分析物，正相无机材料如纯二氧化硅（或化学改性）、氟硅和氧化铝及其混合物，也被证明可以提供满意结果（见综述［15，17，20］，以及其中的参考文献）。然而，由于与反相吸附剂（即吸附与稀释）相比，（极性）分析物在这些材料表面的保留范围较小，一般来说，通常在这些实验中优化洗脱方案的更费力。根据所研究化合物的极性，从非极性到甲醇混合物的溶剂已被用于分析物的洗脱。但是，一般情况下，洗脱方案的复杂性取决于所研究的基质的复杂性，从用单一合适的溶剂[26]直接洗脱到相对复杂的洗脱方案都有不同之处。后来的一个例子提到，在目标极性化合物选择性洗脱前，脂肪样品的处理通常包括一个非极性溶剂（如正己烷）洗脱步骤以消除非极性干扰（见参考文献［34］）。或者，一种更常见的 C_{18} 吸收剂（但不完全是[35]）已被用于共萃物的柱上分馏。其他可能的方法主要是通过与饱和氢氧化钾溶液（在土壤中）或酸性二氧化硅（在含有脂肪的基质中）[12,36]的甲醇溶液混合进行化学消解来消除样品干扰。

最后，值得一提的是，正如 Barker 和同事的早前研究所证明的那样，惰性固体载体，如盐、沙子和硅藻土，也能够破坏样品结构，产生满足色谱分析的干燥的均质粉末。然而在这些混合物中，分析物与固体载体之间相对较弱的相互作用使得 MSPD 过程的选择性本质上取决于分析物在洗脱溶剂中的溶解度。因此，在方法开发中，应仔细优化洗脱方案[33,37]。

12.2.2.2　溶剂选择和洗脱方案

洗脱溶剂选择应尽可能以定量和选择性的方式从 MSPD 柱中提取目标分析物，最好是消耗溶剂最少。然而，选择的最佳方案（如直接洗脱或顺序洗脱）是由所研究基质的复杂性、作为固体载体的吸附剂以及最终仪器分析测定系统提供的选择性和灵敏度决定的。

在直接洗脱中目标分析物用适当的溶剂直接从 MSPD 柱中洗脱，而不需要的基质组分保留在吸附剂表面。当使用反相材料作为分散剂时，这种洗脱方案最常用。在这种情况下，非极性溶剂已被报道可以从基质（如水果、谷物、面粉和果汁）中提供相对浓缩和干净的提取物。然而，含有大量脂质的基质，如杏仁、橄榄油或脂肪食品，通常需要额外的处理以确保在仪器分析之前完全去除脂肪（见文献[15,17,20]及其中的参考资料）。这种处理可以在 MSPD 之前、柱上（添加适当吸收剂[12,35,36]）或离线（通过提取物额外 SPE 纯化/分离收集）进行[12,30]。

关于正相的 MSPD 吸附剂，采用直接分析物洗脱的方法可以使用热水作为萃取剂。许多这些应用使用沙子作为固体支撑，并提供可以直接用于仪器分析的提取物。该方法已经在测定食品[38,39]和动物组织[40]中的抗生素、蔬菜和水果[41]中的农药以及茶叶和咖啡[42]中的咖啡因方面得到了证明。一般来说，在使用 LC-MS 进行最终测定时，可以获得 70%以上的回收率和满意的 RSD。使用助溶剂（如甲醇）并不总是有助于提高方法的性能，正如预期的那样，往往导致大量不受欢迎的基质组分的共萃取[43]。

与直接洗脱方案相比，依序洗脱方法为溶剂选择提供了更高的灵活性，由于最终设定的洗脱步骤的选择性提高，通常会产生更干净、浓缩度高的提取物。一般来说，当使用某种有限保留能力的固体载体（即正相或惰性吸附剂）或处理复杂基质时，特别是目标物是较少组分时采用这种洗脱方案。正如已经解释过的，通过简单地加入先前的洗脱溶剂来引入额外的净化步骤在许多方面是有利的，但如前面讨论的应用案例，也取决于样品的性质和目标分析物的极性。除此之外，使用更苛刻的条件从 MSPD 柱中去除干扰化合物仅受所研究分析物化学稳定性的限制[33,44,45]。

12.3 基质固相分散萃取的发展趋势

最近在 MSPD 的创新主要是开发更具选择性、快速和高效的萃取程序。为此目的，对具有改善吸附能力或特定识别能力的新型吸附剂进行了分析，并评估了与使用定制溶剂相关的好处。

为了提高 MSPD 的效率，减少萃取次数和溶剂消耗，需要将 MSPD 已与现代溶剂萃取技术相结合，特别是将 PLE 和 UAE 技术相结合。最后，与在其他应用领域一样[2]，小型化可以被认为是 MSPD 的一种趋势。在这一领域，对颗粒较小的样品来说，小型化已被认为是一种有价值、简单和绿色的处理方案，而且也是建立完整的联用系统以提高样品通量的必要步骤。表 12.1 所示为这些趋势的一些有代表性的例子。

12.3.1 新型吸附剂

在过去的几十年中，已经合成了许多吸附剂材料，试图避免前面章节中描述的常规固相的主要限制。这些新的吸附剂增强了吸附能力，改善了化学保留的（甚至定制）选择性或替代机制[57,58]。这些改进的特性有时会促进新 MSPD 装置的发展，尤其是小型化方向，但这必须与高灵敏和高选择性的仪器联用来测定最终分析物。

提高碳纳米管（CNTs）的负载能力在 MSPD 中是有利的，因为它有助于减少吸附剂的数量，从而简化样品制备过程[20,21]。在一项有代表性的研究中，Vosough 等[46]从复杂基质城市污水污泥中提取磺胺嘧啶、磺胺嘧啶和磺胺甲恶唑，比较了两种纳米材料（二氧化硅多壁碳纳米管（MWCNT）和纳米二氧化硅-C_{18}）的性能。

MWCNT 方法通过少量分散在 30mg 的 CNT 材料的干燥样品（200mg）和 200mg 二氧化硅作为助吸收剂来准确地测定分析物（回收率在 91%～95%；RSD 低于 10%）。5mL 乙腈足以进行定量和选择性洗脱。这些结果优于以纳米二氧化硅-C_{18} 为分散剂的结果，该方法样品均匀化过程中需要 300mg 吸附剂，7mL 乙酸乙酯作为洗脱剂。尽管用于样品制备的吸附剂数量较大，但与基于 MWCNT 的方法相比，在 LC-DAD 色谱图中可以看到更多的干扰（图 12.4）。

MWCNTs 是目前流行的多功能材料，可用于不同类型的基于 MSPD 的应用研究（表 12.1），但 Shen 等[13]发现它们很难与鱼组织形成均质化混合物。这些作者提到虽然没有形成预期的均匀混合物，但形成了一个黑色的薄膜，不利于从这个脂肪样品中提取磺酰胺。事实上，用 MWCNTs 获得的回收率整体上（和显著地）比用 C_{18}、中性二氧化硅或亲水亲脂平衡（HLB）材料获得的回收率低。HLB 最终被选为研究对象。

表 12.1　基质固相分散萃取最新发展趋势的应用研究

基质（mg）	分析物	分散剂/mg	提取溶剂	回收率（RSD）*/%	检测仪器	参考文献
新型吸附剂						
鱼组织（10）	磺胺类	HLB（20）	甲醇：水：氨（50：49：1）（0.2mL）	71~91（<10）	LC-MS/MS	[13]
城市污水污泥（200）	抗生素	MWCNT（30）+二氧化硅（200）	乙腈（10mL）	91~95（3~10）	LC-DAD	[46]
黄油（500）	激素	石墨 MWCNTs（300）+ MWCNTs（100）	乙酸乙酯（10mL）	85~112（2~9）	—	[47]
薯片，吐司（1000）	丙烯酰胺	壳聚糖-MWCNTs（2000）	n-C_6（2mL）+甲醇：水：乙酸（90：10：2；4mL）	85~94（4）	LC-UV	[48]
土壤，树皮，鱼（100）	PBDEs，MeO-BDEs，OH-BDEs	化学修饰的石墨（10）	n-C_6：DCM（1：1；0.5mL）+丙酮（1mL）	29~116（3~20）	—	[49]
十里香叶子（L.）Jack（25）	聚甲氧基黄酮类	GO 包裹的硅（50）	甲醇（5mL）	93~102（2~5）	UHPLC-UV	[50]
原蜂胶（50）	酚酸和黄酮类	［C_6MIM］Cl-硅（200）	n-C_6（20mL）+甲醇（15mL）	66~92（9~11）	LC-UV	[51]
新型溶剂						
橙子（50）	具有生物活性的黄酮苷	Florisil（150）	［BMIM］BF4（250mol/L，0.4mL）	90~97（1~3）	UPLC-UV	[52]
联用其他萃取技术						
鱼肌肉（2000）	OCPs	Al_2O_3（6000）+Na_2SO_4（10000）+ s-PLE（5min×2；70℃）	n-C_6：DCM（1：1；3.5mL）	91~104（2~5）	GC-MS	[53]
饲料（250）	PCBs 和 PBDEs	Acidic SiO_2（250）+Na_2SO_4（250）+s-PLE（7min ×2；50℃）	n-C_6：DCM（1：1；3.5mL）	86~114（<20）	GC-ITD（MS-MS）/GC-NCI-MS	[54]
水果（200）	OPPs14 和三嗪类	C_8（200）+UAE（1min）	乙酸乙酯（0.7mL）	68~139（2~17）	—	[55]
联用系统						
柑橘，体液（150uL）	黄酮类	C_{18}（200）+UAE（6min）	甲醇（0.5mL，pH，4）	85~101（4~7）	LC-UV	[56]

注：［BMIM］BF4—1-丁基-3-甲基咪唑四氟硼酸酯；［C6MIM］Cl—1-己基-3-甲基咪唑氯化物；ACN—乙腈；DCM—二氯甲烷；EtAcO—乙酸乙酯；HLB—亲水亲脂平衡柱；MeO-BDEs—甲氧基溴代二苯醚；MeOH—甲醇；MWCNTs—多壁碳纳米管；n-C_6—正己烷；OCP—有机氯化农药；OH-BDEs—羟基溴代二苯醚；PBDEs—多溴联苯醚；PCBs—多氯联苯；s-PLE—选择性加压液体萃取。

* 相对标准偏差（RSD）%。

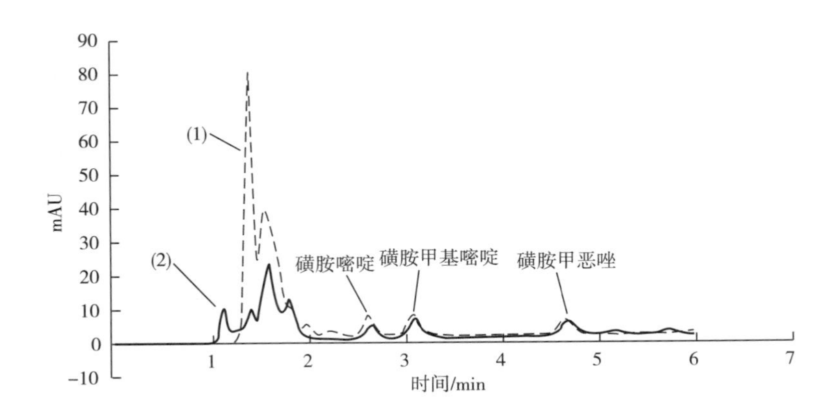

图 12.4　城市污水污泥在经过多壁碳纳米管二氧化硅（1）和纳米二氧化硅-C_{18}（2）基质固相分散后，通过配有紫外可见光检测器的重叠液相色谱检测获得的色谱图[46]

尽管只使用了 10mg 样品，但该吸附剂可以增强吸附能力，再加上最终使用高选择性和高灵敏度的 LC-MS/MS 技术，可以准确测定鱼肉中所研究的 14 种磺酰胺类的绝大部分。在少量的样品中，20mg 的 HLB 足以进行均匀的样品分散。然后将混合物置于柱顶端，仅在 6min 内就能以最少的溶剂完成样品制备（200mL 混合物，甲醇：水：氨，50：49：1，体积比）。方法优化后其性能略优于使用 MSPD 萃取小柱的常规方法（回收率 71%~91%比 66%~87%，RSD 均低于 12%）。基质效应在这两种情况下都是最小的，本质上是相似的。

有趣的是，化学改性的 MWCNTs 在某些应用方面比非改性材料具有更好的结果。通过嫁接壳聚糖的 MWCNTs 分析薯片和烘烤食品中丙烯酰胺时采用了这种方法。在这种情况下，改性的 MWCNTs 比未改性的材料可以更有效地吸附丙烯酰胺（回收率 85%~95%；RSD4%）[48]。

原则上石墨烯具有很大的表面积、化学稳定性、柔韧性和富 π 电子结构，是一种很好的吸附材料。Liu 等[49]一项研究利用这些有利特性，结果表明，这种吸附剂在提取多溴二苯醚及其羟基化（OH-PBDEs）和甲氧基化（MeO-PBDEs）类似物方面比传统材料（C_{18}-二氧化硅、硅酸镁）和 MWCNTs 具有更好的回收率。然而，其他作者也遇到了直接使用这种材料的一些实际困难，即因微小石墨烯片产生玻璃熔块堵塞的风险，不适合有效地破坏样品结构，难以从研钵壁上完全收集小石墨烯，以及通过范德华力相互作用产生不可逆的聚合。将石墨烯固定在合适的硅基吸附剂上，避免了这些问题，同时保持了石墨烯片的有利特性[50]。该方法成功地从中国一种重要的传统草本植物千里香叶中提取多甲氧基黄酮类化合物。本研究通过将氧化石墨烯（GO）的羧基与水溶液中二氧化硅的氨基连接制备了二氧化硅包覆氧化石墨烯。用 50mg 这种吸附剂对 25mg 草本植物叶片进行均质 3min。所得混合物装入 1mL 聚丙烯小柱中，用 5mL 甲醇洗脱。将提取物稀释后不需要纯化，在 3min 内通过超高效液相色谱紫外线检测器测定分析物。可以从图 12.5 中看出，与其他萃取条件苛刻的常规萃取工艺［回流萃取

（RE）或 UAE］相比，优化方法可以明显提高选择性和效率。

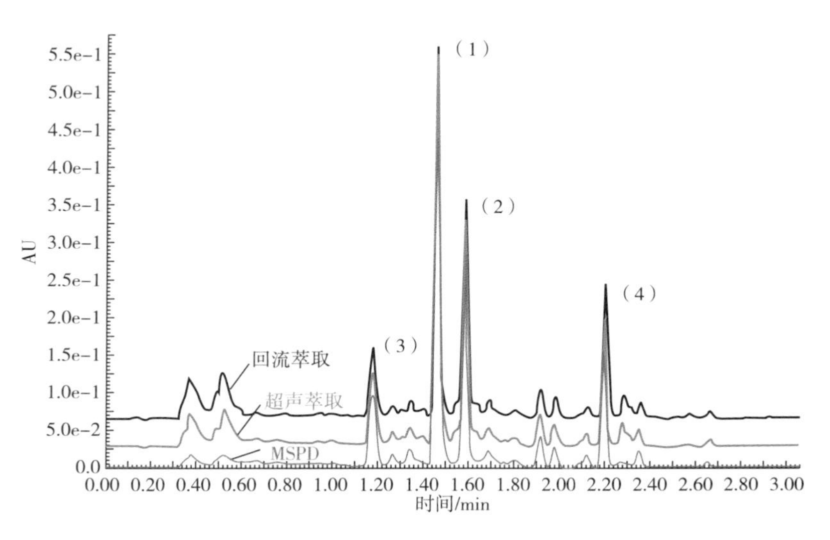

图 12.5 不同的提取程序下从紫菜叶中聚甲氧基化黄酮类化合物的 UPLC-UV 色谱图[50]

新型亲水分子印迹物质（MIPs）的合成促进了它们在 MSPD 中的应用，大多数情况下，这些物质与更粗糙的吸附剂（沙子[59]）相结合，以确保基质适当分解。这些材料除了固有的选择性[58]，其表面亲水性可以促进洗涤步骤并入洗脱过程，有助于分析物洗脱前进一步消除可能存在的基质干扰。因此收集相对浓缩和干净的提取物后，即使没有对其进行额外的纯化，也可以用相对简单的检测器检测其中所研究的化合物。举一个该方法处理复杂亲脂基质的例子，Arabi 等人通过基于功能化二氧化硅纳米粒子的 MIP 从食品饼干和面包中选择性提取和预富集丙烯酰胺[59]。优化后的方法仅需样品 100mg 分散在 150mg 丙烯酰胺-MIP 和 126mg 沙子上就可以实现。脂质和非极性基质组分在提取柱上用 1mL 正己烷就可以被选择性地洗脱，随后用 2.5mL 乙腈和甲醇（50∶50）洗脱液萃取丙烯酰胺。结合脱脂步骤，MIP 改进选择性可以获得非常干净的提取物，甚至可以使用 LC-UV 直接进行测定，检测限（LODs）为 16μg/kg。

最近的一些研究也探讨了在 MSPD 过程中使用离子液体（ILs）的可能性。将二氧化硅吸附剂直接浸泡在 IL 甲醇溶液中，室温下搅拌 12h 后在 150℃ 烘箱中加热除去其中多余溶剂，可以将［C6MIM］Cl 固定在二氧化硅上。所得二氧化硅材料含有 10% 的［C6MIM］Cl，可从 50mg 蜂胶中选择性回收咖啡酸、阿魏酸、莫林、木樨草素、槲皮素、芹菜素、大黄素和山柰素。在吸附剂和样品比为 4∶1 时，用正己烷洗涤后再用 15mL 甲醇选择性洗脱目标分析物。使用 LC-UV 检测时，其 LODs 范围在 5.8～22ng/mL 内。与其他常规提取方法（如 UAE 和 Soxhlet 提取）相比，MSPD 方法的回收率较低（66%～92%），但仍可接受，但所需的溶剂和时间明显减少。

12.3.2 新型溶剂

过去的十年里，在许多研究领域中替代传统挥发性有机溶剂的新型溶剂深受青睐。

在这些溶剂中，ILs 具有化学稳定性、非挥发性，而且通过取代其结构中阳离子或阴离子从而比较容易调整其性能，因而特别具有吸引力[60]。目前为止，在所有关于此方法的文献资料中只能找到一篇基于 MSPD 的研究[52]。在这项工作中，50mg 的干燥酸橙样品和 150mg 硅酸镁在研钵中研磨均质化 1min。将混合物装入聚丙烯 SPE 小柱后，用 0.4mL 的 250mmol/L [BMIM]BF4 洗脱目标分析物（具有生物活性的黄酮苷）。萃取物以 13000r/min 离心 5min 后通过 UPLC-UV 进行分析。作者认为黄酮苷的-OH 基团可以通过氢键与硅酸镁表面连接。然后，[BMIM]BF4 中的氟原子将促进溶剂与目标化合物形成更强氢键，使得洗脱溶剂消耗最少进而获得浓缩提取物。以 60mg/g 的添加水平计算，回收率在 90%~97%的范围内，LODs 在 4~5μg/g 范围内。

12.3.3 溶剂萃取强化联用技术

不论外源化合物和内源性化合物，大多数 MSPD 方法都可以获得满意回收率，但从高吸附性基质中提取分析物往往需要使用相对较多的溶剂[24]。在这些情况下，在萃取柱上应用额外的能量通常有助于提高萃取效率，这反过来又有助于减少定量回收分析物所需的洗脱液体积和总分析时间。到目前为止，最常与 MSPD 联用的强化萃取技术是 PLE 和 UAE。这些组合方法及其主要结果和结论可在以前的论文[1,20,21]中找到。在这里，我们只对最近几项有代表性的研究进行探讨（表 12.1）。

MSPD 和 PLE 技术各自的特点，以及它们组合后的特点，可以从不同的研究小组在含脂复杂基质中测定特定类别的持久性机污染物（即有机氯农药（OCPs）[53]、多氯联苯（PCBs）和多溴二苯）的提取条件中看出[54]。Shen 等人[53]采用 MSPD 方法处理鱼组织时建议使用酸性氧化铝和硫酸钠，并在 60℃下用（两个 5min 循环）正己烷和二氯甲烷（3∶7，体积比）进行 PLE 时使用氧化铝吸收剂来选择性保留共萃取脂肪残留物。同时，Pena-Abaurrea 等[54]通过小型化 PLE 方法中，将 250g 饲料分散在类似数量的酸性二氧化硅和硫酸钠中，通过 2g 酸性二氧化硅进行柱上脂肪保留，该过程在 50℃正己烷和二氯甲烷（1∶1，体积比）的条件下由两个 7min 循环组成。在后者混合物中，二氯甲烷的含量可能较高，这对于用较小溶剂体积（3.5mL）来定量提取高吸附性多溴二苯醚是很有必要的，因此必须使用较多吸收剂，并避免使用与前一种方法相比更高的萃取温度。在任何情况下，这两种程序都产生了可用于直接分析的提取物，定量回收率都超过了 86%（表 12.1）。

同样，所谓的超声辅助 MSPD（US-MSPD）是指将 MSPD 柱浸泡在超声波浴[56]或超声反应器中对其超声[55]，该方法已被证明是提高分析物回收率或加快提取过程的有效方法。Barfi 等[56]指出与传统的 MSPD 相比，使用 US-MSPD 时，柑橘果汁以及人尿和血浆中黄酮类化合物的回收率增加了 10%~25%。同时在小型化 MSPD 方法中，超声反应器在 50%功率和提取 1min 的条件下，从水果中提取农药得到了最好的结果（回收率在 68%以上，重现性在 15%以下）[55]。

12.3.4 小型化发展

与许多其他研究领域一样[2]，MSPD 小型化是一个明显趋势[20,21]。表 12.1 概括的许多应用研究仅使用小于 0.5g 的少量样品，促使每个样品分析所需试剂和溶剂数量显著减少。这样，现有方法的小型化有助于发展更绿色和更具成本和时间效益的分析过程。另一方面，在开发部分集成的特别是联用程序时，简化样品制备步骤是一个基本要求。几年前 Lu 等[23] 和 Gutierrez-Valencia 等[22] 几乎同时证明了后一种方法的可行性。图 12.6 显示了这种联用系统中使用的典型配置，对于这种配置，整个设置的最终复杂性一般取决于所研究基质的性质以及所用仪器的选择性和灵敏度[23]。因此，LC-UV 具有相对较低的灵敏度和选择性，可能需要将提取物进行额外的（在线）SPE 净化[22]，但通过 LC-MS/MS 测定时却不需要这一步[23]。

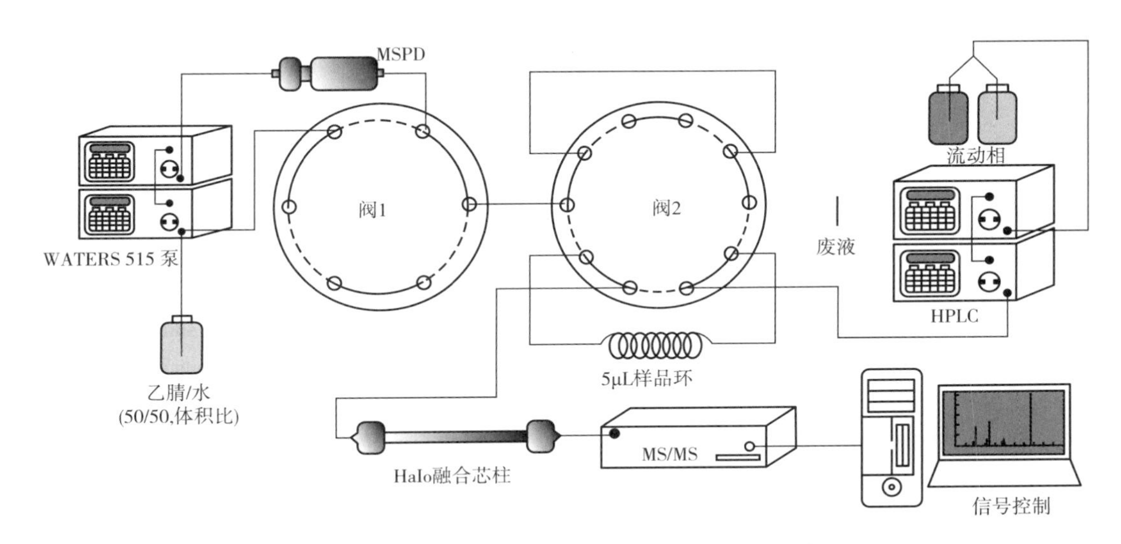

图 12.6 基质固相分散-LC-MS/MS 的设置[23]

12.4 结论

目前 MSPD 是一种公认的技术，在许多分析应用领域已经证明了它的检测能力。该方法相当通用、快速和简单，它不涉及任何特殊设备，而且与其他提取技术相比，样品和试剂的消耗大大减少。设计合理时，方法优化后可以产生直接用于分析的提取物，从而进一步减少样品操作步骤，加快分析过程。此外，对先前的程序进行稍加修改，缩减过程就可以实现小型化，并且许多文献已经证明了后续与色谱检测系统在线联用的可行性。将新设计的纳米结构溶剂和材料纳入 MSPD 处理中，并与增强溶剂萃取技术相结合，促进了新特性方法的发展，激发了新研究的大量涌现，这说明 MSPD 是一个非常活跃的研究领域。

参考文献

[1] L. Ramos, Critical overview of selected contemporary sample preparation techniques, J. Chromatogr. A 1221 (2012) 84-98;

[1a] J. Płotka-Wasylka, N. Szczepańska, K. Owczarek, J. Namieśnik, Miniaturized solid phase

extraction, in: E. Ibanez, A. Cifuentes (Eds.), Green Extraction Techniques: Principles, Advances and Applications, vol. 76, 2017, pp. 279-318;

[1b] J. M. Kokosa, Selecting an appropriate solvent microextraction mode for a green analytical method, in: E. Ibanez, A. Cifuentes (Eds.), Green Extraction Techniques: Principles, Advances and Applications, vol. 76, 2017, pp. 403-425;

[1c] P. Serra-Mora, Y. Moliner-Martinez, C. Molins-Legua, R. Herráez-Hernández, J. Verdú-Andrés, P. Campíns-Falcó, Trends in online in-tube solid phase microextraction, in: E. Ibanez, A. Cifuentes (Eds.), Green Extraction Techniques: Principles, Advances and Applications, vol. 76, 2017, pp. 427-461;

[1d] J. M. Florêncio Nogueira, Stir bar sorptive extraction, in: E. Ibanez, A. Cifuentes (Eds.), Green Extraction Techniques: Principles, Advances and Applications, vol. 76, 2017, pp. 463-481;

[1e] É. A. Souza-Silva, J. Pawliszyn, Recent advances in solid-phase microextraction for contaminant analysis in food matrices, in: E. Ibanez, A. Cifuentes (Eds.), Green Extraction Techniques: Principles, Advances and Applications, vol. 76, 2017, pp. 483-517;

[1f] J. Moreda-Piñeiro, A. Moreda-Piñeiro, Recent advances in the combination of assisted extraction techniques, in: E. Ibanez, A. Cifuentes (Eds.), Green Extraction Techniques: Principles, Advances and Applications, vol. 76, 2017, pp. 519-573.

[2] L. Ramos, B. Richter, Extraction of Micropollutants from Size-Limited Solid Samples, vol. 29, LC-GC Europe, 2016, pp. 558-568.

[3] S. A. Barker, A. R. Long, C. R. Short, Isolation of drug residues from tissues by solid phase dispersion, J. Chromatogr. 475 (1989) 353-361.

[4] S. A. Barker, L. C. Hsieh, C. R. Short, Methodology for the analysis of fenbendazole and its metabolites in plasma, urine, feces, and tissue-homogenates, Anal. Biochem. 155 (1986) 112-118.

[5] S. A. Barker, Matrix solid-phase dispersion, J. Chromatogr. A 885 (2000) 115-127.

[6] S. A. Barker, T. McDowell, B. Charkhian, L. C. Hsieh, C. R. Short, Methodology for the analysis of benzimidazole anthelmintics as drug residues in animal-tissues, J. Assoc. Off. Anal. Chem. 73 (1990) 22-25.

[7] S. A. Barker, Applications of matrix solid-phase dispersion in food analysis, J. Chromatogr. A 880 (2000) 63-68.

[8] S. A. Barker, Matrix solid phase dispersion (MSPD), J. Biochem. Biophys. Methods 70 (2007) 151-162.

[9] A. R. Long, L. C. Hsieh, M. S. Malbrough, C. R. Short, S. A. Barker, Matrix solid-phase dispersion (MSPD) isolation and liquid-chromatographic determination of oxytetracycline, tetracycline, and chlortetracycline in milk, J. Assoc. Off. Anal. Chem. 73 (1990) 379-384.

[10] D. Wianowska, A. L. Dawidowicz, Can matrix solid phase dispersion (MSPD) be more simplified? Application of solventless MSPD sample preparation method for GC-MS and GC-FID analysis of plant essential oil components, Talanta 151 (2016) 179-182.

[11] E. M. Kristenson, E. G. J. Haverkate, C. J. Slooten, L. Ramos, J. J. Vreuls, U. A. T. Brinkman, Miniaturized automated matrix solid-phase dispersion extraction of pesticides in fruit followed by gas chromatographicemass spectrometric analysis, J. Chromatogr. A 917 (2001) 277-286.

[12] J. J. Ramos, M. J. Gonzalez, L. Ramos, Miniaturised sample preparation of fatty foodstuffs for the determination of polychlorinated biphenyls, J. Sep. Sci. 27 (2004) 595-601.

[13] Q. Shen, R. Jin, J. Xue, Y. Lu, Z. Dai, Analysis of trace levels of sulfonamides in fish tissue

using micro-scale pipette tip-matrix solid-phase dispersion and fast liquid chromatography tandem mass spectrometry, Food Chem. 194 (2016) 508-515.

[14] M. E. Hines, A. R. Long, T. G. Snider, S. A. Barker, Lysis and fractionation of mycobacterium-paratuberculosis and escherichia-coli by matrix solid-phase dispersion, Anal. Biochem. 195 (1991) 197-206.

[15] E. M. Kristenson, L. Ramos, U. A. T. Brinkman, Recent advances in matrix solid-phase dispersion, Trends Anal. Chem. 25 (2006) 96-111.

[16] M. Garcia-Lopez, P. Canosa, I. Rodriguez, Trends and recent applications of matrix solid-phase dispersion, Anal. Bioanal. Chem. 391 (2008) 963-974.

[17] A. L. Capriotti, C. Cavaliere, P. Giansanti, R. Gubbiotti, R. Samperi, A. Lagana, Recent developments in matrix solid-phase dispersion extraction, J. Chromatogr. A 1217 (2010) 2521-2532.

[18] L. Ramos, Matrix solid phase dispersion, in: J. Pawliszyn, H. Lord (Eds.), Comprehensive Sampling and Sample Preparation, Extraction Techniques, vol. 2, Elsevier, Academic Press, Oxford, UK, 2012, pp. 299-310.

[19] A. L. Capriotti, C. Cavaliere, A. Lagana, S. Piovesana, R. Samperi, Recent trends in matrix solid-phase dispersion, Trends Anal. Chem. 43 (2013) 53-66.

[20] A. L. Capriotti, C. Cavaliere, P. Foglia, R. Samperi, S. Stampachiacchiere, S. Ventura, A. Lagana, Recent advances and developments in matrix solid-phase dispersion, Trends Anal. Chem. 71 (2015) 186-193.

[21] J. Escobar-Arnanz, L. Ramos, The latest trends in the miniaturized treatment of solid samples, Trends Anal. Chem. 71 (2015) 275-281.

[22] T. M. Gutierrez Valencia, M. Garcia de Llasera, Determination of organophosphorus pesticides in bovine tissue by an on-line coupled matrix solid-phase dispersion-solid phase extraction-high performance liquid chromatography with diode array detection method, J. Chromatogr. A 1218 (2011) 6869-6877.

[23] Y. B. Lu, Q. Shen, Z. Y. Dai, H. Zhang, H. H. Wang, Development of an on-line matrix solid-phase dispersion/fast liquid chromatography/tandem mass spectrometry system for the rapid and simultaneous determination of 13 sulfonamides in grass carp tissues, J. Chromatogr. A 1218 (2011) 929-937.

[24] L. Ramos, E. Eljarrat, L. M. Hernandez, J. Rivera, M. J. Gonzalez, Comparative study of methodologies for the analysis of PCDDs and PCDFs in powdered full-fat milk. PCB, PCDD and PCDF levels in commercial samples from Spain, Chemosphere 38 (1999) 2577-2589.

[25] X. G. Chu, X. Z. Hub, H. Y. Yao, J. Chromatogr. A 1063 (2005) 201-210.

[26] J. Lia, Y. Lic, D. Xub, J. Zhangb, Y. Wangb, C. Luob, Determination of metrafenone in vegetables by matrix solid-phase dispersion and HPLC-UV method, Food Chem. 214 (2017) 77-81.

[27] J. J. Ramos, M. J. González, L. Ramos, Comparison of gas-chromatography-based approaches after fast miniaturised sample preparation for the monitoring of selected pesticide classes, J. Chromatogr. A 1216 (2009) 7307-7313.

[28] P. C. Abhilash, S. Jamil, N. Singh, Matrix solid-phase dispersion extraction versus solidphase extraction in the analysis of combined residues of hexachlorocyclohexane isomers in plant matrices, J. Chromatogr. A 1176 (2007) 43-47.

[29] N. Negreira, I. Rodríguez, E. Rubí, R. Cela, J. Chromatogr. A 1216 (2009) 5895-5902.

[30] R. Cholewa, D. Beutling, J. Budzyk, M. Pietrzak, S. Walorczyk, Persistent organochlorine pesticides in internal organs of coypu, Myocastor coypus, J. Environ. Sci. Heal. B 50 (2015) 590-594.

[31] A. Bacaloni, C. Cavaliere, F. Cucci, P. Foglia, R. Samperi, A. Lagana, J. Chromatogr. A 1179 (2008) 182-189.

[32] A. Argente-García, Y. Moliner-Martínez, P. Campíns-Falcó, J. Verdú-Andrés, R. Herráez-Hernández, Determination of amphetamines in hair by integrating sample disruption, clean-up and solid phase derivatization, J. Chromatogr. A 1447 (2016) 47-56.

[33] R. M. Garcinuño, L. Ramos, P. Fernández-Hernando, C. Cámara, Optimization of a matrix solid-phase dispersion method with sequential clean-up for the determination of ethylene bisdithiocarbamate residues in almond samples, J. Chromatogr. A 1041 (2004) 35-41.

[34] Y. F. Zhu, S. Y. Xie, D. M. Chen, Y. H. Pan, W. Qu, X. Wang, Z. L. Liu, D. P. Peng, L. L. Huang, Y. F. Tao, Targeted analysis and determination of beta-agonists, hormones, glucocorticoid and psychiatric drugs in feed by liquid chromatography with electrospray ionization tandem mass spectrometry, J. Sep. Sci. 13 (2016) 2584-2594.

[35] R. Djatinika, C. C. Hsieh, J. M. Chen, W. H. Ding, Determination of paraben preservatives in seafood using matrix solid-phase dispersion and on-line acetylation gas chromatography-mass spectrometry, J. Chromatogr. B 1036 (2016) 93-99.

[36] J. Zhan, J. Li, D. Liu, C. Liu, G. Yang, Z. Zhou, P. Wang, A simple method for the determination of organochlorine pollutants and the enantiomers in oil seeds based on matrix solid-phase dispersion, Food Chem. 194 (2016) 319-324.

[37] M. Rashidipour, R. Heydari, A. Feizbakhsh, P. Hashemi, Rapid monitoring of carvacrol in plants and herbal medicines using matrix solid-phase dispersion and gas chromatography flame ionisation detector, Nat. Prod. Res. 29 (2015) 621-627.

[38] S. Bogialli, G. D' Ascenzo, A. Di Corcia, A. Lagana, G. Tramontana, Simple assay for monitoring seven quinolone antibacterials in eggs: extraction with hot water and liquid chromatography coupled to tandem mass spectrometry Laboratory validation in line with the European Union Commission Decision 657/2002/EC, J. Chromatogr. A1216 (2009) 794-800.

[39] S. Bogialli, A. Di Corcia, A. Lagana, V. Mastrantoni, M. Sergi, A simple and rapid confirmatory assay for analyzing antibiotic residues of the macrolide class and lincomyc in bovine milk and yoghurt: hot water extraction followed by liquid chromatography/tandem mass spectrometry, Rapid Commun. Mass Spectrom. 21 (2007) 237-246.

[40] S. Bogialli, V. Capitolino, R. Curini, A. Di Corcia, M. Nazzari, M. Sergi, J. Agric. Food Chem. 52 (2004) 3286-3291.

[41] S. Bogialli, R. Curini, A. Di Corcia, M. Nazzari, D. Tamburro, J. Agric. Food Chem. 52 (2004) 665-671.

[42] A. L. Dawidowicz, D. Wianowska, J. Pharm. Biomed. Anal. 37 (2005) 1155-1159.

[43] S. Bogialli, A. Di Corcia, A. Laganá, V. Mastrantoni, M. Sergi, Rapid Commun. Mass Spectrom. 21 (2007) 237-246.

[44] M. T. Pena, M. C. Casais, M. C. Mejuto, R. Cela, J. Chromatogr. A 1165 (2007) 32-38.

[45] S. S. Caldas, C. M. Bolzan, E. J. de Menezes, A. L. Venquiaruti-Escarrone, C. D. G. Martins, A. Bianchini, E. G. Primel, A vortex-assisted MSPD method for the extraction of pesticide residues from fish liver and crab hepatopancreas with determination by GC-MS, Talanta 112 (2013) 63-68.

[46] M. Vosough, M. N. Onilghi, A. Salemi, Optimization of matrix solid-phase dispersion coupled with high performance liquid chromatography for determination of selected antibiotics in municipal sewage

sludge, Anal. Methods 8 (2016) 4853-4860.

[47] R. Su, X. Wang, X. Xu, Z. Wang, D. Li, X. Zhao, X. Li, H. Zhang, A. Yu, Application of multiwall carbon nanotubes-based matrix solid phase dispersion extraction for determination of hormones in butter by gas chromatography mass spectrometry, J. Chromatogr. A 1218 (2011) 5047-5054.

[48] H. Zhao, N. Li, J. Li, X. Qiao, Z. Xu, Preparation and application of chitosan-grafted multiwalled carbon nanotubes in matrix solid-phase dispersion extraction for determination of trace acrylamide in foods through high-performance liquid chromatography, Food Anal. Methods 8 (2015) 1363-1371.

[49] Q. Liu, J. B. Shi, J. T. Sun, T. Wang, L. X. Zeng, N. L. Zhu, G. B. Jiang, Graphene-assisted matrix solid-phase dispersion for extraction of polybrominated diphenyl ethers and their methoxylated and hydroxylated analogs from environmental samples, Anal. Chim. Acta 708 (2011) 61-68.

[50] T. Sun, X. W. Li, J. Yang, L. J. Li, Y. R. Jin, X. L. Shi, Graphene-encapsulated silica as matrix solid-phase dispersion extraction sorbents for the analysis of poly-methoxylated flavonoids in the leaves of *Murraya panaculata* (L.) Jack, J. Sep. Sci. 38 (2015) 2132-2139.

[51] Z. B. Wang, R. Sun, Y. P. Wang, N. Li, L. Lei, X. Yang, A. M. Yu, F. P. Qiu, H. Q. Zhang, Determination of phenolic acids and flavonoids in raw propolis by silica-supported ionic liquid-based matrix solid phase dispersion extraction high performance liquid chromatography-diode array detection, J. Chromatogr. B 969 (2014) 205-212.

[52] J. J. Xu, R. Yang, L. H. Ye, J. Cao, W. Cao, S. S. Hu, L. Q. Peng, Application of ionic liquids for elution of bioactive flavonoid glycosides from lime fruit by miniaturized matrix solid-phase dispersion, Food Chem. 204 (2016) 167-175.

[53] Z. L. Shen, D. Yuan, H. Zhang, M. Hu, J. H. Zhu, X. Q. Zhang, Q. D. Su, Matrix solid phase dispersion-accelerated solvent extraction for determination of OCP residues in fish muscles, J. Chin. Chem. Soc. 58 (2011) 494-502.

[54] M. Pena-Abaurrea, J. J. Ramos, M. J. González, L. Ramos, Miniaturised selective pressurized liquid extraction of polychlorinated biphenyls and polybrominated diphenyl ethers from feedstuffs, J. Chromatogr. A 1273 (2013) 18-25.

[55] J. J. Ramos, R. Rial-Otero, L. Ramos, J. L. Capelo, Ultrasonic-assisted matrix solidphase dispersion as an improved methodology for the determination of pesticides in fruits, J. Chromatogr. A 1212 (2008) 145-149.

[56] B. Barfi, A. Asghari, M. Rajabi, A. Barfi, I. Saeidi, Simplified miniaturized ultrasound-assisted matrix solid phase dispersion extraction and high performance liquid chromatographic determination of seven flavonoids in citrus fruit juice and human fluid samples: hesperetin and naringenin as biomarkers, J. Chromatogr. A 1311 (2013) 30-40.

[57] N. Fontanals, R. M. Marce, F. Borrull, P. A. G. Cormack, Hypercrosslinked materials: preparation, characterisation and applications, Polym. Chem. 6 (2015) 7231-7244.

[58] A. Martin-Esteban, Molecularly-imprinted polymers as a versatile, highly selective tool in sample preparation, Trends Anal. Chem. 45 (2013) 169-181.

[59] M. Arabi, M. Ghaedi, A. Ostovan, Development of dummy molecularly imprinted based on functionalized silica nanoparticles for determination of acrylamide in processed food by matrix solid phase dispersion, Food Chem. 210 (2016) 78-84.

[60] L. Ruiz-Aceituno, M. L. Sanz, L. Ramos, Use of ionic liquids in analytical sample preparation of organic compounds of food and environmental samples, Trends Anal. Chem. 43 (2013) 121-145.

13 绿色分析方法中选择合适的溶剂进行微萃取

John M. Kokosa*
Kettering University, Flint, MI, United States
Mott Community College, Flint, MI, United States
* 通讯作者：E-mail：jmkokosa@ yahoo. com

13.1 引言

术语溶剂微萃取（SME）和液相微萃取（LPME）是此类重要分析技术的起源[1-3]。这两个词语可以互换使用，以描述使用少量（0.5~100μL）溶剂提取和净化多种基质（包括环境、药物、临床、法医和生物学样品）中含有的分析物的过程。在本文中，将首选SME作为该技术的一般描述，因为许多作者还使用LPME一词来指代一种或多种特定的SME模式，而不是通称。此处，LPME仅用于经典中空纤维LPME（HF-LPME），这是此SME模式最常用的描述。为避免混淆，本文中各种SME模式的首字母缩略词使用将受到限制。作为读者的辅助材料，表13.1列出了此处使用的首字母缩写词及其含义。多年来，几乎每个SME研究小组都提出了数百个首字母缩略词来描述他们对SME模式的变化。这导致在研究论文中出现很多困难，试图确定哪种方法真正用于执行提取方法时非常困难。应鼓励读者首先将其SME研究分成此处涵盖的四个主要类别之一。SME已经使用了22年，随着本书的出版，研究、应用和综述类报道的总数可能会超过2000。至少可以说，试图找到一种可能适合分析特定样品中特定化学物质需求的特定方法是一项艰巨的任务。选择一种合适的SME模式的方法并不简单，特别是满足绿色分析要求的方法，实际上可以采取以下这几种方法[4-7]。这里将介绍选择合适的SME模式的主要方法，但是这些方法仅作为指导，而不是分析人员可用的唯一技术。除了考虑用于方法开发的特定研究论文之外，还鼓励读者首先考虑一些非常好的综述论文，这些论文涵盖了各种SME模式，其中一些在13.3中列出，尤其是一些较新的模式。由于大多数新的分析应用程序都需要考虑绿色技术，因此在选择SME模式时也必须考虑这一点[4,5]。与大多数传统的分析提取和纯化技术相比，SME模式是绿色的，具有经济和环保意义，替代了需要大量有毒和昂贵化学品的技术，并且将这些技术视为一般分析方案的一部分。还应记住，SME只是分析人员可以采用的几种绿色方法之一，并且不应将SME视为与这些技术竞争的工具，而应包容它们[8]。实际上，在分析方法开发中，SME已与其他绿色技术一起使用，包括固相萃取和固相微萃取。另外，大多数SME程序可以自动化到某种程度，并且可以在线使用，甚至与微流体设备兼容[9,10]。SME模式可分为两大类：溶剂直接暴露于样品中以及溶剂通过膜与样品分离。这些类别中的每一个又可以分为两个主要的SME类别。直接暴露的SME包含单滴微萃取（SDME）及其变体，如顶空SDME（HS-SDME），分散液液微萃取（DLLME）及其变体。膜保护溶剂微萃取包括HF-LPME及其变体、电膜萃取（EME）。电膜萃取其本身是HF-LPME的变体，但其自身也是很重要的，所以将其单独列出。SME系列中的某些技术，例如浊点提取（CPE）和膜辅助溶剂提取，将不在此处介绍。注意，当看研究论文中提出的方法时，应该记住，大多数论文都经过大学研究小组巨大的努力，目的是降低检测限，通常是使用少量分析物，获得干净的基质，而无需花费精力去验证该过程。因此，当为绿色分析方法选择SME模式时，都需要根据自己感兴趣的样品对任何公开的方法进行完善和验证。

此处选择正确的SME模式所采用的方法涉及多个路线，这些路线可以结合使用也可以独立使用。其中包括评估每种 SME 模式的优缺点；完善已发布的方法，考虑方法要求；以及使用绿色方法。参考文献中提供了对选择合适的 SME 模式所需的规则和方法的详细检查。

表 13.1　　溶剂微萃取模式术语和缩略语

SME 模式	缩略语
Solvent microextraction	SME
Liquid-phase microextraction	LPME
Single drop microextraction	SDME
Direct immersion single drop microextraction	DI-SDME
Headspace single drop microextraction	HS-SDME
Drop-to-drop microextraction	DDME
Directly suspended droplet microextraction	DSDME
Liquid-liquid-liquid microextraction	LLLME
Continuous-flow microextraction	CFME
Dispersive liquid-liquid microextraction	DLLME
Ultrasound-assisted dispersive liquid-liquid microextraction	UA-DLLME
Microwave-assisted dispersive liquideliquid microextraction	MA-DLLME
Vortex-assisted dispersive liquideliquid microextraction	VA-DLLME
Cloud point extraction	CPE
Hollowfibre liquid-phase microextraction	HF-LPME
2-Phase hollowfibre liquid-phase microextraction	HF（2）LPME
3-Phase hollowfibre liquid-phase microextraction	HF（3）LPME
Electromembrane extraction	EME

13.2　溶剂微萃取模式优缺点的评价

如引言中所述，SME 模式可分为两大类：将溶剂直接暴露于样品中以及样品与溶剂通过膜分离[4,11]。从样本基质的角度来看，这些区别很重要。除非事先进行净化，否则肮脏或复杂的基质（例如血液）可能会干扰分析。在这些情况下，通常使用膜保护的 SME 模式，否则通常首选非膜模式。在本节中，将简要概述每种 SME 模式的优缺点。需要指出的是对于那些不熟悉 SME 的人来说，常规使用的常用萃取溶剂（$CHCl_3$、乙醚、乙酸乙酯）不一定适用于该技术，因为它们的溶解度和挥发性都太高了。使用的典型溶剂包括与水不混溶的醇，例如 1-辛醇和 1-十二烷醇；碳氢化合物，例如甲苯，二甲苯和癸烷；离子液体（ILs）和深共熔溶剂（DESs）；在某些情况下，还包括纯天然油和萜烯。SME 几乎替换所有宏观溶剂萃取有机物和金属离子的方法，必要时使用标准络合剂和衍生化技术。提取物的分析仪器部分取决于提取溶剂，但包括所有常规色谱和检测器，包括气相色谱（GC），气相色谱/质谱（GC/MS），高效液相色谱（HPLC），高效液相色谱/质谱（HPLC/MS）和毛细管电泳（CE）。

13.2.1 溶剂微萃取的当前研究方向

正如预期的那样，研究人员致力于开发新的 SME 模式或改进 SME 模式以进行样品分析。为了改进四种 SME 模式以逐步增加检测限，研究人员已经做了很多工作，但很少有人关心方法验证的严格要求。尤其是在过去的 5 年中，已经开展了一些开发工作以使每种 SME 模式自动化，其中一些比较成功，并将绿色化学原理应用于方法开发[5,11-15]。这包括用毒性较小，生物蓄积性较小的化学品代替氯化烃萃取溶剂。还一直在强调对临床和法医应用显示出希望并缩短提取时间的技术。通过百分比模式比较 2010—2014 年发表的 SME 论文（图 13.1）与 2015—2016 年发表的 SME 论文（图 13.2），可以明显地看出 SME 研究的方向。SDME 研究论文的百分比稳步下降，而涉及较新技术的论文（尤其是 DLLME 和 EME）最近占据主导地位，而每年发表的研究论文总数却稳步增长。应该指出的是，图 13.1 中的数据不是将 HF-LPME 和 EME 的单独统计的，而是将它们组合在一起。

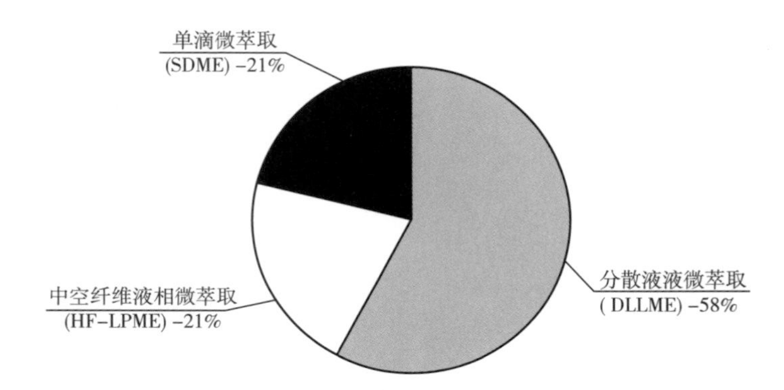

图 13.1 2010—2014 年关于单滴微萃取（SDME）、中空纤维液相微萃取（HF-LPME）/电膜萃取（EME）和分散液液微萃取（DLLME）等 4 种溶剂微萃取研究报道（大约 225 篇）

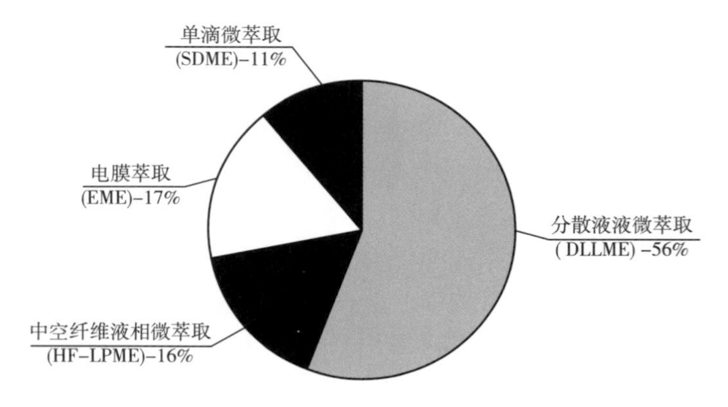

图 13.2 2015—2016 年关于单滴微萃取（SDME）、中空纤维液相微萃取（HF-LPME）/电膜萃取（EME）和分散液液微萃取（DLLME）等 4 种溶剂微萃取研究报道（大约 325 篇）

13.2.2 溶剂微萃取命名法

如引言中所述，多年来，研究人员试图通过为每一种改进的SME赋予唯一的名称和相应的缩写来为该领域做出贡献[16,17]。在某些方面，这使得翻阅文献并找到有用的应用方法很困难。在这里为了避免混淆，将使用最常用的术语和首字母缩写词（表13.1）。术语LPME是一个另外。作为用微升体积的溶剂萃取的通用术语，LPME和SME都是合适的术语。但是，研究人员还使用LPME来指一种或多种特定的SME模式，最常见的是膜保护性溶剂萃取。术语HF-LPME是这些模式的最常用名称，避免将这些技术与其他SME模式混淆。首字母缩略词也将尽可能地限制于四种主要的SME模式和主要变体（使用超声、顶空、直接浸入等）。对于模式变化（使用电磁搅拌器、冷却、加热或蒸馏等）会另外用其他首字母缩写词。

13.2.3 单滴微萃取及其变体（SDME）

13.2.3.1 直接浸入式单滴微萃取（DI-SDME）

SDME是SME模式的一种，其中一滴溶剂（通常为0.5~10μL）暴露于样品，样品可能是固体，液体或气体。SDME本身通常是指浸入液体样品（通常是水）中的注射器针头或导管（例如CE、导管）尖端处的一滴溶剂[4,11,18]。为了区分，SDME通常称为直接浸入式单滴微萃取，指的是液滴浸入液体样品中，此时应使用该术语以避免任何混淆。这是所有SME模式中最简单、最便宜的方法，并且使用的溶剂量最少。这也是最简单的自动化技术，因为它可以使用普通的X-Y自动进样器执行，并且通常使用10μL~10mL的样品量（即使使用1~2mL的样品量）也可以获得足够的灵敏度，以与自动进样器兼容。通常将样品搅拌以加速提取，这是该技术的主要缺点，因为在采样过程中液滴很容易脱落并易于在针头的侧面向上爬行。对于包含固体的样品基质，液滴损失尤其成问题。常用的溶剂包括碳氢化合物，例如甲苯和环己烷，它们的表面张力低，典型的液滴大小为1~4μL。传统的大型萃取溶剂，例如氯仿、二乙醚和乙酸乙酯，由于它们的高挥发性和溶解性而不能单独使用。当使用高表面张力，高黏度溶剂（例如1-辛醇或十二烷）时，可以部分避免这些问题。此外，在过去的几年中，研究人员一直在尝试使用更绿色的替代品来替代标准有机溶剂，例如离子液体（ILs）[19-22]或深共熔溶剂（DESs）[23-26]。较高黏度的溶剂可提供更大的液滴稳定性，液滴大小可增加至10μL。但是，较高黏度的溶剂也需要更长的萃取时间。也应该指出ILs可能不是许多人认可的绿色溶剂，应谨慎使用[27]。IL和DES也可能不是与标准GC仪器一起使用的兼容溶剂，尽管它们与原子吸收、HPLC和CE仪器兼容。典型的提取时间取决于分析物种类和分析物浓度，并且可能在30s到1h之间变化。SDME通常是一个平衡过程。但是，由于可重复性取决于可重复的时间和条件，因此可以缩短时间[4,18]。SDME的一个变体是逐滴SDME（DD-SDME），它几乎没有得到应用。但具有的优势在于，与使用1~10mL样品体积的普通SDME不同，DD-SDME仅使用10μL的样品体积，使用0.3~1.0μL萃取溶剂，无须搅拌，在几分钟内达到平衡，可用于有限的样品量（血液，血清，唾液）。

随着其他技术的流行，SDME 与其他 SME 模式相比使用率有所下降。但是，该技术仍在广泛使用，不应该被淘汰，特别是当从干净的水溶液样品中提取非极性分析物时。SDME 还与在线连续流微萃取和微流体设备兼容，这可能是该技术的未来[9]。

13.2.3.2　顶空单滴微萃取（HS-SDME）

HS-SDME 通常是从含水，油或固体样品中采集挥发性或某些半挥发性分析物的首选[4,11-14]。与标准的顶空进样相比，HS-SDME 的优势在于，通常根据其辛醇/水分配系数（lg K_{ow}）而非其沸点提取分析物[4,11]。例如，使用十四烷作为萃取溶剂从机油中萃取汽油碳氢化合物产生的 GC 色谱图与汽油样品本身相似，而顶空进样产生的 GC 色谱图强调了最易挥发的组分[28]。在上述情况下，将此技术用于沸点低于萃取溶剂的分析物时，必须进行 GC 分流进样才能获得解析峰。对于沸点高于萃取溶剂的分析物，不分流进样效果很好。与 SDME 一样，HS-SDME 也会掉落，但程度要小得多。但是，当使用较高的样品温度时，低沸点萃取溶剂会蒸发并降低顶空/溶剂平衡。已经进行了一些尝试以使用冷却装置降低溶剂温度来控制这些问题，但是对于挥发性分析物，HS-SDME 通常在 20~40℃的温度下进行。

13.2.3.3　直接悬浮液滴微萃取（DSDME）和液液液微萃取（LLLME）

HF-LPME 和 DLLME 的出现，使得直接悬浮液滴微萃取（DSDME）和液-液-液微萃取（LLLME）经常被忽视，这两种方法是非常简单，易于自动化的方法。DSDME 一般是将 5~20μL 低密度溶剂放入搅拌的水性样品（1~20mL）中旋涡[4,11]。相对于 SDME 的优势在于没有滴落损失，并且较大体积的萃取溶剂可萃取更多分析物。当然，缺点是，只有一小部分（1~5μL）很容易被回收用于分析[29]。通过使用熔点低于室温的溶剂（如 1-十二烷醇），在冰中冷却萃取小瓶并取出固化的溶剂进行分析，可以解决此问题。当然，可以使用的溶剂数量是有限的，如果要使用所有溶剂，则 GC 必须进行大体积进样。与 SDME 一样，该技术最适合干净的基质样品。

LLLME 操作类似于 DSDME，用于从水样中提取可电离的分析物[4,11,30,31]。在实践中，将样品（0.5~10mL）搅拌并将低密度萃取溶剂（100~400μL）加入涡旋。萃取一段时间后，将注射器尖端的一滴水放入有机溶剂中以反萃取分析物。对于酸性分析物，将样品设置为低 pH，将水滴设置为高 pH。对于碱性分析物，这是相反的。通常用 CE 或 HPLC 分析样品。该技术还遭受潜在的液滴损失，但由于水具有高的表面张力，因此损失较小。与 DSDME 一样，相对清洁的水样也很重要。DSDME 和 LLLME 在很大程度上已被 HF-LPME 取代，尤其是当样品中含有固体颗粒时。

13.2.4　中空纤维液相微萃取及其变体

HF-LPME 是由 Rasmussen 和 Petersen-Bjergaard 的研究小组在 SDME 之后不久开发的，并且对于包含肮脏或复杂基质的样品（例如未过滤的水样品和生物液体）具有很大的实用性[32]。与 SDME 一样，HF-LPME 方法可以实现自动化，并且与在线和连续流动萃取以及微流体设备兼容[10]。HF-LPME 有两个主要类型：两相和三相。

13.2.4.1　两相中空液相微萃取

在两相 HF-LPME［HF（2）LPME］中，管腔（纤维内部）和纤维孔中都含有与水不混溶的溶剂，例如 1-辛醇[4,33-35]。典型的溶剂体积为 1~20μL，样品体积为 1~20mL。提取纤维（通常为具有 0.6mm 内腔的 Accurel Q3/2 多孔聚丙烯的提取纤维）用于每次提取，价格便宜且可一次性使用，因此该技术不会残留任何分析物。除了用于萃取的溶剂外，还使用 25~50mL 的溶剂（通常是丙酮）清洁纤维。

纤维的一端连接到注射器的针头上，并将纤维浸入搅拌的含水样品中，以将分析物提取到溶剂中。提取后，将溶剂抽回到注射器中，从针头上取下纤维，然后通常通过 GC 分析溶剂。该技术对于包含固体[36]和生物液体（例如血液或血清）[37]的脏基质特别有用。通过保护膜，无须担心液滴损失，可以提高搅拌速度。另外，根据纤维的长度[38,39]，可以使用大量的提取物。该技术也可以用于顶空提取。但是，一个主要的缺点是提取通常要花费一些时间，因为分析物需要穿过纤维壁进入内腔，并且通常需要 1h 的提取时间。与 SDME 一样，这种萃取也是一种平衡萃取，如果所有条件均保持不变，则系统无须获得平衡就可以再现萃取。

13.2.4.2 三相中空纤维液相微萃取

在三相 HF-LPME［HF（3）LPME］中，纤维的多孔壁再次用与水不混溶的溶剂（例如 1-辛醇）浸渍，但是管腔中含有水性反萃取剂。管腔溶剂体积为 1~25μL，样品体积为 1~200mL。与 LLLME 一样，该技术适用于包含可电离的分析物的样品，尤其是血液，血清和唾液等生物流体中的药物和代谢物。但是，该方法还适合在纤维壁和内腔中使用两种不混溶的有机溶剂[40]。使含水样品对酸性分析物呈酸性，以使它们能够通过纤维壁孔内的有机溶剂，并且管腔中的含水萃取剂是碱性的。对于碱性分析物，这是相反的。对于亲水性分析物，也可以使用载体介导的系统[41]。在大多数情况下，提取基本是彻底的。HF（3）LPME 和 HF（2）LPME 具有基本相同的优点和缺点，包括萃取时间长。一种用于缩短提取时间的技术是溶剂棒法，该方法是将纤维两端密封并在搅拌的样品中快速移动[42]。

13.2.5 电膜萃取

开发 EME 是为了解决 HFLPME 的主要缺点：提取时间长。EME 本质上是三相 HF-LPME，在样品溶液中添加了电极，在纤维腔中添加了电极。施加直流电压后，带有正电荷或负电荷的分析物可以迅速地（5min）穿过整个膜并进入管腔内的受体溶液中[43-45]。尽管需要附加设备，但该技术已迅速普及用于可电离的医学，代谢和生物样品的生物学和法医分析[46]。目前，EME 可能是这些类型样品的首选。该技术的另一个优点是，它与微流体设备兼容，可用于临床和法医应用[47,48]。但是，一个潜在的缺点是，分析物在 EME 期间也可能易于发生电解反应[49]。

13.2.6 分散液液微萃取（DLLME）

DLLME 已成为从水性样品中提取分析物的主要方法，其原因有两个：第一，实际提取仅需几秒钟；第二，提取了将近 100%的分析物[35,50-54]。在这两种情况下，这都是正确的，因为分散液产生的溶剂表面积很大，而 SME 理论清楚地表明萃取效率与溶剂表面积直接相关[4,11]。已开发出多种技术来产生分散体，包括溶剂分散，机械分散，超声辅助分散，微波辅助分散和通过化学方法原位形成分散体。就其本质而言，该技术仅用于从液体（通常是水，但有一些从天然油中提取的例子）中提取分析物，因此，必须使用其他方法处理顶空样品和不可溶于水的固体。尽管DLLME比其他SME模式更难以完全自动化，但最近已开发出一些使用X-Y自动进样器进行完全自动化的方法[55-57]。

13.2.6.1　溶剂分散-分散液液微萃取

早期的时候，DLLME涉及将不溶于水的氯代烃（20~100μL）（例如CCl_4或四氯乙烯）溶解在水溶性分散溶剂（100μL~2.5mL）[例如乙腈（ACN），丙酮，甲醇或乙醇]中[4,11,35]。然后用注射器将该溶液快速注入含水样品（5~50mL）中。通过离心或其他分散剂或盐溶液破坏分散液[56]。然后用注射器将提取溶剂取出并进行分析。该过程存在三个问题：首先，含氯萃取溶剂有毒且不环保。其次，溶剂最终在提取容器的底部，从而使自动进样器难以抽出溶剂。再次，分散剂的使用增加了萃取溶剂在水中的溶解度，减少了可用于分析的萃取溶剂的体积。分散溶剂也污染了样品。已经进行了各种研究努力，并获得了一些成功的尝试，即用密度小于水的溶剂代替氯化溶剂，但是这些努力通常涉及特殊的提取瓶，以帮助除去少量的提取溶剂[11,50,51,57]。另一种方法是使用诸如1-十二烷醇之类的溶剂，其熔点刚好低于室温。将提取瓶放在冰水中会使溶剂冻结，可以将其取出以进行分析，该技术类似于LLME中使用的技术[58]。这种技术的有趣逆转使用，也可以冻结含水样品，然后简单地倒出有机萃取物进行分析[59]。

13.2.6.2　机械分散-分散液液微萃取

机械分散涉及将萃取溶剂与含水样品剧烈振摇或涡旋，直到形成分散或伪分散。将样品涡旋后，用于此模式的首字母缩写为VA-DLLME[4,11,35,60]。另一种变体涉及将样品和溶剂快速移入和移出注射器以形成分散液[61]。很难说这些技术实际上不是小规模的液-液萃取。实际上，尽管离心可以加快过程，但混合物在静置时也可能会分离。竖立时分离确实可以使用此技术实现完全自动化的可能性，尤其是在注射器混合方式方面。这些技术还避免了使用分散剂的需要，并且萃取溶剂的体积（20~100μL）要比EPA所谓的微萃取程序（EPA方法为1~5mL而言）要小得多[62-64]。离子液体已成功用于该技术，因此水不溶性DES也应是有效的可以避免使用含氯溶剂。还使用了比水密度小的溶剂，包括接近室温的液体，例如1-十二烷醇。通常，由于很难用这种方法形成真正的分散体，因此提取比DLLME的其他方式需要更多的时间并且效率可能更低。

13.2.6.3　超声辅助分散液液微萃取和微波辅助分散液液微萃取

超声波或微波能非常有效地在萃取溶剂和水相之间形成真正的乳液[65]。这些技术已与添加的乳液溶剂和纯萃取溶剂一起使用。显然，当使用分散溶剂或仅使用超声或微波辅助无法完全分散时，便使用了超声辅助（UA）或微波辅助能量。研究人员对此并不总是很清楚，应当谨慎。其次，不使用分散溶剂的优点是溶剂回收率更高，样品污染更少。也可以使用IL和DES溶剂以及高熔点液体。超声或微波有可能也引发分析物分解或其他化学反应，因此应仔细考虑。

13.2.6.4 原位分散-液液微萃取

IL 或 DES 分散液可通过将单独的组分添加到水性样品中并使其反应形成 IL 或 DES 而形成，如果不溶于水，它们将形成精细的分散液以提取分析物[66]。可以类似的方式，将不溶于水的羧酸盐或胺盐添加到水性样品中，然后将样品制成酸性或碱性，以形成羧酸或胺的精细分散体，以提取分析物[67]。第三种技术涉及向含水样品中加入在室温下不溶但在高温下可溶的提取溶剂。随着样品冷却，再次形成分散体[68]。这项技术被定义为浊点萃取（CPE），但无须添加表面活性剂。这些技术尚未得到广泛使用，但具有的主要优势在于它们避免了使用分散溶剂，超声或微波，从而简化了流程并使其绿色化，特别是在使用天然 DES、酸和碱的情况下。

13.3 现有方法的改进

无论读者是精通 SME 还是本领域的新手，都应参考阅读以下列表中的参考文献，这些文献是近期综述类文献的代表。这些材料包含大量参考资料，为读者根据分析需求对公开的分析方法进行改进提供入门知识。此列表中包含的参考文献是综述文章，以及一本专著，该专著是有关 SME 的最全面的数据收集。本书还详细讨论了 SME 理论，应用以及选择 SME 模式的过程。除了这些综述材料外，还应参考本文引用的其他研究论文。

（1）绿色分析化学（GAC）和溶剂微萃取的 12 条原则[5-7]。

（2）全面的溶剂微萃取理论，发展与应用文本[4]。

（3）全面的溶剂微萃取综述[11-15]。

（4）SDME 和相关技术综述[18]。

（5）HF-LPME 综述[33-35]。

（6）电膜萃取综述[43-45]。

（7）DLLME 综述[35,50-54]。

（8）离子液体环境关注综述[27]。

（9）离子液体综述[19-22]。

（10）深共熔溶剂（DES）综述[23-26]。

（11）将溶剂微萃取与其他微萃取技术相结合[8]。

（12）固化的有机液滴溶剂微萃取[58]。

（13）萃取溶剂温度控制综述[69]。

（14）溶剂微萃取的衍生化技术[70]。

（15）非膜溶剂微萃取的自动化[9,10]。

一旦选择了适当的方法，就不太可能不经改进就使用它，一般需要考虑对其进行改进。即使样品系统和要分析的化学物质是相同的，也不应忘记，在发表之前，原始科学家必须进行许多实验才能获得正确执行分析的技能。也不应忘记，有时关键步骤会以某种方式被遗忘在书面程序中或被忽视。但是，对于所有化学分析方法，甚至对于经过验证的官方标准方法，都是如此，当然，大多数研究论文都不包含验证程序。

13.4 考虑方法的要求

当然，所需的分析方法将影响选择哪种 SME 模式，因为该方法和模式的要求必须匹配。方法要求通常包括样品性质：固体，液体（溶液）还是气体；样本基质以及分析物的性质和浓度。精确度（重复性，坚固性和再现性），准确性，选择性和样品通量也将有期望的要求。这些要求规定了该方法所需的分析仪器的类型，确定了人员要求，包括培训。最后，必须最终验证该方法。特定的 SME 模式是否可以满足所有这些要求，只能通过对满足分析目标的已发布方法进行仔细分析来确定。

13.5 绿色分析化学

在概述了溶剂微萃取的主要模式的基础上，应该对为什么溶剂微萃取模式是绿色的进行更详细的研究。绿色化学起源于为了开发工业规模的有机合成方法。Anastas 和 Warner 提出了绿色化学的 12 条原则，以满足工业的这些需求[71]。Galuszka，Migaszewski 和 Namiesnik[5]最近对 12 条原则进行了修订，以更直接地解决绿色分析化学（GAC）的特定问题。12 条原则如下所示：

（1）应采用直接分析技术以避免样品处理；
（2）最小样本量和最小样本数是目标；
（3）应进行原位测量；
（4）分析过程和操作的集成节省了能源并减少了试剂的使用；
（5）应该选择自动化和小型化的方法；
（6）应避免衍生化；
（7）应避免产生大量分析废物，并应提供对分析废物的适当管理；
（8）与一次使用一种分析物的方法相比，首选多分析物或多参数的方法；
（9）应尽量减少能源的使用；
（10）从可再生资源获得的试剂应该是首选；
（11）有毒试剂应消除或更换；
（12）应提高操作员的安全性。

通过更仔细地查看此列表，作者得出结论，GAC 方法需要解决的关键问题分为四类：

（1）消除或减少化学物质的使用；
（2）最小化能耗；
（3）正确管理分析废物；
（4）增加操作员的安全性。

将 SME 模式整合到绿色分析方法中可以满足所有这些标准。从本质上讲，SME 不会消除溶剂的使用，但从低方面来说，SDME，HF-LPME 和 EME 可能只需要 10μL 溶剂（包括注射器清洁所需的溶剂）。从较高的方面看，DLLME 可能需要 200μL 溶剂，但仍然远远少于传统方法。由于检测限所需的样品量范围从几微升到几毫升，因此额外的试剂要求（缓冲液，衍生试剂）也大大降低了[70]。由于只需要少量的样品和试剂，因此可大大减少浪费。由于试剂量很小，因此操作员的安全性得到了增强，尤其是在采用自动化的 SME 模式时。大多数 SME 程序不需要溶剂浓缩步骤、高操作温度以及昂贵的设备，所以从本质上讲是节能的。此外，SME 通常将提取，浓缩，纯化和直接分析物制备纳入几乎所有类型的仪器分析中，所有这些步骤都可以节省能源，时间和人力。最后，应该指出的是，对于在线，法医或临床需求，SDME，HF-LPME 和 EME 特别适用于大规模常规分析的微流控设备[9,10]。

13.6 为绿色分析化学方法选择合适的溶剂微萃取

没有一种固定的方法可以为特定的绿色分析方法选择适当的 SME 模式。必须考虑许多因素，包括样品的性质（液体、固体、气体、水、土壤、复杂或简单的基质等）；被鉴定和测量的分析物的类型和浓度；分析仪器的可用性；实验室人员培训；时间和通量要求等。不幸的是，目前尚无标准的，公开的使用 SME 的方法，如固相萃取技术。公开的研究方法是最合理的起点，但是有将近 2000 篇 SME 论文可供选择，而且对于一种特定的目标分析物，可能已经使用了一种以上的 SME 模式。一旦读者对 13.2 中的材料有了一般的了解，下一步就是浏览 13.3 中列出的参考材料。完成后，下一步是什么？从图 13.2 的对比来看，似乎 DLLME 是当前最流行的 SME 模式，因此也许这是一个合适的起点。不一定如此。绿色分析化学（GAC）的因素之一是在可能的情况下避免样品处理，并且 DLLME 除了仅适用于水溶液外，还需要相对清洁的基质。因此，DLLME 可能适用于自来水，但对于未经预处理的生物样品可能会出现问题。此外，DLLME 通常用于 5~20mL 的样品量。HF-LPME 或 EME（如果分析物被电离）可以更好地处理样品量有限且含有蛋白质或腐殖质的复杂基质的样品。SDME 是最简单和完全自动化的 SME 模式，可能适用于干净的含水样品，而 HS-SDME（或顶空 HF-LPME）绝对是固体，液体和油性样品中挥发性分析物的首选方法。

虽然没有选择最佳 SME 模式的完美方案，但图 13.3 所示是通用示意图，在这一点上可能会有用。该方案首先考虑分析物的挥发性，然后考虑分析物极性，最后考虑样品基质。Kokosa，Przyjazny 和 Jeannot 在本书中介绍了有关 SME 模式选择方法的其他示意图和详细讨论[4]。本书还包括对 1995—2009 年该领域论文的引用（超过 700 篇），其中包含详细的 SME 理论和 SME 实验程序的示例。

除了选择萃取方式外，还必须谨慎选择溶剂。除了最流行的溶剂 1-辛醇外，许多早期的 SME 程序（有些仍在使用）使用卤化或芳香族萃取溶剂，这些溶剂可能不符合绿色要求。为了避免溶剂由于挥发性，溶解性或液滴不稳定性而损失，离子液体已在所有 SME 模式下获得普及，并认为这些溶剂是绿色的[19-22]。情况可能并非如此，应谨慎使用离子液体[27]。最近，DES 已经可用，并且正在研究将这些材料用于 SME[23-26]。甚至尝试使用天然来源的 DES，以最大程度减少分析废物的产生[72]。因此，建议在可能的情况下，溶剂微萃取考虑使用 DES 溶剂而不是离子液体，尤其是对于大规模使用而言。另一方面，溶剂不仅必须适合于 SME 模式，而且还必须适合仪器分析方法。例如，离子液体（IL）和 DES 可能与 GC 不兼容，但与 HPLC 和 CE 兼容。样品通量也是一个重要的分析要求，并且如前所述，对于 SDME，HF-LPME 和 EME，DES 和 ILs 是高表面张力，高黏度的溶剂，并且与 DLLME 相比，萃取所需的时间比典型的有机萃取溶剂长得多。但是，应该记住，DLLME 模式变体可能很难完全自动化，一旦对其进行了充分验证，就需要进行培训。样品量和萃取溶剂对样品的污染也是重要因素。例如，DLLME 萃取（包括使用 IL 或 DES 时）应尽可能在无分散溶剂的情况下进行。

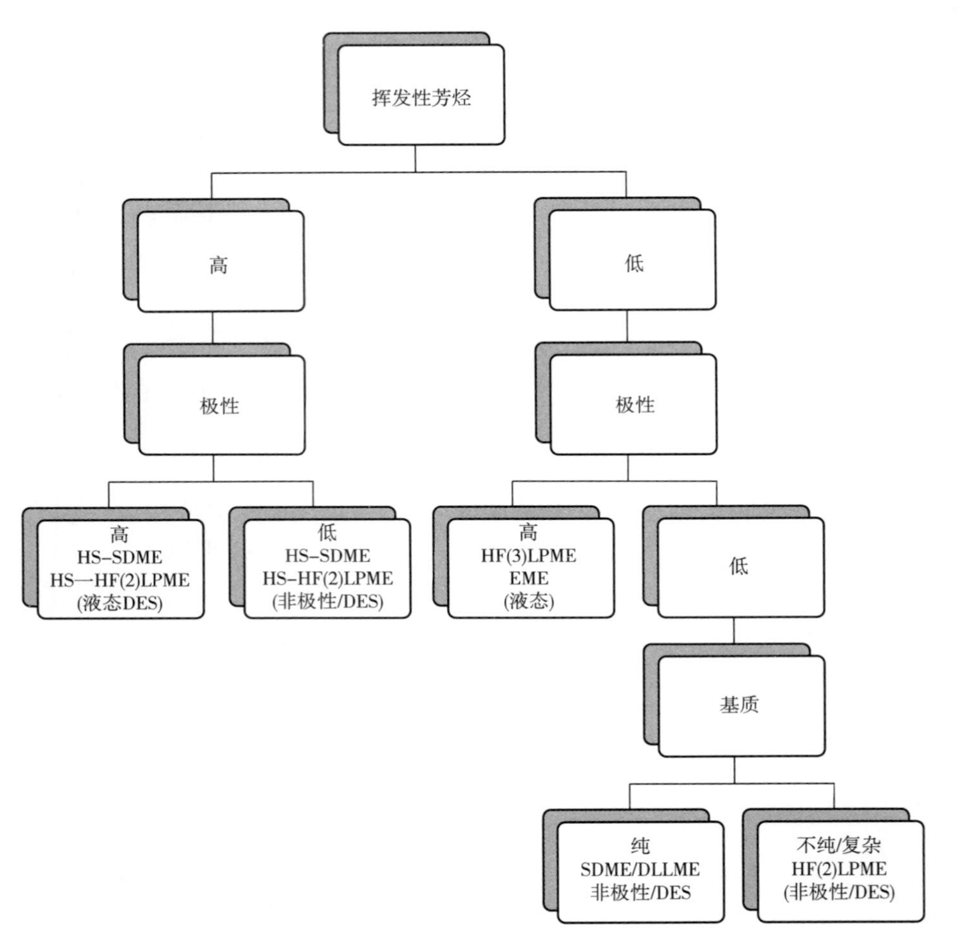

图 13.3　为绿色分析方法选择适当的溶剂微萃取（SME）的通用示意图，包括推荐的溶剂类别

13.7　总结

为一种方法选择合适的溶剂微萃取（SME）模式似乎是一项艰巨的任务。这样做需要对每种模式的优缺点以及对整个方法的要求有基本的了解。最好的入门方式是对相关 SME 的参考资料进行综述。13.3 中列出的参考文献虽然不全面，但却是一个很好

的起点。尤其是如果新方法替代了现有方法，则将可获得有关所需仪器和已知样品的分析要求的信息。然后可以查找已发布的方法并根据需要进行修改。如果 SME 技术似乎符合所提出方法的要求，则最佳建议是简单地尝试重复已发布的方法，以亲自了解它确实按编写的方式工作。但是，应该记住，每一种成功的方法最终都需要经过很多尝试和许多时间才能掌握，而将 SME 技术添加到新的分析程序中时也是如此。

参考文献

[1] M. A. Jeannot, F. F. Cantwell, Solvent microextraction into a single drop, Anal. Chem. 68 (1996) 2236-2240.

[2] M. A. Jeannot, F. F. Cantwell, Mass transfer characteristics of solvent extraction into a single drop at the tip of a syringe needle, Anal. Chem. 69 (1997) 235-239.

[3] Y. He, H. K. Lee, Liquid phase microextraction in a single drop of organic solvent by using a conventional microsyringe, Anal. Chem. 69 (1997) 4634-4640.

[4] J. M. Kokosa, A. Przyjazny, M. A. Jeannot, Solvent Microextraction Theory and Practice, Wiley, Hoboken, NJ, 2009.

[5] A. Galuszka, Z. Migaszewski, J. Namiesnik, The 12 principles of green analytical chemistry and the significance mnemonic of green analytical principles, Trends Anal. Chem. 50 (2013) 78-84.

[6] S. Armenta, S. Garrigues, M. de la Guardia, The role of green extraction techniques in green analytical chemistry, Trends Anal. Chem. 71 (2015) 2-8.

[7] A. Spietelun, L. Marcinkowski, M. del la Guardia, J. Namiesnik, Green aspects, developments and perspectives of liquid phase microextraction techniques, Talanta 119 (2014) 34-45.

[8] J. Moreda-Pineiro, A. Moreda-Pineiro, Recent advances in combining microextraction techniques for sample pre-treatment, Trends Anal. Chem. 71 (2015) 265-274.

[9] M. Alexovic, B. Horskotte, P. Solich, J. Sabo, Automation of static and dynamic nondispersive liquid phase microextraction. Part 1: approaches based on extractant drop-, plug-, film-and microflow-formation, Anal. Chim. Acta 906 (2016) 22-40.

[10] M. Alexovic, B. Horskotte, P. Solich, J. Sabo, Automation of static and dynamic non-dispersive liquid phase microextraction. Part 2: approaches based on impregnated membranes and porous supports, Anal. Chim. Acta 907 (2016) 18-30.

[11] J. M. Kokosa, Solvent microextraction, in: J. Reedijk (Ed.), Elsevier Reference Module in Chemistry, Molecular Sciences and Chemical Engineering, Elsevier, Waltham, MA, 2015, pp. 1-30, http://dx.doi.org/10.1016/B978-0-12-409547-2.2.11640-3.

[12] S. Tang, H. Zhang, H. K. Lee, Advances in sample extraction, Anal. Chem. 88 (2016) 228-249.

[13] J. Plotka-Wasylka, K. Owczarek, J. Namiesnik, Modern solutions in the field of microextraction using liquid as a medium of extraction, Trends Anal. Chem. 85 (2016) 46-64.

[14] M. R. Ganjali, M. Rezapour, P. Norouzi, F. Faridbod, Liquid-phase microextraction, in: Analytical Separation Science, 5 : 9, Wiley, 2015, pp. 1625-1658, http://dx.doi.org/10.1002/9783527678129.assep059.

[15] J. A. Ocana-Gonzalez, R. Fernandez-Torres, M. A. Bello-Lopez, M. Ramos-Payan, New developments in microextraction techniques in bioanalysis. A review, Anal. Chim. Acta 905 (2016) 8-23.

[16] J. Sanrejova, N. Campillo, P. Vinas, V. Andruch, Classification and terminology in dispersive liquid-liquid microextraction, Microchem. J. 127 (2016) 184-186.

[17] J. M. Kokosa, Advances in solvent microextraction techniques, Trends Anal. Chem. 43 (2013) 2-13.

[18] J. M. Kokosa, Recent trends in using single-drop microextraction and related techniques in green analytical methods, Trends Anal. Chem. 71 (2015) 194-204.

[19] D. Han, B. Tang, Y. R. Lee, K. H. Row, Application of ionic liquid in liquid phase microextraction technology, J. Sep. Sci. 35 (2012) 2949-2961.

[20] C. F. Poole, N. Lenca, Green sample-preparation methods using room-temperature ionic liquids for the chromatographic analysis of organic compounds, Trends anal. Chem. 71 (2015) 144-156.

[21] L. Marcinkowski, F. Pena - Pereira, A. Kloskowski, J. Namiesnik, Opportunities and shortcomings of ionic liquids in single-drop microextraction, Trends Anal. Chem. 72 (2015) 153-168.

[22] E. Stanisz, J. Werner, A. Zgola-Grzeskowiak, Liquid-phase microextraction techniques based on ionic liquids for preconcentration and determination of metals, Trends anal. Chem. 61 (2014) 54-66.

[23] Y. Dai, J. van Spronsen, G. -J. Witkamp, R. Verpoorte, Y. H. Choi, Natural deep eutectic solvents as new potential media for green technology, Anal. Chim. Acta 766 (2013) 61-68.

[24] G. Li, Y. Jiang, X. Liu, D. Deng, New levulinic acid-based deep eutectic solvents: synthesis and physicochemical property determination, J. Mol. Liq. 222 (2016) 201-207.

[25] D. J. G. P. van Osch, L. F. Zubar, A. van den Bruinhorst, M. C. Kroon, Hydrophobic deep eutectic solvents as water-immiscible extractants, Green Chem. 17 (2015) 4518-4521.

[26] X. Li, K. H. Row, Development of deep eutectic solvents applied in extraction and separation, J. Sep. Sci. 39 (2016) 3505-3520.

[27] M. Made, J. -F. Liu, L. Pang, Environmental application, fate, effects, and concerns of ionic liquids: a review, Environ. Sci. Technol. 49 (2015) 12611-12627.

[28] J. M. Kokosa, A. Przyjazny, Headspace microdrop analysis-an alternate test method for gasoline diluents and benzene, toluene, ethylbenzene and xylenes in used engine oils, J. Chromatogr. A 983 (2003) 205-214.

[29] M. Asadi, S. Dadfarnia, A. M. H. Shabani, B. Abbasi, Simultaneous extraction and quantification of lamotrigine, Phenobarbital, and phenyltoin in human plasma and urine samples using solidified floating organic drop microextraction and high - performance liquid chromatography, J. Sep. Sci. 38 (2015) 2510-2515.

[30] X. Zhou, M. He, B. Chen, Q. Yang, B. Hu, Membrane supported liquid - liquid - liquid microextraction combined with field - amplified sample injection CE - UV for highsensitivity analysis of six cardiovascular drugs in human urine samples, Electrophoresis 37 (2016) 1201-1211.

[31] V. H. Springer, A. G. Lista, In-line coupled single drop liquid-liquid-liquid microextraction with capillary electrophoresis for determining fluoroquinolones in water samples, Electrophoresis 36 (2015) 1572-1579.

[32] S. Pedersen - Bjergaard, K. E. Rasmussen, Liquid - liquid - liquid microextraction for sample preparation of biological fluids prior to capillary electrophoresis, Anal. Chem. 71 (1999) 2650-2656.

[33] J. A. Jonsson, Membrane-assisted separations, in: Analytical Separation Science, 5 : 5, Wiley, 2015, pp. 1503-1524, http: //dx. doi. org/10. 1002/9783527678129. assep055.

[34] E. Carasek, J. Merib, Membrane-based microextraction techniques in analytical chemistry: a

review, Anal. Chim. Acta 880 (2015) 8-25.

[35] V. Sharifi, A. Abbasi, A. Nosrati, Application of hollow fiber liquid phase microextraction and dispersive liquid-liquid microextraction techniques in analytical toxicology, J. Food Drug Anal. 24 (2016) 264-276.

[36] J. Cai, G. Chen, J. Qiu, R. Jiang, F. Zeng, F. Zhu, G. Ouyang, Hollow fiber based liquid phase microextractionfor the determination of organochlorine pesticides in ecological textiles by gas chromatography-mass spectrometry, Talanta 146 (2016) 375-380.

[37] H. A. Panahi, M. Ejlali, M. Chabouk, Two-phase and three-phase liquid-phase microextraction of hydrochlorothiazide and triamterene in urine samples, Biomed. Chromatogr. 30 (2016) 1022-1028.

[38] A. Dominguez-Tello, A. Arias-Borrego, T. Garcia-Barrera, J. L. Gomez-Aruza, Application of hollow fiber liquid phase microextraction for simultaneous determination of regulated and emerging iodinated trihalomethanes in drinking water, J. Chromatogr. A 1402 (2015) 8-16.

[39] P. -S. Chen, Y. -H. Tseng, Y. -L. Chuang, J. -H. Chen, Determination of volatile organic compounds in water using headspace knotted hollow fiber microextraction, J. Chromatogr. A 1395 (2015) 41-47.

[40] M. Tajik, Y. Yamini, A. Esrafih, B. Ebrahimpour, Automated hollow fiber microextractioni based on two immiscible organic solvents for the extraction of two hormonal drugs, J. Pharmaceut. Biomed. 107 (2015) 24-31.

[41] S. Ncube, A. Poliwoda, H. Tutu, P. Wieczorek, L. Chimuka, Multivariate optimization of the hollow fibre liquid phase microextraction of muscimol in human urine samples, J. Chromatogr. B (2016) 1033-1034, 372-381.

[42] J. J. Pinto, M. Martin, B. Herce - Sesa, J. A. Lopez - Lopez, C. Moreno, Solvent bar microextraction: improving hollow fiber liquid phase applicability in the determination of Ni in seawater samples, Talanta 142 (2015) 84-89.

[43] C. Huang, K. F. Seip, A. Gjelstad, S. Pedersen - Bjergaard, Electromembrane extraction for pharmaceutical and biomedical analysis-Quo vadis, J. Pharmaceut. Biomed. 113 (2015) 97-107.

[44] C. Huang, H. Jensen, K. F. Seip, A. Gjelstad, S. Pedersen - Bjergaard, Mass transfer in electromembrane extraction-the link between theory and experiments, J. Sep. Sci. 39 (2016) 188-197.

[45] A. Oedit, R. Ramautar, T. Hankemeier, P. W. Lindberg, Electroextraction and electromembrane extraction: advances in hyphenation to analytical techniques, Electrophoresis 37 (2016) 1170-1186.

[46] C. Huang, A. Gjelstad, K. F. Seip, H. Jensen, S. Petersen-Bjergaard, Exhaustive and stable electromembrane extraction of acidic drugs from human plasma, J. Chromatogr. A 1425 (2015) 81-87.

[47] Y. A. Asl, Y. Yamini, S. Seidi, B. Ebrahimpour, A new effective on chip electromembrane extraction coupled with high performance liquid chromatography for enhancement of extraction efficiency, Anal. Chim. Acta 898 (2015) 42-49.

[48] Y. A. Asl, Y. Yamini, S. Seidi, M. Rezazadeh, Simultaneous extraction of acidic and basic drugs via on-chip electromembrane extraction, Anal. Chim. Acta 937 (2016) 61-68.

[49] P. Kuban, P. Bocek, The effects of electrolysis on operational solutions in electromembrane extraction: the role of acceptor solution, J. Chromatogr. A 1398 (2015) 11-19.

[50] N. Campillo, P. Vinas, J. Sandrejova, V. Andruch, Ten Years of dispersive liquid - liquid microextraction and derived techniques, Appl. Spectrosc. Rev. (2016) 1 - 149, http://dx.doi.org/10.1080/05704928.2016.1224240.

[51] M. I. Leong, M. R. Fuh, S. D. Huang, Beyond dispersive liquid-liquid microextraction, J. Chromatogr. A 1335 (2014) 2-14.

[52] W. Ahmad, A. A. Al-Sibaai, A. S. Bashammakh, H. Alwael, M. S. El-Shahawi, Recent advances in dispersive liquid-liquid microextraction for pesticide analysis, Trends Anal. Chem. 72 (2015) 181-192.

[53] H. M. Al-Saidi, A. A. A. Emara, The recent developments in dispersive liquid-liquid microextraction for preconcentration and determination of inorganic analytes, J. Saudi Chem. Soc. 18 (2014) 745-761.

[54] R. Jain, R. Singh, Applications of dispersive liquid-liquid micro-extraction in forensic toxicology, Trends Anal. Chem. 75 (2016) 227-237.

[55] L. Guo, H. K. Lee, Automated dispersive liquid-liquid microextraction-gas chromatography-mass spectrometry, Anal. Chem. 86 (2014) 3743-3749.

[56] L. Guo, S. Tan, X. Li, H. K. Lee, Fast automated dual-syringe based dispersive liquid-liquid microextraction coupled with gas chromatography-mass spectrometry for the determination of polycyclic aromatic hydrocarbons in environmental water samples, J. Chromatogr. A 1438 (2016) 1-9.

[57] L. Guo, S. H. Chia, H. K. Lee, Automated agitation-assisted demulsification dispersive liquid-liquid microextraction, Anal. Chem. 88 (2016) 2548-2552.

[58] P. Vinas, N. Campillo, V. Andruch, Recent achievements in solidified floating organic drop microextraction, Trends Anal. Chem. 68 (2015) 48-77.

[59] J. G. March, V. Cerda, A novel procedure for phase separation in dispersive liquid-liquid microextraction based on solidification of the aqueous phase, Talanta (2016) 156-157, 204-208.

[60] C. S. M. Liew, X. Li, H. K. Lee, Miniscale Liquid-liquid extraction coupled with full evaporation dynamic headspace extraction for the gas chromatography/mass spectrometric analysis of polycyclic aromatic hydrocarbons with 4000-to-13000-fold enrichment, Anal. Chem. 88 (2016) 9095-9102.

[61] X. You, Z. Xing, F. Liu, X. Zhang, Air-assisted liquid-liquid microextraction by solidifying the floating organic droplets for the rapid determination of seven fungicide residues in juice samples, Anal. Chim. Acta 875 (2015) 54-60.

[62] J. W. Munch, Analysis of Organohalide Pesticides and Commercial Polychlorinated Biphenyl (PCB) Products in Water by Microextraction and Gas Chromatography, 1995, pp. 1-36. U. S. EPA Method 505 Revision 2. 1.

[63] S. C. Wendelken, M. V. Bassett, T. A. Dattilio, B. V. Pepich, D. J. Munch, Determination of Chlorinated Acids in Drinking Water by Liquid-liquid Microextraction, Derivatization, and Fast Gas Chromatography with Eelectron Capture Detection, 2000, pp. 1-48. U. S. EPA Method 515. 4 Revision 1. 0.

[64] J. W. Hodgeson, A. L. Cohen, Determination of Chlorination Disinfection Byproducts, Chlorinated Solvents, and Halogenated Pesticides/herbicides in Drinking Water by Liquid-liquid Extraction and Gas Chromatography with Electron-capture Detection, 1990, pp. 1-56. U. S. EPA Method 551. 1 Revision 1. 0.

[65] T. Tan, Z. Li, Y. Wan, H. Qiu, Deep eutectic solvent-based liquid-phase microextraction for detection of plant growth regulators in edible vegetable oils, Anal. Methods 17 (2016) 3511-3516.

[66] M. A. Farajzadeh, M. R. Afshar Mogaddam, B. Feriduni, Simultaneous synthesis of a deep eutectic solvent and its application in liquid-liquid microextraction of polycyclic aromatic hydrocarbons from aqueous samples, RSC Adv. 6 (2016) 47990-47996.

[67] M. A. Farajzadeh, B. Feriduni, M. R. Afshar Magaddam, Development of a new version of

homogeneous liquid – liquid extraction based on an acid – base reaction: application for extraction and preconcentration of aryloxyphenoxy–propionate pesticides from fruit juice and vegetable samples, RSC Adv. 6 (2016) 14927–14936.

[68] A. A. A. Nabil, N. Nouri, M. A. Farajzadeh, Determination of three antidepressants in urine using simultaneous derivatization and temperature–assisted dispersive liquid–liquid microextraction followed by gas chromatography–flame ionization detection, Biomed. Chromatogr. 29 (2014) 1094–1102.

[69] A. R. Ghiasvand, S. Hazipour, N. Heidari, Cooling – assisted microextraction: comparison of techniques and applications, Trends Anal. Chem. 77 (2016) 54–65.

[70] H. Lin, J. Wang, L. Zeng, G. Li, Y. Sha, D. Wu, B. Liu, Development of solvent microextraction combined with derivatization, J. Chromatogr. A 1296 (2013) 235–242.

[71] P. T. Anastas, J. C. Warner, Green Chemistry: Theory and Practice, Oxford University Press, Oxford, 1998.

[72] A. V. de Bairros, R. Lanaro, R. M. de Almeida, M. Yonamine, Determination of ketamine, norketamine and dehydronorketamine in urine by hollow–fiber liquid–phase microextraction using an essential oil as supported liquid membrane, Forensic Sci. Int. 243 (2014) 47–54.

14　在线管内固相微萃取技术的发展趋势

Pascual Serra-Mora, Yolanda Moliner-Martínez, Carmen Molins-Legua, Rosa Herráez-Hernández, Jorge Verdú-Andrés and Pilar Campíns-Falcó*
University of Valencia, Valencia, Spain
*通讯作者：E-mail：pilar.campins@uv.es

14.1 引言

虽然在分析仪器技术上取得了许多进步，但是在许多分析中，样品的前处理工作仍然是关键的一步。在检测分析之前，可能需要几个操作来提取、纯化、浓缩或转换目标化合物。样品制备是分析过程中最耗时的阶段（约占总分析时间的60%）。据估计，大约65%的样品分析需要三次或更多的操作，以获得合适的仪器响应。因此，在分析过程产生的误差中，几乎有30%是由样品前处理造成的[1-2]。此外，样品前处理往往需要消耗大量的溶剂、试剂、材料和能源。

在过去的15年里，在分析化学方面的许多研究都集中在自动化样品前处理技术的发展上，利用这种技术可以在更短的时间内提供更精确的结果，最大限度地减少材料的消耗和废物的产生，根据绿色分析化学原理，在整个过程中，利用碳足迹理念进行衡量。在这种情况下，最成功的技术之一是由Arthur和Pawliszyn[3]首次提出的固相微萃取技术（SPME）。SPME最初是使用涂有聚合萃取相（光纤SPME）的二氧化硅纤维开发的，但今天，其他形式的SPME也是可用的，管内固相微萃取（IT-SPME）是其中最受欢迎的一种[4]。IT-SPME被提出作为一种替代方案，以克服纤维SPME的一些局限性，如纤维脆性和低提取能力，也促进了SPME与液相色谱（LC）的耦合。

IT-SPME是基于使用一个在其内表面填充或涂覆有萃取相的毛细管（即毛细管柱）。当样品（液体或气体）通过毛细管时，被分析物发生吸收或吸附，接下来，通过向毛细管中注入适当的溶剂来解吸所提取的分析物，然后收集这些溶剂进行进一步处理（静态解吸），或者同时通过溶剂解吸（动态解吸）并转移到分析仪器上，通常是LC设备的流动相。

虽然萃取毛细管用于样品前处理已与气相色谱相结合使用，但是在以IT-SPME为基础的分析中，LC是迄今为止使用最广泛的技术。最近，人们尝试将IT-SPME技术扩展到与其他仪器技术在线应用[5,6]。另一方面，IT-SPME是一种通用技术，可以使用许多不同的配置实现。特别令人感兴趣的是那些能够实现提取和测量在线偶联的方案，在整个过程可以很容易地实现自动化[7-10]。

IT-SPME效率主要取决于萃取毛细管的相关参数。用于气相色谱（GC）的开放式毛细管柱已广泛用于此目的。因此，在方法开发过程中，必须考虑涂层类型和薄膜厚度以及毛细管尺寸等参数。为了提高该技术的性能（选择性、效率和稳健性），替代材料使用越来越多，包括纳米材料、离子液体和分子印迹聚合物，以及不同的复合材料。

通过施加外力来改善萃取性能的可能性也被探索，促成了新形式的IT-SPME，如磁性IT-SPME，电化学控制IT-SPME和热电控制IT-SPME。分析方法的环境友好性在于保持分析性能参数的可靠性，如灵敏度、精密度和准确度[11,12]的前提下，尽量减少危险溶剂、减少废物、使用节能和成本效益高的程序。在绿色提取过程中，样品的预处理，特别是从感兴趣的基质中提取分析物是最关键的步骤。根据使用的设备，在大多数提取技术中，都使用了试剂和（或）溶剂，并且产生废物和消耗能源。因此，萃取过程的绿色评估主要在于①溶剂/试剂和废物最小化；②提取过程的微型化；③能源消耗的减少，在此过程中要牢记所要解决分析问题的需求。值得注意的是，能源与温室气体（GHG）的排放直接相关。分析过程本身是一个对环境影响有贡献的温室气体排放体，因此，在评价提取技术的环境性能时必须考虑这些参数（图 14.1）。

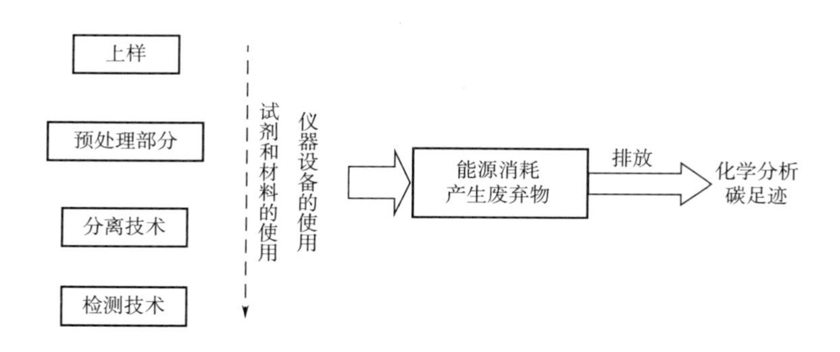

图 14.1　分析过程中碳足迹估计的横向作用

分析过程的环境影响传统上是用定性参数来分类的。在绝大多数这些工作中，能源需求被忽略了。所以，环境影响无法准确估计。于是，必须制订一个定量的环境指标，以反映分析性能和分析方法的可持续性方面的竞争力。因此，碳足迹计算被提议作为一种分析程序[12]的环境绩效指标。

碳足迹被用来反映某一活动产生的温室气体排放量，用 $kgCO_2eq$ 表示。最近，我们团队[12]将其作为定量参数来评价一种分析方法的环境影响。碳足迹的计算基于文件[13]定义的范围 2：电力间接温室气体排放，计算公式如下：

$$kgCO_2eq = \Sigma\ [\text{耗电量}\ (kW \cdot h) \times \text{排放因子}\ (kgCO_2/kW \cdot h)]$$

式中，$kgCO_2eq$ 可以用不同的单位表示，例如时间单位或样本数目。耗电量（$kW \cdot h$）可由电力公司或从设备以及公司的技术资料中取得。排放因子 $kgCO_2/kW \cdot h$ 将电子运动数据转换为排放值。这个因素取决于季节、白昼时间和供应商。因此，涉及的排放因子被用来计算分析过程中[14]的 $kgCO_2eq$。

本章介绍了 IT-SPME 的原理，特别关注了其主要形式及其相应的配置、萃取阶段、相关应用及其碳足迹。

14.2　装置的发展趋势

虽然 IT-SPME 技术可离线应用，但是在在线装置中将分析物的萃取和分离/检测集于一体是该技术最显著的优点之一。到目前为止，大多数报道的 IT-SPME 检测都涉及萃取与液相色谱仪的在线偶联。影响这种偶联的主要方法有两种，分别是吸取/注入（draw/eject）和流动-通过（flow-through）模式。这些模式意味着不同的装置，以及不同水平的仪器要求（图 14.2）。

在 draw/eject 循环系统中，样品通过萃取毛细管连续抽吸和分配，直到样品和萃取相之间的分析物达到分配平衡，或者直到萃取物的量足够。为此，使用了一个经过编

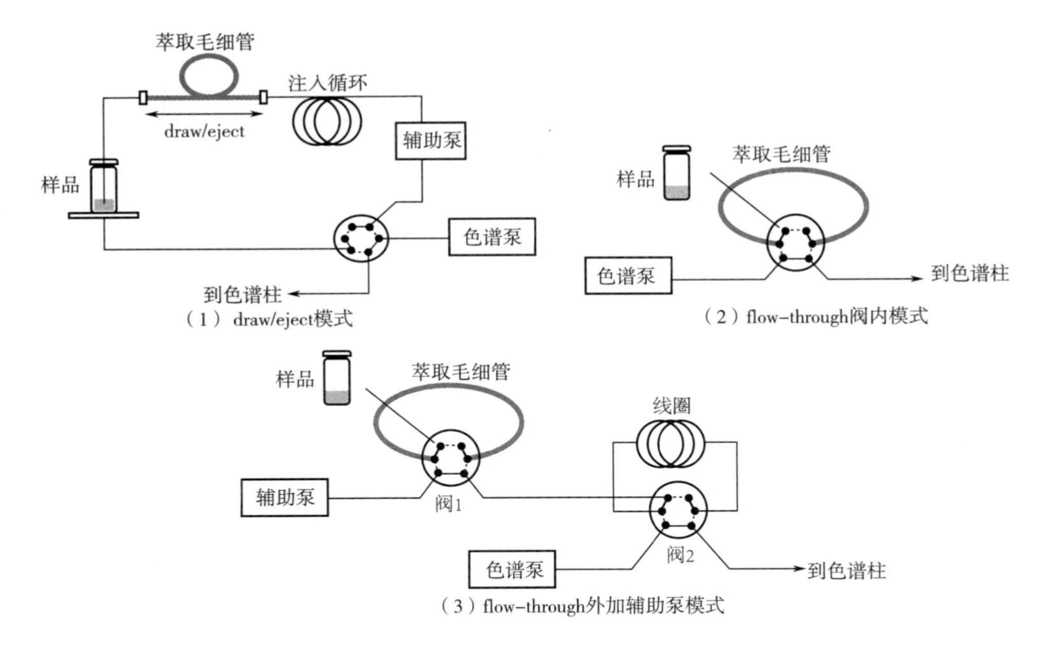

图 14.2 IT-SPME-LC 使用的装置示意图

装置（1）（2）：实线-上样，萃取，净化（静态解吸）；虚线-动态解吸并将分析物转移到色谱柱中。
（3）阀 1 实线-上样，萃取，净化（静态解吸）；阀 1 虚线-动态解吸并将分析物转移到色谱的注入回路中；
阀 2 实线-萃取物注入回路；虚线-阀 2 注射

程的自动采样器。萃取毛细管必须放置在针和自动采样器的线圈之间［图 14.2（1）］。萃取之后，用适当的溶剂填充毛细管进行静态解吸并稍后转移到色谱柱。或者，目标化合物可以在动态模式下通过冲洗毛细管色谱流动相转移。在转移分析物之前，淋洗阶段可以使用清洗溶剂穿过毛细管冲洗（通常是缓冲溶液）。运行时该系统还可以编程来进行净化和清洗毛细管，许多系统包含至少一个额外的泵，以促进输送、清洗和调节操作。

在 flow-through 系统中［图 14.2（2）］，一部分样品通过萃取毛细管（一次），因此有一部分分析物被保留到萃取相，然后通过提供适当的溶剂将保留的分析物解吸并转移到色谱柱，同样是静态或动态的。与 draw/eject 模式不同，毛细管调节、试样加载、清洗和输送操作可以手动或自动完成，可以使用多种不同的方案。两种最常用的设置如图 14.2 所示。最简单的配置是用萃取毛细管替换色谱进样阀上的内环。在加载样品的过程中提取分析物，然后通过转动阀门从 Load 位置到 Inject 位置，将其转移到具有流动相的分析柱中。这种方法被称为插装式 IT-SPME，可以用最少的额外设备来实现，如果要使用开放的毛细管柱进行萃取，这种方法尤其有用。

或者，在萃取后，通过辅助泵将分析物解吸并转移到色谱的进样环［图 14.2（3）］。然后，通过将进样阀从 Load 位置改变到 Inject 位置，将含有分析物的溶剂送入流动相色谱柱。这种结构在色谱分离中提供了更多功能，因为流动相不与萃取相接触，但增加了需要优化和控制的实验变量的数量。

通过增加循环次数，可以提高 draw/eject 法的提取效率，因此，尽管这些技术不是彻底的，而是一种平衡提取方法，但仍可以达到相对较高的回收率。事实上，在 Flow-through 系统中，萃取回收率比平衡法低，但目标化合物的数量可以被萃取，因此分析物的响应，可以通过增加加载到毛细管中的样品体积来增加。由于这些原因，draw/eject 系统更适合于分析复杂基质（如生物液体），此时应限制所处理样品的体积，以防止萃取毛细管变质；样品体积通常在 40～100μL 范围内。Flow-through 更适合分析水样，因为高达 2～4mL 的样品量可以通过萃取毛细管，而不会导致萃取相恶化或堵塞。对环境水体中的许多有机污染物具有良好的敏感性。表 14.1 所示为 2000—2016 年期间 IT-SPME 的相关应用。如前所述，IT-SPME 可以通过不同水平的仪器来实现。在某些形式的 IT-SPME 中，可能需要额外的设备，主要是磁性 IT-SPME 和电控 IT-SPME（见下文）。

将 IT-SPME 的设备与色谱系统在线耦合也值得关注。IT-SPME 可以很容易地在在线模式下与传统 LC 相结合，可以用于多种分析物/基质，色谱条件和检测器（表 14.1）。IT-SPME 与低流动相流速色谱系统如 CapLC 的组合也被广泛记录。在优化条件下，将 IT-SPME 毛细管包含在色谱流路中不会导致额外的带展宽，同时利用了 CapLC 的特性，如提高灵敏度和降低溶剂消耗。然而，在线连接到在较高流速下工作的色谱系统（UPLC）或较低流速下工作的色谱系统（NanoLC）需要一个特定的接口。通过与图 14.2（3）类似的双阀系统，阀内 IT-SPME 首次与 UPLC-MS/MS 偶联，但使用的是带内回路的阀 2[62]。第一个用于阀内 IT-SPME，萃取后，分析物被手动解吸并转移到第二个阀的内部回路。然后，富集分析物的插头从回路转移到色谱系统中，从而避免了在 UPLC 流路中插入萃取毛细管。另一方面，IT-SPME 通过使用两个相互连接的阀门（如常规和微型自动阀）作为注射装置[76]。

在某些应用中，萃取毛细管与微注射器装置组装使用。例如，使用萃取针头和注射器桶作为萃取装置的 IT-SPME 与 MS 的直接偶联已被报道[5]。所述装置用于三嗪类除草剂的富集，然后解吸并在线转移到实时直接分析质谱。

表 14.1　　2000—2016 年 IT-SPME 技术应用的摘要

IT-SPME 模式	基质	分析物	萃取毛细管	分离/检测	检测限/回收率/%	RSD/%	参考文献
draw/eject	尿液，血清	β-阻断剂及其代谢产物	Omegawax 250	LC-ESI-MS	0.1～1.2mg/L	<7.6	[16]
	细胞培养液	亚硝胺及其代谢物	特制 PPY	HPLC-DAD	20～250mg/L	2.9～3.6	[17]
	尿液，唾液	尼古丁，可替宁等生物碱	CP-Pora-Plot	LC-ESI-MS	15～40ng/L	4.7～11.3	[18]
	血浆	干扰素	干扰素 MIP	HPLC-FD	8mg/L	<9.2	[15]

续表

IT-SPME模式	基质	分析物	萃取毛细管	分离/检测	检测限 回收率/%	RSD/%	参考文献
	血浆	干扰素	RAM 蛋白涂层二氧化硅	HPLC-FD	0.06MIU/mL	<8	[19]
	血浆	干扰素	单克隆抗体毛细管柱	HPLC-FD	0.006MIU/mL	<6.2	[20]
	尿液	合成类固醇	Supel-Q PLOT 毛细管柱	LC-MS	9~182 pg/mL；>85	<4	[21]
	玉米	类胡萝卜素	单片 silica-ODS	mCL-UV	7~20ng/mL	4.6	[22]
	血浆	氟西汀和诺氟西汀	聚吡咯（PPY）	LC-FL	10~15ng/mL	13	[23]
	尿液	8-差向前列腺素	羧化 1006 毛细色谱柱	LC-MS/MS	3.3pg/mL；>92	≤8.5	[24]
	尿液	8-羟基-2-脱氧鸟苷	羧化 1006 气相毛细管柱	LC-MS/MS	8.3 pg/mL；92.6~96.4	≤9.6	[25]
	牛乳样品	对羟基苯甲酸酯	在微注射器装置中，一种涂覆分子印记聚合物的熔融二氧化硅毛细管	UPLC-MS/MS	—	2~15	[26]
	饮用水和湖水	苯酚，DMP，苯，DEP，甲苯，萘	PPY 涂层毛细管柱	HPLC-UV-MS	0.1~10mg/L	<7.2	[27]
	海岸水	邻苯二甲酸二正丁酯，邻苯二甲酸二异辛酯	TRB-5（95%聚二甲基硅氧烷-5%聚二苯基硅氧烷	HPLC-DAD	<250μg/L	<20	[28]
	水	氨基甲酸酯类农药	Omegawax 250	HPLC-UV	1μg/L	1.7~5.3	[29]
	海水，污水和废水	雌激素	羊胎素 Q 碳分子筛	HPLC-FD	0.005~0.003；0.01~0.15mg/L	<10	[30]
	地表水	氨基甲酸盐	Omegawax 250	CapLC-DAD	<0.3μg/L	<4.6	[31]
	茶	儿茶酚，咖啡因	PPY	HPLC-MS	0.01μg/L	1.5~5.5	[32]
	果汁，干炸食品	棒曲霉素	CP-Sil 5CB，CP-Sil 19CB，CP-Was 52CB，CP - Pora Plot amine，羊胎素 QPLOT，碳分子筛 1006 PLOT	HPLC-MS	23pg/mL	0.76~4.97	[33]

续表

IT-SPME模式	基质	分析物	萃取毛细管	分离/检测	检测限 回收率/%	RSD/%	参考文献
	茶品，干果	多环芳烃	CP-Sil 19CB	HPLC-FD	0.32~4.63 pg/mL	<7.6	[34]
	坚果和谷物	赭曲霉素A和B	碳分子筛1006 PLOT	LC-MS	0.5~20ng/mL；>88	5.1~7.7	[35]
	坚果，谷物，干果，香料	黄曲霉素（B_1，B_2，G_1，G_2）	羊胎素 QPLOT	LC-MS	0.5~20ng/mL；>80	<11.2	[36]
flow-through	—	磷酸肽	TiO_2-NPs-SiO_2毛细管柱	ESI-QTOF-MS，LC-ESI-MS/MS	—	—	[37]
	血浆，环境水	非甾体抗炎药（NSAIDs）	聚（4-乙烯吡啶-co-乙烯二甲基丙烯酸酯单体）	LC-MS	2.01~4.77ng/mL；88~105	<11	[38]
	猪肝，鸡肉	抗菌药物	OFL-MIP纤维	HPLC-UV-vis	0.016~0.11mg/L	<7.2	[39]
	大鼠血浆	原小檗碱，生物碱	聚（醚醚酮）和整体聚（丙烯胺-乙烯乙二醇二甲基丙烯酸酯）	HPLC-UV	0.01ng/mL；89.8~96.7	—	[40]
In-valve	尿液，血浆	抗抑郁药	单晶硅柱（氰乙基官能团）	LC-MS	0.06~2.84ng/mL；75.2~113	<16.5	[41]
flow-through（离线）	水溶液	β2-微珠蛋白和半胱氨酸蛋白酶抑制剂	熔融二氧化硅毛细管与抗体涂层聚苯乙烯纳米颗粒固定在其内壁	离子增强比浊免疫测定	98.7~101.6	<7.4	[42]
flow-through（In-valve）	人体呼出凝结气	醛	涂有石墨烯/聚苯胺的不锈钢管	LC-UV	0.02~0.04nmol/L；70~120	1.1~19	[43]
flow-through（In-valve）	人体呼出凝结气	醛	涂有石墨烯/PPY的不锈钢管	LC-UV	2.3~3.3nmol/L；85~117	1.8~11.3	[44]
flow-through（In-valve）	尿液	菊糖二羟基的邻苯二甲酸盐	TRB-5	CapLC-DAD	0.05~1.5mg/L；50~90	2~20	[45]
flow-through（In-valve）	大鼠血浆	磺胺类药	用电化学方法填充碳纤维束的PEEK管用聚（3，4-乙二氧噻吩）修饰	LC-UV	0.05~0.1ng/mL；91.7~97.8	≤4.57	[9]
flow-through（In-valve）	口腔液	安非他命，甲基苯丙胺，麻黄素	c-MWCNTs功能化的TRB-5	CapLC-FLD	0.5~0.8 μg/mL	3~5	[46]

续表

IT-SPME模式	基质	分析物	萃取毛细管	分离/检测	检测限 回收率/%	RSD/%	参考文献
flow-through（离线）	血液	己醛和丁酮	聚四氟乙烯管包覆聚多巴胺/双醛淀粉/壳聚糖复合物	LC-DAD	1.6~5.3 nmol/L；70~91	<7.2	[47]
	水	三嗪，有机磷化合物，敌草隆，异丙隆，DEP	TRB-5	CapLC-DAD	5~50ng/L	<19	[48]
	水	有机卤素，有机含氮化合物，多环芳烃，邻苯二甲酸盐，双酚 A	TRB-5	CapLC-DAD	0.008~0.2μg/L	2~6.2	[49]
	双壳类	多环芳烃	TRB-5	HPLC-FD	0.05~0.6ng/g	<10	[50]
	水	脂肪胺	TRB-5	HPLC-FD	0.1~0.4μg/L	<20	[51]
	水	有机磷化合物	TRB-5	CapLC-DAD	0.1~10μg/L	<18	[52]
	水	三嗪	TRB-5	CapLC-DAD	0.1~1μg/L	<25	[53]
	水	苯扎氯胺（BAK）	TRB-5	CapLC-DAD	0.5μg/L	7	[54]
	水	叶绿素 a	TRB-5	CapLC-DAD	0.05μg/L	<7	[55]
	水	甾醇类	TRB-5	CapLC-DAD	1.2~10μg/L	<6	[56]
	水	DEHP 及其降解产物	TRB-5	CapLC-DAD	0.005~1.5μg/L	<9	[57]
	沉积物	DEHP	TRB-5	CapLC-DAD	90μg/kg	<9	[58]
	双壳类	DEHP	TRB-5	CapLC-DAD	170 μg/kg	<10	[59]
	水溶性 PM10	羰基化合物	TRB-5	CapLC-DAD	30~198ng/L	<8	[60]
	水溶性 PM2.5	羰基化合物	TRB-35	CapLC-MS	0.9~8.2ng/L	<25	[61]
	水	三嗪，有机磷化合物，DEHP，敌草隆，异丙隆，氟乐灵	TRB-5	UHPLC-MS	0.025~2.5ng/L	<26	[62]
	水	苯胺化合物	MWNTs - COOH 固化的二氧化硅	HPLC-UV-vis	0.07~0.13μg/L	<9.1	[63]
	水和土壤	多环芳烃	$(FGO-PD)_3$ PTFE	HPLC-FD	0.005~0.1ng/L	<4.7	[64]

续表

IT-SPME模式	基质	分析物	萃取毛细管	分离/检测	检测限 回收率/%	RSD/%	参考文献
	水样	内分泌干扰物，多环芳烃	沉积有十八烷基嫁接的 SiO_2 纳米颗粒	HPLC-UV	0.042~0.78，0.034~0.19μg/L	<4.6	[65]
	水	药物	Fe_3O_4-SiO_2（CTAB）	CapLC-DAD	1.7~5μg/L	<12	[66]
	土壤和灰尘	多环芳烃	不锈钢	HPLC-FD	0.2~2μg/L	<2.9	[67]
flow-through（In-valve）	水	极性三嗪类	TRB-5	CapLC-DAD	0.02~0.1μg/L；98~120	<17	[68]
flow-through（In-valve）	泳池水	氯胺	TRB-5	CapLC-DAD	1.1mg/L；81	<13	[69]
flow-through（In-valve）	水和土壤	多环芳烃	涂有沸石咪唑框架的 PEEK 管	LC-FLD	0.5~5pg/L；82.5~98.6	1~3.6	[70]
flow-through（离线）	自来水，海水和河水	三嗪	涂有氧化石墨烯的熔融石英丝毛细管	LC-MS/MS	0.0005~0.05μg/L；81.6~112	2.7~7.1	[71]
flow-through（In-valve）	河流和雨水，煤灰的水提物	多环芳烃	用铜线填充的铜管	LC-DAD	0.001~0.1μg/L；86.2~115（AR：8.9~83.2）	0.6~3.6	[72]
flow-through（In-valve）	自来水和河水	多环芳烃	纳米银涂层的 PEEK 管	LC-UV	0.15~0.30μg/L；92.3~120	0.7~8.6	[73]
flow-through（In-valve）	水	纳米银离子	TRB-35	CapLC-DAD	0.015nmol/L	7.7	[74]
flow-through（In-valve）	雨水，地下水	多环芳烃	填充碳纤维束的 PEEK 管	LC-UV	0.001~0.1μg/L；92.3~111	0.9~6.8	[75]
多注射 flow-through（In-valve）	药品和河水样本	双氯芬酸	涂覆 SiO_2/PEG-spherical Fe_3O_4 的石英毛细管	Nano-LC-DAD	1ng/mL；50	≤4	[76]
flow-through（In-valve）	雨水和河水	多环芳烃	填充聚酯纤维的 PEEK 管	LC-DAD	0.01~0.03μg/L；91.3~113.4	5.8~6.9	[77]
flow-through（In-valve）	雨水	多环芳烃	TRB-5	LC-DAD-FLD	2.3~28ng/L；72~110（AR：11.0~31.4）	3.4~14.6	[78]
flow-through（In-valve）	瓶装和生活污水	雌激素	用涂有离子液体的铜线填充的铜管	LC-DAD	0.02~0.05μg/L；85~114	1.9~3.0	[79]

续表

IT-SPME 模式	基质	分析物	萃取毛细管	分离/检测	检测限 回收率/%	RSD/%	参考文献
flow-through (In-valve)	海水	防污剂	TRB-35	CapLC-DAD	0.015~0.2μg/L	<3.5	[12]
	动物食物	喹诺酮抗菌药物	聚(甲基丙烯酸-乙二醇二甲基丙烯酸酯)单体	HPLC-QTOF-MS	0.2~30ng/g	<14.5	[81]
flow-through	雪碧	酸性食物添加剂	基于毛细管的1-氨基丙基-3-甲基咪唑氯(液体改性吸附剂)	HPLC-UV	1.2~13.5μg/L;85.4~98.3	<6.9	[82]
	玉米	脂溶性维生素和β-胡萝卜素	单片硅-ODS	CLC-UV/Vis	1.9~173ng/mL;92.7	<5	[83]
FSS 模式分层抽样	花粉,花和种子	24-表油菜素内酯	MIP	DE-FSS-UPLC-FL	0.7ng/L;81.2~116	4.7~9.7	[84]
	植物提取物	28-芸苔素内酯	聚(甲基丙烯酸-co-乙烯二甲基丙烯酸酯)层(ma-co-edma)整体柱	LC-MS	0.101ng/g;80.3~92.1	6.8~9.6	[85]
flow-through	水,牛乳和果汁	苯甲酸酯类	涂有纳米结构聚苯胺复合材料的不锈钢管	LC-UV	0.02~0.04μg/L;80.3~90.2	5.9~7.0	[86]
flow-through	牛奶	环丙氨嗪和三聚氰胺	熔融石英管填充poly(1-vinyl-3-(butyl-4-sulfonate) imidazolium-co-acrylamide-Co-N, N'-methylenebis (acrylamide))单体的纳米颗粒	LC-UV	0.3~21.1ng/mL	≤7.6	[87]
	阳离子聚合物	乙二胺	TRB-5	HPLC-FD	20μg/L	<20	[88]
	清洁剂	苯扎氯铵及其同系物	TRB-35	CapLC-DAD-MS	0.3~0.8ng/mL	<10	[89]
flow-through (In-valve)	抗微生物剂配方	季氨化合物	TRB-35	CapLC-DAD	0.006~0.03μg/mL;117	≤18	[90]
flow-through (In-valve)	河水,香灰的水提取物	多环芳烃	镀金功能化的不锈钢管	LC-DAD	0.05~0.1μg/mL	≤4.2	[91]

续表

IT-SPME 模式	基质	分析物	萃取毛细管	分离/检测	检测限 回收率/%	RSD/%	参考文献
磁性							
磁性 flow-through (In-valve)	尿液	莫西沙星	填充十二烷基硫酸钠涂层四氧化三铁纳米颗粒的不锈钢管	LC-UV	0.03μg/L; 93~97.9	4.5	[92]
磁性 flow-through (In-valve)	水和尿液	氟喹诺酮	填充十二烷基硫酸钠涂层四氧化三铁纳米颗粒的不锈钢管	LC-UV	0.01~0.05μg/L; 98.2~108.2	4.6	[93]
磁性	水	有机磷化合物	Fe_3O_4-SiO_2 (CTAB)	CapLC-DAD	10~50μg/L	<15	[94]
	水	三嗪类	Fe_3O_4-SiO_2 (CTAB)	CapLC-DAD	0.3~0.4 μg/L	<10	[95]
磁性 flow-through (离线)	水样	雌激素	填充3-（三甲氧基硅基）丙基甲基丙烯酸酯整体功能化的四氧化三铁纳米颗粒的熔融石英管	LC-DAD	0.06~25 μg/L; 70~100	2.09~7.2	[96]
磁性 flow-through	农田，湖泊和河流的水	三嗪	Fe_3O_4 纳米颗粒掺杂的单片聚甲基丙烯酸辛酯-二乙基丙烯酸乙二醇酯	LC-DAD	0.074~0.23μg/L; 70.7~119	<10	[97]
电化学							
in valve	尿液	萘普生	填充 PPY 的不锈钢管	LC-UV	0.07μg/L; 92~99	<8	[98]
flow-through (电化学控制)	尿液，血浆和血液	吲哚美辛	填充混有乙二醇二甲基丙烯酸酯分子印记聚合物 PPY 的不锈钢管	LC-UV	0.07~2.0μg/L; 92.6~108.3	4.9~8.3	[99]
flow-through (In-valve) (电化学控制)	水，血浆和尿液	酸性、碱性和中性药物	填充 PPY 和吲哚-2-羧酸共聚物的不锈钢管	LC-DAD	0.05~19ng/L; 35.8~78.6 (AR)	≤6	[100]
flow-through (电化学控制)	自来水，废水和河水	无机硒	填充混有乙二醇二甲基丙烯酸酯聚合物 PPY 的不锈钢管	氰化物发生原子吸收光谱法	0.004μg/L	2.0~2.5	[6]

续表

IT-SPME 模式	基质	分析物	萃取毛细管	分离/检测	检测限 / 回收率/%	RSD/%	参考文献
draw/eject（热控制）	水	普萘洛尔	Rt-U Plot（二乙烯基苯乙二醇/二甲基丙烯酸酯毛细管柱）	LC-UV	6.3ng；6.3	1.2~6.2	[101]
Direct	水，橙汁	三嗪	聚（MAA-EDMA-SWNTs）单体	Direct DART-MS	0.02~0.14μg/L	<8	[5]

最近，利用电化学控制的管内固相微萃取（EC-IT-SPME）和氢气发生系统以及原子吸收光谱仪[6]连接，进行水样品中硒的预浓缩和测定。所述系统能够检测水样品中超痕量浓度的无机硒。

图 14.3 展示在 IT-SPME 应用中采用的毛细管的长度和内径。最佳内径与 LC 模态的偶联有关，而长度并不是一个关键参数。用于 LC 偶联的 IT-SPME 毛细管内径需要比 Nano-LC、Cap-LC 和传统 LC 更小。

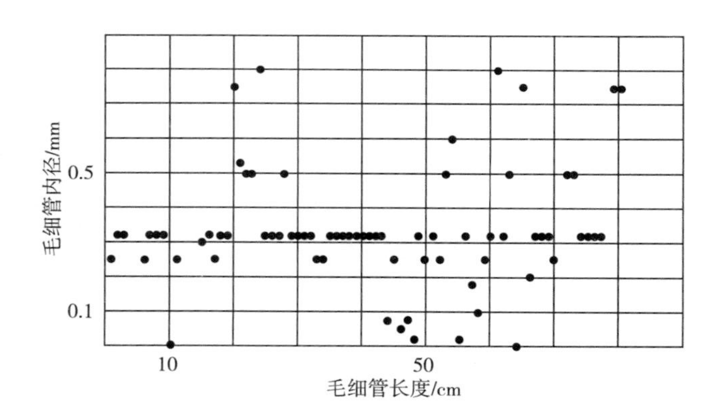

图 14.3　表 14.1 中报道的 IT-SPME 中毛细管的尺寸：内径和长度

14.3　萃取相的发展趋势

利用 IT-SPME 对样品进行萃取、净化和浓缩是依据样品各组分吸附/吸收行为的差异，这直接关系到萃取相的亲和性（平衡和动力学）。因此，吸附剂的吸附/吸收能力是与萃取过程相关的主要性能。通常，商业气相色谱柱已用于 IT-SPME 萃取。传统毛细管柱中主要使用的吸附剂可分为：①硅基萃取相和②碳基萃取相。硅基萃取相被采纳在大多数的 IT-SPME 应用中，77%的研究是使用这些吸附剂进行的。图 14.4（1）所示为研究最广泛的硅基萃取相的应用百分比。可以看出，尽管在文献中也发现了二甲硅氧烷和氰基甲基硅酮，但是二苯基改性聚二甲基硅氧烷的研究较为深入。基于碳的吸附剂主要包括二乙烯基苯，聚乙二醇和碳分子筛。二乙烯基苯和聚乙二醇的研究最为广泛［图 14.4（2）］。表 14.2 所示为毛细管涂层和传统萃取相的主要应用领域。

表 14.2 常规毛细管柱使用的主要毛细管涂层及应用领域

	涂层	应用领域
碳基吸附剂	二乙烯基苯	生物分析
	聚乙二醇	环境分析
	碳分子筛	食品分析
硅基吸附剂	二甲基硅氧烷	环境分析
	二苯基改性聚二甲基硅氧烷	工业产品分析
	腈基甲基硅酮	食品分析

如表 14.2 所示，碳基吸附剂（二乙烯苯、聚乙二醇和碳分子筛）主要用于生物和食品样品的分析。同时，硅基吸附剂的功能主要体现在环境、食品和工业产品分析中。这些吸附剂的萃取能力已被证实在非极性化合物上；然而，若干研究已经显示出了这些吸收剂对极性分析物的效用。表 14.1 所示为这些应用的几个代表性示例。

常规毛细管涂层面临着萃取效率低的主要限制。这种影响可以归因于低的吸附装载能力、稳定性，以及在某些情况下，由于分析物从样品到毛细管涂层的缓慢扩散导致较长的萃取时间。

因此，不断努力寻找更有效的毛细管涂层和探索新的萃取相已成为微萃取技术的一个主要任务，特别是对于 IT-SPME。这些毛细管涂层可以调节相互作用，从而提高灵敏度、选择性、稳定性和萃取时间。主要设计的吸附剂包括聚吡咯(PPY)[32 17,27,32,44]，碳纳米材料吸附剂（碳纳米管、石墨烯和碳纤维）[63、64、68]，磁性纳米颗粒（NPs）(Fe_3O_4）毛细管涂层[66、93、95]，整体毛细管柱[81、87]，分子印迹聚合物(MIPs)[26]和限进性材料（RAM）吸附剂和抗体吸附剂[13、19]。此外，还可以发现其他材料，如分子有机骨架[70]、离子液体改性吸附剂[82]、铜导线[79]和银纳米结构吸附剂[73]。然而，这些涂料的应用仍然有限。图 14.4（3）描述了过去几年提出的主要毛细管涂层的设计。可见，碳（纳米）材料萃取相是使用最广泛的，其次是 Fe_3O_4 吸附剂和整体毛细管柱。一般来说，萃取相的新类别已经在 flow-through 模式下进行了测试。以 Fe_3O_4 吸附剂和整体毛细管柱为例，主要采用 flow-through 模式方式进行研究。与此同时，基于抗体的吸附剂主要是在 draw/eject 模式中描述的。分子印迹聚合物毛细管涂层已用于这两种模式。

在分析方面，所设计的毛细管柱已经用于测定了各种不同性质和极性的化合物。表 14.3 所示为使用不同萃取相在不同基质中分析物的一些代表性例子。可以看出，最常用的毛细管涂层已被用于分析环境、生物和食品样品。值得注意的是，因为应用领域尚未探索，所以用设计的毛细管柱分析工业样品仍然是一个挑战。

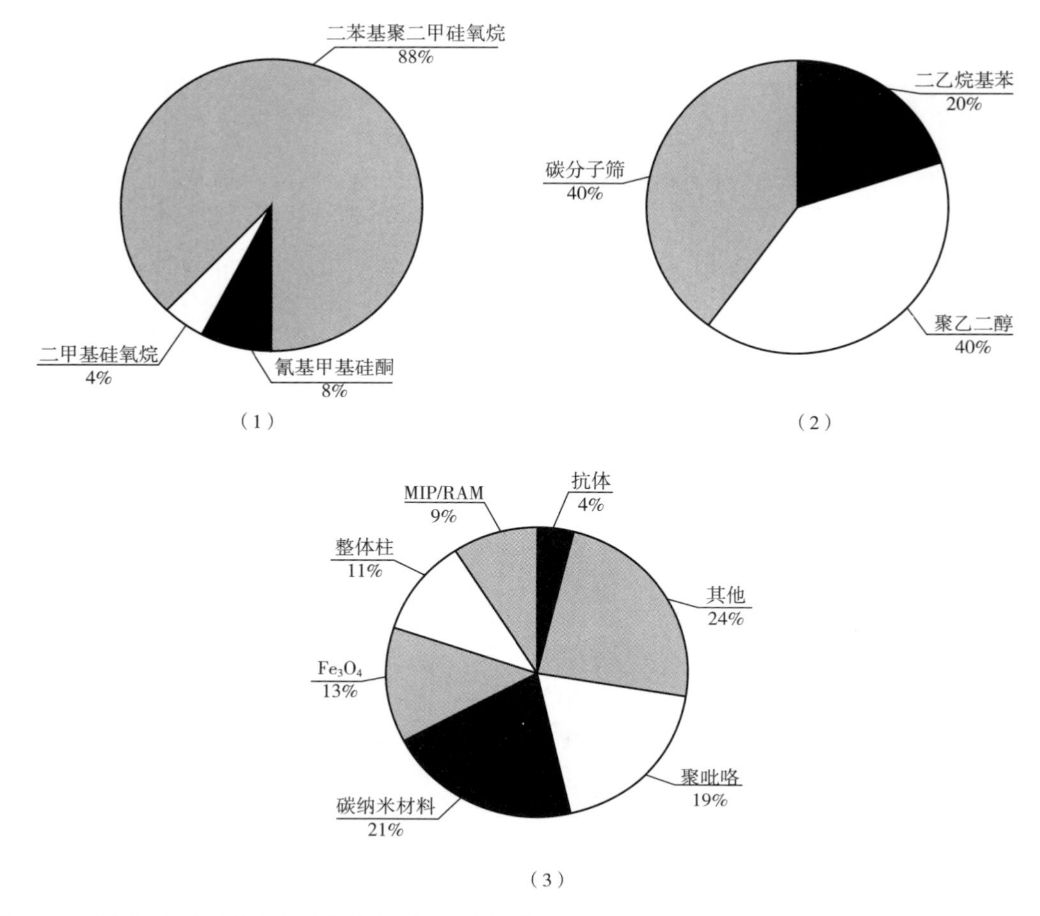

图 14.4 以（1）硅基萃取相（常规）、（2）碳基萃取相（常规）、（3）设计的萃取相作为研究的比例

表 14.3 不同萃取相测定的分析物

毛细管涂层	环境样本	生物样本	食物样本
碳纳米材料	三嗪，有机磷，邻苯二甲酸盐的化合物，苯胺类、多环芳烃	醛，酮，安非他明	三嗪
Fe_3O_4 吸附剂	雌激素，有机磷化合物，药物，三嗪	莫西沙星，氟喹诺酮	—
整体毛细管涂层	三嗪	非甾体抗炎药	喹诺酮类，苯甲酸，羟基苯甲酸，肉桂酸，二氯苯氧基乙酸
PPY 吸附剂	苯基物，苯，甲苯，萘酚	亚硝胺、毒品	苯甲酸酯类、咖啡因，儿茶素
基于抗体的涂层	β-微珠蛋白，半胱氨酸蛋白酶抑制剂	重组人干扰素 α-2a	—

续表

毛细管涂层	环境样本	生物样本	食物样本
基于 MIP/RAM 的吸附剂	对硝基苯酚	苯甲酸酯类、抗生素药物	抗生素药物
基于分子有机骨架的吸附剂	多环芳烃	—	—
基于铜线吸附剂	多环芳烃，雌激素	—	—
离子液体吸附剂	—	—	酸性食品添加剂
硅镁土 NPs 吸附剂	—	—	灭蝇胺，三聚氰胺
二氧化钛-二氧化硅修饰吸附剂	—	磷酸化多肽	—

如前所述，新设计的毛细管柱提高了萃取效率；然而，定量的萃取效率的研究仍在进行中。因此，在评估了这些新萃取相的潜力之后，毛细管涂层与外部源的内在相互作用的组合可以增强分析物对萃取相的亲和力也被研究。外界源，如磁场、电场和热能已经被测试来帮助萃取过程[66,96]。在使用磁场和电场时，这些模式分别被称为磁性 IT-SPME 和电化学控制 IT-SPME。此外，还提出了一种温度响应型涂层来开发 IT-SPME 应用[80]。由于需要额外的设备来提供最容易实现的磁、电或热场，所以这些方法更可能以 flow-through 模式进行，从而在 flow-through 设备中实现。

将固有的相互作用与使用外部资源结合起来的主要优点是提高了萃取效率或绝对回收率。表 14.4 所示为在不同的磁萃取相下，磁性作为外力作用下的绝对回收率。如表 14.4 所示，绝对回收率几乎是量化的。需要指出的是，传统毛细管柱的绝对回收率通常高达 30%。提高萃取效率的机理取决于萃取阶段。对于负载 Fe_3O_4（CTAB）NPs 涂层 SiO_2 和整体功能化的 Fe_3O_4（NPs）涂层，其改善与施加磁场时孤立的超顺磁 Fe_3O_4 NPs 的磁化强度有关。磁场产生具有不同局部磁场的区域。抗磁分析物往往被捕获在磁场梯度最小的区域，从而增加了被 SiO_2 相吸附剂吸附的能力。在解吸过程中，需要通过改变外加磁场的极性来逆转 NPs 的磁化。这种外部刺激在 NPs 附近经历快速变化（例如，一个快速退磁/磁化过程）。磁强度的这些快速变化导致被吸附分析物的解吸[66,96]。填充了 Fe_3O_4-SDS 涂层的不锈钢管萃取效率的提高可以归因于在 Fe_3O_4 NPs 表面形成的混合半胶团[92,93]。

表 14.4　　基于磁性萃取相获得的绝对回收率

磁萃取相	分析物	绝对回收率/%	参考文献
负载 Fe_3O_4（CTAB）NPs 涂层的 SiO_2	三嗪	60~63	[95]
	有机磷	60~84	[94]
	药品	70~100	[66]
整体功能化的 Fe_3O_4（NPs）涂层	雌激素	70~100	[96]
	三嗪	65~100	[97]

电化学和 IT-SPME 的结合也进行了探索。在这种情况下，提高萃取效率是基于导

电聚合物的电化学氧化还原转变。外部电场引起聚合物在导电和非导电阶段之间的氧化还原转换，导致分析物的萃取/解吸。该方法已被用于测定硒、吲哚美辛、阳离子氮化合物和萘普生等分析物。主要应用领域是环境和生物分析[6,99,100]。

此外，还对温度响应聚合物涂层的使用进行了测试。这些聚合物具有亲水性和疏水性随温度而变化的特征。因此，与分析物的相互作用可以在毛细管柱内进行调谐。例如，在食品分析中对雌激素的测定中，热辅助萃取的萃取效率提高了 1.5 倍[80]。

14.4 在线 IT-SPME-LC 的应用

在用于在线进行样品预浓缩和清洗的众多方法中，IT-SPME 由于其通用性和速度快而被广泛接受。表 14.1 对在不同的分析领域如环境、生物分析、食品和工业不同复杂程度基质的应用进行了描述。在本节中，我们将总结这些应用并讨论所使用的方法。

IT-SPME 的大部分研究工作是在环境领域实现的；报道的 2000—2016 年间的文献中，约 57% 属于这一领域。在表 14.1 中可以看到，不同种类的水如废水[6,30,48,49,53-56,62, 79, 94]，河水[6,53,63,71-73,76,77,91,95,97]，海水[12,30,55,68,71]，过度域水[54,68]，海岸水[28,51,54,57,62]，湖水[27,97]，表层水[31,53]，饮用水[27,79]，自来水[6,54,70,71,73]，雨水[72,75,77,78]，地表水[53,75]和泳池水[69]进行了分析。这些应用大多数是在阀模式下进行的，通过毛细管的样品体积通常在 0.2~4mL。一般这些样品可以不经样品预处理直接注入色谱系统，在某些情况下只需要过滤或离心。根据分析物（酸或碱性化合物）情况，可通过改变样品的 pH 来帮助萃取进行。从表 14.1 可以看出，在线 IT-SPME 已经应用于极性和非极性化合物的分析，方法采用了常规的液相色谱、毛细管液相、纳米液相色谱法和二极管阵列检测法（DAD）、荧光检测法（FLD）或质谱检测法。该技术已经应用于不同极性的污染物，如多环芳烃[49,64,65,67,70,72,73,75,77,78,91]，有机卤素[28,49]或杀虫剂如有机磷[48,52,62,94]，三嗪[48,53,62,68,71,95,97]，除草剂[48,62]，防污剂 Irgarol[12]和氨基甲酸盐[28,31]，或内分泌干扰物如邻苯二甲酸盐[28,48,49,57-59,65]，双酚 A[49]，脂肪胺[51]，苯胺化合物[63]和氯胺[69]。其他有机化合物如雌激素[30,79,96]、甾醇[56]、叶绿素[55]、制药化合物等紧急染物（乙酰水杨酸、乙酰氨基酚、阿替洛尔、双氯芬酸、布洛芬、酮洛芬、芬布芬和普萘洛尔[66,76,101]、季铵盐化合物表面活性剂[54]）已通过该萃取技术被测定。除了测定有机化合物的适用性外，值得一提的是它的其他应用。因此，我们的研究小组证明了 IT-SPME 结合 Cap-LC-DAD 用于表征和稳定性研究 AuNPs 的效用[74]。在其他工作中，Asiabi 等[6]使用 EC-IT-SPME 和 HG-AAS 测定了水样中的无机硒。

其他环境样本，如生物群、土壤、沉积物和水中的颗粒物质也被处理。这些样本的复杂性要求提前预处理。土壤[64,67,70]和沉积物[58]中的 PAHs 和 DEPH 被分别测定。土壤样品的预处理采用不同的步骤（超声、离心、蒸发、提取液溶解和稀释）[67]。对于沉积物，先进行基质固相分散处理，时间为 10min；所需的总分析时间为 20min，比文献中描述的其他方法要短。这种组合提取技术也应用于测定双壳类样品中的 DEPH，样品处理时间不到 30min。花粉、花朵、种子[84]和植物[85]分别采用了 MW 萃取法和固相萃取法进行了预处理。对颗粒物 PM2.5 和 PM10 组分中的羰基化合物采用了 IT-

SPME-LC UV[60]或 MS 进行检测[61]，从而避免了离线浓缩步骤。

生物基质如血浆[9,13,19,20,23,38,40,41,99,100]和尿液[16,18,21,24,25,41,45,92,93,98-100]等是 IT-SPME 主要分析的样本。其他样本如血清[16]，血液[47,99]，唾液[18,46]，母乳[26]或呼出气体[43,44]也被分析。通常这些样本（尿液和血清）在进样前只需要过滤、离心或稀释。而血浆、血液和母乳一般需要沉淀和离心。而对于毛发[102]或细胞培养[17]等较为复杂的基质，则采用超声和离心的方法对样品进行预处理。对于分析物，IT-SPME 已被应用于药物（受体阻滞剂和代谢物[16]、抗抑郁药[23,41]、甾体和非甾体抗炎药[38,99]、杀菌剂[26]、滥用药物[18,46,100]、合成类固醇[21]、抗生素[39,92,93]、原小檗碱生物碱[40]和磺胺类[9]）的测定。内源性化合物如 8-异差向前列腺素[24]，8-羟基-2’-脱氧鸟苷[25]，磷酸肽[37]，醛[43,44,47]和酮[47]也被该方法测定。Chaves 和 Queiroz 用不同的萃取相（MIP、RAM 蛋白涂层和免疫亲和管）[15]测定了血浆中的干扰素 $\alpha 2a$。Yammamoto 等[102]和 Jornet-Martinez 等人[45]分别检测了头发和尿液中多环芳烃或未代谢的二烷基邻苯二甲酸酯等污染物。细胞培养[17]中的亚硝胺及其代谢物也被测定。

在食品分析领域，介绍了一些该法在液体和固体基质中的应用。大多数报道的应用都是在液体样品中进行的，如果汁[33,86]、牛乳[86,87]和雪碧[82]。在这些情况下，样品在进样前进行了最低程度的处理（调整 pH、过滤、离心和稀释）[86]。此外，还分析了动物性食品[81]、乳粉[87]、干货食品[33,65]、茶叶[32]、坚果、谷物等样品[35,36]。Saito 等提出了几种测定坚果、谷物、干果和香料中毒素[33,35,36]的方法；分析物通过振荡用甲醇萃取，如果必要的话，用正己烷去除脂肪。最后将提取的 0.1mL 甲醇萃取液用蒸馏水稀释至 1mL。Zheng 等[81]分析测定了动物性食品（牛乳、鸡蛋、鸡和鱼）的喹诺酮类抗菌药物。其他类型的化合物还包括玉米中的脂溶性维生素和 β-胡萝卜素[22,83]，可口可乐和雪碧中的酸性食品添加剂[82]。其他的分析物，如茶[32]中的儿茶素和咖啡因，牛乳和果汁中的苯甲酸酯也已被测定。三嗪[5]或多环芳烃[65]等污染物已分别在橙汁或茶制品中被测定。该法还测定了乳粉中的环丙嗪及其代谢物（三聚氰胺）[87]。

在工业分析方面，Campíns-Falcó 等人报道了几种用于测定洗涤剂和杀菌剂配方中的季氨化合物[89,90]和阳离子聚合物中的乙二胺[88]的方法。最近，Bu 等人提出了一种测定蚊香灰中多环芳烃的方法[91]。

14.5 在线 IT-SPME-LC 的性能表征

表 14.1 所示为所提方法的检测限、回收率和精密度。可以看出，在线 IT-SPME 和 LC 结合可以获得非常高的灵敏度。一般不存在基质效应，回收率接近 100%，精度也较好。如图 14.5 所示，分析的时间也很好，处理样品过程所用的时间大部分不到 30min。

IT-SPME 的环境友好性已从其拟议的碳足迹值中得到了定量论证，其优于 SPE、SBSB、SDE、HS-SPME 和 MF-SPME 等相应的碳足迹值[12,68]。因为进样、萃取、清洗、预浓缩和分离是在一个步骤中完成的，所以它可以被认为是一种高效有力的萃取技术。IT-SPME 的碳足迹取决于操作设备。基本上，需要考虑的参数有：①注入系统（手动或自动）；②萃取装置（单泵或双泵装置）；③分离技术（常规 LC 或小型化 LC

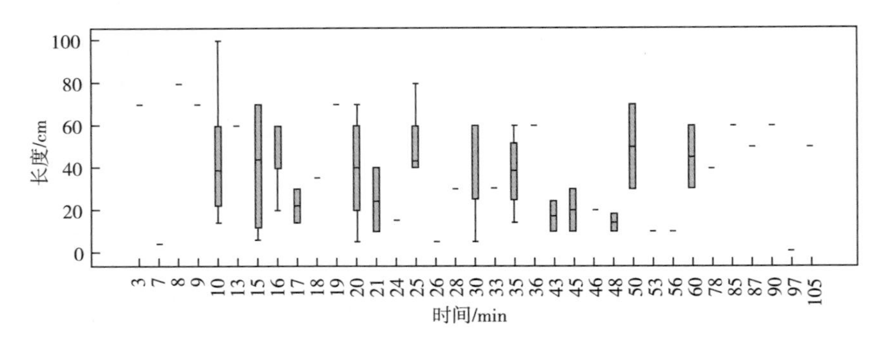

图 14.5　表 14.1 中毛细管长度对样品前处理时间的影响

系统)；④检测器（光学、电化学和质谱）。表 14.5 总结了在线 IT-SPME-LC-DAD 系统中使用的不同操作设备的碳足迹计算，其中目标分析物保留时间 t_r 为 10min，参考排放因子为 0.247kg CO_2/kWh[15]。

表 14.5　在线 IT-SPME-LC-DAD 系统中使用的不同操作设备的碳足迹计算

		碳足迹（kgCO_2eq）
flow-through 模式		
	手动进样-单泵-IT-SPME	1.1
	自动进样-单泵-IT-SPME	1.8
	自动进样-双泵-IT-SPME	3.0
draw-eject 模式		
	无洗涤泵	1.8
	有洗涤泵	3.0

可以看到，不同 IT-SPME 的模式之间有轻微的差异。这些差异可以归因于使用的辅助泵，因为电力消耗较高。这些结果已经证明了 IT-SPME 在绿色分析化学中的潜力。表 14.5 所示的数据是使用 DAD 检测计算出来的。需要指出的是，MS 检测将使这些值增加到 8 kgCO_2eq。

离线萃取方法的碳足迹比 IT-SPME 高一个数量级，这主要是由于使用了额外的仪器，如搅拌器、蒸发器、超声波、热解吸系统等[12]。此外，使用 MS 或 MS/MS 检测可将这些值提高到 250kgCO_2eq。但是，必须指出的是，整个分析方法的选择取决于每个分析问题的需求。因此，它必须被选择为分析性能和环境性能之间的折中解决方案。

14.6　结论

本文总结了 IT-SPME 和 LC 系统之间在线偶联的发展趋势并对 LC 体系进行了总

结。一个趋势是 IT-SPME 和 LC 在线偶联的小型化设备的发展。分析系统的小型化要在尽可能短的时间得到全面和真实的样品信息，在解决问题的同时还要考虑环境适应性（碳足迹工具）和经济性，这是当前和未来的分析化学一个主要目标。

虽然毛细管液相色谱或纳米液相色谱的样品稀释比低于传统液相色谱，但在某些应用中的灵敏度并不理想，这主要是由于使用的样品体积小。此外，有些样本不能直接分析，需要进行处理。为了解决这些问题，IT-SPME 集成了在线提取和预浓缩，是一种很有吸引力的技术。很多工作是需要小型化 LC 和 IT-SPME 在线偶联的。从仪器的角度 IT-SPME-Capillary LC 偶联已经得到了充分的解决，并进一步优化了 IT-SPME-nanoLC。在微纳米流体中，尺寸的设计在很大程度上是至关重要的。

与萃取相相比，IT-SPME 处理的样品的体积或数量是高的，通常，分析物的萃取是不彻底的。目前，大多数研究都集中在开发新的萃取材料，以提高选择性和萃取能力。在其他领域广泛使用的一些材料和纳米材料仍未被探索。因此，未来的应用是可以期待的。我们应该强调的是，IT-SPME 的性能不仅可以通过与萃取相的化学作用来实现，还可以通过先进的（纳米）材料来实现物理响应。很明显，先进萃取相将有助于 IT-SPME 领域的全面发展。

致谢

感谢 Generalitat Valenciana（PROMETEO 2016/109）和西班牙经济与竞争大臣（MINECO）/FEDER（项目 CTQ2014-53916-P）。P. S. -M. 对 MINECO/FEDER 的博士学位授予表示感谢。

参考文献

[1] R. E. Majors, Overview of sample preparation, LCGC 9 (1991) 16-20.

[2] D. E. Raynie, Trends in sample preparation, LCGC North Am. 34 (2016) 174-188.

[3] C. L. Arthur, J. Pawliszyn, Solid-phase microextraction with thermal-desorption using fused silica optical fibers, Anal. Chem. 62 (1990) 2145-2148.

[4] R. Eisert, J. Pawliszyn, Automated in-tube solid phase microextraction coupled to high-performance liquid chromatography, Anal. Chem. 69 (1997) 3140-3147.

[5] X. Wang, X. Li, Z. Li, Y. Zhang, Y. Bai, H. Liu, Online coupling of in-tube solid phase microextraction with direct analysis in real time mass spectrometry for rapid determination of triazine herbicides in water using carbon-nanotubes-incorporated polymer monolith, Anal. Chem. 86 (2014) 4739-4747.

[6] H. Asiabi, Y. Yamini, S. Seidi, M. Shamsayei, M. Safari, F. Rezaei, On-line electrochemically controlled in-tube solid phase microextraction of inorganic selenium followed by hydride generation atomic absorption spectrometry, Anal. Chim. Acta 922 (2016) 37-47.

[7] H. Kataoka, A. Ishizaki, Y. Nonaka, K. Saito, Developments and applications of capillary microextraction techniques: a review, Anal. Chim. Acta 655 (2009) 8-29.

[8] M. E. C. Queiroz, L. P. Melo, Selective capillary coating materials for in - tubesolid phase

microextraction coupled to liquid chromatography to determine drugs and biomarkers in biological samples: a review, Anal. Chim. Acta 826 (2014) 1-11.

[9]X. Ling, W. Zhang, Z. Chen, Electrochemically modified carbon fiber bundles as selective sorbent for online solid-phase microextraction of sulfonamides, Microchim. Acta 183 (2016) 813-820.

[10]Y. Moliner-Martinez, R. Herráez-Hernández, J. Verdú-Andrés, C. Molins-Legua, P. Campíns-Falcó, Recent advances of in-tube solid-phase microextraction, Trends Anal. Chem. 71 (2015) 205-213.

[11]C. Turner, Sustainable analytical chemistry-more than just being green, Pure Appl. Chem. 85 (2013) 2217-2229.

[12]J. Pla-Tolós, P. Serra-Mora, L. Hakobyan, C. Molins-Legua, Y. Moliner-Martinez, P. Campins-Falcó, A sustainable on-line CapLC method for quantifying antifouling agents like irgarol-1051 and diuron in water samples: estimation of the carbon footprint, Sci. Total Environ. 569-570 (2016) 611-618.

[13]S. Schmitz, B. Dawson, M. Spannagle, F. Thomson, J. Koch, R. Eaton, The Greenhouse Gas Protocol-A Corporate Accounting and Reporting Standard, Revised Edition, The GHG Protocol Corporate Accounting and Reporting Standard. 2004, 2004.

[14] L. Jiménez, J. De la Cruz, A. Carballo, J. Domench, Enfoques metodológicos parael cálculo de la Huella de Carbono, Observatorio de la sostenibilidad en España. NIPO: 770-11-252-8, Estudios Gráficos Europeos, S. A.

[15]A. R. Chaves, M. E. C. Queiroz, In-tube solid phase microextraction with molecularly imprinted polymers to determine interferon α2a in plasma samples by high performance liquid chromatography, J. Chromatogr. A 1318 (2013) 43-48.

[16] H. Kataoka, S. Narimatsu, H. L. Lord, J. Pawliszyn, Automated in-tube solid phase microextraction coupled with liquid chromatography/electrospray ionization mass spectrometry for the determination of β-blockers and metabolites in urine and serum samples, Anal. Chem. 71 (1999) 4237-4244.

[17]W. M. Mullet, K. Levsen, J. Borlak, J. C. Wu, J. Pawliszyn, Automated in-tube solidphase microextraction coupled with HPLC for the determination of N-nitrosamines in cell cultures, Anal. Chem. 74 (2002) 1695-1701.

[18]H. Kataoka, R. Inoue, K. Yagi, K. Saito, Determination of nicotine, cotinine and related alkaloids in human urine and saliva by automated in-tube solid phase microextraction coupled with liquid chromatography-mass spectrometry, J. Pharm. Biomed. Anal 49 (2009) 108-114.

[19]A. R. Chaves, B. J. G. Silva, F. M. Lanças, M. E. C. Queiroz J, Biocompatible in-tube solid-phase microextraction coupled with liquid chromatography-fluorescence detection for the determination of interferon α2a in plasma samples, J. Chromatogr. A 1218 (2011) 3376-3381.

[20]A. R. Chaves, M. E. C. Queiroz, Immunoaffinity in-tube solid phase microextraction coupled with liquid chromatography with fluorescence detection for determination of interferon α2a in plasma samples, J. Chromatogr. B 928 (2013) 37-43.

[21]K. Saito, K. Yagi, A. Ishizaki, H. Kataoka, Determination of anabolic steroids in human urine by automated in-tube solid-phase microextraction coupled with liquid chromatography-mass spectrometry, J. Pharm. Biomed. Anal. 52 (2010) 727-733.

[22]S. Zhang, K. Jia, S. Wang, Determination of b-Carotene in Corn by in tube SPME coupled to micro-LC, Chromatographia 72 (11) (2010) 1231-1233.

[23] B. J. Conçalves-Silva, F. Mauro-Lanças, M. E. Costa-Queiroz, Determination of fluoxetine and norfluoxetine enantiomers in human plasma by polypyrrole-coated capillary in tube solid-phase microextraction coupled with liquid chromatographyfluorescence detection, J. Chromatogr. A 1216 (2009) 8590-8597.

[24] K. Mizuno, H. Kataoka, Analysis of urinary 8-isoprostane as an oxidative stress biomarker by stable isotope dilution using automated online in-tube solid-phase microextraction coupled with liquid chromatography-tandem mass spectrometry, J. Pharm. Biomed. Anal. 112 (82015) (2015) 36-42.

[25] H. Kataoka, K. Mizuno, E. Oda, A. Saito, Determination of the oxidative stress biomarker urinary 8-hydroxy-2-deoxyguanosine by automated on-line in-tube solid-phase microextraction coupled with liquid chromatography-tandem mass spectrometry, J. Chromatogr. B 1019 (2016) 140-146.

[26] I. D. Souza, L. P. Melo, I. C. S. F. Jardim, J. C. S. Monteiro, A. M. S. Nakano, M. E. C. Queiroz, Selective molecularly imprinted polymer combined with restricted access material for in-tube SPME/UHPLC-MS/MS of parabens in breast milk samples, Anal. Chim. Acta 932 (2016) 49-59.

[27] J. C. Wu, J. Pawliszyn, Polypyrrole coated capillary coupled to HPLC for in-tube solid phase microextraction and analysis of aromatic compounds in aqueous samples, Anal. Chem. 73 (2001) 55-63.

[28] C. Chafer-Pericás, P. Campíns-Falcó, M. C. Prieto-Blanco, Automatic in-tube SPME and fast chromatography: a cost effective method for the estimation of dibutyl and di-2-ethylhexyl phthalates in environmental water samples, Anal. Chim. Acta 610 (2008) 268-273.

[29] Y. N. Gou, R. Eisert, J. Pawliszyn, Automated in-tube solid phase microextraction-high performance liquid chromatography for carbamate pesticide analysis, J. Chromatogr. A 873 (2000) 137-147.

[30] J. Aufartová, M. E. Torres-Padron, Z. Sosa-Ferrera, P. Solich, J. J. Santana-Rodriguez, Optimization of an in-tube solid-phase microextraction method coupled with HPLC for determination of some oestrogens in environmental liquid samples using different capillary columns, Int. J. Environ. Anal. Chem. 92 (2012) 382-396.

[31] Y. Gou, J. Pawliszyn, In-tube solid-phase microextraction coupled to capillary LC for carbamate analysis in water samples, Anal. Chem. 72 (2000) 2774-2779.

[32] J. C. Wu, W. Xie, J. Pawliszyn, Automated in-tube solid phase microextraction coupled with HPLC-ES-MS for the determination of catechins and caffeine in tea, Analyst 125 (2000) 2216-2222.

[33] H. Kataoka, M. Itano, A. Ishizaki, K. Saito, Determination of patulin in fruit juice and dried fruit samples by in-tube solid phase microextraction coupled with liquid chromatography-mass spectrometry, J. Chromatogr. A 1216 (2009) 3746-3750.

[34] A. Ishizaki, K. Saito, N. Hanioka, S. Nsrimatsu, H. Kataoka, Determination of polycyclic aromatic hydrocarbons in food samples by automated on-line in-tube solid phase microextraction coupled to high-performance liquid chromatographyfluorescence detection, J. Chromatogr. A 1217 (2010) 5555-5563.

[35] K. Saito, R. Ikeuchi, H. Kataoka, Determination of ochratoxins in nuts and grain samples by in-tube solid-phase microextraction coupled with liquid chromatography-mass spectrometry, J. Chromatogr. A 1220 (2012) 1-6.

[36] Y. Nonaka, K. Saito, N. Hanioka, S. Narimatsu, H. Kataoka, Determination of aflatoxins in food samples by automated on-line in-tube solid-phase microextraction coupled with liquid chromatography-mass spectrometry, J. Chromatogr. A 1216 (2009) 4416-4422.

[37] B. Lin, T. Li, Y. Zhao, F.-K. Huang, L. Guo, Y.-Q. Feng, Preparation of a TiO_2 nanoparticle-deposited capillary column by liquid phase deposition and its application to phosphopeptide analysis, J. Chromatogr. A 1192 (2008) 95-102.

[38]Q. W. Yu, X. Wang, Q. Ma, B. F. Yuan, H. B. He, Y. Q. Feng, Automated analysis of non-steroidals anti-inflammatory drugs in human plasma and water samples by in-tube solid-phase microextraction coupled to liquid chromatography-mass spectrometry based on a poly (4-vinylpyridine-co-ethylene dimethacrylate) monolith, Anal. Methods 4 (2012) 1538-1545.

[39]Y. L. Hu, C. Y. Song, G. K. Li, Fiber-in-tube solid-phase microextraction with molecularly imprinted coating for sensitive analysis of antibiotic drugs by high performance liquid chromatography, J. Chromatogr. A 1263 (2012) 21-27.

[40]W. Zhang, Z. Chen, Mussel inspired polydopamine functionalized poly (ether ether ketone) tube for online solid-phase microextraction-high performance liquid chromatography and its applications in analysis of protoberberine alkaloids in rat plasma, J. Chromatogr. A 1278 (2013) 29-36.

[41]M. M. Zheng, S. T. Wang, W. K. Hu, Y. Q. Feng, In tube solid-phase microextraction based on hybrid silica monolith coupled to liquid chromatography-mass spectrometry for automated analysis of ten antidepressants in human urine and plasma, J. Chromatogr. A 1217 (2010) 7493-7501.

[42]B. Xu, S. Cheng, X. Wang, D. Wang, L. Xu, Novel polystyrene/antibody nanoparticle-coated capillary for immunoaffinity in-tube solid-phase microextraction, Anal. Bioanal. Chem. 407 (2015) 2771-2775.

[43]Y. Li, H. Xu, Development of a novel graphene/polyaniline electrodeposited coating for on-line in-tube solid phase microextraction of aldehydes in human exhaled breath condensate, J. Chromatogr. A 1395 (2015) 23-31.

[44]S. Wang, S. Hu, H. Xu, Analysis of aldehydes in human exhaled breath condensates by in-tube SPME-HPLC, Anal. Chim. Acta 900 (2015) 67-75.

[45]N. Jornet-Martínez, C. Antón-Soriano, P. Campíns-Falcó, Estimation of the presence of unmetabolized dialkyl phthalates in untreated human urine by an on-line miniaturized reliable method, Sci. Total Environ. 532 (2015) 239-244.

[46] A. I. Argente-García, Y. Moliner-Martínez, E. López-García, P. Campíns-Falcó, R. Herráez-Hernández, Application of carbon nanotubes modified coatings for the determination of amphetamines by in-tube solid-phase microextraction and capillary liquid chromatography, Separations 3 (2016), http://dx. doi. org/10. 3390/chromatography 30100007.

[47]S. Wu, C. Cai, J. Cheng, M. Cheng, H. Zhou, J. Deng, Polydopamine/dialdehyde starch/chitosan composite coating for in-tube solid-phase microextraction and insitu derivation to analysis of two liver cancer biomarkers in human blood, Anal. Chim. Acta 935 (2016) 113-120.

[48]Y. Moliner-Martinez, C. Molins-legua, J. Verdú-Andrés, R. Herráez-Hernández, P. Campíns-Falcó, Advantages of monolithic over particulate columns for multiresidue analysis of organic pollutants by in-tube solid phase microextraction coupled to capillary liquid chromatography, J. Chromatogr. A 1218 (2011) 6256-6262.

[49]P. Campíns-Falcó, J. Verdú-Andrés, A. Sevillano-Cabeza, R. Herráez-Hernández, C. Molins-Legua, Y. Moliner-Martinez, In-tube solid phase microextraction coupled by in-valve mode to capillary LC-DAD: improving the detectability to multiresidue organic pollutants analysis in several whole waters, J. Chromatogr. A 1217 (2010) 2695-2702.

[50]P. Campíns-Falcó, J. Verdú-Andrés, A. Sevillano-Cabeza, C. Molins-Legua, R. Herráez-Hernández, New micromethod combining miniaturized matrix solidphase dispersion and in-tube-in-valve solid phase microextraction for estimating polycyclic aromatic hydrocarbons in bivalves, J. Chromatogr. A 1211

(2008) 13-21.

[51] P. Campíns-Falcó, R. Herráez-Hernández, J. Verdú-Andrés, C. Cháfer-Pericás, Online determination of aliphatic amines in water using in-tube solid-phase microextraction-assisted derivatization in in-valve mode for processing large sample volumes in LC, Anal. Bioanal. Chem. 394 (2009) 557-565.

[52] C. Cháfer-Pericás, R. Herráez-Hernández, P. Campíns-Falcó, In-tube solid-phase microextraction-capillary liquid chromatography as a solution for the screening analysis of organophosphorus pesticides in untreated environmental water samples, J. Chromatogr. A 1141 (2007) 10-21.

[53] C. Chafer-Pericás, R. Herráez-Hernández, P. Campíns-Falcó, On-fibre solid-phase microextraction coupled to conventional liquid chromatography versus in-tube solid-phase microextraction coupled to capillary liquid chromatography for screening analysis of triazines in water samples, J. Chromatogr. A 1125 (2006) 159-171.

[54] M. C. Prieto-Blanco, Y. Moliner-Martinez, P. Lopez-Mahia, P. Campíns-Falcó, Ionpair in-tube solid-phase microextraction and capillary liquid chromatography using titania-based column: application to the specific lauralkonium chloride determination in water, J. Chromatogr. A 1248 (2012) 55-59.

[55] Y. Vitta, Y. Moliner-Martinez, P. Campíns-Falcó, A. F. Cuervo, An in-tube SPME device for the selective determination of chlorophyll a in aquatic systems, Talanta 82 (2010) 952-956.

[56] Y. Moliner-Martinez, R. Herráez-Hernández, C. Molins-Legua, P. Campíns-Falcó, Improving analysis of apolar organic compounds by the use of a capillary titania-based column: application to the direct determination of faecal sterols, cholesterol and coprostanol in wastewater samples, J. Chromatogr. A 1217 (2010) 4682-4687.

[57] N. Jornet-Martinez, M. Muñoz-Ortuño, Y. Moliner-Martinez, R. Herráez-Hernández, P. Campíns-Falcó, On-line in-tube solid phase microextraction-capillary liquid chromatography method for monitoring degradation products of di- (2-ethylhexyl) phthalate in water, J. Chromatogr. A 1347 (2014) 157-160.

[58] M. Muñoz-Ortuño, A. Argente-garcia, Y. Moliner-Martinez, J. Verdú-Andrés, R. Herráez-Hernández, M. T. Picher, P. Campins-Falcó, A cost-effective method for estimating di- (2-ethylhexyl) phthalate in coastal sediments, J. Chromatogr. A1324 (2014) 57-62.

[59] M. Muñoz-Ortuño, Y. Moliner-Martinez, S. Cogollos-Costa, R. Herráez-Hernández, P. Campíns-Falcó, A miniaturized method for estimating di- (2-ethylhexyl) phthalate in bivalves as bioindicators, J. Chromatogr. A 1260 (2012) 169-173.

[60] M. C. Prieto-Blanco, P. Lopez-Mahia, P. Campíns-Falcó, On-line analysis of carbonyl compounds with derivatization in aqueous extracts of atmospheric particulate PM_{10} by in-tube solid phase microextraction coupled to capillary liquid chromatography, J. Chromatogr. A 1218 (2011) 4834-4839.

[61] M. C. Prieto-Blanco, Y. Moliner-Martinez, P. López-Mahia, P. Campíns-Falcó, Determination of carbonyl compounds in particulate matter $PM_{2.5}$ by in-tube solid phase microextraction coupled to capillary liquid chromatography/massspectrometry, Talanta 115 (2013) 876-880.

[62] A. Masiá, Y. Moliner-Martínez, M. Muñoz-Ortuño, Y. Picó, P. Campíns-Falcó, Multiresidue analysis of organic pollutants by in-tube solid phase microextraction coupled to ultra-high performance liquid-chromatography-electrospray-tandem mass spectrometry, J. Chromatogr. A 1306 (2013) 1-11.

[63] X. -Y. Liu, Y. -S. Ji, H. -X. Zhang, M. -C. Liu, Highly sensitive analysis of substituted aniline compounds in water samples by using oxidized multiwalled carbon nanotubes as an in-tube solid phase microextraction medium, J. Chromatogr. A1212 (2008) 10-15.

[64]W. P. Zhang, J. Zhang, T. Bao, W. Zhou, J. W. Meng, Z. L. Zheng, Universal multilayer assemblies of graphene in chemically resistant microtubes for extraction, Anal. Chem. 85 (2013) 6846-6854.

[65]T. Li, J. Xu, J. -H. Wu, Y. -Q. Feng, Liquid-phase deposition of silica nanoparticles into a capillary for in-tube solid phase microextraction coupled with high-performance liquid chromatography, J. Chromatogr. A 1216 (2009) 2989-2995.

[66]Y. Moliner-Martínez, H. Prima-García, A. Ribera, E. Coronado, P. Campíns-Falcó, Magnetic in-tube solid phase microextraction, Anal. Chem. 84 (2012) 7233-7240.

[67]W. Zhang, Z. Zhang, J. Meng, W. Zhou, Z. Chen, Adsorptive behavior and solidphase microextraction of bare stainless steel sample loop in high performance liquid chromatography, J. Chromatogr. A 1365 (2014) 19-28.

[68]Y. Moliner-Martínez, P. Serra-Mora, J. Verdú-Andrés, R. Herráez-Hernández, P. Campíns-Falcó, Analysis of polar triazines and degradation products in waters by in-tube solid-phase microextraction and capillary chromatography: an environmentally friendly method, Anal. Bioanal. Chem. 407 (2015) 1485-1497.

[69]J. Pla-Tolós, Y. Moliner-Martínez, C. Molins-Legua, R. Herráez-Hernández, J. Verdú-Andrés, P. Campíns-Falcó, Selective and sensitive method based on capillary liquid chromatography with in-tube solid phase microextraction for determination of monochloramine in water, J. Chromatogr. A 1388 (2015) 17-23.

[70]J. Zhang, W. Zhang, T. Bao, Z. Chen, Polydopamine-based immobilization of zeolitic imidazolate framework-8 for in-tube solid-phase microextraction, J. Chromatogr. A 1388 (2015) 9-16.

[71]F. Tan, C. Zhao, L. Li, M. Liu, X. He, J. Gao, Graphene oxide based in-tube solid-phase microextraction combined with liquid chromatography tandem mass spectrometry for the determination of triazine herbicides in water, J. Sep. Sci. 38 (2015) 2312-2319.

[72]M. Sun, J. Feng, Y. Bu, C. Luo, Highly sensitive copper fiber-in-tube solid-phase microextraction for on line selective analysis of polycyclic aromatic hydrocarbons coupled with high performance liquid chromatography, J. Chromatogr. A 1408 (2015) 41-48.

[73]M. Sun, J. Feng, Y. Bu, C. Luo, Nanostructured-silver-coated polyetheretherketone tube for online in-tube solid-phase microextraction coupled with high-performance liquid chromatography, J. Sep. Sci. 38 (2015) 3239-3246.

[74]R. A. Gonzalez-Fuenzalida, Y. Moliner-Martínez, C. Molins-Legua, V. Parada-Artigues, J. Verdu-Andres, P. Campins-Falco, New tools for characterizing metallic nanoparticles: AgNPs, a case study, Anal. Chem. 88 (2016) 1485-1493.

[75]J. Feng, M. Sun, Y. Bu, C. Luo, Development of a cheap and accessible carbon fibersin-poly (ether ether ketone) tube with high stability for online in-tube solid-phase microextraction, Talanta 148 (2016) 313-320.

[76]R. A. González-Fuenzalida, E. López-García, Y. Moliner-Martínez, P. Campíns-Falcó, Adsorbent phases with nanomaterials for in-tube solid-phase microextraction coupled on-line to liquid nanochromatography, J. Chromatogr. A 1432 (2016) 17-25.

[77]Y. Bu, J. Feng, M. Sun, C. Zhou, C. Luo, Facile and efficient poly (ethylene terephthalate) fibers-in-tube for online solid-phase microextraction towards polycyclic aromatic hydrocarbons, Anal. Bioanal. Chem. 408 (2016) 4871-4882.

[78]M. Fernández-Amado, M. C. Prieto-Blanco, P. López-Mahía, S. Muniategui-Lorenzo, D. Prada-Rodríguez, A novel and cost-effective method for the determination of fifteen polycyclic aromatic hydrocarbons in low volume rain water samples, Talanta 155 (2016) 175-184.

[79]M. Sun, J. Feng, Y. Bu, C. Luo, Ionic liquid coated copper wires and tubes for fiberin-tube solid-phase microextraction, J. Chromatogr. A 1458 (2016) 1-8.

[80]Q. -W. Yu, Q. Ma, Y. Q. Feng, Temperature-response polymer coating for in-tube solid phase microextraction coupled to high performance liquid chromatography, Talanta 84 (2011) 1019-1025.

[81]M. -M. Zheng, G. -D. Ruan, Y. -Q. Feng, Evaluating polymer monolith in-tube solid phase microextraction coupled to liquid chromatography/quadrupole time of flight mass spectrometry for reliable quantification and confirmation of quinolone antibacterials in edible animal food, J. Chromatogr. A 1216 (2009) 7510-7519.

[82] T. T. Wang, Y. H. Chen, J. F. Ma, M. J. Hu, U. Li, J. H. Fang, H. Q. Gao, A novel ionic liquid-modified organic-polymer monolith as the adsorbent for in-tube solid phase microextraction of acidic food additives, Anal. Bioanal. Chem. 406 (2014) 4955-4963.

[83]H. Xu, L. Jia, Capillary liquid chromatographic analysis of fat-soluble vitamins and b-carotene in combination with in-tube solid-phase microextraction, J. Chromatogr. B 877 (1—2) (2009) 13-16.

[84]J. Pan, Y. Huang, L. Liu, Y. Hu, G. Li, A novel fractionized sampling and stacking strategy for online hyphenation of solid-phase based extraction to ultra-high performance liquid chromatography for ultrasensitive analysis, J. Chromatogr. A 1316 (2013) 29-36.

[85]X. Wang, Q. Ma, M. L, C. Chang, Y. Bai, Y. Feng, H. Liu, Automated and sensitive analysis of 28-epihomobrassinolide in *Arabidopsis thaliana* by on-line polymer monolith microextraction coupled to liquid chromatography-mass spectrometry, J. Chromatogr. A 1317 (2013) 121-128.

[86]H. Asiabi, Y. Yamini, S. Seidi, A. Esrafili, F. Rezaei, Electroplating of nanostructured polyaniline-polypyrrole composite coating in a stainless-steel tube for on-line in-tube solid phase microextraction, J. Chromatogr. A 1397 (2015) 19-26.

[87]T. Wang, Y. Chen, J. Ma, Q. Qian, Z. Jin, L. Zhang, Y. Zhang, Attapulgite nanoparticles-modified monolithic column for hydrophilic in-tube solid-phase microextraction of cyromazine and melamine, Anal. Chem. 88 (2016) 1535-1541.

[88]M. C. Prieto-Blanco, C. Cháfer-Pericás, P. López-Mahía, P. Campíns-Falcó, Automated on-line in tube solid-phase microextraction-assisted derivatization coupled to liquid chromatography for quantifying residual dimethylamine in cationic polymers, J. Chromatogr. A 1188 (2008) 118-123.

[89] M. C. Prieto-Blanco, Y. Moliner-Martínez, P. Campíns-Falcó, Combining poly (dimethyldiphenylsiloxane) and nitrile phases for improving the separation and quantification of benzalkonium chloride homologues: in-tube solid phase microextraction-capillary liquid chromatography-diode array detection-mass spectrometry for analyzing industrial samples, J. Chromatogr. A 1297 (2013) 226-230.

[90]M. C. Prieto-Blanco, A. Argente-García, P. Campíns-Falcó, A capillary liquid chromatography method for benzalkonium chloride determination as a component or contaminant in mixtures of biocides, J. Chromatogr. A 1431 (2016) 176-183.

[91]Y. Bu, J. Feng, M. Sun, C. Zhou, C. Luo, Gold-functionalized stainless-steel wire and tube for fiber-in-tube solid-phase microextraction coupled to high-performance liquid chromatography for the determination of polycyclic aromatic hydrocarbons, J. Sep. Sci. 39 (2016) 932-938.

[92]A. Manbohi, S. H. Ahmadi, V. Jabbari, On-line microextraction of moxifloxacin using Fe_3O_4

nanoparticle-packed in-tube SPME, RSC Adv. 5 (2015) 57930-57936.

[93] A. Manbohi, S. H. Ahmadi, In-tube magnetic solid phase microextraction of some fluoroquinolones based on the use of sodium dodecyl sulfate coated Fe_3O_4 nanoparticles packed tube, Anal. Chim. Acta 885 (2015) 114-121.

[94] Y. Moliner-Martinez, Y. Vitta, H. Prima-Garcia, R. A. Gonzalez-Fuenzalida, A. Ribera, P. Campíns-Falcó, E. Coronado, Silica supported Fe_3O_4 magnetic nanoparticles for magnetic solid-phase extraction and in-tube solid phase microextraction: application to organophosphorous compounds, Anal. Bioanal. Chem. 406 (2014) 2211-2215.

[95] R. A. González-Fuenzalida, Y. Moliner-Martínez, H. Prima-Garcia, A. Ribera, P. Campins-Falcó, R. J. Zaragozá, Evaluation of superparamagnetic silica nanoparticles for extraction of triazines in magnetic in-tube solid phase microextraction coupled to capillary liquid chromatography, Nanomaterials 4 (2014) 242-255.

[96] M. Mei, X. Huang, Q. Luo, D. Yuan, Magnetism-enhanced monolith-based in-tube solid phase microextraction, Anal. Chem. 88 (2016) 1900-1907.

[97] M. Mei, X. Huang, X. Yang, Q. Luo, Effective extraction of triazines from environmental water samples using magnetism-enhanced monolith-based in-tube solid phase microextraction, Anal. Chim. Acta 937 (2016) 69-79.

[98] S. H. Ahmadi, A. Manbohi, K. T. Heydar, Electrochemically controlled in-tube solid phase microextraction of naproxen from urine samples using an experimental design, Analyst 140 (2015) 497-505.

[99] H. Asiabi, Y. Yamini, S. Seidi, F. Ghahramanifard, Preparation and evaluation of a novel molecularly imprinted polymer coating for selective extraction of indomethacin from biological samples by electrochemically controlled in-tube solid phase microextraction, Anal. Chim. Acta 913 (2016) 76-85.

[100] H. Asiabi, Y. Yamini, S. Seidi, M. Safaria, M. Shamsayei, Evaluation of in-tube solidphase microextraction method for co-extraction of acidic, basic, and neutral drugs, RSC Adv. 6 (2016) 14049-14058.

[101] Y. Yang, A. Rodriguez-Lafuente, J. Pawliszyn, Thermoelectric-based temperature controlling system for in-tube solid-phase microextraction, J. Sep. Sci. 37 (2014) 1617-1621.

[102] Y. Yamamoto, A. Ishizaki, H. Kataoka, Biomonitoring method for the determination of polycyclic aromatic hydrocarbons in hair by online in-tube solid-phase microextraction coupled with high performance liquid chromatography and fluorescence detection, J. Chromatogr. B1 000 (2015) 187-191.

15　搅拌棒吸附萃取

José Manuel Florêncio Nogueira
University of Lisbon, Lisbon, Portugal
E-mail: nogueira@ fc. ul. pt

15.1 引言

在过去三十年中，基于吸附的微萃取技术作为现代样品制备方法在分析化学中发挥了重要作用[1,2]。特别是由 Bzltuseen 等在 1999 年提出的搅拌棒吸附萃取（SBSE），这是一种符合绿色分析化学原理的新型被动无溶剂技术[3]。从那以后，SBSE 技术在世界范围内获得了专利，并注册了“Twister”商标。对于痕量分析，SBSE 已被证明是气体、液体和固体样品中挥发性至半挥发性化合物分析的最佳选择。这种无溶剂分析方法在应用色谱或联用技术之前，被证明是可以提高选择性和灵敏度的高通量方法。SBSE 技术的主要特点和优势在于，只需一步就能完成分析物的萃取和浓缩，操作简单易行，所需样品量少，可重复利用，可以在没有任何特殊要求的情况下过夜操作，易于与现代仪器系统相结合，并已证明具有广泛的应用前景。尽管 SBSE 与固相微萃取（SPME）的原理相同[4]，但其灵敏度更高，因为其使用更多吸附剂相，并且更稳健，已成功应用于监测多个科学领域真实基质中痕量乃至超痕量的重点和新出现的有机化合物。如今，SBSE 是一项成熟技术，在过去 15 年中，已被报道的出版物数量增加到约 1000 篇，在科学界获得了广泛认可。对于初学者来说，可以阅读几篇综述文章[4-14]以进一步了解这种分析技术，包括 SBSE 原理、方法开发和应用。

15.2 理论分析

最开始，这种技术采用聚二甲基硅氧烷（PDMS）作为吸附相，当时市面上已有几种这类商用装置。图 15.1 是 SBSE 分析装置的示意图和实际图像。该装置由一根磁性搅拌棒组成，该搅拌棒被结合到玻璃夹套（长度为 10mm 或 20mm）中，通常涂有 24~126μL（膜厚度为 0.5mm 或 1.0mm）的 PDMS。这种基于硅胶的非极性相特征在于促进与目标分析物的疏水相互作用，尽管 PDMS 中的氧原子也可以形成氢键，但是其保留机制主要利用范德华力。这种聚合物相表现出可观的扩散和热稳定特性，可在广泛的温度范围内操作[15]。因此，该无溶剂技术最开始被认为与气相色谱法（GC）分析后的热解吸（TD）相关。相对于 SPME 纤维体积（膜厚度为 100 μm 时可达 0.5μL），所涉及的 PDMS 体积较大，可实现较低的相比率（$\beta = V_W/V_{PDMS}$，其中 V_W 为水样体积，V_{PDMS} 为 PDMS 体积），从而让 SBSE 具有较好的萃取能力和定量回收率，特别是对于疏水性溶质。与 SPME 相比，这种方法能够将灵敏度提高至 250 倍，将检测限降低数倍至超痕量水平（万亿分之一）。SBSE 是一个平衡过程[4,5,10]，其分析物的回收率与搅拌棒的 PDMS 相和水样（W）之间的分配有关，呈现出与辛醇-水分配系数（$K_{PDMS/W} \approx K_{O/W}$）所描述的分布相似的行为。非极性溶质的 $\lg K_{O/W}$ 值几乎都大于或等于 3，而极性更大的溶质则更低。因此，$K_{O/W}$ 和 β 是以下公式预测理论回收率的重要参数：

$$回收率\% = \frac{K_{O/W}/\beta}{1+K_{O/W}/\beta}$$

由公式可见，目标溶质的疏水性越高或 PDMS 体积越大（β 越小），SBSE 的预期

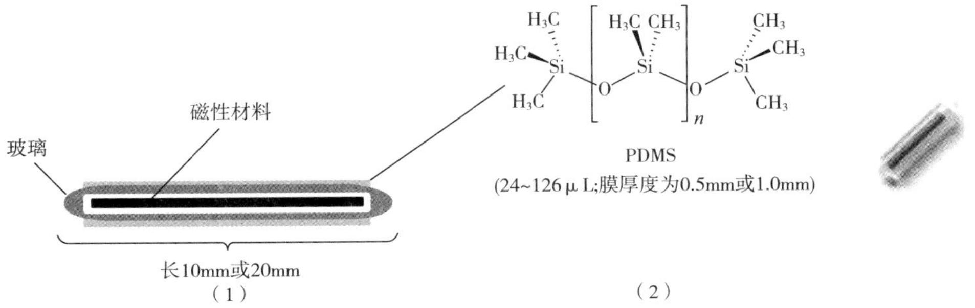

玻璃

磁性材料

长10mm或20mm

（1）

PDMS

(24~126μL;膜厚度为0.5mm或1.0mm)

（2）

图 15.1 搅拌棒吸附萃取装置的示意图（1）和实际图像（2）

萃取收率越高。因此，使用该公式，只要知道 β 和 $K_{O/W}$，我们就可以预测特定化合物的理论回收率，其中 $K_{O/W}$ 可以很容易地从文献中的理论值获得或通过其他可用软件计算。作为实际示例（图 15.2），如果通过 SBSE 分析含有非极性溶质（例如 lg $K_{O/W}$ = 4.0）的 24mL 样品，使用含有 24μL PDMS（即 β = 1000）的搅拌棒，预期平均理论回收率为 90.9%。我们还可以预测，对于 $K_{O/W}/\beta = 1$ 的 SBSE 工艺，预期理论回收率为 50.0%，而对于 lg $K_{O/W}$ 值大于 5 的溶质，萃取将是定量或完全萃取。由于每种分析物的 $K_{O/W}$ 值恒定，因此可针对每种特定应用类型调整 β。然而，如果观察到偏离预期理论数据，则必须优化影响微萃取动力学或热力学的最重要实验参数，以达到稳态条件。图 15.2 描绘了理论效率对 lg $K_{O/W}$ 的曲线图，由图可见目标分析物的疏水性越高，通过 SBSE 预测的萃取收率就越大。还可见 β 对 SBSE 理论回收率的影响，对于较低的 β 值（样品体积小），曲线向左移动，预期具有较高的效率。为更好地理解，图 15.2 还举例说明了三种不同 β 水平对 lg $K_{O/W}$ = 4 的有机化合物在特殊情况下的理论效率的影响。

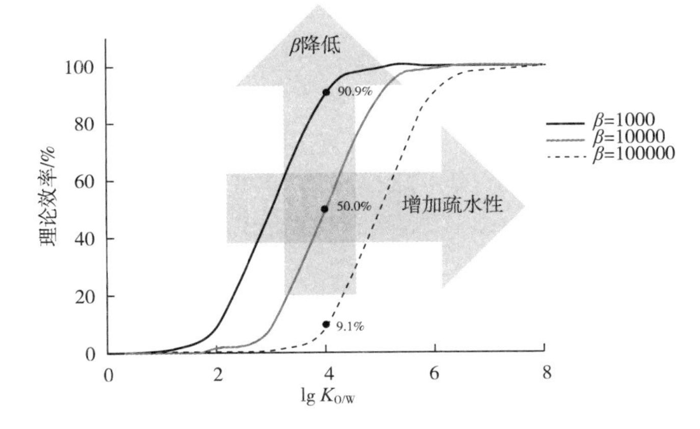

图 15.2 相关相（β）对搅拌棒吸附萃取相对于 lg $K_{O/W}$ 的理论效率的影响，以及在三种不同 β 水平下 lg $K_{O/W}$ = 4 的有机化合物的特殊情况示例。lg $K_{O/W}$ =4 的有机化合物在三种不同 β 水平影响下的理论回收率

15.3 实验分析

与许多其他基于吸附的微萃取技术一样，SBSE 实验分析包括两个主要阶段。在第一阶段，通过静态吸附过程从原样品向 PDMS 相萃取目标分析物。然后，采用 TD 或液体解吸（LD）模式，将溶质从聚合物反相萃取至色谱系统。从实验的角度来看，SBSE 方法开发应始终通过系统的分析，使用单变量或多变量策略进行优化。主要目标是获得影响分析系统的最佳实验参数，从而能够评估不同变量之间的主要相互作用[10,16,17]。在优化评估之后，必须执行验证程序以及应用于真实基质，以评估分析的有效性。

15.3.1 萃取阶段

在萃取过程中，在搅拌样品的同时，采用顶空（HS）或浸没取样模式使搅拌棒与溶质接触。图 15.3 描绘了两种操作模式下的装置示意图和实际图像。在专用于分析挥发性化合物的 HS-SBSE 模式下，将搅拌棒悬挂在顶空瓶顶部进行采样［图 15.3（1）］，此时聚合物与液体或固体样品的蒸汽相保持静态接触，避免被潜在的干扰化合物污染。图 15.3（2）以示意图显示了以浸没模式运行的 SBSE 图像，注意其中最值得关注的分析特征。该阶段通常在稳态条件下进行，以获得最大效率，其中需控制了几个实验参数。在这方面，必须进行系统分析，以优化影响萃取过程的最重要变量，即动力学参数（平衡时间和搅拌速度），其影响分析物向 PDMS 相的传质速度。如果动力学过程缓慢（即达到稳态条件需要更长的平衡时间），则可以使用 SBSE，因为与其他微萃取方法（例如动态顶空技术）相比，SBSE 能够过夜操作，没有任何特殊要求，因而无缺点。基质的酸碱度、极性和离子强度影响萃取热力学过程，影响分析物和 PDMS 相之间的相互作用[4,10,17,18]；基质的酸碱度影响可解离的分析物，其中非解离条件始终是获得更好回收率的必要条件；添加电解质（例如氯化钠）有利于“盐析”或“油效应”现象，因此，可能对富集极性化合物更为有效（$\lg K_{O/W}<3$），特别是 HS 采样模式；最后，添加改性剂（例如甲醇）可最大程度减少疏水性更强的溶质（$\lg K_{O/W}\geqslant 3$）发生“壁效应”现象。其他参数，例如样品和 PDMS 体积、稀释系数、温度等也非常重要，特别是对于要达到超痕量分析所需的灵敏度。PDMS 体积影响回收率，特别是对于极性较大的溶质，而温度对所涉及化合物的蒸气压有很大影响，尤其是在 HS 采样模式下。然而，为了最大限度地减少萃取时间，并在非平衡条件下工作也具有足够的性能，可牺牲灵敏度和精密度。必须强调的是，任何 SBSE 方法的开发都应该从在加标所需浓度的模型或目标化合物的超纯水中进行分析开始，以评估分析效果。一般而言，该技术涉及的实际测定非常容易操作，因为我们只需要使用镊子和简单的磁性夹子来操作搅拌棒。SBSE 技术最值得关注的特点之一是，每个搅拌棒可以使用数百次，而不会使 PDMS 涂层产生任何物理退化。即便如此，也必须避免使用极端条件，例如高离子强度（例如氯化钠），因为它们会破坏 PDMS 相。唯一可行的条件是，在重复使用之前，应使用合适的溶剂（例如，乙腈）或通过 TD 处理（例如 320℃）将搅拌棒清理干净，以尽量减少干扰和记忆效应。

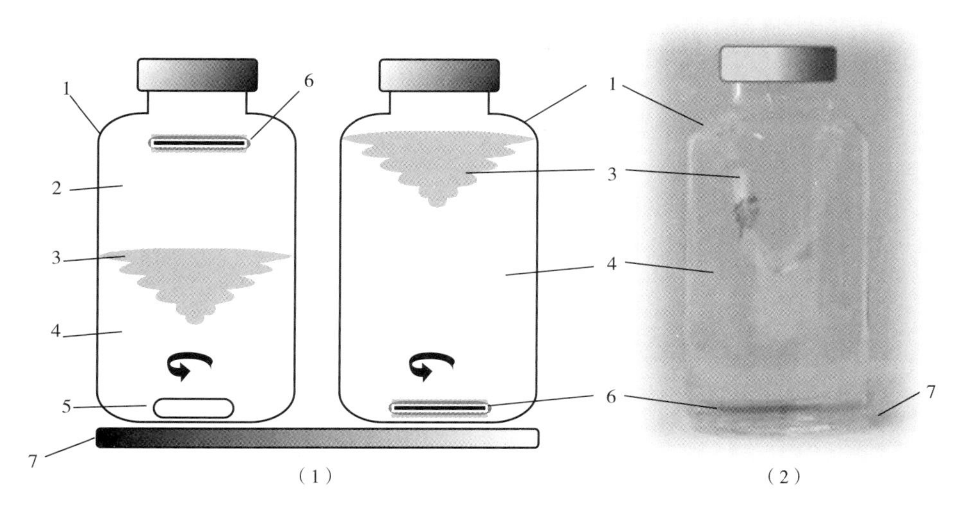

图 15.3 通过顶空（HS）（1）和浸没（2）模式进行搅拌棒吸附萃取（SBSE）操作的示意图和实际图像
1—取样瓶 2—HS 3—样品涡旋 4—样品 5—特氟龙磁力搅拌棒 6—SBSE 装置 7—搅拌板

15.3.2 反萃取阶段

在萃取阶段之后，通常从水性基质中取出搅拌棒，用蒸馏水冲洗以清除任何可能存在的干扰（例如，盐、糖、蛋白质或其他不良成分），用干净的纸巾去除水分，并通过 TD 或 LD 模式进行反萃取过程，如图 15.4 所示。从一开始，SBSE 方法就被精心设计为符合绿色分析化学原理的替代性无溶剂方法，正因为如此，将其开发为与采用 TD 模式的现代仪器系统联用。尽管 TD 是最直接的反萃取方法，但其只能分析热稳定性挥发性至半挥发性溶质，而 LD 应用更广泛，其通常更适用于半挥发性至非挥发性不耐热化合物。TD 模式需要一个玻璃管和一个专门用于烘烤操作的装置（高达 350℃），仅适用于 GC 分析。在 TD 操作期间，应优化几个仪器参数，尤其是解吸温度、吹扫流速以及 GC 进样器温度，其中程序升温汽化（PTV）入口是必要优化参数[17]。PTV 在较低温度下使用氮气作为低温液体对分析物进行低温聚焦，随后通过快速加热色谱柱进行汽化，以避免谱带变宽。图 15.4（1）（I）为 TD 单一模式，分析前每个样品仅使用一个搅拌棒置于玻璃管内。溶剂类型（例如甲醇、乙腈、混合物等）、浸泡时间和解吸步骤数是 LD 模式的重要变量。为此，必须在机械或超声处理下将搅拌棒完全浸没入玻璃瓶或衬管中的溶剂，以提高反萃取效率[16,18,19]。图 15.4（2）为 LD 模式，其使用装满合适溶剂的常规玻璃瓶或衬管。尽管如此 LD 模式也不能完全符合绿色分析化学原理，所涉及的溶剂体积（通常达 1.5mL）较多，因此认为该步骤不环保。更局限的是，LD 在仪器分析（离线方法）之前通常需要蒸发和/或溶剂切换步骤，这使得与 TD 相比，这一阶段有时并不方便使用。另一方面，LD 为样品再分析提供了机会，并且除了非常具有成本效益外，还因为其可与 GC、高效液相色谱（HPLC）[19-21] 甚至毛细管电泳[22-24] 系统相结合而具有通用性。LD 模式是为了克服 TD 分析半挥发性至非挥发性耐热性化合物的局限性，而引入的首个修正模式。相反，TD 是一种在线方法，其能够进行直接和完整的样品分析，使得极高的灵敏度和自动化成为可能（例如，CombiPAL）。但 TD 模式仅限于 GC 系统，并且由于其价格昂贵，仅适合常规分析。

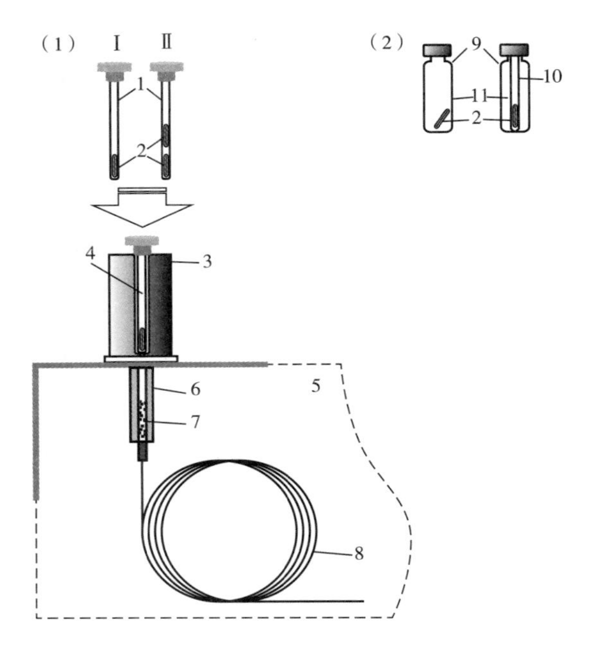

图 15.4 通过热解吸（TD）（1）和液体解吸（LD）（2）模式的搅拌棒吸附萃取（SBSE）反萃取操作示意图

1—TD 玻璃管 2—SBSE 装置 3—TD 装置 4—带有 SBSE 装置的 TD 玻璃管 5—气相色谱仪 6—程序升温蒸发进样器 7—内衬玻璃棉 8—毛细管柱 9—小瓶 10—衬管 11—LD 溶剂 Ⅰ—TD 单模 Ⅱ—TD 多模

15.3.3 验证和潜在应用

验证评估始终是分析开发的重要任务[10,14]。因此，完成 SBSE 优化过程后，将进行验证评估，以评价所开发方法的分析限度、线性动态范围、精密度和准确度等参数。一般而言，验证参数应符合大多数现行国际指南的再现性要求。另外，应评估优化方法在实际样品中的应用，以证明与其他专用方法相比，该方法作为替代技术具有分析优势。SBSE 的有效性可能受到许多实际基质复杂性的严重影响，因为潜在的干扰可能影响回收率。因此，应将经过优化和验证的 SBSE 方法应用于特定的实际样品，以验证分析效果，以及有无可能发生导致基质效应的干扰。但如果出现这种情况，补偿基质效应的最简单方法是采用标准加入法。自从引入 SBSE 以来，可以找到数百种分析应用，其集中了所有优于其他常规技术的优点。一般来说，作者认可的主要观点是，SBSE 是一种无溶剂方法，其实际操作非常简单，作为用于复杂基质的多残留分析方法，其能实现极大的选择性和灵敏度。根据文献报道，大多数 SBSE 应用主要是采取浸没取样模式，该模式涉及对新出现的优先目标进行痕量分析，这些目标可能来自自然或人为来源，主要在重要领域，例如环境、食品、香料和芳香剂、天然产品、药物和个人护理产品、生物医学、法医等。良好的应用实例包括测定气体样品中的挥发性和半挥发性有机化合物（例如香料和芳香剂）、水基质中的持久性有机污染物（例如杀虫剂）、饮料、水果、蔬菜和植物中的污染物和芳香化合物，以及生物体液（例如唾液、尿液样品）和废水样品中的代谢物、药物（例如性激素）和药品等。除了数百篇独立的文章外，在文献［25-69］中还可以找到几篇专门针对 SBSE 应用的推荐报告。简而言之，SBSE 已被证明是一种卓越的基于吸附的微萃取技术，专门用于监测多种真实基质（尤其是水基样品）中从非极性到中等极性超痕量水平的具有挥发性至半挥发性特征的物质，图 15.5 举例说明了 SBSE 的三个成功应用：图 15.5（Ⅰ）为马德拉葡萄酒样品应用 SBSE/TD-GC-MS 的色谱图，其中可见表征由痕量水平的几类有机化合物组成的复杂芳香族化合物达到了极高灵敏度[38]；在图 15.5（Ⅱ）的色谱图中，可见在 SBSE-LD/HPLC-二极管阵列检测（DAD）分析之前通过衍生化获得的几个真实样品中存在乙二醛和甲基乙二醛[46]；图 15.5（Ⅲ）为 SBSE-LD/胶束电动色谱法（MEKC）-DAD 测定河口样品中六种多环芳香烃的电泳图[23]。

15.4 SBSE 主要缺点和解决方案

尽管 SBSE 是一种完全成熟的基于吸附的技术，这种分析工具最初设计用于监测中等极性至非极性化合物（即 lg $K_{O/W} \geqslant 3$），尤其是具有挥发性至半挥发性特征的化合物，因此一般与 GC 系统联用。然而，如果我们要研究具有从中等极性到高极性特征的分析物上（即 lg $K_{O/W} < 3$），其中大部分为非挥发性，则由于涉及的疏水相互作用较弱或不存在，以及 TD 反萃取存在较大限制，SBSE 结合 GC 分析已被证明无效。因此，在过去十年中，已经提出了几种分析策略或新概念来克服这些缺点。其中包括引入多模式分析、衍生步骤、替代吸附剂相和新型被动微萃取技术[10,14,17]。

15.4.1 多模式分析

通常进行多残留分析非常困难，尤其是不同目标或化合物类别的实验条件要求差别很大时。即便如此，在双重或连续模式下使用 SBSE 均可克服这些分析局限性[47,48]。SBSE 双模式，在同一试验中使用多种搅拌棒来提高灵敏度。另一种方案是在相同或不同实验条件下，在不同试验中萃取样品，每次分析使用同一个搅拌棒装置。随后通过 GC-TD 或 LD 法来分析各种富集的搅拌棒。在多次多残留分析过程中，在不同实验条件下可以较好地萃取几种化合物。对于这些特殊情况，可以使用最佳实验参数进行几次分析，然后，搅拌棒可以单独或同时解吸。例如，随着离子强度增强，亲水性更强的溶质的回收率会增加，而疏水性更强的溶质的回收率会降低。因此，对于这些情况，可以用连续 SBSE 模式，其中一个样品通过添加盐进行分析，第二个样品在未改变的条件下使用一个或多个搅拌棒进行分析。当决定使用多模式方法时，可以使用市售磁力夹，借助磁力将一个或多个搅拌棒定位在样品瓶内壁上。图 15.4（1）（Ⅱ）举例说明了 TD 多模式方法，其中两个搅拌棒在相同的玻璃管中，在相同或不同的实验条件下，

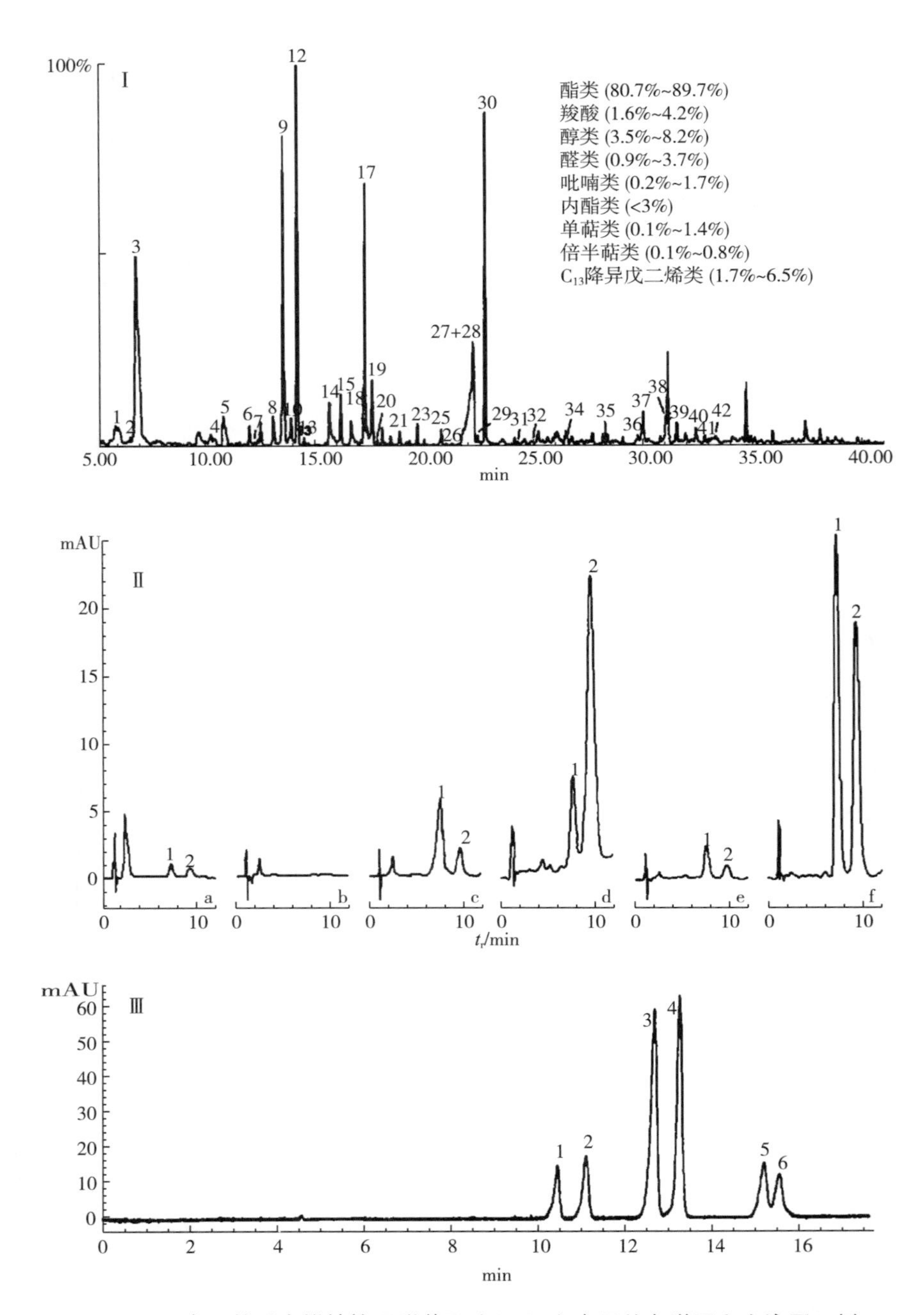

图 15.5　实际基质中搅拌棒吸附萃取（SBSE）应用的色谱图和电泳图示例

（Ⅰ）用 SBSE/TD-GC-MS 法从马德拉葡萄酒样品（Verdelho-99）中萃取芳香物质[38]；（Ⅱ）SBSE（DAN）$_{原位}$-LD/HPLC-DAD 法测定超纯水（0.1μg/L）(a)、自来水（b）、游泳池水（c）样品、啤酒（d）、酵母细胞萃取液（e）和尿液（f）真实样品中的乙二醛（1）和甲基乙二醛（2）[46]；（Ⅲ）用 SBSE-LD/MEKC-DAD 法分析河口沉积物中六种多核芳香烃（1，联苯；2，芴；3，菲 ISD；4，蒽；5，荧蒽；6，芘）[23]。DAD，二极管阵列检测；DAN，2，3-二氨基萘；MEKC，胶束电动色谱；TD-GC-MS，热解吸气相色谱法。

每个或两个搅拌棒都来自相同或不同的分析。

15.4.2 衍生步骤

当回收率非常低或没有回收时，对极性较大、热不稳定的化合物使用衍生化试剂是实施 SBSE 技术的一种可能方法。该程序可降低亲水性更强的溶质的极性，因此可增加对 PDMS 相的亲和力，同时挥发性或发色基团可获得更好的 GC 或 HPLC 响应[46,53]。为了实施这种方法，可以采用不同的衍生策略，包括原位衍生、搅拌棒上衍生或萃取后衍生模式，在这些模式中，可以主要使用烷基化、乙酰化、酰化、酯化和甲硅烷基化试剂等进行多种特定的衍生反应。原位衍生模式是最常见和最简单的方法，在富集阶段之前或与富集阶段同时发生在水基质中，使用浸没或 HS 采样方法。搅拌棒上衍生模式可以采用衍生剂预加载 PDMS 涂层，然后，将分析物并入 PDMS 相中（同时衍生和萃取）。另一种可以采用的方法是将分析物预先富集到 PDMS 相中，然后将搅拌棒暴露于衍生剂的蒸汽中（衍生后萃取）。尽管衍生化方法目前已被用于克服 SBSE 对大多数极性溶质缺乏效率的问题，但根据所涉及的分析物的化学性质，它仅限于少数几种特定的衍生试剂。

15.4.3 替代吸附剂相

由于 PDMS 不适用于极性化合物，因此克服这一局限性的方法是使用具有更好的亲和性和选择性的替代涂层相[57]。2005 年提出的第一个替代相是双相搅拌棒[58]，它结合了两种或多种吸附剂材料，即活性炭（AC）和 PDMS，表现出不同的富集能力。同时，其他材料已证明是在水基质中浓缩分析物的简单且廉价的方法，例如 PDMS 棒和作为聚合物吸收剂的聚丙烯（PP）微孔膜。由于涂层相的物理损伤，提出了溶胶-凝胶技术制备厚膜，以获得具有更高热稳定性和溶剂稳定性、低渗出、良好的可重复性和更长寿命的相，因为涂层和玻璃表面之间通过化学键结合有很强的黏附力[59-61]。通过这种技术，在 PDMS 中引入不同化学基团，包括 β-环糊精、二乙烯基苯和聚乙烯醇[10,11,14,17]。尽管这些新型搅拌棒涂层可提高几种极性分析物的萃取选择性，但仍观察到与聚合物开裂有关的问题，其导致随着时间的推移，涂层会逐渐劣化。最近，PDMS/乙二醇共聚物搅拌棒开始在市场上销售，尽管还没有对此提出有效的应用。随后，制备整体材料，以及其他吸附剂材料，例如聚（酞嗪醚砜酮）、聚吡咯、聚丙烯酸酯等[10,14,17,62,63]也被提出作为 SBSE 吸附剂涂层，其中一些通过浸没沉淀技术制备，表现出热稳定性高、寿命长和良好的富集。聚氨酯泡沫也被提议作为 SBSE 相，其显示出显著的热稳定性、对有机溶剂优异的机械抗性和广泛的适用性[64-67]。在过去几年中，还合成和评估了涂有限制进入材料和分子印迹聚合物的选择性更高的搅拌棒，其显示出极高的选择性和快速吸附平衡，但仅适用于特定系统[68]。尽管有这些努力，在为极性更强的化合物开发新吸附剂方面，PDMS 仍然是 SBSE 最常用的吸附剂涂层，因为它具有很好的稳定性、显著的可重复使用性和更高的稳健性。

15.4.4 新型被动微萃取技术

在过去几年中，已经提出了新的静态微萃取方法（例如，单滴微萃取、中空纤维膜萃取、分散固相萃取和许多其他基于无溶剂微萃取和搅拌的方法)[69]。然而，其中的大部分方法对极性溶质既不具有高效性，也不具备 SBSE 技术所展示的稳健性和广泛的适用性。到目前为止，众所周知，极性溶质由于其所表现出的结构、较大的特定面积和吸附能力，容易保留在分裂的固体材料上。因此，最近引入了条形吸附微萃取(BAμE)[70]，其原理与通常基于吸附剂的方法相同。BAμE 已经成功提出，其纳米结构材料（例如 AC）或聚合物可采用黏合剂薄膜方便地固定在聚丙烯条形基板上。出于静态采样目的，这种新分析方法在流动采样技术下运行，这是一种用于微萃取分析的新型富集概念。这种新富集原理有几个优点，可避免设备与玻璃壁或采样管底部的直接接触。此外，这种新颖的微萃取技术允许为每种特定类型的应用选择最方便的吸附剂相。文献［10，70］中已报道的几个成功应用表明，当 SBSE 存在局限性或缺乏有效性时，这种新型分析技术都是监测实际基质中大量极性溶质的非常有效且互补的替代方法。此外，BAμE 具有成本效益且便于用户操作，在痕量分析中表现出优异的性能、高再现性和显著的灵敏度。

15.5 结论与展望

在过去十年中，SBSE 已成为一种成熟的被动采样工具，其显示出操作简便、容量大、性能卓越，以及对优先和新出现的非极性至中极性有机化合物进行超痕量分析的高灵敏度和选择性。与其他已提出或已商业化的被动微萃取技术相比，SBSE 已经证明具有更高的再现性、稳定性和更广泛的适用性，特别是在涉及复杂基质的情况下。尽管本文讨论了所有优点，但从理论、实践和应用的角度来看，这种卓越的分析方法不能有效地监测大量极性有机化合物。尽管已经提出了一些分析策略或新概念来克服一些局限性，但在未来，将会明确提出新的 SBSE 发展和改进方向，特别是与顶级分析仪器（例如，综合色谱或串联质谱系统）的联用。当然，在保证符合绿色分析化学原理的前提下，还将引入创新的吸附剂相（例如离子液体），以应对分析挑战和克服局限性。

致谢

作者感谢里斯本大学（葡萄牙）科学学院化学和生物化学中心分离科学与技术小组色谱和毛细管电泳实验室的所有研究合作者，以及感谢 Fundação para a Ciência e a Tecnologia 提供的资助（葡萄牙；UID/MULTI/00612/2013）。

参考文献

[1] S. Mitra, Sample Preparation Techniques in Analytical Chemistry, John Wiley & Sons, New Jersey, 2003.

[2] D. E. Raynie, Modern extraction techniques, Anal. Chem. 82 (2010) 4911-4916.

[3] C. L. Arthur, J. Pawlizyn, Solid phase micro-extraction with thermal desorption using fused silica optical fiber, Anal. Chem. 62 (1990) 2145-2148.

[4] E. Baltussen, P. Sandra, F. David, C. A. Cramers, Stir bar sorptive extraction (SBSE), a novel extraction technique for aqueous samples: theory and principles, J. Microcol. Sep. 11 (1999) 737-747.

[5] E. Baltussen, C. A. Cramers, P. J. F. Sandra, Sorptive sample preparation - a review, Anal. Bioanal. Chem. 373 (2002) 3-22.

[6] F. David, P. Sandra, Stir bar sorptive extraction for trace analysis, J. Chromatogr. A 1152 (2007) 54-69.

[7] T. Hyotylainen, M. L. Riekkola, Sorbent - and liquid - phase micro - extraction techniques and membrane-assisted extraction in combination with gas chromatographic analysis: a review, Anal. Chim. Acta 614 (2008) 27-37.

[8] F. Sanchez-Rojas, C. Bosch-Ojeda, J. M. Cano-Pavon, A review of stir bar sorptive extraction, Chromatographia 69 (2009) S79-S94.

[9] F. M. Lanças, M. E. C. Queiroz, P. Grossi, I. R. B. Olivares, Recent developments and applications of stir bar sorptive extraction, J. Sep. Sci. 32 (2009) 813-824.

[10] J. M. F. Nogueira, Novel sorption-based methodologies for static micro-extraction analysis: a review on SBSE and related techniques, Anal. Chim. Acta 757 (2012) 1-10.

[11] I. Rykowska, W. Wasiak, Advances in stir bar sorptive extraction coating: a review, Acta Chromatog. 25 (2013) 27-46.

[12] A. Spietelun, Ł. Marcinkowski, M. Guardia, J. Namiesnik, Recent developments and future trends in solid phase micro-extraction techniques towards green analytical chemistry, J. Chromatogr. A 1321 (2013) 1-13.

[13] M. He, B. Chen, B. Hu, Recent developments in stir bar sorptive extraction, Anal. Bioanal. Chem. 406 (2014) 2001-2026.

[14] J. M. F. Nogueira, Stir - bar sorptive extraction: 15 years making sample preparation more environment-friendly, TrAC 71 (2015) 214-223.

[15] S. Seethapathy, T. Gorecki, Applications of polydimethylsiloxane in analytical chemistry: a review, Anal. Chim. Acta 750 (2012) 48-62.

[16] P. Serôdio, M. S. Cabral, J. M. F. Nogueira, Use of experimental design in the optimization of stir bar sorptive extraction for the determination of polybrominated diphenyl ethers in environmental matrices, J. Chromatogr. A 1141 (2007) 259-270.

[17] K. MacNamara, R. Leardi, F. McGuigan, Comprehensive investigation and optimisation of the main experimental variables in stir - bar sorptive extraction (SBSE) - thermal desorption - capillary gas chromatography (TD-CGC), Anal. Chim. Acta 636 (2009) 190-197.

[18] A. Prieto, O. Basauri, R. Rodil, A. Usobiaga, L. A. Fernúndez, N. Etxebarria, O. Zuloaga, Stir- bar sorptive extraction: a view on method optimisation, novel applications, limitations and potential solutions, J. Chromatogr. A 1217 (2010) 2642-2666.

[19] P. Serôdio, J. M. F. Nogueira, Development of a stir bar sorptive extraction - liquid desorption - large-volume injection capillary gas chromatographic - mass spectrometric method for pyrethroid pesticides in water samples, Anal. Bioanal. Chem. 382 (2005) 1141-1151.

[20] P. Serôdio, J. M. F. Nogueira, Multi-residue screening of endocrine disrupters chemicals in water samples by stir bar sorptive extraction - liquid desorption - capillary gas chromatography - mass spectrometry

detection, Anal. Chim. Acta 517 (2004) 21-32.

[21] P. Popp, C. Bauer, A. Paschke, L. Montero, Application of a polysiloxane-based extraction method combined with column liquid chromatography to determine polycyclic aromatic hydrocarbons in environmental samples, Anal. Chim. Acta 504 (2004) 307-312.

[22] A. De Villiers, G. Vanhoenacker, P. Sandra, Stir bar sorptive extraction-liquid desorption applied to the analysis of hop-derived bitter acids in beer by micellar electro- kinetic chromatography, Electrophoresis 25 (2004) 664-669.

[23] P. M. A. Rosário, J. M. F. Nogueira, Combining stir bar sorptive extraction and MEKC for the determination of polynuclear aromatic hydrocarbons in environmental and biological matrices, Electrophoresis 27 (2006) 4694-4702.

[24] A. Juan-Garcia, Y. Pico, G. Font, Capillary electrophoresis for analyzing pesticides in fruits and vegetables using solid-phase extraction and stir-bar sorptive extraction, J. Chromatog. A 1073 (2005) 229-236.

[25] M. Kawaguchi, R. Ito, K. Saito, H. Nakazawa, Novel stir bar sorptive extraction methods for environmental and biomedical analysis, J. Pharmac. Biomed. Anal. 40 (2006) 500-508.

[26] F. J. Camino-Sanchez, A. Zafra-Gomez, J. P. Perez-Trujillo, J. E. Conde-Gonzalez, J. C. Marques, J. L. Vilchez, Validation of a GC-MS/MS method for simultaneous determination of 86 persistent organic pollutants in marine sediments by pressurized liquid extraction followed by stir bar sorptive extraction, Chemosphere 84 (2011) 869-881.

[27] M. Kawaguchi, R. Ito, H. Nakazawa, A. Takatsu, Applications of stir-bar sorptive extraction to food analysis, TrAC 45 (2013) 280-293.

[28] B. A. Lukman, H. T. Guan, Review of SBSE technique for the analysis of pesticide residues in fruits and vegetables, Chromatographia 77 (2014) 15-24.

[29] F. J. Camino-Sanchez, A. Zafra-Gomez, J. Ruiz-Garcia, J. L. Vilchez, Screening and quantification of 65 organic pollutants in drinking water by stir bar sorptive extraction-gas chromatography-triple quadrupole mass spectrometry, Food Anal. Met. 6 (2013) 854-867.

[30] M. Barriada-Pereira, P. Serôdio, M. J. Gonzalez-Castro, J. M. F. Nogueira, Determination of organochlorine pesticides in vegetable matrices by stir bar sorptive extraction with liquid desorption and large volume injection-gas chromatography-mass spectrometry towards compliance with European Union directives, J. Chromatogr. A 1217 (2010) 119-126.

[31] B. T. Weldegergis, A. M. Crouch, Analysis of volatiles in Pinotage wines by stir bar sorptive extraction and chemometric profiling, J. Agric. Food Chem. 56 (2008) 10225-10236.

[32] P. Serôdio, J. M. F. Nogueira, Considerations on ultra-trace analysis of phthalates in drinking water, Water Res. 40 (2006) 2572-2582.

[33] J. D. Caven-Quantrill, J. Alan, Comparison of volatile constituents extracted from model grape juice and model wine by stir bar sorptive extraction-gas chromatog- raphy-mass spectrometry, J. Chromatogr. A 1218 (2011) 875-881.

[34] R. Perestrelo, J. M. F. Nogueira, J. S. Câmara, Potentialities of two solventless extraction approaches-stir bar sorptive extraction and headspace solid-phase microextraction for determination of higher alcohol acetates, isoamyl esters and ethyl esters in wines, Talanta 80 (2009) 622-630.

[35] E. Coelho, M. A. Coimbra, J. M. F. Nogueira, S. M. Rocha, Quantification approach for assessment of sparkling wine volatiles from different soils, ripening stages, and varieties by stir bar sorptive

extraction with liquid desorption, Anal. Chim. Acta 635 (2009) 214-221.

[36]I. Lavagnini, A. Urbani, F. Magno, Overall calibration procedure via a statistically based matrix-comprehensive approach in the stir bar sorptive extraction - thermal desorption - gas chromatography - mass spectrometry analysis of pesticide residues in fruit-based soft drinks, Talanta 83 (2011) 1754-1762.

[37]E. D. Guerrero, R. C. Mejias, R. N. Marin, C. G. Barroso, Optimization of stir bar sorptive extraction applied to the determination of pesticides in vinegars, J. Chromatogr. A 1165 (2007) 144-150.

[38]R. F. Alves, A. M. D. Nascimento, J. M. F. Nogueira, Characterization of the aroma profile of Madeira wine by sorptive extraction techniques, Anal. Chim. Acta 546 (2005) 11-21.

[39]E. Coelho, R. Perestrelo, N. R. Neng, J. S. Câmara, M. A. Coimbra, J. M. F. Nogueira, S. M. Rocha, Optimisation of stir bar sorptive extraction and liquid desorption combined with large volume injection-gas chromatography-quadrupole mass spectrometry for the determination of volatile compounds in wines, Anal. Chim. Acta 624 (2008) 79-89.

[40]C. Yang, J. Wang, D. Li, Micro - extraction techniques for the determination of volatile and semivolatile organic compounds from plants: a review, Anal. Chim. Acta 799 (2013) 8-22.

[41]R. J. De La Torre - Roche, W. Y. Lee, S. I. Campos - Diaz, Soil - borne polycyclic aromatic hydrocarbons in El Paso, Texas: analysis of a potential problem in the United States/ Mexico border region, J. Haz. Mat. 163 (2009) 946-958.

[42]C. Almeida, J. M. F. Nogueira, Determination of steroid sex hormones in water and urine matrices by stir bar sorptive extraction and liquid chromatography with diode array detection, J. Pharmac. Biom. Anal. 41 (2006) 1303-1311.

[43]A. R. M. Silva, J. M. F. Nogueira, New approach on trace analysis of triclosan in personal care products, biological and environmental matrices, Talanta 74 (2008) 1498-1504.

[44]A. R. M. Silva, J. M. F. Nogueira, Stir bar-sorptive extraction and liquid desorption combined with large-volume injection gas chromatography-mass spectrometry for ultra-trace analysis of musk compounds in environmental water matrices, Anal. Bioanal. Chem. 396 (2010) 1853-1862.

[45]I. Carpinteiro, M. Ramil, I. Rodriguez, J. M. F. Nogueira, Combining stir bar sorptive extraction and large volume injection-gas chromatography-mass spectrometry for the determination of benzotriazole UV stabilizers in wastewater matrices, J. Sep. Sci. 35 (2012) 459-467.

[46]N. R. Neng, C. A. A. Cordeiro, A. P. Freire, J. M. F. Nogueira, Determination of glyoxal and methylglyoxal in environmental and biological matrices by stir bar sorptive extraction with in-situ derivatization, J. Chromatogr. A 1169 (2007) 47-52.

[47]N. Ochiai, K. Sasamoto, H. Kanda, E. Pfannkoch, Sequential stir bar sorptive extraction for uniform enrichment of trace amounts of organic pollutants in water samples, J. Chromatogr. A 1200 (2008) 72-79.

[48]M. C. Sampedro, M. A. Goicolea, N. Unceta, A. S. Ortega, R. J. Barrio, Sequential stir bar extraction, thermal desorption and retention time locked GC-MS for determination of pesticides in water, J. Sep. Sci. 32 (2009) 3449-3456.

[49]J. Vercauteren, C. Peres, C. Devos, P. Sandra, F. Vanhaecke, L. Moens, Stir bar sorptive extraction for the determination of ppq-level traces of organotin compounds in environmental samples with thermal desorption capillary gas chromatography-ICP mass spectrometry, Anal. Chem. 73 (2001) 1509-1514.

[50]A. M. C. Ferreira, M. Möder, M. E. F. Laespada, Stir bar sorptive extraction of parabens, triclosan

and methyl triclosan from soil, sediment and sludge with in situ derivatization and determination by gas chromatography-mass spectrometry, J. Chromatogr. A 1218 (2011) 3837-3844.

[51]A. Stopforth, A. Tredoux, A. Crouch, P. van Helden, P. Sandra, A rapid method of diagnosing pulmonary tuberculosis using stir bar sorptive extraction - thermal desorption - gas chromatography - mass spectrometry, J. Chromatogr. A 1071 (2005) 135-139.

[52]A. Stopforth, C. J. Grobbelaar, A. M. Crouch, P. Sandra, Quantification of testosterone and epitestosterone in human urine samples by stir bar sorptive extraction - thermal desorption - gas chromatography/mass spectrometry: application to HIV-positive urine samples, J. Sep. Sci. 30 (2007) 257-265.

[53]N. Ochiai, K. Sasamoto, S. Daishima, A. C. Heiden, A. Hoffmann, Determination of stale-flavor carbonyl compounds in beer by stir bar sorptive extraction with in-situ derivatization and thermal desorption-gas chromatography-mass spectrometry, J. Chromatogr. A 986 (2003) 101-110.

[54]J. B. Quintana, R. Rodil, S. Muniategui-Lorenzo, P. López-Mahía, D. Prada-Rodríguez, Multiresidue analysis of acidic and polar organic contaminants in water samples by stir-bar sorptive extraction-liquid desorption-gas chromatography-mass spectrometry, J. Chromatogr. A 1174 (2007) 27-39.

[55]M. Kawaguchi, N. Sakui, N. Okanouchi, R. Ito, K. Saito, H. Nakazawa, Stir bar sorptive extraction and trace analysis of alkylphenols in water samples by thermal desorption with in tube silylation and gas chromatography-mass spectrometry, J. Chromatogr. A 1062 (2005) 23-29.

[56]N. R. Neng, R. P. Santalla, J. M. F. Nogueira, Determination of tributyltin in environ- mental water matrices using stir bar sorptive extraction with in-situ derivatisation and large volume injection-gas chromatography-mass spectrometry, Talanta 126 (2014) 8-11.

[57]N. Gilart, R. M. Marcé, F. Borrull, N. Fontanals, New coatings for stir-bar sorptive extraction of polar emerging organic contaminants, TrAC 54 (2014) 11-23.

[58]C. Bicchi, C. Cordero, E. Liberto, P. Rubiolo, B. Sgorbini, F. David, P. Sandra, Dual- phase twisters: a new approach to headspace sorptive extraction and stir bar sorptive extraction, J. Chromatogr. A 1094 (2005) 9-16.

[59]A. Kabir, K. Furton, G. Kenneth, A. Malik, Innovations in sol-gel micro-extraction phases for solvent-free sample preparation in analytical chemistry, TrAC 45 (2013) 197-218.

[60]W. Fan, X. Mao, M. He, B. Chen, B. Hu, Development of novel sol-gel coatings by chemically bonded ionic liquids for stir bar sorptive extraction-application for the determination of NSAIDS in real samples, Anal. Bioanal. Chem. 28 (2014) 7261-7273.

[61]C. Yu, B. Hu, Sol-gel polydimethylsiloxane/poly (vinylalcohol) -coated stir bar sorptive extraction of organophosphorus pesticides in honey and their determination by large volume injection GC, J. Sep. Sci. 32 (2009) 147-153.

[62]J. Y. Barletta, P. C. F. de Lima Gomes, A. J. dos Santos-Neto, F. M. Lanças, Development of a new stir bar sorptive extraction coating and its application for the determination of six pesticides in sugarcane juice, J. Sep. Sci. 34 (2011) 1317-1325.

[63]N. Gilart, N. Miralles, R. M. Marce, F. Borrull, N. Fontanals, Novel coatings for stir bar sorptive extraction to determine pharmaceuticals and personal care products in environmental waters by liquid chromatography and tandem mass spectrometry, Anal. Chim. Acta 774 (2013) 51-60.

[64]N. R. Neng, M. L. Pinto, J. Pires, P. M. Marcos, J. M. F. Nogueira, Development, optimisation and application of polyurethane foams as new polymeric phases for stir bar sorptive extraction, J. Chromatogr.

A 1171 (2007) 8-14.

[65] F. C. M. Portugal, M. L. Pinto, J. M. F. Nogueira, Optimization of polyurethane foams for enhanced stir bar sorptive extraction of triazinic herbicides in water matrices, Talanta 77 (2008) 765-773.

[66] A. R. M. Silva, F. C. M. Portugal, J. M. F. Nogueira, Advances in stir bar sorptive extraction for the determination of acidic pharmaceuticals in environmental water matrices: comparison between polyurethane and polydimethylsiloxane polymeric phases, J. Chromatogr. A 1209 (2008) 10-16.

[67] R. C. P. Sequeiros, N. R. Neng, F. C. M. Portugal, M. L. Pinto, J. Pires, J. M. F. Nogueira, Development and application of stir bar sorptive extraction with polyurethane foams for the determination of testosterone and methenolone in urine matrices, J. Chromatogr. Sci. 49 (2011) 297-302.

[68] A. Martin-Esteban, Molecularly-imprinted polymers as a versatile, highly selective tool in sample preparation, TrAC 45 (2013) 169-181.

[69] R. Lucena, Extraction and stirring integrated techniques: examples and recent advances, Anal. Bioanal. Chem. 403 (2012) 2213-2223.

[70] A. R. M. Silva, F. Portugal, J. M. F. Nogueira, Adsorptive micro-extraction techniques-Novel analytical tools for trace levels of polar solutes in aqueous media, J. Chromatogr. A 1209 (2008) 10-16.

16

16 固相微萃取技术在食品污染物分析中的进展

Érica A. Souza-Silva[1,*], Janusz Pawliszyn[2]
1. Instituto de Química, Universidade Federal do Rio Grande do Sul (UFRGS), Porto Alegre, Brazil
2. University of Waterloo, Waterloo, ON, Canada
*通讯作者，E-mail：ericachemistry@gmail.com

16.1 引言

随着分析仪器现代化的推进，分析前复杂且费力的样品制备方法应得以简化甚至消除。然而，在大多数情况下，样品制备仍然是寻求最佳分析方法的瓶颈。特别是，复杂基质中痕量分析物的测定通常需要在分析之前采用繁杂的样品制备方案。样品制备所涉及的步骤，通常包括从基质中萃取分析物、净化及预浓缩，从而使特定分析方法达到足够的灵敏度。因此，寻求更简单的样品制备方案，不仅是为了减少样品处理所需的时间，也是为了减少与程序的每个步骤相关的误差。因为在统计学上，方法的不确定度与其包含的步骤数直接相关。对包括样品制备在内的整个工作流程的自动化，也提升了方法的再现性，同时减少了样品制备所需的工作量。

因此，样品预处理的主要目标之一是减少多步样品制备技术所涉及的时间和工作量。在 20 世纪 90 年代初，固相微萃取技术（SPME）的引入，解决了传统样品制备的一些难题，成功整合了多种分析步骤，如采样、萃取、预浓缩，以及在气相色谱（GC）应用中为仪器分析导入样品[1]。此外，SPME 是一种简单、灵敏、省时、经济、可靠、易于自动化和便携的样品制备技术，可将溶剂消耗量降至最低。

与常规的完全萃取方法（如 SPE）相比，SPME 技术是一种基于分析物在样品基质和萃取相之间实现分配平衡的不完全萃取技术。在 SPME 中，萃取相可直接暴露在样品基质中（DI）或样品上方的顶空（HS）。当 SPME 的萃取相直接接触样品，达到平衡状态时，萃取的分析物的量（n_e）为[2]：

$$n_e = \frac{K_{fs} \times V_s \times V_f}{K_{fs} \times V_f + V_s} \times C_s \tag{16.1}$$

在式（16.1）中，n_e 与分析物在涂层和样品基质之间的分配系数（K_{fs}）、萃取相的体积（V_f）、样品体积（V_s）、样品基质中的分析物浓度（C_s）成比例。式（16.1）表明萃取到涂层上的分析物的量（n_e）与样品中分析物的浓度（C_s）呈线性关系，这是利用 SPME 进行定量分析的基础。

如果样品体积很大，$V_f K_{fs} << V_s$，式（16.1）可简化为：

$$n_e = K_{fs} \times V_f \times C_s \tag{16.2}$$

实际上，式（16.2）表明无须在分析之前收集既定的样品，因为纤维可以直接暴露在样品基质中，例如在涉及可食用植物的应用中。这对于将 SPME 应用到体内取样（稍后论述）至关重要。

如图 16.1 所示，还可以进行预平衡萃取，以缩短分析所需的时间。尽管未达到萃取平衡，但萃取到纤维上的分析物的量与样品基质中分析物的浓度，仍然具有线性关系。达到完全平衡所需的时间是无限长的；因此，当从样品中萃取出分析物平衡量的 95%时，即认为达到平衡时间[2,3]。

这个解释是对 SPME 技术基本原理的简单概述。为了更全面的探讨 SPME 的理论基础，强烈推荐想深入了解 SPME 理论的读者，参考相关文献［2-7］。

因此，SPME 技术的萃取效率可通过增大 K_{fs}（萃取相的化学性质），和/或增大萃取

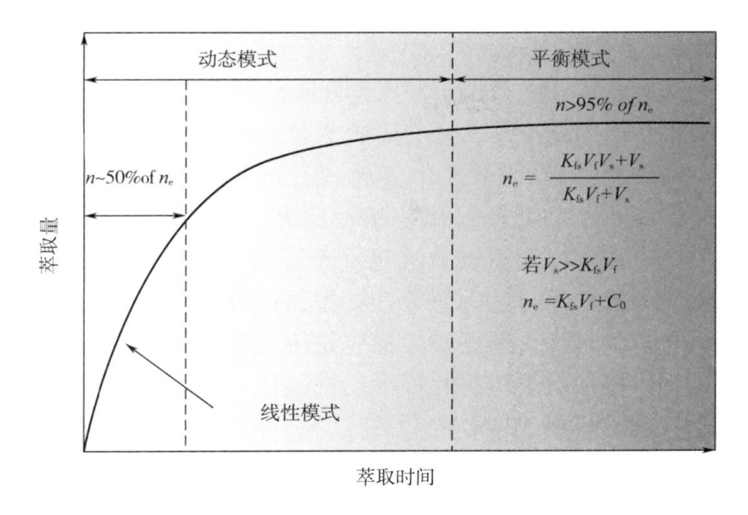

图 16.1　固相微萃取的萃取时间曲线

［资料来源：T. A. Souza-Silva, E. Gionfriddo, J. Pawliszyn, A critical review of the state of the art of solid-phase microextraction of complex matrices II. Food analysis, TrAC -Trends Anal. Chem. 71 (2015) 236-248[8]. 经 Elsevier 许可。］

相的体积或活性表面积（萃取相的几何形态）而提升。

当不便改变萃取相的几何形态时，K_{fs} 值对分析物的回收和分配发挥关键作用。在上述式（16.1）中，为计算平衡条件下所萃取的分析物的量，要考虑实际参与萃取过程的萃取相的量。对于液相涂层，宜考虑参与萃取的萃取相总体积（V_f）。但是对于固体多孔涂层，分析物与萃取相的相互作用，仅限于构成多孔聚合物活性表面积的表面活性位点。因此，为了准确计算固相涂层的平衡萃取量，必须将上述公式中的萃取相体积值（V_f）替换为总表面积（S_a）。这说明固态多孔涂层的萃取量受限于可被用于萃取的活性点位的数量。涂层饱和的一个直接后果是被分析物之间的竞争性吸附。这一现象最初是在双组分体系中进行研究，近期被研究以解释复杂基质中的分析物行为[9,10]。

16.2　SPME 应用于食品污染物的测定

食物代表着最多样化和最复杂的基质成分。正确地说，食物是由一组不同的基质组成的，而不是某一特定的基质。食品（动物的或植物的）的起源，产生了一系列的成分，即蛋白质、碳水化合物、矿物质、维生素等。这种本质上的多样性带来了挑战和困难，需要食品分析人员求助于现有的最佳科学技术。食品分析对于评价与生鲜和加工产品质量有关的属性很重要，如营养价值、风味、掺假、污染等。然而，无论食品分析的目的为何，都必须特别关注所选用的样品制备方法。分析人员面临的挑战是开发更环保、更快速、更精确和准确的方法，以确保食品的安全性、质量、真实性和可追溯性。在这方面，SPME 是食品分析中更有前景的、新型绿色样品制备技术。21 世纪以来，新的萃取相和设备的发展已经扩展了 SPME 在食品研究中的应用范围。

通常，安全性一直是食品分析的主要目标。因此，农药分析一直是重中之重。超过1000种有效成分在世界范围内被用作农药。它们属于许多不同的化学类别，唯一的共同特点是对害虫有效。这意味着需要考虑大量具有不同理化性质的分析物。

传统的测定食品中农药的方法，从费时费力且对环境不友好的方法演变为涵盖更广泛的分析物的简单方法。作为食品中多农残分析的一种更简单的样品制备方法，Lehotay 和 Anastassiades 首先引入了代表快速、简单、廉价、有效、可靠和安全的 QuEChERS 方法[11]。QuEChERS 萃取方法是基于乙腈、乙酸乙酯或丙酮的液液分配，并由硫酸镁或其与其他盐的混合物促进分配，再通过分散固相萃取进行净化。这种方法有效涵盖广泛的分析物范围，因此被广泛应用于水果、蔬菜以及动物食品中的多农残分析。这种方法的主要缺点是对每克样品的富集能力较低，最终提取物必须浓缩至更大浓度以获得适宜的定量限（LOQ）。另外，QuEChERS 是一种多步骤方法，不易实现自动化。样品前处理和仪器导入等步骤的整合不易实现。关于 QuEChERS 的深入讨论参阅本书第 11 章。

理想情况下，用于食品农药分析的方法应当快速、简便，需要最少的化学品（尤其是溶剂）消耗量，具有一定的选择性，且能覆盖广泛的分析物和不同的基质[12]。大多数的样品前处理方法涉及大量繁杂的多步骤处理，增大了分析物损失、样品污染及分析误差的概率。因此，如果采用简单的样品处理方法以减少步骤的数量，可提高方法的再现性（精密度）和准确性。食品基质中农药分析自动化方法的发展呈现出高样品通量的重要优势，整个分析是完全自动化的，从而减少与人为错误相关的误差，并允许高通量的样品进行分析。此外，随着样品制备向小型化、易操作和自动化方向发展，人们对于 SPME 在分析食品中农药和其他污染物方面的应用越来越感兴趣。

SPME 通过将采样、萃取、浓缩和样品导入集成到一个单一的无溶剂步骤，已经能够满足上述大多数要求。SPME 的主要优点是易于小型化、设备自动化以及便于与色谱仪器联用。然而，为了使高质量的分析方法与 SPME 相结合，需要优化影响萃取效率的参数：即萃取相化学性质、萃取方式、搅拌方法、样品修饰（pH、离子强度、有机溶剂含量）、样品温度、萃取时间、解吸条件[13]。在上述参数中，选择适宜的纤维涂层是 SPME 方法发展的最关键步骤之一。纤维涂层对特定分析物的适宜性是由涂层的极性和相对于其他基质组分对分析物的选择性所决定的。由于 SPME 是基于分析物在样品和萃取相之间的平衡分布，吸附剂的性质对其性能起关键作用。

16.3 SPME 设备在食品污染物分析中的进展

16.3.1 溶胶-凝胶

尽管 SPME 有许多优点，但一些关键性的限制阻碍了它在食品分析中的应用，例如：①可商用的吸附剂涂层数量有限；②由于物理结合吸附涂层的热稳定性差，操作温度相对较低；③涂层在有机溶剂中的不稳定和膨胀；④物理结合吸附涂层寿命短。过去几年里，人们在寻找最佳的 SPME 涂层、纤维、器件方面做了大量的研究。从这个意义上说，溶胶凝胶技术极大地促进了新萃取相的发展[14-17]。溶胶凝胶技术无疑是创造定制萃取相的最巧妙方法之一。事实上，自 Malik 和同事引入 SPME 以来，由于溶胶凝胶涂层提高了热稳定性和机械稳定性，溶胶凝胶技术极大地拓宽了 SPME 在食品污染物分析中的应用范围。通过溶胶凝胶技术生产的涂层具有更好的热稳定性，在 GC 应用中可实现更高的解吸温度。此外，吸附剂与基体的共价结合增强了机械稳定性，增强了溶胶凝胶涂层对有机溶剂和酸性/碱性溶液的稳定性，提高了其在基于 LC 的 SPME 方法中的可行性[15-19]。考虑到食物基质中成分的多样性，这一点尤为重要，因此，低挥发性化合物可以成功地从纤维涂层中脱附，从而减少残留问题。在食品分析中，被引用最多的一种用于 SPME 的溶胶基涂层是基于乙烯基冠醚结构的（详见表 16.1）。冠醚是一种具有空腔结构、中等极性和强电负性的杂环化合物。冠醚的极性程度随环数的增加而增大。乙烯基冠醚涂层通过环上的杂原子表现出很强的电负性，这使得纤维涂层对极性化合物具有选择性。此外，这些涂层具有多孔的三维网状结构，为萃取提供了更大的表面积，提供了更快的传质速率。然而，多孔的三维结构可能会对涂层表面造成污染，这是直接浸入食品样品中经常遇到的问题。事实上，下文中介绍的在食品分析中应用多孔溶胶凝胶涂层的研究大多数都是在 HS 模式下进行的[15,20,21]。

Cai 等采用自由基交联乙烯基冠醚溶胶凝胶法制备了三种涂层。将苯并-15-冠-5 涂层应用于 HS-SPME，并与 GC-FPD（火焰光度检测器）联用，以分析苹果、番茄和苹果汁中的有机磷农药（OPPs），可达到 3~90pg/g 的检测限，与商用纤维［85μm PA 和 65μm PDMS-DVB（聚二甲基硅氧烷-二乙烯苯）］相比，对 OPPs 的提取效率和灵敏度更高[21]。通过同样的方式，以冠醚结构为基础合成了不同类型的溶胶凝胶混合涂层，并应用于草莓、青苹果、葡萄、蜂蜜、果汁、柑橘和小白菜中 OPPs 的测定[20,22]。最近，通过电化学沉积法制备了一种涂覆聚吡咯/溶胶凝胶（Ppy/solgel）的 SPME 纤维。这种薄涂层具有多孔的表面，在超过 150 次的萃取/脱附循环中保持稳定。应用 Ppy/solgel 涂层通过 DI-SPME 测定莴苣和黄瓜样品中的 3 个 OPPs，然后进行 GC 和氮磷检测（GC-NPD）[23]。Ppy/solgel 涂层的萃取效率优于聚吡咯、PDMS 和 PDMS/DVB 纤维。Ppy/solgel 涂层对分析物的亲和性归功于涂层中的苯基和亲水性基团，这些基团增强了与所选农药的 π-π 相互作用、氢键和偶极-偶极相互作用[23]。

此外，溶胶凝胶技术通过利用牢固的芯部（如铂、金和不锈钢）解决了纤维涂层在熔融石英毛细管上的机械稳定性差的问题[19,24]。Supelco 还将纤维涂敷在金属芯上进行商业化，尽管这种涂层是物理沉积在基体上，而不是共价结合。

16.3.2 离子液体

与溶胶凝胶技术类似，离子液体（ILs）作为萃取相在 SPME 中的应用在 21 世纪以来有了显著增长[25]。ILs 可以针对特定应用进行结构优化，这一主要特点使其成为开发选择性萃取相的一个非常有吸引力的选择。基于热稳定性、忽略不计的蒸汽压、高黏度以及通过阳离子和阴离子的作用进行疏水性和/或亲水性的结构调整，ILs 也能与 GC 兼容。通过优化所有这些特性，可以设计所需的萃取相，用于对各种环境和食品样品

表 16.1　SPME 在食品污染物分析中的选择应用

基质	分析物	模式/配置	涂层	分离/检测	定量	附注	参考文献
苹果 番茄	农药 （OPPs）	HS-SPME 纤维 （~80μm）	乙烯基冠醚 Solgel（极性）	GC-FPD	基质匹配	纤维萃取能力优于 PDMS-DVB 65μm 和 PA 80μm 纤维	[21]
蜂蜜 柑橘 小白菜	农药 （OPPs）	DI-SPME 纤维 （~40μm）	冠醚 Solgel （混合）	GC-FPD	标准加入法	纤维寿命延长（超过 200 次萃取周期）	[22]
青苹果 草莓 葡萄（水提取物）	农药 （OPPs）	HS-SPME 纤维 （~17μm）	冠醚 Solgel （混合）	GC-FPD	外校法（水）	纤维具有类似或优于 PDMS 100μm 纤维的萃取能力	[20]
生菜 黄瓜	农药 （OPPs）	DI-SPME 纤维 （~18μm）	聚吡咯/solgel	GC-NPD	标准加入法	在不锈钢丝上电化学沉积镀层	[23]
蔬菜（己烷提取物）	农药 （拟除虫菊酯）	DI-SPME 纤维 （15~20μm）	离子液体 $ViHDIm^{+}PF6^{-}$	GC-ECD	外校法（己烷）	与 PDMS 涂层相比，聚合 IL 涂层在非极性有机溶剂中表现出更好的化学稳定性	[26]
洋葱 大米 （水提取物）	农药 （三嗪类）	DI-SPME 纤维 （15~20μm）	MIP （阿特拉津，莠灭净模板）	GC-MS GC-FID	外校法（水）	涂层的特异性使得在脱附前可对纤维进行有效的清洗。不易破损的纤维：MIP-莠灭净涂敷在经阳极氧化的硅烷化铝丝上	[37，84，85]
*番茄酱*辣椒粉 家禽饲料	苏丹染料（I-IV）	DI-SPME 纤维 （~19.8μm） *纤维（~0.55μm）	MIP （苏丹红 I 模板）	HPLC-UV *LC-MS/MS	外校法 （丙酮/水）	（1）不易破损的纤维：MIP 涂覆在不锈钢丝上；（2）涂层批量制备；（3）耐溶剂和 pH；*（4）通过链转移制备超薄涂层	[31，34]
黄瓜，青椒，大白菜，茄子，莴苣	农药 （OPPs）	HS-SPME 纤维 （~50μm）	Solgel MIP PEG 二嗪农模板	GC-NPD	基质匹配	耐水极性涂层，LOQs 在 10^{-12} 级别	[39]

牛乳	多类别的 40 种农药	DI-SPME 纤维	商用 PDMS/DVB（65μm）	LP-GC-MS/MS	基质匹配 内标	稀释牛乳样品的蛋白质变性，单内标（五氯苯）	[62]
植物油（已烷提取物）	PAHs	DI-SPME	商用的 Carbopack Z/PDMS	GC×GC-ToFMS GC-MS	基质匹配（橄榄油） 多内标	纤维在脱附前用己烷冲洗。涂层与平面化合物具有良好的 π-π 相互作用	[63，64]
乳制品	己烯雌酚	中空纤维 HF-SPME	MWCNTs Solgel	HPLC	—	缺乏自动化	[65]
牛乳	四环素	分散 SPME	硅基高分子吸附剂（30mg）	HPLC-DAD	基质匹配	商用 SPE 吸附剂，程序与 QuEChERS 类似（乙腈提取）	[73]
猪肝 鸡 （乙腈提取物）	氟喹诺酮 磺胺类	管内 SPME（填充纤维）	MIP	HPLC-UV	外校法（乙腈）	PEEK 管填充不同功能的 MIP 纤维与 HPLC 联用	[33]
莴苣	多类别的 10 种农药	DI-SPME	商用的 CW/TRP	HPLC-DAD	基质匹配	由于化学和机械稳定性差，该涂层已不再在商用	[59]
茶	多类别的 36 种农药	HS-SPME	商用的 DVB/Car/PDMS	GC×GC-ToFMS	基质匹配	将 SPME 方法与溶剂萃取-凝胶渗透色谱法进行比较	[60]
苹果	农药（氨基甲酸酯）	HF-SPME	CNTs	HPLC-DAD	基质匹配 内标	在浸入苹果样品前，将 CNTs 在 1-辛醇中浸湿	[61]
黄瓜	多类别的 7 种农药	DI-SPME 纤维	PDMS	GC-MS	基质匹配 内标	96 孔板格式（敞口）；气相色谱溶剂解吸	[44]
葡萄 草莓	农药（三唑类）	DI-SPME 纤维	改良的 PDMS PDMS/DVB	GC-ToFMS	基质匹配 内标	直接浸入果肉；无样品预处理；基质相容性涂层	[52，74]
葡萄	多类别的 40 种农药	DI-SPME 纤维	改良的 PDMS PDMS/DVB	GC-ToFMS	基质匹配 多内标	直接浸入果肉；无样品预处理；基质相容性涂层	[54]
康科德葡萄汁	多类别的 11 种污染物	DI-SPME 纤维	改良的 PDMS PDMS/DVB	GC-ITMS	基质匹配	直接浸入；不稀释；在解吸前用去离子水冲洗纤维	[55]
牛油果	多类别的 9 种污染物	DI-SPME 纤维	改良的 PDMS PDMS/DVB	GC-ITMS GC×GC-ToFMS	基质匹配	直接浸入；在解吸前用丙酮：水（9：1，体积比）冲洗纤维	[66]

中不同类别的分析物/污染物进行痕量测定。一种新型的IL SPME 特制涂层，即1-乙烯基-3-十六烷基咪唑六氟磷酸盐（$ViHDIm^+PF6^-$），通过直接浸泡在蔬菜的己烷提取物中，对拟除虫菊酯进行选择性、灵敏性萃取，然后用GC-ECD测定[26]。基于以下几方面考虑，这项工作带来了重要影响：

（1）拟除虫菊酯是全球杀虫剂市场的主要组成部分，因此对拟除虫菊酯的残留分析在农业和环境控制方面具有重要意义；

（2）鉴于拟除虫菊酯的理化性质，这些分析物具有在实验器具表面进行较强的非特异性吸附的能力，因此用于这些分析物样品制备的成熟方法通常是以溶剂为基础的完全萃取方法；

（3）PDMS是萃取拟除虫菊酯的最佳涂层。然而由于上述事实，这需要进行预先萃取步骤，而且PDMS与非极性有机溶剂存在化学不相容性。

最近，一种基质兼容的聚合物IL（PIL）涂层得以合成并被表征，通过DI-SPME与GC[27]和LC[28]平台联用，以应用于咖啡和咖啡粉中丙烯酰胺的测定。

16.3.3 分子印迹聚合物

SPME在食品基质中应用的另一突破涉及分子印迹聚合物（MIP）涂层的开发，以提高方法的灵敏度和选择性。SPME中吸附剂的体积通常比SPE小得多，因此SPME的萃取量很小。此外，在食物基质中，分析物通常与蛋白质和高分子质量碳水化合物等基质成分高度结合。采用特制MIP作为SPME涂层可显著克服吸附剂体积小的限制，使该方法的选择性增强，灵敏度提升[17,24,29-36]。采用单片MIP涂层的DI-SPME方法与GC-MS联用，对洋葱和大米中的三嗪类除草剂进行选择性萃取，结果表明该方法萃取效率高，由于采用直接浸入法，其LODs可达20~88ng/mL，且回收率高。通过在自来水、洋葱、大米等加标样品中进行SPME，对所制备的纤维在实际样品中提取阿特拉津及其他类似物的可靠性进行了研究和验证。值得注意的是，萃取之后，作者在脱附前对纤维进行清洗，以消除可能从基体吸附到涂层表面的有机物。因此在方法中增加了一个由水和甲醇组成的清洗步骤。尽管作者没有提到甲醇清洗可能造成的损失，但可以理解的是，由于涂层与目标分析物的特定亲和力，这种损失不会很严重[37,38]。在另一项研究中，Hu和他的同事介绍了一种高通量方法，可以每批制造20多个不易破损的MIP涂层SPME纤维：它使用苏丹I作为模板，不锈钢纤维作为基体。制得的MIP涂层不锈钢纤维对苏丹I~IV染料（致癌化合物）有较高的萃取能力和特定选择性。采用基于MIP的SPME与高效液相色谱联用，对辣椒粉和家禽饲料中微量苏丹I~IV染料进行快速、选择性的测定，苏丹I-IV染料的检测限在2.5~4.6ng/g[31,34]。

还有一些研究结合溶胶凝胶技术和MIP技术，制备了耐水MIP涂层[39,40]。由于其特殊的吸附性能和粗糙多孔的表面，该涂层比非印迹聚合物和商业纤维的萃取能力更大。此外，该纤维还具有优异的热稳定性和化学稳定性。关于MIP-SPME的更多应用可参阅本书第10章。

16.3.4 高通量技术

SPME的简化特性促进其应用的扩展，这反过来推动了SPME吸附剂、配置和应用等方面新进展的不断增长，并将其应用于更广泛的样品基质和分析物。SPME 可以在不同的配置中使用，如图 16.2 所示。在其最著名的配置中，SPME 装置包括一个涂覆到熔融石英棒上的萃取相。由于 SPME 专用自动进样器的商业化，以及纤维与 GC 应用中的普通注射器的相似性，当纤维式 SPME 与 GC 联用时，使得整个分析工作流程能够完全自动化[2,13]。

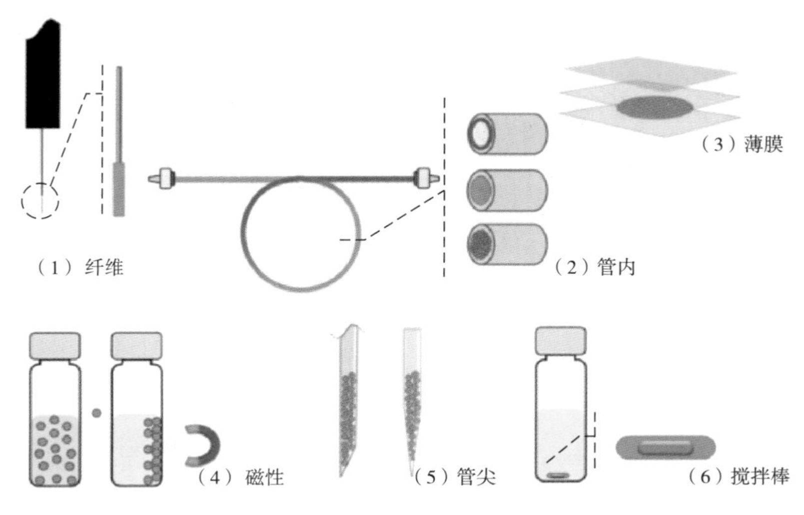

图 16.2　各种 SPME 配置

对于 SPME 的新发展，减少分析时间和提高样品通量是关键问题。在 GC 应用中，SPME 早期的自动化可以归功于 SPME 纤维设计的简易性。多年来，自动进样器已经发展到可以执行完整的 SPME-GC 工作流程，包括纤维更换，而不需要人工干预。

通过开发管内固相微萃取（IT-SPME），首次实现了 SPME 与液相色谱（LC）的自动联用。与纤维 SPME 相比，IT-SPME 由涂覆到熔融石英管内壁的萃取相组成[41]。有关 IT-SPME 发展趋势的更深入讨论，请参阅本书第 14 章。

另一种用于增加萃取相体积的方法是薄膜 SPME（TFME）[42,43]。除了用 SBSE 增加了萃取相体积之外，TFME 还提供了更高的表面/体积比，这不仅提高了灵敏度，而且由于萃取的快速传质，取样时间也更快。随后，TFME 技术与 96 孔板结合，成为一个全自动化、软件操作、离线式机器人样品制备工作站（Concept 96），如图 16.3 所示。

Hagehri 和同事利用 96 孔板型的制备工作站，成功地建立了一种结合气相色谱的自动化纤维 SPME 方法，用于测定黄瓜中的农药残留。在这项工作中，作者使用 PDMS 涂层进行萃取，在乙腈中进行液体解吸，然后进行溶剂蒸发，重新浓缩萃取物至正辛烷中。所建立的方法用于所研究的黄瓜样品中农药的分析，具有通量大、灵敏度高、适宜性好等明显优势。该方法是针对一些半挥发性农药进行的开发和验证。相反地，对于在挥发性较大的农药上的应用，可以预见其在溶剂交换等敞口方法的实施中会有一些限制[44]。

尽管这些技术具备优势，但它们的主要缺点是在这一领域缺乏标准化的程序，导

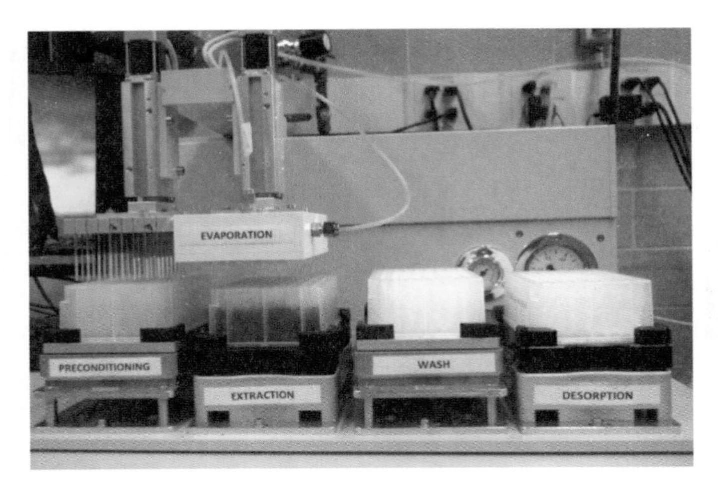

图 16.3　PAS Concept 96 高通量薄膜微萃取（德国马格达拉）

致实验室间的重现性问题，所以在常规应用中是不可行的。因此，尽管商用涂层存在缺陷，但它们仍然是常规用途及实验室间验证的首选。

16.3.5　复杂基质中直接浸入式固相微萃取

虽然大多数食品应用使用 HS-SPME 模式，因为它保护纤维涂层免受损害，但 HS-SPME 并不适用于所有情况；主要的限制在于对挥发性差或极性分析物的低萃取率，以及固体涂层更易出现纤维饱和。在这种情况下，应采用萃取相直接与样品接触的直接浸入式 SPME（DI-SPME），以保证具有较高的灵敏度以及从基质中萃取分析物的良好代表性。尽管关于应用直接浸入式 SPME 方法分析水果和蔬菜等复杂基质中农药的文献很多，但在大多数情况下，样品在进行 SPME 萃取之前要进行一些额外的预处理或净化，如离心、稀释或在有机溶剂中预萃取[45-50]。

Beltran 等发表的一篇综述显示，许多利用 DI-SPME 进行食品基质中农药的分析均使用 PDMS 涂层。PDMS 是一种表面光滑均匀的液体涂层，相比于固体涂层，基质组分造成的不可逆污染效应较小。这一事实使得 PDMS 涂层成为直接分析复杂基质最可靠的选择，因此无论对目标分析物的灵敏度如何，这种涂层都是首选[51]。

有研究介绍了一种不同的 DI-SPME，其中包括 SPME 纤维的浸入，而且在纤维周围有一层薄膜，以阻止大分子质量化合物的污染。然而，薄膜改变了所萃取化合物的平衡速率（即 K_{fs}），减缓了萃取过程[2]。

所有上述事实和局限性促使我们对现有的商用 SPME 纤维涂层进行了修改，使用薄层 PDMS 创建了一种新型 SPME 纤维涂层，实现了基体相容性，同时保持了原有涂层对目标分析物的灵敏度[52]。在该研究中，对一种商用 PDMS/DVB 纤维进行了改性，在其表面添加了一层永久性的 PDMS 薄膜。测试了改性 SPME 纤维涂层直接在葡萄基质下的萃取效率和使用寿命；在整个葡萄浆中连续进行了 130 次直接浸入 SPME 循环。涂层的可靠性通过整个实验中提取分析物的量来衡量，RSDs 低于 20%。考虑到所研究基质的复杂性，这是一个令人印象深刻的成就，其性能远远优于原来的商业纤维。例如，De Jager 等报告说，在 DI-SPME 模式下使用 PDMS/DVB 纤维分析被水稀释的食品样品，在第 10 次萃取时信号下降了 65%[53]。在该研究中，介绍了在复杂食品基质中完全自动化直接浸入 SPME 的可能性，而无须使用任何样品预处理，而在应用 DI-SPME 进行食品分析的文献中，所有其他已有的报道都是在之前附加了样品预处理步骤。图 16.4 所示为 PDMS 改性涂层的 SEM 图像。

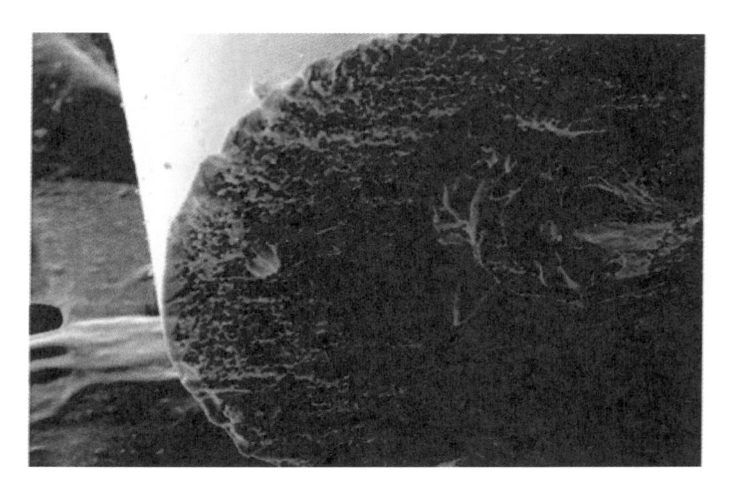

图 16.4 PDMS-DVB/PDMS 涂层的 SEM 图像

SEM 放大倍率 1.38 k。DVB，二乙烯基苯；PDMS，聚二甲基硅氧烷；SEM，扫描电镜

在对食品样品中的污染物进行多类别分析方面，SPME 的应用还没有得到相当程度的确立；然而，未来高灵敏度和基质相容的 SPME 涂层的发展将极有可能促进食品基质中农药高通量分析的全自动方法的发展。采用 PDMS 修饰的 PDMS/DVB 涂层，利用多元方法优化影响 SPME 性能的最重要因素，建立了一种快速、灵敏的 DI-SPME-GC-ToFMS 方法以测定葡萄中的多种残留农药[54]。通过使用自下而上的方法对适宜的内标物进行了全面调查，从而选择了可以充分涵盖属于不同类别的 40 种农药的适当化合物。经验证的方法具有良好的准确度、精密度和灵敏度，定量限≤10 μg/kg。但是，必须强调所提方法的局限性：对于极性较大的农药（如乙酰甲胺磷、氧化乐果、乐果）和高度疏水性的农药，如拟除虫菊酯，DI-SPME 方法并没有提供令人满意的性能。对于高度疏水性的拟除虫菊酯，问题似乎在于它们由于与瓶壁相互作用而损失。在极性较大的农药中，灵敏度的限制，如下节所述，在于这些化合物在水基基质中的高溶解度和它们对萃取相的低亲和力。

考虑到 PDMS 外层在该涂层对极性分析物的限制方面的影响，一项全面的研究使用了 11 种分析物对外包 PDMS 的纤维进行了彻底的评估，涵盖了不同的应用类别（农药、工业化学品和药品）和广泛的 lg *P* 值（从 1.43~6），如图 16.5 中的色谱所示[55,56]。LODs 的范围在 0.2~1.3ng/g[55]。

虽然结果表明 PDMS 外层的厚度并不影响纤维的寿命，但后续的研究表明，PDMS

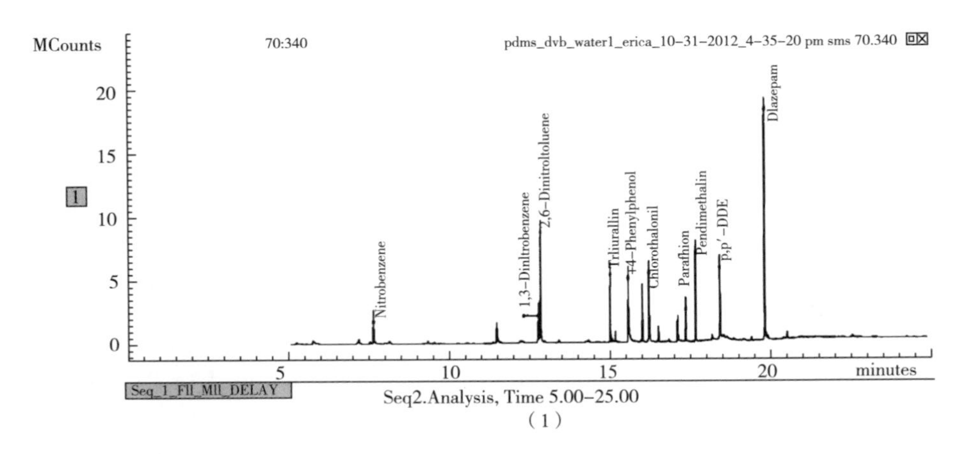

（1）

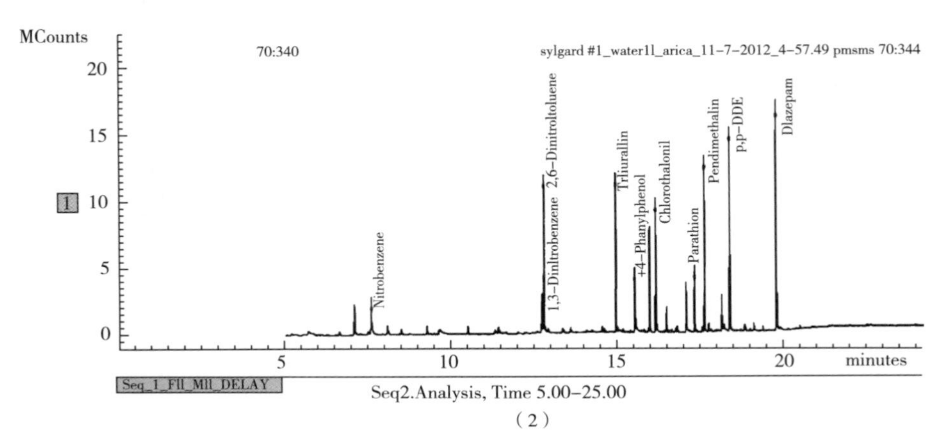

（2）

图 16.5　代表性色谱图：使用商用 PDMS/DVB 纤维（1）和 PDMS 外包覆涂层（2）从水中提取的标准混合物中的分析物

DVB，二乙烯基苯；PDMS，聚二甲基硅氧烷

外层厚度对经 PDMS 修饰的涂层吸附分析物的动力学和热力学具有相当重要的影响。结果可以简化为两种模型：①限速步骤为通过涂层的扩散；②限速步骤为通过水扩散边界层的扩散。如图 16.6 所示，对于极性化合物（如硝基苯），根据理论讨论，限速步骤为通过涂层的扩散；因此，外部的 PDMS 层影响了基体相容涂层的吸收速率。另一方面，对于非极性化合物（如 p，p'-DDE），吸收过程的限速步骤是通过水扩散边界层的扩散；因此，与 DVB/PDMS 纤维相比，外部包覆的 PDMS 不影响基体相容涂层的吸收速率。从热力学角度来看，所计算出的纤维常数进一步证实了这一假设，即：额外的 PDMS 层并不影响萃取相的能力[56]。

16.4　具有挑战性的极性化合物

目前，用于多类别测定的 SPME 方法的主要限制是萃取相的极性，以及他们对极

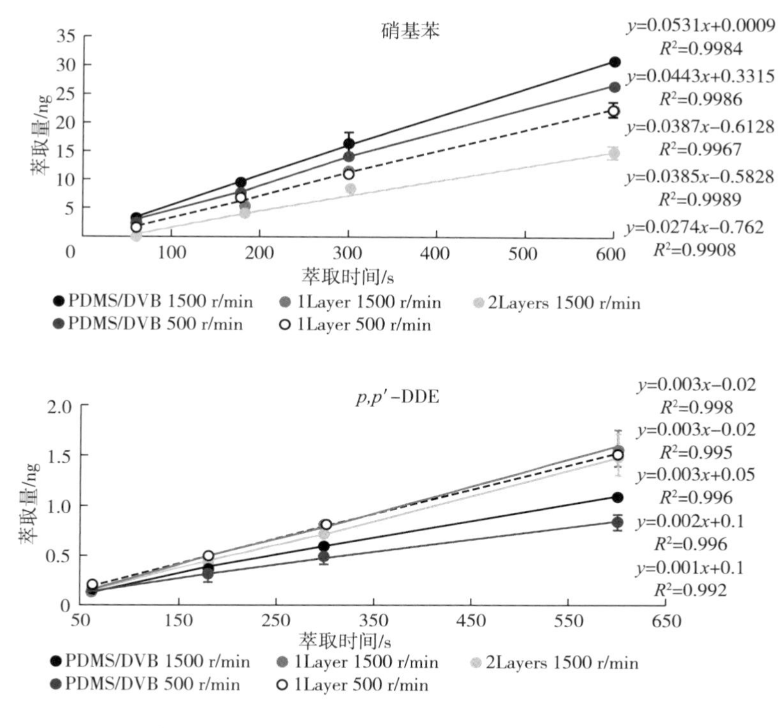

图 16.6　在 30℃的水溶液中使用 DI-SPME 萃取极性（硝基苯）和非极性（*p*，*p*′-DDE）分析物的质量吸收曲线

DI-SPME，直接浸入式固相微萃取

性分析物的萃取能力的限制。

Kloskowski 和同事报道了一个有趣的例子，旨在克服从极性样品中分离极性分析物（如水）的困难[57,58]。在上述研究中，作者制备了一种多层 SPME 纤维，在疏水性 PDMS 层下含有一层高极性萃取相聚乙二醇（PEG）。该体系被命名为膜 SPME（M-SPME），其外疏水层防止了水溶性 PEG 涂层的损失。在这个体系中，双层涂层允许高极性吸附剂作为水基质中的萃取剂，而没有溶解的风险。在 M-SPME 中使用高度极性的吸附剂以及双层涂层的优点，使其能够克服从极性基质中萃取极性分析物的相关问题[15,43]。关于 M-SPME 的更多应用参阅本书第 10 章。

Melo 和同事们在开发一种用 SPME-HPLC-DAD 分析生菜中 10 种农药的方法时，就遇到了这样的问题。尽管作者使用了聚乙二醇/模板化树脂（CW/TPR），但是对于大多数极性分析物（$\lg K_{ow} < 2$），该方法并不令人满意。CW/TRP 是一种极性涂层，由于与水相样品接触时溶胀引起的低化学稳定性而停止了商业化[59]。一种结合了 HS-SPME 与全二维气相色谱联用高速飞行时间质谱（HS-SPME-GC×GC/TOF MS）的方法，可快速、灵敏的检测 36 种可能污染茶叶样品（绿茶、红茶和果茶）的农药。使用商用 DVB/Car/PDMS 涂层的 SPME 方案与使用乙酸乙酯萃取和高性能凝胶渗透色谱（HPGPC）的方法进行比较，得到了类似的结果。尽管如此，与传统的 HPGPC 萃取方

式相比，HS-SPME 程序得到更洁净的提取物，且干扰性的基质成分更少，从而导致了更干净的背景，从化学噪声中更好的识别化合物，并消除积聚在进样器及分离毛细管前端的非挥发性沉积物[60]。利用纳米技术，结合碳纳米管增强中空纤维 SPME（CNTs-HF-SPME）和高效液相色谱-二极管阵列检测器（HPLC-DAD），建立了一种高效、灵敏的苹果中 5 种氨基甲酸酯类农药的检测方法。通过添加表面活性剂，CNTs 被分散在水中，然后在毛细管力和超声作用的支持下保持在 HF 的孔隙中。将 SPME 装置用 1-辛醇浸润，置于搅拌过的苹果样品中提取目标物。提取后，分析物经解吸后用 HPLC-DAD 分析。在优化的提取条件下，富集系数在 49~308 倍，纤维间重复性和批次间重现性好，线性范围好，回收率高。检测限为 0.09~6ng/g。因此，结果表明该新方法是一种用于苹果中氨基甲酸酯类农药痕量测定的有效的预处理及富集方法[61]。

16.5 脂类基质的特别注意事项

在开发 SPME 方法时，脂肪基质特别麻烦。分析物与基体的结合导致分析物的自由浓度非常低。含有中至高疏水性的分析物（$\lg K_{ow} > 3$）在脂肪基质中也显示低亨利定律常数，这对 HS 模式的实施具有消极影响。直接浸入式 SPME 甚至更为复杂，因为它可能会污染萃取相表面。在这种情况下，在 SPME 萃取前对样品进行预处理是不可避免的。

Gonzalez-Rodriguez 和同事开发了一种结合 SPME 和低压气相色谱（LP-GC）分析牛乳中 40 种不同农药残留的方法。作者在不使用溶剂萃取进行样品预处理的情况下取得了成功的结果。取而代之的是，用水稀释牛乳样品，并添加三乙胺以促进蛋白质变性（增加水相中分析物的自由浓度）。SPME 采用商用 PDMS/DVB 涂层在直接浸入模式下进行[62]。

Purcaro 和同事在两项不同的研究中，分别通过 SPME 与快速 GC 或 GC×GC 联用，完成了用 SPME 方法从植物油中提取多环芳香烃（PAHs）的任务。油样或者在乙腈中萃取，并在已烷中共萃取，或者直接在已烷中稀释，然后通过直接浸入已烷萃取物进行 SPME。在两项研究中，都使用了商用 SPME 纤维 Carbopack Z/PDMS。与构成 DVB/Car/PDMS 和 PDMS/Car 的 carboxen 颗粒不同，Carbopack-Z 是一种多孔石墨化炭黑，孔径约为 10nm。这种固体涂层具有较低的微孔率，因此，在萃取高分子质量的分析物方面，这一萃取相不会像受到 carboxen 涂层那样的限制。这种纤维的主要萃取方式是碳表面和分析物之间的 π-π 相互作用。平面化合物，如多环芳烃，可以与碳表面有更大的相互作用并被保留，而其他分析物不被有效保留。这在非极性溶剂基质中尤其重要。在这些条件下，PDMS 的作用被大大降低，但在水基基质中，PDMS 相的吸附作用要大得多。所提出的方法描述了快速、灵敏的解决方案，以减少甘油三酯的干扰，并提高柱寿命，避免频繁清洗质谱仪离子源。此外，与其他方法相比，该方法的溶剂消耗量和样品处理最少[63,64]。

Yang 等利用碳纳米管增强中空纤维 SPME（CNTs-HF-SPME）结合 HPLC 提取并测定乳制品中的己烯雌酚（DES）。利用溶胶凝胶技术将多壁碳纳米管（MWCNTs）填充在中空纤维的壁孔中。使用中空纤维可以限制像蛋白质这样的大分子进入小孔隙。中空纤维壁孔中的 MWCNTs 能够吸附目标分子，从而有效的从乳制品中选择性的提取 DES。此外，在该研究中仅使用了一次 CNTs 以防止可能的残留问题。本研究表明，CNTs-HF-SPME 与 HPLC 联用是一种简单、快速、经济的技术，可用于监测乳制品中二乙基己烯酚的残留。因此，应进一步推广该方法，以扩大牛乳样品中测定分析物的范围。尽管该方法具有良好的灵敏度，但其主要缺点是缺乏自动化和需要多次样品处理，这可能会对该方法增加相当大的误差[65]。

建立了一种在线管内纤维 SPME 方法，对猪肝和鸡肉样品中四种氟喹诺酮类物质进行了分析。该技术包括纵向将分子印迹纤维（MIP-纤维）装入 PEEK 管中作为在线萃取单元。由于涂层体积的增加，降低了背压，加快了动力学速率，并提高了萃取能力。此外，通过使用分子印迹涂层（多种氧氟沙星印迹纤维）对分析物的萃取进行了特别定制，极大地减少了样品基质的干扰。结果灵敏，检出限低至 0.016~0.11μg/L。事实上，MIP 确保了对模板及其结构类似物的特定识别。相反，MIP 涂层的专一性使得它不适合同时萃取多类化合物。由于污染物残留通常需要同时分析，为了解决这种限制并扩大方法的适用性，在 PEEK 管内填充由氧氟沙星和磺胺甲嘧啶分别印迹的两种不同的纤维，以同步萃取这两个类别的抗生素药物。初步结果表明，混合填充方式可以同时富集复杂样品中的目标物。同时还研究了在添加了氟喹诺酮类和磺胺类药物的猪肝样品中应用该方法的可行性[33,35]。

脂类基质的特性在最近的一项研究中得到了较好描述，即采用外包 PDMS 的 PDMS/DVB 涂层测定葡萄汁中的 11 种污染物的方案[55]应用于牛油果（*Persea Americana*）基质，发现明显缺乏再现性[66]。这种基质的含油量是一个很大的挑战。先前为葡萄汁基质（高糖基质）开发的纤维清洗方案，由于非挥发性物质在涂层表面沉积，而不能有效地避免纤维污染。由于非挥发性物质沉积在涂层表面的油类特性，用丙酮：水溶液（9：1，体积比）取代了纯水预冲洗步骤，脱附后清洗步骤的甲醇也被丙酮取代，从而提高涂层对基质污染的处理效率[66]。

16.6 体内固相微萃取

作为一种无溶剂制备样品的微萃取技术，SPME 的小巧及分析物的非完全萃取使得 SPME 成为体内分析和现场分析样品制备方法的理想候选者。此外，根据基本原理，在样品基质中分析物损耗可以忽略的情况下，SPME 萃取量与样品容量无关，这在大样品容量和/或分析物具有较低的纤维涂层/样品基质分布系数的分析中得以实现[2]。在分析之前不需要收集确定的和有代表性的样品量，从而允许在体内进行测量的情况下，这一原则是有利的。尽管 SPME 被广泛认为是一种测定食品质量和真实性的工具[67]，但关于使用体内 DI-SPME 研究可食用植物中污染物的文献资料一直很少[68-70]。

最早的报道之一涉及应用体内 SPME 追踪菜心（*Brassica parachinensis*）和芦荟（*Barbadensis*）中 OPPs 的吸收和消除，以及对倍硫磷代谢的研究，以更好地了解 OPPs 在活体植物中的归趋[68]。本研究突出了体内 SPME 对不同活体可食植物中污染物追踪的应用。如图 16.7 所示，微小的 SPME 探针（PDMS 萃取相）被导入植物的枝叶中。将菜心植株暴露于 OPPs 溶液中，通过吸收曲线追踪根系的吸收情况。对芦荟植株来说，在活体植物上喷洒农药溶液可以在喷洒后 8～12h 观察到最大吸收。在 180h 内，所有 OPPs 均达到基线浓度水平。体内示踪方法简便，可为评价不同农药的风险提供更多的数据，对农业生产安全具有重要意义。

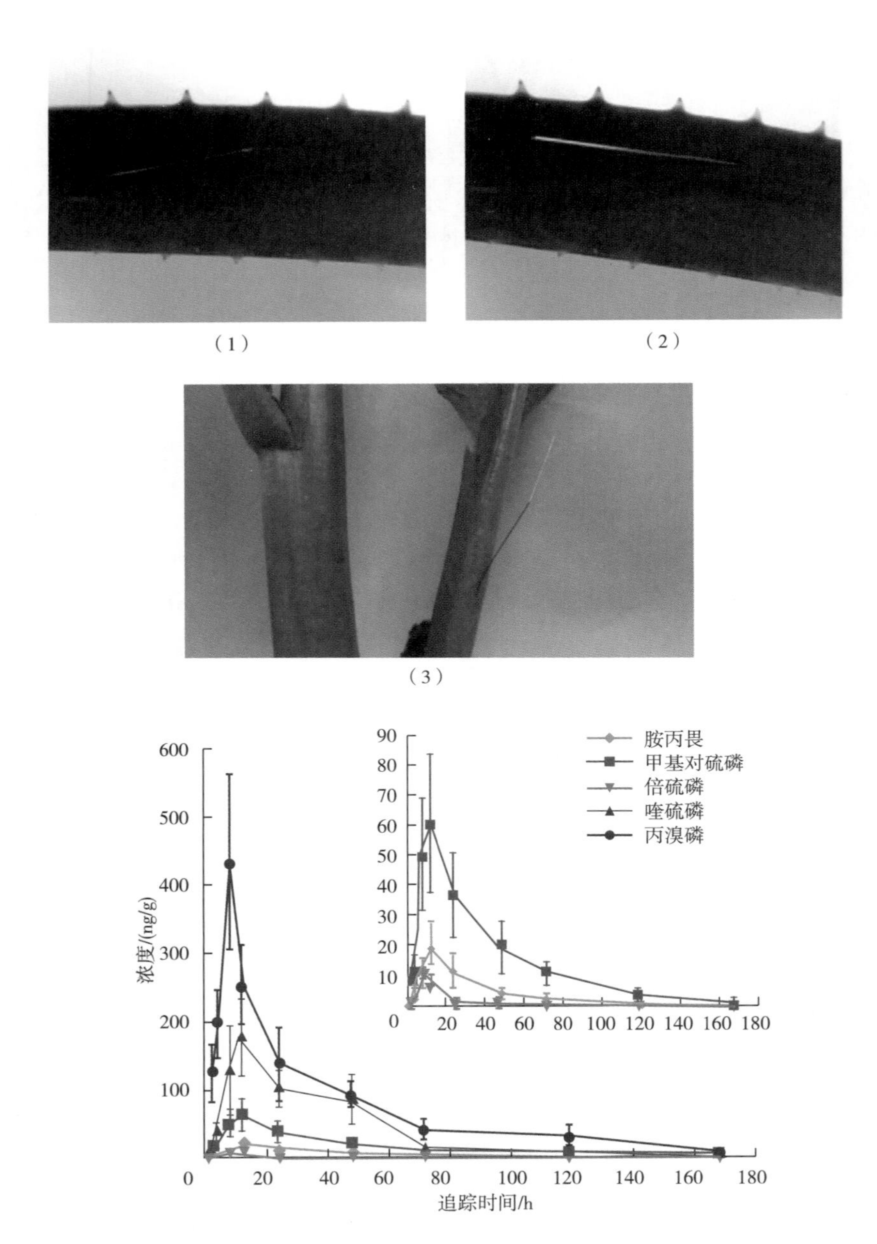

图 16.7　活体植物的体内固相微萃取采样，芦荟植物中有机磷农药的体内示踪

［资料来源：J. Qiu, G. Chen, H. Zhou, J. Xu, F. Wang, F. Zhu, et al., In vivo tracing of organophosphorus pesticides in cabbage (Brassica parachinensis) and aloe (Barbadensis), Sci. Total Environ. 550 (2016) 1134-1140. 经 Elsevier 许可转载。］

16.7 复杂食品基质中的定量固相微萃取

由于微萃取技术的平衡性质，SPME 萃取量与样品中分析物的自由浓度成正比。这一原则经常引起关于 SPME 定量能力的误解，特别是在处理复杂基质时，分析物的绝对回收量可能在总量的一个很小的百分比内。在这种情况下，要实现准确的定量方法，就需要选择适当的校准技术。

水溶液的外部校准很少适用于复杂基质。在基质呈现高含水量和中等至低的复杂性的情况下，可以采用外部校准。在某些情况下，最初的复杂食物基质被大量稀释（通常用水稀释），从而由于基质成分浓度降低，水溶液模型中的外部校准可能是适用的。然而，必须研究每个基质/分析物组合，以确保没有基质效应。必须考虑基质效应，特别是样品基质中含有有机溶剂，如酒精饮料，定量一般需要用基质匹配的标准溶液或标准加入法[2,13,71,72]。

在存在复杂基质的食品分析中，最常用的校准方法是将基质与未知样品匹配[73]。也应考虑使用同位素标记的内标。例如，采用 DI-SPME-GC 对水果中三唑类的定量测定，是通过基质匹配校准曲线来考虑可能的基质效应；此外，还使用了同位素标记的内标物[74]。内标（IS）提高了所开发方法的准确度和精密度，因为它纠正了仪器响应漂移和处理生物基质时经常发生的变异。在完全萃取方法中（主要涉及溶剂萃取），分析物的总量被萃取出，内标主要用于校正仪器漂移和由于最终萃取物中存在的基质成分而可能产生的基质效应。但是，由于 SPME 的非完全萃取性质，在没有考虑到对所研究的基质和分析物的适用性并进行相关的前期研究的情况下，应用这些在完全萃取方法中被广泛使用的内标是不合适的。在 SPME 中，内标应尽可能类似于系统基质/涂层中分析物的表现；或者说，IS 应该能够模拟分析物在萃取相和任何竞争相中的分配。在此基础上，近期的研究对在各种测定农药的方法中被广泛用作内标和代标的 10 种化合物进行了评价，探讨了它们作为 DI-SPME 方法的内标以用于葡萄中多残留农药检测的适用性[54]。最重要的是，研究表明，只用一对内标（对硫磷-d10 或 TPP+五氯硝基苯或 α-HCH-d6）足以确保可靠的校正仪器漂移和处理自然基质时可能发生的变异。此外，选择 TPP/五氯硝基苯作为内标将带来显著的成本节省，因为这些化合物比同位素标记的化合物便宜得多。

当无法获得基质空白样品时，可选择标准加入法。这类方法也适用于样品间差异较大（如 pH、盐含量、糖含量、含水量等）的基质，这些差异会导致显著的基质效应（校准曲线的斜率有统计学差异）[75]。相反地，在使用标准加入法和研究固态或半固态基质时，应该注意所添加的标准物和天然分析物的传质机制可能不同[76,77]。这也被 Mirnaghi 等提及，当采用 SPME 研究浆果中的酚类含量时，作者指出，内源性基质化合物和添加标准物在键合和吸附点位上的差异，导致所研究（经溶剂萃取验证的研究）的复杂基质中的分析物被低估[78]。

除了在食品分析中使用的传统校准方法外，SPME 还提供了其他校准方法。对于 SPME 校准，平衡萃取是最简单的方法。直接应用式（16.1），可以很容易地计算出样品浓度。在 SPME 中实现分配平衡是很重要的，因为在这种情况下，传质的变化不会影响最终的萃取质量。然而，平衡时间可能很长，当使用吸附剂萃取相（如 PDMS/DVB、DVB/Car/PDMS、PDMS/Car）对多组分、复杂样品进行测定时，必须仔细检查分析物间发生取代的情况[10,79,80]。最近，有报道称，在多分析物体系中使用固态的 SPME 涂层时，与 DI 模式相比，这种取代效应更容易发生在 HS 模式中[10]。这主要与分析物在 HS/基质界面处的传质阻力及其亨利定律常数有关。那些对涂层具有较高亲和力和具有较高亨利定律常数的化合物，容易富集在基质上方的顶空，被涂层萃取，并始终占据其活性位点。相反，半挥发性和亲水性分析物在水相介质中溶解度较高，在不同相之间的界面处传质阻力增大，从而以较慢的速度富集顶空。与 HS 取样相比，即使是在预平衡条件下，DI 萃取在萃取分析物方面提供了更均衡的覆盖范围。因为亲水化合物的萃取动力学更快，而疏水化合物的萃取更易受其在萃取相周围的边界层中的扩散所影响。

此外，当采用 SPME 固态涂层分析复杂基质时，基质成分、分析物和内标之间的竞争可能导致极性化合物的线性动态范围变窄，极性化合物可能容易被其他具有更高浓度和具有对萃取相更高亲和力的吸附物取代，或者当涂层发生饱和被取代。然而，在这种情况下，通过限制促进涂层饱和的因素（萃取时间、萃取方式、样品稀释、样品和/或 HS 体积的调整等)，仍然可以获得良好的线性度。或者，对于疏水化合物，线性动态范围可能受到它们在水基样品中的溶解度、与基质成分的结合程度或对瓶壁的竞争性吸附的限制。此外，在灵敏度允许的情况下，可采用多次预平衡萃取；使用标准品对纤维进行预吸附或使用动态校正机制（图 16.1 中，当 $n < 95\% \ n_e$ 时）等程序已得到令人满意的定量结果。

值得注意的是，对于某些食品方面的应用，在纤维上加载标准品时，所发生的分析物迁移/解吸进入固态和半固态基质孔隙的现象会使动态校正失效（图 16.8）。天然基质化合物的解吸速率可能与加载的分析物从涂层到基质的解吸速率不同，导致加载的标准品和内源性分析物之间存在各向同性成为错误假设。

如前所述，考虑到可以延长涂层的使用寿命，因此在复杂基质中，DI-SPME 经常在预平衡状态下进行。然而，当使用预平衡 SPME 时，必须考虑和校正复杂基质对取样动力学的影响。基于此，一种基于非平衡 SPME 的新校正方法被开发[81]，并被验证可用于如牛乳等高蛋白质含量的基质[82]。由于多环芳烃可与蛋白质和脂肪基质紧密结合，因此被选为模型分析物。使用该校正方式获得的牛乳样品中多环芳烃的回收率在 75%~108%[82]。

最后，SPME 是一种平衡萃取方法，在大多数情况下，只有一小部分的分析物被萃取。但是，如果分布常数（K_{fs}）非常大，则 $V_s << K_{fs}V_f$，可以萃取给定分析物的总量；换句话说，可能会发生完全萃取。完全萃取的校正在 SPME 中并不常用，因为它通常只适用于样品体积小、分布系数很大的情况。由于吸附剂的表面积/体积比很大，薄膜几何形状的应用也证实了完全萃取能力[78]。完全萃取的一个特殊应用是多重 SPME，其中样品被重复萃取，即使样品基质中的分析物没有被完全萃取，也可以从少数萃取物中推断出分析物的总量[76,83]。这种方法的主要优点是可以通过计算样品中分析物的

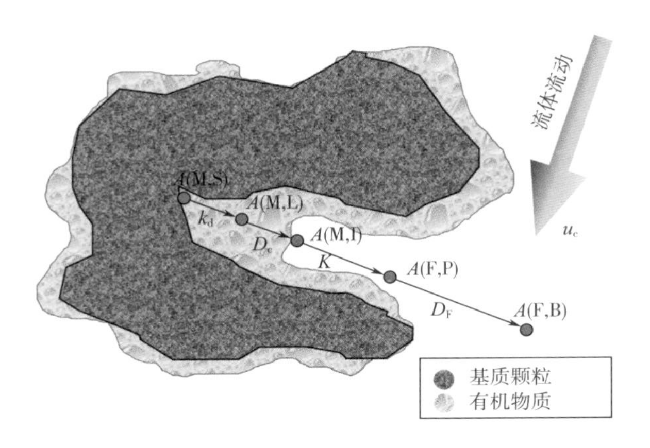

图 16.8　固体颗粒上萃取过程的各个步骤的示意图

A(M，S)：吸附在基质固体表面上的分析物分子；A(F，B)：在流动的流体中的分析物；A(F，P)：溶解于颗粒空隙内的流体中的分析物；A(M，I)：在基质-流体界面处的分析物；A(M，L)：附着于基质颗粒上的有机物质中的分析物；D_c：分析物在膨胀的基质有机成分中的扩散系数；D_F：在流体中的扩散系数；K：分析物的基质流体分布常数；k_d：分析物-基质络合物的解离速率常数。

总量来避免基质效应。但是，萃取量不能忽略，否则连续萃取的区域是相同的区域，该方法将无效。

16.8　结论与展望

尽管仪器有了很大的发展，但在大多数分析应用中，复杂的样品不能直接引入仪器。因此，为了从这些复杂的基质中分离和浓缩目标分析物，样品制备通常是必要的。SPME 是一种不断发展的样品制备方法。将 SPME 方法用于复杂食品基质分析的关键是小型化、进一步自动化、高通量、重现性、可追溯性和便携性。SPME 在复杂样品分析中的应用，在很大程度上依赖于新萃取相化学的发展。萃取相对复杂基质的稳定性和兼容性是获得可靠的定性和定量数据的关键因素。然而，没有一种通用的方法能够处理广泛的分析物/基质组合。需要对所调查的系统进行适当的研究，以选择和优化影响可靠和准确数据的关键因素。自动化提供了高的样品通量，因此，与手工方法相比，在节省时间和提高重现性方面是非常有效的；然而，自动化系统需要一些初始投入。只要采用适当的校正方法，平衡法和预平衡法都可以实现可靠和准确的定量。最后，我们希望本章将作为选择有效的 SPME 策略的指导，特别是在定量测定复杂食品基质中的污染物方面。

参考文献

[1] C. L. Arthur, J. Pawliszyn, Solid phase microextraction withthermal desorption using fused silica optical fibers, Anal. Chem. 62 (1990) 2145-2148.

[2] J. Pawliszyn, Handbook of Solid Phase Microextraction, Chemical Industry Press, Beijing, 2009.

[3] J. Ai, Solid phase microextraction for quantitative analysis in nonequilibrium situations, Anal. Chem. 69 (1997) 1230-1236.

[4] C. L. Arthur, L. M. Killam, K. D. Buchholz, J. Pawliszyn, J. R. Berg, Automation and optimization of solid-phase microextraction, Anal. Chem. 64 (1992) 1960-1966.

[5] J. Pawliszyn, Solid Phase Microextraction: Theory and Practice, first ed., Wiley-VCH, New York, NY, 1997.

[6] J. Pawliszyn, Applications of Solid Phase Microextraction, Cambridge: Royal Society of Chemistry, Cambridge, UK, 1999.

[7] J. Ai, Headspace solid phase microextraction. Dynamics and quantitative analysis before reaching a partition equilibrium, Anal. Chem. 69 (1997) 3260-3266.

[8] T. A. Souza-Silva, E. Gionfriddo, J. Pawliszyn, A critical review of the state of the art of solid-phase microextraction of complex matrices II. Food analysis, TrAC 71 (2015) 236-248.

[9] T. Górecki, X. Yu, J. Pawliszyn, Theory of analyte extraction by selected porous polymer SPME fibresł, Analyst 124 (1999) 643-649.

[10] E. Gionfriddo, E. A. Souza-Silva, J. Pawliszyn, Headspace versus direct immersion solid phase microextraction in complex matrixes: investigation of analyte behavior in multicomponent mixtures, Anal. Chem. 87 (2015) 8448-8456.

[11] M. Anastassiades, S. J. Lehotay, D. Stajnbaher, F. J. Schenck, Fast and easy multiresidue method employing acetonitrile extraction/partitioning and dispersive solid-phase extraction for the determination of pesticide residues in produce, J. AOAC Int. 86 (2003) 412-431.

[12] D. A. Lambropoulou, T. A. Albanis, Methods of sample preparation for determination of pesticide residues in food matrices by chromatography-mass spectrometry-based techniques: a review, Anal. Bioanal. Chem. 389 (2007) 1663-1683.

[13] S. Risticevic, V. H. Niri, D. Vuckovic, J. Pawliszyn, Recent developments in solid-phase microextraction, Anal. Bioanal. Chem. 393 (2009) 781-795.

[14] M. Mclean, A. Malik, S. Florida, Sol-Gel Materials in Analytical Microextraction, Elsevier, 2012.

[15] A. Kabir, K. G. Furton, A. Malik, Innovations in sol-gel microextraction phases for solvent-free sample preparation in analytical chemistry, TrAC 45 (2013) 197-218.

[16] S. L. Chong, D. X. Wang, J. D. Hayes, B. W. Wilhite, A. Malik, Sol-gel coating technology for the preparation of solid-phase microextraction fibers of enhanced thermal stability, Anal. Chem. 69 (1997) 3889-3898.

[17] F. Augusto, E. Carasek, R. G. C. Silva, S. R. Rivellino, A. D. Batista, E. Martendal, et al., New sorbents for extraction and microextraction techniques, J. Chromatogr. A 1217 (2010) 2533-2542.

[18] H. Bagheri, H. Piri-Moghadam, A. Es'haghi, An unbreakable on-line approach towards sol-gel capillary microextraction, J. Chromatogr. A 1218 (2011) 3952-3957.

[19] H. Bagheri, H. Piri-Moghadam, M. Naderi, Towards greater mechanical, thermal and chemical stability in solid-phase microextraction, TrAC 34 (2012) 126-139.

[20] W. A. Wan Ibrahim, H. Farhani, M. M. Sanagi, H. Y. Aboul-Enein, Solid phase microextraction using new sol-gel hybrid polydimethylsiloxane-2-hydroxymethyl-18-crown-6-coated fiber for determination of organophosphorous pesticides, J. Chromatogr. A 1217 (2010) 4890-4897.

[21] L. S. Cai, S. L. Gong, M. Chen, C. Y. Wu, Vinyl crown ether as a novel radical crosslinked sol-

gel SPME fiber for determination of organophosphorus pesticides in food samples, Anal. Chim. Acta 559 (2006) 89-96.

[22] J. Yu, C. Wu, J. Xing, Development of new solid-phase microextraction fibers by sol-gel technology for the determination of organophosphorus pesticide multiresidues in food, J. Chromatogr. A 1036 (2004) 101-111.

[23] M. Saraji, B. Rezaei, M. K. Boroujeni, A. A. H. Bidgoli, Polypyrrole/sol-gel composite as a solid-phase microextraction fiber coating for the determination of organophosphorus pesticides in water and vegetable samples, J. Chromatogr. A 1279 (2013) 20-26.

[24] J. Xu, J. Zheng, J. Tian, F. Zhu, F. Zeng, C. Su, et al., New materials in solid-phase microextraction, TrAC 47 (2013) 68-83.

[25] H. Yu, T. D. Ho, J. L. Anderson, Ionic liquid and polymeric ionic liquid coatings in solid-phase microextraction, TrAC 45 (2013) 219-232.

[26] Y. Zhang, X. Wang, C. Lin, G. Fang, S. Wang, A novel SPME fiber chemically linked with 1-Vinyl-3-hexadecylimidazolium hexafluorophosphate ionic liquid coupled with GC for the simultaneous determination of pyrethroids in vegetables, Chromatographia 75 (2012) 789-797.

[27] C. Cagliero, T. D. Ho, C. Zhang, C. Bicchi, J. L. Anderson, Determination of acrylamide in brewed coffee and coffee powder using polymeric ionic liquid-based sorbent coatings in solid-phase microextraction coupled to gas chromatography-mass spectrometry, J. Chromatogr. A 1449 (2016) 2-7.

[28] C. Cagliero, H. Nan, C. Bicchi, J. L. Anderson, Matrix-compatible sorbent coatings based on structurally-tuned polymeric ionic liquids for the determination of acrylamide in brewed coffee and coffee powder using solid-phase microextraction, J. Chromatogr. A 1459 (2016) 17-23.

[29] J. Feng, H. Qiu, X. Liu, S. Jiang, The development of solid-phase microextraction fibers with metal wires as supporting substrates, TrAC 46 (2013) 44-58.

[30] M. Hu, M. Jiang, P. Wang, S. Mei, Y. Lin, X. Hu, et al., Selective solid-phase extraction of tebuconazole in biological and environmental samples using molecularly imprinted polymers, Anal. Bioanal. Chem. 387 (2007) 1007-1016.

[31] X. Hu, Y. Fan, Y. Zhang, G. Dai, Q. Cai, Y. Cao, et al., Molecularly imprinted polymer coated solid-phase microextraction fiber prepared by surface reversible addition-fragmentation chain transfer polymerization for monitoring of Sudan dyes in chilli tomato sauce and chilli pepper samples, Anal. Chim. Acta 731 (2012) 40-48.

[32] E. Turiel, J. L. Tadeo, A. Martin-Esteban, Molecularly imprinted polymeric fibers for solid-phase microextraction, Anal. Chem. 79 (2007) 3099-3104.

[33] Y. Hu, C. Song, G. Li, Fiber-in-tube solid-phase microextraction with molecularly imprinted coating for sensitive analysis of antibiotic drugs by high performance liquid chromatography, J. Chromatogr. A 1263 (2012) 21-27.

[34] X. Hu, Q. Cai, Y. Fan, T. Ye, Y. Cao, C. Guo, Molecularly imprinted polymer coated solid-phase microextraction fibers for determination of Sudan I-IV dyes in hot chili powder and poultry feed samples, J. Chromatogr. A 1219 (2012) 39-46.

[35] Y. Hu, J. Pan, K. Zhang, H. Lian, G. Li, L. Ramos, et al., Novel applications of molecularly-imprinted polymers in sample preparation, TrAC 43 (2013) 37-52.

[36] D. Siegel, Applications of reversible covalent chemistry in analytical sample preparation, Analyst 137 (2012) 5457-5482.

[37] D. Djozan, B. Ebrahimi, Preparation of new solid phase micro extraction fiber on the basis of atrazine-molecular imprinted polymer: application for GC and GC/MS screening of triazine herbicides in water, rice and onion, Anal. Chim. Acta 616 (2008) 152-159.

[38] D. Djozan, M. Mahkam, B. Ebrahimi, Preparation and binding study of solid - phase microextraction fiber on the basis of ametryn-imprinted polymer: application to the selective extraction of persistent triazine herbicides in tap water, rice, maize and onion, J. Chromatogr. A 1216 (2009) 2211-2219.

[39] Y.-L. Wang, Y.-L. Gao, P.-P. Wang, H. Shang, S.-Y. Pan, X.-J. Li, Sol-gel molecularly imprinted polymer for selective solid phase microextraction of organophosphorous pesticides, Talanta 115 (2013) 920-927.

[40] J.-W. Li, Y.-L. Wang, S. Yan, X.-J. Li, S.-Y. Pan, Molecularly imprinted calixarene fiber for solid-phase microextraction of four organophosphorous pesticides in fruits, Food Chem. 192 (2016) 260-267.

[41] H. Kataoka, Automated sample preparation using in-tube solid-phase microextraction and its application-a review, Anal. Bioanal. Chem. 373 (2002) 31-45.

[42] D. Vuckovic, E. Cudjoe, F. M. Musteata, J. Pawliszyn, Automated solid-phase microextraction and thin-film microextraction for high-throughput analysis of biological fluids and ligand-receptor binding studies, Nat. Protoc. 5 (2010) 140-161.

[43] E. Cudjoe, D. Vuckovic, D. Hein, J. Pawliszyn, Investigation of the effect of the extraction phase geometry on the performance of automated solid-phase microextraction, Anal. Chem. 81 (2009) 4226-4232.

[44] H. Bagheri, A. Es'haghi, A. Es-haghi, N. Mesbahi, A. Es'haghi, A high-throughput approach for the determination of pesticide residues in cucumber samples using solid-phase microextraction on 96-well plate, Anal. Chim. Acta 740 (2012) 36-42.

[45] L. B. Abdulra'uf, W. A. Hammed, G. H. Tan, L. B. Abdulra'uf, SPME fibersfor the analysis of pesticide residues in fruits and vegetables: a review, Crit. Rev. Anal. Chem. 42 (2012) 152-161.

[46] A. Menezes Filho, F. N. dos Santos, P. A. D. P. Pereira, P. A. de Paula Pereira, Development, validation and application of a methodology based on solid - phase microextraction followed by gas chromatography coupled to mass spectrometry (SPME/GC-MS) for the determination of pesticide residues in mangoes, Talanta 81 (2010) 346-354.

[47] S. Cortés-Aguado, N. Sánchez-Morito, F. J. Arrebola, A. G. Frenich, J. L. M. Vidal, S. Cortesaguado, et al., Fast screening of pesticide residues in fruit juice by solid-phase microextraction and gas chromatography-mass spectrometry, Food Chem. 107 (2008) 1314-1325.

[48] N. Aguinaga, N. Campillo, P. Viñas, M. Hernandez-Cordoba, P. Vinas, M. Hernandez-Cordoba, Solid-phase microextraction coupled to gas chromatography-mass spectrometry for the analysis of famoxadone in wines, fruits, and vegetables, Spectrosc. Lett. 42 (2009) 320-326.

[49] S. C. Aguado, N. Sanchez-Morito, A. G. Frenich, J. L. M. Vidal, F. J. Arrebola, Screening method for the determination at parts per trillion levels of pesticide residues in vegetables combining solid-phase microextraction and gas chromatography-tandem mass spectrometry, Anal. Lett. 40 (2007) 2886-2914.

[50] P. Viñas, N. Campillo, N. Martínez-Castillo, M. Hernández-Córdoba, Method development and validation for strobilurin fungicides in baby foods by solid-phase microextraction gas chromatography-mass-spectrometry, J. Chromatogr. A 1216 (2009) 140-146.

[51] J. Beltran, F. J. López, F. Hernández, Solid-phase microextraction in pesticide residue analysis, J. Chromatogr. A 885 (2000) 389-404.

[52] É. A. Souza-Silva, J. Pawliszyn, Optimization of fiber coating structure enables direct immersion solid phase microextraction and high-throughput determination of complex samples, Anal. Chem. 84 (2012) 6933-6938.

[53] L. S. De Jager, G. A Perfetti, G. W. Diachenko, Analysis of tetramethylene disulfotetramine in foods using solid-phase microextraction-gas chromatography-mass-spectrometry, J. Chromatogr. A 1192 (2008) 36-40.

[54] É. A. Souza-Silva, J. Pawliszyn, Direct immersion solid-phase microextraction with matrix-compatible fiber coating for multiresidue pesticide analysis of grapes by gas chromatography-time-of-flight mass spectrometry (DI-SPME-GC-ToFMS), J. Agric. Food Chem. 63 (2015) 4464-4477.

[55] É. A. Souza-Silva, E. Gionfriddo, R. Shirey, L. Sidisky, J. Pawliszyn, Methodical evaluation and improvement of matrix compatible PDMS-overcoated coating for direct immersion solid phase microextraction gas chromatography (DI-SPME-GC) -based applications, Anal. Chim. Acta (2016) .

[56] É. A. Souza-Silva, E. Gionfriddo, M. N. Alam, J. Pawliszyn, Insights into the effect of the PDMS-layer on the kinetics and thermodynamics of analyte sorption onto the matrix-compatible solid phase microextraction coating, Anal. Chem. (2017), http://dx.doi.org/10.1021/acs.analchem.6b04442 (in press) .

[57] A. Kloskowski, M. Pilarczyk, J. Namiesnik, Membrane solid-phase microextraction-a new concept of sorbent preparation, Anal. Chem. 81 (2009) 7363-7367.

[58] A. Spietelun, M. Pilarczyk, A. Kloskowski, J. Namiesnik, Polyethylene glycol-coated solid-phase microextraction fibres for the extraction of polar analytes-a review, Talanta 87 (2011) 1-7.

[59] A. Melo, A. Aguiar, C. Mansilha, O. Pinho, I. M. P. L. V. O. Ferreira, Optimisation of a solid-phase microextraction/HPLC/Diode Array method for multiple pesticide screening in lettuce, Food Chem. 130 (2012) 1090-1097.

[60] J. Schurek, T. Portolés, J. Hajslova, K. Riddellova, F. Hernandez, Application of head-space solid-phase microextraction coupled to comprehensive two-dimensional gas chromatography-time-of-flight mass spectrometry for the determination of multiple pesticide residues in tea samples, Anal. Chim. Acta 611 (2008) 163-172.

[61] X. -Y. Song, Y. -P. Shi, J. Chen, Carbon nanotubes-reinforced hollow fibre solid-phase microextraction coupled with high performance liquid chromatography for the determination of carbamate pesticides in apples, Food Chem. 139 (2013) 246-252.

[62] M. J. González-Rodríguez, F. J. Arrebola Liébanas, A. Garrido Frenich, J. L. Martínez Vidal, F. J. Sánchez López, Determination of pesticides and some metabolites in different kinds of milk by solid-phase microextraction and low-pressure gas chromatography-tandem mass spectrometry, Anal. Bioanal. Chem. 382 (2005) 164-172.

[63] G. Purcaro, M. Picardo, L. Barp, S. Moret, L. S. Conte, Direct-immersion solid-phase microextraction coupled to fast gas chromatography mass spectrometry as a purification step for polycyclic aromatic hydrocarbons determination in olive oil, J. Chromatogr. A1307 (2013) 166-171.

[64] G. Purcaro, P. Morrison, S. Moret, L. S. Conte, P. J. Marriott, Determination of polycyclic aromatic hydrocarbons in vegetable oils using solid-phase microextraction-comprehensive two-dimensional gas chromatography coupled with time-of-flight mass spectrometry, J. Chromatogr. A 1161 (2007) 284-291.

[65] Y. Yang, J. Chen, Y. -P. Shi, Determination of diethylstilbestrol in milk using carbon nanotube-reinforced hollow fiber solid-phase microextraction combined with high-performance liquid chromatography, Talanta 97 (2012) 222-228.

[66] S. De Grazia, E. Gionfriddo, J. Pawliszyn, A new and efficient Solid Phase Microextraction approach for analysis of high fat content food samples using a matrix-compatible coating, Talanta (2017), http://dx.doi.org/10.1016/j.talanta.2017.01.064 (in press).

[67] E. A. Souza Silva, S. Risticevic, J. Pawliszyn, Recent trends in SPME concerning sorbent materials, configurations and in vivo applications, TrAC 43 (2013) 24-36.

[68] J. Qiu, G. Chen, H. Zhou, J. Xu, F. Wang, F. Zhu, et al., In vivo tracing of organophosphorus pesticides in cabbage (*Brassica parachinensis*) and aloe (*Barbadensis*), Sci. Total Environ. 550 (2016) 1134-1140.

[69] J. Qiu, G. Chen, J. Xu, E. Luo, Y. Liu, F. Wang, et al., In vivo tracing of organochloride and organophosphorus pesticides in different organs of hydroponically grown Malabar spinach (*Basella alba* L.), J. Hazard. Mater. 316 (2016) 52-59.

[70] J. Qiu, G. Chen, F. Zhu, G. Ouyang, Sulfonated nanoparticles doped electrospun fibers with bioinspired polynorepinephrine sheath for in vivo solid-phase microextraction of pharmaceuticals in fish and vegetable, J. Chromatogr. A 1455 (2016) 20-27.

[71] S. Risticevic, H. Lord, T. Gorecki, C. L. Arthur, J. Pawliszyn, Protocol for solid-phase microextraction method development, Nat. Protoc. 5 (2010) 122-139.

[72] S. Risticevic, Y. Chen, L. Kudlejova, R. Vatinno, B. Baltensperger, J. R. Stuff, et al., Protocol for the development of automated high-throughput SPME-GC methods for the analysis of volatile and semivolatile constituents in wine samples, Nat. Protoc. 5 (2010) 162-176.

[73] W. -H. Tsai, T. -C. Huang, J. -J. Huang, Y. -H. Hsue, H. -Y. Chuang, Dispersive solid-phase microextraction method for sample extraction in the analysis of four tetracyclines in water and milk samples by high-performance liquid chromatography with diodearray detection, J. Chromatogr. A 1216 (2009) 2263-2269.

[74] É. A. Souza-Silva, V. Lopez-Avila, J. Pawliszyn, Fast and robust direct immersion solid phase microextraction coupled with gas chromatography-time-of-flight mass spectrometry method employing a matrix compatible fiber for determination of triazole fungicides in fruits, J. Chromatogr. A 1313 (2013) 139-146.

[75] F. -F. Lei, J. -Y. Huang, X. -N. Zhang, X. -J. Liu, X. -J. Li, Determination of polycyclic aromatic hydrocarbons in vegetables by headspace SPME-GC, Chromatographia 74 (2011) 99-107.

[76] G. Ouyang, J. Pawliszyn, A critical review in calibration methods for solid-phase microextraction, Anal. Chim. Acta 627 (2008) 184-197.

[77] G. Ouyang, J. Pawliszyn, SPME in environmental analysis, Anal. Bioanal. Chem. 386 (2006) 1059-1073.

[78] F. S. Mirnaghi, F. Mousavi, S. M. Rocha, J. Pawliszyn, Automated determination of phenolic compounds in wine, berry, and grape samples using 96-blade solid phase microextraction system coupled with liquid chromatography-tandem mass spectrometry, J. Chromatogr. A 1276 (2013) 12-19.

[79] O. Lasekan, A. Khatib, H. Juhari, P. Patiram, S. Lasekan, Headspace solid-phase microextraction gas chromatography-mass spectrometry determination of volatile compounds in different varieties of African star apple fruit (*Chrysophillum albidum*), Food Chem. 141 (2013) 2089-2097.

[80] S. Risticevic, J. Pawliszyn, Solid-phase microextraction in targeted and nontargeted analysis:

displacement and desorption effects, Anal. Chem. 85 (2013) 8987-8995.

[81] R. Jiang, J. Xu, W. Lin, S. Wen, F. Zhu, T. Luan, et al. , Investigation of the kinetic process of solid phase microextraction in complex sample, Anal. Chim. Acta 900 (2015) 111-116.

[82] W. Lin, S. Wei, R. Jiang, F. Zhu, G. Ouyang, Calibration of the complex matrix effects on the sampling of polycyclic aromatic hydrocarbons in milk samples using solid phase microextraction, Anal. Chim. Acta 933 (2016) 117-123.

[83] R. Costa, L. Tedone, S. De Grazia, P. Dugo, L. Mondello, Multiple headspace-solid-phase microextraction: an application to quantification of mushroom volatiles, Anal. Chim. Acta 770 (2013) 1-6.

[84] D. Djozan, M. Mahkam, B. Ebrahimi, Preparation and binding study of solid - phase microextraction fiber on the basis of ametryn-imprinted polymer, J. Chromatogr. A1216 (2009) 2211-2219.

[85] D. Djozan, B. Ebrahimi, M. Mahkam, M. A. Farajzadeh, Evaluation of a new method for chemical coating of aluminum wire with molecularly imprinted polymer layer. Application for the fabrication of triazines selective solid-phase microextraction fiber, Anal. Chim. Acta 674 (2010) 40-48.

17

17 辅助萃取偶联技术的进展

Jorge Moreda-Piñeiro[1], Antonio Moreda-Piñeiro[2*]
1. Grupo Química Analítica Aplicada (QANAP), University Institute of Research in Environmental Studies (IUMA), Centro de Investigaciones Científicas Avanzadas (CICA), University of A Coruña, A Coruña, Spain
2. University of Santiago de Compostela, Santiago de Compostela, Spain
＊通讯作者 E-mail：antonio. moreda@ usc. es

17.1 引言

在前面的章节中，由于减少了有机溶剂和高毒性或生态毒性的试剂并防止了废物的产生，微萃取技术（吸附剂微萃取，液相微萃取和膜微萃取）被认为是环保的程序。微萃取技术所使用的萃取溶剂相对于样品体积而言是少量的，这意味着有高富集因子(高样品体积与萃取剂体积比)[1,2]。吸附剂微萃取可以被认为是一种先进的微型固相萃取（SPE）技术。固相微萃取（SPME）和搅拌棒吸附萃取（SBSE）属于此类。类似地，液相微萃取（LPME）和膜微萃取技术可被视为小型化的液液萃取程序。分散液-液微萃取（DLLME）和单滴微萃取（SDME）属于LPME技术。关于膜微萃取，受体相通过合适的聚合物膜（负载的液体膜，SLM）与大体积样品分离，或者萃取剂可被捕获到聚合物膜的孔中［中空纤维（HF）膜］。因此，第三种微萃取技术包括支撑液膜萃取（SLM）和中空纤维液相微萃取（HF-LPME）程序。

微萃取技术有很多，不能一一列举。但当前开发和应用的是这些分析化学中样品预处理技术具有易于操作、小型化、低成本以及对多种样品和分析物的适应性等多种优势。使用偶联的微萃取技术，可以依次（离线偶联）或同时（在线）执行目标物的提取，预浓缩和净化过程，这缩短了整个前处理的时间。这些绿色偶联技术的其他优点包括操作简单、成本低和富集因子高。与SPE结合使用新的吸附剂，例如碳纳米管(CNT)、分子印迹聚合物（MIP）、磁性纳米颗粒（MNP），以及微SPE（μ-SPE）技术，结合现代微萃取技术，可提高选择性，也适用于处理复杂样品。进行基于膜的微萃取技术时，将固体吸附剂用作新的受体相可以消除相渗漏问题。但是，需要进一步的发展来实现在线偶联并扩大偶联技术在复杂样品中的适用性。这一事实将有助于成功地将这些方法应用于常规分析。

在本章中，对如何将不同的微萃取技术以及基于使用非离子和两性离子表面活性剂，微波/超声波能量和加压的其他环保样品预处理技术相偶联的可能性进行了详细的讨论。还讨论了这些技术的最新进展和应用，以及偶联技术的优缺点。

17.2 与浊点萃取技术的偶联

浊点萃取（CPE）方法基于非离子和两性离子表面活性剂的性质，将它们的水溶液加热到一定温度以上时，它们会分成两个液相。微波［CPE—微波辅助提取(MAE)］已用于辅助从胶束相中提取目标物（表17.1)。有文献报道从PVC薄膜的水提取物中分离出己二酸二乙基己酯（DEHA）和柠檬酸乙酰基三丁酯（ATBC)[3]，从水中分离出邻苯二甲酸酯（PEs)[4]。先进的CPE—MAE方法提供了60（DEHA和ATBC)[71]和71~85（PEs)[4]的富集因子。而且定量分析的回收率为89%~100%。这些结果与使用常规溶剂萃取（分析回收率在65%~80%范围内）相比更好[3]。

微波也已用于从土壤，沉积物和污泥样品中提取磺酰胺（SAs）进入胶束相（表17.1)[5]。在最佳条件下，10种磺酰胺的回收率在69.7%~102.7%。

表 17.1 各种浊点萃取实验条件的总结

结合萃取/预富集技术	目标物	样品基质	分析仪器	溶剂（体积）	LOD①	EF②	偶联方式	参考文献
CPE-MAE	DEHA、ATBC	水溶液	GC-FID	异辛烷（150uL）	15 和 19μg/mL	60	离线	[3]
CPE-MAE	PEs	矿泉水	GC-FID	异辛烷（150uL）	11.5~19.3μg/L	71~85	离线	[4]
MAE-CPE	SAs	土壤、沉积物、污泥	HPLC-UV	免溶剂	0.42~0.68ng/g	—③	在线	[5]
CPE-MD-μ-SPE-UAE	BZ3、MBC、EMC、BDM	自来水、湖水	HPLC-UV/PDA	DCM（250μL）	1.43~7.5μg/L	—③	离线	[6]
CPE-MD-μ-SPE-UAE	PEs	自来水、河水、罐装水、井水、瓶装水	HPLC-UV	甲醇（200μL）	0.5~1.0μg/L	14.6~25.4	离线	[7]
CPE-MD-μ-SPE-UAE	DOX、ALF	药物制备、尿液、血浆	FS	乙醇（200μL）	DOX 和 ALF 分别为 0.21，0.16μg/L	—③	离线	[8]
SPE-CPE	Cd	自来水、瓶装水、海水	ETAAS	免溶剂	1.0μg/L	1050	离线	[9]
SPE-CPE	Cu、Fe	井水、泉水、湖水、自来水	FAAS	免溶剂	0.07~0.09μg/L	150	离线	[10]
SPE-CPE	Ni	绿茶、椰子水	紫外-可见光谱	辛醇（1mL）、DMF（2mL）	5.0μg/L	—③	离线	[11]

注：ALF—阿夫唑嗪；ATBC—乙酰柠檬酸三丁酯；BDM—丁基甲氧基二苯甲酰甲烷；BZ3—2-羟基-4-甲氧基二苯甲酮；CPE—浊点提取；DCM—二氯甲烷；DEHA—己二酸二（2-乙基己基）酯；DMF—二甲基甲酰胺；DOX—多沙唑嗪；EMC—4-甲氧基肉桂酸乙基己酯；ETAAS—电热原子吸收光谱法；FAAS—火焰原子吸收光谱法；FID—火焰离子检测器；FS—分子荧光光谱法；GC—气相色谱；HPLC—高效液相色谱法；MAE—微波辅助萃取；MBC—3-（4-甲基苄二烯）-樟脑；MD-μ-SPE—磁分散微固相萃取；PDA—光电二极管阵列；PEs—邻苯二甲酸酯；SAs—磺酰胺；SPE—固相萃取；UAE—超声辅助提取；UV—紫外线。

①检出限。

②富集因子。

③未给出。

CPE 还与磁性分散微萃取（D-μ-SPE）技术结合使用（使用聚硅氧烷涂层的核壳 Fe_2O_3[6] 和硅藻土负载的磁铁矿纳米颗粒[7,8]）。超声辅助进一步从磁性纳米材料中反萃取分离出的目标物[7,8]。该方法已成功应用于从河流和湖泊水中提取紫外线滤光剂（防晒霜）[6]。同时也用于分析自来水、河水、罐装水、井水和瓶装水中的 PE[7]，以及用于分析药物制剂（阿夫唑嗪和多沙唑嗪）[8]。CPE-D-μ-SPE 在磁场的作用下，可以快速收集胶束纳米颗粒，具有实际优势[6-8]。发现所提出的 CPE-D-μ-SPE 方法是准确的（依据定量分析回收率[6-8]）。

CPE 与使用纳米材料的 SPE 结合也用于痕量金属（Cd[9]，Fe 和 Cu[10] 和 Ni[11]）测定。在这些方法中，金属离子首先保留在银纳米颗粒上[9]，用苋菜红修饰的 SnO_2 纳米粉[10] 和 TiO_2/海泡石复合材料[11] 中。然后将它们转移到富含表面活性剂的相中。在最优化条件下，方法回收率接近 100%。

17.3 与基于吸附剂的微萃取技术的偶联

17.3.1 固相微萃取

SPME 是一种可将分析物吸附到涂覆的石英纤维表面上的技术。在此阶段之后，将分析物解吸到气相色谱仪（GC）或高效液相色谱仪（HPLC）中。在操作上，SPME 是不完全的，平衡和非平衡以及间歇性和流动性[1]。

SPME 可以离线或在线与其他环保萃取技术相偶联（如微波辅助萃取（MAE），超声辅助萃取（UAE），超临界流体萃取（SFE），加压液体萃取（PLE）/加压热水萃取（PHWE），基质固相分散萃取（MSPD）。SPME 主要充当目标化合物离线或在线偶联的预浓缩过程。但是，SPME 的在线偶联可以同时进行目标物富集/净化过程，这缩短了整个分析过程。

17.3.1.1 离线固相微萃取的偶联

将 SPME 与其他环保提取技术偶联在一起的大多数方法都是离线的。这些偶联技术（表 17.2）都是基于完全萃取技术的，如 MAE[12-25]，UAE[14,26-28]，SFE[29-31]，PLE/PHWE[32-50]。除了预浓缩能力外，SPME 还可以用作测定前的有效净化程序。在其他情况下，也建议使用基质固相分散萃取（MSPD），然后是常规的固相微萃取[51,52]，或管内（阀内）固相微萃取[53,54]。

（1）离线微波-固相微萃取　在分析药用植物[12] 和有机物含量低的土壤[19] 中的有机氯农药（OCPs）；灰分[15] 和土壤[20] 中的多氯联苯（PCBs）；皮脂中的半挥发性化合物[16]，传统中药（TCM）中的樟脑和冰片[17]，姜黄素，姜黄酮和杀菌酮[18]，丹皮酚[22]；全血中的汞形态（原位丙基化后）[24] 时，离线 MAE-SPME 程序是定量方法。另外，在分析 CRM804-050（土壤）中某些有机氯农药时，微波辅助胶束萃取（MAME）-SPME 方法[25] 已被证明是准确的。然而，在分析茶叶样品[14] 以及有机物含量高的土壤[19] 时，某些有机氯的回收率较低（低于 70%）。还可以通过分析 SRM 1941a（海洋沉积物）中的多环芳烃（PAH）[13]；Metranal 16 和 CNS300-04-100（河

表 17.2　各种 SPME 偶联技术的实验条件的简要概述

偶联萃取/富集技术	目标物	样品基质	检测仪器	溶剂（体积）	LOD①	偶联方式	参考文献
MAE-DI-SPME	OCPs	TCM	GC-ECD	免溶剂	0.05~0.13μg/kg	离线	[12]
MAE-DI-SPME	PAHs	海洋沉积物	GC-MS	免溶剂	0.28~7.66μg/L	离线	[13]
MAE-HS-SPME	OCPs	茶叶	GC-ECD	免溶剂	0.015~0.081ng/L	离线	[14]
MAE-HS-SPME	PCBs	灰	GC-ECD、GC-MS	DMSO（30mL）	0.2~1.5ng/g	离线	[15]
MAE-DI-SPME	SVOC	皮脂	GC-MS	丙酮（6.7mL）、正己烷（3.3mL）	—②	离线	[16]
MAE-HS-SPME	樟脑和冰片	TCM	GC-MS	免溶剂	—②	离线	[17]
MAE-HS-SPME	姜黄醇，姜黄酮和杀菌酮	姜黄根茎	GC-MS		—②	离线	[18]
MAE-HS-SPME	OCPs；PCBs	土壤；土壤和沉积物	GC-MS/MS	丙酮（10mL）、正己烷（10mL）和乙醇（720uL）	0.02~3.6ng/g、0.4~1.0ng/g	离线	[19，20]
MAE-HS-SPME	OCPs	河口沉积物	GC-MS	甲醇（10mL）	0.005~0.11ng/g	离线	[21]
MAE-HS-SPME	丹皮酚	TCM	GC-MS	免溶剂	0.24μg/mL	离线	[22]
MAE-HS-SPME	有机杂质	3，4-亚甲二氧基甲基苯丙胺药丸	GC-MS	免溶剂	—	离线	[23]
MAE-HS-SPME	汞的形态	全血	GC-ICP-MS	免溶剂	30pg/g	离线	[24]
MAE-DI-SPME	OCPs	土壤	HPLC-UV	甲醇（80uL）	—	离线	[25]
UAE-HS-SPME	OCPs	茶叶	GC-ECD	免溶剂	0.015~0.081ng/L	离线	[14]
UAE-DI-SPME	杀菌剂乙烯菌核利和氯硝胺	土壤	GC-MS	丙酮（5mL）	2~13ng/g	离线	[26]
UAE-HS-SPME	OCPs	鸟肝	GC-ECD	丙酮（4mL）、正己烷（16mL）、甲醇（50uL）	0.01~0.60ng/g	离线	[27]
UAE-HS-SPME	PBBs 和 PBDEs	污水污泥，河道沉积物	GC-MS/MS	正己烷（8mL）	0.01~1.20ng/g	离线	[28]
SFE-HS-SPME	OTCs	蛤蜊	GC-MS	免溶剂	—②	离线	[29]
SFE-HS-SPME	PCBs，PBBs 和 PBDEs	鱼饲料	GC-MS/MS	正己烷（2mL）	0.2~8.9ng/kg	离线	[30]

续表

偶联萃取/富集技术	目标物	样品基质	检测仪器	溶剂（体积）	LOD①	偶联方式	参考文献
SFE-HS-SPME	香气化合物	镇江香醋	GC-MS	免溶剂	—②	离线	[31]
PHWE-DI-SPME	PAHs 和 PCBs	土壤，沉积物和颗粒物	GC-MS、Gc-ECD	免溶剂	—②	离线	[32，33]
PHWE-DI-SPME	精油	TCM	GC-MS	免溶剂	—②	离线	[34-37]
PHWE-HS-SPME	脂肪族伯胺	污水污泥	GC-MS/MS	免溶剂	9～135μg/kg	离线	[38，40]
PHWE-HS-SPME	*N*-亚硝胺	污水污泥	GC-MS/MS	免溶剂	0.03～0.14μg/kg	离线	[39]
PHWE-DI-SPME	甲基汞	土壤	GC-MS	免溶剂	1.3～5μg/kg	离线	[41]
PHWE-DI-SPME	CPs	土壤	GC-MS	乙腈（—）	1.1～6.7μg/kg	离线	[42]
PHWE-DI-SPME	阿特拉津	牛肾	GC-MS	乙醇（—）	20μg/kg	离线	[43]
PHWE-DI-SPME	OCPs 和氯苯	水果和蔬菜	GC-MS	丙酮（—）	0.5～40μg/kg	离线	[44，45]
PHWE-HS-SPME	PAHs	海洋沉积物	GC-MS	甲醇（—）	0.4～15μg/kg	离线	[46]
PHWE-HS-SPME	OCPs	海洋沉积物	GC-MS	甲醇（—）	0.11～16μg/kg	离线	[47]
PHWE-DI-SPME	PFRs	鱼	GC-FPD	乙腈（—）	0.010～0.208ng/g	离线	[48]
PLE-DI-SPME	杀虫剂	空气样品	GC-MS	乙腈（—）	2～750pg/m^3	离线	[49]
PLE-DI-SPME	OCPs 和 PCBs	空气样品	GC-ECD	乙腈（—）	0.02～4.9ng/m^3	离线	[50]
MSPD-IT-SPME	PAHs			乙腈（1.2mL）	0.05～0.6μg/kg	离线	[51]
MSPD-DI-SPME	PCPs	化妆品	GC-MS	正己烷（0.5mL）/丙酮（0.5mL）	0.03～0.55ng/mL	离线	[52]
MSPD-DI-SPME	OCPs 和 PBDEs	贻贝和蛤	GC-ECD	乙腈（3.0mL）	0.3～7μg/kg	离线	[53]
MSPD-IT-SPME	邻苯二甲酸二（2-乙基己基）酯	海洋沉积物	CapLC-DAD	乙腈（1.2mL）	90～270μg/kg	离线	[54]
MAE-HS-SPME	敌敌畏	蔬菜	GC-ECD	乙二醇（20mL）	1ug/L	在线	[55]
MAE-HS-SPME	VOCs	TCM	GC-MS	免溶剂	—②	在线	[56]
MAE-HS-SPME	VOCs	肾果小扁豆	GC-MS	免溶剂	—②	在线	[57]

MAE-HS-SPME	EOs	TCM	GC-MS	免溶剂	—②	在线	[58]
MAE-HS-SPME	EOs	TCM	GC-MS	免溶剂	—②	在线	[59]
MAE-HS-SPME	EOs	TCM	GC-MS	免溶剂	—②	在线	[60]
MD-HS-SPME	EOs	TCM		免溶剂	—②	在线	[61]
MAE-HS-SPME	OCPs	水	GC-ECD	免溶剂	0.002~0.070mg/L	在线	[62]
MAE-HS-SPME	SVOCs	河水，自来水	GC-ECD	免溶剂	0.2~10.7ng/L	在线	[63]
MAE-HS-SPME	OPEs	水	GC-MS	免溶剂	0.5 和 4ng/L	在线	[64]
MAE-HS-SPME	PAHs	污染的水	GC-MS	免溶剂	26.8~128ng/L	在线	[65，66]
UMHE-HS-SPME	EOs	TCM	GC-MS	免溶剂	—②	在线	[67]
UMHE-HS-SPME	挥发性有机化合物	烟草	GC-MS	免溶剂	—②	在线	[68]
UAE-HS-SPME	2-AAP	酒	GC-MS	免溶剂	0.01μg/L	在线	[69]
UAE-HS-SPME	挥发性有机化合物	贯叶连翘和贯叶金丝桃	GC-MS	免溶剂	—②	在线	[70]
SFE-HS-SPME	对羟基苯甲酸酯和多酚类抗氧化剂	化妆品	GC-MS	免溶剂	0.5~0.83μg/kg	在线	[71]
SFE-HS-SPME	全氟羧酸	海洋沉积物	GC-MS/MS	丁醇（500uL）	0.39~0.54μg/kg	在线	[72]

注：AAP—2-氨基苯乙酮；ACN—乙腈；CapLC—毛细管液相色谱；CPs—氯酚；DAD—二极管阵列检测器；DI—直接浸入；DMSO—二甲基亚砜；ECD—电子捕获检测；EOs—精油；FPD—火焰光度检测器；GC—气相色谱；HPLC—高效液相色谱法；HS—顶空；ICP-MS：电感耦合等离子体质谱法；MAE—微波辅助萃取；MS—质谱；MS/MS—串联质谱；MSPD—基质固相分散体；OCPs—有机氯农药；OPEs—有机磷酸酯；OTCs—有机锡化合物；PAHs—多环芳烃；PBBs—多溴联苯；PBDEs—多溴联苯醚；PCBs—多氯联苯；PCP—个人护理产品；PFRs—有机磷阻燃剂；PHWE—加压热水提取；PLE—加压液体萃取；SFE—超临界流体萃取；SPME—固相微萃取；SVOCs—半挥发性有机化合物；TCM—中药；UAE—超声辅助提取；UMHE—超声-微波混合辅助提取；UV—紫外线；VOC—挥发性有机化合物。

①检出限。

②未给出。

口沉积物）中的有机氯[21]；以及 SRM 966（级别 2 血液）中的汞形态[24]来证明离线 MAE-SPME 的准确性。微波照射时间从 4min[18,22]到 30min[23]不等，而所需的 SPME 采样时间从 20min[18,22]到 120min[13]。

（2）离线超声波-固相微萃取　据报道，离线 UAE-SPME 从污水污泥样品中提取多溴二苯醚（PBDE）时的超声处理时间为 15min[28]，从茶中提取 OCPs 时的超声处理时间为 60min[14]。顶空（HS）-SPME 采样（时间在 40～60min 范围内）[26-28]给出了污水污泥中的多溴二苯醚[28]，土壤中的乙烯菌核利和氯硝胺[26]以及禽肝中的 OCPs 的定量分析回收率[27]。然而，Cai 等报道茶叶中 OCPs 的回收率低于 40%，并显示 MAE-HS-SPME 具有获得更好的定量结果的便利性[14]。

（3）离线超临界流体-固相微萃取　HS-SPME 已用于超临界流体萃取后衍生化[29]和没有衍生化[30,31]的目标物的预富集。有文献报道从蛤样本中用甲醇改性的二氧化碳（45℃和 30MPa，15min）提取有机锡化合物，然后进行乙基化并同时吸附到聚二甲基硅氧烷（PDMS）/二乙烯基苯（DVB）纤维上（50℃，30min）[29]。该报道未提供有关该方法定量的数据。有文献报道通过 SFE-HS-SPME（对 CRM 准确进行分析测试）对水产养殖鱼类饲料和养殖海洋物种中的多溴联苯（PBB）、多溴二苯醚（PBDE）和多氯联苯（PCB）进行定量提取并进行浓缩，该方法在静态提取 5min 后进行 27min 的动态提取和离线 HS-SPME（PDMS）60min[30]。最后，SFE-HS-SPME 还用于镇江香醋中香气化合物的定性表征[31]。

（4）离线加压液体萃取（PLE）/加压热水萃取（PHWE）-固相微萃取　以水为溶剂的 PHWE 已成功用于从土壤和空气颗粒物中分离出 PAHs，从受污染的土壤中分离出烷基苯和芳香胺[32]，在土壤和沉积物中分离出了多氯联苯[33]，并从鱼类中分离出了有机磷阻燃剂[48]。在 HS-SPME[32,33]之前的 60min[32,33]和 5min[48]的静态时间，或采用直接浸入（DI）-SPME 采样（PDMS）10～40min[48]，可以得到令人满意的精度（通过分析 CRM）[33]和分析回收率[33,48]。PHWE 和 SPME 的组合被广泛用于中药的定性表征[34-37]。据报道进行 5min 加压热水萃取（PHWE）和顶空-固相微萃取（HS-SPME）富集 10min 的时间[34-37]。也使用水（pH 调节至 4.0[38,40]或 7.5[39]）作为加压液体萃取脂肪族伯胺的溶剂（预热时间为 5min，两次静态萃取为 15min）[38,40]和污水污泥中的 *N*-亚硝胺萃取的溶剂（预热时间为 6min，静态提取为 5min 的两个循环）[39]。聚丙烯酸酯（PA）纤维（HS-SPME 采样时间为 20min）用于脂肪族伯胺的预浓缩[38,40]；相反，*N*-亚硝胺被吸附到 DVB/羧基（CAR）/PDMS 纤维上需要 60min[39]。尽管评估了重复性和再现性的数据，但没有给出准确性数据[38-40]。最后，Beichert 等[41]也使用水从土壤中分离出有机汞，然后进行原位乙基化—顶空固相微萃取（HS-SPME）（10min 取样）。作者证明，加压时间为 30min 可确保有机汞形态的完整性。然而，甲基汞的回收率出乎意料的低，进一步的实验表明，不锈钢萃取池的表面是回收率低的原因之一[41]。

将 PHWE 和 SPME 结合在一起的其他方法需要使用可变数量的改性剂[42-47]。使用乙腈（ACN）作为改性剂（95：5，水/ACN 作为溶剂）加压 30min，然后使用 HS-SPME 采样 20min，发现土壤中的氯酚回收率更高（从 42%～82%）[42]。在其他情况下，使用乙醇改性的亚临界水（70∶30，水/乙醇）作为溶剂从牛肉肾脏中提取阿特拉津（预热时间为5min，静态为5min的三个循环）[43]。用碳纤维二乙烯基苯（CW/DVB）作为SPME纤维处理30min[43]，获得了良好的分析回收率（在104%~111%的范围内）[43]。

Wennrich等[44,45]提出了使用水/丙酮混合物（90∶10，体积比）和两个10min（静态时间）的循环从水果和蔬菜中分离OCPs和氯苯的方法。然而，据报道，使用PDMS/DVB纤维采样60min后，回收率较低［从p，p'-DDT的0.6%到六氯环己烷（α-HCH）的13.7%］[45]。用甲醇改性的水用于从海洋沉积物中提取PAH（90∶10，水/甲醇）[46]和OCPs（95∶5，水/甲醇）[47]。通过加压19min（预热9min，静态时间10min）可实现定量的PAHs回收率[46]。但是，提取OCPs时，较短的时间就足够了（预热时间为7min，静态时间为5min）[47]。使用PDMS/DVB纤维进行HS-SPME进行预浓缩，分别对PAH和OCP使用60和45min的采样时间[46,47]，通过分析CRM，证明了所开发方法的准确性。最后，在进行DI-SPME之前，使用大量有机溶剂（例如乙腈）提取空气样品中的农药[49]以及OCPs和PCBs[50]。在45min（3个循环，每次15min）中提取了来自大体积样品的玻璃纤维过滤器和XAD-2树脂，然后将其预浓缩到PA纤维（50℃，55min）[49]和PDMS纤维（80℃，40min）[50]。但是，未测试准确性或分析其回收率。

（5）离线基质固相分散-固相微萃取　MSPD与SPME的结合使用比直接注射MSPD提取物获得的检出限更低[51]。用C_{18}相分散（5min）并通过ACN洗脱后，通过含有弗罗里硅土相作为脂肪保留剂的SPE管，在连续搅拌下用PDMS/DVB纤维进行DI-SPME处理45min[51]。MSPD与SPME的结合也用于测定个人护理产品（PCP）并阐明化妆品防腐剂的可能光转化[52]。回收率实验显示结果在50%~130%范围内[51,52]。MSPD与管内SPME工艺相结合的进一步发展[53,54]包括用SPME毛细管柱代替传统进样阀的不锈钢进样环。然后通过1mL精密注射器将足够体积的标准溶液和提取物加载到SPME毛细管柱中（内部体积约为64μL或32μL），并通过加载50~100μL的过滤后去离子水冲洗过量的加载溶液。最后，旋转进样阀，使分析物从具有流动相的毛细管柱的萃取相中解吸出来，然后泵送到分析柱中进行分离和检测[53,54]。在使用C_{18}作为分散剂，使用Florisil作为净化相并使用乙腈进行洗脱，该方法成功应用于贻贝中PAHs的测定（分析回收率为56%~120%）[53]。另外，对于海岸沉积物中邻苯二甲酸二（2-乙基己基）酯，建议采用C_{18}进行MSPD和用乙腈洗脱（分析回收率为80%~20%）[54]。

17.3.1.2　在线SPME偶联

（1）在线MAE/UAE-SPME　还探讨了MAE-SPME在色谱分析之前一步制备原位样品的可能性（表17.2）。Chen等[55]首先提出了使用自制的MAE-SPME设备（以HS模式运行的SPME）。该系统包括一个改进的家用微波炉，该微波炉配备有冷却冷凝器，该冷凝器具有连续的自来水流以冷凝蒸汽（反复使用溶剂），以及用于密封和引导蒸汽通过SPME纤维的玻璃管。保持顶部空间尽可能小。将置于微波炉腔体外部的玻璃管直接连接到装有样品和提取溶液的烧瓶（烧瓶置于微波炉腔体内部）。发现高挥发性溶剂（例如丙酮和甲醇）（用于离线方法提取农药敌敌畏）已不合适。否则，当从蔬菜中提取农药敌敌畏并在线预浓缩到 SPME 纤维（PDMS）上时，乙二醇水溶液会有优势[55]。在 10min 内同时完成了目标物的定量微波提取和 SPME 采样（10mg/L 敌敌畏的分析回收率为 106%），这大大缩短了离线偶联所需的预处理时间。

一个非常相似但没有有机溶剂的设计被进一步用于从香椿[56]，肾果小扁豆[57]，广玉兰[58]，益智[59]，细叶茯苓[60]中提取/预浓缩挥发性化合物/精油（EO）。该方法使用 PDMS/DVB[56,57]，CAR/PDMS[58,59]和涂有聚噻吩/六角有序二氧化硅纳米复合材料（PT/SBA-15）的纤维。在这些无溶剂方法中，MAE-HE-SPME 后需要添加少量水。在 4~40min 内同时进行微波蒸馏（MD）和 SPME 预浓缩。在 Ye 和 Zheng 提出的无溶剂 MAE-HS-SPME 方法中，使用预先与样品（干燥的紫苏）混合的胺功能化磁性纳米颗粒（AMN），只需要很短的预浓缩时间（2min）[61]。胺功能化磁性纳米颗粒用作微波吸收固体介质，用于在预浓缩（HS-SPME）之前对样品中的精油成分进行干馏。但是，必须指出的是，该方法没有提供有关验证（定量回收率/准确性）的数据[56-61,65,66]。

在线 MAE-SPME 方法还用于协助从液体样品中浓缩/富集挥发性化合物（OCPs[62]，半挥发性有机化合物（SVOC）[63]，有机磷酸酯（OPE）[64]和 PAHs[65,66]）。基于 Li 等人提出的 MAE-HS-SPME 设备[62]，将样品引入烧瓶中，该烧瓶与连接到水冷凝器的 Y 型玻璃相连。尽管已为此目的使用了改进的家用微波炉，但最近的发展涉及使用聚焦微波炉从水中分离 OPEs[64]。将 SPME 纤维（DVB/CAR/PDMS）暴露于 Y 型玻璃管的顶部空间，并用 80% 功率的微波（家用微波炉）辐射 10min，同时在萃取过程中以 300r/min 的速度磁力搅拌样品[62]。作者验证了该方法避免基质效应的作用，这是与 DI-SPME 相比从水中预浓缩 OCPs 的主要优势[62]。其他建议包括使用聚焦的微波炉，该微波炉带有一个螺帽（PTFE 面隔片）小瓶，其中装有放置在微波炉腔中的样品[63]。在选定的微波辐射时间内，PDMS/DVB 纤维暴露在顶部空间中，以吸附从样品中释放的 SVOC[63]。作者得出的结论是，对于所有 OCPs 和硝基苯，MAE-HS-SPME 程序的萃取效率比常规 HS-SPME（水浸法）要高 1.47~3.91 倍[63]。这些差异归因于使用微波辅助时更有效的能量传输过程。最后，Tsao 等人也使用了聚焦微波炉[64]来提取水中磷酸三正丁酯和磷酸三（2-乙基己基）酯（表 17.2）。这种一步式 MAE-HS-SPME 在水中的应用，SVOC 的分析回收率在 94%~117%，OPEs 的分析回收率在 86%~106%[64]。大多数 OCPs（92%~110%）也获得了良好的分析回收率[62]。Feng 等[67]提出了一种在线超声-微波混合辅助提取（UMHE）系统与 HS-SPME 结合，用于从干燥中药中提取精油（EOs）。这种方法利用了超声的振动空化，微波的高能作用以及 HS-SPME 的高富集因子的优势。UMHE-HS-SPME 由改进的家用微波炉组成，带有冷却冷凝器和玻璃管，用于密封和引导蒸汽通过 SPME 纤维，类似于之前的设计[55-61]。另外，使用三颈烧瓶盛装样品和溶剂（10mL 超纯水），并在提取烧瓶内安装玻璃管，超声探头和温度传感器。相似的设计还用于提取/预浓缩烟草中的挥发性化合物[68]。微波辐射 10min 后完成 EOs 的萃取和 SPME 预浓缩，时间比使用常规蒸汽蒸馏所需的时间短（约 6h）。已开发的方法可用于识别中药中的 80 个成分，但缺少有关准确性的数据[67,68]。还发现 UMHE-HS-SPME 方法优于 MAE-HS-SPME，因为检测到更多类型的挥发性成分[68]。

还开发了几种用于色谱分析之前的在线样品制备 UAE-SPME 方法[69,70]。在 UAE-SPME 中，将 SPME 纤维通过硅橡胶隔垫盖引入样品瓶中，然后将设备置于超声浴中（仅将样品暴露于 35kHz 的超声下）。提取后，将纤维引入 GC 进样器进行化合物分析。该方法用于白葡萄酒中 2-氨基苯乙酮（2-AAP）的测定[69]。与使用 DI-SPME 相比，UAE-SPME 提取 2-AAP 的灵敏度较低。但是，UAE-SPME 显示出明显短的提取时间，并且在大约 30 次注射后纤维可能是稳定的。两种提取方法的回收率均无显著差异（74%~86%）[69]。UAE-SPME 还对贯叶连翘和贯叶金丝桃的主要成分进行了鉴定[70]。将样品瓶放入超声仪中，在顶部空间和样品之间达到平衡后，将 SPME 纤维固定器夹在小瓶上方，并将纤维暴露于顶部空间。

（2）在线超临界流体萃取-固相微萃取　与此同时也出现了在线 SFE-SPME 的结合（表 17.2）。来自 SFE 设备的限流器和 SPME 纤维通过恒温小瓶的特富龙隔膜固定。在评估化妆品中对羟基苯甲酸酯时，该方法涉及在样品瓶中直接添加衍生试剂（将 SPME 纤维放置在样品瓶的顶部空间中），限流器和玻璃样品瓶底部之间的端到端距离约为 1 cm[71]。动态提取 15min 后将 SPME 纤维刺穿隔片，到达小瓶内部。该方法的主要优点是目标物衍生化，并且在化合物从超临界流体装置中洗脱的同时会吸附在纤维上[71]。因此，通过使用离线样品预处理，提取时间比其他文献报道的时间短。另外，该方法需要较少量的样品[71]。当分析高浓度（500 和 1000ng/g）加标的乳膏样品时，定量结果得到了报道。但是，在低浓度加标的乳膏样品中，大多数对羟基苯甲酸酯回收率过高（分析回收率在 125%~172%）。

出现了一种类似的装置来测定沉积物中的全氟羧酸（PFCA）[72]，在 SFE 过程中用正丁醇进行了衍生化，并且全氟羧酸丁酯进一步吸附到了 PDMS 纤维（HS-SPME）上。分析回收率范围为 92%~96%（加标实验浓度为 100ng/g）；低浓度（10ng/g）加标时回收率从 85%~88%[72]。

17.3.2 搅拌棒吸附萃取

SBSE 可以视为 SPME 的搅拌器变体，因此，其机理和优点与 SPME 相似。在吸附提取中，将分析物从基质中提取到位于搅拌棒（长度在 1~4cm 范围内）中的相对较厚的一层不可混溶液相（通常为 PDMS）中（0.1~1.0mm）。SBSE 与其他样品预处理（例如，UAE[73]，MAE[74,75] 和 PLE/PHWE[44,45,76-79]）的偶联已在离线模式下进行（表 17.3）。这些欧偶联的主要优点是避免了清理步骤和溶剂（进行 GC 分析时，分析物可以从搅拌棒中热脱附[73,75-77]）。PLE-SBSE 和 PLE-SPME 之间的比较在评估水果和蔬菜中的 OCPs 和氯苯时，可重复性更好，检出限更低[44,45]，并提高了分析回收率[45]。

据报道，即使使用少量样品，使用 PLE-SBSE 仍具有很高的富集能力（测定大气颗粒物中的 SVOC）[74]。使用超声协同萃取（UAE）从城市灰尘中提取 PAHs，然后进行 SBSE，可以使得净化和目标物富集一步完成，这与 USEPA 相比，具有吸引力，适用于 13-A 方法[73]。同样，发现 MAE-SBSE 有利于从软木塞中提取/富集 2，4，6-三氯

表 17.3　各种 SBSE 偶联程序的实验条件的简要概述

偶联萃取/富集技术	目标物	样品基质	检测仪器	溶剂（体积）	LOD①	偶联方式	参考文献
PHWE-SBSE	OCPs 和氯苯	水果和蔬菜	GC-MS	丙酮（—②）	0.5~30μg/kg	离线	[44，45]
UAE-SBSE	PAHs	城市尘埃	GC-MS	丙酮（16mL），正己烷（24mL）	1.73~31.9ng/L	离线	[73]
MAE-SBSE	SVOVs	大气颗粒物	GC-MS	丙酮（24mL）	0.3~8.3pg/m^3	离线	[74]
MAE-SBSE	TCA	软木塞	GC-MS	乙醇（5mL）	0.5ng/L	离线	[75]
PHWE-SBSE	OCPs 和氯苯	土壤	GC-MS	乙腈（—②）	2~1428ng/kg	离线	[76]
PLE-SBSE	PCBs，PAHs，PBDEs，OPPs，OCPs	海洋沉积物	GC-MS/MS	甲醇（—②）	0.00~0.300μg/kg	离线	[77]
PHWE-SBSE	对羟基苯甲酸酯	室内尘埃	GC-MS	免溶剂	1.0~2.1μg/kg	离线	[78]
PHWE-SBSE	PPCPs	海洋沉积物	GC-MS	甲醇（—②）和乙酸乙酯（200μL）	0.01~5μg/kg	离线	[79]
SPE-EH-SBSE	糖苷键合的挥发性前体	黑莓汁	GC-MS	DCM（20mL），甲醇（6mL）	—②	离线	[80]

注：ACN—乙腈；DCM—二氯甲烷；EH—酶促水解；GC—气相色谱；MAE—微波辅助萃取；MS—质谱；MS/MS—串联质谱；OCPs—有机氯农药；OPPs—有机磷农药；PAHs—多环芳烃；PBDEs—多溴联苯醚；PCBs—多氯联苯；PHWE—加压热水提取；PLE—加压液体萃取；PPCPs—药品和个人护理产品；SBSE—搅拌棒吸附萃取；SPE—固相萃取；SVOCs—半挥发性有机化合物；TCA—2，4，6-三氯茴香醚；UAE—超声辅助提取。

①检出限。

②未给出。

茴香醚[75]。紧随 SBSE 之后的 PLE（水作为溶剂，PHWE）也用于测定土壤中的 OCPs 和氯苯（CB）[76]，屋尘中的对羟基苯甲酸酯[78]，以及沉积物样品中的药物和 PCP[79]。此外，基于 PLE（甲醇作为溶剂）-SBSE 的方法学也被用于测定海洋沉积物中的 86 种持久性有机污染物[77]。最后，在文献中也发现了有关使用 SPE 从黑莓汁中预浓缩糖苷，然后通过酶促水解以及进行 SBSE 分离释放的糖苷键合挥发物[80]。从搅拌棒进行的分析物热脱附通常用于 GC 分析[44,45,73-78,80]，尽管也有人提出了在超声（US）辅助下用少量乙酸乙酯解吸附的方法[79]。

尽管对大气颗粒物中的几种 SVOC 进行了足够的准确度分析（SRM 1649a 分析，城市尘埃）[74]，但仍需要进一步的工作来提高 SBSE 程序的准确性。现有数据显示，从土壤（在 4%~60%范围内）提取 OCPs 和 CBs 时，分析回收率较低[76]，海洋沉积物中的持久性有机污染物（<80%）[77]，房屋粉尘中的对羟基苯甲酸酯（从 40%~80%）[78]以及沉积物样品中的药品和五氯苯酚（PPCP）（大多数目标低于 60%，邻苯二甲酸二丁酯和三氯甲基甲烷过高[79]）。但是，MAE/UAE-SBSE 偶联无法使用回收率数据[73,75]。

17.4 液相微萃取技术的偶联

LPME 涵盖了一组简单、小型化、廉价且环保的样品预处理技术。如前所述，这些方法可以分为 DLLME 和 SDME。

17.4.1 分散液液微萃取

在 DLLME 中，样品溶液中的目标物迅速转移到萃取溶剂的细小液滴中。离心后，回收并分析锥形管底部的沉淀相。

分析液体样品时，大多数偶联技术都会在 DLLME 之前使用 SPE 进行目标物预富集和消除干扰物[81-116]。

当前的趋势涉及基于 DLLME 和 D-μ-SPE[117-126]或 DLLME 和 SBSE[127]的两步提取技术。在通过 UAE[110,111]或 MAE[115]提取分析物之后，还将 SPE 与 DLLME 结合使用也适用于固体材料。基于 DLLME 和 D-μ-SPE 的萃取技术使用 MNP（磁性纳米颗粒，Fe_3O_4）[118,120,123,125,126]，壳聚糖-氧化锌 NP[124]，3-氯丙基三乙氧基硅烷改性的磁性纳米颗粒[117]，正辛基三乙氧基硅烷（C_8MNPs）改性的磁性纳米颗粒[119]，S-BaFe 改性的磁性纳米颗粒[121]，聚多巴胺改性的磁性纳米颗粒（Fe-PD/Ag）[122]和油酸改性的磁性纳米颗粒[124]（表 17.4）。可以对 MNP 进行表面修饰，使其具有特定的表面特性，从而可以潜在地用于云层形成后回收各种萃取溶剂。MNP 的使用可以扩展 DLLME 的适用性，因为它允许使用更广泛的溶剂，包括密度小于 1.0g/mL 的溶剂，这些溶剂不适合获得稳定的云层。因此，在这些进展中，已报道了诸如 1-辛醇、1-十一烷醇、1-十二烷和 1-十二烷之类的萃取溶剂[106,117,119]。MNP 的使用避免了与传统 DLLME 相关的烦琐步骤（离心和冷冻以冻结溶剂，手动除去有机溶剂），因为分离是通过施加外部磁场实现的。

表 17.4　各种 DLLME 偶联程序的实验条件的简要概述

偶联萃取/富集技术	目标物	样品基质	检测仪器	溶剂（体积）	LOD①	EF②	偶联方式	参考文献
SPE-DLLME	CPs	井水，自来水，河水	GC-ECD	丙酮（1.0mL），CIBZ（13μL），Ac_2O（50μL）	0.5~100ng/L	4390~1787	离线	[81]
SPE-DLLME	杀菌剂：苯甲酰胺，邻苯二甲酰亚胺，唑类，嗜球果伞素	白酒和红酒	GC-ECD，GC-MS	丙酮（1mL），TCE（100μL）	20~250ng/L③	156~254	离线	[82]
SPE-DLLME	PBDEs	井水，河水，海水，渗滤液和三叶草提取物	GC-ECD	正己烷（2.0mL），乙腈（1.0mL），1，1，2，2-TeCA（22μL）	水样 0.03~0.15ng/L；三叶草提取物 0.05~0.16μg/kg	7169~9405	离线	[83]
SPE-US-DLLME	瘦肉精	猪组织提取物	HPLC-UV	甲醇（4.75mL），1，1，2，2-TeCA（22μL）	0.07μg/kg	62	离线	[84]
SPE-DLLME	OPPs	自来水，井水，灌溉水	GC-MS	丙酮（2mL），TeCM（15μL）	38~230pg/L	~2000~~5000	离线	[85]
SPE-DLLME	PBDE	牛奶提取物	GC-MS	正己烷（2mL），丙酮（2mL）	0.01~0.4μg/L	—②	离线	[86]
SPE-DLLME	MNT	井水，河水，海水	GC-FID	乙腈（1.0mL），PCE（11μL）	0.2μg/L	4247~4570	离线	[87]
MIP-SPE-DLLME	NNAL	人类头发提取物	LC-MS/MS	甲醇（0.3mL），DCM（2.7mL）	0.08ng/g	20	离线	[88]
SPE-DLLME	CBZ	尿液和血浆	HPLC-UV	乙腈（1.5mL），氯仿（60μL），甲醇（30μL）	0.8~1.7μg/L	105~132	离线	[89]
SPE-DLLME	拟除虫菊酯类农药	河水	GC-ECD	丙酮（2.4mL），正己烷（1.6mL），甲醇（0.5mL），PCE（40μL）	0.48~3.81ng/L	1545~2138	离线	[90]
SPE-DLLME	乙酚	红酒	GC-MS	丙酮（2.5mL），AC_2O（50μL），TCE（60μL）	0.3~0.8μg/L③	67~78	离线	[91]
SPE-DLLME	OPPs	井水和农场水	GC-FPD	丙酮（1mL），CIBZ（12μL）	0.2~1.5μg/L	15160~21000	离线	[92]

SPE -DLLME -SFO	PCs	井水，河水，海水和废水	GC-MS	甲醇（2mL），1-十一烷醇（9.0μL）	2~40ng/L	752~3135	离线	[93]
MIP-SPE-DLLME	MAMP，MDMA	尿液	GC-FID	甲醇（1mL），氯甲酸丁酯（30μL）	2~18ng/L	427~285	离线	[94]
SPE-DLLME	氨基甲酸酯农药	自来水	HPLC-UV	乙腈（1mL），三氯甲烷（35μL），甲醇（25μL）	1~10ng/L	5408~7647	离线	[95]
MIP-SPE-DLLME	t，t-MA	尿液	GC-MS	乙醇（4.2mL），吡啶（100μL），ECF（100μL），TCE（80μL）	37μg/L	—④	离线	[96]
MIP-SPE-DLLME	LOT	水、尿液、血浆	HPLC-PDA	甲醇（5mL），TFA（50μL），1-癸醇（150μL）	0.2~0.6μg/L	24~30	离线	[97]
SPE-DLLME	新烟碱类杀虫剂	蜂蜜	LC-DAD，LC-MS/MS	乙腈（1.5mL），氯仿（100μL）	0.02~1.0μg/kg	—④	离线	[98]
SPE-DLLME	叠氮离子	自来水，井水，矿泉水	UV-vis	丙酮（400μL），1，2-二氯甲烷（100μL），乙醇（400μL）	0.05μg/L	250	离线	[99]
SPE-DLLME	苯甲醛	注射用制剂溶液（NaDCF，复合B族维生素和扶他林）	GC-FID	乙腈（1.5mL），1，2-二氯甲烷（55μL）	0.08μg/L	—④	离线	[100]
SPE-DLLME	PPCPs	饮用水，海水，河水和废水	HPLC-MS/MS	甲醇（1.5mL），ACN（1.5mL），DCM（2mL）	0.08~625μg/L	75~1059	离线	[101]
SPE-DLLME	PAHs	熏培根提取物	HPLC-UV-vis	乙腈（1mL），DCM（0.2mL）	0.01~0.05μg/kg	3478~3842	离线	[102]
SPE-DLLME	BZPs	超纯水和自来水，苹果汁，无酒精啤酒和尿液	GC-FID	丙酮（2mL）	0.02~0.2μg/L	838~7222	离线	[103]
SPE -DLLME -SFO	OCPs	河水和井水	GC-ECD	丙酮（2mL），十二烷醇（10μL）	0.10~0.39μg/L	8280~2822	离线	[104]

续表

偶联萃取/富集技术	目标物	样品基质	检测仪器	溶剂（体积）	LOD①	EF②	偶联方式	参考文献
SPE-IL-DLLME	PAHs	井水，自来水，地表水和河水	HPLC-UV-vis	丙酮（1mL）	0.002~0.1μg/L	2768~5409	离线	[105]
SPE-LDS-SD-DLLME	Cd	蒸馏水，自来水，河水，湖水，海水和废水	FAAS	辛醇（50μL），乙腈（500μL），乙醇（~500μL）	0.03μg/L	165	离线	[106]
SPE-DLLME-SFO	As（Ⅲ）As（Ⅴ）	矿泉水，自来水，井水，湖水	ETAAS	丙酮（2mL），1-十一烷醇（60μL）	2.5ng/L	1520	离线	[107]
SPE -DLLME -SFO	Hg（Ⅱ）	河水和井水	ETAAS	1-十一烷醇（40μL）	9ng/L	1540	离线	[108]
SPE-DLLME	抗生素	废水，饮用水，自来水和河水	UHPLC-MS/MS	甲醇（5.6mL），DCM（800μL），乙腈（600μL）	0.08~1.67ng/mL	1763~4990	离线	[109]
UAE-SPE-DLLME	拟除虫菊酯（溴氰菊酯和苄氯菊酯）	蜂蜜	GC-MS	乙腈（10mL），甲醇（1mL），TeCM（30μL）	0.02~0.04ng/g	4925~4955	离线	[110]
UAE-SPE-DLLME	BPA	鱼，蜂蜜，乳粉，汽水，水样	HPLC-FS	乙腈（10mL），乙醇（2.5mL），丙酮（1mL），十一烷醇（30μL）	0.002ng/g	1940	离线	[111]
SPE-DLLME	农药	牛乳，蜂蜜，水果和水样	GC-MS	甲醇（1.5mL），CIBZ（20μL）	0.5~1ng/kg	2362~10593	离线	[112]
SPE-DLLME	农药酚	酒	GC-MS	甲醇（2mL），丙酮（500μL），甲苯（60）	—④	—④	离线	[113]
SPE-DLLME	BZD	人类尿液和血浆	HPLC-UV	乙醇（1.5mL），DCM（120μL）	0.4~0.7μg/L	—④	离线	[114]
MAE-SPE-DLLME	REE	湖水、河水和沉积物	ETV-ICP-MS	TeCM（15μL），乙醇（300μL）	0.003~0.073ng/L	234~566	离线	[115]

UAE-SPE-DLLME	OPPs	土壤	GC-MS	甲醇（2.5mL），乙腈（1mL），甲苯（10μL）	0.002~0.125ng/g	—④	离线	[116]
DLLME-D-μ-SPE	PAHs	河水	GC-MS	1-辛醇（20μL），乙腈（100μL）	11.7~61.4ng/L	110~186	离线	[117]
IL-US-DLLME-D-μ-SPE	BUs	湖水	HPLC-UV-vis	乙腈（300μL），乙腈（50μL）	0.05~0.15μg/L	261~302	离线	[118]
DLLME-D-μ-SPE	4-NP	饮用水和废水	HPLC-FD	1-辛醇（10μL），甲醇（1.5mL）	13.9μg/L	100	离线	[119]
IL-DLLME-D-μ-SPE	铅	湖水，废水，组织和毛发提取物	FAAS	免溶剂	0.57μg/L	160	离线	[120]
IL-US-DLLME-D-μ-SPE	拟除虫菊酯	蜂蜜	HPLC-UV-vis	乙腈（50μL）	30~50ng/L	—	离线	[121]
US-DLLME-D-μ-SPE	PAH	河水	HPLC-FD	1-辛醇（10μL），乙腈（500μL）	2.0~19.5ng/L	—④	离线	[122]
IL-US-DLLME-D-μ-SPE	Cd	湖水，废水，蔬菜和毛发提取物	FAAS	免溶剂	0.40μg/L	100	离线	[123]
DLLME-D-μ-SPE	AF	牛乳样品	FS	1-庚醇（320μL），乙腈（2mL）	13ng/L	—④	离线	[124]
DLLME-D-μ-SPE	ZEN、AFS	谷物和开心果提取物	FS	1-庚醇（310~320μL），乙腈（1~2mL）	21~25ng/L	58.5	离线	[125，126]
SBSE-DLLME-D-SFO	PAHs	自来水，湖泊，地面和井水	HPLC-UV	甲醇（5mL），1-十一烷醇（30μL）	0.0067~0.10μg/L	1630~2637	离线	[127]
SFE-DLLME	PAHs	海底沉积物	GC-FID	乙腈（1mL），CIBZ（16μL）	0.2mg/kg	165~286	离线	[128]
SFE-DLLME	OPPs	土壤和海底沉积物	GC-FID	乙腈（1mL），TeCM（17μL）	1~9μg/kg	67~144	离线	[129]
SFE-DLLME	*m*-MNT、*p*-MNT	土壤	GC-FID	甲醇（1.15mL），TeCM（20μL），乙腈（1mL）	0.12mg/kg	118~122	离线	[130]

续表

偶联萃取/富集技术	目标物	样品基质	检测仪器	溶剂（体积）	LOD①	EF②	偶联方式	参考文献
PHWE-DLLME	OH-PAHs	沉积物	GC-MS	乙腈（—④），ClBZ（100μL），丙酮（100μL）	13.9~233.4	—	离线	[131]
PLE-DLLME	Ts	化妆品	LC-DAD	TeCM（100μL），乙腈（2mL），甲醇（15μL）	0.13~0.29μg/L	—	离线	[132]
PLE-DLLME	Ts、T3s	菠菜，玉米和芒果-苹果汁	LC-FD、LC-MS	甲醇（—④），异丙醇（—④），TeCM（150μL）	0.15~1.1μg/L	—④	离线	[133]
PHWE-DLLME	EDC	沉积物	GC-MS	丙酮（—④），ClBZ（150μL）	0.006~0.639ng/g	—	离线	[134]
PLE-DLLME	碳水化合物	烟草	GC-FID	乙醇（—④），氯仿（44μL）	0.06~0.9μg/mL	—	离线	[135]
PLE-DLLME	EOs	女贞、川穹	GC-MS	乙醇（—④），氯仿（100μL），乙醇（100μL）	—④	—	离线	[136]
MAE-DLLME	NAms	肉产品	GC-MS	TeCM（20μL），甲醇（1.5mL）	0.12~0.56μg/kg	285~340	离线	[137]
MAE-DLLME	卤代苯甲醚和卤代酚	软木塞和橡木桶	GC-ECD	甲醇（35mL），TeCM（110μL），AC2O（65μL）	18~92ng/kg	—	离线	[138]
MAE-DLLME	PAHs	熏鱼	GC-MS	乙醇（6mL），四氯乙烯（100μL），丙酮（500μL）	0.017~0.48μg/kg	244~373	离线	[139]
MAE-DLLME	CPs	土壤和海底沉积物	HPLC-UV	丙酮（1mL），ClBZ（37μL）	0.5~1.2μg/kg	23~29	离线	[140]
MAE-DLLME	PAHs	熏鱼	GC-MS	乙醇（5mL），四氯乙烯（150μL），丙酮（500μL）	—④	—	离线	[141]
MAE-LDS-DLLME	EOs	TCM	GC-MS	甲苯（100μL），甲醇（500μL）	—④	—④	离线	[142]
MAE-DLLME	NAms	肉产品（香肠和萨拉米香肠）	GC-MS	甲醇（2mL），氯仿（100μL），乙醇（500μL）	0.11~0.48ng/g	126~152	离线	[143]
MAE-DLLME	PAHs	烤肉	GC-MS	乙醇（5mL），四氯乙烯（80μL），丙酮（300μL）	0.15~0.3ng/g	—	离线	[144]

MAE-IL-DLLME	活性化合物	TCM	HPLC-UV-vis	乙醇（10mL），甲醇（300μL）	0.28~0.56ng/mL	—	离线	[145]
MAE-LDS-DLLME	PAs	火鸡胸肉	HPLC-UV-vis	乙腈（300μL），1-辛醇（60μL）	0.8~1.4ng/g	—	离线	[146]
MAE-US-DLLME	兽药残留	鱼	UHPLC-MS/MS	乙腈（8.5mL），DCM（500μL）	4.54~101.3pg/kg	—	离线	[147]
MAE-US-IL-DLLME	硒	食品和饮料	ETAAS	免溶剂	12ng/L	—	离线	[148]
UAE-DLLME	OPPs	马铃薯	GC-FPD	丙酮（5mL），ClBZ（60μL）	0.1~0.5μg/kg	—	离线	[149]
UAE-US-DLLME	EOs	橄榄	GC-MS	乙腈（500μL），ClBZ（100μL）	0.2~29μg/L	—	离线	[150]
UAE-DLLME	调味料	烟草	GC-MS	乙醇（40mL），甲醇（500μL），氯仿（70μL）	40~240ng/L	140~208	离线	[151]
UAE-DLLME	挥发性化合物	绿茶，黑茶，乌龙茶和白茶	GC-MS	甲醇（1mL），氯仿（27μL）	—④	—	离线	[152]
UAE-DLLME	EOs	TCM	GC-MS	丙酮（1mL），甲苯（100μL）	—④	—	离线	[153]
UAE-DLLME	挥发性化合物	藏红花	GC-MS	甲醇（0.32mL），乙腈（0.68mL），氯仿（26μL）	6~123mg/L	3.6~41.3	离线	[154]
UAE-DLLME	挥发性化合物	普洱茶	GC-MS	甲醇（2.5mL），氯仿（25μL）	6.3~8.2ng/mL	—	离线	[155]
UAE-DLLME	AP	清洁产品	LC-MS/MS	乙腈（2mL），TeCM（100μL）	0.2~2.5ng/mL	—	离线	[156]
UAE-DLLME	PE	PCPs 和清洁产品	LC-MS/MS	乙腈（2mL），TeCM（100μL）	0.04~0.45ng/mL	—	离线	[157]

续表

偶联萃取/富集技术	目标物	样品基质	检测仪器	溶剂（体积）	LOD①	EF②	偶联方式	参考文献
UAE-DLLME	NP	土壤		甲醇（3mL），DCM（300μL）	0.4~0.8ng/g	—	离线	[158]
MIM-MSPD-US-DLLME	苏丹染料	蛋黄	HPLC-UV	丙酮（2.85mL），四氯乙烯（100μL），甲醇（49μL）	2.3~6.1μg/kg	—④	离线	[159]
MSPD-US-DLLME	拟除虫菊酯	土壤	GC-ECD	丙酮（3.0mL），四氯乙烯（50μL），正己烷（20μL）	0.45~1.53μg/kg	128~138	离线	[160]
MSPD-MIL-DLLME	三嗪类除草剂	油籽	UFLC-UV	正己烷（6.5mL），乙酸乙酯（1.5mL），乙腈（100μL）	1.20~2.72ng/g	—	离线	[161]

注：1，1，2，2-TeCA—1，1，2，2-四氯乙烷；4-NP—壬基苯酚；ACN—乙腈；AC_2O—乙酸酐；AF—黄曲霉毒素；APs—烷基酚；BPA—双酚 A；BU—苯甲酰脲类杀虫剂；BZPs—苯并二氮杂；CBZ—卡马西平；ClBz—氯苯；CP—氯酚；D-μ-SPE—分散微固相萃取；DAD—二极管阵列检测器；DCM—二氯甲烷；DLLME—分散液-液微萃取；ECD—电子捕获检测器；EDC—破坏内分泌的化学物质；EOs—精油；ETAAS—电热原子吸收光谱法；ETV—电热汽化；FAAS—火焰原子吸收分光光度法；FD—荧光检测；FID—火焰离子检测器；FPD—火焰光度检测器；FS—分子荧光光谱法；GC—气相色谱；HPLC—高压液相色谱；ICP-MS—电感耦合等离子体质谱法；IL—离子液体；LC—液相色谱；LDS—低密度溶剂；LOT—氯雷他定；MAE—微波辅助萃取；MAMP—甲基苯丙胺；MDMA—3，4-亚甲二氧基甲基苯丙胺；MIL—磁性离子液体；MIM—分子印迹；微球 MIP—分子印迹聚合物；MNT——硝基甲苯；MS—质谱；MS/MS—串联质谱；MSPD—基质固相分散体；NAms—N-亚硝胺；NNAL—4-（甲基亚硝氨基）-1-（3-吡啶基）-1-丁醇；NP—硝基酚；OCPs—有机氯农药；OH-PAHs—羟基化多环芳烃；OPPs—有机磷农药；PAs—多胺；PAHs—多环芳烃；PBDEs—多溴联苯醚；PCE—四氯乙烯；PDA—光电二极管阵列；PEs—邻苯二甲酸酯；PHWE—加压热水提取；PLE—加压液体萃取；PPCP—药品和个人护理产品；RREs—稀土元素；SBSE—搅拌棒吸附萃取；SD—基于溶剂的反乳化；SFE—超临界流体萃取；SFO—固化悬浮的有机液滴；SPE—固相萃取；TCE—1，1，1-三氯乙烷；TCM—中药；TeCM—四氯甲烷；TFA—三氟乙酸；Ts—生育酚；T3s—生育三烯酚；UAE—超声辅助提取；UFLC—超快速液相色谱；UHPLC—超高压液相色谱；US—超声；UV—紫外线；vis—可见光；ZEN—玉米赤霉烯酮。

①检出限。

②富集因子。

③定量限。

④未给出。

其中一些与 SPE 的偶联涉及 MIP[88,93,96,97]或多壁 CNT（MWCNT）[99]，以实现高选择性的方法，以提高灵敏度并改善富集和净化效果。据报道有应用该方法测定 4-（甲基亚硝胺基）-1-（3-吡啶基）-1-丁醇，甲基苯丙胺，3，4-亚甲基二氧基甲基苯丙胺，叔丁康酸，氯雷他定，邻苯二甲酸酯和三环类抗抑郁药的方法，具有良好的分析回收率[117-125]。另外，对于镉的测定，其准确性（CRM 分析）也是令人满意的[123]。

DLLME 通常与 SFE[128-130]，PLE/PHWE[131-136]，MAE[137-148]，水浴超声提取[149-155]，超声探针萃取[156-158]相结合。在这些方法中，DLLME 用于从固体样品中提取目标物后进行净化和目标物富集（表 17.4）。这些组合的主要缺点是完成分析物的提取，净化和富集需要很长时间。

通过 SFE 提取的分析物通常在动态提取期间收集在置于冰浴中的小瓶中，以提高 DLLME 之前的收集效率[128-130]。使用两级 PHWE/PLE-DLLME[131-133]的优势在于，PHWE 的有机改性剂[131,134,135]和 PLE 提取的有机馏分[132,133]可以充当 DLLME 的分散剂，因此这种组合更直接。据报道，在土壤和沉积物中提取和预浓缩多环芳烃[128]，有机磷农药（OPPs）[129]和硝基甲苯[130]时，两步 SFE-DLLME 方法具有良好的准确性。同样，PHWE/PLE 和 DLLME 也显示出很好的准确度（婴儿/成人营养配方 SRM 1849a 的分析，全脂乳粉 ERM-BD600 得出的结果表明，α-生育酚与经认证的含量非常吻合[132,133]）。另外，据报道烟草样品中的生育酚[132,133]和碳水化合物[135]回收率良好。相反，当从海洋沉积物中分离出一些羟基化的 PAH（OH-PAH）时，回收率较低（从 57.6%~91.1%）[131]。PLE-DLLME 还是一种可用于破坏沉积物中内分泌化学物质的方法，其回收率从 42.3%（二烯雌酚）~131.3%（4，5a-二氢睾酮）不等。但是，己烯雌酚（15.0%）和壬基酚（29.8%）的分析回收率较低[134]。

因为用于 MAE 的大多数萃取剂是水溶液[137-142,146]，所以执行 DLLME 阶段需要将分散剂溶剂与有机萃取剂一起使用。类似地，在某些使用水[150]和乙醇[151]作为萃取剂的 UAE-DLLME 开发中，还需要在 DLLME 中添加分散剂。否则，在 DLLME 步骤中，用于 MAE[138]和 UAE[149,152-158]的萃取剂将充当分散剂。MAE 与 DLLME 结合使用均有良好的准确性。在分析肉类产品中的亚硝胺（回收率> 83.9%）[137,143]，软木塞和橡木桶中卤代苯甲醚和卤代酚（回收率> 86.0%）[138]，熏制鱼和鱼类中的多环芳烃（回收率从 82.1%~105.5%）[139,144]，土壤和海洋沉积物中氯酚（CP）（分析物的回收率和富集系数分别在 68.0%~82.0%和 25~30）[140]，中成药的活性化合物（回收率从 80.1%~92.9%）[145]，火鸡胸肉中的多胺（回收率从 95%~105%）[146]和鱼类中兽药残留（回收率> 87%）[147]时回收率均较好。通过分析 SRM 1573a 番茄叶片证实了 MAE 和 DLLME 结合提取食品和饮料中硒的准确性[148]。但是，其他一些 MAE-DLLME 应用程序尚未通过回收率研究或 CRM 分析进行验证[141,142]。当从番茄中提取 OPP[149]，从烟草中提取调味化合物[151]，从清洁产品中提取烷基酚[156]，从 PCP 和清洁产品中提取 PE[157]，从普洱茶中提取挥发性化合物时[155]和土壤中的硝基酚（NP）[158]，UAE-DLLME 准确性也很高。但是，未验证 UAE-DLLME 方法提取植物[150,154]，鲜花[153]和茶（绿色、黑色、乌龙茶和白色品种）[152]中的 EO 和挥发性成分的准确性。

一些报道涉及 MSPD 和 DLLME 的偶联[159,160]。但是，已报道的应用程序显示

MSPD-DLLME 偶联并不简单，整个过程涉及许多其他步骤（主要是用于净化的 SPE）用于制备 DLLME 所用的 MSPD 提取物。尽管 MSPD-DLLME 偶联有多步操作，但是在测定蛋黄中的苏丹红时，据报道回收率良好（三个加标浓度的回收率在 87.2%~103.5%范围内）[159]；在测定土壤中拟除虫菊酯时（平均回收率从 83.6%~98.5%）[160]，在测定油料种子中的三嗪类除草剂时（平均回收率从 82.9%~113.7%）[161]。

必须提及的是，某些报道的 DLLME 组合在 SPE[105]，D-μ-SPE[118,120]，UAE-SPE[121,123]，MAE-SPE[148] 和 SMPD-SPE[161] 之后使用离子液体（IL）进行分散萃取（表 17.4）。IL 表现出优异的性能（低挥发性，以及良好的化学和热稳定性），并且已逐渐取代 DLLME 中的经典有毒有机溶剂[105,118,120,121,123,148,161]。

最后，Ciucanu 等[162,163]已将 SPME 与非分散 LPME（称为基于液滴的液-液微萃取）结合使用，用于萃取水中样品中的挥发性有机化合物（VOC）和二氯吡啶酸。在这种组合中，通过固定的 PDMS 纤维在适当的有机溶剂液滴的辅助下，从水性样品中准确地提取了分析物（分析回收率在 94%~101.2%范围内）（表 17.4）。在搅拌下，有机溶剂在水性基质中变成液滴，并同时开始用有机溶剂和纤维进行微萃取。作者得出结论，添加分散剂（DLLME-SPME）会降低极性分析物在萃取溶剂中的溶解度，并降低萃取溶剂与 PDMS 纤维之间的亲和力[163]。

17.4.2 单滴微萃取（SDME）

SDME 是使用与水不混溶剂作为微滴暴露于水溶液样品中（直接浸入式 SDME 或连续流动微萃取模式）。在处理挥发性或半挥发性分析物时，可以通过将液滴暴露于样品上方的顶部空间来实现浓缩。这种情况通常称为顶空单滴微萃取（HS-SDME）。SDME 并非可以萃取所有物质，只有一小部分分析物被提取/预浓缩。文献中很少有单滴微萃取的偶联技术（离线和在线偶联）。即 SDME 与微波[164-168]，US[169-171]，SPE[172,173]，PHWE[174,175] 和 DLLME[176] 的偶联（表 17.5）。

17.4.2.1 离线 SDME 偶联

HS-SDME 模式[164,165,172-175]的应用大多都是与 MAE 偶联，如结合 MAE 从中药中分离丹皮酚[164]，以及从大豆异黄酮膳食补充剂中提取掺入的雌激素[165]。此外，HS-SDME 还与 SPE 结合用于从水样中分离 CP[172] 和 VOC（苯、甲苯、乙基苯、间二甲苯和邻二甲苯）[173]，与 PHWE 结合用于从中药中分离人参三醇和 EO[174,175]。在其他情况下，还有直接浸入模式与 UAE[169] 和 DLLME[176] 结合使用，以从鱼样品[169]中提取 OCP 和水中提取 SA[176]。在 DLLME-SDME 偶联中，在形成浊点（添加 200μL 的 1-辛醇和 750μL 的甲醇）之后，注入 600μL 的 ACN（破乳剂）以分解乳液。将体积为 3μL 的受体溶液（0.1mol/L NaOH）向前推至浸入有机相上层的微针的末端[176]。在多个环境样品中获得了良好的分析回收率（88%[164]，70%[165]，82.1%~95.3%[169] 和 90%~113%）[172,173,176]。

17.4.2.2 在线 SDME 的偶联

尽管 SDME 方法的操作模式不连续，但已经开发了一些在线方法。即 SDME 与微波[166-168]和超声（US）[170,171]的组合（表 17.5）。MAE/UAE-SDME 具有实际优势，因为目标物提取，纯化和富集可以一步进行。

MAE（正式为 MD）和同步 HS-SDME 用于从中药中分离 EO[166-168]。这些方法包括使用一个装有样品的玻璃瓶[166,168]，或者将样品和 MNP 的混合物[167]放入微波炉中。在后一种情况下，MNP（Fe_2O_3）增强了样品对微波的吸收。玻璃瓶连接到外部冷却系统（放置在烤箱中的冷凝器），该系统可降低微注射器的针尖（放置在冷凝器中）的温度。在建议的方法中，将一滴有机溶剂悬浮在微注射器的针尖上，通过微波加热在短时间内将样品中的挥发性成分转移到 HS 中，同时将 HS 中的分析物浓缩到悬浮的微滴溶剂中。

同样，UAE 与 HS-SDME 在线联用，可从中药中提取 EO[170]，而 DI-SDME 可从植物油中提取 Cd[171]。在 HS-SDME 方法中，将与样品混合的水（溶剂）放入萃取容器中，并使微注射器针头穿过塞子并固定在固定位置[170]。将超声设备切换到运行状态后（> 17MHz）后，将分析物从固体样品中提取到溶剂中，然后转移到顶部空间。关闭超声仪时，将 2μL 的庚烷液滴悬浮在微注射器的针尖上，并使分析物与悬浮的液滴相互作用[170]。在 DI-SDME 方法中，将样品（放入玻璃烧杯中）浸入超声波水浴中，将注射器（装有 0.1mol/L HNO_3）夹在烧杯上方，并将针头浸入样品中[171]。按下柱塞，在大体积样品中形成 5μL 液滴的萃取液。超声处理（15min）后，将液滴吸入注射器并直接注入石墨平台[171]。

这些在线方法的优势主要是完成分析物提取、净化和富集所需的时间很短。从植物油中提取 Cd 可获得良好的分析回收率[171]。但是，这些程序的准确性并未针对从中药中提取 EO 进行测试[166-168,170]。

17.5　膜微萃取技术的偶联

膜微萃取技术基于聚合物膜（无孔疏水材料）的使用，可将样品与萃取剂分离。微孔膜中的孔充满了有机溶剂。然而，分析物必须在扩散至无孔膜中的受体相之前扩散通过膜。使用的最典型配置是 HF-LPME，溶剂棒微萃取（SBME），SLM 萃取，微孔膜液-液萃取（MMLLE），膜辅助溶剂萃取，膜保护的微型 SPE（MP-μ-SPE）和电膜萃取（EME）。

很少有将 HF-LPME 与其他技术结合使用的例子。例如 HF-LPME 与 PHWE 结合[177,178]可以从沉积物和土壤中提取 PAH，从污水污泥中提取非甾体类消炎药。UAE[179,180]用于从鱼类组织和土壤中提取 OPP 和苯并咪唑农药；MAE[181-184]用于提取水样品中的 OCP，PCB，CP，HCH 和二氯二苯基三氯乙烷（DDT）及其代谢物[181,183,184]（表 17.6）。这些方法的优势主要是在通过 PHWE，UAE 和 MAE 分离靶标物后，HF-LPME 固有的高富集因子和选择性（纯化和靶标物富集）。但是，与 SPME 相比，主要缺点是每次提取需要使用新鲜的纤维。尽管有报道使用 PHWE 和 UAE 时的过程可以在脱机模式下完成（两阶段提取过程[178-180]），但已经有人进行了一些尝试来执行一步 MAE-HF-LPME 提取过程[183,184]。该设计意味着使用外部冷却系统，在该系统中，将安装在微型注射器针尖上的中空纤维段放置在了微波炉外部。尖端连接到装有样品和

表 17.5　各种 SDME 偶联技术的实验条件的简要概述

偶联萃取/富集技术	目标物	样品基质	检测仪器	溶剂（体积）	LOD①	EF②	偶联方式	参考文献
基于液滴的 DI-SPME	VOCs	自来水，废水	GC-FID、GC-MS	DCM（200μL）	0.02~0.65μg/L	7.1~32.4	二氯吡啶酸在线	[162]
基于液滴的 DI-SPME	二氯吡啶酸	地下水，河水	GC-MS	DCM（200μL）	0.02μg/L	200	在线	[163]
MAE-HS-SDME	丹皮酚	TCM	GC-MS	1-辛醇（1.0μL）	2μg/g	—	离线	[164]
MAE-HS-SDME	掺雌激素	饮食大豆异黄酮	HPLC-UV-vis	甲醇（20mL）1-辛醇（6.0μL）	1.1~1.2μg/L	247~335	离线	[165]
MAE-HS-SDME	EOs	TCM	GC-MS	十二烷（2μL）	—③	—	在线	[166]
MNP-MA-HS-SDME	EOs	TCM	GC-MS	十二烷（2μL）	—③	—	在线	[167]
MA-HS-SDME	EOs	TCM	GC-MS	十七烷（2μL）	—③	—	在线	[168]
UAE-DI-SDME	OCPs	鱼	GC-MS	甲醇（1mL），甲苯（0.6μL）	0.5ng/g	—	离线	[169]
UNE-HS-SDME	EOs	TCM	GC-MS	十七烷（2.0μL）	—③	—	在线	[170]
UAE-DI-SDME	Cd	菜油	HR-CS-ETAAS	免溶剂	2ng/kg	12	在线	[171]
SPE-DI-SDME	CPs	自来水，河水，湖水	HPLC-DAD	乙酸乙酯（200μL），1-丁醇（10μL）	0.04~0.08μg/L	—③	离线	[172]
SPE-HS-SDME	VOCs	河水和稻田水	GC-MS	甲醇（1mL），1-辛醇（2μL）	0.03~0.1μg/L	118~504	离线	[173]
PHWE-HS-SDME	人参三醇	TCM	GC-MS	环己烷（2.0μL）	2μg/g	—	离线	[174]
PHWE-HS-SDME	EOs	TCM	GC-MS	环己烷（2.0μL）	—③	—	离线	[175]
DLLME-DI-SDME	SAs	湖泊，渔业和废水	HPLC-UV-vis	1-辛醇（200μL），甲醇（750μL），ACN（600μL）	0.22~1.92μg/L	6~91	离线	[176]

注：ACN—乙腈；CPs—氯酚；DAD—二极管阵列检测器；DCM—二氯甲烷；DI—直接浸入；DLLME—分散液液微萃取；EOs—精油；FID—火焰离子检测器；GC—气相色谱；HPLC—高压液相色谱法；HR-CS-ETAAS—高分辨率连续谱源电热原子吸收光谱法；HS—顶空；MA—微波辅助；MAE—微波辅助萃取；MNP—磁性纳米粒子；MS—质谱；OCPs—有机氯农药；PHWE—加压热水提取；SAs—磺酰胺；SDME—单滴微萃取；SPE—固相萃取；SPME—固相微萃取；TCM—中药；UAE—超声辅助提取；UNE—超声雾化提取；UV—紫外线；vis—可见；VOCs—挥发性有机化合物。

①检出限。

②富集因子。

③未给出。

表 17.6　各种膜微萃取（HF-LPME、SBME、MP-uSPE、MMLLME、SLM、MASE、EME）偶联技术的实验条件的简要概述

偶联萃取/富集技术	目标物	样品基质	检测仪器	溶剂（体积）	LOD①	EF②	偶联方式	参考文献
PHWE-HF-LPME	PAHs	土壤和沉积物	GC-MS	环己烷（—③）	0.11~1.22μg/g④	9~55	在线	[177]
PHWE-HF-LPME	NSAIDs	下水道污泥	LC-MS	免溶剂	0.4~3.7ng/g	947~1213	离线	[178]
UAE-HF-LPME	OPPs	鱼组织	GC-MS	丙酮（15mL），甲醇（400μL），二甲苯（30μL）	2~4.5ng/g	—	离线	[179]
UAE-HF-LPME	苯并咪唑类农药	土壤	HPLC-FD	甲醇（20mL），辛醇（20μL），乙腈（11μL）	0.001~6.94ng/g	—	离线	[180]
MAE-HF-LPME	OCPs，PCBs	海洋沉积物	GC-MS	辛醇（5μL）	0.07~0.7μg/kg	—	离线	[181]
MAE-HF-LPME	CPs	农田水	GC-ECD	辛醇（3μL）	0.04~0.7μg/L	75~515	在线	[182]
MAE-HF-LPME	HCHs	农田水和河水	GC-ECD	辛醇（4μL）	0.03~0.4μg/L	88~143	在线	[183]
MAE-HF-LPME	DDT 及其衍生物	农田水，河水，海水	GC-ECD	辛醇（4μL）	20~30ng/L	92~299	在线	[184]
MAE-SBME	PAHs	尿液	GC-MS	甲苯（1μL）	30~220ng/kg	—	在线	[185]
MAE-MP-μ-SPE	POPs	人类组织	GC-MS	正己烷（100μL）	2~9ng/kg	—	在线	[186]
MAE-MP-μ-SPE	PAHs	土壤	GC-FID，GC-MS	ACN（150μL）	1.7~5.7ng/kg	—	在线	[187]
MAE-MP-μ-SPE	OPPs	水果和蔬菜	GC-MS	乙酸乙酯（2.5mL），正己烷（8mL）	0.06~0.23μg/kg	—	在线	[188]
IL-US-DLLME-MP-μ-SPE	TCAs	运河水	HPLC-UV-vis	甲醇（200μL），甲醇（20μL）	0.3~1.0μg/L	17~43	离线	[189]
VA-MP-μ-SPE-LDS-DLLME	PEs	河水	GC-MS	ACN（350μL），正己烷（30μL），ACN（350μL）	6~20ng/L	—③	离线	[190]
MMLLME-SPE	OPEs	血浆	GC-NPD，GC-MS	正己烷（—③），丙酮（4.1mL）	0.06~0.9μg/kg	—③	离线	[191]
MMLLME-MIP-SPE	17β-雌二醇	废水	HPLC-UV	正己烷/乙酸乙酯 3∶2 体积比（2.5mL），ACN（3mL）	31μg/L	—③	离线	[192]

续表

偶联萃取/富集技术	目标物	样品基质	检测仪器	溶剂（体积）	LOD①	EF②	偶联方式	参考文献
PHWE-MMLLME	PAHs	海洋沉积物和土壤	GC-MS	环己烷（—③）	0.11~1.22μg/g④	9~55	在线	[177]
PHWE-MMLLME	PAHs	土壤	GC-MS	异辛烷（—③）	0.05~0.13μg/kg	18.9~82.3	在线	[193]
PHWE-MMLLME	OCPs，PCBs	葡萄	GC-MS	异辛烷（150μL）	0.3~1.8μg/kg④	24~69	在线	[194]
SLM-SPE	三嗪类除草剂	水果汁	CE	甲醇（10mL）	30μg/L	—③	离线	[195]
SLM-MIP-SPE	三嗪类除草剂	水果和蔬菜	HPLC-UV	甲苯（2.5mL），DCM（2mL）	22~38μg/kg	40~63	离线	[196]
SLM-SPE	赭曲霉素 A	红酒	CE	甲醇（100μL）	30μg/L	200	离线	[197]
MASE-MIP-SPE	三嗪类除草剂	豆，小玉米和植物提取物	HPLC-UV	甲苯（1mL），甲醇（3.5ml）	30μg/kg	—	离线	[198]
EME-LDS-US-DLLME	CPs	污水	GC-MS	甲苯（30μL）	5~20ng/L	1450~2158	离线	[199]
EME-DLLME	吡啶衍生物	尿液	GC-MS	氯仿（20μL），甲醇（50μL）	0.25~2ng/mL	40~202	离线	[200]
SPE-EME	氯苯氧酸类除草剂	河水，海水	CE	甲醇（3.68mL），1-辛醇（20μL）	0.3~0.5μg/L	1950~2000	离线	[201]
MAE-EME	高氯酸盐	海产食品	LC-CD	正己烷（—③）	0.04μg/g	—	离线	[202]

注：ACN—乙腈；CD—电导率检测仪；CE—毛细管电泳；CP—氯酚；DCM—二氯甲烷；DDT—二氯二苯基三氯乙烷；DLLME—分散液微萃取；ECD—电子捕获检测器；EME—电膜萃取；FD—荧光检测器；FID—火焰离子检测器；GC—气相色谱；HCH—六氯环己烷；HF—中空纤维；HPLC—高压液相色谱法；IL—离子液体；LC—液相色谱；LDS—低密度溶剂；LPME—液相微萃取；MAE—微波辅助萃取；MASE—膜辅助溶剂萃取；MIP—分子印迹聚合物；MMLLME—微孔膜液-液萃取；MP-μ-SPE—膜保护的微固相萃取；MS—质谱；NPD—氮磷检测器；NSAIDs—非甾体类抗炎药；OCPs—有机氯农药；OPEs—有机磷酸酯；OPPs—有机磷农药；PAHs—多环芳烃；PCBs—多氯联苯；PEs—邻苯二甲酸酯；PHWE—加压热水提取；POPs—持久性有机污染物；SBME—溶剂棒微萃取；SLM—支撑液膜萃取；SPE—固相萃取；TCAs—三环抗抑郁药；UAE—超声辅助提取；US—超声；UV—紫外线；vis—可见。

①检出限。

②富集因子。

③未给出。

④定量限。

萃取剂的玻璃瓶（放入微波炉中）。外部冷却系统使采样点的温度保持恒定，从而防止了 LPME 萃取溶剂的蒸发[182]。此外，外部冷却系统还有助于在顶部空间中形成密集的分析物-水蒸气云。

据报道的大多数 MAE-HF-LPME 方法具有高度可重复性，并且还获得了良好的回收率。鱼组织中八种 OPPs（三个加标水平）的分析回收率在 73.8%~101.8%[179]。在森林土壤中，苯并咪唑农药的回收率在 85%~117%范围内，在火山土壤中为 86%~115%[180]。据报道，OCPs[181]，PCBs[181]，CPs[182]，HCH[183] 和 DDT[184] 的回收率也很高。

MAE 已经与其他基于膜的微萃取技术相结合，例如 SBME[185] 和 MP-μ-SPE[186-188]（表 17.6）。这些组合比 MAE-HF-LPME 的开发更具优势，因为目标萃取（MAE）和目标富集（SBME 和 MP-μ-SPE）同时发生（含有萃取溶剂或固体吸附剂的中空纤维与样品和微波萃取溶剂一起在微波容器内）。实际上，MAE-MP-μ-SPE 优于 MAE-SBME，因为使用合适的溶剂可以轻松地将分析物从微固体材料中解吸，而在 SBME 方法中，中空膜的一端必须用尖锐的刀头切掉，将富含分析物的受体溶剂小心地抽入微注射器中。如表 17.6 所示，已经报道了 MAE-SBME 结合用于从土壤中定量提取 PAHs（富含分析物的溶剂可直接进行分析，而无须额外的净化过程）[185]。关于 MAE-MP-μ-SPE 的应用，集中于从沉积物和土壤中提取多环芳烃（石墨纤维作为微固体吸附剂）[187]，从水果和蔬菜中提取 OPPs（HayeSep A 和 C_{18}）[188] 和从人体组织中提取持久性有机污染物（POPs）（活性炭）[186]。据报道，人体组织中的 POPs[186] 以及水果和蔬菜中的 OPPs 的回收率高（68.2%~95.3%）[93.5%~104.6%][188]。MP-μ-SPE 还与 DLLME 结合使用，以增强环境水中抗抑郁药和 PE 的富集因子[189,190]。在这些方法中，首先通过 MP-μ-SPE［使用咪唑啉沸石骨架 4（5mg）[189] 和 MWCNT[190]（4mg）］把分析物萃取和预浓缩在多孔聚丙烯膜袋中，然后通过超声解吸到甲醇[189] 或乙腈[190] 中，然后采用 DLLME 进行进一步预浓缩。抗抑郁药的分析回收率在 94.3%~114.7%[189] 范围内，PE 的分析回收率在 89.8%~104.1%范围内[190]。

MMLLE 已与其他样品预处理技术相结合，如与 SPE 结合从血浆中提取 OPE[191]，从废水中提取 17β-雌二醇[192]。并与 PHWE 一起用于从土壤，海洋沉积物和葡萄样品中提取多环芳烃和农药[177,193,194]（表 17.6）。据报道，前一个固相萃取阶段可以提高选择性，并在处理复杂的液体样品（例如血浆和废水）时可以进一步净化萃取物[191,192]。Nemulenzi 等人在评估废水中的 17β-雌二醇时[192] 发现通过使用 MIP 可以增强该方法的选择性。因为 MIPs 也形成了受体相的一部分。其他发展意味着在线 PHWE-MMLLE 组合（表 17.6）可将目标物从固体基质中分离出来，并同时进行净化和预浓缩（MMLLE 在 PHWE 之后用作捕集装置）。实验室制造的 PHWE-MMLLE-GC 装置[177,193,194] 已用于多环芳烃以及土壤，沉积物和葡萄样品中的农药测定。将来自 PHWE 设备的水（流速为 0.5~1mL/min）引导至膜单元的供体侧，并将分析物提取至膜另一侧的受体溶液（异辛烷或环己烷）中。在萃取过程中，MMLLE 中的有机受体停滞，并且堵塞了受体侧的出口管。提取完成后，将 PHWE 烤箱冷却，并关闭水流。最后，来自受体侧的提取物被洗脱到 GC 的样品定量环中[177,193,194]。在实验室中进行这种组合比较困难（仪器无法在市场上买到），而且准确度也很差（回收率低于 26%）[194]。相反，当使用 PHWE-MMLLE-GC 时，土壤和沉积物样品中 PAHs 的平均回收率良好[177]。

SLM[195-197]和 MASE[198]与 SPE 和 MIP-SPE 结合使用，可增加富集因子，降低检测限和减少大量干扰物质。并为测定食物和酒等复杂样品中的三嗪除草剂[195,196,198]和曲霉毒素 A[197]提供干净的提取物（表 17.6）。SLM 通过用 H_2SO_4（0.5mol/L）[195]，甲苯[196]和辛醇[197]浸渍氟多孔聚四氟乙烯多孔膜来构建。SLM 和 SPE 组合技术的回收率研究表明，果汁类水果中三嗪类除草剂的分析回收率低于 76%[195]。然而，莴苣、苹果果实（79%~98%）[196]和酒（90%~110%）[197]中的三嗪类化合物的分析回收率很高。

在 MASE-MIP-SPE 组合中，将 MIP 颗粒（100mg）和甲苯（1mL）放入膜袋中。提取（120min）后，将受体内容物转移到 SPE 柱中[198]。提取效率达到 60%~80%[198]。

最后，EME 还与 DLLME[199,200]，SPE[201]和 MAE[202]结合用于环境和生物样品中的 CP、吡啶衍生物、酸性除草剂和高氯酸盐的提取和净化（表 17.6）。在第一种偶联中（EME 与 DLLME 结合使用），首先通过 EME 提取目标，然后通过 DLLME 进一步浓缩分析物。由于 EME 中的膜可保护受体溶液免受干扰物质的影响，因此无须进行额外的净化[199]。在第二种偶联中，将 EME 之前采用 SPE 步骤（石墨烯作为吸附剂）预浓缩和净化[201]。在第三种偶联中，通过 SLM（1-己醇）从海鲜的酸消化液中电动提取高氯酸根离子[202]。与这种组合技术相关的缺点来自以下几个阶段：EME 萃取，然后通过微注射器收集受体溶液，调节 pH，并添加超纯水以增加用于执行第二步萃取步骤的溶液量（DLLME 和 SPE）。据报道，EME-DLLME 方法[200]以及 EME-SPE[201]和 MAE-EME[202]都具有良好的准确性。

17.6 展望

偶联的微萃取技术是吸引人的方法，可以应对复杂的材料同时进行处理。可以达到分析物的分离，预浓缩和分析物满足某些仪器技术要求的程度，从而缩短了整个样品预处理的时间。然而，改进的目的必须是开发在线偶联，以及开发用于微萃取程序的选择性吸附剂。还应努力使这些技术适应复杂的样品，主要是固体样品。这些进展将有助于使偶联微萃取技术成为常规分析的有吸引力的方法。

参考文献

[1] H. Lord, J. Pawliszyn, Microextraction of drugs, J. Chromatogr. A 902 (2000) 17-63.

[2] C. Nerín, J. Salafranca, M. Aznar, R. Batlle, Critical review on recent developments in solventless techniques for extraction of analytes, Anal. Bioanal. Chem. 393 (2009) 809-833.

[3] P. D. Zygoura, E. K. Paleologos, K. A. Riganakos, M. G. Kontominas, Determination of diethylhexyladipate and acetyltributylcitrate in aqueous extracts after cloud point extraction coupled with microwave assisted back extraction and gas chromatographic separation, J. Chromatogr. A 1093 (2005) 29-35.

[4] Y. -K. Lv, W. Zhang, M. -M. Guo, F. -F. Zhao, X. -X. Du, Centrifugal microextraction tube-cloud point extraction coupled with gas chromatography for simultaneous determination of six phthalate esters in

mineral water, Anal. Methods 7 (2015) 560–565.

[5] H. Wang, J. Ding, L. Ding, N. Ren, Analysis of sulfonamides in soil, sediment, and sludge based on dynamic microwave–assisted micellar extraction, Environ. Sci. Pollut. Res. 23 (2016) 12954–12965.

[6] D. L. Giokas, Q. Zhu, Q. Pan, A. Chisvert, Cloud point–dispersive μ–solid phase extraction of hydrophobic organic compounds onto highly hydrophobic core–shell Fe_2O_3@C magnetic nanoparticles, J. Chromatogr. A 1251 (2012) 33–39.

[7] H. Wu, H. Tian, M.–F. Chen, J.–C. You, L.–M. Du, Y.–L. Fu, Anionic surfactant micelle–mediated extraction coupled with dispersive magnetic microextraction for the determination of phthalate esters, J. Agric. Food Chem. 62 (2014) 7682–7689.

[8] N. Gao, H. Wu, Y. Chang, X. Guo, L. Zhang, L. Du, Y. Fu, Mixed micelle cloud point–magnetic dispersive μ–solid phase extraction of doxazosin and alfuzosin, Spectrochim. Acta A 134 (2015) 10–16.

[9] I. López–García, Y. Vicente–Martínez, M. Hernández–Córdoba, Cloud point extraction assisted by silver nanoparticles for the determination of traces of cadmium using electrothermal atomic absorption spectrometry, J. Anal. At. Spectrom. 30 (2015) 375–380.

[10] S. Nekouei, F. Nekouei, H. I. Ulusoy, H. Noorizadeh, Simultaneous application of cloud point and solid–phase extraction for determination of Fe (Ⅲ) and Cu (Ⅱ) ions by using SnO_2 nanopowder in micellar medium, Desalin. Water Treat. 57 (2016) 12653–12662.

[11] C. Fan, Q. Pan, Q. Li, L. Wang, Cloud point–TiO_2/sepiolite composites extraction for simultaneous preconcentration and determination of nickel in green tea and coconut water, J. Iran Chem. Soc. 13 (2016) 331–337.

[12] W.–H. Ho, S.–J. Hsieh, Solid phase microextraction associated with microwave assisted extraction of organochlorine pesticides in medicinal plants, Anal. Chim. Acta 428 (2001) 111–120.

[13] V. Pino, J. H. Ayala, A. M. Afonso, V. González, Micellar microwave–assisted extraction combined with solid–phase microextraction for the determination of polycyclic aromatic hydrocarbons in a certified marine sediment, Anal. Chim. Acta 477 (2003) 81–91.

[14] L. Cai, J. Xing, L. Dong, C. Wu, Application of polyphenylmethylsiloxane coated fiber for solid–phase microextraction combined with microwave–assisted extraction for the determination of organochlorine pesticides in Chinese teas, J. Chromatogr. A 1015 (2003) 11–21.

[15] M. Ramil–Criado, I. Rodríguez–Pereiro, R. Cela–Torrijos, Determination of polychlorinated biphenyls in ash using dimethylsulfoxide microwave assisted extraction followed by solid–phase microextraction, Talanta 63 (2004) 533–540.

[16] L. M. Díaz–Vázquez, O. García, Z. Velázquez, I. Marrero, O. Rosario, Optimization of microwave–assisted extraction followed by solid phase micro extraction and gas chromatography–mass spectrometry detection for the assay of some semi volatile organic pollutants in sebum, J. Chromatogr. B 825 (2005) 11–20.

[17] C. Deng, Y. Mao, N. Yao, X. Zhang, Development of microwave–assisted extraction followed by headspace solid–phase microextraction and gas chromatography–mass spectrometry for quantification of camphor and borneol in *Flos Chrysanthemi Indici*, Anal. Chim. Acta 575 (2006) 120–125.

[18] C. Deng, J. Ji, N. Li, Y. Yu, G. Duan, X. Zhang, Fast determination of curcumol, curdione and germacrone in three species of Curcuma rhizomes by microwave–assisted extraction followed by headspace solid–phase microextraction and gas chromatography–mass spectrometry, J. Chromatogr. A 1117 (2006)

115-120.

[19] P. Herbert, S. Morais, P. Paíga, A. Alves, L. Santos, Development and validation of a novel method for the analysis of chlorinated pesticides in soils using microwave-assisted extraction-headspace solid phase microextraction and gas chromatography-tandem mass spectrometry, Anal. Bioanal. Chem. 384 (2006) 810-816.

[20] P. Herbert, S. Morais, P. Paíga, A. Alves, L. Santos, Analysis of PCBs in soils and sediments by microwave-assisted extraction, headspace-SPME and high resolution gas chromatography with ion-trap tandem mass spectrometry, Intern. J. Environ. Anal. Chem. 86 (2006) 391-400.

[21] P. N. Carvalho, P. Nuno, R. Rodrigues, F. Alves, R. Evangelista, M. C. P. Basto, M. T. S. D. Vasconcelos, An expeditious method for the determination of organochlorine pesticides residues in estuarine sediments using microwave assisted pre-extraction and automated headspace solid-phase microextraction coupled to gas chromatography-mass spectrometry, Talanta 76 (2008) 1124-1129.

[22] Q. Ye, D. Zheng, Z. Chen, Rapid determination of paeonol in traditional Chinese medicinal preparations by microwave-assisted extraction followed by headspace solid-phase microextraction and gas chromatography-mass spectrometry, J. Anal. Chem. 66 (2011) 285-289.

[23] P. J. Giebink, R. W. Smith, Development of microwave-assisted extraction procedure for organic impurity profiling of seized 3, 4-methylenedioxymethamphetamine (MDMA), J. Forensic Sci. 56 (2011) 1483-1492.

[24] G. M. Mizanur-Rahman, M. Mulugeta-Wolle, T. Fahrenholz, H. M. Skip Kingston, M. Pamuku, Measurement of mercury species in whole blood using speciated isotope dilution methodology integrated with microwave-enhanced solubilization and spike equilibration, headspace-solid-phase microextraction, and GC-ICP-MS analysis, Anal. Chem. 86 (2014) 6130-6137.

[25] D. Vega Moreno, Z. Sosa Ferrera, J. J. Santana Rodríguez, Sample extraction method combining micellar extraction-SPME and HPLC for the determination of organochlorine pesticides in agricultural soils, J. Agric. Food Chem. 54 (2006) 7747-7752.

[26] D. A. Lambropoulou, T. A. Albanis, Determination of the fungicides vinclozolin and dicloran in soils using ultrasonic extraction coupled with solid-phase microextraction, Anal. Chim. Acta 514 (2004) 125-130.

[27] D. A. Lambropoulou, I. K. Konstantinou, T. A. Albanis, Coupling of headspace solid phase microextraction with ultrasonic extraction for the determination of chlorinated pesticides in bird livers using gas chromatography, Anal. Chim. Acta 573-574 (2006) 223-230.

[28] C. Salgado-Petinal, M. Llompart, C. García-Jares, M. García-Chao, R. Cela, Simple approach for the determination of brominated flame retardants in environmental solid samples based on solvent extraction and solid-phase microextraction followed by gas chromatography-tandem mass spectrometry, J. Chromatogr. A 1124 (2006) 139-147.

[29] T. Liu, S. Li, S. Liu, G. Lv, Optimization of supercritical fluid extraction/headspace solid-phase microextraction and gas chromatography-mass spectrometry method for determinating organotin compounds in clam samples, J. Food Process. Eng. 34 (2011) 1125-1143.

[30] R. Rodil, A. M. Carro, R. A. Lorenzo, R. Cela-Torrijos, Selective extraction of trace levels of polychlorinated and polybrominated contaminants by supercritical fluidsolid-phase microextraction and determination by gas chromatography/mass spectrometry. Application to aquaculture fish feed and cultured marine species, Anal. Chem. 77 (2005) 2259-2265.

[31] Z. -M. Lu, W. Xu, N. -H. Yu, T. Zhou, G. -Q. Li, J. -S. Shi, Z. -H. Xu, Recovery of aroma compounds from Zhenjiang aromatic vinegar by supercritical fluid extraction, Int. J. Food Sci. Technol. 45 (2011) 1508-1514.

[32] K. J. Hageman, L. Mazeas, C. B. Grabanski, D. J. Miller, S. B. Hawthorne, Coupled subcritical water extraction with solid-phase microextraction for determining semivolatile organics in environmental solids, Anal. Chem. 68 (1996) 3892-3898.

[33] S. B. Hawthorne, C. B. Grabanski, K. J. Hageman, D. J. Miller, Simple method for estimating polychlorinated biphenyl concentrations on soils and sediments using subcritical water extraction coupled with solid-phase microextraction, J. Chromatogr. A 814 (1998) 151-160.

[34] C. Deng, N. Li, X. Zhang, Rapid determination of essential oil in *Acorus tatarinowii Schott.* by pressurized hot water extraction followed by solid-phase microextraction and gas chromatography-mass spectrometry, J. Chromatogr. A 1059 (2004) 149-155.

[35] C. Deng, A. Wang, S. Shen, D. Fu, J. Chen, X. Zhang, Rapid analysis of essential oil from *Fructus Amomi* by pressurized hot water extraction followed by solid-phase microextraction and gas chromatography-mass spectrometry, J. Pharm. Biomed. Anal. 38 (2005) 326-331.

[36] C. Deng, J. Ji, X. Wang, X. Zhang, Development of pressurized hot water extraction followed by headspace solid-phase microextraction and gas chromatography-mass spectrometry for determination of ligustilides in *Ligusticum chuanxiong* and *Angelica sinensis*, J. Sep. Sci. 28 (2005) 1237-1243.

[37] L. Dong, J. Wang, C. Deng, X. Shen, Gas chromatography-mass spectrometry following pressurized hot water extraction and solid-phase microextraction for quantification of eucalyptol, camphor, and borneol in *Chrysanthemum* flowers, J. Sep. Sci. 30 (2007) 86-89.

[38] A. Llop, E. Pocurull, F. Borrull, Automated on-fiber derivatization with headspace SPME-GC-MS-MS for the determination of primary amines in sewage sludge using pressurized hot water extraction, J. Sep. Sci. 34 (2011) 1531-1537.

[39] A. Llop, F. Borrull, E. Pocurull, Pressurised hot water extraction followed by headspace solid-phase microextraction and gas chromatography-tandem mass spectrometry for the determination of N-nitrosamines in sewage sludge, Talanta 88 (2012) 284-289.

[40] A. Llop, F. Borrull, E. Pocurull, Pressurised hot water extraction followed by simultaneous derivatization and headspace solid-phase microextraction and gas chromatography-tandem mass spectrometry for the determination of aliphatic primary amines in sewage sludge, Anal. Chim. Acta 665 (2010) 231-236.

[41] A. Beichert, S. Padberg, B. W. Wenclawiak, Selective determination of alkylmercury compounds in solid matrices after subcritical water extraction, followed by solid-phase microextraction and GC-MS, Appl. Organometal. Chem. 14 (2000) 493-498.

[42] L. Wennrich, P. Popp, M. Möder, Determination of chlorophenols in soils using accelerated solvent extraction combined with solid-phase microextraction, Anal. Chem. 72 (2000) 546-551.

[43] M. S. S. Curren, J. W. King, Ethanol-modified subcritical water extraction combined with solid-phase microextraction for determining atrazine in beef kidney, J. Agric. Food Chem. 49 (2001) 2175-2180.

[44] L. Wennrich, P. Popp, J. Breuste, Determination of organochlorine pesticides and chlorobenzenes in fruit and vegetables using subcritical water extraction combined with sorptive enrichment and CGC-MS, Chromatographia 53 (2001) S380-S386.

[45] L. Wennrich, P. Popp, G. Köller, J. Breuste, Determination of organochlorine pesticides and chlorobenzenes in strawberries by using accelerated solvent extraction combined with sorptive enrichment and

gas chromatography/mass spectrometry, J. AOAC Int. 84 (2001) 1194-1201.

[46] V. Fernández-González, E. Concha-Graña, S. Muniategui-Lorenzo, P. López-Mahía, D. Prada-Rodríguez, Pressurized hot water extraction coupled to solid-phase microextraction-gas chromatography-mass spectrometry for the analysis of polycyclic aromatic hydrocarbons in sediments, J. Chromatogr. A 1196-1197 (2008) 65-72.

[47] E. Concha-Graña, V. Fernández-González, G. Grueiro-Noche, S. Muniategui—Lorenzo, P. López-Mahía, E. Fernández-Fernández, D. Prada-Rodríguez, Development of an environmental friendly method for the analysis of organochlorine pesticides in sediments, Chemosphere 79 (2010) 698-705.

[48] Z. Gao, Y. Deng, W. Yuan, H. He, S. Yang, C. Sun, Determination of organophosphorus flame retardants in fish by pressurized liquid extraction using aqueous solutions and solid-phase microextraction coupled with gas chromatography-flame photometric detector, J. Chromatogr. A 1366 (2014) 31-37.

[49] C. Raeppel, M. Fabritius, M. Nief, B. M. R. Appenzeller, M. Millet, Coupling ASE, sylilation and SPME-GC/MS for the analysis of current-used pesticides in atmosphere, Talanta 121 (2014) 24-29.

[50] H. Mokbe, E. J. Al Dine, A. Elmoll, C. Liaud, M. Millet, Simultaneous analysis of organochlorine pesticides and polychlorinated biphenyls in air samples by using accelerated solvent extraction (ASE) and solid-phase micro-extraction (SPME) coupled to gas chromatography dual electron capture detection, Environ. Sci. Pollut. Res. 23 (2016) 8053-8063.

[51] Y. Moliner-Martinez, P. Campíns-Falco, C. Molins-Legua, L. Segovia-Martínez, A. Seco-Torrecillas, Miniaturized matrix solid phase dispersion procedure and solid phase microextraction for the analysis of organochlorinated pesticides and polybrominated diphenylethers in biota samples by gas chromatography electron capture detection, J. Chromatogr. A 1216 (2009) 6741-6745.

[52] G. Álvarez-Rivera, M. Llompart, C. García-Jares, M. Lores, Identification of unwanted photoproducts of cosmetic preservatives in personal care products under ultraviolet-light using solid-phase microextraction and micro-matrix solid-phase dispersion, J. Chromatogr. A 1390 (2015) 1-12.

[53] P. Campíns-Falcó, J. Verdú-Andrés, A. Sevillano-Cabeza, C. Molins-Legua, R. Herráez-Hernández, New micromethod combining miniaturized matrix solidphase dispersion and in-tube in-valve solid-phase microextraction for estimating polycyclic aromatic hydrocarbons in bivalves, J. Chromatogr. A 1211 (2008) 13-21.

[54] M. Muñoz-Ortuño, A. Argente-García, Y. Moliner-Martínez, J. Verdú-Andrés, R. Herráez-Hernández, M. T. Picher, P. Campíns-Falcó, A cost-effective method for estimating di (2-ethylhexyl) phthalate in coastal sediments, J. Chromatogr. A 1324 (2014) 57-62.

[55] Y. -I. Chen, Y. -S. Su, J. -F. Jen, Determination of dichlorvos by on-line microwaveassisted extraction coupled to headspace solid-phase microextraction and gas chromatography-electron-capture detection, J. Chromatogr. A 976 (2002) 349-355.

[56] R. Mu, X. Wang, S. Liu, X. Yuan, S. Wang, Z. Fan, Rapid determination of volatile compounds in *Toona sinensis* (A. Juss.) Roem. by MAE-HS-SPME followed by GC-MS, Chromatographia 65 (2007) 463-467.

[57] Z. Yang, H. Mao, C. Long, C. Sun, Z. Guo, Rapid determination of volatile composition from *Polygala furcata* Royle by MAE-HS-SPME followed by GC-MS, Eur. Food Res. Technol. 230 (2010) 779-784.

[58] Z. Q. Fan, S. B. Wang, R. M. Mu, X. R. Wang, S. X. Liu, X. L. Yuan, A simple, fast, solvent-free method for the determination of volatile compounds in *Magnolia grandi-flora* Linn, J. Anal.

Chem. 64 (2009) 289-294.

[59] D. Wu, K. Chen, M. Xu, G. Song, Y. Hu, H. Cheng, Rapid analysis of essential oils in fruits of *Alpinia oxyphylla Miq.* by microwave distillation and simultaneous headspace solid-phase microextraction coupled with gas chromatography-mass spectrometry, Anal. Methods 6 (2014) 9718-9724.

[60] M. Piryaei, M.M. Abolghasemi, H. Nazemiyeh, Rapid analysis of *Achillea tenuifolia Lam* essential oils by polythiophene/hexagonally ordered silica nanocomposite coating as a solid - phase microextraction fibre, Nat. Prod. Res. 29 (2015) 1789-1792.

[61] Q. Ye, D. Zheng, Rapid analysis of the essential oil components of dried *Perilla frutescens* (L.) by magnetic nanoparticle - assisted microwave distillation and simultaneous headspace solid - phase microextraction followed by gas chromatography-mass spectrometry, Anal. Methods 1 (2009) 39-44.

[62] H.-P. Li, G.-C. Li, J.-F. Jen, Determination of organochlorine pesticides in water using microwave assisted headspace solid-phase microextraction and gas chromatography, J. Chromatogr. A 1012 (2003) 129-137.

[63] Y. Huang, Y.-C. Yang, Y.Y. Shu, Analysis of semi-volatile organic compounds in aqueous samples by microwave-assisted headspace solid-phase microextraction coupled with gas chromatography-electron capture detection, J. Chromatogr. A 1140 (2007) 35-43.

[64] Y.-C. Tsao, Y.-C. Wang, S.-F. Wu, W.-H. Ding, Microwave-assisted headspace solid-phase microextraction for the rapid determination of organophosphate esters in aqueous samples by gas chromatography-mass spectrometry, Talanta 84 (2011) 406-410.

[65] P.-C. Hsieh, J.-F. Jen, C.-L. Lee, K.-C. Chang, Determination of polycyclic aromatic hydrocarbons in environmental water samples by microwave-assisted headspace solidphase microextraction, Environ. Eng. Sci. 32 (2015) 301-309.

[66] P.-C. Hsieh, C.-L. Lee, J.-F. Jen, K.-C. Chang, Complexation-flocculation combined with microwave-assisted headspace solid-phase microextraction in determining the binding constants of hydrophobic organic pollutants to dissolved humic substances, Analyst 140 (2015) 1275-1280.

[67] X.-F. Feng, N. Jing, Z.-G. Li, D. Wei, M.-R. Lee, Ultrasound-microwave hybridassisted extraction coupled to headspace solid phase microextraction for fast analysis of essential oil in dry traditional Chinese medicine by GC-MS, Chromatographia 77 (2014) 619-628.

[68] Y. Yang, G. Chu, G. Zhou, J. Jiang, K. Yuan, Y. Pan, Z. Song, Z. Li, Q. Xia, X. Lu, W. Xiao, Rapid determination of the volatile components in tobacco by ultrasoundmicrowave synergistic extraction coupled to headspace solid-phase microextraction with gas chromatography-mass spectrometry, J. Sep. Sci. 39 (2016) 1173-1181.

[69] M.M. Žulj, L. Maslov, I. Tomaz, A. Jeromel, Determination of 2-aminoacetophenone in white wines using ultrasound assisted SPME coupled with GC-MS, J. Chromatogr. Sci. 54 (2016) 264-270.

[70] A. Ghiasvand, S. Shadabi, S. Hajipour, A. Nasirian, M. Borzouei, E. Hassani-Moghadam, P. Hashemi, Comparison of ultrasound-assisted headspace solid-phase microextraction and hydrodistillation for the identification of major constituents in two species of hypericum, J. Chromatogr. Sci. 54 (2016) 264-270.

[71] T.-J. Yang, F.-J. Tsai, C.-Y. Chen, T.C.-C. Yang, M.-R. Lee, Determination of additives in cosmetics by supercritical fluid extraction on-line headspace solid-phase microextraction combined with gas chromatography-mass spectrometry, Anal. Chim. Acta 668 (2010) 188-194.

[72] W.-L. Liu, B.-H. Hwang, Z.-G. Li, J.-F. Jen, M.-R. Lee, Headspace solid phase microextraction in-situ supercritical fluid extraction coupled to gas chromatography-tandem mass spectrometry

for simultaneous determination of perfluorocarboxylic acids in sediments, J. Chromatogr. A 1218 (2011) 7857–7863.

[73] C. Yamaguchi, W. -Y. Lee, A cost effective, sensitive, and environmentally friendly sample preparation method for determination of polycyclic aromatic hydrocarbons in solid samples, J. Chromatogr. A 1217 (2010) 6816–6823.

[74] O. Alvarez-Avilés, L. Cuadra-Rodríguez, F. González-Illán, J. Quiñones-González, O. Rosario, Optimization of a novel method for the organic chemical characterization of atmospheric aerosols based on microwave-assisted extraction combined with stir bar sorptive extraction, Anal. Chim. Acta 597 (2007) 273–281.

[75] J. Vestner, S. Fritsch, D. Rauhut, Development of a microwave assisted extraction method for the analysis of 2, 4, 6-trichloroanisole in cork stoppers by SIDA-SBSE-GC-MS, Anal. Chim. Acta 660 (2010) 76–80.

[76] R. Rodil, P. Popp, Development of pressurized subcritical water extraction combined with stir bar sorptive extraction for the analysis of organochlorine pesticides and chlorobenzenes in soils, J. Chromatogr. A 1124 (2006) 82–90.

[77] F. J. Camino-Sánchez, A. Zafra-Gómez, J. P. Pérez-Trujillo, J. E. Conde-González, J. C. Marques, J. L. Vílchez, Validation of a GC-MS/MS method for simultaneous determination of 86 persistent organic pollutants in marine sediments by pressurized liquid extraction followed by stir bar sorptive extraction, Chemosphere 84 (2011) 869–881.

[78] N. Ramírez, R. M. Marcé, F. Borrull, Determination of parabens in house dust by pressurised hot water extraction followed by stir bar sorptive extraction and thermal desorption-gas chromatography-mass spectrometry, J. Chromatogr. A 1218 (2011) 6226–6231.

[79] M. G. Pintado-Herrera, E. González-Mazo, P. A. Lara-Martín, Environmentally friendly analysis of emerging contaminants by pressurized hot water extraction-stir bar sorptive extraction-derivatization and gas chromatography-mass spectrometry, Anal. Bioanal. Chem. 405 (2013) 401–411.

[80] X. Du, C. E. Finn, M. C. Qian, Bound volatile precursors in genotypes in the pedigree of 'Marion' blackberry (*Rubus* Sp.), J. Agric. Food Chem. 58 (2010) 3694–3699.

[81] N. Fattahi, S. Samadi, Y. Assadi, M. R. M. Hosseini, Solid-phase extraction combined with dispersive liquid-liquid microextraction-ultra preconcentration of chlorophenols in aqueous samples, J. Chromatogr. A 1169 (2007) 63–69.

[82] R. Montes, I. Rodríguez, M. Ramil, E. Rubí, R. Cela, Solid-phase extraction followed by dispersive liquid-liquid microextraction for the sensitive determination of selected fungicides in wine, J. Chromatogr. A 1216 (2009) 5459–5466.

[83] X. Liu, J. Li, Z. Zhao, W. Zhang, K. Lin, C. Huang, X. Wang, Solid-phase extraction combined with dispersive liquid-liquid microextraction for the determination for polybrominated diphenyl ethers in different environmental matrices, J. Chromatogr. A 1216 (2009) 2220–2226.

[84] B. Liu, H. Yan, F. Qiao, Y. Geng, Determination of clenbuterol in porcine tissues using solid-phase extraction combined with ultrasound-assisted dispersive liquide-liquid microextraction and HPLC-UV detection, J. Chromatogr. B 879 (2011) 90–94.

[85] A. C. H. Alves, M. M. P. B. Gonçalves, M. M. S. Bernardo, B. S. Mendes, Determination of organophosphorous pesticides in the ppq range using a simple solid-phase extraction method combined with dispersive liquid-liquid microextraction, J. Sep. Sci. 34 (2011) 2475–2481.

[86] X. Liu, A. Zhao, A. Zhang, H. Liu, W. Xiao, C. Wang, X. Wang, Dispersive liquid-liquid microextraction and gas chromatography-mass spectrometry determination of polychlorinated biphenyls and polybrominated diphenyl ethers in milk, J. Sep. Sci. 34 (2011) 1084-1090.

[87] H. R. Sobhi, H. Farahani, A. Kashtiaray, M. R. Farahani, Tandem use of solid-phase extraction and dispersive liquid-liquid microextraction for the determination of mononitrotoluenes in aquatic environment, J. Sep. Sci. 34 (2011) 1035-1040.

[88] L. Yao, J. Yang, Y. -F. Guan, B. -Z. Liu, S. -J. Zheng, W. -M. Wang, X. -L. Zhu, Z. -D. Zhang, Development, validation, and application of a liquid chromatography-tandem mass spectrometry method for the determination of 4- (methylnitrosamino) -1- (3-pyridyl) -1-butanol in human hair, Anal. Bioanal. Chem. 404 (2012) 2259-2266.

[89] M. Rezaee, H. A. Mashayekhi, Solid-phase extraction combined with dispersive liquid-liquid microextraction as an efficient and simple method for the determination of carbamazepine in biological samples, Anal. Methods 4 (2012) 2887-2892.

[90] H. Yan, Y. Han, J. Du, Combination of solid-phase extraction and dispersive liquide-liquid microextraction for detection of cypermethrin and permethrin in environmental water, Anal. Methods 4 (2012) 3002-3006.

[91] I. Carpinteiro, B. Abuín, I. Rodríguez, M. Ramil, R. Cela, Mixed-mode solid-phase extraction followed by dispersive liquid-liquid microextraction for the sensitive determination of ethylphenols in red wines, J. Chromatogr. A 1229 (2012) 79-85.

[92] S. Samadi, H. Sereshti, Y. Assadi, Ultra-preconcentration and determination of thirteen organophosphorus pesticides in water samples using solid-phase extraction followed by dispersive liquid-liquid microextraction and gas chromatography with flame photometric detection, J. Chromatogr. A 1219 (2012) 61-65.

[93] H. Faraji, S. W. Husain, M. Helalizadeh, Determination of phenolic compounds in environmental water samples after solid-phase extraction with β-cyclodextrin-bonded silica particles coupled with a novel liquid-phase microextraction followed by gas chromatography-mass spectrometry, J. Sep. Sci. 35 (2012) 107-113.

[94] D. Djozan, M. A. Farajzadeh, S. M. Sorouraddin, T. Baheri, Molecularly imprintedsolid phase extraction combined with simultaneous derivatization and dispersive liquid-liquid microextraction for selective extraction and preconcentration of methamphetamine and ecstasy from urine samples followed by gas chromatography, J. Chromatogr. A 1248 (2012) 24-31.

[95] S. Zhou, B. Wu, C. Ma, Y. Ye, H. Chen, Solid-phase extraction followed by dispersive liquid-liquid microextraction for the sensitive determination of carbamates in environmental water by high-performance liquid chromatography, J. Liq. Chromatogr. Rel. Tech. 35 (2012) 2860-2872.

[96] M. K. R. Mudiam, A. Chauhan, K. P. Singh, S. K. Gupta, R. Jain, Ch Ratnasekhar, R. C. Murthy, Determination of t, t-muconic acid in urine samples using a molecular imprinted polymer combined with simultaneous ethyl chloroformate derivatization and pre-concentration by dispersive liquid-liquid microextraction, Anal. Bioanal. Chem. 405 (2013) 341-349.

[97] H. Ebrahimzadeh, K. Molaei, A. A. Asgharinezhad, N. Shekari, Z. Dehghani, Molecularly imprinted nanoparticles combined with miniaturized homogenous liquid-liquid extraction for the selective extraction of loratadine in plasma and urine samples followed by high performance liquid chromatography-photo diode array detection, Anal. Chim. Acta 767 (2013) 155-162.

[98] N. Campillo, P. Viñeas, G. Ferez-Melgarejo, M. Hernandez-Cordoba, Liquid chromatography with diode array detection and tandem mass spectrometry for the determination of neonicotinoid insecticides in honey samples using dispersive liquid-liquid microextraction, J. Agric. Food Chem. 61 (2013) 4799-4805.

[99] A. R. Zarei, R. Hajiaghabozorgy, Solid-phase extraction combined with dispersive liquid-liquid microextraction for the spectrophotometric determination of ultra-trace amounts of azide ion in water samples, Anal. Methods 6 (2014) 5784-5791.

[100] H. Haddadi, M. Rezaee, A. Semnani, H. A. Mashayekhi, A. Hosseinian, Application of solid-phase extraction coupled with dispersive liquid-liquid microextraction for the determination of benzaldehyde in injectable formulation solutions, Chromatographia 77 (2014) 951-955.

[101] R. Celano, A. L. Piccinelli, L. Campone, L. Rastrelli, Ultra - preconcentration and determination of selected pharmaceutical and personal care products in different water matrices by solid-phase extraction combined with dispersive liquid - liquid microextraction prior to ultra high pressure liquid chromatography tandem mass spectrometry analysis, J. Chromatogr. A 1355 (2014) 26-35.

[102] X. -F. Liu, S. Zhou, Q. -F. Zhu, Y. Ye, H. -X. Chen, Ultra preconcentration of polycyclic aromatic hydrocarbons in smoked bacon by a combination of SPE and DLLME, J. Chromatogr. Sci. 52 (2014) 932-937.

[103] M. Ghobadi, Y. Yamini, B. Ebrahimpour, SPE coupled with dispersive liquid - liquid microextraction followed by GC with flame ionization detection for the determination of ultra-trace amounts of benzodiazepines, J. Sep. Sci. 37 (2014) 287-294.

[104] M. Mirzaei, M. Rakh, Preconcentration of organochlorine pesticides in aqueous samples by dispersive liquid - liquid microextraction based on solidification of floating organic drop after SPE with multiwalled carbon nanotubes, J. Sep. Sci. 37 (2014) 114-119.

[105] L. Liu, L. He, X. Jiang, W. Zhao, G. Xiang, J. L. Anderson, Macrocyclic polyamine-functionalized silica as a solid-phase extraction material coupled with ionic liquid dispersive liquid-liquid extraction for the enrichment of polycyclic aromatic hydrocarbons, J. Sep. Sci. 37 (2014) 1004-1011.

[106] M. Behbahani, A. Esrafili, S. Bagheri, S. Radfar, M. K. Bojdi, A. Bagheri, Modified nanoporous carbon as a novel sorbent before solvent - based de - emulsification dispersive liquid - liquid microextraction for ultra - trace detection of cadmium by flame atomic absorption spectrophotometry, Measurement 51 (2014) 174-181.

[107] M. Shamsipur, N. Fattahi, Y. Assadi, M. Sadeghi, K. Sharafi, Speciation of As (Ⅲ) and As (Ⅴ) in water samples by graphite furnace atomic absorption spectrometry after solid phase extraction combined with dispersive liquid-liquid microextraction based on the solidification of floating organic drop, Talanta 130 (2014) 26-32.

[108] M. Sadeghi, Z. Nematifar, M. Irandoust, N. Fattahi, P. Hamzei, A. Barati, M. Ramezani, M. Shamsipur, Efficient and selective extraction and determination of ultra trace amounts of Hg^{2+} using solid phase extraction combined with ion pair based surfactant-assisted dispersive liquid-liquid microextraction, RSC Adv. 5 (2015) 100511-100521.

[109] N. Liang, P. Huang, X. Hou, Z. Li, L. Tao, L. Zhao, Solid-phase extraction in combination with dispersive liquid-liquid microextraction and ultra-high performance liquid chromatography-tandem mass spectrometry analysis: the ultra-trace determination of 10 antibiotics in water samples, Anal. Bioanal. Chem. 408 (2016) 1701-1713.

[110] M. Shirani, H. Haddadi, M. Rezaee, A. Semnani, S. Habibollahi, Solid-phase extraction

combined with dispersive liquid-liquid microextraction for the simultaneous determination of deltamethrin and permethrin in honey by gas chromatography-mass spectrometry, Food Anal. Methods 9 (2016) 2613-2620.

[111] M. Sadeghi, Z. Nematifar, N. Fattahi, M. Pirsaheb, M. Shamsipur, Determination of bisphenol A in food and environmental samples using combined solid-phase extraction-dispersive liquid-liquid microextraction with solidification of floating organic drop followed by HPLC, Food Anal. Methods 9 (2016) 1814-1824.

[112] M. Shamsipur, N. Yazdanfar, M. Ghambarian, Combination of solid-phase extraction with dispersive liquid-liquid microextraction followed by GC-MS for determination of pesticide residues from water, milk, honey and fruit juice, Food Chem. 204 (2016) 289-297.

[113] T. Rodríguez-Cabo, I. Rodríguez, M. Ramil, A. Silva, R. Cela, Multiclass semivolatile compounds determination in wine by gas chromatography accurate time-of-flight mass spectrometry, J. Chromatogr. A 1442 (2016) 107-117.

[114] H. A. Mashayekhi, F. Khalilian, Development of solid-phase extraction coupled with dispersive liquid-liquid microextraction method for the simultaneous determination of three benzodiazepines in human urine and plasma, J. Chromatogr. Sci. 54 (2016) 1068-1073.

[115] X. Q. Guo, X. T. Tang, M. He, B. B. Chen, K. Nan, Q. Y. Zhang, B. Hu, Dual dispersive extraction combined with electrothermal vaporization inductively coupled plasma mass spectrometry for determination of trace REEs in water and sediment samples, RSC Adv. 4 (2014) 19960-19969.

[116] M. Mohammadi, H. Tavakoli, Y. Abdollahzadeh, A. Khosravi, R. Torkaman, A. Mashayekhia, Ultra-preconcentration and determination of organophosphorus pesticides in soil samples by a combination of ultrasound assisted leaching-solid phase extraction and low-density solvent based dispersive liquid-liquid microextraction, RSC Adv. 5 (2015) 75174-75181.

[117] Z-g. Shi, H. K. Lee, Dispersive liquid-liquid microextraction coupled with dispersive μ-solid-phase extraction for the fast determination of polycyclic aromatic hydrocarbons in environmental water samples, Anal. Chem. 82 (2010) 1540-1545.

[118] J. Zhang, M. Li, M. Yang, B. Peng, Y. Li, W. Zhou, H. Gao, R. Lu, Magnetic retrieval of ionic liquids: fast dispersive liquid-liquid microextraction for the determination of benzoylurea insecticides in environmental water samples, J. Chromatogr. A 1254 (2012) 23-29.

[119] K. S. Tay, N. A. Rahman, M. R. B. Abas, Magnetic nanoparticle assisted dispersive Liquid-liquid microextraction for the determination of 4-n-nonylphenol in water, Anal. Methods 5 (2013) 2933-2938.

[120] E. Yilmaz, M. Soylak, Ionic liquid-linked dual magnetic microextraction of lead (II) from environmental samples prior to its micro-sampling flame atomic absorption spectrometric determination, Talanta 116 (2013) 882-886.

[121] M. Li, J. Zhang, Y. Li, B. Peng, W. Zhou, H. Gao, Ionic liquid-linked dual magnetic microextraction: a novel and facile procedure for the determination of pyrethroids in honey samples, Talanta 107 (2013) 81-87.

[122] N. Wang, R. Shen, Z. Yan, H. Feng, Q. Cai, S. Yao, Magnetic retrieval of an extractant: fast ultrasound assisted emulsification liquid-liquid microextraction for the determination of polycyclic aromatic hydrocarbons in environmental water samples, Anal. Methods 5 (2013) 3999-4004.

[123] S. Khan, T. G. Kazi, M. Soylak, Rapid ionic liquid-based ultrasound assisted dual magnetic microextraction to preconcentrate and separate cadmium-4-(2-thiazolylazo)-resorcinol complex from

environmental and biological samples, Spectrochim. Acta A 123 (2014) 194-199.

[124] M. Amoli-Diva, Z. Taherimaslak, M. Allahyari, K. Pourghazi, M. H. Manafi, Application of dispersive liquid-liquid microextraction coupled with vortex-assisted hydrophobic magnetic nanoparticles based solid-phase extraction for determination of aflatoxin M1 in milk samples by sensitive micelle enhanced spectrofluorimetry, Talanta 134 (2015) 98-104.

[125] M. Hashemi, Z. Taherimaslak, S. Parvizi, M. Torkejokar, Spectrofluorimetric determination of zearalenone using dispersive liquid-liquid microextraction coupled to micro-solid phase extraction onto magnetic nanoparticles, RSC Adv. 4 (2014) 45065-45073.

[126] Z. Taherimaslak, M. Amoli-Diva, M. Allahyari, K. Pourghazi, M. H. Manafi, Low density solvent based dispersive liquid-liquid microextraction followed by vortexassisted magnetic nanoparticle based solid-phase extraction and surfactant enhanced spectrofluorimetric detection for the determination of aflatoxins in pistachio nuts, RSC Adv. 5 (2015) 12747-12754.

[127] M. Shamsipur, B. Hashemi, Extraction and determination of polycyclic aromatic hydrocarbons in water samples using stir bar sorptive extraction (SBSE) combined with dispersive liquid-liquid microextraction based on the solidification of floating organic drop (DLLMESFO) followed by HPLC-UV, RSC Adv. 5 (2015) 20339-20345.

[128] M. Rezaee, Y. Yamini, M. Moradi, A. Saleh, M. Faraji, M. H. Naeeni, Supercritical fluid extraction combined with dispersive liquid-liquid microextraction as a sensitive and efficient sample preparation method for determination of organic compounds in solid samples, J. Supercrit. Fluids 55 (2010) 161-168.

[129] M. H. Naeeni, Y. Yamini, M. Rezaee, Combination of supercritical fluid extraction with dispersive liquid-liquid microextraction for extraction of organophosphorus pesticides from soil and marine sediment samples, J. Supercrit. Fluids 57 (2011) 219-226.

[130] M. Jowkarderis, F. Raofie, Optimization of supercritical fluid extraction combined with dispersive liquid-liquid microextraction as an efficient sample preparation method for determination of 4-nitrotoluene and 3-nitrotoluene in a complex matrix, Talanta 88 (2012) 50-53.

[131] X. Wang, L. Lin, T. Luan, L. Yang, N. F. Y. Tam, Determination of hydroxylated metabolites of polycyclic aromatic hydrocarbons in sediment samples by combining subcritical water extraction and dispersive liquid-liquid microextraction with derivatization, Anal. Chim. Acta 753 (2012) 57-63.

[132] P. Viñas, M. Pastor-Belda, N. Campillo, M. Bravo-Bravo, M. Hernández-Córdoba, Capillary liquid chromatography combined with pressurized liquid extraction and dispersive liquid-liquid microextraction for the determination of vitamin E in cosmetic products, J. Pharm. Biomed. Anal. 94 (2014) 173-179.

[133] P. Viñas, M. Bravo-Bravo, I. L ópez-García, M. Pastor-Belda, M. HernándezCórdoba, Pressurized liquid extraction and dispersive liquid-liquid microextraction for determination of tocopherols and tocotrienols in plant foods by liquid chromatography with fluorescence and atmospheric pressure chemical ionization-mass spectrometry detection, Talanta 119 (2014) 98-104.

[134] K. Yuan, H. Kang, Z. Yue, L. Yang, L. Lin, X. Wang, T. Luan, Determination of 13 endocrine disrupting chemicals in sediments by gas chromatography-mass spectrometry using subcritical water extraction coupled with dispersed liquid-liquid microextraction and derivatization, Anal. Chim. Acta 866 (2015) 41-47.

[135] K. Cai, D. Hu, B. Lei, H. Zhao, W. Pan, B. Song, Determination of carbohydrates in tobacco by pressurized liquid extraction combined with a novel ultrasound-assisted dispersive liquid-liquid microextraction method, Anal. Chim. Acta 882 (2015) 90-100.

[136] G. Yang, Q. Sun, Z. Hu, H. Liu, T. Zhou, G. Fan, Optimization of an accelerated solvent extraction dispersive liquid-liquid microextraction method for the separation and determination of essential oil from *Ligusticum chuanxiong* Hort by gas chromatography with mass spectrometry, J. Sep. Sci. 38 (2015) 3588-3598.

[137] N. Campillo, P. Viñas, N. Martínez-Castillo, M. Hernández-Córdoba, Determination of volatile nitrosamines in meat products by microwave-assisted extraction and dispersive liquid-liquid microextraction coupled to gas chromatography-mass spectrometry, J. Chromatogr. A 1218 (2011) 1815-1821.

[138] C. Pizarro, C. Sáenz-González, N. Pérez-del-Notario, J. M. González-Sáiz, Microwave assisted extraction combined with dispersive liquid-liquid microextraction as a sensitive sample preparation method for the determination of haloanisoles and halophenols in cork stoppers and oak barrel sawdust, Food Chem. 132 (2012) 2202-2210.

[139] V. Ghasemzadeh-Mohammadi, A. Mohammadi, M. Hashemi, R. Khaksar, P. Haratian, Microwave-assisted extraction and dispersive liquid-liquid microextraction followed by gas chromatography-mass spectrometry for isolation and determination of polycyclic aromatic hydrocarbons in smoked fish, J. Chromatogr. A 1237 (2012) 30-36.

[140] M. H. Naeeni, Y. Yamini, M. Rezaee, S. Seidi, Microwave-assisted extraction combined with dispersive liquid-liquid microextraction as a new approach to determination of chlorophenols in soil and sediments, J. Sep. Sci. 35 (2012) 2469-2475.

[141] A. Mohammadi, V. Ghasemzadeh-Mohammadi, P. Haratian, R. Khaksar, M. Chaichi, Determination of polycyclic aromatic hydrocarbons in smoked fish samples by a new microextraction technique and method optimization using response surface methodology, Food Chem. 141 (2013) 2459-2465.

[142] Z. -Y. Ye, Z. -G. Li, D. Wei, M. -R. Lee, Microwave-assisted extraction/dispersive Liquid-liquid microextraction coupled with DSI-GC-it/MS for analysis of essential oil from three species of cardamom, Chromatographia 77 (2014) 347-358.

[143] H. Ramezani, H. Hosseini, M. Kamankesh, V. Ghasemzadeh-Mohammadi, A. Mohammadi, Rapid determination of nitrosamines in sausage and salami using microwave-assisted extraction and dispersive liquid-liquid microextraction followed by gas chromatography-mass spectrometry, Eur. Food Res. Technol. 240 (2015) 441-450.

[144] M. Kamankesh, A. Mohammadi, H. Hosseini, Z. M. Tehrani, Rapid determination of polycyclic aromatic hydrocarbons in grilled meat using microwave-assisted extraction and dispersive liquid-liquid microextraction coupled to gas chromatography-mass spectrometry, Meat Sci. 103 (2015) 61-67.

[145] J. Li, Q. Shi, Y. Jiang, J. Gao, M. Li, H. Ma, Microwave-assisted extraction and ionic liquid based dispersive liquid-liquid microextraction followed by HPLC for the determination of trace components in a Langdu preparation, Anal. Methods 8 (2016) 5218-5227.

[146] M. Bashiry, A. Mohammadi, H. Hosseini, M. Kamankesh, S. Aeenehvand, Z. Mohammadi, Application and optimization of microwave-assisted extraction and dispersive liquid-liquid microextraction followed by high-performance liquid chromatography for sensitive determination of polyamines in Turkey breast meat samples, Food Chem. 190 (2016) 1168-1173.

[147] P. Huang, P. Zhao, X. Dai, X. Hou, L. Zhao, N. Liang, Trace determination of antibacterial pharmaceuticals in fishes by microwave-assisted extraction and solid-phase purification combined with dispersive liquid-liquid microextraction followed by ultra-high performance liquid chromatography-tandem mass spectrometry, J. Chromatogr. B 1011 (2016) 136-144.

[148] M. Tuzen, O. Z. Pekiner, Ultrasound - assisted ionic liquid dispersive liquid - liquid microextraction combined with graphite furnace atomic absorption spectrometric for selenium speciation in foods and beverages, Food Chem. 188 (2015) 619-624.

[149] A. Bidari, M. R. Ganjali, P. Norouzi, M. R. M. Hosseini, Y. Assadi, Sample preparation method for the analysis of some organophosphorus pesticides residues in tomato by ultrasound-assisted solvent extraction followed by dispersive liquid-liquid microextraction, Food Chem. 126 (2011) 1840-1844.

[150] H. Sereshti, Y. Izadmanesh, S. Samadi, Optimized ultrasonic assisted extraction - dispersive liquid-liquid microextraction coupled with gas chromatography for determination of essential oil of *Oliveria decumbens* Vent. J. Chromatogr. A 1218 (2011) 4593-4598.

[151] P. Li, X. Zhu, S. Hong, Z. Tian, J. Yang, Ultrasound - assisted extraction followed by dispersive liquid - liquid microextraction before gas chromatography - mass spectrometry for the simultaneous determination of flavouring compounds in tobacco additives, Anal. Methods 4 (2012) 995-1000.

[152] H. Sereshti, S. Samadi, M. Jalali-Heravi, Determination of volatile components of green, black, oolong and white tea by optimized ultrasound - assisted extraction - dispersive liquid - liquid microextraction coupled with gas chromatography, J. Chromatogr. A 1280 (2013) 1-8.

[153] Y. Wen, J. Nie, Z. -G. Li, X. -Y. Xu, D. Wei, M. -R. Lee, The development of ultrasound-assisted extraction/dispersive liquid - liquid microextraction coupled with DSI - GC - IT/MS for analysis of essential oil from fresh flowers of *Edgeworthia chrysantha* Lindl. Anal. Methods 6 (2014) 3345-3352.

[154] H. Sereshti, R. Heidari, S. Samadi, Determination of volatile components of saffron by optimized ultrasound- assisted extraction in tandem with dispersive liquid - liquid microextraction followed by gas chromatography-mass spectrometry, Food Chem. 143 (2014) 499-505.

[155] J. Ye, W. Wang, C. Ho, J. Li, X. Guo, M. Zhao, Y. Jiang, P. Tu, Differentiation of two types of pu-erh teas by using an electronic nose and ultrasound-assisted extraction-dispersive liquid-liquid microextraction-gas chromatography-mass spectrometry, Anal. Methods 8 (2016) 593-604.

[156] P. Viñas, N. Campillo, M. Pastor - Belda, M. Hernández - Córdoba, Ultrasound assisted extraction and dispersive liquid-liquid microextraction with liquid chromatography-tandem mass spectrometry for determination of alkylphenol levels in cleaning products, Anal. Methods 7 (2015) 6718-6725.

[157] P. Viñas, N. Campillo, M. Pastor-Belda, A. Oller, M. Hernández-Córdoba, Determination of phthalate esters in cleaning and personal care products by dispersive liquid-liquid microextraction and liquid chromatography-tandem mass spectrometry, J. Chromatogr. A 1376 (2015) 18-25.

[158] J. I. Cacho, N. Campillo, P. Viñas, M. Hernández - Córdoba, Dispersive liquid - liquid microextraction for the determination of nitrophenols in soils by microvial insert large volume injection - gas chromatography-mass spectrometry, J. Chromatogr. A 1456 (2016) 27-33.

[159] H. Yan, H. Wang, J. Qiao, G. Yang, Molecularly imprinted matrix solid - phase dispersion combined with dispersive liquid-liquid microextraction for the determination of four Sudan dyes in egg yolk, J. Chromatogr. A 1218 (2011) 2182-2188.

[160] H. Wang, H. Yan, J. Qiao, Miniaturized matrix solid - phase dispersion combined with ultrasound-assisted dispersive liquid-liquid microextraction for the determination of three pyrethroids in soil, J. Sep. Sci. 35 (2012) 292-298.

[161] Y. Wang, Y. Sun, B. Xu, X. Li, X. Wang, H. Zhang, D. Song, Matrix solid - phase dispersion coupled with magnetic ionic liquid dispersive liquid-liquid microextraction for the determination of triazine herbicides in oilseeds, Anal. Chim. Acta 888 (2015) 67-74.

[162] I. Ciucanu, V. Agotici, Solid phase microextraction assisted by droplet-based liquid-liquid microextraction for analysis of volatile aromatic hydrocarbons in water by gas chromatography, J. Sep. Sci. 35 (2012) 1651-1658.

[163] I. Ciucanu, P. S. Popa, V. Agotici, G. Dehelean, Droplets-based microextraction assisted SPME for GC-MS analysis of polar compounds such as clopyralid in water, Anal. Methods 6 (2014) 6571-6576.

[164] C. Deng, N. Yao, B. Wang, X. Zhang, Development of microwave-assisted extraction followed by headspace single-drop microextraction for fast determination of paeonol in traditional Chinese medicines, J. Chromatogr. A 1103 (2006) 15-21.

[165] X. Xiao, Y. Yin, Y. Hu, G. Li, Microwave-assisted extraction coupled with single drop microextraction and high-performance column liquid chromatography for the determination of trace estrogen adulterants in soybean isoflavone dietary supplements, J. AOAC Int. 93 (2010) 849-854.

[166] C. Deng, Y. Mao, F. Hu, X. Zhang, Development of gas chromatography-mass spectrometry following microwave distillation and simultaneous headspace singledrop microextraction for fast determination of volatile fraction in Chinese herb, J. Chromatogr. A 1152 (2007) 193-198.

[167] Q. Ye, Rapid analysis of essential oil components of dried *Zanthoxylum bungeanum* Maxim by Fe_2O_3-magnetic-microsphere-assisted microwave distillation and simultaneous headspace single drop microextraction followed by GC-MS, J. Sep. Sci. 36 (2013) 2028-2034.

[168] M. Piryaei, H. Nazemiyeh, Fast analysis of volatile components of *Achillea tenuifolia* Lam with microwave distillation followed by headspace single-drop microextraction coupled to gas chromatography mass spectrometry (GC-MS), Nat. Prod. Res. 30 (2016) 991-994.

[169] K. Shrivas, H.-F. Wu, Ultrasonication followed by single-drop microextraction combined with GC/MS for rapid determination of organochlorine pesticides from fish, J. Sep. Sci. 31 (2008) 380-386.

[170] L. Wang, Z. Wang, H. Zhang, X. Li, H. Zhang, Ultrasonic nebulization extraction coupled with headspace single drop microextraction and gas chromatography-mass spectrometry for analysis of the essential oil in *Cuminum cyminum* L. Anal. Chim. Acta 647 (2009) 72-77.

[171] J. S. Almeida, T. A. Anunciação, G. C. Brandão, A. F. Dantas, V. A. Lemos, L. S. G. Teixeira, Ultrasound-assisted single-drop microextraction for the determination of cadmium in vegetable oils using high-resolution continuum source electrothermal atomic absorption spectrometry, Spectrochim. Acta B 107 (2015) 159-163.

[172] N. Sharma, A. Jain, V. K. Singh, K. K. Verma, Solid-phase extraction combined with headspace single-drop microextraction of chlorophenols as their methyl ethers and analysis by high-performance liquid chromatography-diode array detection, Talanta 83 (2011) 994-999.

[173] H. Bagheri, M. Zare, H. Piri-Moghadam, A. Es-haghi, A combined micro-solid phase-single drop microextraction approach for trace enrichment of volatile organic compounds, Anal. Methods 7 (2015) 6514-6519.

[174] C. Deng, X. Yang, X. Zhang, Rapid determination of panaxynol in a traditional Chinese medicine of *Saposhnikovia divaricata* by pressurized hot water extraction followed by liquid-phase microextraction and gas chromatography-mass spectrometry, Talanta 68 (2005) 6-11.

[175] C. Deng, N. Yao, A. Wang, X. Zhang, Determination of essential oil in a traditional Chinese medicine, *Fructus amomi* by pressurized hot water extraction followed by liquid-phase microextraction and gas chromatography-mass spectrometry, Anal. Chim. Acta 536 (2005) 237-244.

[176] X. Li, Q. Li, A. Xue, H. Chen, S. Li, Dispersive liquid-liquid microextraction coupled with single-drop microextraction for the fast determination of sulfonamides in environmental water samples by high performance liquid chromatography-ultraviolet detection, Anal. Methods 8 (2016) 517-525.

[177] K. Kuosmanen, T. Hyötyläinen, K. Hartonen, M. -L. Riekkola, Analysis of polycyclic aromatic hydrocarbons in soil and sediment with on-line coupled pressurised hot water extraction, hollow fibre microporous membrane liquid-liquid extraction and gas chromatography, Analyst 128 (2003) 434-439.

[178] A. Saleh, E. Larsson, Y. Yamini, J. Å. Jönsson, Hollow fiber liquid phase microextraction as a preconcentration and clean-up step after pressurized hot water extraction for the determination of non-steroidal anti-inflammatory drugs in sewage sludge, J. Chromatogr. A 1218 (2011) 1331-1339.

[179] X. Sun, F. Zhu, J. Xi, T. Lu, H. Liu, Y. Tong, G. Ouyang, Hollow fiber liquid-phase microextraction as clean-up step for the determination of organophosphorus pesticides residues in fish tissue by gas chromatography coupled with mass spectrometry, Mar. Pollut. Bull. 63 (2011) 102-107.

[180] M. Asensio-Ramos, J. Hernández-Borges, G. González-Hernández, M. A. Rodríguez-Delgado, Hollow-fiber liquid-phase microextraction for the determination of pesticides and metabolites in soils and water samples using HPLC and fluorescence detection, Electrophoresis 33 (2012) 2184-2191.

[181] C. Basheer, J. P. Obbard, H. K. Lee, Analysis of persistent organic pollutants in marine sediments using a novel microwave assisted solvent extraction and liquid-phase microextraction technique, J. Chromatogr. A 1068 (2005) 221-228.

[182] Y. -A. Shi, M. -Z. Chen, S. Muniraj, J. -F. Jen, Microwave-assisted headspace controlled temperature liquid-phase microextraction of chlorophenols from aqueous samples for gas chromatography-electron capture detection, J. Chromatogr. A 1207 (2008) 130-135.

[183] M. -Y. Tsai, P. V. Kumar, H. -P. Li, J. -F. Jen, Analysis of hexachlorocyclohexanes in aquatic samples by one-step microwave-assisted headspace controlled-temperature liquid-phase microextraction and gas chromatography with electron capture detection, J. Chromatogr. A 1217 (2010) 1891-1897.

[184] P. V. Kumar, J. -F. Jen, Rapid determination of dichlorodiphenyltrichloroethane and its main metabolites in aqueous samples by one-step microwave-assisted headspace controlled-temperature liquid-phase microextraction and gas chromatography with electron capture detection, Chemosphere 83 (2011) 200-207.

[185] L. Guo, H. K. Lee, Microwave assisted extraction combined with solvent bar microextraction for one-step solvent-minimized extraction, cleanup and preconcentration of polycyclic aromatic hydrocarbons in soil samples, J. Chromatogr. A 1286 (2013) 9-15.

[186] C. Basheer, K. Narasimhan, M. Yin, C. Zhao, M. Choolani, H. K. Lee, Application of micro-solid-phase extraction for the determination of persistent organic pollutants in tissue samples, J. Chromatogr. A 1186 (2008) 358-364.

[187] L. Xu, H. K. Lee, Novel approach to microwave-assisted extraction and micro-solid—phase extraction from soil using graphite fibers as sorbent, J. Chromatogr. A 1192 (2008) 203-208.

[188] Z. Wang, X. Zhao, X. Xu, L. Wu, R. Su, Y. Zhao, C. Jiang, H. Zhang, Q. Ma, C. Lu, D. Dong, An absorbing microwave micro-solid-phase extraction device used in nonpolar solvent microwave-assisted extraction for the determination of organophosphorus pesticides, Anal. Chim. Acta 760 (2013) 60-68.

[189] D. Ge, H. K. Lee, Ionic liquid based dispersive liquid-liquid microextraction coupled with

micro-solid phase extraction of antidepressant drugs from environmental water samples, J. Chromatogr. A 1317 (2013) 217-222.

[190] L. Guo, H. K. Lee, Vortex-assisted micro-solid-phase extraction followed by low—density solvent based dispersive liquid-liquid microextraction for the fast and efficient determination of phthalate esters in river water samples, J. Chromatogr. A 1300 (2013) 24-30.

[191] O. B. Jonsson, E. Dyremark, U. L. Nilsson, Development of a microporous membrane liquid-liquid extractor for organophosphate esters in human blood plasma: identification of triphenyl phosphate and octyl diphenyl phosphate in donor plasma, J. Chromatogr. B 755 (2001) 157-164.

[192] O. Nemulenzi, B. Mhaka, E. Cukrowska, O. Ramström, H. Tutu, L. Chimuka, Potential of combining of liquid membranes and molecularly imprinted polymers in extraction of 17 b-estradiol from aqueous samples, J. Sep. Sci. 32 (2009) 1941-1948.

[193] K. Kuosmanen, T. Hyötyläinen, K. Hartonen, J. Å. Jönsson, M. -L. Riekkola, Analysis of PAH compounds in soil with on-line coupled pressurised hot water extraction-microporous membrane liquid-liquid extraction-gas chromatography, Anal. Bioanal. Chem. 375 (2003) 389-399.

[194] K. Lüthje, T. Hyötyläinen, M. Rautiainen-Rämä, M. -L. Riekkola, Pressurised hot water extraction-microporous membrane liquid-liquid extraction coupled on-line with gas chromatography-mass spectrometry in the analysis of pesticides in grapes, Analyst 130 (2005) 52-58.

[195] M. Khrolenko, P. Dżygiel, P. Wieczorek, Combination of supported liquid membrane and solid-phase extraction for sample pretreatment of triazine herbicides in juice prior to capillary electrophoresis determination, J. Chromatogr. A 975 (2002) 219-227.

[196] B. Mhaka, E. Cukrowska, B. T. S. Bui, O. Ramstrom, K. Haupt, H. Tutu, L. Chimuka, Selective extraction of triazine herbicides from food samples based on a combination of a liquid membrane and molecularly imprinted polymers, J. Chromatogr. A 1216 (2009) 6796-6801.

[197] S. Almeda, L. Arce, M. Valcárcel, Combined use of supported liquid membrane and solid-phase extraction to enhance selectivity and sensitivity in capillary electrophoresis for the determination of ochratoxin A in wine, Electrophoresis 29 (2008) 1573-1581.

[198] L. Chimuka, M. van Pinxteren, J. Billing, E. Yilmaz, J. Å. Jönsson, Selective extraction of triazine herbicides based on a combination of membrane assisted solvent extraction and molecularly imprinted solid phase extraction, J. Chromatogr. A 1218 (2011) 647-653.

[199] L. Guo, H. K. Lee, Electro membrane extraction followed by low-density solvent based ultrasound-assisted emulsification microextraction combined with derivatization for determining chlorophenols and analysis by gas chromatography-mass spectrometry, J. Chromatogr. A 1243 (2012) 14-22.

[200] L. Arjomandi-Behzad, Y. Yamini, M. Rezazadeh, Extraction of pyridine derivatives from human urine using electromembrane extraction coupled to dispersive liquid-liquid microextraction followed by gas chromatography determination, Talanta 126 (2014) 73-81.

[201] H. Tabani, A. R. Fakhari, A. Shahsavani, M. Behbahani, M. Salarian, A. Bagheri, S. Nojavan, Combination of graphene oxide-based solid phase extraction and electro membrane extraction for the preconcentration of chlorophenoxy acid herbicides in environmental samples, J. Chromatogr. A 1300 (2013) 227-235.

[202] H. Nsubuga, C. Basheer, M. M. Bushra, M. H. Essa, M. H. Omar, A. M. Shemsi, Microwave-assisted digestion followed by parallel electromembrane extraction for trace level perchlorate detection in biological samples, J. Chromatogr. B 1012-1013 (2016) 1-7.

18 绿色样品制备技术中的新材料：研究进展与发展趋势

Meire R. da Silva[1], Bruno H. Fumes[1], Carlos E. D. Nazario[2] and Fernando M. Lancas[1,*]

1. University of São Paulo, Institute of Chemistry at São Carlos, São Carlos, Brazil

2. Federal University of Mato Grosso do Sul, Institute of Chemistry, Campo Grande, Brazil

*通讯作者：E-mail：flancas@ iqsc. usp. br

18.1 引言

传统的样品制备技术主要局限于使用液液萃取（LLE）方法，或是使用开放管状玻璃柱的低压液相色谱。虽然上述方法操作简单，且对昂贵仪器的需求低，但两者都技术存在几个缺点，最重要的是要使用大量有毒溶剂，给环境及健康带来一些影响，而这正是新材料和替代技术需要克服的问题。最近，微型样品制备技术的发展引起了人们的特别关注，其在本质上减少甚至是不使用有毒溶剂。即使那些仍在使用（微量）溶剂的技术，也根本无法与过去（二十世纪九十年代）需要消耗大量溶剂的技术相提并论。

在本章中，我们将探讨更多的选择性材料的制备及其在绿色样品制备中的应用，并介绍离子液体（ILs）、分子印迹聚合物（MIPs）及其相关材料、限进材料（RAM）、石墨烯衍生物等材料，讨论它们的优点和主要应用。此外，还将讨论合成这些材料的最新方法，包括溶胶-凝胶和分子印迹技术及其在绿色样品制备技术中的主要应用。

最后，在本章末尾的表中，列举了这些材料在绿色样品制备技术方面的主要应用，并提供了相关参考文献，以供感兴趣的读者参考。

18.2 硅基吸附剂

二氧化硅（silico oxide）是目前分离科学中应用最多的无机聚合物。通过调节聚合物的形貌或改变其极性，从而实现样品的萃取和预富集，是一种非常有意思的方法。自固相萃取（SPE）技术发展以来，二氧化硅作为制样中的吸附剂一直备受关注[1]。此后，二氧化硅被广泛应用于现代及微量样品制备技术。

硅基材料的特性极大地依赖于其制备方法。在合成二氧化硅粒子的方法中，我们要重点关注烷氧基硅烷的水解、硅酸盐盐溶液的凝胶化以及稳定溶胶的可控聚集[2]。前两种方法包括了溶胶-凝胶技术，它们已被广泛用于新型吸附剂的合成中。

溶胶一词指的是胶体粒子的分散，而凝胶则特指由胶粒或聚合物链的刚性结构形成的系统，液相被锁定在系统的间隙当中。溶胶凝胶过程分为两个不同的阶段：①水解；②二氧化硅前驱体在催化剂和溶剂存在下的缩合反应[3]。在水解阶段，前驱体反应形成硅醇基（Si-OH）。然后，部分水解前体之间的缩合反应形成胶体溶液。通过胶体颗粒的聚集，聚合物链生长形成一种交联的三维宏观结构，称为凝胶。

溶胶-凝胶技术是一种通用而受到广泛重视的技术，通过调节反应条件（前体的性质、类型和浓度、反应温度、溶剂、催化剂、pH、反应时间、老化时间和干燥温度），即可实现无机和杂化（有机-无机）吸附剂的可控制备[4]。催化介质可以影响吸附剂的最终特性。例如，碱性 pH 条件容易生成高交联度的聚合物，而酸性 pH 条件往往生成低交联网状物[5]。

由于成本较低，硅酸盐溶液的凝胶化往往成为合成硅基吸附剂的基础。虽然这种方法会含有大量的金属（金属>0.3%），但用阳离子树脂和酸洗进行预处理可以减少杂质并提高吸附剂的质量[6,7]。硅酸盐溶液在高 pH 下是稳定的，只有在反应介质中加入酸才会形成聚合物网络结构。值得注意的是，其他低成本硅源也可用于制备吸附剂，包括沙子、黏土和稻壳[8]。

烷氧基硅烷成为溶胶凝胶工艺中最常用的二氧化硅前驱体，是因为能制备出的低金属含量的聚合物网络。四甲氧基硅烷（TMOS）和正硅酸乙酯（TEOS）是制备二氧化硅吸附剂最常用的试剂。此外，其他二氧化硅前驱体也可用于制备混合吸附剂，以及改变聚合物表面的极性和选择性。因此，一些功能基团，如氨基（3-氨基丙基三乙氧基硅烷-APTES）、氰基（3-氰基丙基三乙氧基硅烷-CNPrTEOS）、氯取代基（3-氯丙基三乙氧基硅烷-CPTMS）、巯基（3-巯基丙基三甲氧基硅烷-MPTS），环氧基（3-缩水甘油氧基丙基三甲氧基硅烷-GPTMS）、烯烃（烯丙基氧基三甲基硅烷-ATMS）、丙烯酸酯（3-丙烯酰氧基丙基三甲氧基硅烷-AcPTMS）、甲基丙烯酸酯（γ-甲基丙烯酸氧基丙基三甲氧基硅烷-γ-MPTS），以及二胺基［3-（2-氨基氨乙基）丙基三甲氧基硅烷-AATMS］可以容易地并入二氧化硅网络中[9]。必须强调的是，使用这类混合前体通常有两个目的，一是对二氧化硅聚合物表面进行包覆，二是形成聚合物网络。

二氧化硅网络中存在的一些反应性基团（如 NH_2、SH、丙烯酸酯和环氧基团）极大地延展了吸附剂改性的可能性。这一方法可以合成基于下列材料的新型吸附剂，包括碳材料［碳纳米管（CNTs）、石墨烯及其衍生物］、离子液体（ILs）、分子印迹聚合物、免疫亲和性、有机聚合物、整体材料或其他材料等。例如，Zhu 等[10]制备了具有弱阳离子交换特性的硅基吸附剂（图 18.1）。首先，由 CPTMS 制备叠氮基团，然后，通过点击反应将羧基和烷基固定在二氧化硅微球上。该混合型硅基吸附剂可用于从环境水样中提取芳香伯胺。

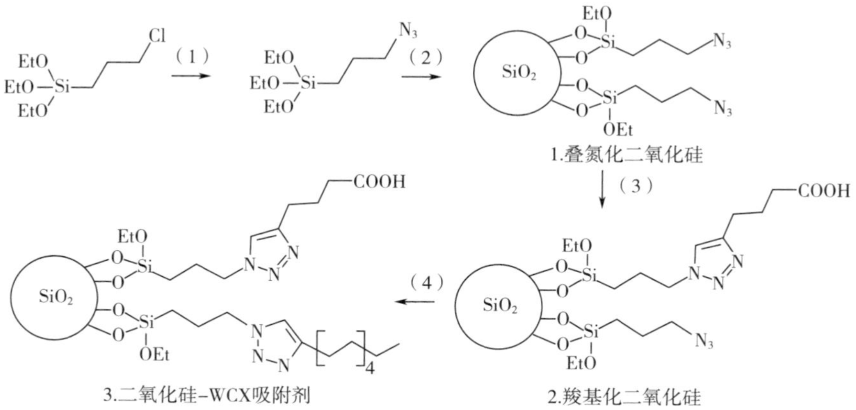

图 18.1 混合型硅基吸附剂的合成路线：（1）叠氮化物改性二氧化硅前驱体；（2）叠氮化物前驱体对硅球进行功能化；（3 和 4）弱阳离子交换，以及通过 Cu（I）催化叠氮炔烃环加成点击反应，制备疏水官能基团

［资料来源：Y. Zhu，W. Zhang，L. Li，C. Wu，J. Xing，Preparation of a mixed-mode silica-based sorbent by click reaction and its application in the determination of primary aromatic amines in environmental water samples，Anal. Methods，6（2014）2102-2111.］

二氧化硅良好的机械强度、热稳定性、化学惰性、比表面积和孔结构可控、价格低廉、与有机和水介质相容性好等优点，使其成为制样步骤用吸附剂而备受人们的关注。此外，二氧化硅吸附剂的合成条件温和，且吸附剂的选择性可通过调整前体（结构基团）或对聚合物表面的合成后改性来进行相应地设计。

硅基吸附剂可制备成颗粒、整体材料、薄膜、纤维或涂层形式。这一特性使其可用在多种制样技术当中，如固相萃取（SPE）、固相微萃取（SPME）、搅拌棒吸附萃取（SBSE）、填充吸附微萃取（MEPS）、中空纤维液相微萃取及各种衍化技术。例如，二氧化硅前驱体被广泛用作制备 SPME 的新型纤维，或者 SBSE 涂层的关键材料。因此，与商业纤维和涂层相比，更高化学稳定性和热稳定性的聚合物在提取各种分析物方面显示出了诱人的前景。硅基吸附剂在食品、饮料、环境和生物样品分析中的应用也有较多文献报道，主要用于金属和有机化合物的萃取、净化及预浓缩[11-14]。

18.3 离子液体

离子液体是室温（25℃）下的液态盐，其熔点低于 100°[15]，于 1976 年由 Atwood 等首次确定[16]。然而，直到 1992 年才合成了第一种在空气和水条件下稳定的离子液体，1-乙基-3-甲基咪唑[15]。

离子液体具有独特的物理化学性质，在不同领域得到了广泛的应用。离子液体由两部分组成，一是来源于路易斯碱的有机阳离子（如咪唑啉、吡咯烷、吡啶、四烷基铵、四烷基膦和硫鎓离子），二是含有不同无机或有机结构的多原子阴离子（如咪唑啉、吡咯烷、吡啶、四烷基铵、四烷基膦和硫鎓离子）组成。含咪唑阳离子的离子液体是目前使用最多的离子液体。

由于阴、阳离子之间可以进行不同的配对组合，离子液体又被称为“可设计溶剂”，它们的黏度允许其在水和有机溶剂中混溶，从而扩大了它们在不同领域中的应用[9]。

最近，基于离子液体的材料已被应用到高效液相色谱（HPLC）和气相色谱（GC）中的固定相，SPME 纤维上的涂层，LLE 及分散液液微萃取中的溶剂萃取器，修饰电极和 SPE 上使用不同载体（如二氧化硅）的吸附剂，多壁碳纳米管、石墨烯（G）以及在其他材料上面的应用[15,17]。

基于离子液体的材料可以制备于某个基体之上，而离子液体则作为功能性基团，通过化学键合或固定在基体上面。通常，基材为二氧化硅颗粒、聚合物、石墨烯、碳纳米管、熔融石英及磁性纳米颗粒。离子液体与不同基体的结合可以让材料同时具有二者的特性。

如图 18.2 所示，可以采用不同的方法将离子液体键合到二氧化硅表面上。在所有的步骤当中，第一步都是使用强酸来活化二氧化硅颗粒。常见的制备方法是将活化后的二氧化硅粒子与一种硅烷偶联剂反应得到氯丙基硅，然后氯丙基硅与咪唑或烷基支链咪唑衍生物反应，剩余的氯再与其他阴离子进行离子交换，最终得到所需的离子液体［图 18.2（1）］[18]。

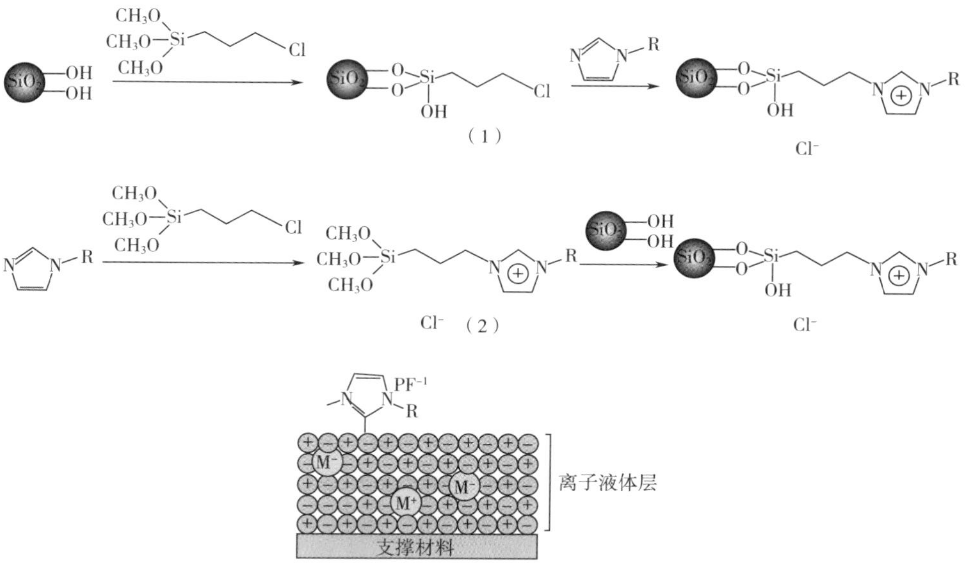

图 18.2　在二氧化硅表面键合离子液体的合成方法：
（1）活性二氧化硅粒子与硅烷偶联剂的反应，之后再与咪唑或烷基支链咪唑衍生物反应；
（2）硅烷偶联剂之间的反应；（3）不同基体上的多层反应

［资料来源：N. Fontanals, F. Borrull, R. M. Marcé, Ionic liquids in solid-phase extraction, TrAC Trends Anal. Chem. 41（2012）15-26.］

还有其他的一些方法，如硅烷偶联剂、咪唑（或咪唑衍生物）与活性硅粒子进行反应［图 18.2（2）］。还有第三类方法，可能是在基体上形成多层结构，如图 18.2（3）[18]所示。

离子液体附加到基体上的困难之处在于其附着性差，形成的键较弱，带来的后果就是离子液体的流失。为了避免离子液体流失，可以采用共价键结合的方式，将聚合物与离子液体嫁接到材料的表面[19]。

利用 ILs 制备的 SPME 纤维，既具有较高的热稳定性，又具有理想的选择性，从而受到了特别的关注。Anderson 等人将基于 ILs 的新型纤维应用于不同的领域，并研究了不同的合成路径，以避免纤维涂层的流失[20,21]。

ILs 和聚合物 ILs（PILs）有许多不同的涂层固定方法，包括使用烷氧基硅烷功能化 IL 对基体进行硅烷化改性、将 IL/PILs 键合的二氧化硅粒子黏合到纤维基体上、离子液体掺杂聚苯胺复合物的电化学沉积、溶胶凝胶技术，衍生化底物上的表面自由基链转移反应和衍生化底物上的自由基共聚[21]。

PILs 作为 SPME 的涂层是由 Anderson 等人开发的[22]，最近，该方法已被用于 SPME 直接浸入的模式，与 HPLC 进行联用。一些研究表明，基于 PIL 的 SPME 纤维可用于 GC 和 HPLC，在有机溶剂中表现出较好的稳定性，并且延长了纤维的使用寿命。

由于 ILs 具有良好的化学稳定性、热稳定性和较高的萃取效率等优点，近年来人们对 ILs 吸附剂的应用进行了大量的研究。此外，这些材料已广泛应用于不同领域的半极性和非极性化合物的测定[23-30]。

不同基体与 ILs 的结合赋予了材料的多功能化，其用于小型制样技术当中，显示出了较好的选择性，预浓缩因子也得以改善，并且环境友好。关于 ILs 在样品制备中的更多应用，读者可参考 Lucena 和 Cardenas[30a]的“离子液体在样品制备中的应用”一章。

18.4 分子印迹技术

印迹技术包括分子印迹（MIPs）、分子印迹整体材料和离子印迹聚合物。基于印迹技术的材料对目标分子/离子或类似分析物具有特殊识别位点，从而具有高的选择性[31]。

第一批关于印迹聚合物合成的报道见于 20 世纪中叶，由 Polyakov[32]和 Dickey[33]使用二氧化硅来构造聚合物网络。然而，直到 1972 年，Wulff 和 Sarhan 才将有机聚合物用于分子识别[34]。研究人员用共价的方法合成了一种对甘油酸具有选择性位点的有机聚合物。九年后，Arshady 和 Mosbach[35]采用非共价的方法制备出了 MIP。

提取生物基质中有机化合物的研究表明，MIP 可以作为提取材料来使用[36,37]。在验证了印迹技术的提取能力之后，许多文献报道了如何提高样品制备过程中的选择性[38]。

一般来说，MIPs 的合成是通过前体（由催化剂引发）在分散于溶剂中的模板存在下的聚合反应来实现。在模板周围形成三维聚合物网络后，后者被移除，从而形成具有特定相互作用位点的选择性空腔（图 18.3）[39]。

制备选择性空腔的方法有三种：共价、半共价和非共价。合成选择性好的 MIPs 主要是基于共价的方法，包括聚合物和模板之间化学键的形成/裂解。其缺点是可用于该合成路线的前体较少。在半共价法合成过程中，聚合物与模板之间形成化学键；然而，在萃取步骤，有可能产生非共价键，进而导致选择性降低。非共价的方法可以拥有大量的聚合物前体供选择，后期分析物从选择性空腔中的脱出条件也较为温和。虽然这是最常用的合成方法，但由于功能基团是游离于聚合物网络之中，从而导致了较低选择性[38,40]。

印迹聚合物可分为有机、无机或杂化材料。对于有机 MIPs，通常是将功能单体（甲基丙烯酸、4-乙烯基吡啶、丙烯酰胺、甲基丙烯酸羟乙基酯）、交联剂（乙烯基二甲基丙烯酸乙酯、三羟甲基丙三烯甲基丙烯酸和二乙烯基苯）、模板和自由基引发剂（双异丁腈）的混合物溶解在溶剂中来进行制备。在制备有机 MIPs 的方法中，本体聚合、沉淀聚合或悬浮聚合是特别值得注意的制备方法[40]。

无机和杂化印迹聚合物则采用溶胶-凝胶法来制备。分子印迹溶胶-凝胶材料一般是基于二氧化硅、二氧化锆或二氧化钛的聚合物。二氧化硅的合成路线首选 TEOS 或 TMOS 的水解缩合反应。分子识别主要来自于聚合物中的硅醇基团与模板之间的非共价相互作用（氢键）[41]。

其他二氧化硅前驱体，如 APTES、MPTS 和 CNPrTEOS，已用于合成杂化聚合物，

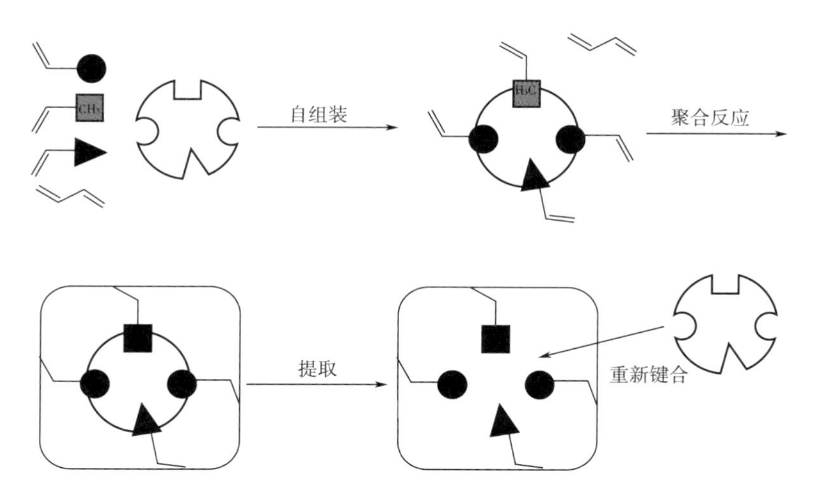

图 18.3　印迹技术示意图

[资料来源：S. Eppler, M. Stegmaier, F. Meier, B. Mizaikoff, A novel extraction device for efficient clean-up of molecularly imprinted polymers, Anal. Methods 4 (2012) 2296-2299.]

并扩展了相互作用的基本原理[42]。另一个值得关注的方法是有机单体与溶胶凝胶前驱体之间的结合。有机单体（γ-MPTS、ATMS 和 AcPTMS）进入二氧化硅网络后，有机单体与交联剂发生反应并形成一层有机层。杂化 MIP 则具有有机聚合物和无机聚合物双重特性的优点。此外，需要强调的是，二氧化硅网络中参加反应的功能基团给新型选择性材料的制备提供了更多的可能性，这主要通过设计不同的结构单元来实现[43]。

在样品制备过程中，如何选择合适的位点，主要是基于在不损失目标分析物的情况下，如何更有效地去除干扰。因而，在方法验证中，基体效应、回收率、精密度和准确度可以获得了较好的结果。此外，MIP 技术还具有合成成本低、物理强度高、热稳定性好、可重复使用等优点。

前驱体的选择是制备印迹材料时需要明确的一个重要参数（应考虑模板和前驱体之间相互作用的强度和性质）。此外，为了提高分子识别率和避免聚合物溶胀，萃取介质通常与合成 MIP 时使用的溶剂相似。

除了溶胀问题外，MIP 合成后的模板去除可能很耗时，甚至有时候还无法完全去除（模板渗出）。解决这种情况的方法是使用与目标分析物结构相似的伪模板[44]。

印迹聚合物主要用于 SPE（离线与在线），但在 SPME、管内 SPME、SBSE、MEPS 和膜萃取等其他制样技术方面，印迹聚合物也表现出了很高的性能，用于不同基质中药物、个人护理产品、农药、食品添加剂和金属离子的预富集和净化[25,31,45]。

18.5　限进材料吸附剂

Desilets 等在 1991 年引入了一种新的固定相，能够从生物基质中排除大分子物质（内源性化合物），称之为限进材料[46]。

RAM 结构的特点是在载体颗粒的外表面覆盖亲水基团，亲水基团起到亲水屏障的作用，使小分子能够通过固定相的疏水部分进行渗透，而大分子则被物理或化学机制，或其联合机制排除在外[47,48]。

事实上，RAM 吸附剂已经引起了相当大的兴趣，因为它们的亲水表面阻止了蛋白质的不可逆吸附，直接允许从富含蛋白质的基质中提取小分子。包括对复杂样本（如血清、血液、尿液和细胞培养物等）中药物、肽和内源性物质的测定[49]。

在制备 RAM 吸附剂的各种方法中，重点应关注原位制备，其包括两个简单步骤。首先，用载体来填充柱子，然后再将功能材料固定在颗粒表面。一般用于制备 RAM 吸附剂的载体包括多孔二氧化硅或多孔聚合物，而功能材料具有非吸附性和亲水性表面。

RAM 按其颗粒结构可分为内表面反相、屏蔽疏水相（SHP）、半渗透表面（SPSs）、双区相和混合功能相，如图 18.4 所示。

市面上有很多种商业化的 RAM 吸附剂，包括甘氨酸-L-苯丙氨酸-L-苯丙氨酸、烷基二醇二氧化硅、SPS 相、SHPs、α_1-酸性糖蛋白、复合配体和混合功能材料[46]。此外，各研究小组还采用了不同的功能材料来作为 RAM 吸附剂，如牛血清白蛋白、单甲基丙烯酸甘油酯、二甲基丙烯酸甘油酯、甲基丙烯酸缩水甘油酯和糖[50-52]。

基于 RAM 吸附剂的新材料的应用，如分子印迹聚合物（MIPs）与 RAM 的复合物、整体材料与 RAM 的复合物、磁性材料与 RAM 的复合物等，提高了萃取方法的选择性。

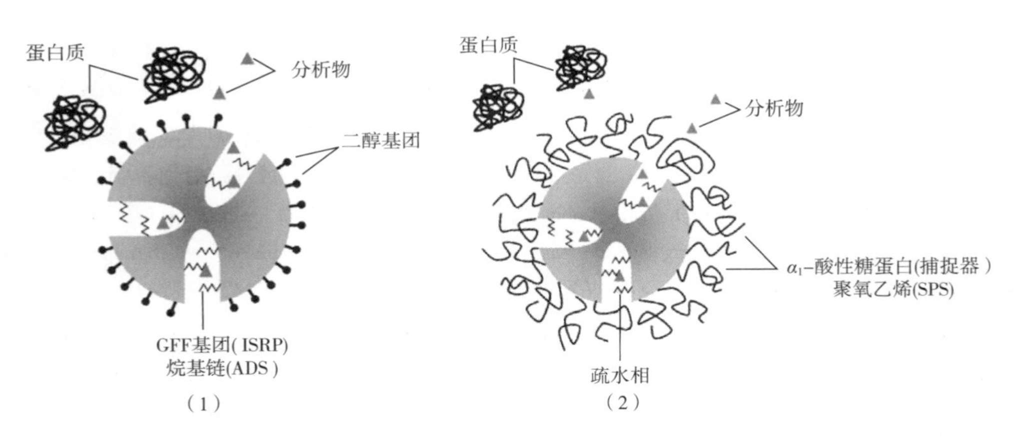

图 18.4　限进材料粒子结构示意图（1）具有甘氨酸-L-苯丙氨酸-L-苯丙氨酸基团的内表面反相（ISRP）和带有烷基链的烷基二醇二氧化硅（ADS）。（2）半透膜表面（SPS）和蛋白涂层二氧化硅支架

［资料来源：S. Souverain, S. Rudaz, J. L. Veuthey, Restricted access materials and large particle supports for on-line sample preparation: an attractive approach for biological fluids analysis, J. Chromatogr. B 801（2004）141-156.］

RAM 改性 MIP（RAM-MIP）材料是一种新型的高分子材料，它能有效地排除高分子，保留低分子量化合物。这些改性方法是将亲水单体键合在 MIP 表面上。

有文献报道了一种由疏水性内表面和亲水性外表面组成的新型限进材料-杂化整体柱。大孔结构可以使用较高的流速来避免堵塞柱子[53]。

限进材料磁性吸附剂则是另一种发展趋势，其利用纳米颗粒或微球来作为载体，

包括 $Fe_3O_4@SiO_2$、$\gamma-Fe_3O_4$ 或 $\gamma-Fe_2O_3$ 等。这种材料的主要优点是选择性高，该材料的亲水外表面可阻止蛋白质的不可逆吸附和变性，从而成为快速而简便的生物样品提取方式[54]。

还可以利用多孔二氧化硅颗粒作为载体，以亲水性聚合物（聚乙烯醇、牛血清白蛋白等）为载体，内部键合上手性选择剂（b-CD、糖肽抗生素等），从而制备了限进手性固定相或吸附剂[55]。

RAM 材料具有广泛的应用前景，它们与其他材料的结合更能提高萃取方法的选择性。因为多孔载体的存在，可允许使用较高的流速从而避免柱子堵塞。比较有意思的是将这些材料应用于小型制样，除了对环境友好外，对于样品量有限的生物样品（富含蛋白质的基质）来说，它还是一种可行的替代品。

18.6 石墨烯材料

自 2004 年发现石墨烯[56]以来，人们就开始关注石墨烯在样品制备及相关技术方面的应用。与碳纳米管和富勒烯一样，石墨烯也是石墨的同素异形体，但由于石墨烯的纳米片结构，因此，作为吸附剂，它在制样技术中还具有另外的优势。当分子试图进入其内壁时，富勒烯和碳可能存在空间位阻，而石墨烯则不会存在这种情况，因为石墨烯的两个表面都可以进行分子吸附[57]。图 18.5 显示了不同结构中碳的 sp^2 键合形式[58]。

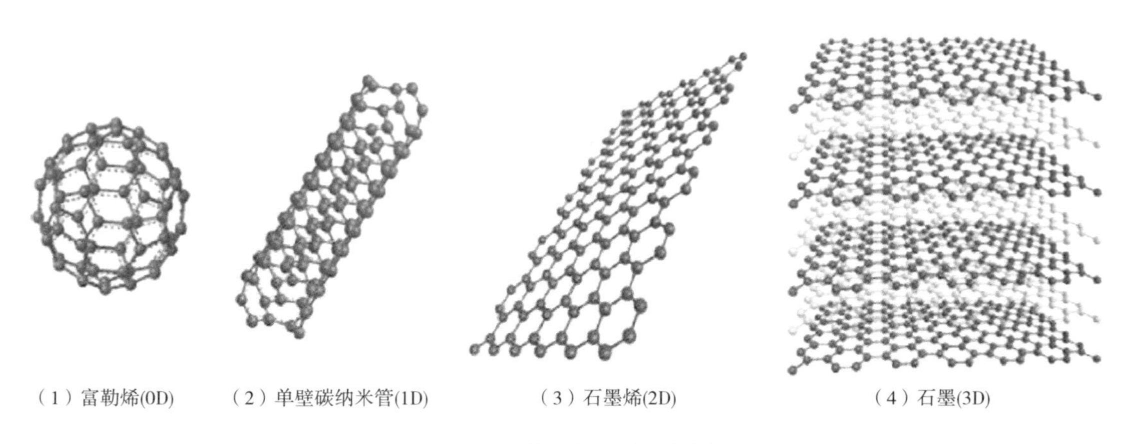

图 18.5 碳的各种 sp^2 杂化形态

［资料来源：Elsevier S. Basu，P. Bhattacharyya，Recent developments on graphene and graphene oxide based solid state gas sensors，Sens. Actuators B Chem. 173（2012）1-21.］

石墨烯的特性还包括在制样方面的应用优势，如热稳定性、机械稳定性、高比表面积和离域 π 电子系统，该系统与芳香环的分子［如杀虫剂、医药和兽药，和多环芳烃（PHAs）等］有很好的亲和力。此外，对于不含芳环的分析物，可使用氧化石墨烯（即石墨烯前体）来探索其可用性。石墨烯和氧化石墨烯的主要区别在于极性的大小，因为氧化石墨烯比石墨烯具有更多的羧基和羟基。

一些研究者将氧化石墨烯还原后得到的石墨烯命名为还原氧化石墨烯、化学还原氧化石墨烯甚至化学转化石墨烯（如图 18.6 所示）；原因之一是纯的石墨烯不表现出含有羟基或羧基[59]。在关于样品制备的文献中，以上几个名称都曾被人们所使用。出于教学的目的，我们使用石墨烯这一术语，是考虑到合成石墨烯的方法，这一术语很容易理解。图 18.6 显示了从石墨开始制备石墨烯的方法[60]。

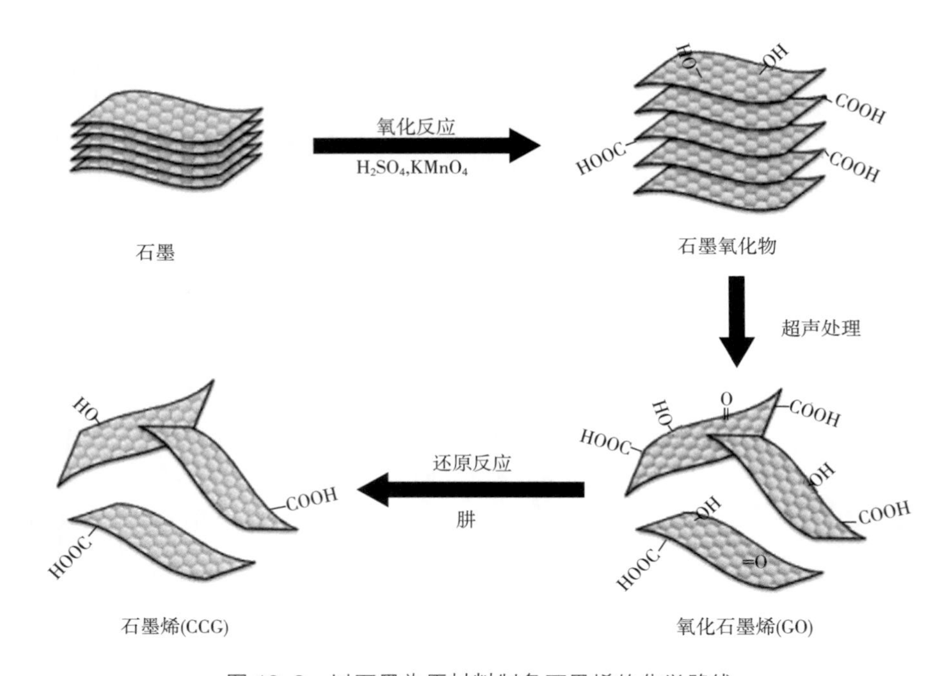

图 18.6　以石墨为原材料制备石墨烯的化学路线

［资料来源：Elsevier Q. Liu，J. Shi，G. Jiang，Application of graphene in analytical sample preparation，TrAC，Trends Anal. Chem. 37（2012）1-11.］

石墨烯很容易从石墨中获得，不需要昂贵的试剂和仪器设备。这个过程是从氧化石墨来获得石墨氧化物开始，通常采用最早由 Hummers 提出的制备方法[61]。然后，石墨氧化物经过一个剥离（超声波处理）的过程，该过程包括制备石墨氧化物溶液（通常为 1mg/L），并置于超声波中 1h。溶液被冻干后，形成一种褐色的粉末材料（氧化石墨烯），最后可以用肼将其还原为石墨烯。另一种方法是将这种氧化石墨烯键合到另一种材料上（例如，通常采用氨丙基硅的方法），以获得氧化石墨烯，其上含有共价方式结合的氨丙基硅，而氨丙基硅在后续中可以被还原[57,62]。

在样品制备过程中使用石墨烯甚至氧化石墨烯，优点之一是它们符合绿色化学的原则。较高的比表面积使得这两种材料可以用在只需少量吸附剂的小型制样技术中，如 SPME、SBSE 和分散 SPE。但是，这一特性可能会在其他的技术中产生问题。例如，在 MEPS 和 SPE 中，可能会产生熔块阻塞甚至高背压，从而使得提取困难或不可能实现。为了克服这一困难，氧化石墨烯必须预先铺设在另一个表面上，在这种情况下，最常用的是氨丙基硅的方法。

石墨烯或氧化石墨烯在制样中的应用主要与 SPME 有关。既然这样，材料在纤维上的固定方式，就可以通过共价键的方式结合到预处理过的石英毛细管上[63]，或通过黏在不锈钢丝上[64]来完成。第一种方式中，必须使用氧化石墨烯将其结合到预处理过的石英毛细管壁上，例如利用 APTES 上的 NH_2 与氧化石墨烯上的—COOH 反应[63]。类似的，也可将氧化石墨烯与氨丙基硅相结合。尽管 SPE[62] 中已经使用了这种方法，但目前未见其在 MEPS 中使用的报道。这种方法可以将石墨烯或氧化石墨烯铺设到球形甚至不规则形状的二氧化硅或改性二氧化硅表面。在 SBSE 中，石墨烯可以用作聚合物复合材料，也可以像在 SPME、SPE 中采用的共价结合方式。

在大多数情况下，使用石墨烯和氧化石墨烯吸附剂，通常会使目标化合物在其结构中呈现出芳香环（例如，PHAs、磺胺类、拟除虫菊酯、氯酚类、氟喹诺酮类等）。在环境和食物样品中都检出了这些化合物，这表明很大可能在样品制备中使用了石墨烯材料作为吸附剂，这一原因尚未得到明确的解释[15]。

18.7 磁性材料

磁性材料的应用是在 1973 年由 Robinson 等首次报道[65]，用经溴化氰活化过的氧化铁来实现对酶的固定。从那时起，磁性材料开始在制样技术中脱颖而出，主要是因为这些材料可能与本章已讨论过的其他吸附材料结合在一起。这种磁性材料可引入含有目标化合物的样品中，通过施加磁场（例如，使用磁棒）进行分离。这种材料能够克服纳米材料在基于吸附剂的萃取技术中的限制，由于吸附剂和样品之间高度接触，通常能够提供非常好的回收率。

在这种方法中使用的所有材料都具有磁性，因此当外部施加磁场时，它们会受到强烈的吸引。当磁性材料用在样品制备中时，所采用的方法类似于分散固相萃取（SPE），因此被称为磁性固相萃取（MSPE）。在这种方式下，一般将材料分散到溶液中，并保持一段时间，即所谓培养时间。然后施加磁场并除去样品溶液。下一步是清洗材料以消除可能的干扰；再利用磁场来来分离洗涤液。最后一步是将吸附在磁性材料表面的目标化合物洗脱，并将提取物注入色谱系统进行分析。图 18.7 显示了使用磁性材料作为吸附剂进行提取的简要过程[66]，该过程并没有洗涤步骤。

在所有情况下，磁性材料都是经涂膜或与其他材料复合后来使用[67]。通常使用磁性氧化铁颗粒（Fe_3O_4 和 γ-Fe_2O_3）、铁（Fe）、钴（Co）或镍（Ni）来制备磁性材料。这些金属或金属氧化物负责提供磁性；它们可以和先前相关的材料（通过溶胶凝胶工艺获得的二氧化硅、ILs、石墨烯或氧化石墨烯、MIP、RAM 及其他基于溶胶凝胶技术的材料）一起回收利用。磁性吸附剂具有容易制备、易表面改性、操作简单、回收率高、材料可重复使用、萃取过程简单等优点。例如，在分散 SPE 中，当使用纳米材料时，通常需要额外的离心分离来分离剩余的固体材料。离心步骤中的高转速可能会导致一些干扰物的沉淀，并且一些目标化合物也可能会共沉淀，特别是生物样品[68]。在这种情况下，使用磁性材料有助于克服这一限制，其很容易被磁场从溶液中分离出来。

这些材料已应用于食品、生物或环境基质中的几种污染物残留分析，例如农药、

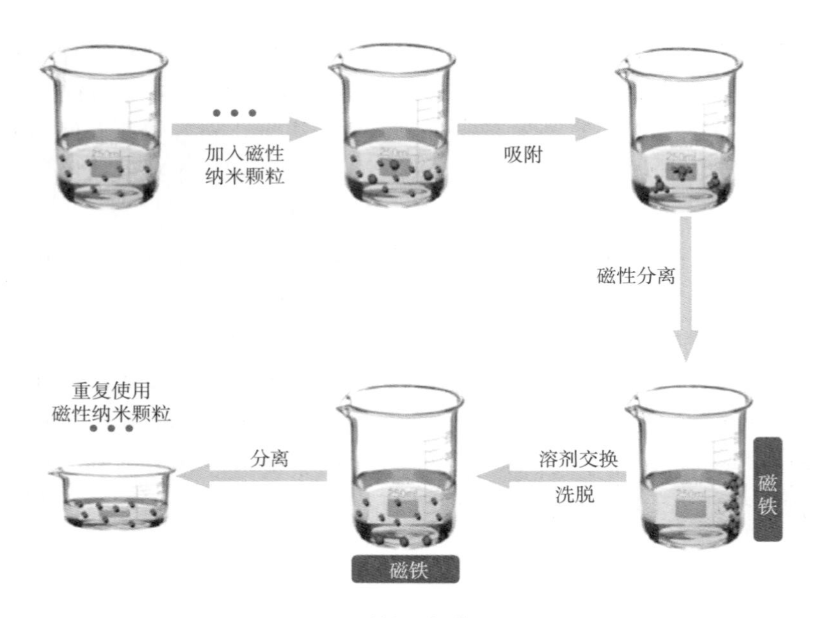

图 18.7　磁性固相萃取的步骤

［资料来源：J. Płotka-Wasylka, N. Szczepanska, M. de la Guardia, J. Namiesnik, Miniaturized solid-phase extraction techniques, TrAC, Trends Anal. Chem. 73（2015）19-38.］

药物等的检测和定量分析[68,69]。固定在磁性粒子表面的材料必须进行仔细挑选，充分考虑目标化合物的性质和可能存在的样品干扰。理想情况下，它必须与目标化合物有亲和力，而与干扰物的亲和力较差。前面第一部分中讨论的材料均可以成为具有磁性的复合材料的原材料。

18.8　结论与展望

最近的文献中出现了大量属于不同化学和物理类别的新化合物，旨在制备选择性更强、效率更高、更环保的样品制备材料。这使得开发更有效的合成方法成为可能，包括溶胶凝胶和分子印迹技术等，从而允许制备更具选择性的材料。最后，对绿色制样技术感兴趣的分析人员可以收获几种新的样品制备方法、技术和材料。这些可能性均可在表 18.1 中找到，其中列出了这些材料在现代制样技术中的部分应用及其相关参考信息。

如表 18.1 所示，在绿色样品制备中对这些材料的使用，有两种方法可以认为是该领域的一个发展趋势：样品制备（SPME、SBSE、MEPS 等）的微型化，以及在线处理与色谱联用技术，或是两者的结合。预计在未来几年，这两种方法的使用将更加普遍。

表 18.1　　新材料在制样技术中的应用

材料	化合物	基质	制样技术	参考文献
正硅酸乙酯/三异丁基硫化膦	汞	沉积物	SPE	[70]

续表

材料	化合物	基质	制样技术	参考文献
正硅酸乙酯/多壁碳纳米管/K_7［$PW_{11}NiO_{40}$］	萘普生	毛发	HF-μ-SPE	［71］
聚四氢呋喃（溶胶-凝胶法制备）	雌激素	尿液	膜分离	［72］
聚吡咯/β-萘磺酸（溶胶-凝胶法制备）	挥发性有机化合物	红头蚁和芫荽	SPME	［11］
PDMS/PTMS/β-CD（溶胶-凝胶法制备）	类固醇	自来水和污水	SBSE	［73］
IL-修饰磁性高分子微球	磺胺类	尿液	MDSPE	［10］
IL-MIP	硫铬烷酮衍生物	尿液	PT-SPE	［12］
IL-浸渍 SPME 纤维	甲基苯丙胺和苯丙胺	人类尿液	SPME	［29］
N-甲基咪唑	除草剂	水和土壤	SPE-LC-UV	［74］
1-丁基-3-甲基咪唑六氟磷酸	镉	水	μ-SPE-LC-UV	［75］
TMOS/MPTS（IIP）	银	头发和指甲	SBSE	［76］
MAA/EGDMA（MIP）	邻苯二甲酸二甲酯	瓶装饮料	SPME	［77］
MAA/EGDMA（MIP）	三嗪	玉米	MEPS	［78］
4-VP/EGDMA（MIP）	倍硫磷	鱼，鸡肉和蜂蜜	SPE	［79］
吡咯/EGDMA（MIP）	吲哚美辛	血液、血浆和尿液	管内 SPME	［80］
ADS/C_{18}	他汀类	人类血浆	在线 SPE	［51］
PHEMA/β-CD	手性药物	血浆	直列式 SPE	［81］
ADS/C_{18}，SPS/C_{18}，BSA/C_{18}	抗抑郁药、驱虫药	尿液和血浆	在线 μ-SPE	［82］
m-APBA/二醇/环氧丙基/Fe_3O_4@SiO_2	多巴胺，肾上腺素，去甲肾上腺素	尿液	离线 MSPE	［83］
ODS-IIP	铜	尿液和血浆	在线 SPE	［84］
氧化石墨烯	三环类抗抑郁药	血浆	顶空 SPME	［85］
石墨烯/聚苯胺	醛类	人类呼气冷凝液	管内 SPME	［86］
石墨烯	氯酚	河水	SPE	［87］
还原石墨烯@硅	氟喹醇	水	SPE	［88］
石墨烯	磺胺类	牛乳	PT-SPE	［89］

续表

材料	化合物	基质	制样技术	参考文献
磁性 MIP	可卡因	血浆	MSPE	[90]
磁性 MWCNT-IL	溶菌酶	蛋白质	MSPE	[69]
Fe_3O_4@SiO_2-石墨烯	多环芳烃	环境水	MSPE	[69]
磁性材料	氨基甲酸酯	水	MSPE	[91]

注：IIP—离子印迹聚合物；IL—离子液体；LC—液相色谱；MWCNT—多壁碳纳米管；MSPE—磁固相萃取；SPE—固相萃取；SPME—固相微萃取。

致谢

作者非常感谢圣保罗研究基金会（FAPESP）基金#2015/15462-5、#2014/07347-9、#2015/12306-2、#2012/22055-9，以及国家科学技术发展理事会（CNPq）基金#307293/2014-9 提供的经费支持。

参考文献

[1] I. Liška, Fifty years of solid-phase extraction in water analysis-historical development and overview, J. Chromatogr. A 885 (2000) 3-16.

[2] J. Nawrocki, C. Dunlap, A. McCormick, P. W. Carr, Part I. Chromatography using ultra-stable metal oxide-based stationary phases for HPLC, J. Chromatogr. A 1028 (2004) 1-30.

[3] L. L. Hench, J. K. West, The sol-gel process, Chem. Rev. 90 (1990) 33-72.

[4] G. Kickelbick, Hybrid Materials, first ed., Wiley-VCH, Weinheim, 2006.

[5] A. Kloskowski, M. Pilarczyk, W. Chrzanowski, J. Namiesnik, Sol-gel technique-a versatile tool for adsorbent preparation, Crit. Rev. Anal. Chem. 40 (2010) 172-186.

[6] D. A. Barrett, V. A. Brown, R. C. Watson, M. C. Davies, P. N. Shaw, H. J. Ritchie, P. Ross, Effects of acid treatment on the trace metal content of chromatographic silica: bulk analysis, surface analysis and chromatographic performance of bonded phases, J. Chromatogr. A 905 (2001) 69-83.

[7] C. E. D. Nazario, P. C. F. de Lima Gomes, F. M. Lancas, Analysis of fluoxetine and norfluoxetine in human plasma by HPLC-UV using a high purity C_{18} silica-based SPE sorbent, Anal. Methods 6 (2014) 4181-4187.

[8] E. R. Essien, O. A. Olaniy, L. A. Adams, R. O. Shaibu, Sol-gel-derived porous silica: economic synthesis and characterization, J. Miner. Mater. Charact. Eng. 11 (2012) 976-981.

[9] R. Liu, J. Liu, Y. Yin, X. Hu, G. Jiang, Ionic liquids in sample preparation, Anal. Bioanal. Chem. 393 (2009) 871-883.

[10] Y. Zhu, W. Zhang, L. Li, C. Wu, J. Xing, Preparation of a mixed-mode silica-based sorbent by click reaction and its application in the determination of primary aromatic amines in environmental water samples, Anal. Methods 6 (2014) 2102-2111.

[11] Z. Zhang, L. Zhu, Y. Ma, Y. Huang, G. Li, Preparation of polypyrrole composite solid-phase

microextraction fiber coatings by sol-gel technique for the trace analysis of polar biological volatile organic compounds, Analyst 138 (2013) 1156-1166.

[12] X. Mao, B. Hu, M. He, B. Chen, High polar organic-inorganic hybrid coating stir bar sorptive extraction combined with high performance liquid chromatography - inductively coupled plasma mass spectrometry for the speciation of seleno-amino acids and seleno-oligopeptides in biological samples, J. Chromatogr. A 1256 (2012) 32-39.

[13] W. Zhang, J. Du, C. Su, L. Zhu, Z. Chen, Development of B-cyclodextrin-modified silica and polyporous polymer particles for solid-phase extraction of methyl jasmonate in aqueous and plant samples, Anal. Lett. 46 (2013) 900-911.

[14] C. Basheer, W. Wong, A. Makahleh, A. A. Tameem, A. Salhin, B. Saad, H. K. Lee, Hydrazone-based ligands for micro-solid phase extraction-high performance liquid chromatographic determination of biogenic amines in orange juice, J. Chromatogr. A 1218 (2011) 4332-4339.

[15] B. H. Fumes, M. R. Silva, F. N. Andrade, C. E. D. Nazario, F. M. Lanças, Recent advances and future trends in new materials for sample preparation, TrAC, Trends Anal. Chem. 71 (2015) 9-25, http://dx.doi.org/10.1016/j.trac.2015.04.011.

[16] J. D. Atwood, J. L. Atwood, Inorganic compounds with unusual properties, in: Advances in Chemistry Series, American Chemical Society, Washington, 1976, pp. 112-122.

[17] A. Berthod, L. He, D. W. Armstrong, Ionic liquids as stationary phase solvents for methylated cyclodextrins in gas chromatography, Chromatographia 53 (2001) 63-68.

[18] N. Fontanals, F. Borrull, R. M. Marcé, Ionic liquids in solid-phase extraction, TrAC, Trends Anal. Chem. 41 (2012) 15-26.

[19] G. Zhao, S. Chen, J. Yue, R. He, Imidazolium-modified sulfonated polyetheretherketone for selective isolation of hemoglobin, Anal. Methods 5 (2013) 5425-5430.

[20] E. Wanigasekara, S. Perera, J. A. Crank, L. Sidisky, R. Shirey, A. Berthod, D. W. Armstrong, Bonded ionic liquid polymeric material for solid-phase microextraction GC analysis, Anal. Bioanal. Chem. 396 (2010) 511-524.

[21] H. Yu, T. D. Ho, J. L. Anderson, Ionic liquid and polymeric ionic liquid coatings in solid-phase microextraction, TrAC, Trends Anal. Chem. 45 (2013) 219-232.

[22] I. Pacheco-Fernández, A. Najafi, V. Pino, J. L. Anderson, J. H. Ayala, A. M. Afonso, Utilization of highly robust and selective crosslinked polymeric ionic liquid-based sorbent coatings in direct-immersion solid-phase microextraction and high-performance liquid chromatography for determining polar organic pollutants in waters, Talanta 158 (2016) 125-133.

[23] H. Yan, M. Gao, C. Yang, M. Qiu, Ionic liquid-modified magnetic polymeric microspheres as dispersive solid phase extraction adsorbent: a separation strategy applied to the screening of sulfamonomethoxine and sulfachloropyrazine from urine, Anal. Bioanal. Chem. 406 (2014) 2669-2677.

[24] X. Ding, Y. Wang, Y. Wang, Q. Pan, J. Chen, Y. Huang, K. Xu, Preparation of magnetic chitosan and graphene oxide-functional guanidinium ionic liquid composite for the solid-phase extraction of protein, Anal. Chim. Acta. 861 (2015) 36-46.

[25] Y. Yuan, S. Liang, H. Yan, Z. Ma, Y. Liu, Ionic liquid-molecularly imprinted polymers for pipette tip solid-phase extraction of (Z) -3- (chloromethylene) -6-flourothiochroman-4-one in urine, J. Chromatogr. A 1408 (2015) 49-55.

[26] H. Yan, S. Liu, M. Gao, N. Sun, Ionic liquids modified dummy molecularly imprinted

microspheres as solid phase extraction materials for the determination of clenbuterol and clorprenaline in urine, J. Chromatogr. A 1294 (2013) 10-16.

[27] K. Klotz, J. Angerer, Quantification of naphthoquinone mercapturic acids in urine as biomarkers of naphthalene exposure, J. Chromatogr. B 1012-1013 (2016) 89-96.

[28] W. Fan, X. Mao, M. He, B. Chen, B. Hu, Development of novel sol-gel coatings by chemically bonded ionic liquids for stir bar sorptive extraction - application for the determination of NSAIDS in real samples, Anal. Bioanal. Chem. 406 (2014) 7261-7273.

[29] Y. He, J. Pohl, R. Engel, L. Rothman, M. Thomas, Preparation of ionic liquid based solid-phase microextraction fiber and its application to forensic determination of methamphetamine and amphetamine in human urine, J. Chromatogr. A 1216 (2009) 4824-4830.

[30] M. Ebrahimi, Z. Es'haghi, F. Samadi, M. S. Hosseini, Ionic liquid mediated sol-gel sorbents for hollow fiber solid-phase microextraction of pesticide residues in water and hair samples, J. Chromatogr. A 1218 (2011) 8313-8321;

[30a] R. Lucena, S. Cárdenas, Ionic liquid in sample preparation, in: E. Ibanez, A. Cifuentes (Eds.), Green Extraction Techniques: Principles, Advances and Applications, vol. 76, 2017, pp. 203-224.

[31] K. Tang, X. Gu, Q. Luo, S. Chen, L. Wu, J. Xiong, Preparation of molecularly imprinted polymer for use as SPE adsorbent for the simultaneous determination of five sulphonylurea herbicides by HPLC, Food Chem. 150 (2014) 106-112.

[32] M. V. Polyakov, Adsorption properties and structure of silica gel, Zhur. Fiz. Khim. 2 (1931) 799-805.

[33] F. H. Dickey, The preparation of specific adsorbents, Proc. Natl. Acad. Sci. 35 (1949) 227-229.

[34] G. Wulff, A. Sarhan, Use of polymers with enzyme - analogous structures for resolution of racemates, Angew. Chem. Int. Ed. 11 (1972) 341-343.

[35] R. Arshady, K. Mosbach, Synthesis of substrate-selective polymers by host-guest polymerization, Die Makromol. Chem. 182 (1981) 687-692.

[36] G. Vlatakis, L. I. Andersson, R. Muller, K. Mosbach, Drug assay using antibody mimics made by molecular imprinting, Nature 361 (1993) 645-647.

[37] B. Sellergren, Direct drug determination by selective sample enrichment on an imprinted polymer, Anal. Chem. 66 (1994) 1578-1582.

[38] A. Martín-Esteban, Molecularly-imprinted polymers as a versatile, highly selective tool in sample preparation, TrAC, Trends Anal. Chem. 45 (2013) 169-181.

[39] S. Eppler, M. Stegmaier, F. Meier, B. Mizaikoff, A novel extraction device for efficient clean-up of molecularly imprinted polymers, Anal. Methods 4 (2012) 2296-2299.

[40] G. Vasapollo, R. Del Sole, L. Mergola, M. R. Lazzoi, A. Scardino, S. Scorrano, G. Mele, Molecularly imprinted polymers: present and future prospective, Int. J. Mol. Sci. 12 (2011) 5908-5945.

[41] J. Ou, Z. Liu, H. Wang, H. Lin, J. Dong, H. Zou, Recent development of hybrid organic-silica monolithic columns in CEC and capillary LC, Electrophoresis 36 (2015) 62-75.

[42] B. Du, T. Qu, Z. Chen, X. Cao, S. Han, G. Shen, L. Wang, A novel restricted access material combined to molecularly imprinted polymers for selective solid-phase extraction and high performance liquid chromatography determination of 2-methoxyestradiol in plasma samples, Talanta 129 (2014) 465-472.

[43] C. I. Lin, A. K. Joseph, C. K. Chang, Y. C. Wang, Y. Der Lee, Synthesis of molecular imprinted organic-inorganic hybrid polymer binding caffeine, Anal. Chim. Acta. 481 (2003) 175-180.

[44] E. Turiel, A. Martín-Esteban, Molecularly imprinted polymers for sample preparation: a review, Anal. Chim. Acta. 668 (2010) 87-99.

[45] X. Zhang, S. Xu, J. -M. Lim, Y. -I. Lee, Molecularly imprinted solid phase microextraction fiber for trace analysis of catecholamines in urine and serum samples by capillary electrophoresis, Talanta 99 (2012) 270-276.

[46] S. Souverain, S. Rudaz, J. L. Veuthey, Restricted access materials and large particle supports for on- line sample preparation: an attractive approach for biological fluids analysis, J. Chromatogr. B 801 (2004) 141-156.

[47] M. Rogeberg, H. Malerod, H. Roberg-Larsen, C. Aass, S. R. Wilson, On-line solid phase extraction-liquid chromatography, with emphasis on modern bioanalysis and miniaturized systems, J. Pharm. Biomed. Anal. 87 (2014) 120-129.

[48] S. H. Yang, H. Fan, R. J. Classon, K. A. Schug, Restricted access media as a streamlined approach toward on-line sample preparation: recent advancements and applications, J. Sep. Sci. 36 (2013) 2922-2938.

[49] N. M. Cassiano, J. C. Barreiro, M. C. Moraes, R. V. Oliveira, Q. B. Cass, Restrictedaccess media supports for direct high-throughput analysis of biological fluid samples: review of recent applications, Bioanalysis 1 (2009) 577-594.

[50] K. Hua, L. Zhang, Z. Zhang, Y. Guo, T. Guo, Surface hydrophilic modification with a sugar moiety for a uniform-sized polymer molecularly imprinted for phenobarbital in serum, Acta Biomater. 7 (2011) 3086-3093.

[51] V. F. Fagundes, C. P. Leite, G. A. Pianetti, C. Fernandes, Rapid and direct analysis of statins in human plasma by column-switching liquid chromatography with restricted access material, J. Chromatogr. B 947e948 (2014) 8-16.

[52] H. Sambe, K. Hoshina, K. Hosoya, J. Haginaka, Direct injection analysis of bisphenol A in serum by combination of isotope imprinting with liquid chromatography-mass spectrometry, Analyst 130 (2005) 38-40.

[53] Y. -K. Lv, Z. -Y. Guo, J. -Z. Wang, M. -M. Guo, L. -K. Yu, H. Fang, Preparation of a restricted access material-macroporous hybrid monolithic column-for on-line solid-phase extraction of the sulfonamides residues from honey, Anal. Methods 7 (2015) 1563-1571.

[54] L. Ye, Q. Wang, J. Xu, Z. -G. Shi, L. Xu, Restricted-access nanoparticles for magnetic solid-phase extraction of steroid hormones from environmental and biological samples, J. Chromatogr. A 1244 (2012) 46-54.

[55] Q. B. Cass, T. Ferreira Galatti, A method for determination of the plasma levels of modafinil enantiomers, (±) -modafinic acid and modafinil sulphone by direct human plasma injection and bidimensional achiral-chiral chromatography, J. Pharm. Biomed. Anal. 46 (2008) 937-944.

[56] K. S. Novoselov, Electric field effect in atomically thin carbon films, Science 306 (2004) 666-669.

[57] R. Sitko, B. Zawisza, E. Malicka, Graphene as a new sorbent in analytical chemistry, TrAC, Trends Anal. Chem. 51 (2013) 33-43.

[58] S. Basu, P. Bhattacharyya, Recent developments on graphene and graphene oxide based solid state

gas sensors, Sens. Actuators B Chem. 173 (2012) 1-21.

[59] D. R. Dreyer, S. Park, C. W. Bielawski, R. S. Ruoff, The chemistry of graphene oxide, Chem. Soc. Rev. 39 (2010) 228-240.

[60] Q. Liu, J. Shi, G. Jiang, Application of graphene in analytical sample preparation, TrAC, Trends Anal. Chem. 37 (2012) 1-11.

[61] W. S. Hummers, R. E. Offeman, Preparation of graphitic oxide, J. Am. Chem. Soc 80 (1958) 1339.

[62] Q. Liu, J. Shi, J. Sun, T. Wang, L. Zeng, G. Jiang, Graphene and graphene oxide sheets supported on silica as versatile and high-performance adsorbents for solid-phase extraction, Angew. Chem. Int. Ed. 50 (2011) 5913-5917.

[63] S. Zhang, Z. Du, G. Li, Layer-by-layer fabrication of chemical-bonded grapheme coating for solid-phase microextraction, Anal. Chem. 83 (2011) 7531-7541.

[64] Y. Wang, X. Wang, Z. Guo, Y. Chen, Ultrafast coating procedure for graphene on solid-phase microextraction fibers, Talanta 119 (2014) 517-523.

[65] P. J. Robinson, P. Dunnill, M. D. Lilly, The properties of magnetic supports in relation to immobilized enzyme reactors, Biotechnol. Bioeng. 15 (1973) 603-606.

[66] J. Płotka-Wasylka, N. Szczepańska, M. de la Guardia, J. Namiesnik, Miniaturized solidphase extraction techniques, TrAC, Trends Anal. Chem. 73 (2015) 19-38.

[67] Q. Liu, Q. Zhou, G. Jiang, Nanomaterials for analysis and monitoring of emerging chemical pollutants, TrAC, Trends Anal. Chem. 58 (2014) 10-22.

[68] C. Herrero-Latorre, J. Barciela-García, S. García-Martín, R. M. Peña-Crecente, J. Otarola-Jiménez, Magnetic solid-phase extraction using carbon nanotubes as sorbents: a review, Anal. Chim. Acta 892 (2015) 10-26.

[69] W. Wang, R. Ma, Q. Wu, C. Wang, Z. Wang, Magnetic microsphere-confined grapheme for the extraction of polycyclic aromatic hydrocarbons from environmental water samples coupled with high performance liquid chromatography-fluorescence analysis, J. Chromatogr. A 1293 (2013) 20-27.

[70] F. Mercader-Trejo, R. Herrera-Basurto, E. R. de San Miguel, J. de Gyves, Mercury determination in sediments by CVAAS after on line preconcentration by solid phase extraction with a sol-gel sorbent containing CYANEX 471X®, Int. J. Environ. Anal. Chem. 91 (2011) 1062-1076.

[71] E. Naddaf, M. Ebrahimi, Z. Es'haghi, F. F. Bamoharram, Application of carbon nanotubes modified with a Keggin polyoxometalate as a new sorbent for the hollow-fiber micro-solid-phase extraction of trace naproxen in hair samples with fluorescence spectrophotometry using factorial experimental design, J. Sep. Sci. 38 (2015) 2348-2356.

[72] R. Kumar, Gaurav, Heena, A. K. Malik, A. Kabir, K. G. Furton, Efficient analysis of selected estrogens using fabric phase sorptive extraction and high performance liquid chromatography-fluorescence detection, J. Chromatogr. A 1359 (2014) 16-25.

[73] W. A. Wan Ibrahim, K. V. Veloo, M. M. Sanagi, Novel sol-gel hybrid methyltrimethoxysilane-tetraethoxysilane as solid phase extraction sorbent for organophosphorus pesticides, J. Chromatogr. A 1229 (2012) 55-62.

[74] G. Fang, J. Chen, J. Wang, J. He, S. Wang, N-methylimidazolium ionic liquid functionalized silica as a sorbent for selective solid-phase extraction of 12 sulfonylurea herbicides in environmental water and soil samples, J. Chromatogr. A 1217 (2010) 1567-1574.

[75] P. Liang, L. Peng, Ionic liquid-modified silica as sorbent for preconcentration of cadmium prior to its determination by flame atomic absorption spectrometry in water samples, Talanta 81 (2010) 673-677.

[76] E. Kazemi, S. Dadfarnia, A. M. Haji Shabani, Development and evaluation of selective coating for stir bar sorptive extraction of silver using sol-gel technique in combination with double-imprinting concept, J. Iran. Chem. Soc. 12 (2015) 929-936.

[77] Y.-F. Jin, Y.-P. Zhang, M.-X. Huang, L.-Y. Bai, M. L. Lee, A novel method to prepare monolithic molecular imprinted polymer fiber for solid-phase microextraction by microwave irradiation, J. Sep. Sci. 36 (2013) 1429-1436.

[78] F. N. Andrade, Á. J. Santos-Neto, F. M. Lanças, Microextraction by packed sorbent liquid chromatography with time-of-flight mass spectrometry of triazines employing a molecularly imprinted polymer, J. Sep. Sci. 37 (2014) 3150-3156.

[79] S. Sadeghi, M. Jahani, Selective solid-phase extraction using molecular imprinted polymer sorbent for the analysis of Florfenicol in food samples, Food Chem. 141 (2013) 1242-1251.

[80] H. Asiabi, Y. Yamini, S. Seidi, F. Ghahramanifard, Preparation and evaluation of a novel molecularly imprinted polymer coating for selective extraction of indomethacin from biological samples by electrochemically controlled in-tube solid phase microextraction, Anal. Chim. Acta. 913 (2016) 76-85.

[81] W. J. Song, J. P. Wei, S. Y. Wang, H. S. Wang, Restricted access chiral stationary phase synthesized via reversible addition-fragmentation chain-transfer polymerization for direct analysis of biological samples by high performance liquid chromatography, Anal. Chim. Acta. 832 (2014) 58-64.

[82] A. J. Santos-Neto, K. E. Markides, P. J. R. Sjoberg, J. Bergquist, F. M. Lancas, Capillary column switching restricted-access media-liquid chromatography-electrospray ionization-tandem mass spectrometry system for simultaneous and direct analysis of drugs in biofluids, Anal. Chem. 79 (2007) 6359-6367.

[83] D. Xiao, S. Liu, L. Liang, Y. Bi, Magnetic restricted-access microspheres for extraction of adrenaline, dopamine and noradrenaline from biological samples, Microchim. Acta 183 (2016) 1417-1423.

[84] C. Cui, B. Hu, B. Chen, M. He, Restricted accessed material-Cu (II) ion imprinted polymer solid phase extraction combined with inductively coupled plasma-optical emission spectrometry for the determination of free Cu (II) in urine and serum samples, J. Anal. At. Spectrom. 28 (2013) 1110-1117.

[85] M. H. Banitaba, S. S. H. Davarani, H. Ahmar, S. K. Movahed, Application of a new fiber coating based on electrochemically reduced graphene oxide for the cold-fiber headspace solid-phase microextraction of tricyclic antidepressants, J. Sep. Sci. 37 (2014) 1162-1169.

[86] Y. Li, H. Xu, Development of a novel graphene/polyaniline electrodeposited coating for on-line in-tube solid phase microextraction of aldehydes in human exhaled breath condensate, J. Chromatogr. A 1395 (2015) 23-31.

[87] Q. Liu, J. Shi, L. Zeng, T. Wang, Y. Cai, G. Jiang, Evaluation of graphene as an advantageous adsorbent for solid-phase extraction with chlorophenols as model analytes, J. Chromatogr. A 1218 (2011) 197-204.

[88] A. Speltini, M. Sturini, F. Maraschi, L. Consoli, A. Zeffiro, A. Profumo, Graphene—derivatized silica as an efficient solid-phase extraction sorbent for pre-concentration of fluoroquinolones from water followed by liquid-chromatography fluorescence detection, J. Chromatogr. A 1379 (2015) 9-15.

[89] H. Yan, N. Sun, S. Liu, K. H. Row, Y. Song, Miniaturized graphene-based pipette tip extraction coupled with liquid chromatography for the determination of sulfonamide residues in bovine milk,

Food Chem. 158 (2014) 239–244.

[90] J. Sánchez–González, T. Barreiro–Grille, P. Cabarcos, M. J. Tabernero, P. Bermejo–Barrera, A. Moreda–Piñeiro, Magnetic molecularly imprinted polymer based–micro–solid phase extraction of cocaine and metabolites in plasma followed by high performance liquid chromatography–tandem mass spectrometry, Microchem. J. 127 (2016) 206–212.

[91] Q. Wu, G. Zhao, C. Feng, C. Wang, Z. Wang, Preparation of a graphene–based magnetic nanocomposite for the extraction of carbamate pesticides from environmental water samples, J. Chromatogr. A 1218 (2011) 7936–7942.

19　全二维色谱中的绿色样品制备技术

Francesco Cacciola[1], Mariarosa Maimone[1], Paola Dugo[1,2], Luigi Mondello[1,2,*]

1. University of Messina, Messina, Italy

2. University Campus Bio-Medico of Rome, Rome, Italy

*通讯作者：E-mail：lmondello@ unime. it

19.1 引言

绿色分析化学的统一定义是在技术和方法中：①减少或不使用对人体健康和环境有害的化学物质；②在不影响性能的前提下实现更快、更高效的分析[1]。

气相色谱法（GC）是公认的一种相对绿色的分析技术，而液相色谱（LC）本身并不是“绿色”的，只有减少溶剂量（如纳米或毛细管 LC）或使用更环保的友好溶剂（如过热水），才可称之为“绿色”。

在进行“绿色分析”时，最关键的步骤是样品制备，这通常是一个相当消耗溶剂的过程。样品制备有两个主要目标，即富集分析物和去除干扰成分。

真正的绿色样品制备技术在过去的二十年里受到了极大的关注，旨在利用固相微萃取（SPME）、搅拌棒吸附萃取法（SBSE）、固相萃取法（SPE）等手段，对样品进行提取、纯化和富集。与固相微萃取、搅拌棒吸附萃取、固相萃取不同的是，其他微萃取方式，如单液滴微萃取和微固相萃取，在全二维气相色谱（GC×GC）中的研究较少。其他方法，如超临界流体萃取（SFE）、加压液体萃取（PLE）、基质固相分散萃取（MSPD）、超声辅助提取（UAE）、微波辅助萃取（MAE）和分散液相微萃取（DLLME），尽管被认为比传统的索氏提取更环保，但仍需要使用溶剂，将只在本章作简单介绍[2-4]。

值得注意的是，由于先进质谱（MS）技术的快速发展，例如，三重四极杆，具有高选择性和灵敏度的高分辨率质谱，样品制备开始变得简单和小型化。此外，使用高分辨率色谱技术，如 GC×GC，还可以扩大分离空间，从而减少样品制备的时间。

固相微萃取（SPME）技术的发展，成为样品制备领域重要发展的代表，该技术由 Arthur 和 Pawliszyn 于 1990 年提出[5]。SPME 是目前最流行的绿色样品制备技术，已拥有多种商业化产品（依据不同的选择性和灵敏度）。SPME 最具突破性的方面，是在很多情况下完全不需要有机溶剂[5-7]。在 Liu 和 Phillips 提出仅仅一年之后，GC×GC 即被认为是当今最强大的 GC 方法之一[8]。与传统的 1D-GC 相比，GC×GC 有很多优点，主要表现为更高的分离能力、选择性、灵敏度，以及群组化学图案的形成。值得关注的是，这一创新技术，是否有可能减少样品制备的时间[9,10]。

1994 年，第一篇结合了绿色样品制备步骤与 GC×GC 的论文公开发表，是基于 SFE 的人类血清的检测[11]：将 10mL 掺有农药的人类血清通过 C_{18} SPE 柱，然后用氮气吹干。之后，用超临界 CO_2 处理 SPE 的填料 70min，并用三甲基硅烷基咪唑作为改性剂。用火焰离子化检测器（FID）来进行检测，这是在 GC×GC 应用的前 10 年中最常用的检测器。在早期的关于 GC×GC 的论文中，作者把注意力更多地集中在分离步骤上，而不是样品处理上：事实上，提取液收集在 10mL 二氯甲烷中，已经丧失了实验的“绿色”性质。

4 年后，第二篇结合了绿色样品制备步骤和 GC×GC 论文发表，其采用 SPME 进行前处理，然后用 GC×GC-FID 来分析地下水中的苯、甲苯、乙苯和二甲苯异构体（BTEX）以及甲基叔丁基醚和乙基丁基醚[12]。在该研究[12]中，评价了两种不同的纤维涂层，即 100μm 的聚二甲基硅氧烷（PDMS）和 75μm 的碳分子筛（CAR）/PDMS，后者具有较高的灵敏度。所有的 SPME 步骤都是手动进行，而“目标”分析物在异于常规的组合柱上进行分离（第一维度，^{1}D：非极性 2.0 m×0.10mm I.D.×5μm；第二维度，^{2}D：极性 1.0 m×0.10mm I.D.×0.14μm）。分析物在两个维度之间的转移是通过热吹扫器来进行。方法的检出限在 0.36~0.63μg/L（体积比）范围内，精密度数据令人满意，相对标准偏差（RSD）小于 10.2%。不同于第一篇论文的工作，现在整个 SPME 处理过程都是以自动化的方式进行，并且纤维的可用性更广，市场上还可以找到更多具有不同选择性的纤维。现在 GC×GC 装置中使用的调节器更加耐用，一般使用低温流体进行操作。此外，GC×GC 与 MS 联用技术自 1999 年首次应用以来，就获得了广泛的应用[13]。

除 SPME 和 SPE 外，只有少数几种绿色样品制备技术应用于 GC×GC，凸显了这两个研究领域之间的差距。在过去的 12 年里（1994 年至 2016 年 10 月），绿色样品制备技术与 GC×GC 分离相结合的论文总数是相当高的（207 次引用），如果将采用了一种以上的绿色样品制备技术的论文统计在内的话，这个数字还会有所增加（218 次引用）。最受瞩目的是 SPME 的应用，它是迄今为止 GC×GC 领域应用最多的绿色样品制备技术（超过总量的 50%）。

一般来说，GC×GC 方法（无论前面有或没有绿色样品制备技术）是为“非靶向”或（前或后）“靶向”分析而开发的。前一种应用是“非靶向”的，旨在识别样本中尽可能多的化合物，例如食品的香气，“预靶向”实验是基于对有限数量的“已知”化合物的测定，例如地下水中的石油污染物，而“后靶向”分析则是在随后进行的，通过对采集到的数据进行新的调查分析，并基于“后靶向”数据以“靶向”的方式进行后分析[14]。

绿色样品制备技术、GC×GC 以及 MS 的组合，可能会产生高灵敏度、高选择性和高分离度的分析方法。GC×GC 分析前的绿色样品制备步骤通常基于低温调制，主要基于 MS 检测。大多数应用都是在食品研究领域，例如饮料和香料（超过 40%），其次是环境及植物调查。

很早以前就有人提出缩小色谱柱内径（I.D.）的好处[15-17]。当然，低 I.D. 柱的主要优点是能够在低流速下使用较少体积的样品。因此，纳米或毛细管液相色谱比传统液相色谱具有更高的质量灵敏度。降低柱内径带来的其他好处是洗脱所需的溶剂和添加剂体积最小，并且由于低内径色谱柱上快速有效的热传递，易于对分离温度进行控制。另一个小型化的关键参数是液相色谱柱吸附剂颗粒的直径。随着吸附剂粒径的逐渐减小，可以获得极高的分离效率和峰值容量，这与气相色谱法相类似[18]。然而，要付出的代价是整个塔的压降增加，这与吸附剂颗粒直径的平方成反比。另一个与柱和吸附剂尺寸不断小型化有关的问题是额外的柱带加宽，事实上，从实际角度来看，这意味着所有通过空间的流量（毛细管连接、检测器单元体积）必须小型化，数据采集率应该最大化[19]。一涉及小型 LC 柱的命名，情况就非常复杂，在本章中，仅使用“纳米”和“毛细管”等术语：作为柱名称，后者的典型内径在 0.1~0.5mm 范围内，而前者在 0.01~0.1mm 范围内[20]。当分析人员使用有限数量的有机溶剂时，环境友好的样品制备以及纳米/毛细管液相色谱和质谱联用可被视为一种高灵敏度的方法，尤其是一种“绿色分析技术”。

19.2 全二维气相色谱分析前的绿色样品制备技术

19.2.1 固相微萃取分析

一般来说，非靶向代谢组学包括分析生物系统中所有或尽可能多的代谢物，是SPME GC×GC-MS分析的“理想”样品。后者可以被认为是一种非常令人兴奋的“绿色”方法，能够提供非常高的灵敏度、选择性、分离能力和识别潜力[21-29]。另一方面，高分辨率气相色谱法的要求在“靶向”分析中大大降低，例如，现实生活食品样品中的农药。在后一种情况下，主要是质谱的选择性，例如萃取离子色谱（EIC）、选择离子监测（SIM）、MS/MS流程、准确的MS数据，这些都可以减少或消除背景噪声，并解决目标分析物和基质组分之间的共溶情况[30-35]。然而，值得注意的是，由于干扰、非挥发性成分和副产品仍然存在于分析系统中，因此，必须提前进行充分的样品制备。

19.2.1.1 非靶向固相微萃取

Chaintreau等在关于烤箱中烤牛肉香气中的硫化物的论文[36]中，首次详述了GC×GC分析前的“非靶向”SPME步骤。通过将SPME（PDMS纤维）与GC×GC联用飞行时间质谱（ToF-MS）结合，共发现了70多种硫化物，其中有50种被准确识别。为了克服数据库中缺乏多个保留指数的问题，他们采用多元线性回归法对缺失值进行模拟，来帮助峰的识别。通过气相色谱和气味测定法，即GC-“SNIF”技术，筛选出了最重要的硫黄味物质。并首次在牛肉香气中检测到7种化合物，其中只有一种在自然界中被发现。之后，Tranchida[37]和Corderoet等[38]报道了采用二乙烯基苯（DVB）/CAR/PDMS相的顶空（HS）-SPME-GC×GC分离技术，结合单重四极杆质谱进行分析，分别测定了烘焙咖啡和烤榛子的挥发性成分。在第一项工作[37]中，首次将高分辨率的0.05mm内径毛细管以及0.25mm内径毛细管用作二维柱，对GC×GC仪器进行优化。第二项工作[38]采用了一个类似的装置，旨在建立不同品种、不同地理来源的烤榛子的GC×GC指纹图谱。Bean[39]等人报道了采用HS-SPME-GC×GC和低分辨率（LR）ToF-MS联用技术来提高灵敏度、选择性和分离能力的典型实例。特别是对细菌（铜绿假单胞菌）在生长24h后释放的挥发物的定性特征进行了研究。用于提取细菌挥发物的纤维与文献［37，38］中报道的在50℃下操作10min的纤维相同。GC×GC方法可以检测大约500个峰，信噪比（s/n）高于250。对峰进行解卷积后，通过质谱数据库检索，初步鉴定出了大约40种化合物，分别属于不同的化学类别：脂肪族和芳烃、醇、醛、酮、酸、酯和杂环芳烃，并利用特殊的GC×GC软件功能，对样品进行了详细的鉴别。Purcaro[25]等人也采用了类似的方法，将HS-SPME-GC×GC数据与初榨橄榄油的感官质量相关联。在这项研究中，结合GC×GC特殊的软件功能与统计分析，作者推断出了与橄榄油缺陷组合有关的“化学蓝图”。食品的化学蓝图将特定的挥发物与特定的食品特性联系起来，也就是与感官品质进行了关联。Risticevic[21,22,27]等人进行了非常有意思的代谢组学相关研究。在他们的前期工作[21]中，采用了一种优化的 HS-SPME-GC×GC 与 ToF-MS 联用分析方法，用于分析苹果代谢物。该研究涉及 7 种纤维的测试，即 PDMS、聚丙烯酸酯（PA）、聚乙二醇、DVB/CAR/PDMS、PDMS/DVB、CAR/PDMS 和 Carbopack Z/PDMS，分别检测到 549、977、897、1163、1053、1167 和 745 种代谢物（最小信噪比和数据匹配度：50 和 750）。通过选择 800 为最小匹配度对峰进行后处理，从而减少对峰的误判。在对纤维涂层的选择性进行了非常详细的描述之后，作者指出，DVB/CAR/PDMS 阶段提供了最平衡的分析物覆盖率和最多的提取代谢物数量（830）。一些非常密集的峰尖和色谱图得以显现出来，进一步凸显了 GC×GC 分离的必要性。此外，由于 GC×GC 可以获得非常仔细的分离，因此该技术能够提供更全面的关于 SPME 分析物覆盖率、解吸效率、动态范围和置换效应的信息[22]。在他们最近的研究[27]中，首次以苹果（Malus×domestica Borkh.）为模型系统，采用直接浸入式固相微萃取（DI-SPME）的活体采样模式，获得了活体植物样本的代谢组。从苹果组织中提取代谢物，通过热脱附引入 GC×GC-ToF-MS。在代谢物全面分析中，基于微萃取的原理（包括可忽略的游离分析物浓度损耗、无溶剂采样和样品制备以及现场相容性），确定了该采样方法的可行性。本研究的目的不是采用传统的样品制备方法，对新陈代谢进行抑制和费劲去制备样品，而是在体内直接采集代谢组分，并采用不同的 SPME 采样模式，评估该方法能够提供公正的提取覆盖率的可行性，并比较该分析方法的精密度。通过评估果实成熟过程中代谢指纹的变化，评估了活体 DI-SPME 在植物代谢组学中的潜力。最近，Carlin 等利用 HS-SPME-GC×GC-ToF-MS 和多元分析，对意大利起泡酒的两个主产区中 48 个酒庄和 6 个年份的 70 种葡萄酒中的挥发性化合物进行了全面的分析[28]。最后的研究范围是描述这些葡萄酒的代谢组学空间，并验证葡萄品种特征、产区的土壤气候影响和复杂的工艺是否在最终产品中能够被检测。葡萄酒色谱图提供了丰富的信息，共发现 1695 种化合物。为了了解新酒与陈年葡萄酒共同的代谢变化，采用基于相关性的网络分析（CNA），对 α-异佛尔酮和藏红花醛进行了仔细的研究，如图 19.1 所示，凸显了当化合物受到外界/环境影响时，CNA 是如何作为一种工具去检测其行为差异性的。

19.2.1.2 靶向固相微萃取

SPME-GC×GC-MS 方法已被提议用于各种样品的“靶向”分析。2008 年，Schurek 等用 HS-SPME-GC×GCToF-MS 法测定了绿茶、红茶和果茶中的 36 种农药[30]。采用了非极性（40m×0.18mm 内径 I.D. ×0.18μm df）柱作为气相色谱的^{1}D 柱，而用中等极性（2.5m×0.1mm 内径 I.D. × 0.1μm df）柱作为^{2}D 柱。作者对使用 1D 和 2D 气相色谱法得到的结果进行了比较。当使用 HS-SPME-GC×GC-ToF MS 时，LoQ 值在 1~28μg/kg 范围内，正如预期的那样，始终低于通过 HS-SPME-GC-ToF-MS 获得的 LoQ 值。最后，将 HS-SPME 与传统的样品制备技术（乙酸乙酯萃取法和高效凝胶渗透色谱法）进行了比较。一般来说，HS-SPME 在灵敏度和减少基体干扰方面具有更好的性能，而传统的样品制备技术在线性（超过分析浓度范围）和重复性方面更为可取。一年后，Purcaro[31]等人将 SPME（使用 100 μm PDMS 纤维）与 GC×GC 及快速扫描 qMS 结合起来，用于测定饮用水中的农药，有效提高了 SPME 的富集能力。方法的定

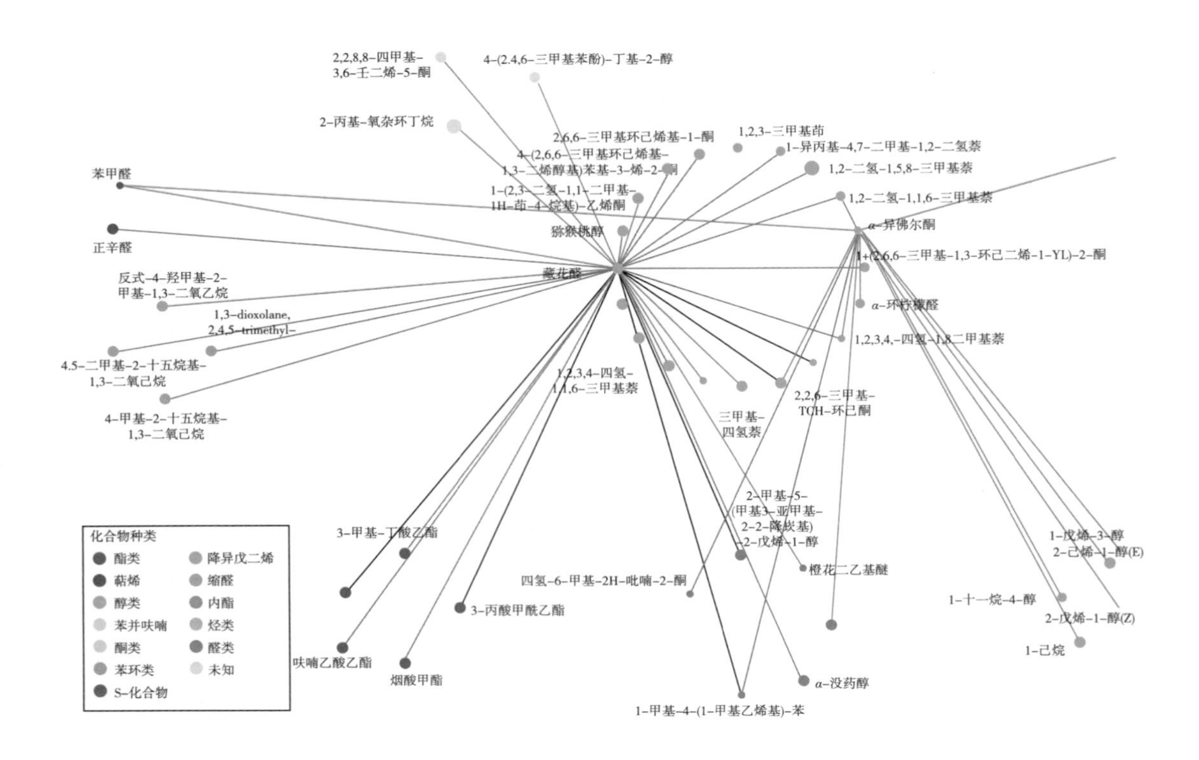

图 19.1　网络分析：α-异佛尔酮和藏红花醛的对称差加交集网络，
新的和陈年的特伦托多克起泡酒之间的比较

[资料来源：S. Carlin，U. Vrhovsek，P. Franceschi，C. Lotti，L. Bontempo，F. Camin，D. Toubian，F. Zottele，G. Toller，A. Fait，F. Mattivi，Regional features of northern Italian sparkling wines，identified using solid-phase micro extraction and comprehensive two-dimensional gas chromatography coupled with time-of-flight mass spectrometry，Food Chem. 208（2016）68-80. 经 Elsevier 许可。]

量限（LOQ）在 3ng/mL～0.084μg/mL 范围内，并且在任何情况下始终低于欧洲法律规定的最大残留限量。Sulej-Suchomska[35] 等人最近报道了一项非常有意思的工作：机场径流的污染治理。多环芳烃（PAHs）是一类重要的外源化合物，广泛存在于机场径流中。不管机场位于何处，用 HSGC-GC-ToF-MS 方法分析的 34 个样品（1.8～26.3μg/L）中，菌、菲和芘始终是含量最高的 PAH 化合物。这一方法可用于跟踪环境中的多环芳烃，并评估机场对环境的影响。

19.2.2　搅拌棒吸附萃取

基于与 SPME 相同的概念，SBSE 于 1999 年首次报告[40]，其主要区别在于存在大量的绿色样品制备技术，这也使得灵敏度高。然而，与 SPME 不同的是，SBSE 吸附剂（PDMS 和 PDMS/乙二醇）的种类要有限得多，SBSE 热脱附过程需要专用的 GC 注射器。SBSE 与 GC×GC 联用的报道较少，主要用于高度稀释样品的靶向模式。“非靶向”模式主要是由于存在大量的固定相，不利于萃取过量的溶质，特别是在二维色谱柱中，会产生“超载”现象。Ochiai[41]等人使用了SBSE-GC ×GC技术，联合HR-ToF-MS系统，分析了河水中的“预靶向”（有机氯农药）和“非靶向”污染物。搅拌棒的特征是含有20mm长、0.5mm厚的PDMS层（相体积=47μL）。该方法的灵敏度高，检出限在10~44pg/L。作者还进行了“非靶向”的检索，并在河水中发现了另外20种污染物。在对河水和废水的后一个“预靶向”GC×GC研究中，利用LR ToF-MS系统，提高了SBSE对13种苯环类化合物（PCPs）、15种多环芳烃和27种农药的萃取效率[42]。方法检出限为0.01~2.15ng/L（废水在0.02~2.5ng/L范围内）。

19.2.3 固相萃取

SPE是继SPME之后应用最广泛的样品制备技术，可作为从液体样品中分离分析物的富集和净化方法。根据分析物的不同化学性质，有几种吸附剂可担当此项任务，例如C_{18}、离子交换、分子印迹聚合物、免疫吸附剂等[42]。第一次应用是由Matamoros[43]等人于2010年发布的，其研究了SPE-GC×GC与LR ToF-MS联用测定河水中97种污染物的方法。SPE过程中使用的滤芯为苯乙烯-二乙烯基苯等吸附剂。

方法对狄氏剂、异狄氏剂、阿特拉津、菲、艾氏剂和加拉唑内酯的检出限分别为2、3、3、3、5和6ng/L，97种污染物的检出限范围是0.5~100ng/L。一年后，Weldegergis等人采用SPE-GC×GC-LR-ToF-MS对南非红酒中的挥发物进行了详细的研究[44]。研究发现，反相（RP）-SPE（使用苯乙烯-二乙烯基苯吸附剂）比作者在之前的研究中采用的HS-SPME更具有优势[18]，由于还提取出了少量的极性化合物，如酸和醇，这些都是导致色谱分析结果不好的原因，即过载、拖尾、缠绕、与其他化合物共溶。SPME纤维由悬浮在液态聚合物（CAR/PDMS）中的微孔碳组成，在极性基础上具有良好的分析物覆盖率，但对高分子质量代谢物表现出高度的辨别力[45]。所提出的SPE-GC×GC飞行时间质谱联用方法被证明有利于检测萜烯类、酚类、内酯类及含硫溶质，其在葡萄酒中的含量都处于μg/L的含量水平。总的来说，三种葡萄酒中共鉴定出276种分析物，其中许多分析物都是首次通过质谱数据库匹配、线性保留指数和标准化合物（如可用）[44]比对识别出来的。最近，Dimitriou Christidis等[46]报道了一个非常有意思的应用方向，即用于污水处理厂进水、出水、一级污泥和二级污泥基质（包括液相和粒相）中非极性卤化微污染物的测定。所研究的GC×GC微电极捕捉检测（μECD）方法使他们能够测定多达59种目标分析物，例如毒杀芬类、多氯萘类、有机氯农药、多氯联苯、多溴联苯醚以及新兴的持久性和生物累积性化学品。通过RP-SPE提取的大多数分析物中，回收率都在70%~130%。主要研究发现，溶解性有机碳的吸附对污水处理厂中疏水性微污染物的表观固-液分布有重要影响。

以动态针捕集装置（Needle trap，NTs）为代表的小型化设备介于SPE和SPME之间，它们是通过在针状物内填充吸附材料来起作用。NTs用于分析物的萃取、富集和热脱附释放。与传统的捕集相比，NTs在现场取样方面具有很大的优势。典型的NT应用与挥发性有机化合物的分离有关，如烷烃混合物（C_6~C_{15}）[47]和人类呼吸[48]。

19.2.4 其他绿色样品制备技术

除了被视为“绿色”的SPME、SBSE和SPE外，其他样品制备技术运用了一定量的溶剂进行萃取，因此被认为比前一种方法“绿色化程度低”，但肯定比传统方法（如索氏试剂）更“绿色”，其中包括 SFE、PLE、MSPD、UAE、MAE、分散液液微萃取（DLLME）。

由于不产生有害废弃物，SFE 是传统液相萃取技术的理想替代品。由于 CO_2 的特殊特性，即临界温度和压力为 31℃、7200kPa，以及低的毒性和萃取的快速性，大多数 SFE 工艺都是以 CO_2 为基础的。然而，超临界 CO_2 萃取也存在一系列的局限性，例如对极性化合物的选择性较低（使用少量有机改性剂可提高选择性），以及经常需要净化[49]。在 GC×GC 分析之前，SFE 的使用已在“非靶向”分析中有所报道。例如，Pripdeevech[50] 等人使用 GC×GC-FID 和 GC×GC-qMS 对通过 SFE、MAE、同时蒸馏和萃取（SDE）和溶剂萃取（SE）获得的香根草油样品进行了表征。对于这四种技术，使用了不同数量的溶剂（SDE，150mL 二氯甲烷；SFE，10%二氯甲烷，10%甲醇和 20%甲苯；MAE，二氯甲烷，甲醇和甲苯各 30mL；SE，二氯甲烷，甲醇和甲苯各 500mL）。SFE 是最快速的方法，而且在提取能力方面也是最有效的方法，与 SE 和 MAE 相比，由于非挥发物的大量减少，SFE 提供了较为纯净的提取物。对所有提取物而言，GC×GC-qMS 分析能够识别 245 种化合物。

PLE 是基于在高温高压下使用液态溶剂的原理。在这种条件下，萃取过程以更快的方式进行，且使用的溶剂量更少。通过向萃取池中添加特定的吸附剂，使得溶剂渗透到基质中的能力得以改善，最终让净化步骤成为可选项[49]。Ong[51] 等人使用 PLE-GC×GC-FID 对土壤中 24 种多环芳烃进行筛选（其使用了三种土壤萃取方法），均在 150℃、14MPa 下进行，即①溶剂：己烷/丙酮（1：1，体积比）；采取萃取后色谱净化的步骤；②溶剂：己烷；采取细胞内净化；③与方法 2 相同，但采用不同的萃取溶剂混合物［己烷/二氯甲烷（3：1，体积比］。所有提取物的 GC×GC 色谱图都很相似，因此细胞内和细胞外净化方法之间没有显著性差异。GC×GC 所提供的扩展分离空间，有效地避免了干扰物与目标分析物的峰重叠。方法 2 和方法 3 测定的土壤中多环芳烃浓度（mg/kg）低于方法 1。很有可能分析物没有从用于细胞内净化过程的吸附剂中完全洗脱出来。

MSPD 通过使用少量的样品，且只需一个步骤，就让简单、快速、低成本的提取和净化成为可能[52]。Barker 等于 1989 年在一项研究工作中首次报道了这种情况，该研究工作侧重于对动物组织中的残留药物进行分离[52]。2009 年，Ramos 等用 MSPD-GC×GC-μECD 测定了不同水果（橘子、梨、葡萄、苹果）中的多种有机磷农药、三嗪类和拟除虫菊酯[53]。MSPD 方法包括使用 100mg 均质果皮（除总量中分析的葡萄外），与相同数量的 C_8-键合二氧化硅混合，然后封装在 3mL SPE 柱中；然后用 15mL 水清洗 SPE 柱，再用 0. 7mL 乙酸乙酯进行洗脱。方法的灵敏度较好，且检出限小于 1μg/kg。

UAE 利用低频声波产生的机械能来加速提取，通常产量很高。UAE 的两种主要方法基于水浴或探头的存在。尽管前一种选择是最受欢迎的一种，但超声波探头可以直接与样品接触，或者可以更好地将其能量集中到样品上，从而大大缩短了 UAE 时间[54]。有时该过程需要采取额外的净化步骤。Morales-Muñoz[55] 等人借助动态 UAE 与 GC×GC-LRToF-MS 相结合，分析了海洋沉积物中的污染物。UAE 的特点是在距萃取室表面1mm处放置一个探针，在将萃取物收集到小瓶之前，用6mL溶剂（正己烷）前后各泵入萃取室15min。而传统的SE则需要更高的溶剂用量（80mL己烷）和更长的萃取周期（24 h）。从两种方法的比较来看，UAE回收率和重复性值优于传统的SE。

MAE是基于利用微波产生的能量从基质中溶解提取分析物。MAE的主要优点是样品的溶剂混合物加热迅速，萃取时间短，有机溶剂消耗相对较低。当MAE在密闭容器中进行时，萃取温度可超过沸点，例如在PLE中，从而可以进一步提高萃取效率。此外，MAE可同时接受多个样品，因此提高了样品的处理量[56]。Kristenson[57]等对MAE、PLE和UAE进行了比较，并对于污泥中多氯联苯（PCBs）进行GC×GC-μECD分析：这三种提取物在气相色谱分析之前都用铜处理，以减少含硫化合物的干扰。与MAE相比，尽管UAE具有最佳回收率（60%~100%）和精密度（RSDs=2%~11%），但是提取周期较长；另一方面，PLE的回收率最低（45%~75%），与MAE（2）相比，可能与较高的溶剂-样品比（3.3）和比UAE低的提取时间有关。

DLLME是在2006年由Rezaee[58]等首次提出，用于水中有机化合物的测定，其主要是在水样中使用小体积的萃取和分散溶剂。据我们所知，唯一的DLLME-GC×GC研究工作是由Tugizimana等报道，其用于监测麦角甾醇诱导烟草细胞代谢产物的变化[59]。DLLME-GC与LRToF-MS联用技术能够让对照细胞与经麦角甾醇处理后的细胞之间的分化更明显。

19.3　毛细管全二维液相色谱法

与传统的1D-LC的相比，LC×LC以其不容忽视的优点，在过去的二十年里引起了广泛的关注[60-62]。LC×LC在纳米/毛细管尺度下开展工作，可大大提高整个方法的灵敏度，并且当样品制备和分离只涉及有限量的有机溶剂（<10mL）时，可被视为“绿色”的分析工具。与传统的1D-LC相比，LC×LC还使MS的联用变得更为有用，迄今为止，大多数应用都集中在蛋白质组学、临床生物化学、环境和食品等领域。

Nagele[63]等在LC×LC装置上分析酵母蛋白质组，LC×LC装置包括^{1}D的毛细管强阳离子交换柱［35mm×0.3mm（内径）］和^{2}D中的纳米C_{18}柱［150mm×0.75mm（内径）］。在切换配置中，使用带有两个C_{18}捕集柱（5mm×0.3mm）的6端口和10端口开关阀来转移和预浓缩来自^{1}D的分析物，然后在^{2}D中进行RP-LC分离。Wang[64]等利用类似的设置，对肝细胞癌（HCC）的蛋白质组进行了分析，在分析的组织中鉴定出了229个蛋白质，其中一些蛋白质被发现与HCC疾病的发展有关。Haunet[65]等人采用了一维纳米液相色谱柱和二维毛细管柱相结合的典型“连续流动”液相色谱排列（图19.2）。10端口2位开关阀有两个回路，交替地由1D的流出物以纳米流速填充：在1min长的调制周期中，两个回路交替切换到2D毛细管流。Pinkse[66]等人使用了自动化纳米LC×LC-MS用于分析^{1}D内TiO_2-C_{18}预柱上分馏的肽。磷酸肽保留在TiO_2预柱上，而非磷酸肽被保留在C_{18}相。首先，用水-乙腈梯度洗脱^{1}D中的非磷酸肽，并使其在纳米C_{18}柱上的^{2}D分离成为可能。磷酸肽通过碳酸氢铵溶液从TiO_2预柱中洗脱，并在第二次梯度洗脱中分离。这种结构就像一个通风柱，其中一个六通阀打开或关闭排气口，关闭或打开分流管。最近的其他的研究工作也基于相同的原理，在通风柱装置中使用了 2D 预柱[67-69]。Boersema[70] 和 Di Palma[71] 采用了基于亲水作用液相色谱和 RP-LC 相结合的离线蛋白质分析方法。这些装置采用 0.1mm 的预柱和纳米柱（内径 0.075mm）。这些组分被收集在 96 孔板上，然后用 RP-LC 进行分析。最后，Sommella[72] 等人于 2012 年首次开发了一种全自动毛细管 LC×LC 方法，该方法基于分别在^{1}D 和^{2}D 中使用高 pH RP 和低 pH RP 分离的方法，进行蛋白质组学的分析。通过这种方法，尽管使用相同的固定相，但获得了较高的峰容量，而流动相消耗则可忽略不计，灵敏度也得以提高。

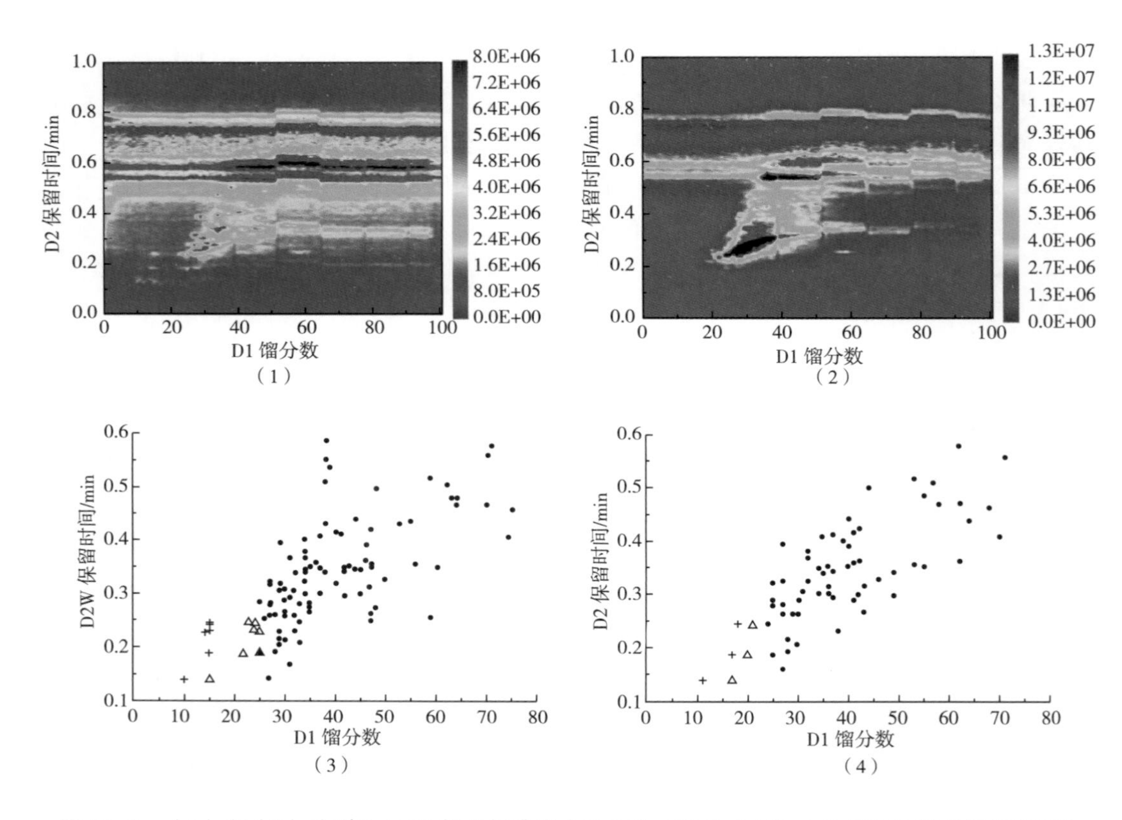

图 19.2　（1）和（2）分别显示 99 组分标准混合物和废水样品的 LC×LC 分离的等高线总离子流色谱图，（3）和（4）表示标准混合物和废水样品中目标物的分析物图

分别基于分析物的峰值最大值，用●表示。^{1}D 双峰用+和 Δ 突出显示，表示第一个和第二个最大值

［资料来源：J. Haun, J. Leonhardt, C. Portner, T. Hetzel, J. Tuerk, T. Teutenberg, T. C. Schmidt, Online and splitless nano-LC×capillary LC with quadrupole/time-of-flight mass spectrometric detection for comprehensive screening analysis of complex samples, Anal. Chem. Anal. Chem. 85（2013）10083-10090。版权所有 2013 美国化学学会。］

19.4　结论

SPME 和 SPE 结合 GC×GC 分离无疑是最成熟的分离方法（约占文献中全部工作的

50%)，在 GC×GC 分析之前，仅有限的部分出版著作涉及使用其他的绿色样品制备技术。SPME-GC×GC 的应用，提供了非常高的灵敏度、选择性、分离能力和鉴定潜力，并且大多数是在代谢组学研究中以“非靶向”模式进行的；然而，绿色样品制备与 GC 相结合的应用数量远远高于 GC×GC 应用的数量，由此可以得出结论，目前绿色样品制备研究最多的是 GC 领域。除此之外，考虑到 GC×GC 转移装置，应强调低温调制 GC×GC 不能真正视为一种绿色方法，因为它的高运行成本（主要是由于低温流体的消耗）；另外，环境友好型 GC×GC，例如流量调节系统，可能是一个可行的替代方案。另一方面，在纳米/毛细管尺度下运行的 LC×LC 成为一种有价值的“绿色”分析工具，由于在样品制备过程中大量减少有机溶剂的使用量（<10mL），因而可以极大地提高整个方法的灵敏度。可以预见，在纳米/毛细管尺度下进行的 GC×GC 和 LC×LC 应用，将非常适合与 MS 方法进行直接联用。

参考文献

[1] P. T. Anastas, J. C. Warner, Green Chemistry: Theory and Practice, Oxford University Press, New York, 1998.

[2] S. Armenta, S. Garrigues, M. de la Guardia, Green analytical chemistry, Trends Anal. Chem. 27 (2008) 497-511.

[3] M. Farré, S. Pérez, C. Gonçalves, M. F. Alpendurada, D. Barcelo, Green analytical chemistry in the determination of organic pollutants in the aquatic environment, Trends Anal. Chem. 29 (2010) 1347-1362.

[4] L. Ramos, Critical overview of selected contemporary sample preparation techniques, J. Chromatogr. A 1221 (2012) 84-98.

[5] C. L. Arthur, J. Pawliszyn, Solid phase microextraction with thermal desorption using fused silica optical fibers, Anal. Chem. 62 (1990) 2145-2148.

[6] E. A. Souza Silva, S. Risticevic, J. Pawliszyn, Recent trends in SPME concerning sorbent materials, configurations and in vivo applications, Trends Anal. Chem. 43 (2013) 24-36.

[7] H. Yu, T. D. Ho, J. L. Anderson, Ionic liquid and polymeric ionic liquid coatings in solid-phase microextraction, Trends Anal. Chem. 45 (2013) 219-232.

[8] Z. Liu, J. B. Phillips, Comprehensive two-dimensional gas chromatography using an on—column thermal modulator interface, J. Chromatogr. Sci. 29 (1991) 227-231.

[9] P. J. Marriott, S. -T. Chin, B. Maikhunthod, H. -G. Schmarr, S. Bieri, Multidimensional gas chromatography, Trends Anal. Chem. 34 (2012) 1-21.

[10] P. Q. Tranchida, G. Purcaro, P. Dugo, L. Mondello, Modulators for comprehensive two-dimensional gas chromatography, Trends Anal. Chem. 30 (2011) 1437-1461.

[11] Z. Liu, S. R. Sirimanne, D. G. Patterson Jr., L. L. Needham, J. B. Phillips, Comprehensive two-dimensional gas chromatography for the fast separation and determination of pesticides extracted from human serum, Anal. Chem. 66 (1994) 3086-3092.

[12] R. B. Gaines, E. B. Ledford Jr., J. D. Stuart, Analysis of water samples for trace levels of oxygenate and aromatic compounds using headspace solid-phase microextraction and comprehensive two-dimensional gas chromatography, J. Microcolumn Sep. 10 (1998) 597-604.

[13] G. S. Frysinger, R. B. Gaines, Comprehensive two-dimensional gas chromatography with mass spectrometric detection (GC×GC/MS) applied to the analysis of petroleum, J. High Resol. Chromatogr. 22 (1999) 251-255.

[14] P. Q. Tranchida, M. Maimone, G. Purcaro, P. Dugo, L. Mondello, The penetration of green sample-preparation techniques in comprehensive two-dimensional gas chromatography, TRAC-Trend Anal. Chem. 71 (2015) 74-84.

[15] P. Kucera, Microcolumn High-performance Liquid Chromatography, Elsevier, 1984.

[16] M. V. Novotny, D. Ishii, Microcolumn Separations: Columns, Instrumentation and Ancillary Techniques, Elsevier, 1985.

[17] M. Krejci, Trace Analysis with Microcolumn Liquid Chromatography, Marcel Dekker, 1992.

[18] J. C. Medina, N. J. Wu, M. L. Lee, Comparison of empirical peak capacities for high efficiency capillary chromatographic techniques, Anal. Chem. 73 (2001) 1301-1306.

[19] F. Gritti, G. Guiochon, The current revolution in column technology: how it began, where is it going? J. Chromatogr. A 1228 (2012) 2-19.

[20] Y. Saito, K. Jinno, T. Greibrokk, Capillary columns in liquid chromatography: between conventional columns and microchips, J. Sep. Sci. 27 (2004) 1379-1390.

[21] S. Risticevic, J. R. DeEll, J. Pawliszyn, Solid phase microextraction coupled with comprehensive two-dimensional gas chromatography-time-of-flight mass spectrometry for high-resolution metabolite profiling in apples: implementation of structured separations for optimization of sample preparation procedure in complex samples, J. Chromatogr. A 1251 (2012) 208-218.

[22] S. Risticevic, J. Pawliszyn, Solid-phase microextraction in targeted and nontargeted analysis: displacement and desorption effects, Anal. Chem. 85 (2013) 8987-8995.

[23] C. Cordero, C. Cagliaro, E. Liberto, L. Nicolotti, P. Rubiolo, B. Sgorbini, C. Bicchi, High concentration capacity sample preparation techniques to improve the informative potential of two-dimensional comprehensive gas chromatography-mass spectrometry: application to sensomics, J. Chromatogr. A 1318 (2013) 1-11.

[24] S. R. Rivellino, L. W. Hantao, S. Risticevic, E. Carasek, J. Pawliszyn, F. Augusto, Detection of extraction artifacts in the analysis of honey volatiles using comprehensive two-dimensional gas chromatography, Food Chem. 141 (2013) 1828-1833.

[25] G. Purcaro, C. Cordero, E. Liberto, C. Bicchi, L. S. Conte, Toward a definition of blueprint of virgin olive oil by comprehensive two-dimensional gas chromatography, J. Chromatogr. A 1334 (2014) 101-111.

[26] M. Jiang, C. Kulsing, Y. Nolvachai, P. J. Marriott, Two-dimensional retention indices improve component identification in comprehensive two-dimensional gas chromatography of saffron, Anal. Chem. 87 (2015) 5753-5761.

[27] S. Risticevic, E. A. Souza-Silva, J. R. DeEll, J. W. Cochran, J. Pawliszyn, Capturing plant metabolome with direct-immersion in vivo solid phase microextraction of plant tissues, Anal. Chem. 88 (2016) 1266-1274.

[28] S. Carlin, U. Vrhovsek, P. Franceschi, C. Lotti, L. Bontempo, F. Camin, D. Toubian, F. Zottele, G. Toller, A. Fait, F. Mattivi, Regional features of northern Italian sparkling wines, identified using solid-phase micro extraction and comprehensive two-dimensional gas chromatography coupled with time-of-flight mass spectrometry, Food Chem. 208 (2016) 68-80.

[29] T. Dymerski, J. Namiesnik, H. Leontowicz, M. Leontowicz, K. Vearasilp, A. L. Martinez-Ayala, G. A. Gonzalez-Aguilar, M. Robles-Sánchez, S. Gorinstein, Chemistry and biological properties of berry volatiles by two-dimensional chromatography, fluorescence and Fourier transform infrared spectroscopy techniques, Food Res. Int. 83 (2016) 74-86.

[30] J. Schurek, T. Portolés, J. Hajslova, K. Riddellova, F. Hernández, Application of headspace solid-phase microextraction coupled to comprehensive two-dimensional gas chromatography-time-of-flight mass spectrometry for the determination of multiple pesticide residues in tea samples, Anal. Chim. Acta 611 (2008) 163-172.

[31] G. Purcaro, P. Q. Tranchida, L. Conte, A. Obiedzinska, P. Dugo, G. Dugo, L. Mondello, Performance evaluation of a rapid-scanning quadrupole mass spectrometer in the comprehensive two-dimensional gas chromatography analysis of pesticides in water, J. Sep. Sci. 34 (2011) 2411-2417.

[32] C. C. Loureiro, A. S. Oliveira, M. Santos, A. Rudnitskaya, A. Todo-Bom, J. Bousquet, S. M. Rocha, Urinary metabolomic profiling of asthmatics can be related to clinical characteristics, Allergy 71 (2016) 1362-1365.

[33] S.-T. Chin, G. T. Eyres, P. J. Marriott, Application of integrated comprehensive/multidimensional gas chromatography with mass spectrometry and olfactometry for aroma analysis in wine and coffee, Food Chem. 185 (2015) 355-361.

[34] L. Rust, K. D. Nizio, S. L. Forbes, The influence of ageing and surface type on the odour profile of blood-detection dog training aids, Anal. Bioanal. Chem. 408 (2016) 6349-6360.

[35] A. M. Sulej-Suchomska, Ż. Polkowska, T. Chmiela, T. Dymerskia, Z. Kokot, J. Namieśnik, Solid phase microextraction-comprehensive two-dimensional gas chromatography-time-of-flight mass spectrometry: a new tool for determining polycyclic aromatic hydrocarbons in airport runoff water samples, Anal. Methods 8 (2016) 4509-4520.

[36] S. Rochat, J.-Y. de Saint Laumer, A. Chaintreau, Analysis of sulfur compounds from the in-oven roast beef aroma by comprehensive two-dimensional gas chromatography, J. Chromatogr. A 1147 (2007) 85-94.

[37] P. Q. Tranchida, G. Purcaro, L. Conte, P. Dugo, G. Dugo, L. Mondello, Enhanced resolution comprehensive two-dimensional gas chromatography applied to the analysis of roasted coffee volatiles, J. Chromatogr. A 1216 (2009) 7301-7306.

[38] C. Cordero, E. Liberto, C. Bicchi, P. Rubiolo, P. Schieberle, S. E. Reichenbach, Q. Tao, Profiling food volatiles by comprehensive two-dimensional gas chromatography coupled with mass spectrometry: advanced fingerprinting approaches for comparative analysis of the volatile fraction of roasted hazelnuts (*Corylus avellana* L.) from different origins, J. Chromatogr. A 1217 (2010) 5848-5858.

[39] H. D. Bean, J.-M. D. Dimandja, J. E. Hill, Bacterial volatile discovery using solid phase microextraction and comprehensive two-dimensional gas chromatography-time-of-flight mass spectrometry, J. Chromatogr. B 901 (2012) 41-46.

[40] E. Baltussen, P. Sandra, F. David, C. Cramers, Stir bar sorptive extraction (SBSE), a novel extraction technique for aqueous samples: theory and principles, J. Microcolumn Sep. 11 (1999) 737-747.

[41] N. Ochiai, T. Ieda, K. Sasamoto, Y. Takazawa, S. Hashimoto, A. Fushimi, K. Tanabe, Stir bar sorptive extraction and comprehensive two-dimensional gas chromatography coupled to high-resolution time-of-flight mass spectrometry for ultra-trace analysis of organochlorine pesticides in river water, J. Chromatogr. A 1218 (2011) 6851-6860.

[42] M. J. Gómez, S. Herrera, D. Solé, E. García - Calvo, A. R. Fernandez - Alba, Automatic searching and evaluation of priority and emerging contaminants in wastewater and river water by stir bar sorptive extraction followed by comprehensive two-dimensional gas chromatography-time-of-flight mass spectrometry, Anal. Chem. 83 (2011) 2638-2647.

[43] V. Matamoros, E. Jover, J. M. Bayona, Part - per - trillion determination of pharmaceuticals, pesticides, and related organic contaminants in river water by solid-phase extraction followed by comprehensive two-dimensional gas chromatography time-of-flight mass spectrometry, Anal. Chem. 82 (2010) 699-706.

[44] B. T. Weldegergis, A. M. Crouch, T. Górecki, A. de Villiers, Solid phase extraction in combination with comprehensive two - dimensional gas chromatography coupled to time - of - flight mass spectrometry for the detailed investigation of volatiles in South African red wines, Anal. Chim. Acta 701 (2011) 98-111.

[45] B. T. Weldegergis, A. de Villiers, C. McNeish, S. Seethapathy, A. Mostafa, T. Górecki, A. M. Crouch, Characterisation of volatile components of pinotage wines using comprehensive two-dimensional gas chromatography coupled to time-of-flight mass spectrometry (GC×GC-TOF-MS), Food Chem. 129 (2011) 188-199.

[46] P. Dimitriou-Christidis, A. Bonvin, S. Samanipour, J. Hollender, R. Rutler, J. Westphale, J. Gros, J. S. Arey, GCXGC quantification of priority and emerging nonpolar halogenated micropollutants in all types of wastewater matrices: analysis methodology, chemical occurrence, and partitioning, Environ. Sci. Technol. 49 (2015) 7914-7925.

[47] I. -Y. Eom, J. Pawliszyn, Simple sample transfer technique by internally expanded desorptive flow for needle trap devices, J. Sep. Sci. 31 (2008) 2283-2287.

[48] M. Mieth, J. K. Schubert, T. Gröger, B. Sabel, S. Kischkel, P. Fuchs, D. Hein, R. Zimmermann, W. Miekisch, Automated needle trap heart-cut GC/MS and needle trap comprehensive two-dimensional GC/TOF-MS for breath gas analysis in the clinical environment, Anal. Chem. 82 (2010) 2541-2551.

[49] M. Herrero, M. Castro-Puyana, J. A. Mendiola, E. Ibañez, Compressed fluids for the extraction of bioactive compounds, Trends Anal. Chem. 43 (2013) 67-83.

[50] P. Pripdeevech, S. Wongpornchai, P. J. Marriott, Comprehensive two - dimensional gas chromatography-mass spectrometry analysis of volatile constituents in Thai vetiver root oils obtained by using different extraction methods, Phytochem. Anal. 21 (2010) 163-173.

[51] R. Ong, S. Lundstedt, P. Haglund, P. Marriott, Pressurised liquid extraction - comprehensive two - dimensional gas chromatography for fast - screening of polycyclic aromatic hydrocarbons in soil, J. Chromatogr. A 1019 (2003) 221-232.

[52] S. Barker, A. Long, C. Short, Isolation of drug residues from tissues by solid phase dispersion, J. Chromatogr. 475 (1989) 353-361.

[53] J. J. Ramos, M. J. Gonzalez, L. Ramos, Comparison of gas chromatography - based approaches after fast miniaturised sample preparation for the monitoring of selected pesticide classes in fruits, J. Chromatogr. A 1216 (2009) 7307-7313.

[54] Y. Picó, Ultrasound-assisted extraction for food and environmental samples, Trends Anal. Chem. 43 (2013) 84-99.

[55] S. Morales - Muñoz, R. J. J. Vreuls, M. D. Luque de Castro, Dynamic ultrasound - assisted extraction of environmental pollutants from marine sediments for comprehensive two - dimensional gas

chromatography with time-of-flight mass spectrometric detection, J. Chromatogr. A 1086 (2005) 122-127.

[56] C. Sparr Eskilsson, E. Björklund, Analytical - scale microwave - assisted extraction, J. Chromatogr. A 902 (2000) 227-250.

[57] E. M. Kristenson, H. C. Neidig, R. J. J. Vreuls, U. A. Th Brinkman, Fast miniaturized sample preparation for the screening and comprehensive two - dimensional gas chromatographic determination of polychlorinated biphenyls in sludge, J. Sep. Sci. 28 (2005) 1121-1128.

[58] M. Rezaee, Y. Assadi, M. -R. M. Hosseini, E. Aghaee, F. Ahmadi, S. Berijani, Determination of organic compounds in water using dispersive liquid-liquid microextraction, J. Chromatogr. A 1116 (2006) 1-9.

[59] F. Tugizimana, P. A. Steenkamp, L. A. Piater, I. A. Dubery, Multi - platform metabolomic analyses of ergosterol-induced dynamic changes in *nicotiana tabacum* cells, PLoS One 9 (2014) e-87846.

[60] I. François, K. Sandra, P. Sandra, Comprehensive liquid chromatography: fundamental aspects and practical considerations-a review, Anal. Chim. Acta 641 (2009) 14-31.

[61] P. Donato, F. Cacciola, P. Q. Tranchida, P. Dugo, L. Mondello, Mass spectrometry detection in comprehensive liquid chromatography: basic concepts, instrumental aspects, applications and trends, Mass Spectrom. Rev. 31 (2012) 523-559.

[62] P. Q. Tranchida, P. Donato, F. Cacciola, M. Beccaria, P. Dugo, L. Mondello, Potential of comprehensive chromatography in food analysis, TRAC-Trend Anal. Chem. 52 (2013) 186-205.

[63] E. Nagele, M. Vollmer, P. Horth, Improved 2D nano-LC/MS for proteomics applications: a comparative analysis using yeast proteome, J. Biomol. Tech. 15 (2004) 134-143.

[64] Y. Wang, J. Zhang, C. L. Liu, X. Gu, X. M. Zhang, Nano-flow multidimensional liquid chromatography with electrospray ionization time-of-flight mass spec-trometry for proteome analysis of hepatocellular carcinoma, Anal. Chim. Acta 530 (2005) 227-235.

[65] J. Haun, J. Leonhardt, C. Portner, T. Hetzel, J. Tuerk, T. Teutenberg, T. C. Schmidt, Online and splitless nanoLC × capillary LC with quadrupole/time - of - flight mass spectrometric detection for comprehensive screening analysis of complex samples, Anal. Chem. 85 (2013) 10083-10090.

[66] M. W. H. Pinkse, P. M. Uitto, M. J. Hilhorst, B. Ooms, A. J. R. Heck, Selective iso-lation at the femtomole level of phosphopeptides from proteolytic digests using digestsusing 2D-nanoLC-ESI-MS/MS and titanium oxide pre-columns, Anal. Chem. 76 (2004) 3935-3943.

[67] P. Taylor, P. A. Nielsen, M. B. Trelle, O. B. Horning, M. B. Andersen, O. Vorm, M. F. Moran, T. Kislinger, Automated 2D peptide separation on a 1D nano-LC—MSsystem, J. Proteome Res. 8 (2009) 1610-1616.

[68] S. Xia, D. Tao, H. Yuan, Y. Zhou, Z. Liang, L. Zhang, Y. Zhang, Nano - flow multidimensional liquid chromatography platform integrated with combination of protein and peptide separation for proteome analysis, J. Sep. Sci. 35 (2012) 1764-1770.

[69] T. Rajesh, H. -Y. Park, E. Song, C. Sung, S. -H. Park, J. -H. Lee, D. Yoo, Y. -G. Kim, J. -M. Jeon, B. -G. Kim, Y. -H. Yang, A new flow path design for multidimensional protein identification technology using nano-liquid chromatography electrospray ionization mass spectrometry, Korean J. Chem. Eng. 30 (2013) 417-421.

[70] P. J. Boersema, N. Divecha, A. J. R. Heck, S. Mohammed, Evaluation and optimization of ZIC-HILIC-RP as an alternative MudPIT strategy, J. Proteome Res. 6 (2007) 937-946.

[71] S. Di Palma, S. Mohammed, A. J. R. Heck, ZIC-cHILIC as a fractionation method for sensitive

and powerful shotgun proteomics, Nature Protoc. 7 (2012) 2041-2055.

[72] E. Sommella, F. Cacciola, P. Donato, P. Dugo, P. Campiglia, L. Mondello, Development of an online capillary comprehensive 2D-LC system for the analysis of proteome samples, J. Sep. Sci. 35 (2012) 530-533.

延伸阅读

[1] T. Hyötyläinen, Critical evaluation of sample pretreatment techniques, Anal. Bioanal. Chem. 394 (2009) 743-758.

[2] B. Xu, P. Li, F. Ma, X. Wang, B. Matthaus, R. Chen, Q. Yang, W. Zhang, Q. Zhang, Detection of virgin coconut oil adulteration with animal fats using quantitative cholesterol by GC×GC-TOF/MS analysis, Food Chem. 178 (2015) 128-135.

[3] D. Megson, J.-F. Focant, D.G. Patterson, M. Robson, M.C. Lohan, P.J. Worsfold, S. Comber, R.K.E. Reiner, G. O'Sullivan, Can polychlorinated biphenyl (PCB) signatures and enantiomer fractions be used for source identification and to age date occupational exposure? Environ. Int. 81 (2015) 56-63.

[4] C. Li, D. Wang, N. Li, Q. Luo, X. Xu, Z. Wang, Identifying unknown by-products in drinking water using comprehensive two-dimensional gas chromatography-quadrupole mass spectrometry and in silico toxicity assessment, Chemosphere163 (2016) 535-543.